AF771483

Gas Well Testing Handbook

Gas Well Testing Handbook

Amanat U. Chaudhry

Advanced TWPSOM Petroleum Systems, Inc.
Houston, Texas

ELSEVIER

AMSTERDAM • BOSTON • HEIDELBERG • LONDON
NEW YORK • OXFORD • PARIS • SAN DIEGO
SAN FRANCISCO • SINGAPORE • SYDNEY • TOKYO

Gulf Professional Publishing is an imprint of Elsevier.

G|P
P|

Gulf Professional Publishing is an imprint of Elsevier Science.

Copyright © 2003 by Elsevier Science. All rights reserved.

No part of this publication may be reproduced, stored in a retrieval system, or transmitted in any form or by any means, electronic, mechanical, photocopying, recording, or otherwise, without the prior written permission of the publisher.

∞ Recognizing the importance of preserving what has been written, Elsevier Science prints its books on acid-free paper whenever possible.

Library of Congress Cataloging-in-Publication Data
Chaudhry, Amanat U.
Gas well testing handbook / by Amanat U. Chaudhry.
p.cm.
Includes bibliographical references and index.
ISBN 0-7506-7705-8 (acid-free paper)
1. Gas wells—Testing—Handbooks, manuals, etc. I. Title.

TN871.C453 2003
622′.3385′0287–dc21 2003048310

British Library Cataloguing-in-Publication Data
A catalogue record for this book is available from the British Library.

The publisher offers special discounts on bulk orders of this book.
For information, please contact:

Manager of Special Sales
Elsevier Science
200 Wheeler Road
Burlington, MA 01803
Tel: 781-313-4700
Fax: 781-313-4882

For information on all Gulf Professional Publishing publications available, contact our World Wide Web home page at: http://www.gulfpp.com

10 9 8 7 6 5 4 3 2 1

Printed in the United States of America

Dedication

This book is dedicated to my son Alhaj U. Chaudhry

I am grateful to my parents for providing me with education, inspiration, and confidence. I am also indebted to my ex-wife, Nuraini Smith, who provided the encouragement, fortitude, and extraordinary understanding, which enabled me to steal many hours from my family while writing this book.

Contents

CHAPTER 3

Well Testing Techniques in Horizontal Gas Wells 84

CHAPTER 4

Deliverability Testing and Well Production Potential Analysis Methods 140

CHAPTER 5

Fundamentals of Drawdown Test Analysis Methods .. 237

Foreword

Although elements of gas well testing methods have been practical almost since gas reservoirs were first recognized, the concept of gas well testing techniques has taken form only within the past three decades. Many individual monographs and at least one manual on the subject have been published in the open literature, and it is probable that proprietary presentations of gas well testing concepts are to be found within the internal libraries of some oil- and gas-producing companies. In the present volume, the author presents a treatment of the subject to be published in book form.

The roots of gas well testing are to be found in reservoir engineering taken in its broadest sense as the technology that deals with the well/reservoir behavior through the measuring and analysis of deliverability test, flow, and transient pressure responses in unfractured and fractured gas wells. The concepts related to gas well test data acquisition and interpretation are presented from a practical viewpoint. These concepts are emphasized throughout the book by means of examples and field case studies.

In *Gas Well Testing Handbook*, the author has presented a comprehensive study of the measuring and analysis of deliverability tests, flow, and transient pressure responses in gas wells. The basic principles are reviewed and the applicability and limitations of the various testing techniques are critically discussed and illustrated with actual field examples. The material is presented in a form that will allow engineers directly involved in well deliverability, pressure build-up, and flow testing to re-educate themselves on the subject. At the same time, with its up-to-date review of the literature and extensive bibliography, the book will serve as a useful guide and reference to engineers directly engaged in well pressure behavior work. The author has accomplished the intended objectives of the book in a thorough and excellent manner.

The author has illustrated by field application examples and field case studies to describe the types of wells and reservoir behavior encountered in modern production practice. The source, nature, and precision of the data and studies upon which the calculations and analysis are based are discussed subordinately.

Numerous examples are provided to help the reader develop an understanding of the principles and limitations of applied gas well testing methods.

The book is essential and important to engineers concerned with evaluating well/reservoir systems and the pressure performance of gas wells. The author has extensive experience in this field and is most qualified to treat the subject. It is a timely addition to the literature of petroleum technology.

Dilip Borthaker
Head of Gas Engineering Department
Gulf Indonesia Resources

Preface

The major purpose of writing this book is to provide a practical reference source for knowledge regarding state-of-the-art gas well testing technology. The book presents the use of gas well testing techniques and analysis methods for evaluation of well conditions and reservoir characteristics. All techniques and data described in this book are "field-tested" and are published here for the first time. For example, this book contains new tables and comparisons of the various methods of well test analysis. Most of these techniques and applications are clearly illustrated in worked examples of the actual field data. Several actual field example calculations and field case studies are included for illustration purposes.

This text is a must for reservoir engineers, simulation engineers, practicing petroleum engineers, and professional geologists, geophysicists, and technical managers. It helps engineering professors better acquaint their students with "real-life" solution problems. This instructive text includes practical examples that readers should find easy to understand and reproduce.

Fundamental concepts related to well test data acquisition and interpretation are presented from a practical viewpoint. Furthermore, a brief summary of the advances in this area is presented. Emphasis is given to the most common interpretation methods used at present. The main emphasis is on practical solutions and field application. More than 129 field examples are presented to illustrate effective gas well testing practices, most analysis techniques, and their application.

Many solutions that are presented are based upon the author's experience dealing with various well testing techniques and interpretation around the world. I am very thankful to the many companies with whom I had the opportunity to work in well test analysis for many years.

A properly designed, executed, and analyzed well test can provide information about formation permeability, reservoir initial or average pressure, sand-face condition (well damage or stimulation), volume of drainage area, boundary and discontinuities, reservoir heterogeneity, distance or extension of the fracture induced, validation of geological model, and system identification (type of reservoir and mathematical model).

Further, it is important to determine the ability of a formation to produce reservoir fluids and the underlying reason for a well's productivity. These data, when combined with hydrocarbon production data and with laboratory data on fluid and rock properties, afford the means to estimate the original hydrocarbon in-place and the recovery that may be expected from the reservoir under various modes of exploitation. In addition, well test data and IPR well performance equations, combined with production data, help in designing, analyzing, and optimizing a total well production system or in production optimization.

The rigorous discussions, practical examples, and easy-to-read manner make this a valuable addition to every petroleum professional's library. Our colleagues' discussions and their suggestions were very valuable in making this book useful to a practicing engineer. Most users of this book will find it logically organized and readily applicable to many well testing problem solutions and field applications.

Acknowledgments

I would also like to thank Dr. Furlow Fulton, Head of the Petroleum Engineering Department at the University of Pittsburgh, for educating me in reservoir engineering. I have also been privileged to work with many professionals in the oil industry who have taught me many things and helped me grow and develop as an engineer. I am also thankful to A.C. Carnes, Jr., General Manager, Integrated Technology Petroleum Consulting Services, Houston, who oriented me in the areas of reservoir simulation and well test analysis during my career. I am also thankful to many companies who were generous in providing the field histories and data that were used in the book. I am also thankful to Ambar Sudiono,* General Manager of State Owned Oil Company of Indonesia, who was kind enough to read chapters 3, 4, 5, 7, 10 and provided many valuable suggestions.

Mr. Dilip Borthakur has reviewed the material presented in this book.** He has spent hundreds of hours reading, checking, and critically commenting on all aspects of the material and its presentation. There is no doubt that the book is a much better volume that it would have been without his aid.

Ms. Faiza Azam typed the many versions of the book required to reach the final form. Her technical skills and command of the English language have enabled preparation of this volume to proceed smoothly and on schedule. She also prepared all the original illustrations; she redrew many illustrations taken from the references to provide a consistent nomenclature and format. Her artist's viewpoint, her skill, and her highly accurate work have added substantially to this book. I would also like to acknowledge the support of editorial staff of Elsevier for their patience and hard work in producing this book. Last but not the least, I owe sincere appreciation and thanks to Kyle Sarofeen, Phil Carmical and Christine Kloiber for their contributions to this book.

* Presently General Manager with State Owned Oil Company of Indonesia, Jakarta.

** Presently Head of Gas Engineering Department with Gulf Indonesia Resources.

Amanat Chaudhry

Chapter 1

Introduction

1.1 Role of Gas Well Tests and Information in Petroleum Engineering

Gas well test analysis is a branch of reservoir engineering. Information derived from flow and pressure transient tests about *in-situ* reservoir conditions is important in many phases of petroleum engineering. The reservoir engineer must have sufficient information about the reservoir/well condition and characteristics to adequately analyze reservoir performance and forecast future production under various modes of operation. The production engineer must know the condition of production and injection wells to persuade the best possible performance from the reservoir.

Pressures are most valuable and useful data in reservoir engineering. Directly or indirectly, they enter into all phases of reservoir engineering calculations. Therefore accurate determination of reservoir parameters is very important. In general, gas well test analysis is conducted to meet the following objectives:

- To obtain reservoir parameters.
- To determine whether all the drilled length of gas well is also a producing zone.
- To estimate skin factor or drilling and completion related damage to a gas well. Based upon magnitude of the damage a decision regarding well stimulation can be made.

1.2 History of Gas Well Testing

The first analysis was based on the empirical method applicable to very porous and permeable reservoirs developed by Schellherdt and Rawlins,[1] "Back-Pressure Data on Natural Gas Wells and Their Application to Production Practices." Monograph 7, U.S.B.M. This method today is known as the four-point (sometimes as the one-point) method. The [$(\bar{p}_R^2 - p_{wf}^2)$ versus q_{sc}] square of the average reservoir pressure minus the square of the flowing sand-face pressure is plotted versus the respective flow rates on log-log paper. The maximum rate is read at the pressure equal to the average reservoir pressure after a straight line is drawn through test points for four semi-stabilized flow rates. Later, more practical methods of testing were developed. These included the isochronal test and the modified isochronal test. Such tests have been used extensively by the gas industry.

Most recently flow and pressure transient tests have been developed and used to determine the flow characteristics of gas wells. Development of even tighter gas wells was common during the late 1950s and fracturing with large amounts of sand was routine. Pressure difference across the drainage area often was great. By 1966, a group of engineers working with Russell, Shell Oil, published articles using basic flow equations applicable to all gas wells, regardless of the permeability and fractures used by the operators. The state of the art was summarized in 1967 in "Pressure Buildup and Flow Tests in Wells" by Matthews and Russell,[2] SPE Monograph 1, Henry L. Doherty Series. Earlougher[4] again reviewed the state of the art in 1977 in "Advances in Well Test Analysis" in SPE Monograph 5. One book[5] was published in 1975 covering different aspects of flow and pressure transient analysis.

The analysis of pressure data for fractured gas wells has deserved special attention because of the number of wells that have been stimulated by hydraulic fracturing techniques. References 4 through 7 have presented a summary of the work done on flow toward fractured wells in 1962 and 1978.

1.3 Gas Well Test Data Acquisition, Analysis, and Management

Throughout the life of a gas well, from exploration to abandonment, enough well test data are collected to describe well condition and behavior. It is emphasized that the multidisciplinary professionals need to work as an integrated team to develop and implement well test data management programs.

Efficient Gas Well Test Analysis Programs

Initial bottom hole pressure measurements should be made, preferably at each well and at a selected "Key Gas Well" periodically. According to Woods and Abib, key gas wells represent 25% of the total wells.[2] Also, they found it is beneficial to measure pressure in all wells at least every 2 to 3 years to aid in calibrating reservoir models. It is essential to establish the specification of what and how much well test data need to be gathered and the procedure and frequency to be followed. A logical, methodical, and sequential well test data acquisition and analysis program is shown in Figure 1–1.

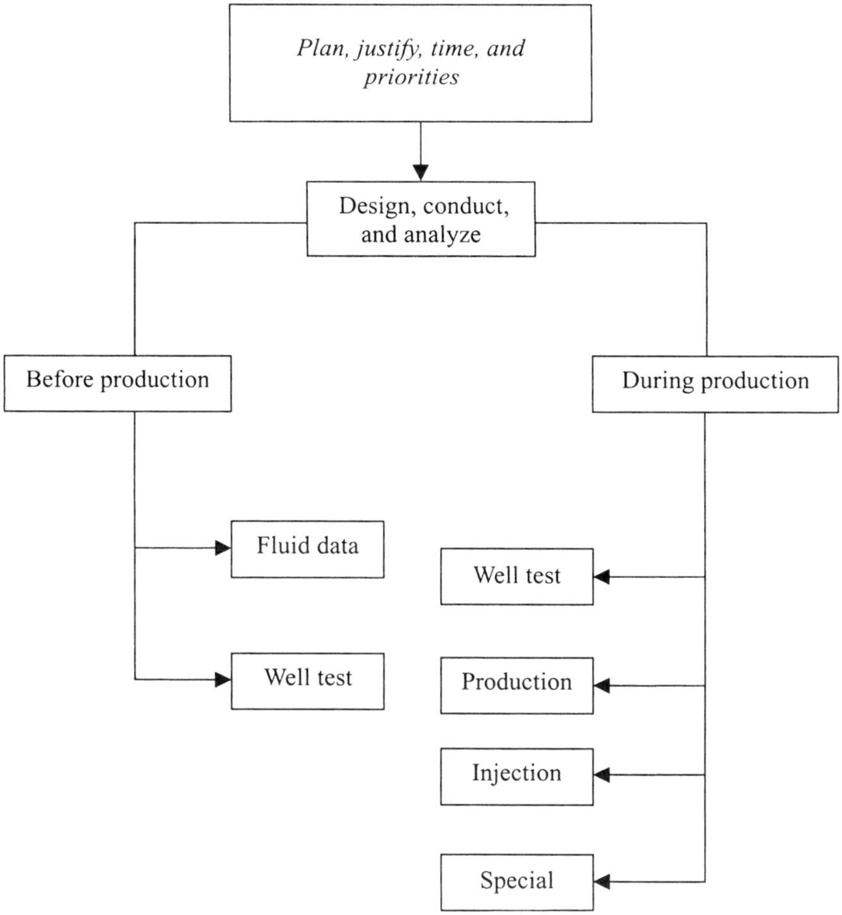

Figure 1–1. Logical well test data acquisition and analysis program.

1.4 Selecting Gas Wells for Optimum Stimulation Treatment

The key to determining whether or not a well is a good candidate for stimulation treatment is diagnosing the well to find the cause for its low productivity. Buildup, drawdown, or drill-stem tests, core analyses, offset well data, and other information can be used to accomplish this. After diagnosis, the optimum well stimulation treatment, either small or massive hydraulic fracturing, can be designed for the well. Figure 1–2 shows several sets of calculations designed to evaluate well/reservoir behavior and evaluate reservoir parameters,

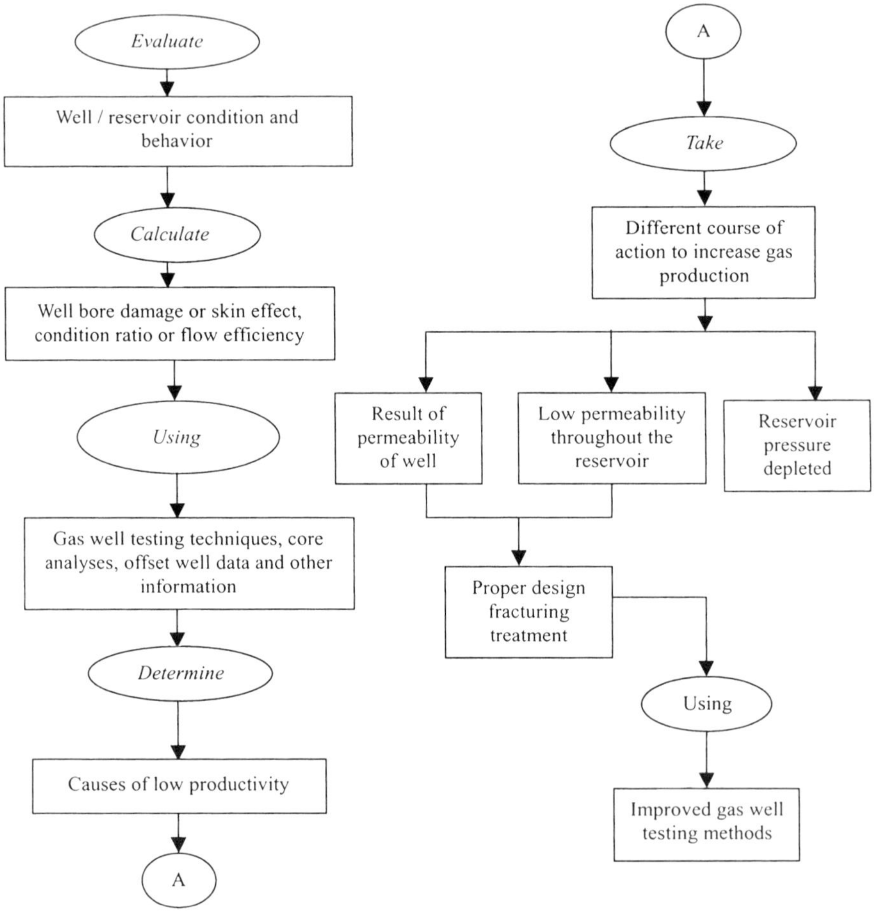

Figure 1–2. Shows selection of gas wells for optimum treatment.

quality, and stimulation efforts to optimize completion methods for enhancing hydrocarbon gas recovery and maximizing profitability.

1.5 Reservoir System Characterization Process

An efficient gas well test data acquisition and analysis program requires careful planning, designing, conducting, and evaluation and well-coordinated team efforts through an integrated approach. Figures 1–3 and 1–4 indicate general activities in reservoir description and inputs from various engineering disciplines (integrated approach). Core analysis measurements of samples selected by the geologist provide data for the preliminary identification of

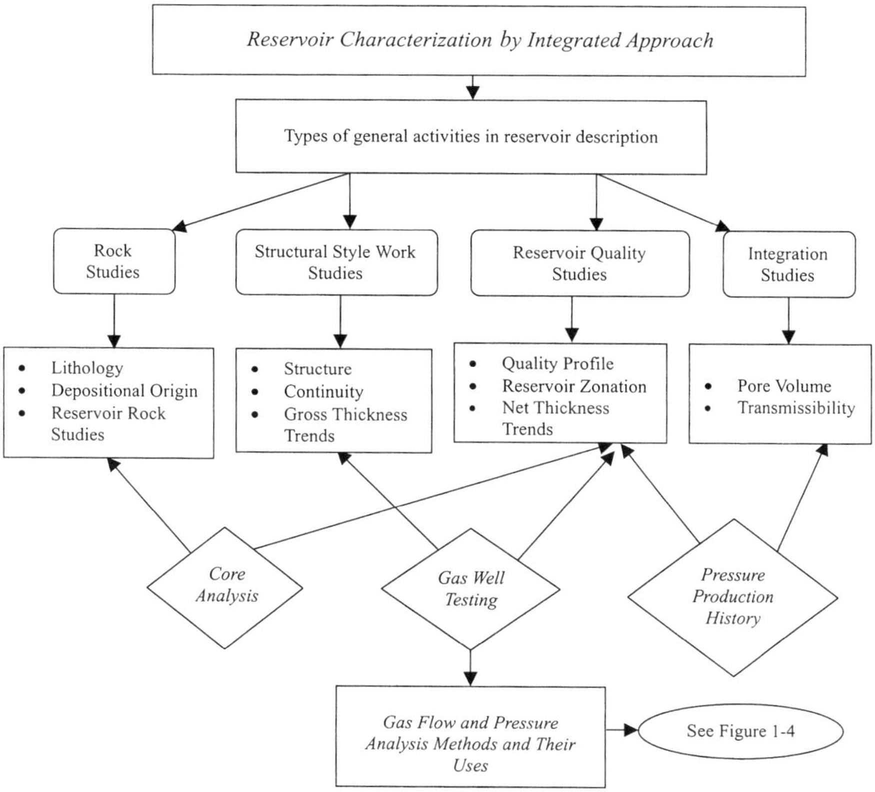

Figure 1–3. Reservoir system characterization flow chart process using integrated approach.

reservoir rock types. Well test results using various techniques were reasonable when compared with known geologic and core data. Well test studies aid in recognizing flow barriers, fractures, and variations in permeability. Various simulation studies can be used to test the physical model against pressure production. Performance adjustments are made to the model until a match is achieved. The major goal is optimization of gas recovery through characterization of the reservoir system.

Most Common Gas Well Test Interpretation Methods

Figure 1–4 shows gas flow and pressure analysis methods. Theory and example applications to illustrate effective well test analysis practices can be found and are discussed in the following chapters.

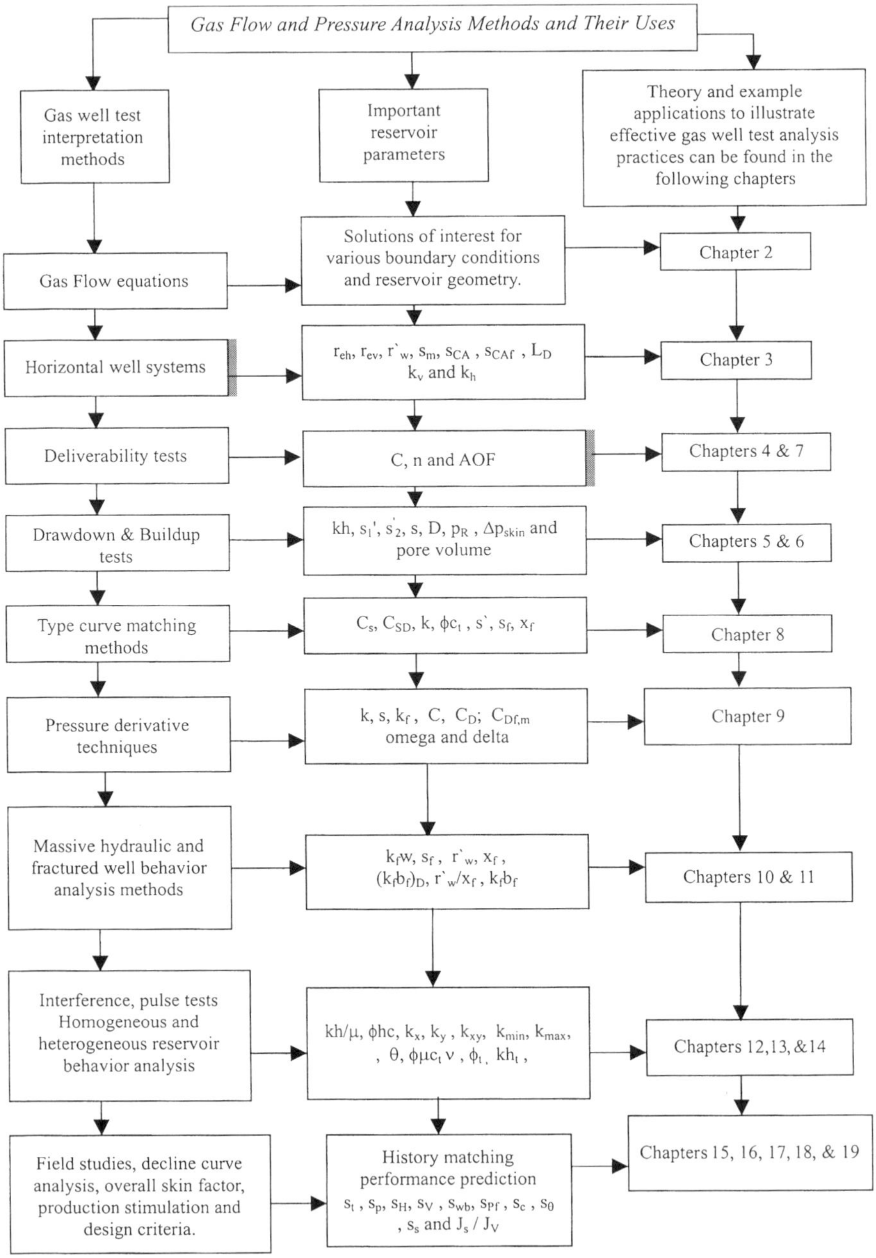

Figure 1–4. Gas flow and pressure analysis methods.

1.6 Scopes and Objective

This book is very important to professional petroleum engineers, teachers, graduate students, and those concerned with evaluating reservoir systems and the pressure performance of gas wells. The data in this book should enable petroleum professionals to design, conduct, and analyze pressure transient tests to obtain reliable information about reservoir and well conditions.

1.7 Organization

The book presents the following:

- Sound fundamental concepts/methodology related to gas well test data acquisition and interpretation from a practical viewpoint
- Modern gas well testing methods and pressure transient test analysis techniques
- Examples illustrating effective well test analysis techniques
- An excellent practical reference source related to pressure transient analysis techniques and their interpretations
- Theory and practices of testing methods and their roles in reservoir engineering management
- Practical examples showing step-by-step solutions to problems
- Various charts, formulae, and tables included for ready reference and quick solutions for gas well testing and analyses

This chapter is an overview of gas well testing and analysis techniques. It also includes a short discussion of unit conversion factors and the SI (metric) unit system. Appendix A provides a list of conversion factors.

Details and supporting materials are presented in the appendices for the benefit of those who would like to learn more.

1.8 Unit Systems and Conversions

In any book of this nature, it is worthwhile to include a comprehensive list of unit conversion factors, since data are often reported in units different from those used in the equations. Such factors are presented in Appendix A. Because of the possibility of eventual conversion of engineering calculations to a metric standard, I also include information about the "SI" system of weights and measures. Finally, I compare some important units and equations in five different unit systems. The calculation procedure is illustrated in following example.

Example 1–1 *Converting Factors and Arithmetic to English Oilfield Units*
The equation of interest, expressed in Darcy units, is

$$\bar{p}_R - p_{pf} = \frac{q\mu_g}{2\pi kh}\left[In\left(\frac{r_e}{r_w}\right) - 0.75 + s \right] \tag{1–1}$$

Equation 1–1, when expressed in field units, becomes

$$(\bar{p} - p_{wf})\text{psi}\left|\frac{\text{atm}}{\text{psi}}\right| = \frac{q\ \text{Mscf/d}\left|\frac{\text{s.cc/sec}}{\text{Mscf/d}}\right|\left|\frac{\text{r.cc/sec}}{\text{s.cc/sec}}\right|}{2\pi k\ \text{mD}\left|\frac{\text{D}}{\text{mD}}\right| h\ \text{ft}\left|\frac{\text{cm}}{\text{ft}}\right|}\mu_g \times \left[In\left(\frac{r_e}{r_w}\right) - 0.75 + s \right] \tag{1–2}$$

In this conversion the ratio is

$$\left|\frac{\text{r.cc/sec}}{\text{s.cc/sec}}\right| = \left|\frac{\text{reservoir cc/sec}}{\text{standard cc/sec}}\right| = \frac{1}{\beta_g} = \frac{1}{\text{Gas formation volume factor}}$$

and in field units,

$$\beta_g = \frac{T_{sc}\,\bar{p}}{p_{sc} zT} = \frac{520\,\bar{p}}{14.70\,zT} = 35.37\frac{\bar{p}}{zT} \tag{1–3}$$

$\bar{p}$, the pressure at which β_g is evaluated, is as yet undefined. The full conversion of the rate term in Eq. 1–1 can be expressed as

$$q_{sc}\ \text{Mscf/d}\left|\frac{\text{Mstb/d}}{\text{Mscf/d}}\right|\left|\frac{\text{stb/d}}{\text{Mstb/d}}\right|\left|\frac{\text{s.cc/sec}}{\text{stb/d}}\right|\frac{1}{\beta_g} = q_{sc}\ \text{r.cc/sec}$$

$$q_{sc}\ \text{Mscf/d}\left|\frac{1}{5.615}\right| |1000|\ |1.84|\ \frac{zT}{35.37\bar{p}} = q_{sc}\ \text{r.cc/sec}$$

$$9.2647\frac{q_{sc}zT}{\bar{p}}\ \text{Mscf/d} = q_{sc}\ \text{r.cc/sec}$$

Including the remaining conversion factors in Equation 1–1 yields

$$\bar{p} - p_{wf} = \frac{711 q_{sc} zT\mu_g}{kh\bar{p}} \cdot \left[In\left(\frac{r_e}{r_w}\right) - 0.75 + s \right] \tag{1–4}$$

Fetkovich[11] has compared Eq. 1–4 with the numerical simulation. This author found that for the same reservoir and flow condition the two were in close agreement provided that the pressure $\bar{p}$ at which the gas formation volume factor was evaluated was set equal to the average of the current, average reservoir pressure, and the bottom hole flowing pressure, i.e.,

$$\tilde{p} = \frac{\bar{p}_R + p_{wf}}{2} \tag{1–5}$$

Furthermore, both μ_g and z should be evaluated at this same pressure so that

$$\mu_g = \mu_g \left[\frac{\bar{p}_R - p_{wf}}{2} \right] \quad \text{and} \quad z = z \left[\frac{\bar{p}_R - p_{wf}}{2} \right] \tag{1–6}$$

and substituting these values of $\tilde{p}$, μ_g, and z in Eq. 1–4 gives

$$\bar{p}_R^2 - p_{wf}^2 = \frac{1.422 \times 10^6 q_{sc} \mu_g z T}{kh} \cdot \left[In\left(\frac{r_e}{r_w}\right) - 0.75 + s \right] \tag{1–7}$$

Equation 1–7 is in pressure-squared form. The Equation 1–7 can be written in pseudo pressure form:

$$\psi(\bar{p}_R) - \psi(p_{wf}) = \frac{1.422 \times 10^6 q_{sc} T}{kh} \cdot \left[In\left(\frac{r_e}{r_w}\right) - 0.75 + s \right] \tag{1–8}$$

Equations 1–7 and 1–8 are the same as Eqs. 4–4 and 4–3 in Chapter 4. Note that the gas flow rate q_{sc} is in mmscfd.

References and Additional Reading

1. Schellhardt, M. A., and Rawlins, E. L., *Back-Pressure Data on Natural Gas Wells and Their Application to Production Practice*, Bureau of Mines, Monograph 7 (1936).
2. Matthews, C. S., and Russell, D. G., *Pressure Buildup and Flow Tests in Wells*, SPE Monograph Series, No. 1, SPE, Dallas, TX (1967).
3. Ramey, H. J., Jr., Kumar, A., and Gulati, M. S., *Gas Well Test Analysis under Water-Drive Conditions*. AGA, Arlington, VA (1973).
4. Earlougher, R. C., Jr., *Advances in Well Test Analysis*, SPE Monograph Series, No. 5, SPE, Dallas (1977).
5. *The Theory and Practice of Testing of Gas Wells*, 3rd ed. Energy Resources Conservation Board, Calgary, Alta. (1975).
6. Ramey, H. J., Jr., *Practical Use of Modern Well Test Analysis*, paper SPE 5878 presented at the SPE 46 Annual California Regional Meeting, Long Beach, CA, April 8–9, 1976.
7. Raghavan, R., "Pressure Behavior of Wells Intercepting Fractures," Proc; Invitational Well-Testing Symposium, Berkeley, CA, Oct. 19–21, 1977.
8. Prats, M., Hazebrock, P., and Sticker, W. R., "Effect of Vertical Fractures on Reservoir Behavior—Compressible Fluid Case," *Soc. Petroleum Eng. J.* (June 1962), 87–94; *Trans. AIME*, 225.

9. Gringarten, A. C., Ramey, H. J., and Raghavan, R., "Applied Pressure Analysis for Fractured Wells," *J. Petroleum Technol.* (July 1975), 887–892; *Trans. AIME*, 259.
10. Cullender, M. H., "The Isochronal Performance Method of Determining the Flow Characteristics of Gas Wells," *J. Petroleum Technol.* (Sept. 1953), 137.
11. Fetkovich, M. J., "Multipoint Testing of Gas Wells," paper presented at the SPE-AIME Mid-Continent Section Continuing Education Course on Well Test Analysis (March 1975).
12. Russell, D. G., Goodrich, J., Perry G. E., and Bruskotter, J. F., 1966. "Methods of Predicting Gas Well Performance," *J. Petroleum Technol.* (January) 99, 108. *Trans. AIME.*

Chapter 2

Application of Fluid Flow Equations to Gas Systems

2.1 Introduction

The aim of this chapter is to develop and present the fundamental equations for flow of gases through porous media, along with solutions of interest for various boundary conditions and reservoir geometries. These solutions are required in the design and interpretation of flow and pressure tests.

To simplify the solutions and application of the solutions, dimensionless terms are used. Assumptions and approximations necessary for defining the system and solving the differential equations are clearly stated. The principle of superposition is applied to solve problems involving interference between wells, variables flow rates, and wells located in noncircular reservoirs. The use of analytical and numerical solutions of the flow equations is also discussed. Formation damage or stimulation, turbulence, and wellbore storage or unloading are given due consideration. This chapter applies in general to laminar, single, and multiphase flow, but deviations due to inertial and turbulent effects are considered. For well testing purposes two-phase flow in the reservoir is treated analytically by the use of an equivalent single-phase mobility.

The equations of continuity, Darcy's law, and the gas equation of state are presented and combined to develop a differential equation for flow of gases through porous media. This equation, in generalized coordinate notation, can be expressed in rectangular, cylindrical, or spherical coordinates and is solved by suitable techniques. The next subsections describe steady-state, pseudo-steady-state, and unsteady-state flow equations including the gas radial diffusivity equation, basic gas flow equations, solutions, and one-, two-, and three-dimensional coordinate systems.

2.2 Steady-State Laminar Flow

Darcy's law for flow in a porous medium is

$$v = \frac{k}{\mu_g}\frac{dp}{dx} \quad \text{or} \quad q = vA = \frac{kA}{\mu_g}\frac{dp}{dx} \tag{2–1}$$

where

v = gas viscosity; q = volumetric flow rate; k = effective permeability; μ_g = gas viscosity; and $\frac{dp}{dx}$ = pressure gradient in the direction of flow

For radial flow, Eq. 2–1 becomes

$$q = \frac{k(2\pi rh)}{\mu_g}\frac{dp}{dx} \tag{2–2}$$

where r is radial distance and h is reservoir thickness,

Equation 2–2 is a differential equation and must be integrated for application. Before integration the flow equation must be combined with an equation of state and the continuity equation. The continuity equation is

$$\rho_1 q_1 = \rho_2 q_2 = \text{constant} \tag{2–3}$$

The equation of state for a real gas is

$$\rho = \frac{pM}{ZRT} \tag{2–4}$$

The flow rate of a gas is usually desired at some standard conditions of pressure and temperature, p_{sc} and T_{sc}. Using these conditions in Eq. 2–3 and combining Eqs. 2–3 and 2–4, we get

$$\rho q = \rho_{sc} q_{sc},$$

or

$$q\frac{pM}{zRT} = q_{sc}\frac{p_{sc}M}{z_{sc}RT_{sc}}$$

Solving for q_{sc} and expressing q_{sc} with Eq. 2–2 gives

$$q_{sc} = \frac{pT_{sc}}{p_{sc}zT}\frac{2\pi rhk}{\mu}\frac{dp}{dr}$$

The variables in this equation are p and r. Separating the variables and integrating:

$$\int_{p_w}^{\bar{p}} p dp = \frac{q_{sc} p_{sc} T \bar{\mu}_g \bar{z}}{T_{sc}\ 2\pi kh} \int_{r_w}^{r_c} \frac{dr}{r}$$

$$\frac{\bar{p}^2 - p_w^2}{2} = \frac{q_{sc} p_{sc} T \bar{\mu} \bar{z}}{T_{sc} 2\pi kh} \ln\left(\frac{r_e}{r_w}\right)$$

$$\text{or} \quad q_{sc} = \frac{\pi kh T_{sc}\left(\bar{p}^2 - p_w^2\right)}{p_{sc} T \bar{\mu}_g \bar{z} \ln\left(\frac{r_e}{r_w}\right)} \tag{2–5}$$

In this derivative it was assumed that μ_g and z were independent of pressure. They may be evaluated at reservoir temperature and average pressure in the drainage area such as

$$\bar{P} = \frac{P_e - P_w}{2}$$

In gasfield units, Eq. 2–5 becomes

$$q_{sc} = \frac{0.007027kh\left(\bar{P}^2 - P_w^2\right)}{\mu_g \bar{z} T \log\left(\frac{r_e}{r_w}\right)} \tag{2–6}$$

$$q_{sc} = \frac{0.000305kh\left(\bar{P}^2 - P_w^2\right)}{\mu_g \bar{z} T \ \ln\left(\frac{r_e}{r_w}\right)} \tag{2–7}$$

Where q_{sc} = mscf/d; k = permeability in mD; h = formation thickness in feet; p_e = reservoir pressure, psi, p_w = well bore pressure, psia, T = reservoir temperature, °R; r_e = drainage radius, ft; r_w = well bore radius, ft; $\bar{z}$ = average compressibility factor, dimensionless; and $\bar{\mu}_g$ = gas viscosity, cP.

This equation incorporates the following values for standard pressure and temperature:

$$p_{sc} = 14.7 \text{ psia},$$

$$T_{sc} = 60°\text{F} = 520°\text{R}$$

The gas flow rate is directly proportional to the pseudopressures. The pseudopressure is defined as

$$\psi(p) = 2\int_{p_{ref}}^{\bar{p}} \frac{p}{\mu z}dp \tag{2–8}$$

In Eq. 2–8, p_{ref} is a reference pressure. At the reference pressure, pseudopressure is assigned a datum value of zero. The Eqs. 2–6 and 2–7 in terms of pseudopressure become

$$q_{sc} = \frac{0.0007027kh[\psi(\bar{p}) - \psi(p_w)]}{T\ \ln\left(\frac{r_e}{r_w}\right)} \tag{2–9}$$

$$q_{sc} = \frac{0.000305kh[\psi(\bar{p}) - \psi(p_w)]}{T\ \log\left(\frac{r_e}{r_w}\right)} \tag{2–10}$$

p^2 and $\psi(p)$ have identical values up to 2500 psia. Above 2500 psia, p^2 and $\psi(p)$ exhibit different values. Thus, below 2500 psia, either p^2 or $\psi(p)$ can be used. Above 2500 psia, $\psi(p)$ should be used. Gas pseudopressure, $\psi(p)$, which is defined by Eq. 2–8, is considered, i.e.,

$$\psi(\bar{p}) - \psi(p_w) = 2\int_{p_{ref}}^{\bar{p}} \frac{pdp}{\mu_g z} - 2\int_{p_{ref}}^{p_w} \frac{pdp}{\mu_g z}$$

It is more difficult and generally engineers feel more comfortable dealing with pressure squared, p^2, rather than an integral transformation. Therefore, it is worthwhile, at this stage, to examine the ease with which these functions can be generated and used. We evaluate the integral in Eq. 2–8 numerically, using values for μ_g and z for the specific gas under consideration, evaluated at reservoir temperature. An example will illustrate this calculation.

Example 2–1 *Calculating Gas Pseudopressure*

Calculate the gas pseudopressure $\psi(p)$ for a reservoir containing 0.732 gravity gas at 250°F as a function of pressure in the range 400 to 4000 psia. Gas properties as functions of pressure are given in Table 2–1.

Solution For $p = 400$ psia:

$$\psi(400) = 2\int_{p_{ref}}^{p} \frac{p}{\mu_g z}dp$$

Table 2–1
Generation of Gas Pseudopressure as a Function of the Actual Pressure

Pressure, p (psia)	μ_g (cP)	Z –	$P/\mu_g z$ (psia/cP)	$\psi(P)$ (mm psia^2/cP)
400	0.014337	0.9733	28.665	11.47
800	0.014932	0.9503	56,378	45.48
1200	0.015723	0.9319	81,899	100.83
1600	0.016681	0.9189	104,383	175.33
2000	0.017784	0.9120	123,312	266.41
2400	0.019008	0.9113	138,552	371.18
2800	0.020329	0.9169	150,217	486.72
3200	0.021721	0.9282	158,719	610.28
3600	0.023151	0.9445	164,638	739.56
4000	0.024580	0.9647	168,689	872.92

$$= 2\frac{\left[\left(\frac{p}{\mu_g z}\right)_0 + \left(\frac{p}{\mu_g z}\right)_{400}\right]}{2}$$

$$= 2\left(\frac{0 + 28,665}{2}\right)(400 - 0)$$

$$= 11.466 \times 10^6 \text{ psia}^2/\text{cp}$$

For $p = 800$ psia:

$$\psi(800) = 11.466 \times 10^6 + 2\left(\frac{28{,}665 + 56{,}378}{2}\right)(800 - 400)$$

$$= 11.466 \times 10^6 + 34.017 \times 10^6$$

$$= 45.483 \times 10^6 \text{ psia}^2/\text{cp}$$

Proceeding in a similar way, we can construct Table 2–1. These results are plotted in Figure 2–1. This plot is used in the gas well test analysis, in which it is assumed that for high pressure, in excess of 2800 psia, the function is almost linear and can be described by

$$\psi(p) = [0.3218p - 416.85] \text{ mm psia}^2/\text{cp}$$

For low pressure, less than 2800 psia, the function is described by a polynomial equation of the form

$$\psi(p) = A + Bp + Cp^2 + Dp^3 + Ep^4 + Fp^5$$

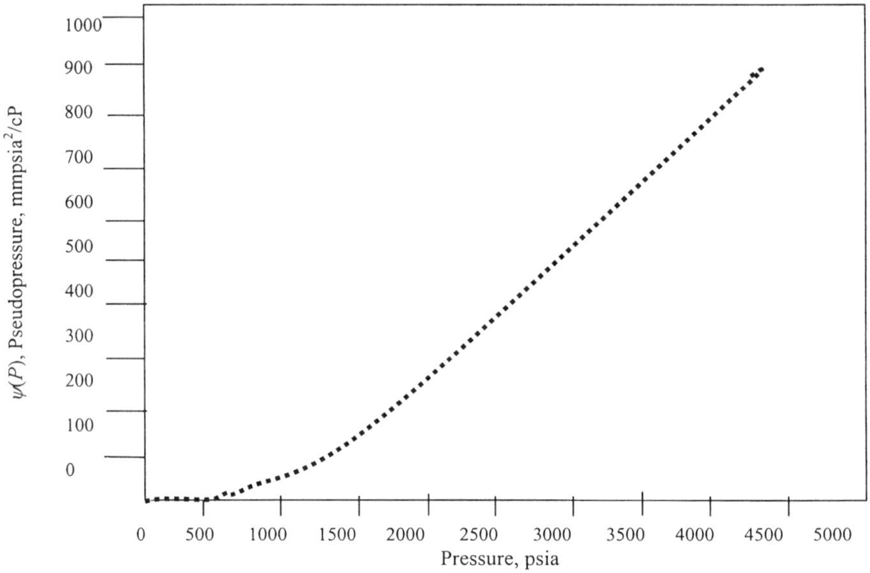

Figure 2–1. Gas pseudopressure $\psi(P)$ versus pressure, psia.

where A, B, C, D, E, and F are polynomial coefficients whose values are

$$A = 39{,}453;\ B = -222.976;\ C = 72.0827$$
$$D = 5.287041\text{E-}04;\ E = -1.993697\text{E-}06;\ \text{and } F = 1.92384\text{E-}10$$

These relationships and the plot can be used to convert from real to pseudopressure and vice versa.

Example 2–2 *Determining Wellbore Pressure Assuming Steady-State Flow Conditions*

Perform this calculation given the following data:

$k = 1.50$ mD, $h = 39$ ft, $q_{sc} = 3900$ mscfd, $p_e = 4625$ psia, $T = 712°$ R, $r_e = 550$ ft, $r_w = 0.333$, $\bar{\mu} = 0.02695$cp, $\gamma_g = 0.759$, $T_{sc} = 520°$R, $P_{sc} = 14.7$ psia.

Solution The solution is iterative since $\bar{z} = f(\bar{p})$, where $\bar{p} = (p_e + p_w)/2$, and p_w is the unknown. As a first estimate, assume $\bar{z} = 1.0$.

First trial using Eq. 2–6:

$$p_w^2 = p_e^2 - \frac{\bar{\mu} T \ln(r_e/r_w) q_{sc} \bar{z}}{0.0007027kh}$$

$$= 4625^2 - \frac{(.02695)(712)(550/.333)(3900) \times \bar{z}}{.0007027(1.5(30)}$$

$$= 2.139 \times 10^7 - 1.756 \times 10^7 (1.0)$$

$$= 3.83 \times 10^6$$

or $p_w = 1957$ psia.

Second trial:

$$\bar{p} = \frac{4625 + 1957}{2} = 3291 \text{ psia}, \ \bar{z} \text{ at } 3291 \text{ psia} = 0.88$$

$$p_w^2 = 2.139 \times 10^7 - 1.756 \times 10^7 (0.88)$$

$$= 5.937 \times 10^6$$

or $p_w = 2436$ psia.

Third trial:

$$\bar{p} = \frac{4625 + 2436}{2} = 3530 \text{ psia}, \ \bar{z} \text{ at } 3530 \text{ psia} = 0.890$$

$$= 2.139 \times 10^7 - 1.756 \times 10^7 (0.89)$$

$$= 5.762 \times 10^6$$

or $p_w = 2400$ psia.

$$\bar{p} = \frac{4625 + 2400}{2} = 3512 \text{ psia and } \bar{z} \text{ at } 3512 \text{ psia} = 0.890$$

Since the value for $\bar{z}$ is the same as for second trial, the solution has converged and the required wellbore pressure is 2400 psia. The solution would have been more complicated if a constant value for μ had not been assumed. The above treatment of steady-state flow assumes no turbulence flow in the formation and no formation or skin damage around the wellbore.

2.3 Steady-State Turbulence Flow

The above treatment of steady-state flow assumes no turbulent flow in the formation and no skin damage around the wellbore. The pressure squared and pseudopressure representations of the steady-state equations including turbulence are

$$p_e^2 - p_w^2 = \frac{50.3 \times 10^6 \mu_g z T P_{sc} q_{sc}}{k h T_{sc}} \left[\ln \frac{r_e}{r_w} + s + D q_{sc} \right] \quad (2\text{–}11)$$

$$\psi(\bar{p}) - \psi(p_w) = \frac{1.422 \times 10^3 T q_{sc}}{kh} \left[\ln \frac{r_e}{r_w} - 0.5 + s + D q_{sc} \right] \quad (2\text{–}12)$$

where Dq_{sc} is interpreted as the rate-dependent skin factor, and

$$D = \frac{5.18 \times 10^{-5} \gamma_g}{\bar{\mu} h r_w k^{0.2}} \beta \quad (2\text{–}13)$$

Expression D is the non-Darcy flow coefficient in psia^2/cP/(mscf/d)^2 and is calculated from Eq. 2–13
where

$$\beta = \frac{2.33 \times 10^3}{k^{1.201}}, \ 1/\text{ft} \quad (2\text{–}14a)$$

or

$$\beta = \frac{2.73 \times 10^{10}}{k^{1.1045}}, \ 1/\text{ft} \quad (2\text{–}14b)$$

where k is the permeability near the wellbore region in mD. Values of the velocity coefficient β for various permeability and porosity can be obtained from Ref. 1 or calculated from Eq. 2–14a or 2–14b. The foregoing equations 2–11 and 2–12 have the forms

$$p_e^2 - p_w^2 = AA' q_{sc} + BB' q_{sc}^2 \quad (2\text{–}11a)$$

where

$$AA' = 50.3 \times 10^3 \frac{\mu_g z T P_{sc}}{k h T_{sc}} [\ln(r_e/r_w) - 0.75 + s] \quad (2\text{–}11b)$$

$$BB' = 50.3 \times 10^3 \frac{\mu_g z T P_{sc}}{k h T_{sc}} D \quad (2\text{–}11c)$$

$$\psi(\bar{p}) - \psi(p_w) = AA q_{sc} + Bb q_{sc}^2 \quad (2\text{–}12a)$$

where

$$AA = \frac{1.422 \times 10^3}{kh}[\ln(r_e/r_w) - .75 + s] \tag{2–12b}$$

$$BB = \frac{1.422 \times 10^3 T}{kh} D \tag{2–12c}$$

Example 2–3 *Calculating Influence of Turbulence in a Vertical Well Using Steady-State Flow Equation*

A vertical gas well is drilled in a 45-ft-thick sandstone reservoir with permeability of 12 mD. The initial reservoir pressure is 2150 psia and well spacing is 640 acres. The well could be operated with a minimum bottomhole pressure of 350 psia. The other data are $T = 590°R$, $\mu_g = 0.02$ cP, $z = 0.90$, $\gamma_g = 0.70$, $r_w = 0.29$ ft, $s' = 0$, perforated length $h_p = 45$ ft.

Use the p^2 equation to calculate the flow rate.

Solution To solve this problem, the Eq. 2–11a has the form

$$p_e^2 - p_w^2 = AA' q_{sc} + BB' q_{sc}^2$$

where

$$AA' = 50.3 \times 10^6 \frac{\mu_g z T P_{sc}}{kh T_{sc}}\left[\ln\left(\frac{r_e}{r_w}\right) - 0.75 + s\right]$$

$$BB' = 50.3 \times 10^6 \frac{\mu_g z T P_{sc}}{kh T_{sc}} D$$

Substituting these parameters in the above equations, we have

$$AA' = 50.3 \times 10^6 \frac{(.02)(.9)(130)(590)(14.7)}{12(45)(520)}\left[\ln\left(\frac{2978}{.29}\right) - 0.75 + 0\right]$$
$$= 237.34$$

The value of BB' can be calculated using the preceding equation:

$$BB' = 50.3 \times 10^6 \frac{(0.02)(0.9)(590)(14.7)}{12(45)(520)} D$$
$$= 0.027965 \times 10^6 D$$

where

$$D = \frac{2.222 \times 10^{-15} \gamma_g kh\beta}{\mu_g r_w h_p^2}$$

and

$$\beta = 2.73 \times 10^{10}\, k^{-1.1045},\ 1/\text{ft}$$

$$= 2.73 \times 10^{10}(12)^{-1.1045} = 1.7547 \times 10^{9}\, 1/\text{ft}$$

$$\therefore D = \frac{2.222 \times 10^{-15}(0.7)(12)(45)}{(0.02)(0.29)(45)(45)}(1.7547 \times 10^{9})$$

$$= 1.255 \times 10^{-4},\ 1/\text{mscfd}$$

Substituting value of D into Eq. 2–11a, BB' is calculated as

$$BB' = 0.027965 \times 10^{3}(1.255 \times 10^{-4}) = 0.351\ 1/\text{mscfd}^2$$

Substituting values of AA' and BB' into Eq. 2–11a:

$$p_e^2 - p_w^2 = 237.34 q_{sc} + 0.351 q_{sc}^2$$

This quadratic equation is rearranged as

$$0.351 q_{sc}^2 + 237.34 q_{sc} - \left(p_e^2 - p_w^2\right) = 0$$

By solving the above quadratic equation the value of q_{sc} is calculated as

$$q_{sc} = \frac{-237.34 + \sqrt{(237.34)^2 + 4(0.351)(p_e^2 - p_w^2)}}{2(0.351)}$$

$$= \frac{-237.34 + \sqrt{56{,}330.271 + 1.404(p_e^2 - p_w^2)}}{0.7020}$$

Calculated values of q_{sc}, both with and without turbulence for various values of p_w, are summarized in Table 2–2. This table indicates a significant effect of turbulence on well productivity.

Table 2–2
Effect of Turbulence on Vertical Well Productivity

P_w (psia)	$p_e^2 - p_w^2$ (psia^2)	No turbulence, $D = 0$ q (mmscfd)	With turbulence q (mmscfd)
1800	138×10^4	5.816	1.673
1400	266×10^4	11.208	2.435
1000	362×10^4	15.252	2.891
500	437×10^4	18.412	3.207

2.4 Pseudo-Steady-State (Finite) Flow

The equations for pseudo-steady-state flow in terms of pressure squared and pseudopressure are:

In terms of pressure-squared treatment:

$$q_{sc} = \frac{0.0007027kh\left(\bar{p}_R^2 - p_w^2\right)}{T\bar{\mu}_g\bar{z}\ \ln(0.472r_e/r_w)} \qquad (2\text{–}15)$$

The effects of skin damage and turbulence are included in Eq. 2–15 as follows:

$$q_{sc} = \frac{0.0007027kh\left(\bar{p}_R^2 - p_w^2\right)}{T\bar{\mu}_g\bar{z}[\ln(0.472r_e/r_w) + s + Dq_{sc}]} \qquad (2\text{–}16)$$

It is frequently necessary to solve Eq. 2–16 for pressure or pressure drop for a known flow rate, q_{sc}.

$$p_R^2 - p_w^2 = \frac{1.422 \times 10^3 T\bar{\mu}_g\bar{z}q_{sc}}{kh}[\ln(0.472r_e/r_w) + s + Dq_{sc}] \qquad (2\text{–}17)$$

Equation 2–17 may be written as follows:

$$\bar{p}_R^2 - p_w^2 = Aq_{sc} + Bq_{sc}^2 \qquad (2\text{–}17a)$$

where

$$A = \frac{1.422 \times 10^3 \bar{\mu}_g\bar{z}T}{kh}\left[\ln\left(\frac{0.472r_e}{r_w}\right) + s\right]$$

and

$$B = \frac{1.422 \times 10^3 \bar{\mu}_g\bar{z}T}{kh}D$$

It is sometimes convenient to establish a relationship between the two parameters that indicate the degree of turbulence occurring in a gas reservoir. These parameters are the velocity coefficient β and the turbulence coefficient D. Equation 2–17a can be written for pseudo-steady-state flow as

$$\bar{p}_R^2 - p_w^2 = 1.422 \times 10^3 \bar{\mu}_g\bar{z}T\left(\ln\frac{0.472r_e}{r_w} + s\right)q_{sc} + \frac{3.161 \times 10^{-12}\gamma_g\bar{z}T\beta}{r_w h^2}q_{sc}^2 \qquad (2\text{–}17b)$$

This form of the equation includes the assumption that $r_e \gg r_w$. Equating the terms and multiplying q_{sc}^2 in Eqs. 2–17a and 2–17b yields

$$\frac{1.422 \times 10^3 \bar{\mu}_g \bar{z} T}{kh} D = \frac{3.161 \times 10^{-12} \gamma_g \bar{z} T}{r_w h^2} \beta$$

or

$$D = \frac{2.22 \times 10^{-15} \gamma_g k}{\bar{\mu}_g h r_w} \beta$$

Expressing β in terms of permeability from Eq. 2–14b, the preceding expression becomes

$$D = \frac{5.18 \times 10^{-5} \gamma_g}{\bar{\mu}_g h r_w k^{0.2}} \tag{2–17c}$$

In terms of pseudopressure treatment:

$$\psi(\bar{p}_R) - \psi(p_w) = A' q_{sc} + B' q_{sc}^2 \tag{2–17d}$$

where

$$A' = \frac{1.422 \times 10^3 T}{kh} \left[\ln\left(\frac{0.472 r_e}{r_w} \right) + s \right]$$

and

$$B' = \frac{1.422 \times 10^3 T}{kh} D$$

It is sometimes convenient to establish a relationship between the two parameters that indicate the degree of turbulence occurring in a gas reservoir. These parameters are the velocity coefficient β and the turbulence coefficient D. Equation 2–17d can be written for pseudo-steady-state flow as

$$\psi(\bar{p}_R) - \psi(p_w) = 1.422 \times 10^3 T \left(\ln \frac{0.472 r_e}{r_w} + s \right) q_{sc} + \frac{3.161 \times 10^{-12} \gamma_g T \beta}{r_w h^2} q_{sc}^2 \tag{2–17e}$$

This form of the equation includes the assumption that $r_e \gg r_w$. Equating the terms and multiplying q_{sc}^2 in Eqs. 2–17d and 2–17e yields

$$\frac{1.422 \times 10^3 \bar{\mu}_g \bar{z} T}{kh} D = \frac{3.161 \times 10^{-12} \gamma_g \bar{z} T}{r_w h^2} \beta$$

or

$$D = \frac{2.22 \times 10^{-15} \gamma_g k}{h r_w} \beta$$

Expressing β in terms of permeability from Eq. 2–14b, the preceding expression becomes

$$D = \frac{5.18 \times 10^{-5} \gamma_g}{h r_w k^{0.2}} \tag{2–17f}$$

2.5 Unsteady-State (Transient) Flow

A well flows in the unsteady-state or transient regime until the pressure disturbance reaches a reservoir boundary or until interference from other wells takes effect. Although the flow capacity of a well is desired for pseudosteady-state or stabilized conditions, much useful information can be obtained from transient tests. This information includes permeability, skin factor, turbulence coefficient, and average reservoir pressure. The procedures are developed on transient testing and the relationship among flow rate, pressure, and time will be presented in this section for various conditions of well performance and reservoir types.

2.6 Gas Radial Diffusivity Equation

By combining an unsteady-state continuity equation with Darcy's law and the gas equation of state, one can derive the diffusivity equation. The equation is

$$\frac{\partial}{\partial x}\left(\frac{k_x \rho}{\mu} \frac{\partial p}{\partial x}\right) = \frac{\partial}{\partial t}(\phi \rho) \tag{2–18}$$

Equation 2–18 can be written in three-dimensional form:

$$\frac{\partial}{\partial x}\left(\frac{k_x \rho}{\mu} \frac{\partial p}{\partial x}\right) + \frac{\partial}{\partial y}\left(\frac{k_y \rho}{\mu} \frac{\partial p}{\partial y}\right) + \frac{\partial}{\partial z}\left(\frac{k_z \rho}{\mu}\left(\frac{\partial p}{\partial z} + \rho\right)\right) = \frac{\partial}{\partial t}(\phi \rho) \tag{2–19}$$

Equation 2–19 represents a general form for the combination of the continuity equation and Darcy's law. The final differential equation, which will result from this equation, depends on the fluid and the equation of state of interest.

For the radial flow case we obtain in a similar manner

$$\frac{1}{r}\frac{\partial}{\partial r}\left(\frac{r \rho k_r}{\mu} \frac{\partial p}{\partial r}\right) = \frac{\partial}{\partial t}(\phi \rho) \tag{2–20}$$

In the case of flow of a nonideal gas, the gas deviation factor z_g is introduced into the equation of state to give

$$\rho = \frac{M}{RT}\frac{\rho}{z_g} \tag{2–21}$$

If we assume laminar flow, neglect gravity, and assume constant rock properties, then Eq. 2–19 becomes, for isothermal conditions,

$$\frac{\partial}{\partial x}\left(\frac{p}{\mu z_g}\frac{\partial p}{\partial x}\right)+\frac{\partial}{\partial y}\left(\frac{p}{\mu z_g}\frac{\partial p}{\partial y}\right)+\frac{\partial}{\partial z}\left(\frac{p}{\mu z_g}\frac{\partial p}{\partial z}\right)=\frac{\phi}{k}\frac{\partial}{\partial t}\left(\frac{p}{z_g}\right) \tag{2–22}$$

For radial flow Eq. 2–22 can be expressed as

$$\frac{1}{r}\frac{\partial}{\partial r}\left(\frac{p}{\mu z_g}r\frac{\partial p}{\partial r}\right)=\frac{\phi}{k}\frac{\partial}{\partial t}\left(\frac{p}{z_g}\right) \tag{2–23}$$

Equation 2–23 in gasfield units is

$$\frac{1}{r}\frac{\partial}{\partial r}\left(\frac{p}{\mu z}r\frac{\partial p}{\partial r}\right)=\frac{\phi}{0.000264}\frac{\partial}{\partial t}\left(\frac{p}{z}\right) \tag{2–24}$$

Equation 2–24 can be modified to account for simultaneous flow of gas, oil, and water; the equation is

$$\frac{1}{r}\frac{\partial}{\partial r}\left(r\frac{\partial p}{\partial z}\right)=\frac{\phi c_t}{0.000264\lambda_t}\frac{\partial p}{\partial t} \tag{2–25}$$

where

z = gas deviation factor
c_t = total system isothermal compressibility, psi^{-1}
λ_t = total mobility

$$c_t = c_g s_g + c_o s_o + c_w s_w c_f \tag{2–26}$$

$$\lambda_t = \frac{k_g}{\mu_g}+\frac{k_o}{\mu_o}+\frac{k_w}{\mu_w} \tag{2–27}$$

2.7 Basic Gas Flow Equations

Gas flow is characterized by Darcy's law and for a gas described by the equation of state:

$$\rho = \frac{M}{RT}\frac{p}{z} \tag{2–28}$$

Equation 2–19 becomes, for constant ϕ and k and negligible gravitational forces,

$$\frac{\partial}{\partial x}\left(\frac{\rho}{\mu z_g}\frac{\partial p}{\partial x}\right)+\frac{\partial}{\partial y}\left(\frac{\rho}{\mu z_g}\frac{\partial p}{\partial y}\right)+\frac{\partial}{z}\left(\frac{\rho}{\mu z_g}\frac{\partial p}{\partial z}\right)$$

$$=\frac{\phi}{0.000264k}\frac{\partial}{\partial t}\left(\frac{p}{z_g}\right) \tag{2–29}$$

Equation 2–29 has a form similar to the following equation:

$$\frac{\partial^2 p}{\partial x^2}+\frac{\partial^2 p}{\partial y^2}+\frac{\partial^2 p}{\partial z^2}=\frac{\phi\mu c}{0.000264k}\frac{\partial p}{\partial t} \tag{2–30}$$

For radial flow, the corresponding equation is

$$\frac{1}{r}\frac{\partial}{\partial r}\left(r\frac{\partial p}{\partial r}\right)=\frac{\phi\mu c}{0.000264k}\frac{\partial p}{\partial t} \tag{2–31}$$

We define a pseudopressure,[1] $\Psi(p)$, as follows:

$$\psi(p)=2\int_{p_0}^{p}\frac{p}{\mu z_g}dp \tag{2–32}$$

where p_0 is a low base pressure, now:

$$\frac{\partial}{\partial t}\left(\frac{p}{z_g}\right)=\frac{d\left(\frac{p}{z_g}\right)}{dp}\frac{\partial p}{\partial t}=\frac{c_g p}{z_g}\frac{\partial p}{\partial t}$$

because

$$c_g=\frac{1}{\rho}\frac{d\rho}{d\rho}=\frac{z_g}{p}\frac{d\left(\frac{p}{z_g}\right)}{dp}$$

and also

$$\frac{\partial\psi}{\partial t}=\frac{\partial\psi}{\partial p}\frac{\partial p}{\partial t}\frac{\partial p}{\partial x}$$

Similar expressions apply for $\frac{\partial\psi}{\partial y}$ and $\frac{\partial\psi}{\partial z}$. Thus, Eq. 2–29 becomes

$$\frac{\partial}{\partial x}\left(\frac{\partial\psi}{\partial x}\right)+\frac{\partial}{\partial y}\left(\frac{\partial\psi}{\partial y}\right)+\frac{\partial}{\partial z}\left(\frac{\partial\psi}{\partial z}\right)=\frac{\phi\mu c_g}{0.000264k}\frac{\partial\psi}{\partial t} \tag{2–33}$$

For radial flow, the equivalent of Eq. 2–33 is

$$\frac{1}{r}\frac{\partial}{\partial r}\left(r\frac{\partial\psi}{\partial r}\right)=\frac{\phi\mu c_g}{0.000264k}\frac{\partial\psi}{\partial t} \tag{2–34}$$

2.8 One-Dimensional Coordinate Systems

Equation 2–29 may be expressed in terms of rectangular, cylindrical, or spherical coordinates:

$$\nabla^2 p=\frac{\phi\mu c}{k}\frac{\partial p}{\partial t} \tag{2–35}$$

where $\nabla^2 p$ is the Laplacian of p. The expression "one-dimensional" refers to a specified coordinate system. For example, one-dimensional flow in the x-direction in rectangular coordinates may be expressed in cylindrical coordinates.

Linear Flow

Flow lines are parallel, and the cross-sectional area of flow is constant and is represented by Eq. 2–36, which is in the rectangular coordinate system and is the one-dimensional form of Eq. 2–35:

$$\frac{\partial^2 p}{\partial x^2}=\frac{\phi\mu c}{k}\frac{\partial p}{\partial t} \tag{2–36}$$

Fractures often exist naturally in the reservoir, and the flow toward the fracture is linear.

Radial Cylindrical Flow

In petroleum engineering the reservoir is often considered to be circular and of constant thickness h, with a well opened over the entire thickness. The flow takes place in the radial direction only. The flow lines converge toward a central point in each point, and the cross-sectional area of flow decreases as the center is approached. Thus flow is directed toward a central line referred to as a line-sink (or line-source in the case of an injection well). In the petroleum literature it is often simply called radial flow in the cylindrical coordinate system and is given by one-dimensional form of Eq. 2–35:

$$\frac{\partial}{r}\frac{\partial}{\partial r}\left(r\frac{\partial p}{\partial r}\right)=\frac{\phi\mu c}{k}\frac{\partial p}{\partial t} \tag{2–37}$$

Radial Spherical Flow

If the well is not opened to the entire production formation because of a thick reservoir (h is very large), then to measure vertical permeability, the one-dimensional form of Eq. 2–35, in the spherical coordinate system, is of interest. It is known as the radial–spherical flow equation and is given by

$$\frac{\partial}{r^2}\frac{\partial}{\partial r}\left(r\frac{\partial p}{\partial r}\right) = \frac{\phi\mu c}{k}\frac{\partial p}{\partial t} \tag{2–38}$$

2.9 Radial Gas Flow Equations in Dimensionless Variables and Groups

Equation 2–35 and the relevant boundary conditions in dimensionless terms are:

$$\nabla^2(\Delta p_D) = \frac{\partial}{\partial t_D}(\Delta p_D) \tag{2–39}$$

where the subscript D means dimensionless, and the dimensionless terms are defined in the next section for various modes of flow.

Pressure Treatment

The pressure case will be considered along with the boundary and initial conditions. Assuming a well is producing at a constant rate q_g from an infinite reservoir, the equation governing flow is

$$\frac{\partial}{r}\frac{\partial}{\partial r}\left(r\frac{\partial p}{\partial r}\right) = \frac{\phi\mu c}{k}\frac{\partial p}{\partial t} \tag{2–40}$$

with the following boundary and initial conditions:

Inner Boundary Condition:

Assuming at the wellbore, the flow rate is constant and from Darcy's law,

$$\left.\frac{q}{2\pi rh}\right|_{well} = \left.\frac{k}{\mu}\frac{\partial p}{\partial r}\right|_{well} \quad \text{for } t > 0 \tag{2–41}$$

That is,

$$\left. r\frac{\partial p}{\partial r}\right|_{well} = \frac{q\mu}{2\pi kh} \tag{2–42}$$

and in terms of standard conditions,

$$r\frac{\partial p}{\partial r}\bigg|_{well} = \frac{q_{sc}\mu}{2\pi kh}\frac{P_{sc}T\bar{z}}{\bar{p}T_{sc}} \tag{2–43}$$

Outer Boundary Condition:

At all times, the pressure at the outer boundary (radius = infinity) is the same as the initial pressure, p_i, that is,

$$p \rightarrow p_i \quad as\, r \rightarrow \infty$$

for all t.

Initial Condition

Initially, the pressure throughout the reservoir is constant, that is,

$$p = p_i \quad at\, t = 0$$

for all t.

At this stage, the variables which affect the solution of Eq. 2–40 are p, p_I, r, r_w, q_{sc}, μ_g, k, h, ϕ, c, and t. Let

$$\Delta p = p_i - p$$
$$r_D = \frac{r}{r_w}\text{(dimensionless)}$$

$$\Delta p'_D = \frac{p_i - p}{p_i}$$

Then Eq. 2–43 becomes

$$r_D\frac{\partial}{\partial r_D}(\Delta p'_D)\bigg|_{r_D=1} = \frac{-q_{sc}\mu_g P_{sc}T\bar{z}}{p_i 2\pi kh\bar{p}}T_{sc} \tag{2–44}$$

Let the dimensionless flow rate be

$$q_D = \frac{q_{sc}\mu P_{sc}T\bar{z}}{p_i 2\pi kh\bar{p}T_{sc}}$$

Equation 2–44 becomes

$$r_D\frac{\partial}{\partial r_D}\left[\frac{(\Delta p'_D)}{q_D}\right]\bigg|_{r_D=1} = -1 \tag{2–45}$$

Let the dimensionless pressure drop be

$$\Delta p_D = \frac{(\Delta p'_D)}{q_D} = \frac{p_i - p}{p_i q_D}$$

Then Eq. 2–45 becomes

$$r_D \frac{\partial}{\partial r_D}(\Delta p_D)\bigg|_{r_D=1} = -1$$

Equation 2–37 becomes

$$\frac{1}{r_D}\frac{\partial}{\partial r_D}\left[r_D \frac{\partial}{\partial r_D}(\Delta p_D)\right] = \frac{\phi \mu c r_w^2}{k}\frac{\partial}{\partial t}(\Delta p_D) \tag{2–46}$$

Let dimensionless time be

$$t_D = \frac{kt}{\phi \mu c r_w^2}$$

Equation 2–37, the radial cylindrical flow equation, may now be expressed in dimensionless terms by

$$\frac{1}{r_D}\frac{\partial}{\partial r_D}\left[r_D \frac{\partial}{\partial r_D}(\Delta p_D)\right] = \frac{\partial}{\partial t_D}(\Delta p_D) \tag{2–47}$$

with the boundary and initial conditions as follows:

1. $r_D \frac{\partial}{\partial r_D}(\Delta p_D)\Big|_{r_D=1} = -1 \quad \text{for } t_D > 0$
2. $\Delta p_D \rightarrow 0 \quad as\ r_D \rightarrow \infty \quad \text{for all } t_D$
3. $\Delta p_D = 0 \quad at\ t_D = 0 \quad \text{for all } r_D$

The solution of Eq. 2–47, which is the dimensionless form of Eq. 2–40, now involves only Δp_D, t_D, and r_D. The dimensionless terms in terms of pressure treatment case are defined in gasfield units as follows:

$$t_D = \frac{0.0002637kt}{\phi \bar{\mu}_g \bar{c} r_w^2} \tag{2–48}$$

$$\Delta p_D = \frac{p_i - p}{p_i q_D}, \tag{2–49}$$

and

$$q_D = \frac{7.085 \times 10^5 q_{sc} \bar{\mu}_g T \bar{z}}{\bar{p} k h p_i} \tag{2–50}$$

where k = formation permeability, mD; t = time, hours; ϕ = porosity, fraction; $\bar{\mu}_g$ = average gas viscosity, cP; T = reservoir temperature, °R; $\bar{z}$ = gas compressibility factor at average pressure; ΔP_D = dimensionless average reservoir pressure, psia; p_i = initial reservoir pressure, psia; h = reservoir thickness, ft; q_{sc} = gas flow rate, mmscfd; T_{sc} = base temperature, °R; P_{sc} = base pressure, psia; and $\bar{c}$ = gas compressibility, psi^{-1}.

Pressure Squared Treatment

Dimensionless variables in terms of pressure squared treatment are defined in gasfield units as follows:

$$t_D = \frac{0.0002637kt}{\phi\bar{\mu}_g\bar{c}r_w^2} \tag{2–51}$$

$$p_D = \frac{p_i^2 - p^2}{p_i^2 q_D} \tag{2–52}$$

and

$$q_D = \frac{1.417 \times 10^6\ \bar{z}Tq_{sc}\bar{\mu}_g}{khp_i^2} \tag{2–53}$$

Pseudopressure Treatment

Dimensionless variables in terms of pseudopressure treatment are defined in gasfield units as follows:

$$t_D = \frac{0.0002637kt}{\phi\bar{\mu}_g\bar{c}r_w^2} \tag{2–54}$$

$$\Delta p_D = \frac{\psi_i - \psi_{wf}}{\psi_i q_D} \tag{2–55}$$

and

$$q_D = \frac{1.417 \times 10^6 Tq_{sc}}{kh\psi_i} \tag{2–56}$$

Example 2–4 *Calculating Dimensionless Quantities Using p, p^2, and $\psi(p)$ Treatment*

A gas reservoir was produced at a constant rate q_{sc} of 6.5 mmscfd for a time, t, of 36 hours. The sandface pressure, p_{wf}, at that time was 1750 psia. General data are as follows:

$\bar{p} = 1925$ psia, $p_i = 2100$ psia, $z_I = 0.842$, $z_i = 0.849$, $z_{1750} = 0.855$, $c_i = 0.000525$ psi^{-1}, $c_{1750} = 0.000571$ psi^{-1}, $\bar{c} = 0.000548$ psi^{-1}, $k =$ 18.85 mD, $T = 595°$R, $r_w = 0.39$ ft, $\mu_i = 0.01495$ cp, $\bar{\mu} = 0.01430$ cp, $\mu_{1,750} = 0.01365$ cp, $h = 40$ ft, and $\phi = 0.138$ fraction.

Calculate the dimensionless quantities t_D, P_D, and q_D using the p, p^2, and ψ treatments.

Solution Pressure treatment, p, from Eq. 2–48:

$$t_D = \frac{0.0002637kt}{\phi\bar{\mu}\bar{c}r_w^2}$$

$$\therefore t_D = \frac{0.0002637(18.85)(36)}{(0.138)(0.01430)(0.000548)(0.39)^2} = 1{,}087{,}925$$

From Eq. 2–50:

$$q_D = \frac{7.085 \times 10^5 q_{sc}\bar{\mu}T\bar{z}}{\bar{p}khp_i}$$

$$\therefore q_D = \frac{7.085 \times 10^5(6.5)(0.0143)(595)(0.849)}{(1925)(18.85)(40)(2100)} = 0.010914$$

From Eq. 2–49:

$$\therefore \Delta p_D = \frac{p_i - p}{p_i q_D} = \frac{2100 - 1750}{2100(0.010914)} = \frac{350}{22.92} = 15.27$$

Pressure-squared treatment, p^2, from Eq. 2–51:

$$t_D = \frac{0.0002637kt}{\phi\bar{\mu}\bar{c}r_w^2}$$

$$\therefore t_D = \frac{0.0002637(18.85)(36)}{(0.138)(0.01430)(0.000548)(0.39)^2} = 1{,}087{,}925$$

From Eq. 2–53:

$$q_D = \frac{1.417 \times 10^6 \bar{z}Tq_{sc}\bar{\mu}}{khp_i^2}$$

$$= \frac{1.417 \times 10^6(0.849)(595)(6.5)(0.0143)}{(18.85)(40)(2100)^2} = 0.020010$$

From Eq. 2–52:

$$\Delta p_D = \frac{p_i^2 - p^2}{p_i^2 q_D}$$

$$= \frac{2100^2 - 1.750^2}{2100^2(0.020010)} = 15.27$$

Pseudopressure treatment, ψ, from Eq. 2–54:

$$t_D = \frac{0.0002637kt}{\phi\bar{\mu}\bar{c}r_w^2}$$

$$\therefore t_D = \frac{0.0002637(18.85)(36)}{(0.138)(0.01430)(0.000548)(0.39)^2} = 1{,}087{,}925$$

From Eq. 2–56:

$$q_D = \frac{1.417 \times 10^6 T q_{sc}}{kh\psi_i}$$

$$p_I = 2100 \text{ psia} \leftrightarrow \psi_i = 335 \text{ mmpsia}^2/\text{cp}$$

$$\therefore q_D = \frac{1.417 \times 10^6(595)(6.5)}{(18.85)(40)(335 \times 10^6)} = 0.021696$$

From Eq. 2–55:

$$\Delta p_D = \frac{\psi_i - \psi_{wf}}{\psi_i q_D}$$

$$p = 1{,}750 \text{ psia} \leftrightarrow \psi(p) = 223 \text{ mmpsia}^2/\text{cp}$$

$$\therefore \Delta p_D = \frac{(335 - 223)10^6}{335 \times 10^6(0.021696)} = 15.41$$

Calculating Gas-Pseudopressure $\psi(p)$ Function

Accuracy of gas well test analysis can be improved in some cases if the pseudopressure $\psi(p)$ is used instead of approximations written in terms of pressure or pressure squared. In this section, we discuss the calculations of pseudopressure. Detailed discussion, including systematic development of working equations and application to drawdown, buildup, and deliverability tests, is provided in Ref. 2. The applications of real gas pseudopressure $\psi(p)$ to flow in gas wells under practical conditions are as follows:

1. When turbulence is not present, the drawdown test provides accurate results. When turbulence is significant, the drawdown test can be misleading.
2. The buildup test can be interpreted accurately even with extreme turbulence.
3. The use of a p^2 well-test plot is usually equivalent to the $\Delta(p)$ method, when well pressures are below 2000 psi.
4. Flow capacity can be determined accurately from $(p)^2$ or p well-test plots if point values, rather than average values, are used for slopes and gas properties.

Calculation of Pseudopressure

Gas pseudopressure, $\psi(p)$, is defined by the integral

$$\psi(p) = 2 \int_{p_{BASE=0}}^{P} \frac{p}{\mu z} dp \qquad (2\text{–}57)$$

An example will illustrate this calculation.

Example 2–5 *Calculating Gas Pseudopressure*

Given data are gas gravity = 0.7, T = 200°F. Gas properties as functions of pressure are given in Table 2–3.

Solution Use the trapezoidal rule for numerical integration.

For p = 150 psia,

$$\psi(150) = 2 \int_{p_{base}}^{p} \frac{p}{\mu z} dp = 2 \frac{\left[\left(\frac{p}{\mu z}\right)_0 + \left(\frac{p}{\mu z}\right)_{150}\right]}{2} (150 - 0)$$

$$= 2 \frac{[0 + 12{,}290]}{2} (150) = 1.844 \times 10^6 \text{ psia}^2/\text{cp}$$

Table 2–3
Gas Properties as Functions of Pressure

Pressure P (psia)	Gas viscosity (cP)	Compressibility factor z	$p/\mu z$ (psia/cP)
150	0.01238	0.9856	12,290
300	0.01254	0.9717	24,620
450	0.01274	0.9582	36,860

For $p = 300$ psia,

$$\psi(300) = 1.844 \times 10^6 + 2\frac{\left[\left(\frac{p}{\mu z}\right)_{150} + \left(\frac{p}{\mu z}\right)_{300}\right]}{2}(300 - 150)$$

$$= 1.844 \times 10^6 + 2\frac{(12{,}290 + 24{,}620)}{2}(300 - 150)$$

$$= 7.381 \times 10^6 \text{ psia}^2/\text{cp}$$

2.10 Analytical Solutions of Gas Flow Equations

Radial flow geometry is of greatest interest in gas well testing. This radial flow equation was developed in terms of dimensionless variables in previous sections. It is Eq. 2–47 and is repeated below.

$$\frac{1}{r_D}\frac{\partial}{\partial r_D}\left[r_D\frac{\partial}{\partial r_D}(\Delta p_D)\right] = \frac{\partial}{\partial t_D}(\Delta p_D) \tag{2–58}$$

Equation 2–58 can be solved for pressure as a function of flow rate and time. Solutions to Eq. 2–47 depend on the reservoir type, the boundary and initial conditions. Direct analytical solutions will be presented in this section.

Constant Production Rate, Radial Cylindrical Flow, Infinite-Acting Reservoir (Transient)

The Eq. 2–58 is reduced to an ordinary differential equation by applying the Boltzmann transformation $X = r_D^2/(4t_D)$. This is then solved by separating the variables and integrating with the above three conditions. The equation form of the solution is

$$\Delta p_D = -0.5E_i\left(-\frac{r_D^2}{4t_D}\right) \tag{2–59}$$

or

$$\Delta p_D = -0.5E_i\left(-\frac{\phi\mu c r^2}{0.0002637kt}\right) \tag{2–60}$$

Values of Δp_D versus t_D can be found in Ref. 5 for various reservoir sizes, that is, for various values of r_D. E_i is the exponential integral and is defined by

$$E_i(-x) = \int_x^\infty \frac{e^{-u}du}{u} = \ln(1.781) - \frac{x}{1!} + \frac{x^2}{2 \times 2!} - \frac{x^3}{3 \times 3!} + \frac{x^4}{4 \times 4!} \cdots + \frac{(-x)^n}{n \times n!} \qquad (2\text{–}61)$$

For values of x less than 0.02, Eq. 2–62 can approximate the exponential integral with an error of less than 0.6:

$$E_i(-x) = \ln(1.781x) \quad \text{for } x < 0.02 \qquad (2\text{–}62)$$

For computing pressures at the borehole such as drawdown pressures or buildup pressures Eq. 2–61 may be used. However, if practical units are used and logarithms to the base 10 are used, constants for Eq. 2–62 must be evaluated. Darcy units apply to Eq. 2–62. Table 2–4 lists Darcy units and practical units.

For $x \geq 10.9$ the exponential integral is closely approximated by zero. To evaluate the E_i function, we can use Table 2–5 for $0.02 < x$ 10.9.

Thus Eq. 2–59 becomes

$$p_D = 0.5 \ \ln\left(\frac{4t_D}{1.781r_D^2}\right) \quad \text{for } \frac{4t_D}{r_D^2} > 100 \qquad (2\text{–}59\text{a})$$

$$p_D = 0.5\left[\ln\left(\frac{t_D}{r_D^2}\right) + 0.80907\right] \quad \text{for } \frac{t_D}{r_D^2} > 25 \qquad (2\text{–}63)$$

Table 2–4
Darcy and Practical Units for Parameters in the Exponential Solution of the Diffusivity Equation

Parameter or variables	Darcy units	Practical units
C	vol/vol/atm	vol/vol/psi
ϕ	Porosity	Porosity
h	cm	ft
K	Darcy	Millidarcies
μ	Centipoise	Centipoise

Table 2–5
Values of the Exponential Integral, $-E_i(-x)$ (after Lee, © SPE, *Well Testing*, 1982)[5]

$-E_i(-x), 0.000 < 0.209$, interval $-$ 0.001

X	0	1	2	3	4	5	6	7	8	9
0.00	∞	6.332	5.639	5.235	4.948	4.726	4.545	4.392	4.259	4.142
0.01	4.038	3.944	3.858	3.779	3.705	3.637	3.574	3.514	3.458	3.405
0.02	3.355	3.307	3.261	3.218	3.176	3.137	3.098	3.062	3.026	2.992
0.03	2.959	2.927	2.897	2.867	2.838	2.810	2.783	2.756	2.731	2.706
0.04	2.681	2.658	2.634	2.612	2.590	2.568	2.547	2.527	2.507	2.487
0.05	2.468	2.449	2.431	2.413	2.395	2.378	2.360	2.344	2.327	2.311
0.06	2.295	2.280	2.265	2.249	2.235	2.220	2.206	2.192	2.178	2.164
0.07	2.251	2.138	2.125	2.112	2.099	2.087	2.074	2.062	2.050	2.039
0.08	2.027	2.016	2.004	1.993	1.982	1.971	1.960	1.950	1.939	1.929
0.09	1.919	1.909	1.899	1.889	1.879	1.870	1.860	1.851	1.841	1.832
0.10	1.823	1.814	1.805	1.796	1.788	1.770	1.770	1.762	1.754	1.745
0.11	1.737	1.729	1.721	1.713	1.705	1.697	1.690	1.682	1.675	1.667
0.12	1.660	1.652	1.645	1.638	1.631	1.623	1.616	1.609	1.603	1.696
0.13	1.589	1.582	1.576	1.569	1.562	1.556	1.549	1.543	1.537	1.530
0.14	1.524	1.518	1.512	1.506	1.500	1.494	1.488	1.482	1.476	1.470
0.15	1.465	1.459	1.453	1.448	1.442	1.436	1.431	1.425	1.420	1.415
0.16	1.409	1.404	1.399	1.393	1.388	1.383	1.378	1.373	1.368	1.363
0.17	1.358	1.353	1.348	1.343	1.338	1.333	1.329	1.324	1.319	1.315
0.18	1.310	1.305	1.301	1.296	1.292	1.287	1.283	1.278	1.274	1.269
0.19	1.265	1.261	1.256	1.252	1.248	1.244	1.239	1.235	1.231	1.227
0.20	1.223	1.219	1.215	1.211	1.207	1.203	1.199	1.195	1.191	1.187

$-E_i(-x), 0.00 < x < 2.09$, interval $=$ 0.01

0.0	∞	4.0380	3.3548	2.9592	2.6813	2.4680	2.2954	2.1509	2.0270	1.9188
0.1	1.8230	1.7372	1.6596	1.5890	1.5242	1.4645	1.4092	1.3578	1.3099	1.2649
0.2	1.2227	1.1830	1.1454	1.1099	1.0763	1.0443	1.0139	0.9850	0.9574	0.9310
0.3	0.9057	0.8816	0.8584	0.8362	0.8148	0.7943	0.7745	0.7555	0.7372	0.7195
0.4	0.7024	0.6860	0.6701	0.6547	0.6398	0.6354	0.6114	0.5979	0.5848	0.5721
0.5	0.5598	0.5479	0.5363	0.5350	0.5141	0.5034	0.4931	0.4830	0.4732	0.5721
0.6	0.4544	0.4454	0.4366	0.4281	0.4197	0.4116	0.4036	0.3959	0.3884	0.3810
0.7	0.3738	0.3668	0.3600	0.3533	0.3468	0.3404	0.3342	0.3281	0.3221	0.3163
0.8	0.3107	0.3051	0.2997	0.2944	0.2892	0.2841	0.2791	0.2742	0.2695	0.2648
0.9	0.2602	0.2558	0.2514	0.2471	0.2429	0.2388	0.2348	0.2308	0.2270	0.2232
1.0	0.2194	0.2158	0.2122	0.2087	0.2053	0.2019	0.1986	0.1954	0.1922	0.1891
1.1	0.1861	0.1831	0.1801	0.1772	0.1744	0.1716	0.1689	0.1662	0.1636	0.1610
1.2	0.1585	0.1560	0.1536	0.1512	0.1488	0.1465	0.1442	0.1420	0.1398	0.1377
1.3	0.1355	0.1335	0.1314	0.1294	0.1274	0.1255	0.1236	0.1217	0.1199	0.1181
1.4	0.1163	0.1146	0.1129	0.1112	0.1095	0.1079	0.1063	0.1047	0.1032	0.1016

Table 2–5 (Continued)

1.5	0.1002	0.0987	0.0972	0.0958	0.0944	0.0930	0.0917	0.0904	0.0890	0.0878
1.6	0.0865	0.0852	0.0840	0.0828	0.0816	0.0805	0.0793	0.0782	0.0771	0.0760
1.7	0.0749	0.0738	0.0728	0.0718	0.0708	0.0698	0.0679	0.0669	0.0669	0.0660
1.8	0.0651	0.0642	0.0633	0.0624	0.0616	0.0607	0.0599	0.0591	0.0583	0/0575
1.9	0.0567	0.0559	0.0552	0.0545	0.0537	0.0530	0.0523	0.0516	0.0509	0.0503
2.0	0.0496	0.0490	0.0483	0.0477	0.0471	0.0465	0.0459	0.0453	0.0448	0.0432
$2.0 < x < 10.9$, interval = 0.1										
2	4.89	4.26	3.72	3.25	2.84	2.49	2.19	1.92	1.69	1.48
3	1.30	1.15	1.01	8.94	7.89	6.87	6.16	5.45	4.82	4.27
4	3.78	3.35	2.97	2.64	2.34	2.07	1.84	1.64	1.45	1.29
5	1.15	1.02	9.08	8.09	7.19	6.41	5.71	5.09	4.53	4.04
6	3.60	3.21	2.86	2.55	2.28	2.03	1.82	1.62	1.45	1.29
7	1.15	1.03	9.22	8.24	7.36	6.58	5.89	5.26	4.71	4.21
8	3.77	3.37	3.02	2.70	2.42	2.16	1.94	1.73	1.55	1.39
9	1.24	1.11	9.99	8.95	8.02	7.18	6.44	5.77	5.17	4.64
10	4.15	3.73	3.34	3.00	2.68	2.41	2.16	1.94	1.74	1.56×60^{-6}

Δp_D varies with the boundary conditions, but for the case of constant productivity rate from an infinite-acting reservoir, Δp_D is given by

$$\Delta p_D = -0.5E_i\left(-\frac{1}{4t_D}\right) \tag{2–64}$$

When $r = r_w$, $r_D = 1$. In terms of the logarithmic approximation, from Eq. 2–63

$$\Delta p_D = 0.5\,(\ln t_D + 0.809) \quad \text{for } t_D > 25 \tag{2–65}$$

It is evident that p_D for an infinite-acting reservoir is identical to the $r_D = 1$ curve for p_D, is expressed in dimensionless terms, and is the value at the well without inertial-turbulent and skin effects.[1] The effects of skin inertial-turbulent flow are treated earlier.

Example 2–6 *Calculating Flowing Pressure at the Well due to Laminar Flow in an Infinite-Acting Reservoir Using p, p^2, and Pseudopressure Treatments.*

Using the following data, calculate the pressure at the well after a flowing time of 24 hours using p, p^2, and ψ treatment. Given data are $h = 40$ ft, $k = 20$ mD, $p_i = 2000$ psia, $r_w = 0.399$ ft, $T = 580°$R, $q_{sc} = 7.0$ mmscfd, $\phi = 0.16$, $\bar{z} = 0.850$, $\bar{\mu} = 0.0152$ cP, $\bar{c} = 0.00061$ psi^{-1}.

Solution Pressure treatment:

From Eq. 2–54:

$$t_D = \frac{0.0002637kt}{\phi\bar{\mu}\bar{c}r_w^2}$$

$$= \frac{0.0002637(20)(24)}{0.16(0.0152)(0.00061)(0.399)^2} = 535{,}935$$

From Eq. 2–65, since $t_D > 25$:

$$\therefore \Delta p_D = 0.5(\ln t_D + 0.809)$$

$$= 0.5(\ln(535{,}935) + 0.809)) = 7.00$$

The value of Δp_D can also be obtained from Ref. 5, $r_D = 1.0$ curve.

First trial:

Assume

$$\bar{p} = p_i = 2000 \text{ psia}$$

From Eq. 2–50:

$$q_D = \frac{7.085 \times 10^5 \bar{z} T q_{sc}\bar{\mu}}{\bar{p}khp_i}$$

$$= \frac{7.085 \times 10^5(0.85)(580)(7.0)(0.0152)}{(2000)(20)(40)(2000)} = 0.01161$$

Using Eq. 2–49:

$$\Delta p_D = \frac{p_i - p}{p_i q_D}$$

$$p = p_i - p_i \Delta p_D q_D = 2000 - 2000(0.01161)(7.00)$$

$$= 2000 - 163 = 1837 \text{ psia}$$

Second trial:

Assume

$$\bar{p} = \frac{p_i + p}{2} = \frac{2000 + 1837}{2} = 1919 \text{ psia}$$

From Eq. 2–50:

$$q_D = \frac{7.085 \times 10^5(0.85)(580)(7.0)(0.0152)}{1919(20)(40)(2000)} = 0.01210$$

or

$$p = 2000 - 2000(7.0)(0.01210) = 1831 \text{ psia}$$

Third trial:
Assume

$$\bar{p} = \frac{p_i + p}{2} = \frac{2000 + 1831}{2} = 1916 \text{ psia}$$

$$q_D = \frac{7.085 \times 10^5 (0.85)(0.0152)(580)(7.0)}{1916(20)(40)(2000)} = 0.01212$$

or

$$p = 2000 - 2000(7.0)(0.01212) = 1830 \text{ psia}$$

Pressure-squared treatment:
Assuming $\bar{\mu}$, $\bar{z}$, and $\bar{c}$ are constants, therefore, using Eqs. 2–65 and 2–53:
From Eq. 2–53:

$$q_D = \frac{1.417 \times 10^6 \bar{z} T q_{sc} \bar{\mu}}{khp_i^2}$$

$$= \frac{1.417 \times 10^6 (0.85)(580)(7.00)(0.0152)}{(20)(40)(2000)^2} = 0.02323$$

From Eq. 2–52:

$$\Delta p_D = \frac{p_i^2 - p^2}{p_i^2 q_D}$$

or

$$p = \sqrt{p_i^2 - p_i^2 \Delta p_D q_D} = [2000^2 - 2000^2 (7.00)(0.02323)]^{0.5}$$

$$= 1830 \text{ psia}$$

(the same as the results from the pressure treatment).

Pseudopressure treatment:

The values of z_i, μ, and c_i are calculated at p_i; therefore

$$\psi_i = 329.6 \text{ mmpsia}^2/\text{cP}, \quad z_i = 0.84, \quad \mu_i = 0.0156 \text{ cP},$$
$$c_i = 0.00058 \text{ psi}^{-1}$$

From Eq. 2–54:

$$t_D = \frac{0.0002637kt}{\phi\mu_i c_i r_w^2}$$

$$= \frac{0.0002637(20)(24)}{(0.16)(0.0156)(0.00058)(0.399)^2} = 549{,}203$$

Since $t_D > 25$ and using Eq. 2–65:

$$\Delta p_D = 0.5(\ln t_D + 0.809)$$

$$= 0.5[\ln(549{,}203) + 0.809] = 7.013$$

From Eq. 2–56:

$$\Delta p_D = \frac{1.417 \times 10^6\, T q_{sc}}{kh\psi_i}$$

$$= \frac{1.417 \times 10^6(580)(7.0)}{(20)(40)(329.6 \times 10^6)} = 0.02182$$

From Eq. 2–55:

$$\Delta p_D = \frac{\psi_i - \psi_{wf}}{\psi_i q_D}$$

Therefore:

$$\psi_{wf} = \psi_i - \psi_i q_D \Delta p_D$$

$$= 329.6 \times 10^6 - 329.6 \times 10^6(0.02182)(7.013)$$

$$= 279.16 \text{ mmpsia}^2/\text{cP} = 1818 \text{ psia}$$

The values of p_{wf} calculated by the p, p^2, and ψ treatments are 1830, 1830, and 1818 psi respectively.

Example 2–7 *Calculating Flowing Pressure away from the Well due to Laminar Flow in an Infinite-Acting Reservoir Using p, p^2, and Pseudopressure Treatments*

A gas well is situated in an infinite-acting reservoir. Calculate the flowing pressure, due to laminar flow, at a radius of 100 feet from the well, after 24 hours of production. Reservoir and well data are as follows:

$p_i = 2000$ psia, $\psi_i = 329.6$ mmpsia^2/cP, $z_i = 0.835$, $\mu_i = 0.0159$ cP, $c_i = 0.00055$ psia^{-1}, $r = 50$ ft, $r_w = 0.33$ ft, $\phi = 0.15$, $k = 20$ mD, $t = 24$ hours, $q_{sc} = 7.50$ mmscfd, $T = 580°$R, $h = 40$ ft.

Solution From Eq. 2–54:

$$t_D = \frac{0.0002637kt}{\phi\mu_i c_i r_w^2}$$

$$= \frac{0.0002637(20)(24)}{(0.15)(0.0159)(0.00055)(0.33)^2} = 886{,}079$$

$$r_D = \frac{r}{r_w} = \frac{50}{.33} = 152$$

$$\therefore \frac{t_D}{r_D^2} = \frac{886{,}079}{152^2} = 38.35$$

Since $\frac{t_D}{r_D^2} > 25$ and using Eq. 2–63:

$$\Delta p_D = 0.5\left[\ln\left(\frac{t_D}{r_D^2}\right) + 0.80907\right]$$

$$= 0.5[\ln(38.35) + 0.80907] = 2.228$$

From Eq. 2–56:

$$q_D = \frac{1.417 \times 10^6 T q_{sc}}{kh\psi_i}$$

$$= \frac{1.417 \times 10^6(580)(7.50)}{(20)(40)(329.6 \times 10^6)} = 0.02338$$

From Eq. 2–55:

$$\Delta p_D = \frac{\psi_i - \psi_{wf}}{\psi_i - q_D}$$

$$\therefore \psi_{wf} = \psi_i - \psi_i \Delta p_D q_D$$

$$= 329.6 \times 10^6 - 329.6 \times 10^6(2.228)(0.02338)$$

$$= 327.88 \text{ mmpsia}^2/\text{cP}$$

Using the ψ-p curve, $p_{wf} = 1942$ psia.

Radial–Cylindrical Flow, Finite Reservoir, Constant Production Rate, with No Flow at Outer Boundary (Pseudo-Steady-State)

Equation 2–58 can be written as follows:

$$\frac{\partial^2}{\partial r_D^2}(\Delta p_D) + \frac{1}{r_D}\frac{\partial}{\partial r_D}(\Delta p_D) = \frac{\partial}{\partial t_D}(\Delta p_D) \tag{2–66}$$

Using Laplace transform[15] and Bessel functions, Δp_D, which is the solution at the well, is obtained as follows.

For values of $t_D < 0.25\ r_{eD}^2$:

$$\Delta p_D = 0.5\ \ln(t_D + 0.80907) \tag{2–67}$$

For $\frac{t_D}{r_{eD}^2} > 0.25$: the equation of the form solution is

$$\Delta p_D = \frac{2t_D}{r_{eD}^2} + \ln(0.472\ r_{eD}) \tag{2–68}$$

where

$$r_{eD} = \frac{r_e}{r_w}$$

Values of Δp_D versus t_D can be found in Ref. 5 for various reservoir sizes. At early times the solution is represented by Eq. 2–61 and for large times and where $r_w \ll r_e$, the solution at the well is given by Eq. 2–68. The transition from infinite to finite behavior occurs at

$$t_D \approx 0.25\ r_{eD}^2 \tag{2–68a}$$

Example 2–8 *Calculating Flowing Sandface Pressure in Finite-Acting (Closed) Reservoir*

A gas well in a finite-acting (closed) reservoir ($r_e = 1850$ ft) was produced at a constant rate of 7.5 mmscfd. Assuming gas composition, reservoir, and well data pertinent to the test are the same as in Example 2–1, calculate the flowing sandface pressure, p_{wf}, after 80 days of production.

Solution Since the gas is the same as that of Example 2–1, the ψ–p curve already constructed for Figure 2–1 is applicable to this problem.

$$t = 80 \times 24 = 1920 \text{ hours}$$

From Eq. 2–54:

$$t_D = \frac{0.0002637kt}{\phi\mu_i c_i r_w^2}$$

$$= \frac{0.0002637(20)(1{,}920)}{(0.15)(0.0159)(0.00055)(0.33)^2} = 70{,}886{,}315$$

From Eq. 2–56:

$$q_D = \frac{1.417 \times 10^6 T q_{sc}}{kh\psi_i}$$

$$= \frac{1.417 \times 10^6(580)(7.5)}{(20)(40)(329.6 \times 10^6)} = 0.02338$$

$$r_{eD} = \frac{r_e}{r_w} = \frac{1850}{.33} = 5606$$

$$r_{eD}^2 = 5606^2 = 31{,}427{,}236$$

$$\therefore t_D/r_{eD}^2 = \frac{70{,}886{,}315}{31{,}427{,}236} = 2.256$$

Since $t_D/r_{eD}^2 > 0.25$, Δp_D is given by Eq. 2–68:

$$\Delta p_D = \frac{2t_D}{r_{eD}^2} + \ln(0.472\, r_{eD})$$

$$= 2(2.256) + \ln(0.472 \times 5606)$$

$$= 12.392$$

$$\therefore \psi_{wf} = \psi_i - \psi_i \Delta p_D q_D$$

$$= 329.6 \times 10^6 - 329.6 \times 10^6(12.392)(0.02338)$$

$$= 234.11 \text{ mmpsia}^2/\text{cP}$$

$$p_{wf} = 1790 \text{ psia} \quad \text{(Figure 2–1)}$$

The transition from infinite to finite behavior occurs at

$$t = \frac{0.25 r_{eD}^2 \phi\mu_i c_i r_w^2}{0.0002637k}$$

$$= \frac{0.25(31{,}427{,}236)(0.15)(0.0159)(0.00055)(0.33)^2}{0.0002637(20)} = 212.8 \text{ hours}$$

Radial–Cylindrical Flow, Finite Circular Reservoir, Constant Production Rate with Constant Pressure at Outer Boundary (Steady-State Conditions)

The conditions for this situation are:

1. Flow rate at the well is constant
2. The pressure at the boundary is constant at all times, $p_e = p_i$ for all t
3. Initially the pressure throughout the reservoir is uniform

By the use of the Laplace transform, Bessel functions,[3] and the above boundary conditions, the solution of the Eq. 2–66 is found to be (Carslaw and Jaeger, 1959, p. 334)[20]

$$\Delta p_D = \ln r_{eD} \quad \text{for } t_D > 1.0\, r_{eD}^2 \quad \text{(approximately)} \tag{2–69}$$

This equation may also be derived directly by integration of Darcy's law for a radial flow. Equation 2–69 represents the steady-state condition. Values of Δp_D versus t_D can be found in Ref. 5 for various reservoir sizes, which are for various values of r_D.

Example 2–9 *Calculation of Flowing Bottom-Hole Pressure Assuming Steady-State Conditions*

Rework Example 2–9, assuming a steady-state condition is achieved after long producing time. Calculate the flowing bottom hole pressure, p_{wf}, after 1920 hours of production.

Solution From Example 2–8, we have $t_D = 70{,}886{,}315$, $q_D = 0.02338$, $r_{eD} = 5606$, $(r_{eD})^2 = 31{,}427{,}236$. Since $t_D > 1.0r_{eD}^2$, Δp_D is given by Eq. 2–69,

$$\Delta p_D = \ln(r_{eD}) = \ln(5606) = 8.632$$

From Eq. 2–55:

$$\psi_{wf} = \psi_i - \psi_i \Delta p_D q_D$$

$$= 329.6 \times 10^6 - 329.6 \times 10^6(8.632)(0.02338)$$

$$= 263.8 \text{ mmpsia}^2/\text{cP}$$

From Figure 2–1, $p_{wf} = 1970$ psia.

Radial–Cylindrical Flow, Infinite and Finite Circular Reservoir, Constant Production Rate, Solution at the Well

The Δp_D functions may also be expressed in steady-state form by introducing the idea of an effective drainage radius. This concept, along with the concepts of radius of investigation and time to stabilization, is discussed in detail hereafter. Possible expressions for the effective drainage radius for various systems are as follows.

Infinite reservoir:

$$\ln\left(\frac{r_d}{r_w}\right) = \frac{1}{2}(\ln t_D + 0.809) \quad \text{for } t_D > 25. \tag{2–70}$$

Closed outer boundary:

$$r_d = 0.472 r_e \quad \text{for } t_D > 0.25 r_{eD}^2 \tag{2–70a}$$

Constant-pressure outer boundary:

$$r_d = r_e \quad \text{for } r_d = r_e$$

In terms of pressure treatment:

$$\Delta p_D = \frac{\bar{p}_R - p_{wf}}{p_i q_D} = \ln\left(\frac{r_d}{r_w}\right) \tag{2–71}$$

In terms of pressure-squared:

$$\frac{\bar{p}_R^2 - p_{wf}^2}{p_i^2 q_D} = \ln\left(\frac{r_d}{r_w}\right) \tag{2–72}$$

In terms of pseudopressure:

$$\frac{\bar{\psi}_R - \psi_{wf}}{\psi_i q_D} = \ln\left(\frac{r_d}{r_w}\right) \tag{2–73}$$

Radial–Cylindrical Flow, Constant Well Pressure, Infinite and Finite Circular Reservoir

When the well is producing at a constant pressure, the flow rate is not constant but declines continuously. The cumulative production is given by Katz et al. (1959, p. 414)[21] and may be written as

$$G_p = 2\pi \phi c r_w^2 h \frac{T_{sc}}{T} \frac{p_i}{P_{sc}} (p_i - p_{wf}) Q_{pD} \tag{2–74}$$

where

Gp = cumulative gas produced, and
Q_{pD} = dimensionless total production number which has been tabulated for certain boundary conditions, and can be found in Ref. 5.

For $t_D < 0.01$:

$$Q_{pD} = \left(\frac{t_D}{\pi}\right)^{0.5} \tag{2–75}$$

For $t_D \geq 200$ or

$$t_D \propto Q_{pD} = \frac{-4.29881 + 2.02566 t_D}{\ln t_D} : \tag{2–76}$$

$$t_D = \frac{0.0002637kt}{\phi \bar{\mu} \bar{c} r_w^2}$$

In terms of pressure-squared treatment:

$$G_p = \frac{0.111 \phi h r_w^2 c \left(p_i^2 - p_{wf}^2\right)}{\bar{z} T} Q_{pD} \tag{2–77}$$

where

G_p = cumulative gas produced, mscf, and
$r_D = r/r_w$

Values of Q_{pD} as a function of dimensionless time t_D and dimensionless radius can be found in tabular form in Ref. 5.

Linear Flow, Constant Production Rate, Infinite Reservoir

When flow is in the vicinity of a fracture (of length x_f), the flow will be linear and the pressure at any distance x from the sandface ($x \neq 0$) is given by Katz *et al.* (1959, p. 411)[21] as

$$\Delta p_D = \frac{2}{\sqrt{\pi}} \left(\frac{t_D}{x_D^2}\right)^{0.5} \exp\left(-\frac{x_D^2}{4t_D}\right) - \operatorname{erfc}\left[0.5\left(\frac{x_D^2}{t_D}\right)^{0.5}\right] \tag{2–78}$$

where

$$t_D = \frac{0.0002637kt}{\phi \bar{\mu} \bar{c} x_f^2} \tag{2–79}$$

x_f is half fracture length, ft

$$x_D = \frac{x}{x_f}$$

In terms of pressure treatment:

$$q_D = \frac{4.467 \times 10^6 \bar{z} T q_{sc} \bar{\mu}}{\bar{p} k h p_i} \tag{2–80}$$

In terms of pressure-squared treatment:

$$q_D = \frac{8.933 \times 10^6 \bar{z} T q_{sc} \bar{\mu}}{k h p_i^2} \tag{2–81}$$

In terms of pseudopressure treatment:

$$q_D = \frac{8.933 \times 10^6 T q_{sc}}{k h \psi_i} \tag{2–82}$$

and erf is the error function defined as

$$\operatorname{erf} x = \frac{2}{\sqrt{\pi}} \int_0^x e^{-t^2} dt \tag{2–83}$$

$$\operatorname{erf} x = \frac{2}{\sqrt{\pi}} \left(x - \frac{x^2}{1!3} + \frac{x^5}{2!5} - \frac{x^7}{3!7} + - \cdots \right) \tag{2–83a}$$

$\operatorname{erf}(\infty) = 1$, the complementary error function, and is defined by

$$\operatorname{erfc} x = 1 - \operatorname{erf} x = \frac{2}{\sqrt{\pi}} \int_x^\infty e^{-t^2} dt \tag{2–83b}$$

The values of error and complementary functions are given in Table 2–6.

Radial–Spherical Flow, Constant Production Rate, Infinite Reservoir

The dimensionless Δp_D, at any radius r, is given by (Carslaw and Jaeger, 1959, p. 261)[20]

$$\Delta p_D = \frac{1}{2} \operatorname{erfc} \left(\frac{r_D^2}{4 t_D} \right)^{0.5} \tag{2–84}$$

Table 2–6
Complementary Error Function (after Katz *et al.*, 1959, © McGraw-Hill)[21]

X	erf *x*	erfc *x* = 1 − erf *x*
0.0	0.0000	1.0000
0.1	0.1114	0.8887
0.2	0.2227	0.7773
0.3	0.3256	0.6745
0.4	0.4284	0.5716
0.5	0.5162	0.4839
0.6	0.6039	0.3961
0.7	0.6730	0.3268
0.8	0.7421	0.2579
0.9	0.7924	0.2076
1.0	0.8427	0.1573
1.1	0.8765	0.1235
1.2	0.9103	0.0897
1.3	0.9313	0.0687
1.4	0.9523	0.0477
1.5	0.9643	0.0356
1.6	0.9763	0.0237
1.7	0.9827	0.0173
1.8	0.9891	0.0109
1.9	0.9922	0.0078
2.0	0.9953	0.0047
2.1	0.9967	0.0033
2.2	0.9981	0.0019
2.3	0.9987	0.0013
2.4	0.9993	0.0007
2.5	0.9996	0.0005
2.6	0.9998	0.0002
2.7	0.9999	0.0001
2.8	0.9999	0.0001
2.9	1.0000	0.0000
3.0	1.0000	0.0000
3.1	1.0000	0.0000
3.2	1.0000	0.0
3.3	1.0000	0.0
3.4	1.0000	0.0
3.5	1.0000	0.0
3.6	1.0000	0.0
3.7	1.0000	0.0
3.8	1.0000	0.0
3.9	1.0000	0.0
4.0	1.0000	0.0

where

$$t_D = \frac{0.0002637kt}{\phi\bar{\mu}\bar{c}r_w^2} \tag{2–85}$$

In terms of pressure treatment:

$$q_D = \frac{7.110 \times 10^5 \bar{z} T q_{sc} \bar{\mu}}{\bar{p}krp_i} \tag{2–86}$$

In terms of pressure-squared treatment:

$$q_D = \frac{1.422 \times 10^6 \bar{z} T q_{sc} \bar{\mu}}{krp_i^2} \tag{2–87}$$

In terms of pseudopressure treatment:

$$q_D = \frac{1.422 \times 10^6 T q_{sc}}{kr\psi_i} \tag{2–88}$$

In thick formations, radial–spherical flow may exist in the vicinity of the well when only a limited portion of the formation is opened to flow.

2.11 Application of Superposition Techniques

Superposition may be considered to be a problem-solving technique in which the pressure behavior at any point at any time is the sum of the histories of each of the effects that may be considered to affect the solution at that point. Particular applications of superposition, which are important in the analysis of pressure test data, are discussed in the following section.

Investigating for Rate Change Effects

The following example will illustrate the principle of superposition as applied to the pressure drawdown due to two different flow rates. The method may be extended to any number of changing flow rates. Thus the total pressure drop for the well would be

$$(\Delta\psi)_{total} = |\psi_i \Delta p_{D1} q_{D1}|_{q1} + |\psi_i \Delta p_{D2} q_{D2}|_{q2-q1} + |\psi_i \Delta p_{D3} q_{D3}|_{q3-q2} + \cdots. \tag{2–89}$$

$$\psi_{mf} = \psi_i - (\Delta\psi)_{total} \tag{2–90}$$

The variable-rate production history is illustrated in Figure 2–2.

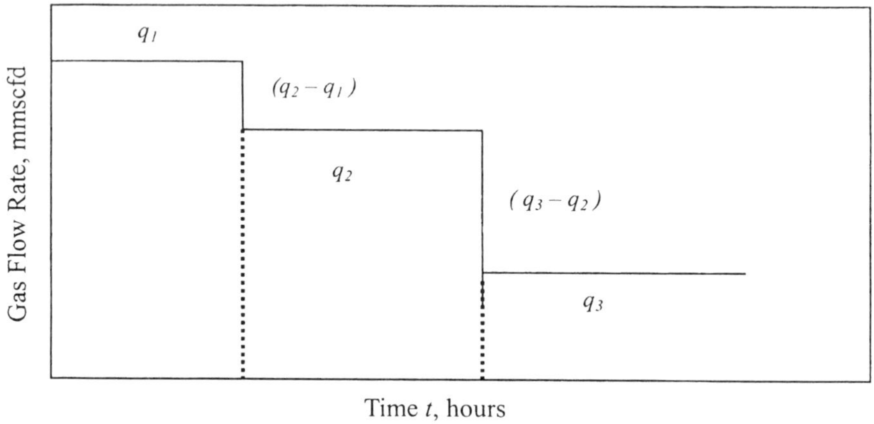

Figure 2–2. Variable-rate production of a gas well.

Example 2–10 *Calculating Flowing Sandface Pressure Accounting for Rate Change Effects*

A well situated in an infinite-acting reservoir was produced at constant rate of 5 mmscfd for 55 hours, at which time the flow rate was changed to 15 mmscfd. The stabilized shut-in pressure, $\bar{p}_R$, prior to the test was 2100 psia. General data pertinent to the test are as follows: $k = 25$ mD, $T = 600°$R, $r_w = 0.35$ ft, $h = 35$ ft, $\phi = 0.16$, $c_i = 0.00053$ psi^{-1}, $\mu_i = 0.0147$ cP, $\psi_i = 320$ mmpsia^2/cP, $t_1 = 45$ hours, $t_2 = 70$ hours, $q_1 = 5$ mmscfd, $q_2 = 15$ mmscfd.

Using the principle of superposition, calculate the flowing sandface pressure, p_{wf}, after 40 hours of production at the increased flow rate.

Solution Total production time $= t_1 + t_2 = 45 + 70 = 115$ hours.

From Eq. 2–54:

$$t_D = \frac{0.0002637kt}{\phi \mu_i c_i r_w^2}$$

$$t_{D1} = \frac{0.0002637(25)(115)}{(0.16)(0.0147)(0.00053)(0.35)^2} = 4{,}964{,}765$$

$$t_{D1} = \frac{0.0002637(25)(115 - 45)}{(0.16)(0.0147)(0.00053)(0.35)^2} = 3{,}022{,}031$$

From Eq. 2–56:

$$q_D = \frac{1427 \times 10^3 T q_{sc}}{kh\psi_i}$$

$$q_{D1} = \frac{1427 \times 10^3(600)(5)}{(25)(35)(320 \times 10^6)} = 0.01518$$

$$q_{D2} = \frac{1427 \times 10^6(600)(10)}{(25)(35)(320 \times 10^6)} = 0.03036$$

Since the reservoir is infinite-acting, Eq. 2–65 applies, so that

$$\Delta p_D = 0.5\,[\ln t_D + 0.809]$$

$$\Delta p_{D1} = 0.5\,[\ln(4{,}964{,}765) + 0.809] = 8.1134$$

$$\Delta p_{D2} = 0.5\,[\ln(3{,}022{,}031) + 0.809] = 7.86522$$

$$(\Delta\psi)_{total} = \psi_i \Delta p_{D1} q_{D1} + \psi_i \Delta p_{D2} q_{D2}$$

$$= 320 \times 10^6(8.1134)(0.01518) + 320 \times 10^6(7.86522)(0.03036)$$

$$= 115.82 \text{ mmpsia}^2/\text{cP}$$

$$\psi_{wf} = \psi_i - (\Delta\psi)_{total}$$

$$= 320 \times 10^6 - 115.82 \times 10^6 = 204.18 \text{ mmpsia}^2/\text{cP}$$

from Figure 2–1; $\therefore\ p_{wf} = 1604$ psia.

Estimating for Effects of More Than One Well

In some cases more than one well is producing from a common reservoir. As an example, consider three wells A, B, and C that start to produce at the same time, from an infinite-acting reservoir, the pressure at a point C in the producing wells (see Figure 2–3). Thus the pressure at a point C in the reservoir is obtained by superposing (adding) the solution at point C due to well A to that at point C due to well B. Each of these solutions is independent of the other and, to obtain it, the pressure behavior at any point r in the reservoir is required: that is, the general solution of the partial differential equation and

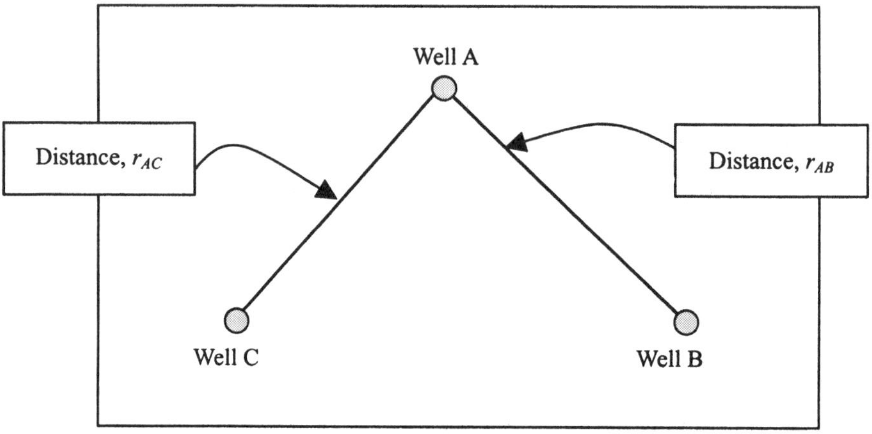

Figure 2–3. Three wells in an infinite reservoir.

not just the solution at the well. Thus

$$\Delta p|_{PointC} = p_i q_{AD}\left[-0.5E_i\left(\frac{r_{AD}^2}{4t_D}\right)\right] + p_i q_{BD}\left[-0.5E_i\left(\frac{r_{BD}^2}{4t_D}\right)\right] \quad (2\text{–}91)$$

where

r_A = distance from C to well A.
$r_{AD} = r_A/r_w$
r_B = distance from C to well B
$r_{BD} = r_B/r_w$

This is the basis of "interference" type tests used to determine reservoir characteristics. In such a test, point C is really an observation well and the interference of other producing wells is measured at C. Figure 2–3 illustrates this concept.

Example 2–11 *Accounting for the Effects of More Than One Well*

Consider the three wells in Figure 2–4. Well B is put on production at rate of 3.0 mmscfd after well A has produced for 2 months at a rate of 5.2 mmscfd. After well A has produced 3 months, what is the pressure at well C, where a well C is to be drilled? Rock and fluid properties are as follows:

p_i = 3700 psia, ψ_i = 772.56 mmpsia²/cP, c_i = 0.00023 psi^{-1}, μ_i = 0.0235 cP, ϕ = 0.1007 fraction, r_w = 0.4271 ft, T = 710°R, h = 41 ft, k = 8.5 mD.

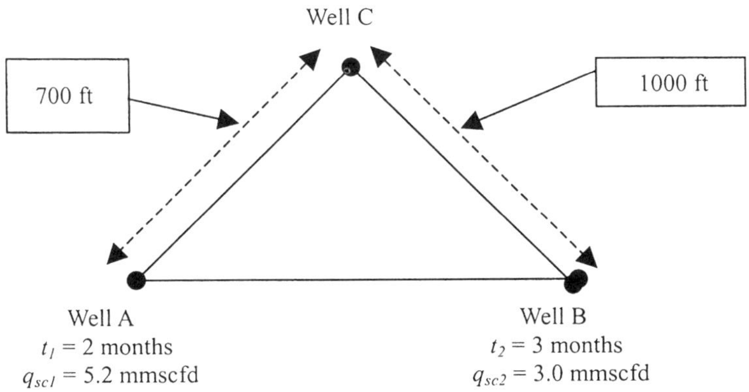

Figure 2–4. Illustration of three wells in infinite system.

Solution From Eq. 2–51:

$$t_D = \frac{0.0002637kt}{\phi\mu_i c_i r_w^2}$$

$$t_{DA} = \frac{0.0002637 \times 8.5 \times 2 \times 30.5 \times 24}{0.1007(0.0235)(0.00023)(0.4271)^2} = 33{,}051{,}092.58$$

$$t_{DB} = \frac{0.0002637 \times 8.5 \times 3 \times 30.5 \times 24}{0.1007(0.0235)(0.00023)(0.4271)^2} = 49{,}576{,}638.87$$

From Eq. 2–56:

$$q_D = \frac{1427 \times 10^3 T q_{sc}}{kh\psi_i}$$

$$q_{DA} = \frac{1427 \times 10^3(710)(5.2)}{(8.5)(41)(772.56 \times 10^6)} = 0.019568$$

$$q_{DB} = \frac{1427 \times 10^3(710)(3.0)}{(8.5)(41)(772.56 \times 10^6)} = 0.011289$$

r_A = distance from well C to well A = 700 ft

$$r_{AD} = \frac{r_A}{r_w} = \frac{700}{0.4271} = 1638.96$$

r_B = distance from well C to well B = 1000 ft

$$r_{BD} = \frac{1000}{0.4271} = 2{,}341.37$$

Using Eq. 2–91:

$$\Delta p|_{well\,C} = p_i(q_{AD})\left[0.5E_i\left(\frac{r_{DA}^2}{4t_{DA}}\right)\right] + p_i(q_{DB})\left[0.5E_i\left(\frac{r_{DB}^2}{4t_{DB}}\right)\right]$$

$$= 3700(0.019568)\left[0.5E_i\left[\frac{(1{,}638.96)^2}{4(33{,}051{,}092.58)}\right]\right]$$

$$+ 3{,}700(0.011289)\left[0.5E_i\left(\frac{(2{,}341.37)^2}{4(49{,}576{,}638.87}\right)\right]$$

$$= 72.4016[0.5E_i(0.020318)] + 41.7693[0.5E_i(0.027644)]$$

From Table 2–5, $E_i(0.020318) = 3.355$ and $E_i(0.027644) = 3.062$

$$\therefore\ \Delta p|_{well\,C} = 72.4016[0.5(3.355)] + 41.7693[0.5(3.062)]$$

$$= 185.40 \text{ psia}$$

Pressure at well C = 3700 − 185.50 = 3515 psia.

Determining Pressure Change Effects

Superposition is also used in applying the constant pressure-rate case. In cases where two pressure changes have occurred, the constant-pressure solution will be applied to each individual pressure change. This means that in this particular case we have to use Eq. 2–92 two times. The following generalized form of Eq. 2–92 will be used in applying the principle of superposition to pressure changes in the constant-pressure case:

$$G_p = \frac{0.111\phi h r_w^2 c}{T}\sum_{j=1}^{j=m}\left(\frac{\Delta p_j^2}{\bar{z}}\right)Q_{pD} \qquad (2\text{–}92)$$

$$\Delta p_j^2 = p_{old}^2 - p_{new}^2$$

and

$\bar{z}$ is calculated at $\left(\dfrac{p_{old} + p_{new}}{2}\right)$

For illustration, let us assume that a well has experienced the pressure history shown in Figure 2–5.

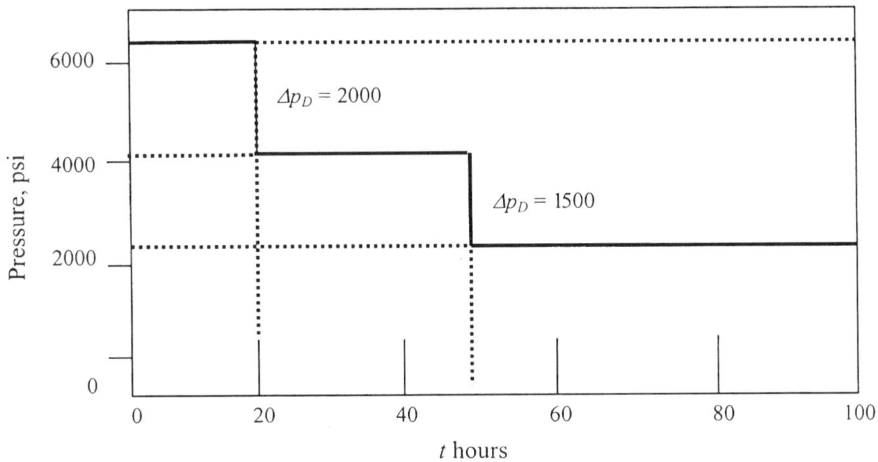

Figure 2–5. Variable pressure history of a gas well.

Simulating Boundary Effects

The principle of superposition concept can be applied to infinite-acting solutions to reservoirs that are limited in one or more direction, i.e., pressure behavior in bounded fault. Figure 2–6 shows a well, A, located at a distance $L/2$ from a no-flow barrier and producing at a constant rate. This system can be treated by replacing the barrier by an imaging well A′ identical to the real well but situated at a distance L from it. Thus the pressure history of the well will be that of an infinite-acting well at A, plus the effect at point A′ of an infinite-acting well at A′, that is,

$$\Delta p_D|_{well} = p_i q_D\left[-0.5E_i\left(-\frac{\phi\mu c r_w^2}{0.00105kt}\right)\right]$$

← caused by A →

$$+\left[-0.5E_i\left(-\frac{\phi\mu c L^2}{0.00105kt}\right)\right] \tag{2–93}$$

→ effect of A′ at A →

Equation 2–67 may approximate the first E_i term because the agreement is usually less than 0.01 for all practical times. However the second E_i term is not true because of the presence of L^2 (usually a large number) in the argument.

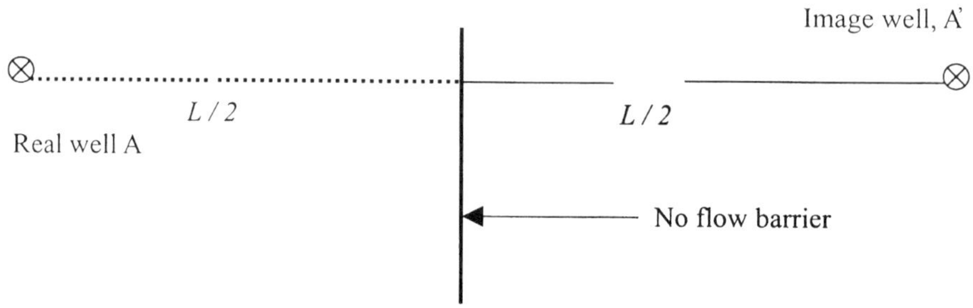

Figure 2–6. Well near no-flow boundary illustrating use of imaging.

Therefore:

$$\Delta p_D = p_i q_D \left[0.5(\ln t_D + 0.809) - 0.5E_i\left(-\frac{\phi \mu c L^2}{0.00105kt}\right)\right] \quad (2\text{–}94)$$

The following example will illustrate the principle of superposition applied to the simulation of no-flow barriers within a reservoir.

Example 2–12 *Simulating No-Flow Boundaries within a Reservoir*

In an infinite-acting gas reservoir, a well is situated 150 ft from a barrier and produced at a constant rate of 5 mmscfd for 36 hours. The stabilized shut-in reservoir pressure, p_R, prior to the test was 2100 psia. Calculate the flowing bottom hole pressure. Other data are as follows:

$k = 25$ mD, $T = 580°$R, $h = 41$ ft, $r_w = 0.35$ ft, $\phi = 0.16$, $\mu_i = 0.0157$ cP, $c_i = 0.00059$ psi^{-1}, $p_i = 2{,}100$ psia, $\psi_i = 320$ mmpsia^2/cP.

Solution From Eq. 2–51:

$$t_D = \frac{0.0002637kt}{\phi \mu_i c_i r_w^2}$$

$$= \frac{0.0002637(25)(36)}{(0.16)(0.0157)(0.00059)(0.35)^2} = 1{,}307{,}209$$

From Eq. 2–56:

$$q_D = \frac{1417 \times 10^3 T q_{sc}}{kh\psi_i}$$

$$= \frac{1417 \times 10^3(580)(5)}{(25)(41)(320 \times 10^6)} = 0.01253$$

Equation 2–55 may be written in terms of pseudopressure as

$$\psi_{wf} = \psi_i - \psi_i \Delta p_D q_D$$

where

$$\Delta p_D = 0.5(\ln t_D + 0.809) - 0.5E_i\left(-\frac{\phi \mu_i c_i L^2}{0.00105kt}\right)$$

$$= 0.5(\ln 1{,}307{,}209 + 0.809)$$

$$-0.5\ E_i\left(-\frac{(0.16)(0.0157)(0.00059)(150)^2}{(0.00105)(25)(36)}\right)$$

$$= 7.446 - 0.5E_i(-0.353) = 7.447 - 0.5(2.75) = 6.07$$

Therefore

$$\psi_{wf} = 320 \times 10^6 - 320 \times 10^6(6.07)(0.01253)$$

$$= 320 \times 10^6 - 24.34 \times 10^6 = 295.66 \text{ mmpsia}^2\text{cP}$$

from $\therefore p_{wf} = 1865$ psia.

Use of Horner's Approximation

In 1951, Horner[11] introduced an approximation that could be used in many cases to avoid the use of the tedious superposition principle as applied to model production history of a variable-rate well instead of using the sequence of E_i functions, i.e., one E_i function for each rate change. With the help of this approximation, we are able to use one equation with one single producing rate and one single producing time.

Thus, mathematically,

$$t_p = \frac{24G_p}{q_{last}} \tag{2–95}$$

where

G_p = cumulative production, mmscf, and
q_{last} = constant-rate just before shut-in, mmscfd.

Accounting for Different Reservoir Geometry

Ramey[22] has presented models of pseudo-steady-state flow in more general reservoir shapes. For practical applications, the concept of the shape factor, C_A, which depends on the shape of the area and the well position, is quite useful. Defining a dimensionless time based on drainage area, A, as

$$t_D = \frac{0.0002637kt}{\phi\mu c A} \tag{2–51}$$

$$t_{DA} = t_D \frac{r_w^2}{A} \tag{2–96}$$

$$p_i - p_{wf} = p_i q_D \frac{1}{2}\left[\ln\left(\frac{2.2458 A t_{DA}}{r_w^2}\right) + 4\pi t_{DA} - F\right] \tag{2–97}$$

where dimensionless pressure Δp_D is

$$\Delta p_D = \frac{1}{2}\left[\ln\left(\frac{2.2458 A t_{DA}}{r_w^2}\right) + 4\pi t_{DA} - F\right] \tag{2–98}$$

and F is the Matthews, Brons, and Hazebroek[23] dimensionless pressure function that has been evaluated for various reservoir shapes and well locations. For small values of t_{DA}, that is, the transient region of flow, the well is infinite-acting and

$$F = 4\pi t_{DA} \tag{2–99}$$

and

$$\Delta p_D = 0.5 \ln\left(\frac{2.2458 A t_{DA}}{r_w^2}\right) \tag{2–100}$$

For large values of t_{DA}, when all the boundaries have been felt, that is, at pseudo-steady state,

$$F = \ln(C_A t_{DA}) \tag{2–101}$$

and

$$\Delta p_D = 0.5 \ln\left(\frac{2.2458 A}{r_w^2 C_A}\right) + 2\pi t_{DA} \tag{2–102}$$

The late transient between transient and pseudo-steady-state varies with each situation. During this period, the pressure drop function may be obtained from

$$\Delta p_D = 0.5\left[\ln\left(\frac{2.2458 A t_{DA}}{r_w^2}\right) + 4\pi t_{DA} - F\right] \tag{2–103}$$

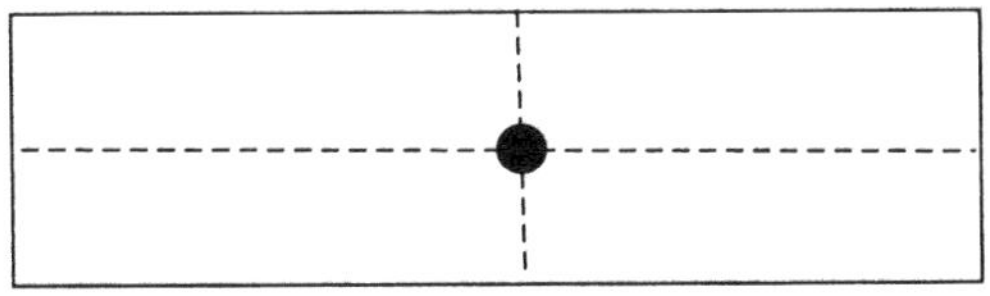

Figure 2–7. Gas well is situated in the center of a rectangle.

Dimensionless pressure function i is obtained from Table B–1[23] or graphically[23] from Figures B–1 through B–7. Shape factors C_A for various drainage shapes and well locations can be found from Table B–1.[13]

Example 2–13 *Accounting for Different Reservoir Geometry*

A gas well is situated in the center of a rectangle, as shown in Figure 2–7, having closed no-flow boundaries and an area A of 8×10^6 sq ft, was produced at a constant rate of 5 mmscfd. The stabilized shut-in reservoir pressure, $\bar{p}_R$, prior to the test was 2100 psia. Use gas composition given in Example 2–1. Other data are as follows: $k = 25$ mD, $T = 580°$R, $h = 41$ ft, $r_w = 0.35$ ft, $\phi = 0.16$, $\mu_i = 0.0157$ cP, $c_i = 0.0059$ psi^{-1}, $\bar{p}_R = 2100$ psia, $\bar{\psi}_R = 320$ mmpsia^2/cP.

Calculate flowing pressure, p_{wf}, after 40 and 2000 hours of production.

Solution Since the gas is the same as that of Example 2–1, the $\psi - p$ curve already constructed (Figure 2–1) is applicable to the problem.

$t = 40$ hours:

From Eq. 2–51:

$$t_{DA} = \frac{0.0002637kt}{\phi\mu_i c_i A}$$

$$= \frac{0.0002637(25)(40)}{(0.16)(0.0157)(0.00059)\ (8 \times 10^6)} = 0.02224$$

From Eq. 2–56:

$$q_D = \frac{1417 \times 10^3 T q_{sc}}{kh\psi_i}$$

$$= \frac{1417 \times 10^3(580)(5)}{(25)((41)(320 \times 10^6)} = 0.01253$$

Calculate F from Table B–1:[23] $F = 0.2806$.

From Eq. 2–103:

$$\Delta p_D = 0.5\left[\ln\frac{2.2458At_{DA}}{r_w^2} + 4\pi t_{DA} - F\right]$$

$$= 0.5\left[\ln\frac{2.2458(8\times 10^6)(0.02224)}{(0.35)^2} + 4(22/7)(0.01253) - 0.2806\right]$$

$$= 7.29$$

Also,

$$\Delta p_D = \frac{\psi_i - \psi_{mf}}{\psi_i q_D}$$

After rearranging:

$$\psi_{wf} = \psi_i - \psi_i \Delta p_D q_D$$

$$= 320\times 10^6 - 320\times 10^6(7.29)(0.01253) = 290.77 \text{ mmpsia}^2/\text{cP}$$

From the $\psi - p$ curve (Figure 2–1), $P_{wf} = 1845$ psia.

$t = 2000$ hours:

From Eq. 2–51:

$$t_{DA} = \frac{0.0002637kt}{\phi\mu_i c_i A}$$

$$= \frac{0.0002637(25)(2000)}{(0.16)(0.0157)(0.00059)(8\times 10^6)} = 1.1120$$

From Eq. 2–56:

$$q_D = \frac{1417\times 10^3 T q_{sc}}{kh\psi_i}$$

$$= \frac{1417\times 10^3(580)(5)}{(25)(41)(320\times 10^6)} = 0.01253$$

Calculate F from Table B–1:[23] $F = 3.2000$

From Eq. 2–103:

$$\Delta p_D = 0.5\left[\ln\frac{2.2458 A t_{DA}}{r_w^2} + 4\pi t_{DA} - F\right]$$

$$= 0.5\left[\ln\frac{2.2458(8 \times 10^6)(1.1120)}{(0.35)^2} + 4(22/7)(0.01253) - 3.200\right]$$

$$= 14.84$$

Also,

$$\Delta p_D = \frac{\psi_i - \psi_{mf}}{\psi_i q_D}$$

After rearranging the preceding equation:

$$\psi_{wf} = \psi_i - \psi_i \Delta p_D q_D$$

$$= 320 \times 10^6 - 320 \times 10^6(14.84)(0.01253) = 260.50 \text{ mmpsia}^2/\text{cP}$$

From the $\psi - p$ curve (Figure 2–1), $P_{wf} = 1746$ psia.

Alternatively, from Table B–2,[13] t_{DA} required for stabilization equals 0.15 and $C_A = 21.8369$. Because t_{DA} at 2000 hours = 1.1120 > 0.15, Eq. 2–102 can be used to evaluate Δp_D.

From Eq. 2–102:

$$\Delta p_D = 0.5 \ln\left(\frac{2.2458A}{r_w^2 C_A}\right) + 2\pi t_{DA}$$

$$= 0.5 \ln\left(\frac{2.2458(8 \times 10^6)}{(0.35)^2(21.8369)}\right) + 2(22/7)(1.1120) = 14.84$$

Therefore,

$$\psi_{wf} = \psi_i - \psi_i \Delta p_D q_D$$

$$= 320 \times 10^6 - 320 \times 10^6(14.84)(0.01253) = 260.50 \text{ mmpsia}^2/\text{cP}$$

From the $\psi - p$ curve (Figure 2–1), $P_{wf} = 1746$ psia.

2.12 Choice of Equation for Gas Flow Testing and Analysis

This section will discuss correlation of the gas flow solutions in terms of the pressure; pressure squared, and real-gas pseudopressure approaches. An analysis of these approaches has been conducted by Aziz, Mattar, Ko, and Brar.[7] They consider the analytical solution at the well for an infinite reservoir given by Eq. 2–104:

$$\Delta p_D = -0.5E_i\left(-\frac{1}{4t_D}\right) \tag{2–104}$$

Calculate the sandface pressure from this equation, using different approaches.

Pressure Case

For pressure >3000 psi the simpler form is in terms of pressure, p. The differential equation is

$$\frac{1}{r}\frac{\partial}{\partial r}\left(r\frac{\partial}{\partial r}\right) = \frac{\phi\mu c}{0.0002637k}\frac{\partial p}{\partial t} \tag{2–105}$$

The diffusivity equation in dimensionless variables becomes

$$\frac{1}{r_D}\frac{\partial}{\partial r_D}\left[r_D\frac{\partial}{\partial r_D}(\Delta p_D)\right] = \frac{\partial}{\partial t_D}(\Delta p_D) \tag{2–106}$$

The dimensionless time, t_d, in Eq. 2–106 is defined by

$$t_D = \frac{0.0002637kt}{\phi r_w^2}\left(\frac{1}{\mu c}\right) \tag{2–107}$$

The definition of Δp_D, however, is different for this approach. For the pressure case,

$$\Delta p_D = \frac{p_i - p}{\frac{70.85\times 10^4 T q_{sc}}{kh}\left(\frac{\mu z}{p}\right)} \tag{2–108}$$

Both quantities $(\frac{1}{\mu c})$ and $(\frac{\mu c}{p})$ in Eqs. 2–107 and 2–108 are evaluated at $(p_i + p)/2$.

Pressure-Squared Case

For pressure <2000 psi a simple form in terms of p^2 is more generally applicable.

$$\frac{\partial^2 p^2}{\partial r^2} + \frac{1}{r}\frac{\partial p^2}{\partial r} = \frac{\phi \mu c}{0.0002637k}\frac{\partial p^2}{\partial t} \tag{2–109}$$

The diffusivity equation in dimensionless variables becomes

$$\frac{\partial^2 \Delta p_D}{\partial r_D^2} + \frac{1}{r_D}\frac{\partial \Delta p_D}{\partial r_D} = \frac{\partial}{\partial t_D}(\Delta p_D) \tag{2–110}$$

The definition of Δp_D, however, is different for this approach. For the pressure-squared case,

$$\Delta p_D = \frac{p_i^2 - p^2}{\frac{1{,}417\times10^3 T q_{sc}}{kh}(\mu z)} \tag{2–111}$$

The quantities $(\frac{1}{\mu c})$ and (μz) in Eqs. 2–107 and 2–108 are evaluated at p_i.

Pseudopressure Case

For both low and high pressures the equation in terms of pseudopressure is best fitted to this role, is denoted by $\psi(p)$, and is defined by the integral[10]

$$\psi(p) = 2\int_{p_{base}}^{p} \frac{p}{\mu z}\,dp \tag{2–112}$$

The differential equation in terms of this approach is

$$\frac{1}{r}\frac{\partial}{\partial r}\left(r\frac{\partial \psi}{\partial r}\right) = \frac{\phi \mu c_g}{0.0002637k}\frac{\partial \psi}{\partial t} \tag{2–113}$$

The diffusivity equation in dimensionless variables becomes

$$\frac{1}{r_D}\frac{\partial}{\partial r_D}\left(r_D\frac{\partial \Delta\psi_D}{\partial r_D}\right) = \frac{\partial \Delta\psi_D}{\partial t_D} \tag{2–114}$$

The definition of $\Delta\psi_D$ is

$$\Delta\psi_D = \frac{\psi_i - \psi}{\frac{1{,}417\times10^3 T q_{sc}}{kh}} \tag{2–115}$$

The properties are evaluated at initial conditions.

2.13 Skin, IT Flow, and Wellbore Storage Effects

In the derivation of the equations it was assumed that the porous medium was homogeneous and isotropic and that flow was single-phase and obeyed Darcy's law. It was also supposed that opening and shut-in of the well was done at the sandface. In actual fact these idealizations are not realistic, and derivations from the ideal model are too frequent and important to be ignored. Ways of accounting for skin effects; IT flow, and wellbore storage will be treated in the following sections.

Accounting for Effects of Formation Damage

The permeability of the formation immediately around the well can be damaged by the well drilling process or improved by fracturing or acidizing the well on completion. To account for this altered permeability a skin factor was defined by Van Everdingen[8] as

$$(\Delta p_D)_{skin} = s, \text{ a constant} \tag{2–116}$$

so that

$$\Delta p_D|_{well} \text{ (including skin)} = p_D + s \tag{2–117}$$

This essentially states that there will be an added pressure difference due to the skin effect given by Eq. 2–117. A position value of s indicates a damaged well, and a negative value, an improved well. Hawkins[9] proposed that the skin be treated as a region of radius r_{skin} with permeability k_{skin}, with the skin factor given by

$$s = \left(\frac{k}{k_{skin}} - 1\right) \ln \frac{r_{skin}}{r_w} \tag{2–118}$$

Equation 2–118 is valid for both positive skin ($k_{skin} < k$) and negative skin ($k_{skin} > k$) but there is no unique set of values of k_{skin} and r_{skin} for a particular s.

An alternative treatment of the skin effect is that of an "effective wellbore radius" (Matthews and Russell, 1967, p. 21),[15] defined as that radius which makes the pressure drop in an ideal reservoir equal to that in an actual reservoir with skin. Thus:

$$r_w \text{(effective)} = r_w e^{-s} \tag{2–119}$$

For positive skin, r_w (effective) $< r_w$, that is, the fluid must travel through additional formation to cause the observed pressure drop, Δp. For negative skin, r_w (effective) $> r_w$. This is a useful concept in hydraulically fractured wells.

Accounting for Effects of Turbulence

For gas flow, however, inertial and/or turbulent (IT) flow effects, not accounted for by Darcy's law, are frequently of significance and should not be ignored. IT flow is most pronounced near the well and results in an additional pressure drop similar to the skin effect, except that it is not a constant but varies directly with flow rate.[24] Smith[25] confirmed with actual test results and with numerical solutions that IT flow could be treated as an additional, rate-dependent skin effect.

$$(\Delta p_D)_{IT} = Dq_{sc} \tag{2–120}$$

Where D = IT flow factor for the system, the pressure at the well is given by

$$\Delta p_D|_{well} = p_D + s + Dq_{sc} \tag{2–121}$$

or

$$s' = (\Delta p_D)_{skin} + (\Delta p_D)_{IT} = s + Dq_{sc} \tag{2–122}$$

The following example will show how pressure drop is attributed to laminar flow, skin, and IT flow effects. It assumes negligible effects of viscosity on turbulence.

Example 2–14 *Calculating Pressure Drop due to Laminar Skin and IT Flow Effects*

In an infinite-acting gas reservoir, a well was produced at a constant rate, q_{sc1}, of 8 mmscfd for a period of 35 hours. The flowing bottom hole pressure, p_{wf1}, at that time was 1550 psia. The same well was produced at a constant rate, q_{sc2}, of 11 mmscfd for a time of 25 hours. The flowing bottom hole pressure, p_{wf2}, at that time was 1300 psia. The stabilized shut-in pressure, $\bar{p}_R$, prior to each of the two flowing periods, was 2100 psia. Other data pertinent to the test are given below:

$k = 25$ mD, $r_w = 0.35$ ft, $h = 35$ ft, $T = 600°$R,
$\phi = 0.16$, $\mu_i = 0.0147$ cP, $c_i = .00053$ psi^{-1}, $\psi_i = 320.00$ mmpsia^2/cP
$t_1 = 35$ hours, $q_{sc1} = 8$ mmscfd, $p_{wf1} = 1550$ psia
$t_2 = 25$ hours, $q_{sc2} = 11$ mmscfd, $p_{wf2} = 1300$ psia

Calculate the skin and IT flow effects, s and D, respectively. Also calculate, for the second flow rate, using the same gas composition given in Example 2–2:

(a) the pressure drop due to the laminar flow effect
(b) the pressure drop due to skin effects

(c) the pressure drop due to IT flow effects
(d) total pressure drop

Solution From Eq. 2–54:

$$t_D = \frac{0.0002637kt}{\phi \mu_i c_i r_w^2}$$

Therefore

$$t_{D1} = \frac{0.0002637(25)(35)}{(0.16)(0.0147)(0.00053)(0.35)^2} = 1{,}511{,}015$$

and

$$t_{D2} = \frac{0.0002637(25)(25)}{(0.16)(0.0147)(0.00053)(.35)^2} = 1{,}077{,}296$$

From Eq. 2–56:

$$q_d = \frac{1417 \times 10^3 T q_{sc}}{kh\psi_i}$$

Therefore

$$q_{D1} = \frac{1417 \times 10^3(600)(8)}{(25)(35)(320 \times 10^6)} = 0.02429$$

$$q_{D2} = \frac{1417 \times 10^3(600)(11)}{(25)(35)(320 \times 10^6)} = 0.03340$$

Since the reservoir is infinite-acting, Eq. 2–65 applies, so that

$$p_t = p_D = 0.5\,[\ln t_D + 0.809]$$

Therefore,

$$p_{t1} = p_{D1} = 0.5\,[\ln(1{,}511{,}015) + 0.809] = 7.519$$

$$p_{t2} = p_{D2} = 0.5\,[\ln(1{,}079{,}296) + 0.809] = 7.351$$

From Eq. 2–55:

$$\Delta p_D = \frac{\psi_i - \psi_{wf}}{\psi_i q_D}$$

From the $\psi - p$ curve, $P_{wf1} = 1550$ psia $\leftrightarrow$ $\psi_{wf1} = 207 \times 10^6$ psia^2/cP

$p_{wf2} = 1300$ psia $\leftrightarrow$ $\psi_{wf2} = 145 \times 10^6$ psia^2/cP

Therefore,

$$\Delta p_{D1} = \frac{320 \times 10^6 - 207 \times 10^6}{320 \times 10^6 (0.02429)} = 14.54$$

$$\Delta p_{D2} = \frac{320 \times 10^6 - 145 \times 10^6}{320 \times 10^6 (0.03340)} = 16.37$$

From Eq. 2–121:

$$\Delta p_D = p_D \quad \text{or} \quad p_t = s + Dq_{sc}$$

Substituting the calculated values of Δp_D, p_D, or p_t and q_{sc} in the above equation gives

$$14.54 = 7.519 + s + 8D$$
$$16.37 = 7.351 + s + 11D$$

Solving these equations simultaneously gives

$$D = \frac{(16.37 - 14.54) - (7.351 - 7.519)}{(11 - 8)} = 0.666$$

$$s = 14.54 - 7.519(8)(0.666) = 1.69$$

For the second production rate, q_{sc2} is as follows:

(a) Pressure drop due to laminar flow effects is given by

$$p_{t2} = \frac{\psi_i - \psi}{\psi_i q_{D2}}$$

Therefore

$$\begin{aligned} \psi &= \psi_i - \psi_i p_{t2} q_{D2} \\ &= 320 \times 10^6 - 320 \times 10^6 (7.351)(0.3340) \\ &= 241.43 \text{ mmpsia}^2/\text{cP} \\ &= 1720 \text{ psia (from } \psi - p \text{ curve)} \end{aligned}$$

and $\Delta p_{laminar\,flow} = p_i - p = 2100 - 1720 = 380$ psia.

(b) Pressure drop due to skin effects is given by

$$s = \frac{\psi_i - \psi}{\psi_i q_{D2}}$$

$$\begin{aligned} \therefore \psi &= \psi_i - \psi_i s q_{D2} = 320 \times 10^6 - 320 \times 10^6 \times 1.69 \times 0.03340 \\ &= 302 \text{ mmpsia/cP} \leftrightarrow p = 1910 \text{ psia} \end{aligned}$$

$$\Delta p_{skin} = p_i - p = 2100 - 1910 = 190 \text{ psia}$$

(c) Pressure drop due to IT flow effects is given by

$$
\begin{aligned}
Dq_{sc2} &= \frac{\psi_i - \psi}{\psi_i q_{D2}} \\
\therefore \psi &= \psi_i - \psi_i D q_{sc2}\, q_{D2} \\
&= 320 \times 10^6 - 320 \times 10^6 \times 0.666 \times 11 \times 0.03440 \\
&= 239.35 \text{ mmpsia}^2/\text{cP} \leftrightarrow p = 1690 \text{ psia} \\
\therefore \Delta p_{IT\,flow} &= p_i - p = 2100 - 1690 = 410 \text{ psia}
\end{aligned}
$$

(d) Total pressure drop $= \Delta p_{laminar\,flow} + \Delta p_{skin} + \Delta p_{IT\,flow} = 380 + 190 + 410 = 980$ psia.

Wellbore Storage Effects

Wellbore storage effects are associated with a continuously varying flow rate in the formation. One solution[8] is to assume that the rate of unloading of, or storage in, the wellbore per unit pressure difference is constant. This constant is known as the wellbore storage constant, C_S, and is given by

$$C_S = V_{WS} \times C_{WS} \tag{2–123}$$

where

V_{WS} = Volume of the wellbore tubing (and annulus, if there is no packer) ft^3
$V_{WS} = \pi r_w^2 L$, ft^3
L = well depth, ft
C_{WS} = compressibility of the wellbore fluid evaluated at the mean wellbore pressure and temperature, psi^{-1}

The wellbore storage constant may be expressed in a dimensionless term as

$$C_{SD} = \frac{0.159 C_S}{\phi h C r_w^2} \tag{2–124}$$

The rate of flow of fluid from the formation may then be obtained from

$$q = q_{sc}\left[1.0 - C_{SD}\frac{\partial}{\partial t_D}(\Delta p_D)\bigg|_{wellbore}\right] \tag{2–125}$$

The time for which wellbore storage effects are significant is given by

$$t_{WSS} = 60C_{SD} \tag{2–126}$$

The time at which wellbore storage effects become negligible is given by

$$t_{WS} = \frac{36{,}177\mu C_S}{kh}, \text{ hours} \tag{2–127}$$

Example 2–15 *Finding the End of Wellbore Storage Effects*

The following characteristics are given: well depth = 5500 ft, $r_w = 0.39$ ft., $C_{WS} = 0.000595$ psi^{-1}, $h = 5$ ft, $k = 25$ mD, $\mu = 0.0175$ cP. Assume there is no bottomhole packer. Calculate the time required for wellbore storage effects to become negligible.

Solution From Eq. 2–123:

$$V_{WS} = \pi r_w^2 L = 22/7(0.39)^2(5500) = 2629 \text{ ft}^3$$

From Eq. 2–123: $C_S = C_{WS}V_{WS} = 0.000595 \times 1629 = 1.565$ ft^3/psi^{-1}
From Eq. 2–127:

$$t_{WS} = \frac{36{,}177(0.0175)(1.565)}{25(45)} = 0.88 \text{ hours}$$

After a time of 0.88 hours, wellbore storage effects become negligible and the analytical solutions for transient flow apply.

Radius of Investigation

The radius of investigation has several uses in pressure transient test analysis and design:

1. Provides a guide for well test design
2. Estimates the time required to test the desired depth in the formation
3. Provides a means of estimating the length of time required to achieve "stabilized" flow (i.e., the time required for a pressure transient to reach the boundaries of a tested reservoir)

An infinite reservoir may be considered to be a limited reservoir with a closed outer boundary at r, provided r is allowed to increase with t_D. This changing value of r is defined as the radius of investigation, r_{inv}, that is,

$$t_D = 0.25r_{eD}^2$$

or (2–128)

$$r_{eD}^2 = 4t_D$$

$$\left(\frac{r_{inv}}{r_w}\right)^2 = 4t_D \tag{2–128a}$$

$$r_{inv} = \left(\frac{0.00105kt}{\phi\mu c_t}\right)^{0.5}, \text{ ft}, \quad \text{for } r_{inv} \le r_e \tag{2–128b}$$

If the value of r_{inv} obtained from Eq. 2–128a is greater than r_e, then the radius of investigation is taken to be r_e.

Time of Stabilization

If a well is centered in a cylindrical drainage area of radius r_e, then setting $r_{inv} = r_e$, the time required for stabilization, t_S, is defined as follows:

$$t_D = 0.25r_{eD}^2$$
$$= \frac{1}{4}r_{eD}^2$$

or

$$t_S = \frac{1}{4} \cdot \frac{\phi\mu C r_e^2}{0.0002637k}$$
$$= \frac{948\phi\mu C r_e^2}{k}, \text{ hours} \tag{2–129}$$

Example 2–16 *Estimating Radius of Investigation*

We want to conduct a flow test on an exploratory gas well for a long enough time to ensure that the well will drain a radius of more than 1500 ft. Well and fluid data are as follows: $\phi = 0.18$ fraction, $k = 9.0$ mD, $r_i = 1500$ ft, $\mu_i = 0.0156$ cP, $C_{ti} = 2.2 \times 10^{-4}$ psi^{-1}. What length of flow test appears advisable? What flow rate do you suggest?

Solution From Eq. 2–128a, the time required is

$$r_{inv} = \left(\frac{0.00105kt}{\phi\mu c_t}\right)^{0.5}, \text{ ft}, \qquad \text{for } r_{inv} \le r_e$$

In principle, any flow rate would sufficient required to achieve a particular radius of investigation is dependent of flow rate.

2.14 Numerical Solutions of Partial Differential Equations

Numerical methods must be used for cases where the partial differential equation and its boundary conditions cannot be linearized, where the reservoir shape is irregular, or when the reservoir is heterogeneous. In some complex situations, analytical solutions may be so difficult to apply that numerical methods are preferred. In this section a brief discussion of the numerical approach is presented including difference equations.

Three-Dimensional Models

Gas flow equations are different from those for liquid flow in that the equations of state that are used are quite different in functional form from those for liquids. The ideal gas law gives the equation of state for an ideal gas:

$$pV = \frac{m}{M}RT \quad \text{and} \quad \frac{m}{V} = \frac{M}{RT}P = \rho$$

where ρ is the density.

In the case of flow of a nonideal gas, the gas deviation factor z_g is introduced into the equation of state to give

$$\rho = \frac{M}{RP}\frac{\rho}{z_g} \tag{2–130}$$

If we assume laminar flow, neglect gravity effects, and assume constant rock properties, Eq. 2–130 becomes

$$\frac{\partial}{\partial x}\left(\frac{p}{\mu z_g}\frac{\partial p}{\partial x}\right) + \frac{\partial}{\partial y}\left(\frac{p}{\mu z_g}\frac{\partial p}{\partial y}\right) + \frac{\partial}{\partial z}\left(\frac{p}{\mu z_g \frac{\partial p}{\partial z}}\right) = \frac{\phi}{k}\frac{\partial}{\partial t}\left(\frac{p}{z_g}\right) \tag{2–131}$$

In field units Eq. 2–131 can be written as

$$\frac{\partial}{\partial x}\left(\frac{p}{\mu z_g}\frac{\partial p}{\partial x}\right) + \frac{\partial}{\partial y}\left(\frac{p}{\mu z_g}\frac{\partial p}{\partial y}\right) + \frac{\partial}{\partial z}\left(\frac{p}{\mu z_g \frac{\partial p}{\partial z}}\right) = \frac{\phi}{0.000264k}\frac{\partial}{\partial t}\left(\frac{p}{z_g}\right) \tag{2–132}$$

In terms of pseudopressure, $\psi(p)$, the equation can be written as follows:

$$\psi(p) = 2\int_{p_0}^{p} \frac{p}{\mu z_g}\, dp \tag{2–133}$$

where p_0 is a low base pressure. Now,

$$\frac{\partial}{\partial t}\left(\frac{p}{z_g}\right) = \frac{d\left(\frac{p}{z_g}\right)}{dp}\frac{\partial p}{\partial t} = \frac{c_g p}{z_g}\frac{\partial p}{\partial t},$$

because

$$c_g = \frac{1}{\rho}\frac{d\rho}{dp} = \frac{z_g}{p}\frac{d\left(\frac{p}{z_g}\right)}{dp}$$

Also note that

$$\frac{\partial \psi}{\partial t} = \frac{\partial \psi}{\partial p}\frac{\partial p}{\partial t} = \frac{2p}{\mu z_g}\frac{\partial p}{\partial t}$$

and

$$\frac{\partial \psi}{\partial x} = \frac{2p}{\mu z_g}\frac{\partial p}{\partial x}$$

Similar expressions apply for $\frac{\partial \psi}{\partial y}$ and $\frac{\partial \psi}{\partial z}$. Thus Eq. 2–131 becomes

$$\frac{\partial}{\partial x}\left(\frac{\partial \psi}{\partial x}\right) + \frac{\partial}{\partial y}\left(\frac{\partial \psi}{\partial y}\right) + \frac{\partial}{\partial z}\left(\frac{\partial \psi}{\partial z}\right) = \frac{\phi \mu c_g}{0.000264k}\frac{\partial \psi}{\partial t} \tag{2–134}$$

Equations 2–131 and 2–134 are in three-dimensional form for single-phase flows and can be used for the study of completely heterogeneous reservoirs.

Radial One-Dimensional Model

For radial flow, the equivalent of Eq. 2–131 is

$$\frac{1}{r}\frac{\partial}{\partial r}\left(\frac{p}{\mu z_g} r \frac{\partial p}{\partial r}\right) = \frac{\phi}{0.000264k}\frac{\partial}{\partial t}\left(\frac{p}{z_g}\right) \tag{2–135}$$

In terms of pseudopressure, $\Psi(p)$ is

$$\frac{1}{r}\frac{\partial}{\partial r}\left(r\frac{\partial \psi}{\partial r}\right) = \frac{\phi}{0.000264k}\frac{\partial \psi}{\partial t} \tag{2–136}$$

For single-well problems, the use of the cylindrical coordinates provides greater accuracy than other coordinate systems. For the study of multiwell systems it is usually necessary to use rectangular coordinates with closely spaced grid points near the well.

Radial Two-Dimensional Coning Model

Where vertical flow is important, a two-dimensional radial model must be considered. The equation to be solved in this case is

$$\frac{1}{r}\frac{\partial}{\partial r}\left(\frac{p}{\mu z_g}r\frac{\partial p}{\partial r}\right)+\frac{\partial}{\partial z}\left(\frac{p}{\mu z_g}\frac{\partial p}{\partial z}\right)=\frac{\phi}{0.000264k}\frac{\partial}{\partial t}\left(\frac{p}{z_g}\right) \tag{2–137}$$

In terms of pseudopressure, $\Psi(p)$ is

$$\frac{1}{r}\frac{\partial}{\partial r}\left(r\frac{\partial \psi}{\partial r}\right)+\frac{\partial}{\partial z}\left(\frac{\partial \psi}{\partial z}\right)=\frac{\phi\mu c_g}{0.000264k}\frac{\partial}{\partial t}\left(\frac{p}{z_g}\right) \tag{2–138}$$

Models of this type can be used to study the effects of anisotropy on the transient pressure analysis of buildup and drawdown tests.

Areal Two-Dimensional Models

Multiwell problems can be solved through the solution of Eq. 2–139:

$$\frac{\partial}{\partial x}\left(\frac{p}{z_g}\frac{hk_x}{\mu}\frac{\partial p}{\partial x}\right)+\frac{\partial}{\partial y}\left(\frac{p}{z_g}\frac{hk_y}{\mu}\frac{\partial p}{\partial y}\right)=\frac{\partial}{\partial t}\left(\frac{\phi hp}{z_g}\right)+q(x,y,t) \tag{2–139}$$

The injection or production from different wells is accounted for by the q term. The reservoir shape may be completely arbitrary and there may be different types of boundary conditions such as no-flow or constant pressure. This model can also be used for interference test analysis.

Studies of this type for Darcy's flow have been reported in the literature, for example, by Carter.[12]

Multiphase (Gas-Condensate Flow) Model

In this section we outline a detailed derivation of an equation describing radial, and a multiphase mixture of gas, condensate, and water. We assume that a porous medium contains gas condensate and water, and that each phase has saturation-dependent effective permeability (k_g, k_o, and k_w); time-dependent saturation (S_g, S_o, and S_w); and pressure-dependent viscosity (μ_g, μ_o, and μ_w). When gravitational forces and capillary pressures are negligible, the differential equation describing this type of flow is

$$\frac{1}{r}\frac{\partial}{\partial r}\left(r\frac{\partial \psi}{\partial r}\right)=\frac{\phi_t c_t}{0.000264\lambda_t}\frac{\partial \psi}{\partial t} \tag{2–140}$$

where

$$c_t = S_g c_g + S_o c_o + S_w c_w + c_f \tag{2–141}$$

c_t is the effective total compressibility and is the sum of the fractional compressibilities. The fractional compressibility of a fluid is its compressibility multiplied by the fraction of the pore space that it occupies (that is, its saturation). The effective total mobility, $(k/\mu)_t$, is given in terms of the in situ permeability to each of the phases by

$$\lambda_t = \left(\frac{k}{\mu}\right)_t = \frac{k_g}{\mu_g} + \frac{k_o}{\mu_o} + \frac{k_w}{\mu_w} \tag{2–142}$$

The in situ permeability to each phase is the product of the permeability of the formation and the relative permeability to that phase. This latter factor depends on the prevailing saturation conditions. The effective total production rate is simply the sum of the individual fluid flow rates.

$$q_t = q_g + q_o + q_w \tag{2–143}$$

Substituting these effective total properties and the total porosity, ϕ_t, for their single-phase equivalents in Eq. 2–108 makes it possible to use the solutions of this equation for multiphase (gas-condensate flow) problems.

Compositional (Multicomponent) Model

In a reservoir system there are generally several species of chemical compounds. These components vary in composition in different phases, and each phase flows at a different rate. Therefore a mass balance must be made on every flowing fraction instead of each phase. Figure 2–8 shows compositional mass balance on element. Detailed discussion and numerical equations can be found in Refs. 16 and 17.

Compositional Mass Balance on Element

There are N species of chemical compounds flowing into the reservoir element in three phases. With the element there are changes due to either or all of the following:

1. Pressure change
2. Production
3. Injection

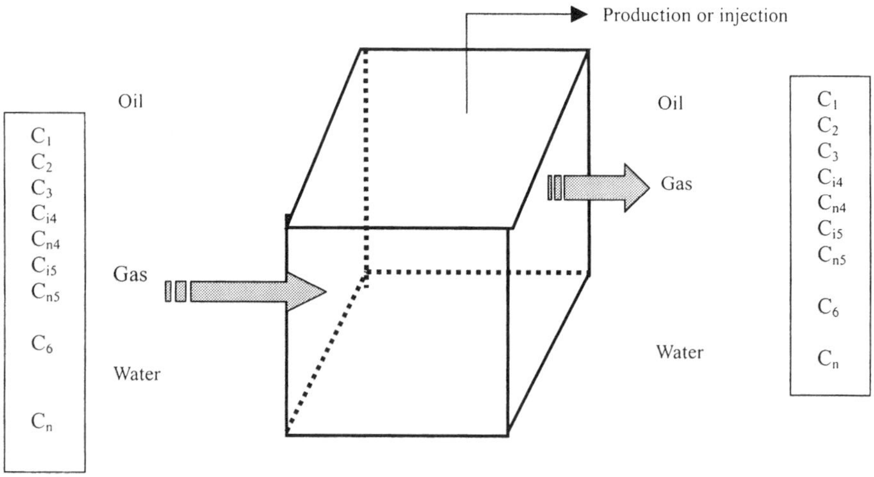

Figure 2–8. Composition mass balance on element (after Roebuck *et al.* © SPE, AIME 1969).[16]

Then we can write

$$\frac{\partial}{\partial x}\left(\frac{k_o\rho_o}{\mu_o}C_{Moj}\frac{\partial p_o}{\partial x}+\frac{k_g\rho_g}{\mu_g}C_{Mgj}\frac{\partial p_g}{\partial x}+\frac{k_w\rho_w}{\mu_w}C_{Mwj}\frac{\partial p_w}{\partial x}\right)$$
$$=\frac{\partial}{\partial t}\left(\phi S_o\rho_o C_{Moj}+\phi S_g\rho_g C_{Mgj}+\phi S_w\rho_w C_{Mwj}\right) \tag{2–144}$$

Consider the conservation of mass applied to one compound. Let

C_{Moj} = mass fraction of jth component in oil
C_{Mgj} = mass fraction of jth component in gas
C_{Mwj} = mass fraction of jth component in water

Equation 2–117 describes the flow of a single component, e.g., CH_4 in a linear system without any sources or sinks. Equation 2–117 also shows that each term on the left represents the mass flux of the jth component in each phase, which is simply derived by the following:

Total mass flux = Density × Volumetric rate

$$=\rho_o q_o=\frac{k_o\rho_o}{\mu_o}\frac{\partial p_o}{\partial x} \tag{2–145}$$

$$\text{Component mass flux}=C_{Moj}\frac{k_o\rho_o}{\mu_o}\frac{\partial p_o}{\partial x} \tag{2–146}$$

Table 2–7

Unknown	Number
C_{mij}	$3N$
p_i	3
S_i	3
ρ_I	3
μ_I	3
k_I	3
	$3N + 15$

Note: $C_{mij} = 1, 2, 3$ $j = 1, \cdots, N$; total $= 3N$

Similarly, the accumulation term embodies the changes in each phase of the specific component:

$$\text{Mass rate of change} = \frac{\text{Mass at time } (t + \Delta t) - \text{Mass at time } t}{\Delta t}$$

A general equation for the N species under observation will be of the form

$$\frac{\partial}{\partial x}\left(\sum_{i=1}^{3} \frac{k_i \rho_i}{\mu_i} C_{Mij} \frac{\partial p_i}{\partial x}\right) = \frac{\partial}{\partial t}\left(\sum_{i=1}^{3} \phi S_i \rho_i C_{Mij}\right), \quad j = 1, \ldots, N \tag{2–147}$$

where

i = represents the phases and
j = the number of components.

We must determine the number of independent variables in the system. These data are listed in Table 2–7 for an N-component system.

In order to solve the system we must have $3N+15$ independent relationships. These relationships come from several sources:

1. Differential equations
2. Phase equilibrium
3. PVT data
4. Relative permeability data
5. Conservation principles
6. Capillary data

Relationship Development

Develop the necessary relationships as follows:

1. Write one partial differential equation for each component in the system, thus providing N relationships.
2. Since the pore space is always fluid-filled, the fluid phase saturations must always sum to unity:

$$S_o + S_g + S_w = 1 \tag{2–148}$$

   This is one relationship.
3. The mass fraction of each component in each fluid phase must sum to unity, since mass conservation of each component is required. Thus:

$$\sum_{j=1}^{N} C_{Moj} = 1$$
$$\sum_{j=1}^{N} C_{Mgj} = 1 \tag{2–149}$$
$$\sum_{j=1}^{N} C_{Mwj} = 1$$

   This provides three relationships.
4. The following can be obtained from the PVT data.

$$\begin{aligned} \mu_o &= \int (p_o, C_{Moj}) \\ \mu_g &= \int (p_g, C_{Mgj}) \\ \mu_w &= \int (p_w, C_{Mwj}) \end{aligned} \tag{2–150}$$

$$\begin{aligned} \rho_o &= \int (p_o, C_{Moj}) \\ \rho_g &= \int (p_g, C_{Mgj}) \\ \rho_w &= \int (p_w, C_{Mwj}) \end{aligned} \tag{2–151}$$

   Note: These provide six more relationships. Viscosity and density are computed experimentally or from well-known correlations, which relate these parameters to compositions and pressures.
5. For mobility calculations, we need relative permeability data:

$$\begin{aligned} k_o &= \int (S_g, S_o, S_w) \\ k_g &= \int (S_g, S_o, S_w) \\ k_w &= \int (S_g, S_o, S_w) \end{aligned} \tag{2–152}$$

   This provides three more relationships.

6. For distribution of a component between its liquid and gaseous states, the equilibrium constant can be derived from thermodynamic principles. For example,

$$\begin{aligned} \frac{C_{Mgj}}{C_{Moj}} &= K_{jgo} \\ \frac{C_{Mgj}}{C_{Mwj}} &= K_{jgw} \end{aligned} \tag{2–153}$$

These equilibrium constants are a function of several variables:

$$\begin{aligned} K_{jgo} &= \int (p, T, C_{ij}) \\ K_{jgw} &= \int (p, T, C_{ij}) \end{aligned} \tag{2–154}$$

from which

$$\frac{K_{jo}}{K_{jw}} = \frac{K_{jgw}}{K_{jgo}} = K_{gow} \tag{2–155}$$

Equations 2–154 and 2–155 provide an independent relationship when written for each component in the system.

7. Capillary pressure provides the remaining relationship:

$$\begin{aligned} p_g - p_o &= p_{cgo} = \int (S_g, S_o, S_w) \\ p_o - p_w &= p_{cow} = \int (S_g, S_o, S_w) \end{aligned} \tag{2–156}$$

These relationships are summarized in Table 2–8.

Therefore, according to Table 2–8, we have $3N + 15$ independent unknown and $3N + 15$ independent relationships that can be used to solve the system.

Assumptions

Several simplifying assumptions are usually made to make the problem more amenable to solution:

Table 2–8

Relationship	Unknown	Equations
Differential equation	N	2–147
Phase equilibrium	$2N$	2–153
PVT data	6	2–150 and 2–151
Relative permeability	3	2–152
$\sum$ Mass fraction	3	2–149
$\sum$ Saturation	1	2–148
Capillary pressure	2	2–156

1. Capillary pressure between oil and gas is generally neglected.
2. Several components are grouped together, e.g., a system containing the following nine components will be grouped as shown below:

C_1 Component 1

$$\left.\begin{array}{l} C_2 \\ C_3 \\ C_{i4} \\ C_{in} \\ C_{i5} \\ C_{n5} \\ C_6 \end{array}\right| \text{Component 2}$$

C_{7+} Component 3

3. The mass fraction of components present in the water is so small that the C_{Mwj} terms are also zero. This means that oil and gas are the only phases in which mass transfer occurs. The equation for the water present is still needed.

Sources and Sinks

Sources and sinks can be included in Eq. 2–139 by the addition of a term representing the source or sink:

$$\frac{\partial}{\partial x}\left(\sum_{i=1}^{3}\frac{K_i\rho_i}{\mu_i}C_{Mij}\frac{\partial p_i}{\partial x}\right)-\sum_{i=1}^{3}q_i\alpha_{ij}\delta(x)=\frac{\partial}{\partial t}\left(\sum_{i=1}^{3}\phi_i S_i\rho_i C_{Mij}\right) \tag{2–157}$$

where

q_i = Mass injection rate of phase in suitable units
α_{ij} = Mass fraction of jth component in ith phase
$\delta(x)$ = Delta function

The delta function $\delta(x)$ is defined as follows:

Production or injection in all at $x : \delta(x) = 1$
No production or injection in all at $x : \delta(x) = 0$

The locations of these wells are shown in Figure 2–9.

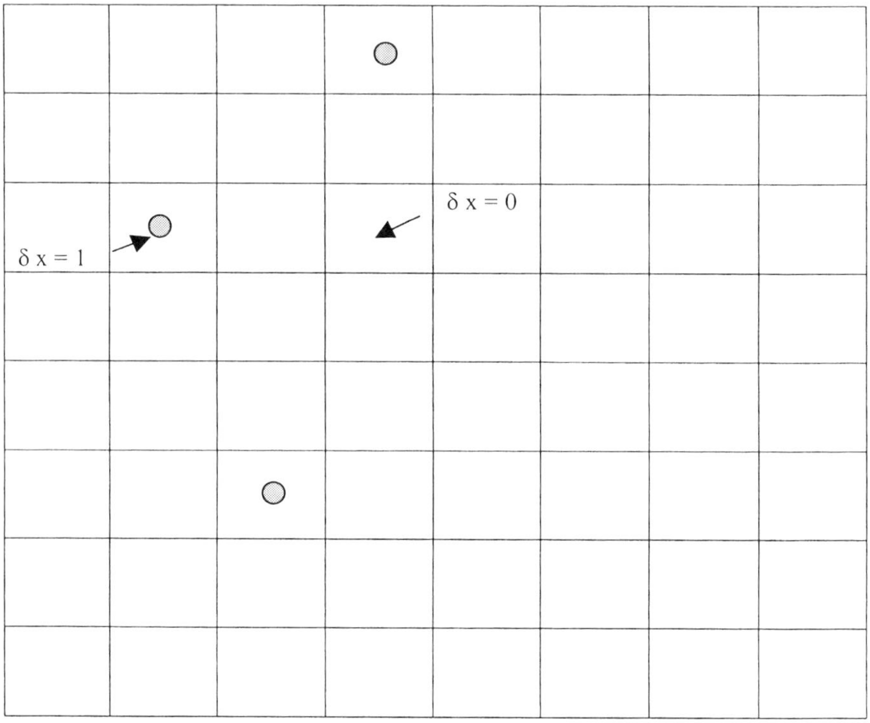

Figure 2–9. Well locations.

Procedure Outline for Solution of Flow Equations

The solution of the compositional model is an iterative one. The process indicated in Figure 2–10 is essentially the solution outline.

2.15 Summary

Chapter 2 provides the basic flow theory for gas well testing and analysis techniques. General equations are used for transient pressure behavior with dimensionless pressure solutions desired. Some important dimensionless pressure functions are presented in this chapter and references to others are provided. The dimensionless pressure approach provides a way to calculate pressure response and to devise techniques for analyzing transient tests in a variety of systems. Sections covering turbulence, wellbore storage effects, wellbore damage, and improvement are included, since the effects have a significant influence on transient well response.

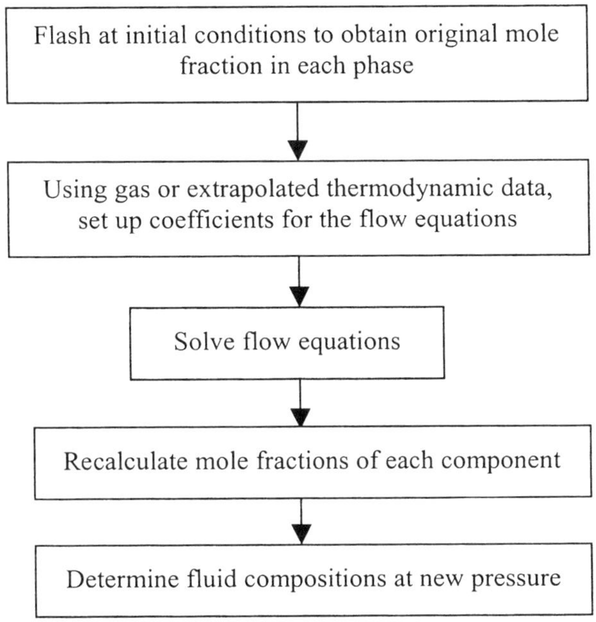

Figure 2–10. Solution Outline.

References and Additional Reading

1. Firoozabadi, A., and Katz, D. L., "An Analysis of High Velocity Gas Flow Through Porous Media." *J. Petroleum Technol.* (Feb. 1979), 221.
2. Al-Hussainy, R, Ramey, H. J., Jr., and Crawford, P. B., "The Flow of Real Gases through Porous Media," *J. Petroleum Technol.* (May 1966), 624–636; Trans. AIME, 237.
3. Watson, E. J., *Laplace Transforms and Applications*, van Nostrand Reinhold Company, New York.
4. Al-Hussainy, R., "Transient Flow of Ideal and Real Gases through Porous Media," Ph.D. Thesis, Texas A&M. University, 1967.
5. Lee, J., *Well Testing*, Vol. 1, SPE, Textbook Series, Society of Petroleum Engineers of AIME, Dallas, TX, 1982.
6. Al-Hussainy, R., "The Flow of Real Gases through Porous Media," M.Sc. Thesis, Texas A&M. University, 1965.
7. Aziz, K., Mattar, L., Ko. S., and Brar, G. S., "Use of Pressure, Pressure-Squared or Pseudo-Pressure in the Analysis of Gas Well Data," submitted for publication, 1975.
8. Van Everdingen, A. F., "The Skin Effect and Its Influence on the Productive Capacity of a Well," *Trans. AIME* (1953) 198, 171–176.

9. Hawkins, M. F., Jr., "A Note on the Skin Effect," Trans. AIME (1956) 207, 356–357.
10. Kirchhoff, H., *Vorlesungen iiber die Theorie der Warme*, Barth, Leipzig, 1894.
11. Horner, D. R., "Pressure Buildup in Wells," Proc. Third World Petroleum Conference; The Hague, Sec. II, 1951, 503–523.
12. Carter, R. D., "Performance Predictions for Gas Reservoirs Considering Two-Dimensional Unsteady-State Flow," *Soc. Petroleum Engineers J.* (1966) 6, 35–43.
13. Dietz, D. N., "Determination of Average Reservoir Pressure from Buildup Surveys," *J. Petroleum Technol.* (August 1965), 955–959.
14. Saidikowski, R. M., "Numerical Simulation of the Combined Effects of Wellbore Damage and Partial Penetration," paper SPE 8204, Sept. 23–26, 1979.
15. Matthews, C. S., and Russell, D. G., *Pressure Buildup and Flow Tests in Wells*, AIME Monograph, Vol. 1, SPE-AIME, New York, 1967.
16. Roebuck, I. F., Henderson, G. E., Douglas, J., Jr., and Ford, W. T., "The Compositional Reservoir Simulator: The Linear Model," *Trans. AIME* (1969) 246, 115.
17. Abel, W., Jackson, R. F., and Wattenbarger, R. A., "Simulation of a Partial Pressure Maintenance Gas Cycling Project with a Compositional Model, Carson Creek Field, Alberta," *J. Petroleum Technol.* (Jan. 1970) 38–46.
18. Van Everdingen, A. F., "The Skin Effect and Its Influence on the Productive Capacity of a Well," *Trans. AIME* (1953) 198, 171–176.
19. Van Poollen, H. K., "Radius of Investigation and Stabilization Time Equations," *Oil Gas J.* (1964) 63(51), 71–75.
20. Carslaw, H. S., and Jaeger, J. C., *Conduction of Heat in Solids*, Oxford University Press, London, 1959.
21. Katz, D. L., Cornell, D., Kobayashi, R., Poettmann, F. H., Vary, J. A., Elenbaas, J. R., and Weinaug. C. F., *Handbook of Natural Gas Engineering*, McGraw-Hill, New York, 1959.
22. Ramey, H. J., Jr., "Application of the Line Source Solution to Flow in Porous Media—A Review," *Producers Monthly* (1967) 31(5), 4–7 and 25–27.
23. Matthew, C. S., Brons, F., and Hazebroek, P., "A Method for Determination of Average Pressure in a Bounded Reservoir," *Trans. AIME* (1954) 201, 182–191.
24. Houpeurt, A., "On the Flow of Gases on Porous Media," *Revue de L' Institut Francais du Petrole* (1959). XIV(11), 1468–1684.
25. Smith, R. V., "Unsteady-State Gas Flow into Gas Wells," *J. Petroleum Technol.* (1961) 13, 1151–1159.

26. Bruce, G. H., Peacemen, D. W., Rachford, H. H., Jr., and Rice, J. D., "Calculation of Unsteady State Gas Flow Through Porous Media," *Trans. AIME* (1953) 198, 79–92.
27. Carter, R. D., "Solutions of Unsteady-State Radial Gas Flow," *J. Petroleum Technol.* (1962) 14, 549–554.
28. Collins, R. E., *Flow of Fluids through Porous Materials*, Reinhold Publishing Corporation; New York, 1961.
29. De Wiest, R. J. M. (ed.), *Flow through Porous Media*, Academic Press, New York, 1969.
30. Earlougher, R. C., Jr., Ramey, H. J., Jr., Miller, F. G., and Mueller, T. D., "Pressure Distributions in Rectangular Reservoirs," *J. Petroleum Technol.* (1968) 20, 199–208.
31. Derradii, S., "Bessel Functions, Laplace Transforms and Their Application," M.S. Report, University of Tulsa, Tulsa, OK, 1983.
32. Van Everdingen, A. F., and Hurst, W., "The Application of Laplace Transformation to Flow Problems in Reservoirs," *Trans. AIME* (1949) 186, 305–324.
33. Abramowitz, M., and Stegun, I. A. (eds.), *Handbook of Mathematical Functions with Formulas, Graphs and Mathematical Tables*, National Bureau of Standards Applied Mathematics Series 55 (June 1964) 227–253.
34. Chatas, A. T., "A Practical Treatment of Non-Steady State Flow Problems in Reservoirs Systems."
35. Watson, G. N., *Theory of Bessel Functions*, Cambridge University Press, London, 1944.

Chapter 3

Well Testing Techniques in Horizontal Gas Wells

3.1 Introduction

This chapter includes steady- and pseudo-steady-state equations for gas flow through a reservoir. For each subsection, flow equations for vertical wells are described first, followed by mathematical equations for horizontal wells. The mathematical determinations of the equations are avoided; this role is filled much better by other publications.[1–5] Field examples are included to provide hands-on understanding of various solution techniques and their applications.

3.2 Steady-State Gas Flow

Steady-state equations for gas flow through a reservoir are given below. Gas flow rate is proportional to the pressure-squared terms. This is generally employed when reservoir pressures are less than 2500 psia. In terms of gas pseudopressure, the gas flow rate is directly proportional to pseudopressure. The pseudopressure is defined as

$$\psi(p) = 2\int_{p_{\mathrm{Ref}}}^{p} \frac{p}{\mu z}\,dp \tag{3–1}$$

$$p_e^2 - p_{wf}^2 = \frac{50.337 \times 10^3 q_g \mu z T P_{sc}}{khT_{sc}} \ln\left(\frac{r_e}{r'_w}\right) \tag{3–2}$$

or

$$p_e^2 - p_{wf}^2 = \frac{115.981 \times 10^3 q_g \mu z T P_{sc}}{khT_{sc}} \log\left(\frac{r_e}{r'_w}\right) \tag{3–3}$$

Also,

$$\psi(p_e) - \psi(p_{wf}) = \frac{50.337 \times 10^3 q_g T P_{sc}}{khT_{sc}} \ln\left(\frac{r_e}{r'_w}\right) \quad (3\text{–}4)$$

$$\psi(p_e) - \psi(p_{wf}) = \frac{115.981 \times 10^3 q_g T P_{sc}}{khT_{sc}} \log\left(\frac{r_e}{r'_w}\right) \quad (3\text{–}5)$$

where

q_g = gas flow rate, mscfd
p_e = pressure at external radius, r_e, psia
p_{wf} = wellbore flowing, psia
$\bar{\mu}$ = average viscosity, cP
$\bar{z}$ = average gas compressibility, dimensionless
T = reservoir temperature, °R
r_e = drainage radius, ft
r'_w = effective wellbore radius, ft
k = permeability, mD
h = thickness, ft
P_{sc} = base pressure, psia
T_{sc} = base temperature, °R

The correct value of $(1/\mu z)_{avg}$ to use in the equation for gas is $(2\int_{p_{wf}}^{p} p/\mu z)/(p_e^2 - p_{wf}^2)$. It has been found that for most natural gases a value of $(1/\mu z)_{avg}$ evaluated at the arithmetic average pressure $(p_e - p_{wf})/2$ will be reasonably accurate. Equations 3–2 through 3–5 are for vertical wells; they can be used to calculate steady-state gas production rate from a horizontal well. This is accomplished by substituting effective wellbore radius r'_w of a horizontal well in the foregoing equations. The following equations can be used to calculate effective wellbore radius r'_w for a horizontal gas well. Figure 3–1 shows major and minor axes of a horizontal well. Assume elliptical drainage area:

$$\text{Drainage area} = \pi r_{eh}^2 = \pi a' b'$$

where

a' = half the major axis of an ellipse $= L/2 + r_{ev}$
b' = half the minor axis of an ellipse $= r_{ev}$
L = length of wellbore, ft
$r_{ev} = \left(\frac{\text{acres} \times 43{,}560}{\pi}\right)^{0.5}$, ft

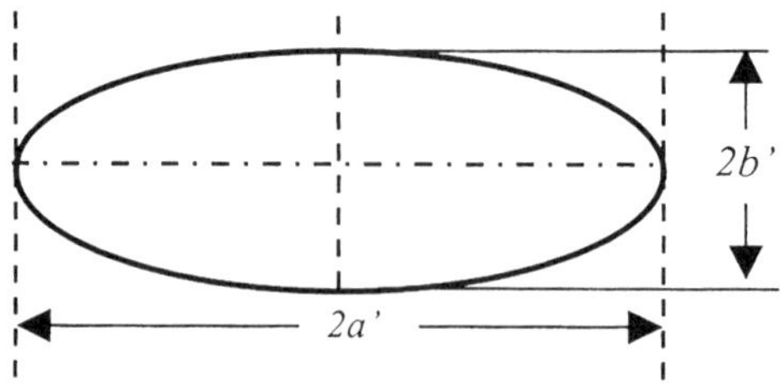

Figure 3–1. Major and minor axes of elliptical drainage area of horizontal gas well.

Radius of Vertical Well

The effective wellbore radius, r'_w of a horizontal well is

$$r_{eh} = [(L/2 + r_{ev})(r_{ev})]^{0.5}, \text{ ft} \tag{3–6}$$

$$r'_w = \frac{r_{eh}(L/2)}{a[1+\sqrt{1-(L/2a)^2}][\beta h/(2r_w)]^{h/L}} \tag{3–7}$$

where

$$a = 0.5 \times L[0.5 + (0.25 + (2r_{eh}/L)^4)^{0.5}]^{0.5} \tag{3–8}$$

$$\beta = (k_h/k_v)^{0.5} \tag{3–8a}$$

Example 3–1 *Steady-State Gas-Flow Rate Calculations (Infinite-Conductivity Fracture)*

Estimate steady-state gas flow rate from a well with a 120-ft-long infinite conductivity fracture. The 6-ft-thick reservoir has a pressure of 2000 psia. The wellbore diameter is $9\frac{1}{2}$ in. Given data are: Base pressure $P_{sc} = 14.7$ psia; base temperature $T_{sc} = 520°$R; reservoir temperature $T = 110°$F; gas compressibility $z = 0.9500$; gas viscosity $\mu = 0.0250$ cP; reservoir permeability $k_h = 0.1$ mD; vertical permeability $k_v = 0.1$ mD; well flowing pressure $p_{wf} = 400$ psia; and drainage area $A = 40$ acres.

Solution Drainage radius, $r_e = \sqrt{\frac{40 \times 43{,}560}{22/7}} = 745$ ft

$$\beta = (0.1/0.1)^{0.5} = 1$$

Equation 3–2 can be rearranged as

$$q_g = \frac{0.019866 \times 10^{-3} khT_{sc}\left(p_e^2 - p_{wf}^2\right)}{\bar{\mu}\bar{z}TP_{sc}\ln(r_e/r'_w)} \tag{3–2a}$$

For an infinite-conductivity fracture of total length 120 ft, half-length x_f is 60 ft. The effective wellbore radius r'_w of an infinite-conductivity fracture is

$$r'_w = x_f/2 = 60/2 = 30 \text{ ft}$$

Substituting the value of r'_w in Eq. 3–2:

$$q_g = \frac{0.019866 \times 10^{-3} \times 0.1 \times 60 \times 520(2000^2 - 400^2)}{0.0250 \times 0.950 \times (110 + 460) \times 14.7 \ln(745/60)}$$

$$= 1196.03/\ln(12.4167) = 1196.03/2.519 = 474.8 \text{ mscfd}$$

Example 3–2 *Calculating Flow Rate for Horizontal Well Assuming Steady-State Conditions*

Two horizontal wells are 1000 and 2000 ft long and drain 100 and 150 acres, respectively. All reservoir properties and drainage areas are the same as those noted in Example 3–1.

Solution The steady-state gas flow rate can be calculated using Eq. 3–2:

$$q_g = \frac{0.019866 \times 10^{-3} khT_{sc}\left(p_e^2 - p_w^2\right)}{\bar{\mu}\bar{z}TP_{sc}\ln(r_e/r_w)}$$

Using the same reservoir properties as Example 3–1, we have

$$q_g = 1196.03/\ln(r_e/r'_w), \text{ mscfd}$$

For a horizontal gas well, r'_w is calculated from Eq. 3–7:

$$r'_w = \frac{r_{eh}(L/2)}{a[1 + \sqrt{1 - (L/2a)^2}][\beta h/(2r_w)]^{h/L}}$$

For 100 acres,

$$r_{ev} = \left(\frac{100 \times 43{,}560}{22/7}\right)^{0.5} = 1177.29 \text{ ft}$$

Therefore for a 1000-ft-long well from Eq. 3–6:

$$r_{eh} = \left[\left(\frac{1000}{2} + 1177.29\right)(1177.29)\right]^{0.5} = 1405 \text{ ft}$$

and from Eq. 3–8:

$$a = 0.5 \times 1000[0.5 + (0.25 + (2 \times 1405/1000)^4)^{0.5}]^{0.5}$$
$$= 500[0.5 + (0.25 + 62.35)^{0.5}]^{0.5} = 500[0.5 + 7.92]^{0.5} = 1450 \text{ ft}$$

Substituting this into Eq. 3–7:

$$r'_w = \frac{1405(1000/2)}{1450[1+\sqrt{1-(1000/(2\times 1450))^2}][1\times 60/0.792]^{60/1000}}$$

$$= \frac{702{,}500}{1450[1+\sqrt{1-0.344829}][75.7576]^{0.06}}$$

$$= \frac{702{,}500}{1450[1+0.65517(1.296479)]} = \frac{702{,}500}{2681.651} = 261.97 \text{ ft}$$

Therefore

$$q_g = 1196.03/\ln\left(\frac{1405}{261.97}\right) = 1196.03/1.67956 = 712.108 \text{ mscfd}$$

For 150 acres,

$$r_{eh} = \sqrt{\frac{150\times 43{,}560}{22/7}} = 1442 \text{ ft}$$

For a 2000-ft-long well, with 150-acre well spacing, using the procedure shown above:

$$a = 1624 \text{ ft}$$
$$r'_w = 520.20 \text{ ft}$$
$$q_g = 1173.070 \text{ mscfd}$$

Thus, comparing the results of Examples 3–1 and 3–2, one can see that for a steady state, a fractured vertical well with 40-acre spacing can produce at a rate of 475 mscfd. A 2000-ft-long horizontal well at 100-acre spacing can produce 712.108 mscfd. A 2000-ft-long horizontal well at 150-acre well spacing can produce 1173.070 mscfd.

3.3 Pressure Transient Characteristics in Horizontal Gas Wells

Horizontal wells have high potential in gas reservoirs. They are suitable and applicable in low- as well as high-permeability gas formations. In low-permeability formations, they drain larger volumes than vertical wells and provide an alternative to achieve long penetration lengths to the formations. They reduce near-wellbore turbulence and enhance well deliverability in high-permeability gas formations. Figure 3–1a illustrates the character of the log-log plot of the pressure—time data in conjunction with the derivative curve. This figure also shows the effects of wellbore damage associated with the minimum

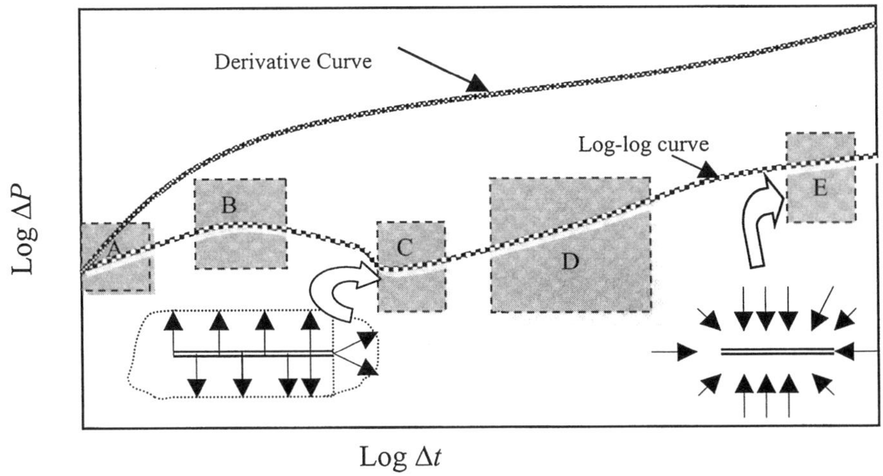

A → Duration of storage effects
B → Wellbore damage
C → Portion to estimate vertical permeability (Early radial flow)
D → Effective length of horizontal well
E → Portion to estimate horizontal permeability and reservoir pressure. (Pseudo radial flow)

Figure 3–1a. Pressure transient responses and major flow regimes in horizontal gas well (after Daviau *et al.*).[7]

and maximum flexures (i.e., zero curves on the derivative curve). Each of the flow regimes, which ideally establish a unique shape on the log-log plot, provides an opportunity for estimating particular reservoir parameters that are much more difficult to determine during other flow regimes[6] (see Figure 3–1a).

Wellbore Storage Effects

Horizontal wellbore storage can have serious consequences on the effectiveness of a pressure transient test, even when the measurement tool is located below a down-hole shut-in device. Reference 7 has shown that the first semilog straight line associated with early-time radial flow almost always disappears because of the effects of wellbore storage. Wellbore storage effects lasts longer than for a vertical well in the same formation, because of greater wellbore volume and also because anisotropy reduces the effective permeability for a horizontal well. At the present time, for horizontal wells, methods are not available to estimate the precise time required to end the wellbore storage flow regime.

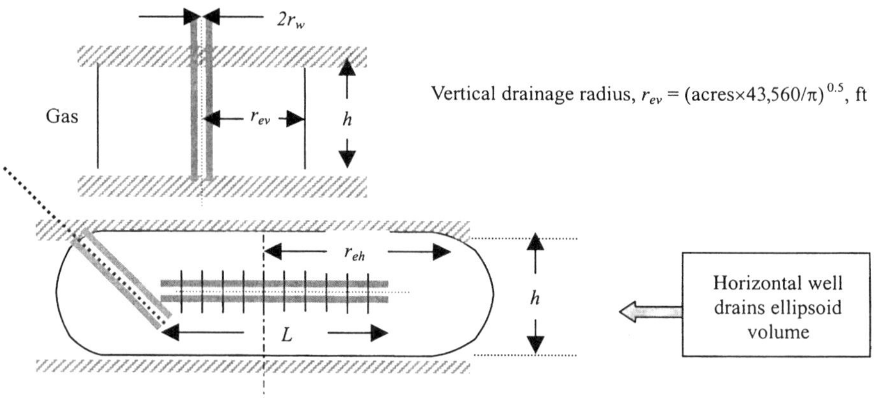

Figure 3–2. Horizontal and vertical well drainage area.

Horizontal Well Drainage Area Concepts

In general, a horizontal well drains a larger reservoir volume than a vertical well. A horizontal well drains an ellipsoid volume whereas a vertical well drains a cylindrical volume. As a rule of thumb, a 1000-ft-long horizontal well can drain two times the area of a vertical well while a 2000-ft-long well can drain three times the area of a vertical well in a given time. Figures 3–2 through 3–4 show drainage areas for vertical and a horizontal wells. The area drained can be calculated by two methods. The first method assumes that a horizontal well drains an ellipse with minor axis a′ and major axis b′. The second method is represented as two half circles of radius r_{ev} at each end and a rectangle, of dimensions $L \times 2r_{ev}$, in the center. Tables 3–1 and 3–2 show the calculations for the first method.

Concept of Skin Factor in Horizontal Wells

Normally, skin factors are determined using drill stem testing (DST) or pressure buildup testing. If the skin factor is known, the pressure drops across the damaged zones can be estimated by the following equations:

$$|\Delta p_{skin}|_{Vertical} = \left[\frac{141.2\mu_g\beta_g}{k}\right] \cdot \frac{q_g}{h}(s) \tag{3–9}$$

$$|\Delta p_{skin}|_{Horizontal} = \left[\frac{141.2\mu_g\beta_g}{\sqrt{k_v k_h}}\right] \cdot \frac{q_g}{L}(s) \tag{3–10}$$

Equations 3–9 and 3–10 indicate that for the same positive skin factor, s, the pressure loss in the skin region is smaller than that in a vertical well (see Figure 3–5).

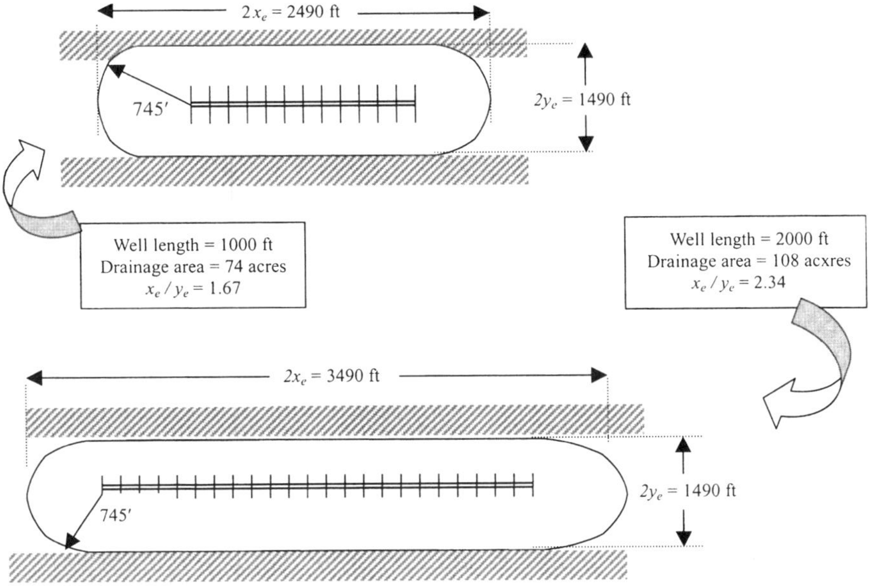

Figure 3–3. Drainage areas of 1000- and 2000-ft-long horizontal wells.

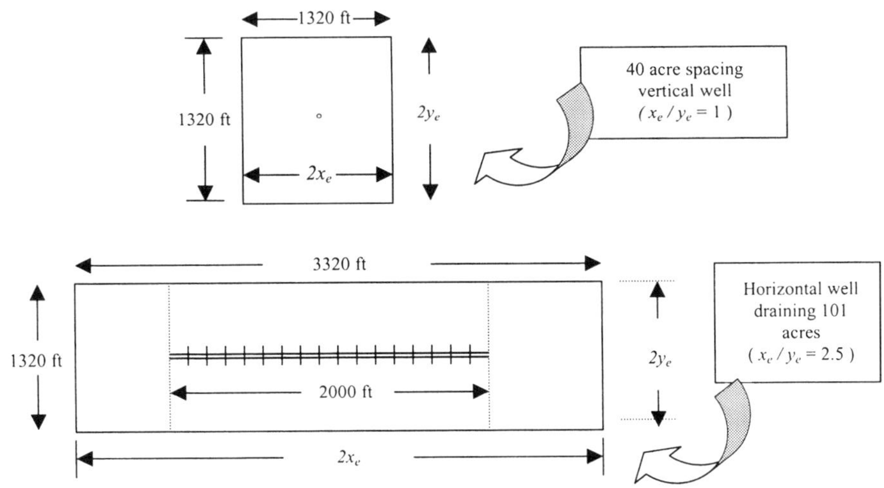

Figure 3–4. Horizontal and vertical well drainage areas for a given time.

Table 3–1
Various Horizontal Well Lengths Draining 400-Acre Lease

Horizontal well length, L (ft)	Half minor axis b'(ft) r_{ev}	Half major axis a' (ft) $L/2+r_{ev}$	Drainage area A, acre $\pi a'b'/43{,}560$	r_{eh}, (ft) $(A\times 43{,}560/\pi)^{0.5}$	Number of wells for 400-acre field $400/A$
500	745	995	53.5	861	7
1000	745	1245	66.9	963	6
2000	745	1745	93.8	1140	4

Table 3–2
Various Horizontal Well Lengths Draining 600-Acre Lease

Horizontal well length L (ft)	Half minor axis b' (ft) r_{ev}	Half major axis a' (ft) $L/2+r_{ev}$	Drainage area A (acre) $\pi a'b'/43{,}560$	r_{eh} (ft)	Number of wells for 600-acre field $600/A$
500	912	1162	76.4	1029	8
1000	912	1412	92.9	1135	6
2000	912	1912	125.8	1320	4 or 5

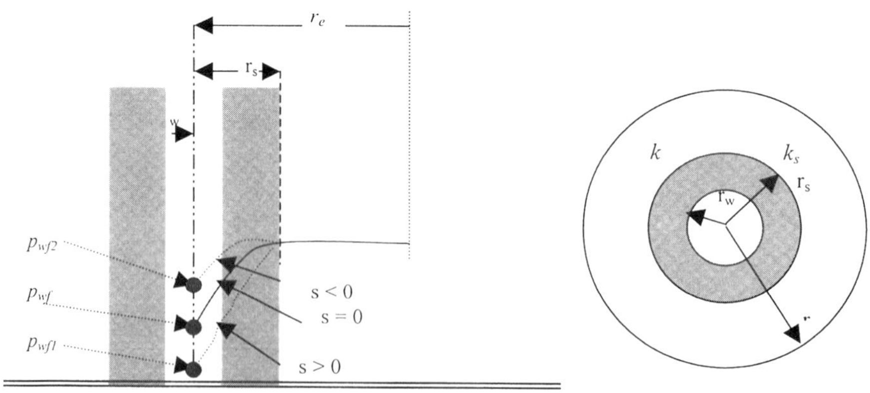

Figure 3–5. Skin factor effects on well flow performance.

Relationships between Effective Wellbore Radius and Skin Factor

The equations listed below along with their practical applications are presented in subsequent sections.

$$s = 1.151\left[\frac{\psi(p_i) - \psi(p_{wf})}{m_h} - \log\left(\frac{k_h}{\phi\mu_g c_t r_w^2}\right) + 3.23\right] \quad (3\text{–}11)$$

$$[(\Delta\psi)_{skin}]_{Horizontal} = 0.869\, m_h\, s \quad (3\text{–}12)$$

$$s = -\ln(r'_w/r_w) \quad \text{or} \quad r'_w = r_w e^{(-s)} \quad (3\text{–}13)$$

$$r'_w = \frac{r_{eh}(L/2)}{a[1 + \sqrt{1 - [L/(2a)]^2}][h/(2r_w)]^{h/L}} \quad (3\text{–}14)$$

where

$$a = L/2[0.5 + \sqrt{0.25 + (2r_{eh}/L)^4}]^{0.5} \quad (3\text{–}15)$$

$$r_{eh} = [(L/2 + r_{ev})(r_{ev})]^{0.5}, \text{ ft} \quad (3\text{–}6)$$

where

Acre spacing $= \pi a' b'/43{,}560$
$a' =$ half major axis $= L/2 + r_{ev}$
$b' =$ half minor axis $= r_{ev}$
Drainage area of vertical well in acres $= A$
$r_{ev} = (A \times 43{,}560/\pi)^{0.5}$
A is the drainage area of vertical well in acres

3.4 Pseudo-Steady-State Gas Flow

The pseudo-steady-state equations for vertical wells, fracture vertical wells and horizontal gas wells on the basis of circular drainage area are

$$\bar{p}_R^2 - p_{wf}^2 = \frac{50.335 \times 10^{-3} q_g \bar{\mu}\bar{z} T P_{sc}}{k h T_{sc}} \times [\ln(r_e/r_w) - AC' + s + s_m + s_{CA} - c' + Dq_g] \quad (3\text{–}16)$$

$$\psi(\bar{p}_R) - \psi(p_{wf}) = \frac{50.337 \times 10^3 q_g T P_{sc}}{k h T_{sc}} \times [\ln(r_e/r_w) - AC' + s + s_m + s_{CA} - c' + Dq_g] \quad (3\text{–}17)$$

where

$$D = \frac{2.222 \times 10^{-15} \times \gamma_g k_a h \beta'}{\mu_{pwf} r_w h_p^2} \tag{3–18}$$

$$\beta' = 2.73 \times 10^{10} k_a^{-1.1045} \tag{3–19}$$

or

$$\beta' = 2.33 \times 10^{10} k_a^{-1.201} \tag{3–20}$$

where

s = equivalent negative skin factor due to either well stimulation or horizontal well
s_m = mechanical skin factor, dimensionless
s_{CA} = shape-related skin factor, dimensionless
c' = shape factor conversion constant, dimensionless
k = permeability, mD
h = reservoir height, ft
p_R = average reservoir pressure, psia
p_{wf} = well flowing pressure, psia
q_g = gas flow rate, mscfd
T = reservoir temperature, °R
$\bar{\mu}$ = gas viscosity evaluated at some average pressure between $\bar{p}_R$ and p_{wf}, cP
μ_g = gas viscosity at well flowing conditions, cP
$\bar{z}$ = gas compressibility factor evaluated at some average pressure between $\bar{p}_R$ and p_{wf}
β' = high velocity flow coefficient, 1/ft
γ_g = gas gravity, dimensionless
r_w = wellbore radius, ft
h_p = perforated interval, ft
k_a = permeability in the near wellbore region, mD
AC′ = 0.75 for a circular drainage area and AC′ = 0.738 for a rectangular drainage area

Equation 3–19 for β' is given in Ref. 6 while Eq. 3–20 is given in Refs. 2 and 8. Depending upon the β' definition used a somewhat different answer will be obtained. We can also write similar equations on the basis of square drainage area:

$$\bar{p}_R^2 - p_{wf}^2 = \frac{50.335 \times 10^{-3} q_g \bar{\mu} \bar{z} T P_{sc}}{k h T_{sc}} \times [\ln(r_e/r_w) - 0.738 + s + s_m + s_{CA} - c' + D q_g] \tag{3–21}$$

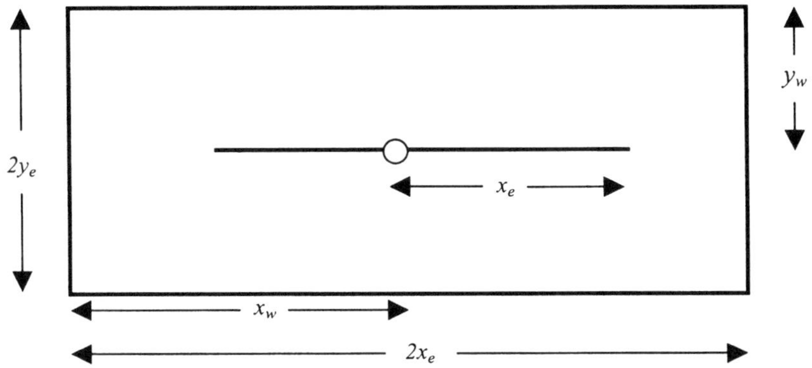

Figure 3–6. An areal view of a fractured gas vertical well.

$$\psi(\bar{p}_R) - \psi(p_{wf}) = \frac{50.337 \times 10^3 q_g T P_{sc}}{k h T_{sc}} [\ln(r_e/r_w) - 0.738 + s + s_m + s_{CA} - c' + D q_g] \quad (3\text{–}22)$$

Shape-Related Skin Factors for Vertical and Fractured Gas Wells

In the preceding equations, definitions of s_{CA} and c'depend upon the type of well. For a vertical well,

$c' = 0$, s_{CA} from Table 3–4

For a fractured vertical well (Figure 3–6),

$c' = 1.386$ and $s_{CA} = s_{CA,f}$

$s_{CA,f} = \ln \sqrt{30.88/c_f}$

c_f is obtained from Table 3–3.

Shape Factors for Horizontal Gas Wells

For a horizontal well (Figure 3–7),

$c' = 1.386$, and $s_{CA} = s_{CA,h}$

$s_{CA,h}$ is obtained from Figures 3–8 through 3–11 or Table 3–4.

Dq_g is a turbulence term and is called turbulence skin or rate dependent skin factor.[2,6] Equations 3–21 and 3–22 can be used for different well flowing

Table 3–3
Shape Factors C_f for Fractured Vertical Wells in a Square Drainage Area[3]

x_f/x_e	Shape factor, C_f
0.1	2.6541
0.2	2.0348
0.3	1.9986
0.5	1.6620
0.7	1.3127
1.0	0.7887

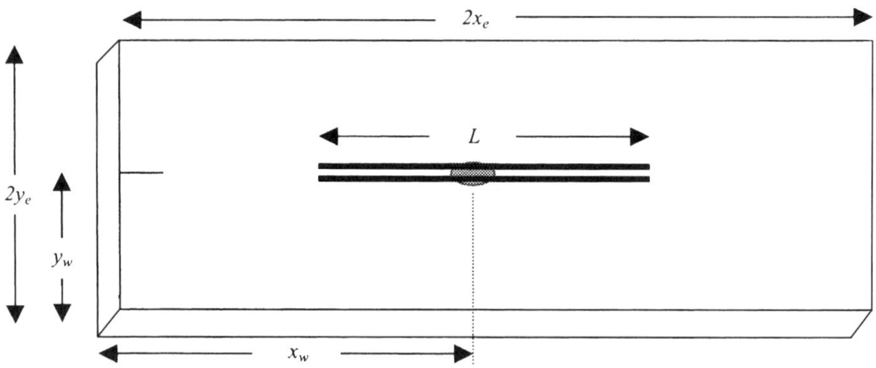

Figure 3–7. A schematic of a horizontal well located in a rectangular drainage volume.

pressures to calculate gas flow rates. Equations 3–21 and 3–22 are rewritten as

$$q_g = \frac{0.019866 \times 10^{-3} khT_{sc}\left(\bar{p}_R^2 - p_{wf}^2\right)}{\bar{\mu}\bar{z}TP_{sc}[\ln(r_e/r_w) - 0.75 + s + s_m + s_{CA} + c' + Dq_g]} = \frac{CC}{BB + Dq_g} \tag{3–23}$$

$$q_g = \frac{0.019866 \times 10^{-3} khT_{sc}(\psi(\bar{p}_R) - \psi(p_{wf}))}{TP_{sc}\left[\ln\left(\frac{r_e}{r_w}\right) - 0.75 + s + s_m + s_{CA} - c' + Dq_g\right]} = \frac{CC'}{BB' + Dq_g} \tag{3–24}$$

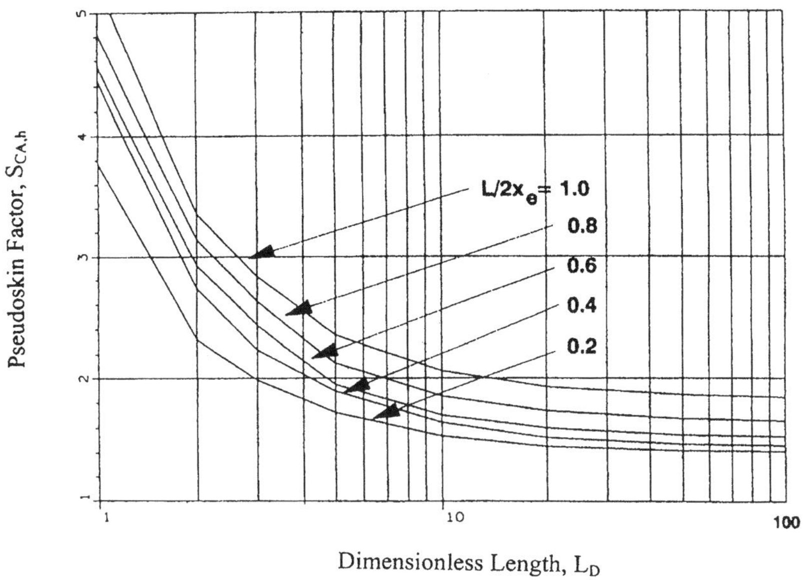

Figure 3–8. Shape related skin factor $S_{CA,h}$ for a horizontal well in a square drainage area ($x_e/y_e = 1$).[8]

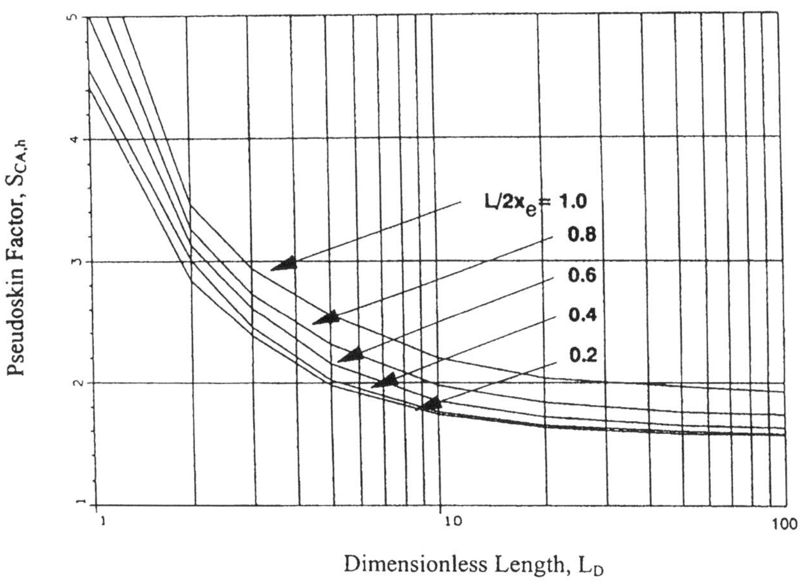

Figure 3–9. Shape related skin factor $S_{CA,h}$, for a horizontal well located in a rectangular drainage area ($x_e/y_e = 2$).[8]

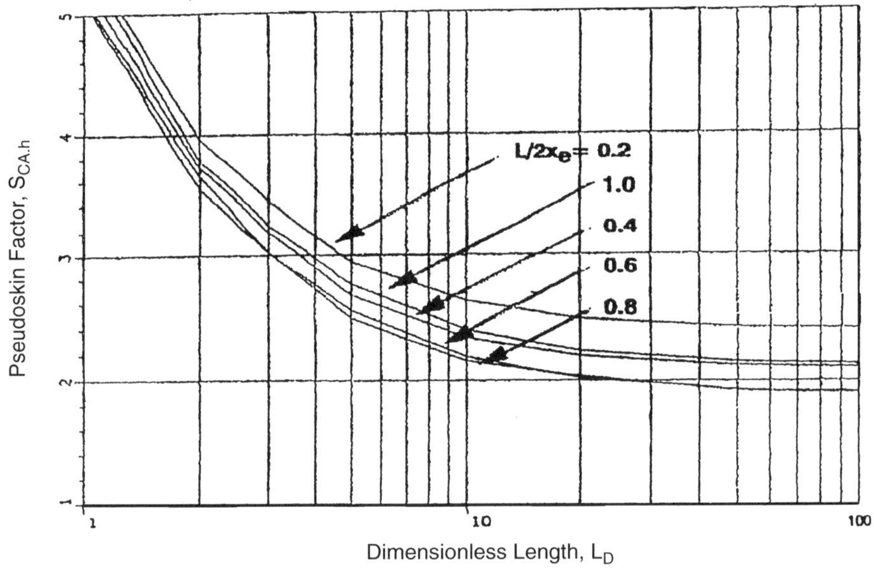

Figure 3–10. Shape related skin factor $S_{CA,h}$ for a horizontal well located in a rectangular drainage area ($x_e/y_e = 5$).[8]

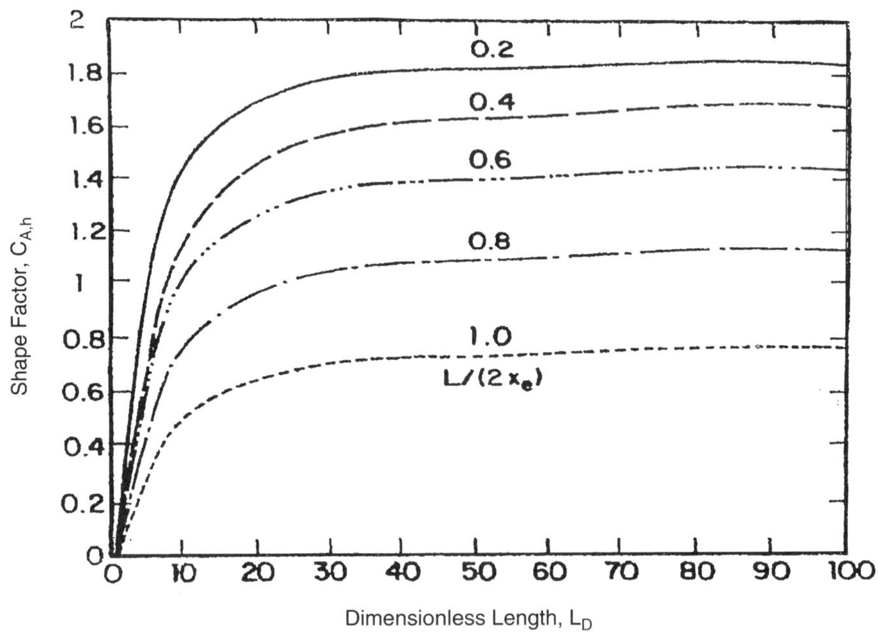

Figure 3–11. Shape factor $C_{A,h}$ for a horizontal well located in a square drainage area for different dimensionless length.[8]

Table 3–4
Shape Factor Dependent Skin Factors, S_{CA}, for Vertical Wells (after Fetkovich and Vienot)[4]

Geometry	C_A	S'_{CA}	t_{Dapss}
	31.62	0.000	0.1
	30.88	0.012	0.1
	31.60	0.000	0.1
	27.60	0.068	0.2
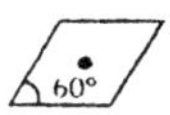	27.10	0.077	0.2
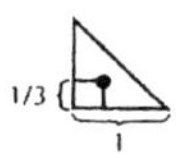	21.90	0.184	0.4
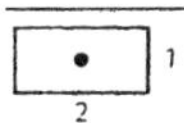	21.84	0.185	0.3
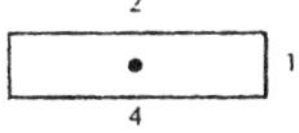	5.379	0.886	0.8
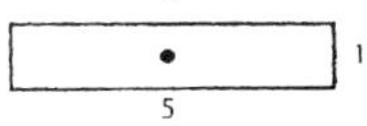	2.361	1.298	1.0
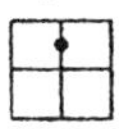	12.98	0.445	0.7
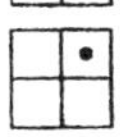	4.513	0.973	0.6
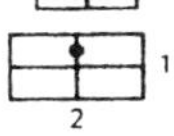	10.84	0.535	0.4
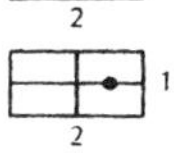	4.514	0.973	2.5
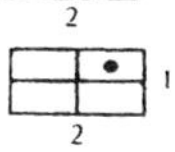	2.077	1.362	1.7
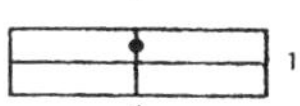	2.690	1.232	0.8
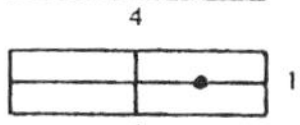	0.232	2.458	4.0
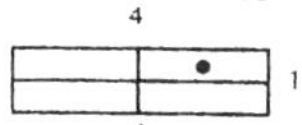	0.115	2.806	4.0

Equations 3–23 and 3–24 are quadratic equations and can be written as

$$Dq_g^2 + BBq_g - CC = 0 \tag{3–25}$$

$$Dq_g^2 + BB'q_g - CC' = 0 \tag{3–26}$$

and

$$q_g = \frac{-BB + \sqrt{(BB)^2 + 4D(CC)}}{2D} \tag{3–27}$$

$$q_g = \frac{-BB' + \sqrt{(BB')^2 + 4D(CC')}}{2D} \tag{3–28}$$

To solve Eqs. 3–27 and 3–28 we need to calculate turbulence factor D using Eqs. 3–18 and 3–19 and assuming $k = k_a$.

Calculation of Skin Factor for Horizontal Gas Well

$$r'_w = L/4 \tag{3–29}$$

$$s = -\ln(r'_w/r_w) \tag{3–30}$$

Figure 3–12 shows a graphic correlation for shape factor C_f.[10] To convert this shape factor C'_f to C_f, the following equation can be used: $C_f = 0.25C'_f$. Instead of calculating shape factors, one can adjust effective wellbore radius of a fractured vertical well to account for both fracture length and shape factor.[10] Figure 3–13 shows a plot of effective wellbore radius for vertical wells with uniform flux and infinite conductivity fractures for different fracture penetration.[7] The effective wellbore radius, r'_w, calculated from Figure 3–13 can be directly substituted in place of r_w in the following equations to calculate gas flow rate in fractured vertical wells, where the vertical well is located centrally in square drainage area. These results can also be extended to rectangular drainage boundaries for varying $2x_e/(2y_e)$ ratio by replacing (x_f/x_e) with $(2x_f/\sqrt{A})$ on the x-axis or Figure 3–13.

$$q_g = \frac{0.0007027kh(\bar{p}_R^2 - p_{wf}^2)}{T\bar{z}\bar{\mu}[\ln(r_e/r_w) - 0.75]}$$

and

$$q_g = \frac{0.0007027kh[\psi(\bar{p}_R) - \psi(p_{wf})]}{T[\ln(r_e/r_w) - 0.75]}$$

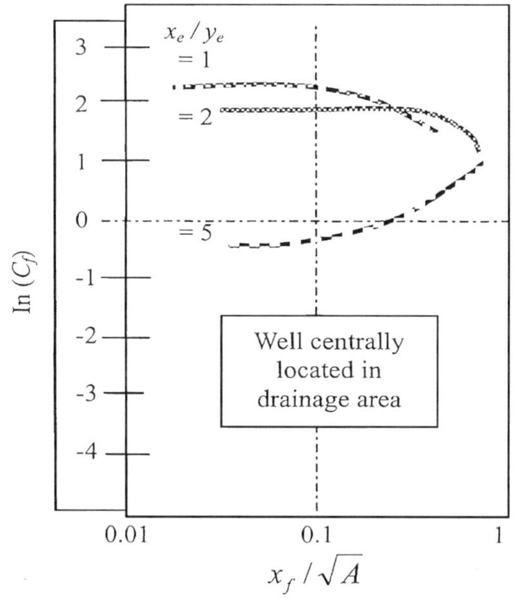

Figure 3–12. Shape factors for fractured vertical wells for different fractured penetration (after Gringarten).[10]

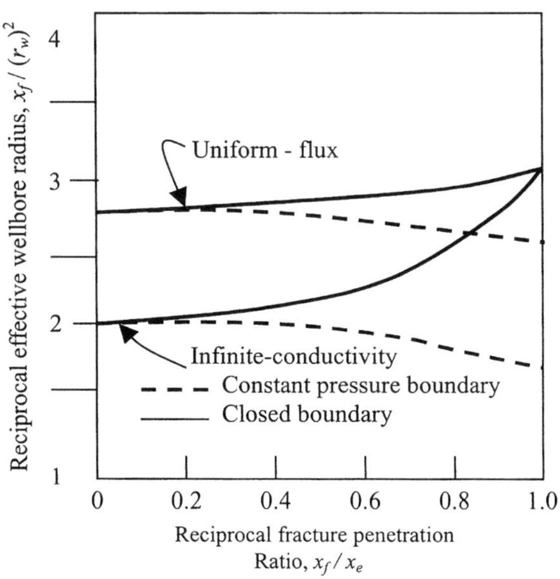

Figure 3–13. Effective wellbore radius for fractured vertical wells for different fracture penetration (after Gringarten).[11]

Shape factors C_f for off-centered-fractured vertical wells[9] are shown in Table 3–5. Shape-related skin factors S_{Cah} for horizontal wells for various well penetrations and different rectangular drainage areas are shown in Table 3–6.

Shape factors, C_f, for fractured vertical wells located centrally in the rectangular drainage area[9] are shown in Table 3–7.

Calculation of s_{CA}

Assuming a square drainage area with each side being $2x_e$ for 640 acres, we have

$$2x_e = \sqrt{\text{acre} \times 43{,}560}, \text{ ft} \tag{3–31}$$

Find $L/(2x_e)$, k_v/k_h and

$$L_D = (L/(2h))\sqrt{k_v/k_h} \tag{3–32}$$

From Figure 3–8 or Table 3–6, corresponding, to L_D and $L/2x_e$, find s_{CA}.

3.5 Horizontal Transient Well Testing Techniques

Possible Flow Regimes and Analytical Solutions

Figure 3–15 shows possible four transient flow regimes depending on the well length relative to the reservoir thickness and drainage area.[6,13] Under certain circumstances, permeability, k anisotropy, and skin factors can be estimated by analyzing these transient flow pressure data. Time and pressure response equations relating to each of the flow regimes to solve for specific reservoir parameters for drawdown and buildup tests can be found in the next sections. Figure 3–14 illustrates the horizontal model.

Well Test Planning Using Flow Time Equations

There are four transient flow regimes that are theoretically possible with a buildup or drawdown test in a horizontal well. In chronological order of development, they are:

Early-Time Radial Flow. The flow is radial and is equivalent to a fully penetrating vertical well in an infinite reservoir (see Figure 3–16).

Intermediate-Time Linear Flow Equation. A horizontal well will generally be long compared to the formation thickness; a period of linear flow may develop once the pressure transient reaches the upper and lower boundaries (see Figure 3–17).

Table 3–5
Shape Factors C_f for Off-Centered-Fractured Vertical Wells[9]

←Influence of y_w/y_e^*→				
y_w/y_e	x_f/x_e	**0.25**	**0.50**	**1.00**
$x_e/y_e = 1$				
	0.1	0.2240	0.8522	2.0200
	0.3	0.2365	0.7880	1.8220
	0.5	0.2401	0.7165	1.6040
	0.7	0.2004	0.5278	1.3170
	1.0	0.1451	0.3606	0.7909
$x_e/y_e = 2$				
	0.1	0.2272	0.7140	1.4100
	0.3	0.3355	0.7700	1.3610
	0.5	0.4325	0.8120	1.2890
	0.7	0.4431	0.7460	1.1105
	1.0	0.2754	0.4499	0.6600
$x_e/y_e = 5$				
	0.1	0.0375	0.09185	0.2110
	0.3	0.1271	0.20320	0.2864
	0.5	0.2758	0.38110	0.4841
	0.7	0.3851	0.49400	0.5960
	1.0	0.2557	0.31120	0.3642
←Influence of x_w/x_e→				
x_w/x_e	x_f/x_e	**0.50**	**0.75**	**1.00**
$x_e/y_e = 1$				
	0.1	0.9694	1.7440	2.0200
	0.3	1.1260	1.7800	1.8200
	0.5	1.2708	1.7800	1.6000
$x_e/y_e = 2$				
	0.1	0.3679	1.0680	1.4098
	0.3	0.5630	1.2980	1.3611
	0.5	0.8451	1.5470	2.2890
$x_e/y_e = 5$				
	0.1	0.0058	0.0828	0.2110
	0.3	0.0317	0.2540	0.2864
	0.5	0.1690	0.7634	0.6050

*x_w and y_w represent the distance of the fracture center from the nearest y and x boundary, respectively.

Table 3–6
Shape-Related Skin Factors S_{Cah} for Horizontal Wells for Various Well Penetrations and Different Rectangular Drainage Areas[9]

			L/(2x_e)		
t_D	0.2	0.4	0.6	0.8	0.10
(1) $x_e/y_e = 1$					
1	3.772	4.439	4.557	4.819	5.250
2	2.231	2.732	2.927	3.141	3.354
3	1.983	2.240	2.437	2.626	2.832
5	1.724	1.891	1.948	2.125	2.356
10	1.536	1.644	1.703	1.851	2.061
20	1.452	1.526	1.598	1.733	1.930
50	1.420	1.471	1.546	1.672	1.863
100	1.412	1.458	1.533	1.656	1.845
(2) $x_e/y_e = 2$					
1	4.425	4.578	5.025	5.420	5.860
2	2.840	3.010	3.130	3.260	3.460
3	2.380	2.459	2.610	2.730	2.940
5	1.982	2.020	2.150	2.310	2.545
10	1.740	1.763	1.850	1.983	2.198
20	1.635	1.651	1.720	1.839	2.040
50	1.584	1.596	1.650	1.762	1.959
100	1.572	1.582	1.632	1.740	1.935
(3) $x_e/y_e = 5$					
1	5.500	5.270	5.110	5.140	5.440
2	3.960	3.720	3.540	3.650	3.780
3	3.440	3.190	3.020	3.020	3.250
5	2.942	2.667	2.554	2.493	2.758
10	2.629	2.343	2.289	2.155	2.399
20	2.491	2.196	2.022	2.044	2.236
50	2.420	2.120	1.934	1.925	2.150
100	2.408	2.100	1.909	1.903	2.136

Late-Time Radial Flow Equation. If the horizontal well length is sufficiently small compared to the reservoir size, a second radial flow known as pseudoradial flow will develop at late times (see Figure 3–18).

Late-Time Linear Flow Equation. This flow period occurs when the pressure transient reaches the lateral extremities of the reservoir. This second and final linear-flow period develops only for reservoirs of finite width. The identification of these flow regimes is critical to the proper interpretation of a horizontal well test (see Figure 3–19).

Table 3–7
Shape Factors, C_f, for Fractured Vertical Wells Located Centrally in a Rectangular Drainage Area

C_f x_f/x_e	1	2	3	5	10	20
0.1	2.020	1.4100	0.751	0.2110	0.0026	0.000005
0.3	1.820	1.3611	0.836	0.2860	0.0205	0.000140
0.5	1.600	1.2890	0.924	0.6050	0.1179	0.010550
0.7	1.320	1.1100	0.880	0.5960	0.3000	0.122600
1.0	0.791	0.6662	0.528	0.3640	0.2010	0.106300

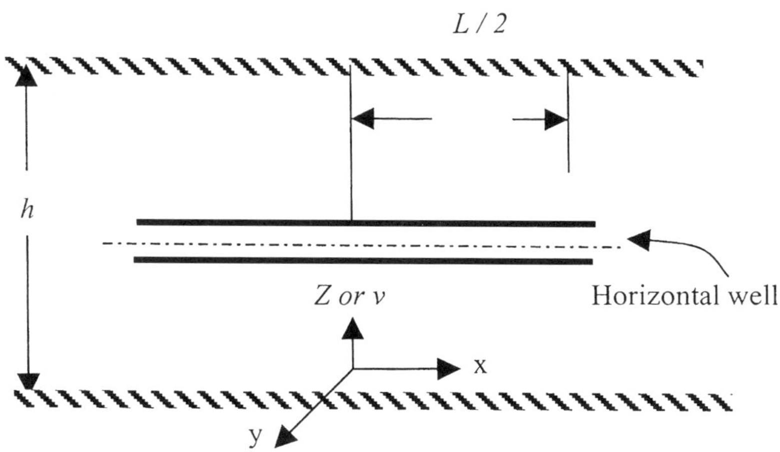

Figure 3–14. Horizontal well model.

Flow Time Equations and Solutions

These sets of equations are presented here for estimating for the various flow regimes based on the concepts of Daviau *et al.*, Kuchuk *et al.*, Odeh *et al.*, and Clonts and Ramey respectively.[7,12–14]

Method 1: Goode and R. K. M. Thambynayagam's Equations[6]

Early-Time Radial Flow. The early-time radial flow period ends at

$$t_{e1} = \frac{190.0\, d_z^{2.095} r_w^{-0.095} \phi \mu_o c_t}{k_v} \tag{3–33}$$

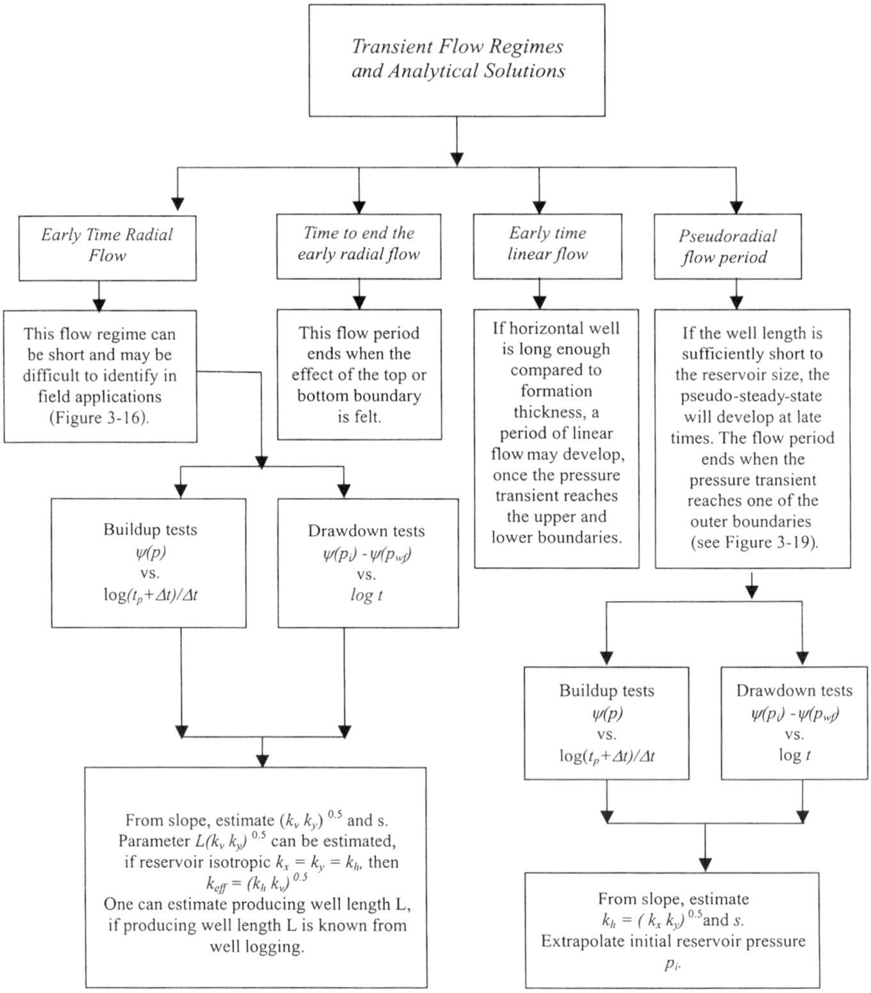

Figure 3–15. Flow regimes and horizontal wellbore pressure responses during flow period.

Intermediate-Time Linear Flow. Intermediate-time linear flow is estimated to end at

$$t_{e2} = \frac{20.8\,\phi\mu_o c_t L^2}{k_v} \tag{3–34}$$

The intermediate-time linear flow may not develop, if the time estimated from Eq. 3–34 is less than the time calculated for the early-time radial flow to end (Eq. 3–33).

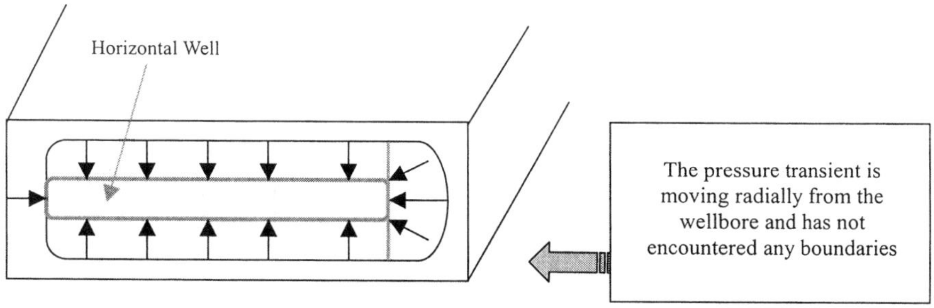

Figure 3–16. Early-time radial flow.

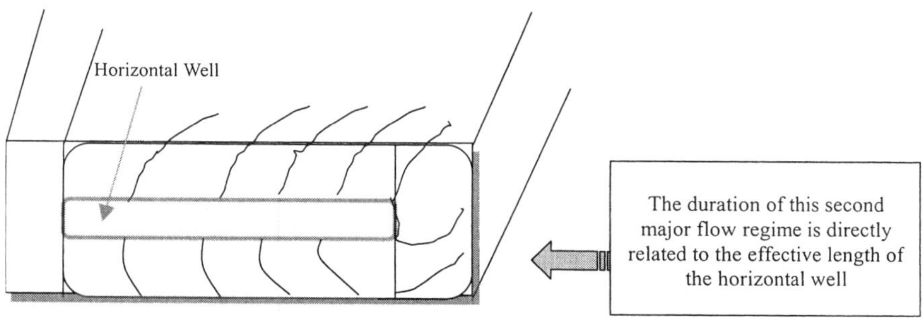

Figure 3–17. Intermediate-time linear flow.

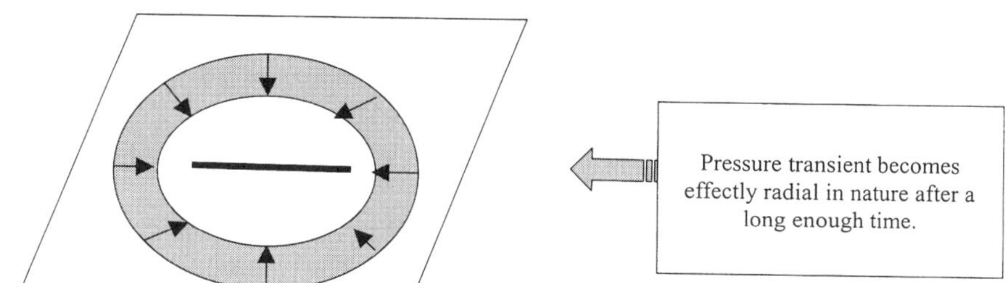

Figure 3–18. Late-time radial flow (pseudoradial flow).

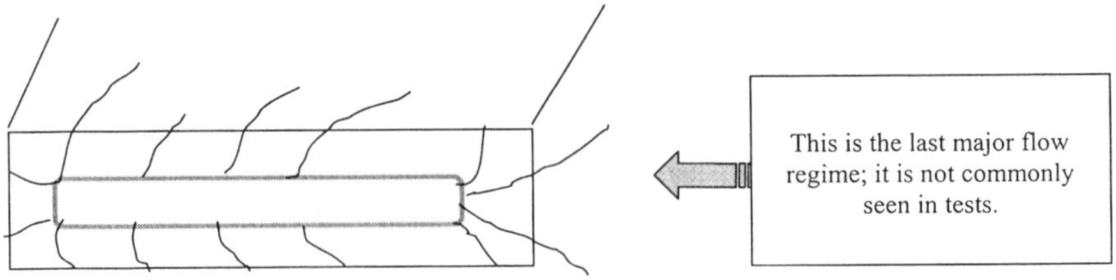

Figure 3–19. Late-time linear flow (pseudo-steady-state).

Late-Time Radial Flow or Pseudoradial Flow. If this flow develops, it will begin at approximately

$$t_{e3} = \frac{1230.0L^2\phi\mu_g c_t}{k_v} \tag{3–35}$$

Reference 9 suggested the following equation to determine the beginning of pseudoradial flow:

$$t_{e3} = \frac{1480L^2\phi\mu_g c_t}{k_x} \tag{3–36}$$

For a reservoir of finite width, pseudoradial flow would end at

$$t_{e3} = \frac{297.0(L_{x1} + L_{xd})^{2.095}L^{-0.095}\phi\mu_g c_t}{k_x} \tag{3–37}$$

This radial flow period will not develop if the estimated time to the end of late-time radial flow (Eq. 3–37) is less than the calculated beginning of pseudoradial flow (Eq. 3–35). It also means that the reservoir is smaller than anticipated. A plot of p_{wf} versus $\sqrt{t}$ can be used to calculate L_{x1} and L_{xd} in Eq. 3–37,

where

d_z = distance from the upper reservoir boundary to the center of the horizontal well, ft
k_v = permeability in vertical-direction, mD
L = effective length of horizontal well, ft
k_x = permeability in x-direction, mD
L_{x1} = distance in x-direction to beginning of horizontal wellbore, ft
L_{xD} = distance in x-direction to end of horizontal wellbore, ft

Method 2: A. S. Odeh and D. K. Babu's Equations [13]

Early-Time Radial Flow. The duration of this period may be approximated by the minimum of the following two terms:

$$t_{e1} = \frac{1800\, d_z^2 \phi \mu_g c_t}{k_v} \tag{3–38}$$

or

$$t_{e1} = \frac{125 L^2 \phi \mu_g c_t}{k_v} \tag{3–39}$$

Intermediate-Time Linear Flow. Time duration for the start and end of linear flow can be found by

$$t_{e2} = \frac{1800 D_z^2 \phi \mu_g c_t}{k_v} \tag{3–40}$$

and

$$t_{e2} = \frac{160\, \phi \mu_g c_t L^2}{k_x} \tag{3–41}$$

Late-Time Radial Flow. This flow period starts at

$$t_{e3} = \frac{1480 L^2 \phi \mu_g c_t}{k_x} \tag{3–42}$$

and ends at the minimum of

$$t_{e3} = \frac{2000\, \phi \mu_g c_t \left(d_x + \frac{L}{4}\right)^2}{k_x} \tag{3–43}$$

or

$$t_{e3} = \frac{1650\, \phi \mu_g c_t d_y^2}{k_y} \tag{3–44}$$

Late-Time Linear Flow. The flow ends at the larger of

$$t_{e4} = \frac{4800\, \phi \mu_g c_t D_x^2}{k_x} \tag{3–45}$$

or

$$t_{e4} = \frac{1800 D_z^2 \phi \mu_g c_t}{k_z} \tag{3–46}$$

where

d_z is the shortest distance between the well and the z-boundary, ft
D_z is the longest distance between the well and the z-boundary, ft
$= (h - d_z)$ and h is the reservoir height
k_y = permeability in y-direction, mD
d_x is the shortest distance between the well and the x-boundary, ft
D_x is the longest distance between the well and the x-boundary, ft

Method 3: Ozham, Raghavan, and Joshi's Equations[15]

Early-Time Radial Flow

$$L_D = \frac{L}{2h}\sqrt{k_v/k_h} \tag{3–47}$$

$$r_{wD} = (2r_w/L)\sqrt{k_h/k_y} \tag{3–48}$$

and assuming isotropic reservoir, i.e., $k_x = k_y$, Eq. 3–48 reduces to

$$r_{wD} = \frac{2r_w}{L} \tag{3–49}$$

Figure 3–20 tell us that horizontal well pressure response depends upon dimensionless well length L_D and dimensionless wellbore radius r_{WD}. The time between dashed lines AA and BB represents transitional flow period from early-time radial flow (vertical radial flow) to pseudoradial flow.

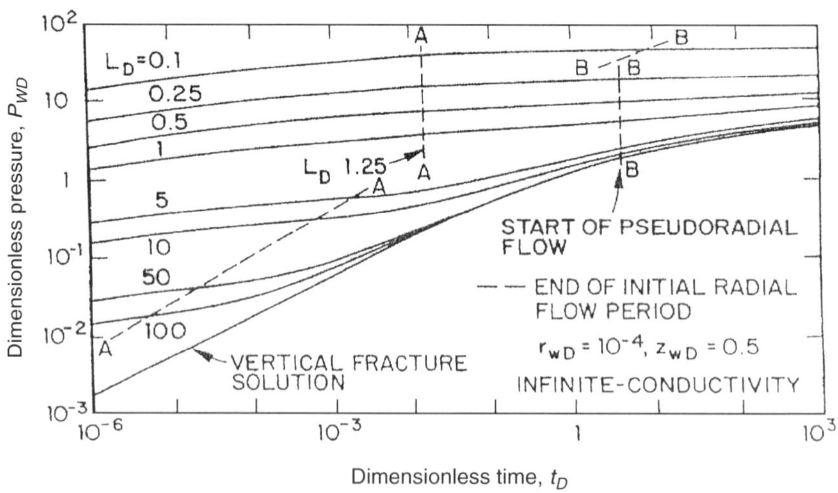

Figure 3–20. Pressure response of horizontal wells (after Ozkan *et al.*).[15]

(Note that the intermediate linear flow regime is not included in Figure 3–20.) As shown in Figure 3–20, once the pseudoradial flow starts, the horizontal well solutions for $L_D \geq 10$ are practically the same as the vertically fractured well solution. After estimating the value of L_D and r_{wD}, and using Figure 3–20, one can find t_D, (dashed line A-A) and the duration of the early-time radial flow, which is given by

$$t_{el} = \frac{t_D \phi \mu_g c_t L_w^2}{0.001055 k_h} \tag{3–50}$$

Late-Time Radial Flow. The start of this radial (pseudoradial) flow can be calculated by using Eqs. 3–47 and 3–48, finding t_D from Figure 3–20, dashed line B-B, and then using Eq. 3–50,

where:

L_D = dimensionless length
h = reservoir thickness, ft
r_{wD} = dimensionless radius
t_D = dimensionless time

Example 3–3 *Calculating the Time Required to End of Early-Time Radial Flow*

A horizontal oil well is 2000 ft long and is drilled in a reservoir with the following characteristics: $h = 120$ ft, $r_w = 0.354$ ft, $\phi = 15.0\%$, $\mu_g = 0.02952$ cP, $c_t = 4.25 \times 10^{-5}$ psi^{-1}, $k_h = 0.8$ mD (from well test data), $k_v = 0.2$ mD (from core data). The well is in the central elevation of the reservoir and the distance from the upper reservoir boundary to the center of the horizontal well is 20 ft. Estimate the time required to end initial radial flow.

Solution $d_z = (120/2 - 20) = 40$ ft

Method 1, using Eq. 3–33:

$$t_{e1} = \frac{190.0 d_z^{2.095} r_w^{-0.095} \phi \mu_g c_t}{k_v}$$

$$= \frac{190.0 \times 40^{2.095} \times 0.354^{-.095} \times 0.15 \times 0.02952 \times 4.25 \times 10^{-5}}{.2}$$

$$= 0.448 \text{ hours}$$

Method 2, using Eq. 3–40:

$$t_{e1} = \frac{1800\, d_z^2 \phi \mu_g c_t}{k_v}$$

$$= \frac{1800 \times 40^2 \times 0.15 \times 0.02952 \times 4.25 \times 10^{-5}}{0.2} = 2.71 \text{ hours}$$

$$t_{el} = \frac{125L^2\phi\mu_g c_t}{k_y} = \frac{125 \times 2000^2 \times 0.15 \times 0.02952 \times 4.25 \times 10^{-5}}{0.8}$$
$$= 117.62 \text{ hours}$$

The minimum of these two values is 0.448 hours. Thus, the initial radial flow period will end in 0.448 hours.

Method 3, using Figure 3–20:

$L_D = 2000/(2 \times 120)\sqrt{0.2/0.8} = 4.17$ (Eq. 3–47)

Assuming it is an isotropic reservoir, and therefore $k_x = k_y = k_h$:

$$r_{wD} = \frac{2r_w}{L}\sqrt{k_h/k_y} = \frac{2 \times 0.354}{2000} \times 1 = 3.54 \times 10^{-4} \quad \text{(Eq. 3–48)}$$

With L_D, r_{WD} known, the time to the end of the initial radial flow period is given by dotted line A-A ($t_D = 1.5 \times 10^{-3}$), and after rearranging and substituting these values in Eq. 3–50,

$$t_{el} = \frac{\phi\mu_g c_t L^2}{.001055k_h}, \quad t_D = \frac{0.15 \times 0.02952 \times 4.25 \times 10^{-5} \times 2000^2}{0.001055 \times 0.8}$$
$$\times (1.5 \times 10^{-3}) = 1.34 \text{ hours}$$

Therefore the initial time required to end initial radial flow would last between 0.448 and 1.34 hours. The reservoir engineers will have to use downhole shut-in devices to enhance the chances of measuring the early radial flow regime.

Example 3–4 *Calculating the Time to Start and Time to End of Early-Time Linear Flow*

For the well described in Example 3–3, assuming $k_x = k_y = 0.8$ mD, calculate the time to start and time to end of early-time linear flow.

Solution The maximum distance of well from either top or bottom boundary is:

$D_z = h - d_z = 120 - 40 = 80$ ft

From Eq. 3–38:

$$t_{e2} = \frac{1800\, D_z^2\phi\mu c_t}{k_v} = \frac{1800 \times 80^2 \times 0.15 \times 0.02952 \times 4.25 \times 10^{-5}}{0.2}$$
$$= 10.84 \text{ hours}$$

From Eq. 3–41:

$$t_{e2} = \frac{160\,\phi\mu c_t L^2}{k_x} = \frac{160 \times 0.15 \times 0.02952 \times 4.25 \times 10^{-5} \times 2000^2}{0.8}$$
$$= 150.55 \text{ hours}$$

From Eq. 3–34:

$$t_{e2} = \frac{20.8\,\phi\mu c_t L^2}{k_x} = \frac{20.8 \times 0.15 \times 0.02952 \times 4.25 \times 10^{-5} \times 2000^2}{0.8}$$
$$= 19.57 \text{ hours}$$

Thus, this flow period will end in about 20 to 150 hours. This indicates that the current well is sufficiently long compared to the reservoir height. Therefore, it is possible to analyze pressure data of the flow period.

Example 3–5 *Calculating the Time Required to Start a Pseudoradial Flow*
For the well/reservoir data given in Example 3–4, calculate time to start and time to end early-time linear flow.

Solution

Method 1, from Eq. 3–35:

$$t_{e2} = \frac{1230\phi\mu c_t L^2}{k_x} = \frac{1230 \times 0.15 \times 0.02952 \times 4.25 \times 10^{-5} \times 2000^2}{0.8}$$
$$= 1157.4 \text{ hours} = 38 \text{ days}$$

Method 2, from Eq. 3–36:

$$t_{e2} = \frac{1480\phi\mu c_t L^2}{k_x} = \frac{1480 \times 0.15 \times 0.02952 \times 4.25 \times 10^{-5} \times 2000^2}{0.8}$$
$$= 1392.6 \text{ hours} = 46 \text{ days}$$

Method 3, from Figure 3–20 Time to start of pseudoradial flow (dashed line B-B) is $t_D = 3.0$; thus from Eq. 3–50,

$$t_{e1} = \frac{\phi\mu c_t L^2}{0.001055 k_h} t_D = \frac{0.15 \times 0.02952 \times 4.25 \times 10^{-5} \times 2000^2}{0.001055 \times 0.8}(3.0)$$
$$= 2675.7 \text{ hours} = 88 \text{ days}$$

It will take 38 to 88 days to reach pseudoradial flow. It will be economically difficult to shut in a well for such a long time. In this case one will have to obtain the necessary information from early-time radial or linear flow periods.

Solution under Pressure Drawdown Tests

Certain reservoir parameters can only be approximated during particular flow regimes; therefore it is important to calculate the times relating to each of the flow regimes. Goode *et al.*[6] developed pressure response function at the horizontal wellbore for conditions of both pressure drawdown and pressure buildup. These pressure response equations, published in 1987, assumed an effective pressure point along the horizontal wellbore. Later work by Kuchuk *et al.*[12] was based on pressure averaging under conditions of pressure drawdown. These equations provide pressure response during each flow regime. In Eq. 3–51b the q_g is conveniently expressed in mmscfd, and the gas formation volume factor, β_g, is then expressed in reservoir barrels per mmscfd, so that the product $q_g\beta_g$ is in reservoir barrels per day (rb/d) as in the analogous equation for slightly compressible liquids. All gas properties (β_g, μ_g, and C_g) are evaluated at original reservoir pressure, p_i:

$$\beta_g = \frac{0.1781 z_i T p_{sc}}{p_i T_{sc}} \tag{3–50a}$$

Early-Time Radial Flow. The wellbore pressure response during this flow period is given by the following equations:

Pseudopressure case:

$$\psi(p_i) - \psi(p_{wf}) = \frac{57.920 \times 10^6 q_g T p_{sc}}{\sqrt{k_v k_y} L T_{sc}} \times \left[\log\left(\frac{\sqrt{k_y k_v}\, t}{\phi \mu_g c_t r_w^2}\right) - 3.23 + 0.868s\right] \tag{3–51}$$

Pressure-squared case:

$$p_i^2 - p_{wf}^2 = \frac{57.920 \times 10^6 q_g T \bar{\mu}_g \bar{z} p_{sc}}{\sqrt{k_v k_y} L T_{sc}} \times \left[\log\left(\frac{\sqrt{k_y k_v}\, t}{\phi \mu_g c_t r_w^2}\right) - 3.23 + 0.868s\right] \tag{3–51a}$$

Pressure case:

$$p_i - p_{wf} = \frac{162.6\, q_g \beta_g \mu_g}{\sqrt{k_v k_y} L}\left[\log\left(\frac{\sqrt{k_y k_v}\, t}{\phi \mu_g c_t r_w^2}\right) - 3.23 + 0.868s\right] \tag{3–51b}$$

where s is mechanical skin damage due to drilling and completion. Equation 3–51b indicates that a plot of wellbore pressure p_{wf} or $(p_i - p_{wf})$

versus $\log t$ will exhibit a semilog straight line with slope given by

$$m_1 = \frac{162.6\, q_g \beta_g \mu_g}{\sqrt{k_v k_y} L} \tag{3–52}$$

The equivalent permeability in a vertical plane around the wellbore can be calculated as

$$\sqrt{k_v k_y} = \frac{162.6 q_g \beta_g \mu_g}{m_1 L} \tag{3–53}$$

Extrapolating the semilog straight line to $t = 1$ hour, the following equation is obtained:

$$p_i - p_{1,hr} = m\left[\log\left(\frac{k_v k_y}{\phi \mu_g c_t r_w^2}\right) - 3.23 + 0.868 s\right] \tag{3–54}$$

where p_i is initial reservoir pressure, and $p_{1,hr}$ is the pressure obtained at $t = 1$ hour. Rearranging Eq. 3–54 gives

$$s = 1.151\left[\frac{p_i - p_{1,hr}}{m_1} - \log\left(\frac{\sqrt{k_v k_y}}{\phi \mu_g c_t r_w^2} + 3.23\right)\right] \tag{3–55}$$

Using Eq. 3–55 one can estimate the skin factor s. If the reservoir is areal isotropic ($k_x = k_y = k_h$), then using Eq. 3–52, we have

$$L\sqrt{k_v k_y} = L k_{eff} = \frac{162.6\, q_g \beta_g \mu_g}{m_1} \tag{3–56}$$

where k_{eff} ($= \sqrt{k_h k_v}$) is the effective reservoir permeability. Thus, if k_{eff} is known, one can estimate producing well length L_w. Conversely, if producing well length L is known by well logging, then one can calculate the effective reservoir permeability. As mentioned earlier, this flow regime can be short in duration and may be difficult to identify in field applications.

Intermediate-Time Linear Flow. Pressure response during this flow period is given by

$$p_i - p_{wf} = \frac{8.128 q_g \beta_g}{Lh}\sqrt{\frac{t}{k_y \phi \mu_g c_t}} + \frac{141.2 q_g \mu_g \beta_g}{L\sqrt{k_y k_v}}(s_z + s) \tag{3–57}$$

where s_z is the pseudoskin factor caused by partial penetration in the vertical direction and is given by[9]

$$s_z = \ln\left(\frac{h}{r_w}\right) + 0.25 \ln\left(\frac{k_y}{k_v}\right) - \ln\left(\sin \frac{180^\circ z_w}{h}\right) - 1.838 \tag{3–58}$$

where z_w = vertical location of well, ft, and h = reservoir height, ft. Equation 3–57 indicates that a plot of $\triangle p = p_i - p_{wf}$ versus $\sqrt{t}$ will exhibit a straight line with slope given by

$$m_2 = \frac{8.128 q_g \beta_g}{Lh} \sqrt{\frac{\mu_g}{\phi c_t k_y}} \tag{3–59}$$

Hence the product of the producing well length square L^2 and permeability k_y can be obtained from the slope

$$L^2 k_y = \left(\frac{8.128 q_g \beta_g}{h m_2} \right)^2 \frac{\mu_g}{\phi c_t} \tag{3–60}$$

Additionally, extrapolating the straight line to $\sqrt{t} = 0$ gives

$$\triangle p\big|_{t=0} = \frac{141.2\, q_g \beta_g \mu_g}{L\sqrt{k_y k_v}} (s_z + s) \tag{3–61}$$

where

$$s = \frac{0.058}{h} \sqrt{\frac{k_v}{\phi \mu_g c_t}} \left(\frac{p_i - p_{wf}(0 = \text{hr})}{m_2} \right) - s_z \tag{3–62}$$

Late-Time Radial Flow. Pressure response during this radial (pseudoradial) flow period is given by[4]

$$p_i - p_{wf} = \frac{162.6\, q_g \beta_g \mu_g}{k_x k_y h} \left[\log\left(\frac{k_x t}{\phi \mu_g c_t L^2} \right) - 2.023 \right] + \frac{141.2\, q_g \mu_g \beta_g}{L\sqrt{k_y k_v}} (s_z + s) \tag{3–63}$$

Equation 3–63 indicates that a plot of p_{wf} or $(p_i - p_{wf})$ versus $\log t$ will exhibit a semilog straight line of slope m_3:

$$m_3 = \frac{162.6\, q_g \beta_g \mu_g}{\sqrt{k_x k_y} h} \tag{3–64}$$

The equivalent horizontal permeability $\sqrt{k_x k_y}$ can be estimated as

$$k_h = \sqrt{k_x k_y} = \frac{162.6\, q_g \beta_g \mu_g}{m_3 h} \tag{3–65}$$

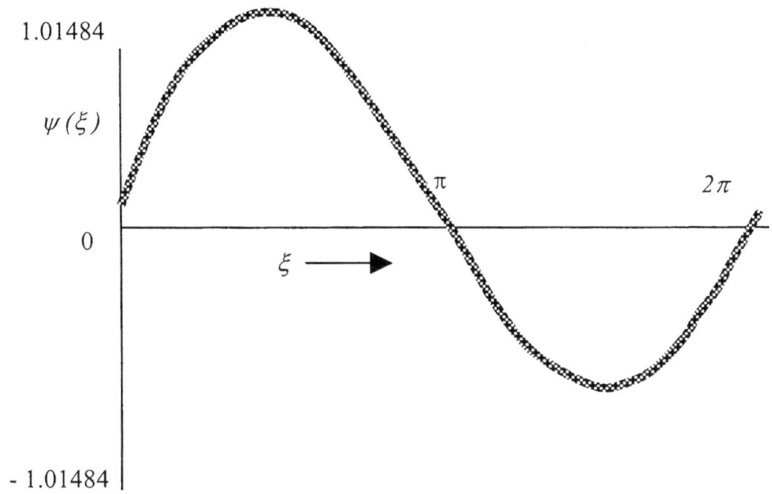

Figure 3–21. Spencer function (after Spencer, *SPE Formation Evaluation*, Dec. 1987).[6]

The skin factor can also be obtained by

$$s = \frac{1.151L}{h}\sqrt{\frac{k_v}{k_x}}\left[\frac{p_i - p_{1hr}}{m_3} - \log\left(\frac{k_x}{\phi\mu_g c_t L_w^2}\right) + 2.023\right] - s_z \quad (3\text{–}66)$$

where p_{1hr} is obtained by extrapolating the late-time radial flow semilog straight line to $t = 1$ hour. Equation 3–58 gives pseudoskin factor. s_z can also be estimated using the Spencer function[6] (Figure 3–21):

$$s_z = \frac{0.07958h_z}{r_{wa}}[\psi(\xi_1) + \psi(\xi_2) - \psi(\xi_3) - \psi(\xi_4)] \quad (3\text{–}67)$$

where

$$\xi_1 = \frac{0.52\pi r_{wa}}{h_z}$$

$$\xi_2 = \frac{\pi}{h_z}(2h_s + 3.48r_{wa})$$

$$\xi_3 = \frac{3.38\pi r_{wa}}{h_z}$$

and

$$\xi_4 = \frac{\pi}{h_z}(2h_s = 0.52r_{wa})$$

where

r_{wa} = apparent wellbore radius, ft
h_s = horizontal well in center of reservoir, ft
h_z = partial penetration in vertical direction, ft

If the bottom and top boundaries are maintained at constant pressure then the pseudoradial or late-time radial flow period will not develop,[6] and there will be steady-state flow at the late time.

Late-Time Linear Flow. Pressure response during this period, *r* also known as pseudo-steady-state is given by[4]

$$p_i - p_{wf} = \frac{8.128 q_g \beta_g}{2x_e h}\sqrt{\frac{\mu_g t}{k_y \phi c_t}} + \frac{141.2 q_g \beta_g \mu_g}{L\sqrt{k_y k_v}}(s_x + s_z + s) \quad (3\text{–}68)$$

where

$2x_e$ = width of reservoir, ft
s_z = pseudoskin factor due to partial penetration in a vertical direction (Eq. 3–58 or 3–67)
s_x = pseudoskin factor due to partial penetration in the x-direction.

Reference 4 gives an expression for s_x:

$$s_x = \frac{0.6366 h_x^2 L_w}{h\sqrt{k_y/k_x}} \sum_{n=1}^{\infty} \frac{\Xi n^2}{n} \quad (3\text{–}69)$$

where

h = reservoir height, ft
h_x = height between the well and the x-boundary, ft

Equation 3–68 indicates that a plot of $p_{wf}(p_i - p_{wf})$ versus $\sqrt{t}$ will exhibit a straight line of slope m_4:

$$m_4 = \frac{8.128\, q_g \beta_g \mu_g}{h_x h\sqrt{\phi c_t k_y}} \quad (3\text{–}70)$$

Reservoir parameter h_x or $\sqrt{\frac{\pi c_t k_y}{\mu_g}}$ can be obtained as

$$h_x = \frac{8.128 q_g \beta_g \mu_g}{m_4 h\sqrt{\phi c_t k_y}} \quad (3\text{–}71)$$

or

$$\sqrt{\frac{\phi c_t k_y}{\mu_g}} = \frac{8.128 q_g \beta_g \mu_g}{m_4 h \sqrt{\phi c_t k_y}}$$

The skin factor s can be found from

$$s = \frac{0.058L}{hh_x}\sqrt{\frac{k_v}{\phi\mu_g c_t}}\left(\frac{p_i - p_{wf(0=hr)}}{m_4}\right) - (s_x + s_z) \tag{3–72}$$

where p_{wf} (0 hr) is the pressure obtained at $t = 0$ hour.

Solution under Pressure Buildup Tests

Early-Time Radial Flow. Pressure buildup response during this flow period is given by the following equations:

Infinite reservoir:

$$p_i - p_{ws} = \frac{162.6\, q_g \beta_g \mu_g}{\sqrt{k_z k_y} L_w}\left[\log\left(\frac{t_p + \Delta t}{\Delta t}\right) + \gamma_1\right] \tag{3–73}$$

where

$$\gamma_1 = \frac{L}{h}\sqrt{\frac{k_v}{k_x}}\left[\log\left(\frac{k_x t}{\phi\mu_g c_t L_w^2}\right) - 2.023\right] - \log(t) - \log\left(\frac{\sqrt{k_v k_y}}{\phi\mu_g c_t r_w^2}\right) + 3.227 + 0.869 s_z$$

Equation 3–73 indicates that a plot of Δp versus $\log(\frac{t_p+\Delta t}{\Delta t})$ will exhibit a semilog straight line with slope, m_{1r}, given by

$$m_{1r} = \frac{162.6\, q_g \beta_g \mu_g}{\sqrt{k_v k_y} L} \tag{3–74}$$

The equivalent permeability in the vertical plane can be estimated by

$$k_v k_y = \left(\frac{162.6\, q_g \mu_g \beta_g}{m_{1r} L}\right)^2 \tag{3–75}$$

Extrapolating the semilog straight line to $t = 1$ hour, the following equation is obtained to estimate s:

$$s = 1.151\left[\frac{p_{1hr} - p_{wf}}{m_{1r}} - \log\left(\frac{k_v k_y}{\phi\mu_g c_t r_w^2}\right) + 3.23\right] \tag{3–76}$$

Finite (bounded) reservoir:

$$p_i - p_{ws} = \frac{162.6\, q_g \beta_g \mu_g}{\sqrt{k_y k_v} L} \left[\log\left(\frac{t_p + \Delta t}{\Delta t} \right) + \gamma_2 \right] \tag{3–77}$$

where

$$\gamma_2 = \frac{0.05L}{h h_x} \sqrt{\frac{k_v t}{\phi \mu_g c_t}} - \log\left(\frac{\sqrt{k_y k_v t}}{\phi \mu_g c_t r_w^2} \right) + 3.227 + 0.868(s_x + s_z)$$

Equation 3–76 gives s.

Intermediate-Time Linear Flow. Pressure buildup response during this flow period is given by the following equations:

Infinite reservoir, first linear flow:

$$p_i - p_{ws} = \frac{8.128\, q_g \beta_g}{hL} \sqrt{\frac{\mu_g \Delta t}{k_y \phi c_t}} + \gamma_3 \tag{3–78}$$

where

$$\gamma_3 = \frac{162.6\, q_g \beta_g \mu_g}{h\sqrt{k_x k_y}} \left[\log\left(\frac{k_x t}{\phi \mu_g c_t L^2} \right) - 2.023 \right]$$

A plot of Δp versus $\sqrt{\Delta t}$ will exhibit a slope given by

$$m_{1l} = \frac{8.128\, q_g \beta_g}{h\sqrt{k_v k_y}} \tag{3–79}$$

The equivalent permeability in the vertical plane can be calculated by

$$k_v k_y = \left(\frac{8.128\, q_g \beta_g}{m_{1l} h} \right)^2 \tag{3–80}$$

The skin factor s is given by

$$s = \frac{0.058}{h} \sqrt{\frac{k_v}{\phi \mu_g c_t}} \left[\frac{p_{1hr} - p_{wf}}{m_{1l}} \right] - s_z \tag{3–81}$$

Finite (bounded) reservoir:

$$p_i - p_{ws} = \frac{8.128\, q_g \beta_g}{hL} \sqrt{\frac{\mu_g}{k_y \phi c_t}} \left(\sqrt{\Delta t} - \frac{L\sqrt{t}}{h_x} \right) + \frac{141.2\, q_g \beta_g \mu_g}{L\sqrt{k_y k_x}} s_x \tag{3–82}$$

Using Eq. 3–81 gives s_x:

Late-Time Radial Flow. Pressure buildup response during this flow period is given by the following equations:

Infinite reservoir:

$$p_i - p_{ws} = \frac{162.6\, q_g \beta_g \mu_g}{h\sqrt{k_x k_y}} \left[\log\left(\frac{t_p + \Delta t}{\Delta t} \right) \right] \tag{3–83}$$

A plot of p_{ws} versus $\log(\frac{t_p+\Delta t}{\Delta t})$ will exhibit a straight line with slope given by

$$m_{2r} = \frac{162.6\, q_g \beta_g \mu_g}{h\sqrt{k_v k_y}} \tag{3–84}$$

The skin factor s is given by

$$s = \frac{1.151L}{h} \sqrt{\frac{k_v}{k_x}} \left[\frac{p_{1hr} - p_{wf}}{m_{2r}} - \log\left(\frac{k_v}{\phi \mu_g c_t L^2} \right) + 2.023 \right] - s_z \tag{3–85}$$

Finite (bounded) reservoir:

$$p_i - p_{ws} = \frac{162.6\, q_g \beta_g \mu_g}{h\sqrt{k_x k_y}} \left[\log\left(\frac{t_p + \Delta t}{\Delta t} \right) + \gamma_4 \right] \tag{3–86}$$

where

$$\gamma_4 = \frac{0.05}{h} \sqrt{\frac{k_x t}{\phi_t \mu_g c_t}} - \log\left(\frac{k_x t}{\phi \mu_g c_t L^2} \right) + 2.023 + 0.868 s_z$$

Calculate s using Eq. 3–85. Generally, only the initial part of the Horner plot generated by Eq. 3–86 will be a straight line.

Late-Time Linear Flow. During this flow period the infinite reservoir case does not exist (pseudoradial flow) and the pressure buildup response for a finite (bounded) reservoir is given by

$$p_i - p_{ws} = \frac{8.128 q_g \beta_g}{h h_x} \sqrt{\frac{\mu_g}{k_y \phi c_t}} (\sqrt{t} - \sqrt{\Delta t}) \tag{3–87}$$

A plot of p_{ws} versus $(\sqrt{t} - \sqrt{\Delta t})$ will produce a straight line that will extrapolate to p_i while a plot of Δp versus $(\sqrt{t} - \sqrt{\Delta t})$ will exhibit a straight line of slope m_{4l}:

$$m_{4l} = \frac{8.128 q_g \beta_g}{h_z h_x \sqrt{\frac{\phi c_t k_y}{\mu_o}}} \tag{3–88}$$

or

$$\sqrt{\frac{\phi c_t k_y}{\mu_o}} = \frac{8.128 q_g \beta_g}{m_{4l} h h_x} \tag{3–89}$$

The skin factor s is given by

$$s = \frac{0.058L}{hk_x} \sqrt{\frac{k_v}{\phi \mu_g c_t}} \left(\frac{p_{1hr} - p_{wf}}{m_{4l}} \right) - (s_z + s_x) \tag{3–90}$$

3.6 Problems in Testing Horizontal Wells

Figure 3–22 illustrates horizontal well test configuration and Figure 3–23 shows possible parameters affecting the transient response and careful test design procedures.

3.7 Horizontal Well Application in Tight Gas Reservoirs

In tight gas reservoirs, the time to start of the pseudo-steady-state can be very large. In such cases vertical gas wells can be drilled at close spacing to effectively drain the reservoir. This will require a large number of vertical wells. In many cases it is difficult to create long fractures in tight reservoirs,

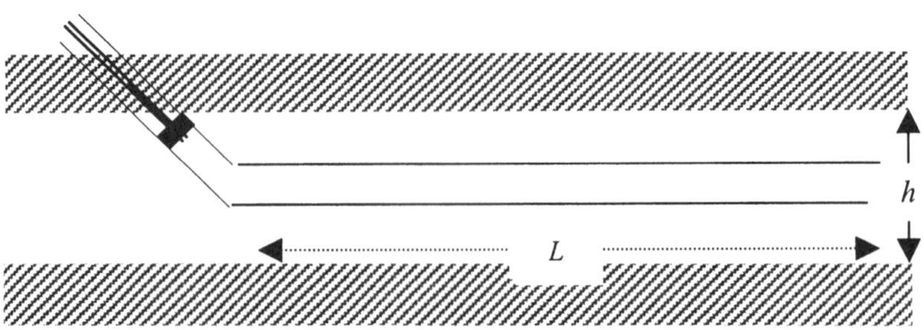

Figure 3–22. Horizontal well test configurations.

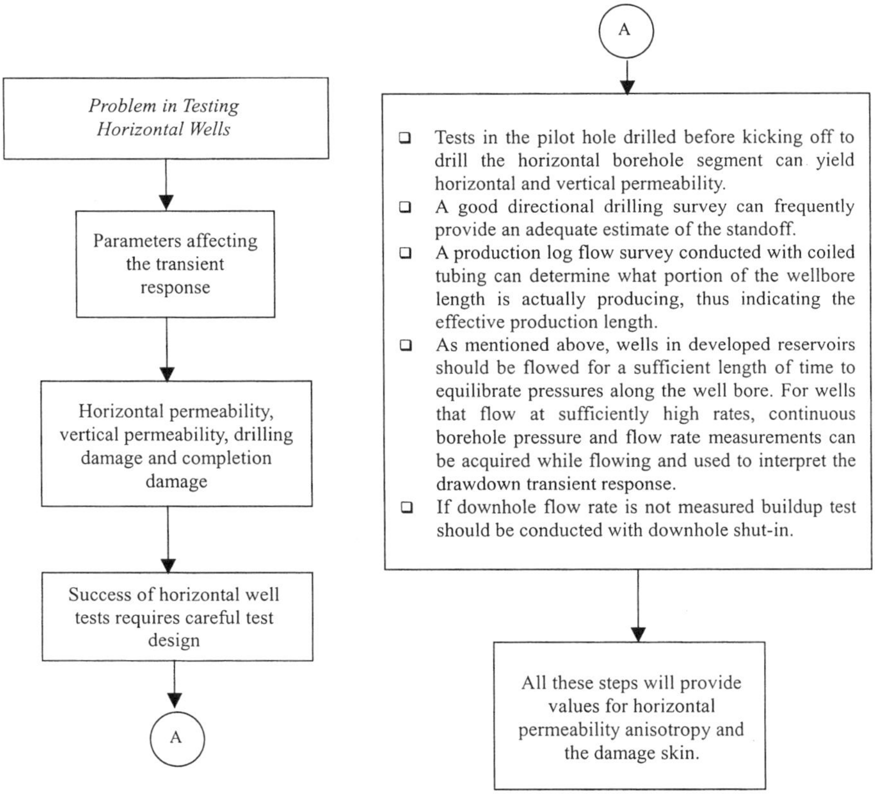

Figure 3–23. Problems in testing horizontal wells.

especially if the reservoir is overlain or underlain by a weak cap or base rock. During the fracture jobs, when pumping pressure exceeds the formation parting pressure for an effective proppant placement, the high pumping pressure may open up a large portion of the weak cap or base rock, resulting in an excessive fracture height and fracture growth in unproductive zones.

The excessive fracture height results in a less-than-desired fracture extension in the reservoir. In such reservoirs, horizontal wells provide an alternative to obtain long fracture extension, since horizontal wells represent a long fracture with a height equal to the wellbore diameter. Vertical wells located near faults can also screen out because of excessive fluid loss, giving short fracture penetration. A long horizontal well can drain larger volumes than a single vertical well in the same time interval. This has an impact on field economics, because one can develop the field using fewer horizontal wells than vertical wells.

3.8 Influence of Turbulence in High-Permeability Gas Wells

Horizontal wells are also useful in high-permeability gas wells, especially in those wells when near-wellbore turbulence is very high. The near-wellbore turbulence is inversely proportional to the well's perforated interval. By drilling a horizontal well, one can increase productive length and therefore decrease the near-wellbore turbulence and enhance well productivity.

Turbulent Flow

Darcy's law of flow through porous media is valid only for laminar flow through the reservoir:

$$dp/dr = av \tag{3–91}$$

where v is velocity, a is a constant, and dp/dr is pressure gradient. Forchheimer modified Darcy's law to account for turbulence effects:

$$dp/dr = av + bv^2 \tag{3–92}$$

Equation 3–92 can be rewritten as either:

$$\bar{p}_R^2 - p_{wf}^2 = aq_g + bq_g^2 \tag{3–93}$$

or

$$\psi(\bar{p}_R) - \psi(p_{wf}) = a'q_g + b'q_g^2 \tag{3–94}$$

where

$$a = \frac{50.337 \times 10^3 \bar{\mu}\bar{z}TP_{sc}}{khT_{sc}}\left[\ln\left(\frac{r_e}{r_w}\right) - 0.75 + s + s_{CA} - c'\right] \tag{3–95}$$

$$b = \frac{50.337 \times 10^3 \bar{\mu}\bar{z}TP_{sc}}{khT_{sc}} D \tag{3–96}$$

$$a' = \frac{50.337 \times 10^3 TP_{sc}}{khT_{sc}}\left[\ln\left(\frac{r_e}{r_w}\right) - 0.75 + s + s_{CA} - c'\right] \tag{3–97}$$

$$b' = \frac{50.337 \times 10^3 TP_{sc}}{khT_{sc}} D \tag{3–98}$$

In the foregoing equations, definitions of s_{CA} and c' depend upon the type of the well as defined earlier. D is the turbulence factor in units of 1/mscfd and is defined in Eq. 3–18. Equations 3–93 and 3–94 tell us that the influence of turbulence is to increase the pressure drop or pressure drawdown required to produce the given gas production rate. Thus, the presence of turbulence reduces net production from a well. Reducing fluid velocity near the wellbore can minimize the turbulence effect. If there is no turbulence near the wellbore, there will be no turbulence in the reservoir. The fluid velocity near the wellbore can be minimized by increasing perforated producing length h_P. A horizontal well provides a means to significantly enhance perforated interval and reduce near-wellbore turbulence.

3.9 Turbulence Identification

A multirate test can be used to confirm the existence of high-velocity effects in a well. Flow tests are conducted with different surface pressures. At each pressure a stabilized gas flow rate q_{sc} is recorded. Based on surface pressure and flow rate, downhole well flowing pressure p_{wf} is estimated. The data are correlated as

$$q_{sc} = C\left(\bar{p}_R^2 - p_{wf}^2\right)^n \tag{3–99}$$

where C is a constant and n is a dimensionless constant $(1/2 \leq n \leq 1)$. Equation 3–99 is rewritten as

$$\ln q_{sc} = \ln C + \ln\left(\bar{p}_R^2 - p_{wf}^2\right)^n \tag{3–100}$$

If slope $n = 1$, then there is no turbulence. However, if $n < 1$, turbulence does exist. The lower the value of n, the higher is the turbulence effect, and when $n = 0.5$ turbulence is dominant. It is important to use pseudopressure ($\psi(p)$ values) when pressure is above 2500 psia.

3.10 Inflow Performance Responses in Vertical and Horizontal Gas Wells

The following example will illustrate the inflow performance responses in both vertical and horizontal gas wells.

Example 3–6 *Inflow Performance Responses for Vertical and Horizontal Wells*

An engineer suggested drilling a 2000-ft-long horizontal gas well not only to reduce near-wellbore turbulence but also to ensure against water coning. Develop inflow performance curves for vertical and horizontal wells. Assume that the gas reservoir is not in communication with the bottom water zone. The following reservoir and gas properties are given:

reservoir = sandstone; reservoir temperature = 226°F; depth = 9011 ft; gas gravity (air = 1.0) = 0.681; reservoir pressure = 1660 psia; pseudocritical temperature = 370.010°R; reservoir thickness = 69 ft; pseudocritical

Table 3–8
Calculated Gas PVT Properties

Pressure (psia)	Compressibility factor, z	Gas viscosity (cP)	Real gas pseudopressure $\psi(p)$ (mmpsia^2/cP)
4000	0.9598	0.023689	903.57
3750	0.9470	0.022859	816.26
3500	0.9354	0.022018	730.52
3250	0.9251	0.021176	646.66
3000	0.9127	0.020345	565.11
2750	0.9119	0.019533	486.41
2500	0.9085	0.018748	411.18
2250	0.9074	0.017997	340.12
2000	0.9089	0.017285	273.93
1750	0.9128	0.016618	213.36
1500	0.9192	0.016002	159.12
1250	0.9279	0.015441	111.91
1000	0.9389	0.014940	72.35
750	0.9518	0.014507	41.00
500	0.9665	0.014147	18.31
250	0.9825	0.013868	4.60
140.65	0.9985	0.013687	0.53

pressure = 650.59 psia; average reservoir permeability = 10 mD; base temperature = 520°R; vertical permeability (assume) = 1.0 mD; base pressure = 14.65 psia; well spacing = 320 acres; and average porosity = 0.146 fraction.

Solution The reservoir has a permeability of 10 mD, and hence a well drilled at 320-acre well spacing will begin pseudo-steady-state in about 30 days. Therefore, the initial transient flow portion is ignored in the following calculations. The inflow performance curve is based upon a pseudo-steady-state solution, i.e., Eq. 3–17. Since the vertical well is centrally located in the drainage plane, $S_{CA} = 0$. For turbulence calculations, Eqs. 3–18 through 3–20 are used. For horizontal well turbulence, the perforated length h_P is simply replaced by well length L in Eq. 3–18.

Calculation of Inflow Performance Curve

The pseudo-steady-state equation for a gas well, Eq. 3–17, is rewritten as

$$q_g = \frac{khT_{sc}[\psi(\bar{p}_R) - \psi(p_{wf})]}{50.337 \times 10^3 T P_{sc}[\ln(r_e/r_w) - 0.75 + s + s_m + s_{CA} - c' - Dq_g]} \tag{3–101}$$

Table 3–9
IPR Calculations for Vertical Well (Example 3–6)

$H =$ p_{wf} (psia)	$\psi(p_{wf})$ (mmpsia^2/cP)	$\Delta\psi = \psi(\bar{p}_R) - \psi(p_{wf})$ (mmpsia^2/cP)	$\mu_{p_{wf}}$ (cP)	D (1/mscfd $\times 10^{-4}$)	q_g (mmscfd)	
					No turb.	With turb.
250	4.60	184.40	0.013868	0.5431	16.965	15.237
500	18.31	170.69	0.014147	0.5325	15.703	14.220
750	41.00	148.00	0.014507	0.5192	13.616	12.489
1000	72.35	116.65	0.014940	0.5042	10.732	9.981
1250	111.91	77.09	0.015441	0.4878	7.092	6.762
1500	159.12	29.88	0.016002	0.4707	2.749	2.688

For a well drainage of 320 acres:

$$\text{Drainage area } = \pi r_e^2 = 320 \times 43{,}560 \frac{\text{ft}^2}{\text{acre}}$$

$$r_e = 2106 \text{ ft}$$

To develop the nonturbulence IPR curve for a vertical well, the turbulence term in Eq. 3–18 is ignored, i.e., $D = 0$. Then Eq. 3–17 can be solved explicitly as

$$q_g = \frac{10 \times 69 \times 520[189 \times 10^6 - \psi(p_{wf})]}{50.335 \times 10^3 \times (226 + 460) \times 14.65[\ln(2106/0.4271) - 0.75]} \tag{3–102}$$

$$q_g = 0.000092[189 \times 10^6 - \psi(p_{wf})] \tag{3–103}$$

Equation 3–103 can be used for different well flowing pressure to calculate gas flow rates. The results are shown in Table 3–9 and Figure 3–24. For turbulent flow, Eq. 3–102 is written as:

$$q_g = \frac{10 \times 69 \times 520[189 \times 10^6 - \psi(p_{wf})]}{50.335 \times 10^3 \times (226 + 460)[\ln(2106/0.4271) - 0.75 + Dq_g]}$$

$$= \frac{CC'}{BB' + Dq_g} \tag{3–104}$$

This is a quadratic equation:

$$Dq_g^2 + BB'q_g - CC' = 0 \tag{3–105}$$

and

$$q_g = \frac{-BB' + \sqrt{(BB')^2 + 4 \times D \times CC'}}{2D} \tag{3–106}$$

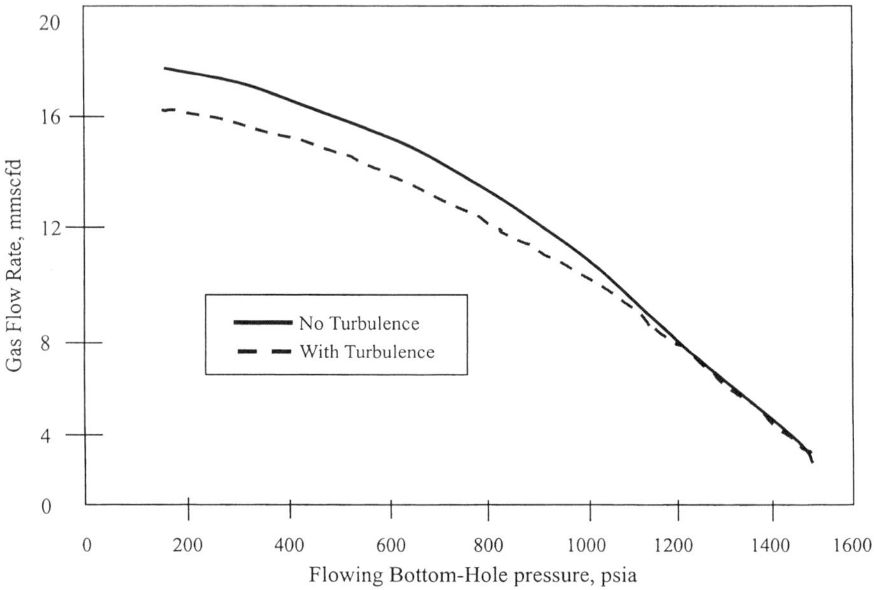

Figure 3–24. IPR performance showing effect of turbulence vertical gas well.

$$q_g = \frac{-7.7533 + \sqrt{(7.7533)^2 + 4 \times D \times 0.000709[189 \times 10^6 - \psi(p_{wf})]}}{2D} \tag{3–107}$$

To solve Eq. 3–107 we need to calculate turbulence factor D using Eqs. 3–18 and 3–20 and assuming $k = k_a$:

$$\beta' = \frac{2.33 \times 10^{10}}{k_a^{1.201}} = \frac{2.33 \times 10^{10}}{10^{1.201}} = 1.467 \times 10^9$$

$$D = \frac{2.222 \times 10^{-15} \times 0.681 \times 10 \times 69 \times 1.467 \times 10^9}{\mu_{p_{wf}} \times 0.4271 \times 69 \times 69} = \frac{7.5326 \times 10^{-7}}{\mu_{p_{wf}}}$$

For horizontal well calculations, Eq. 3–102 is written as

$$q_g = \frac{khT_{sc}[189 \times 10^6 - \psi(p_{wf})]}{50.335 \times 10^3 TP_{sc}[\ln(2106/0.4271) - 0.75 + s + s_m + s_{CA} + Dq_g - c']} \tag{3–108}$$

For a 2000-ft-long horizontal well, skin factor s and shape-related skin factor s_{CA} need to be calculated.

Calculation of Skin Factor s for a Horizontal Well

$$r'_w = L/4 = 2000/4 = 500 \text{ ft}$$
$$r_w = 0.4271 \text{ ft}$$
$$s = -\ln\left(\frac{r'_w}{r_w}\right) = -\ln\left(\frac{500}{0.4271}\right) = -7.07$$

The flow rate results are summarized in Table 3–9 and Figure 3–24. Table 3–9 shows an effect of turbulence on well flow rates.

$$q_g = \frac{khT_{sc}[189 \times 10^6 - \psi(p_{wf})]}{50.335\times 10^3 TP_{sc}[\ln(2106/0.4271) - 0.75 + s + s_m + s_{CA} + Dq_g - c']} \tag{3–108}$$

For a 2000-ft-long horizontal well, skin factor s and shape-related skin factor s_{CA} need to be calculated.

Calculation of s_{CA}

Assuming a square drainage area with each side being $2x_e$ for 320 acres, we have

$$2x_e = \sqrt{320 \times 43560} = 3734 \text{ ft}$$
$$L/(2x_e) = 2000/3734 = 0.536$$
$$k_v/k_h = 1/10 = 0.1$$
$$L_D = (L/(2h))\sqrt{\frac{k_v}{k_h}} = 2000/(2 \times 69)\sqrt{0.1} = 4.58$$

From Figure 3–10, corresponding to $L_D = 4.58$ and $L/(2x_e) = 0.536$, $s_{CA} = 2.75$.

We rewrite Eq. 3–108 to calculate gas flow rate by the following equation:

$$q_g = \frac{0.000709[189 \times 10^6 - \psi(p_{wf})]}{7.7533 + (-7.07) + 0 + 2.75 + Dq_g - 1.386} \tag{3–109}$$

Equation 3–109 assumes that the mechanical skin factor $s_m = 0$. For a nonturbulence case, $D = 0$ and Eq. 3–109 is solved explicitly for various values of p_{wf}. The final results are summarized in Table 3–10 and Figure 3–25. For a turbulence case, D is calculated by substituting 2000 ft as perforated length instead of 69 ft (for a vertical well) as used in Eq. 3–18.

$$D = \frac{2.222 \times 10^{-15} \times 0.681 \times 10 \times 69 \times 1.467 \times 10^9}{\mu_{p_{wf}} \times 0.4271 \times 2000 \times 2000}$$
$$= \frac{8.97 \times 10^{-10}}{\mu_{p_{wf}}} \tag{3–110}$$

Table 3–10
IPR Calculations for Horizontal Well (Example 3–6)

$H =$ p_{wf} (psia)	$\psi(p_{wf})$ (mmpsia^2/cP)	$\Delta\psi = \psi(\bar{p}_R) - \psi(p_{wf})$ (mmpsia^2/cP)	Viscosity $\mu_{p_{wf}}$ (cP)	Turbulence factor D (1/mscfd $\times 10^{-7}$)	q_g (mmscfd) No turb.	With turb.
250	4.60	184.40	0.013868	0.6468	63.863	63.390
500	18.31	170.69	0.014147	0.6341	59.112	59.011
750	41.00	148.00	0.014507	0.6183	51.129	51.000
1000	72.35	116.65	0.014940	0.6004	40.397	40.360
1250	111.91	77.09	0.015441	0.5809	26.697	26.120
1500	159.12	29.88	0.016002	0.5606	10.347	10.327

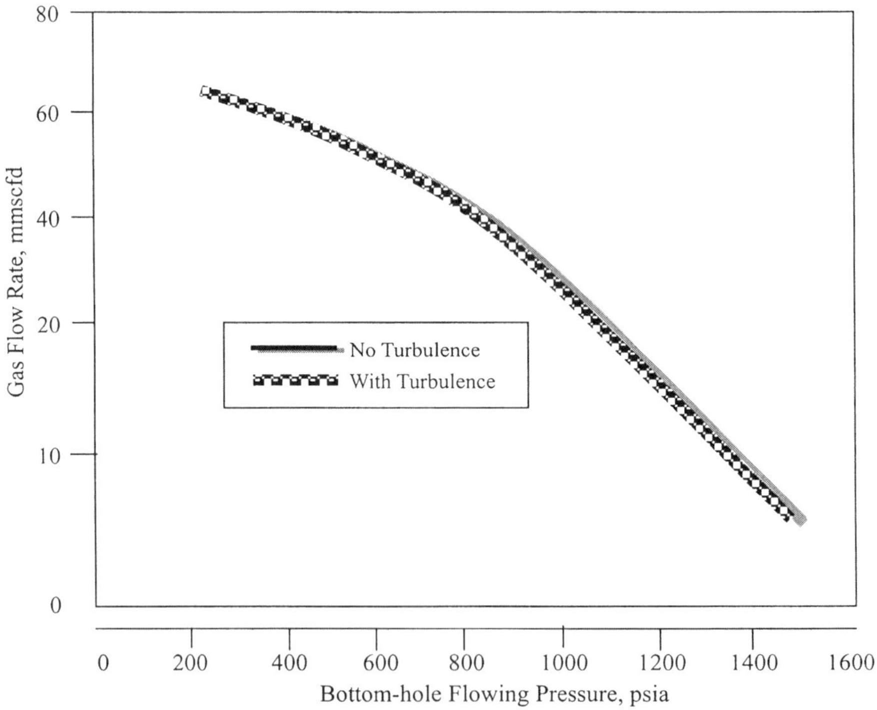

Figure 3–25. IPR performance showing effect of turbulence horizontal gas well.

For turbulence calculations, the value of D from Eq. 3–110 is substituted into Eq. 3–109:

$$q_g = \frac{0.000709[189 \times 10 - \psi(p_{wf})]}{2.047 + Dq_g} = \frac{CC'}{BB' + Dq_g} \tag{3–111}$$

Equation 3–111 is a quadratic equation, which can be solved as

$$q_g = \frac{-BB' + \sqrt{(BB')^2 + 4 \times D \times CC'}}{2D} \tag{3–112}$$

and

$$q_g = \frac{-2.047 + \sqrt{(2.047)^2 + 4 \times D \times 0.000709(189 \times 10^6 - \psi(p_{wf})}}{2D} \tag{3–113}$$

By substituting an appropriate D value, Eq. 3–113 is solved for various well flowing pressure, $\psi(p_{wf})$, values. In Figure 3–25 inflow performance relationships (IPR curves) for horizontal wells are shown. The plots are for two cases: (1) without turbulence and (2) with turbulence.

Celier *et al.*[16] presented the following Eq. 3–114 to calculate the ratio of pressure drop due to turbulence, i.e., non-Darcy flow, in horizontal and vertical wells:

$$\frac{(\Delta p)_{h,t}}{(\Delta p)_{v,t}} = \frac{2\beta^2}{(1+\beta)}\left[\frac{h}{L}\right]^2 \tag{3–114}$$

where

$\beta = \sqrt{k_h/k_v}$
h = reservoir height, ft
L = well length, ft

It is important to note that the preceding equation assumes that h ft of vertical well is perforated in an h-ft-thick reservoir; similarly, L ft of horizontal well is open to flow.

Example 3–7 *Calculating Reduction in Turbulence-Related Pressure Drop*
Determine reduction in turbulence related pressure drop in a 45-ft-thick, 10.0 mD reservoir by drilling a 2000-ft-long horizontal well. What would be the pressure drop ratio if $k_v/k_h = 0.01$, 0.1, and 1.0?

Solution A ratio of horizontal and vertical well pressure drops due to non-Darcy flow is given in Eq. 3–115:

$$\frac{(\Delta p)_{h,t}}{(\Delta p)_{v,t}} = \frac{2\beta^2}{(1+\beta)}g\left[\frac{h}{L}g\right]^2 \tag{3–115}$$

If $k_v/k_h = 0.01$,

$$\beta = \sqrt{\frac{k_h}{k_v}} = \sqrt{100} = 10$$

If $k_v/k_h = 0.1$,

$$\beta = \sqrt{\frac{k_h}{k_v}} = \sqrt{10} = 3.16$$

If $k_v/k_h = 1.0$,

$$\beta = \sqrt{\frac{k_h}{k_v}} = 1.0$$

If $k_v/k_h = 0.01$,

$$\frac{(\Delta p)_{h,t}}{(\Delta p)_{v,t}} = \frac{2 \times (10)^2}{1 + 10}(0.000506) = 0.00920$$

If $k_v/k_h = 0.1$,

$$\frac{(\Delta p)_{h,t}}{(\Delta p)_{v,t}} = \frac{2 \times (3.16)^2}{1 + 3.16}(0.000506) = 0.00243$$

If $k_v/k_h = 1.0$,

$$\frac{(\Delta p)_{h,t}}{(\Delta p)_{v,t}} = \frac{2 \times 1}{1 + 1}(0.000506) = 0.000506$$

3.11 Estimating Reservoir Properties from Production Histories

Homogeneous Isotropic Systems

Production type curves suitable for practical applications are based on the work by Duda[17] and Aminian and Ameri.[18] These type curves can be used either for production feasibility or for estimating reservoir properties from the production histories. These type curves are for two dimensionless wellbore radii, namely, $r_{wD} = 2.5 \times 10^{-4}$ and $r_{wD} = 5.0 \times 10^{-4}$,

where

$$r_{wD} = \sqrt{4r_w^2/L^2} \tag{3–116}$$

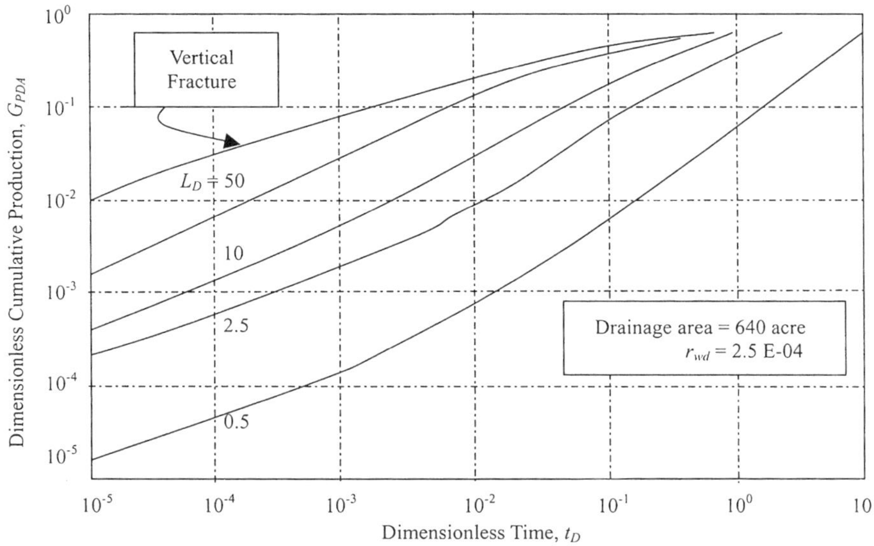

Figure 3–26. Production type curves for horizontal gas wells in a 640-acre drainage area.[17]

The type curves shown in Figures 3–26 through 3–29 are for production in 320- and 640-acre reservoirs, and Figure 3–31 is for production in infinite reservoirs with no flow across the drainage boundary.

Homogeneous Anisotropic Systems

Figure 3–30 shows a production-type curve in an anisotropic reservoir, where a horizontal well is drilled along the x direction. The figure clearly indicates the benefits of drilling a horizontal well along the low-permeability direction. The curves also note loss in production by drilling a horizontal well along the high-permeability direction. The type curves shown in Figure 3–32 also have been developed for rectangular drainage area.[18]

Square Drainage Area

Duda[17] type curves can be used for either production forecasting or estimating reservoir properties from the production histories. In these type curves various dimensionless terms are defined as

$$L_D = [L/(2h)](k_v/k_h)^{0.5} \tag{3–117}$$

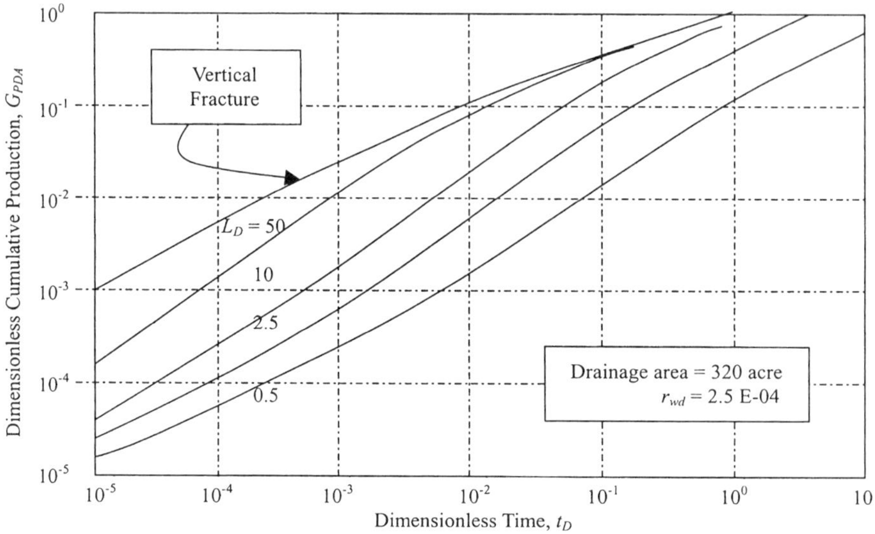

Figure 3–27. Production type curves for horizontal gas wells in a 320-acre drainage area.[17]

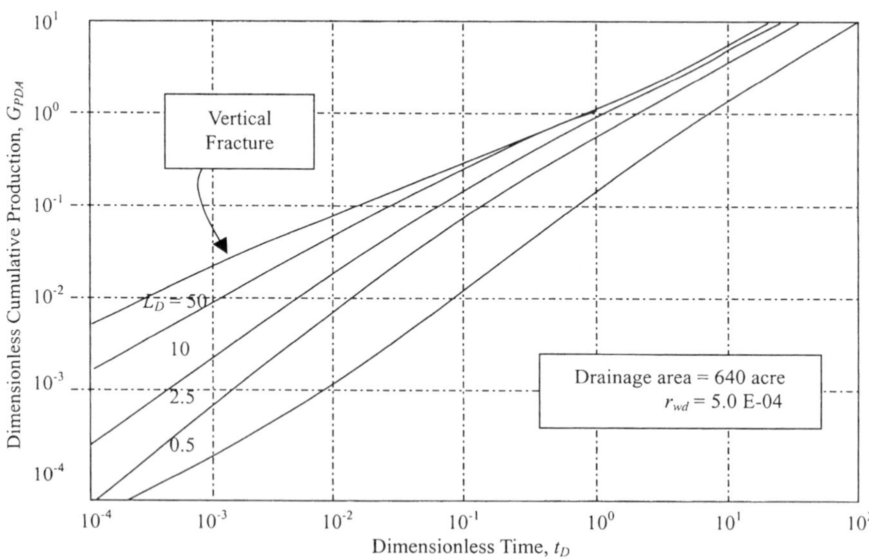

Figure 3–28. Production type curves for horizontal gas wells in a 640-acre drainage area.[17]

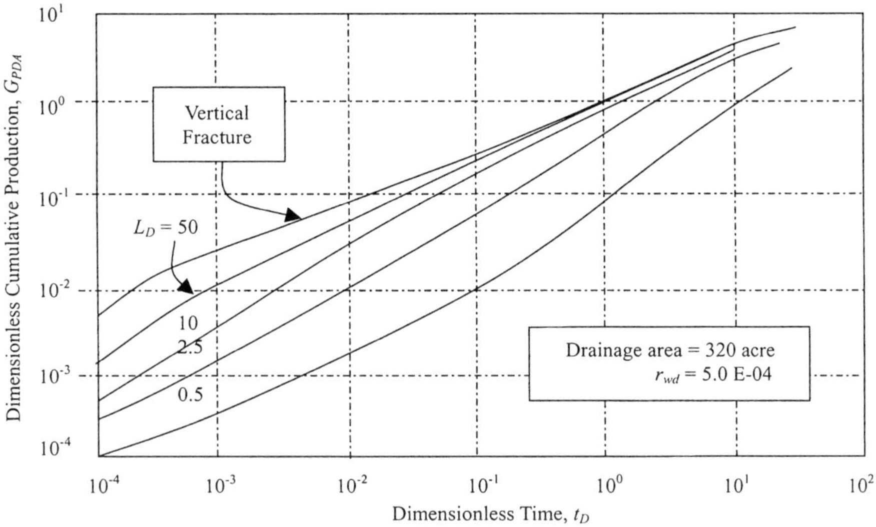

Figure 3–29. Production type curves for horizontal gas wells in a 320-acre drainage area.[17]

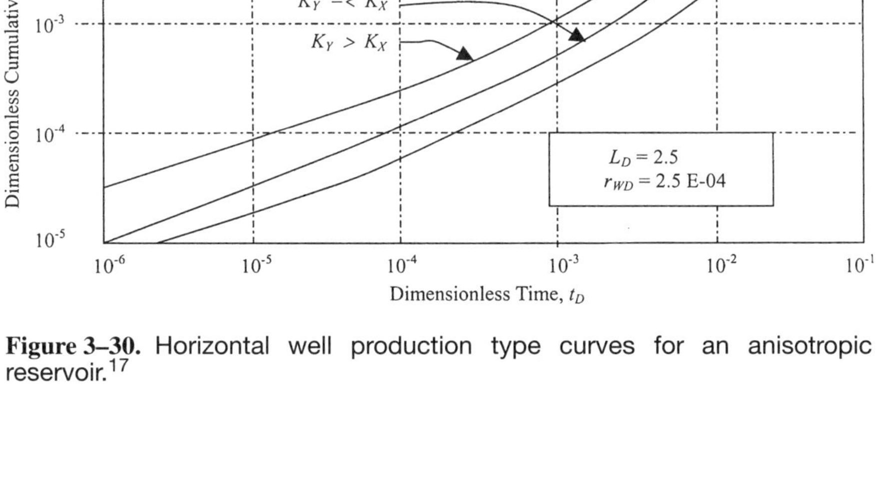

Figure 3–30. Horizontal well production type curves for an anisotropic reservoir.[17]

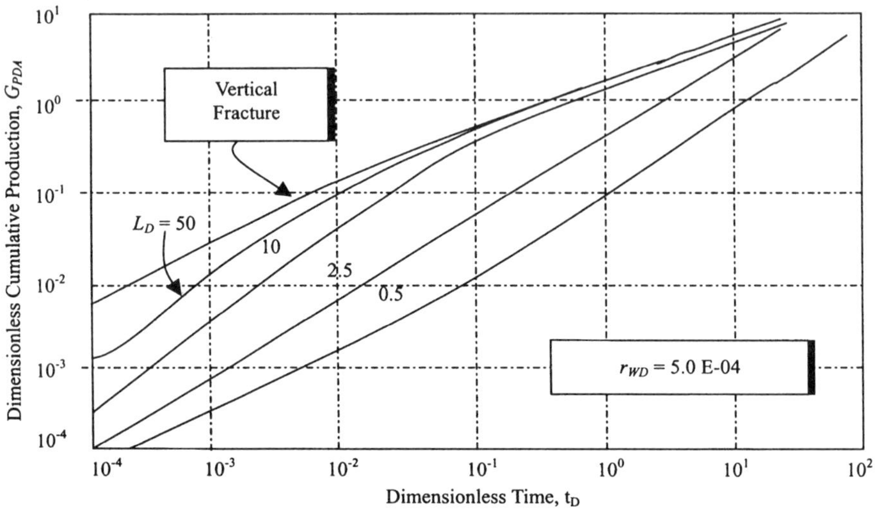

Figure 3–31. Production type curves for horizontal gas wells in an infinite reservoir.[17]

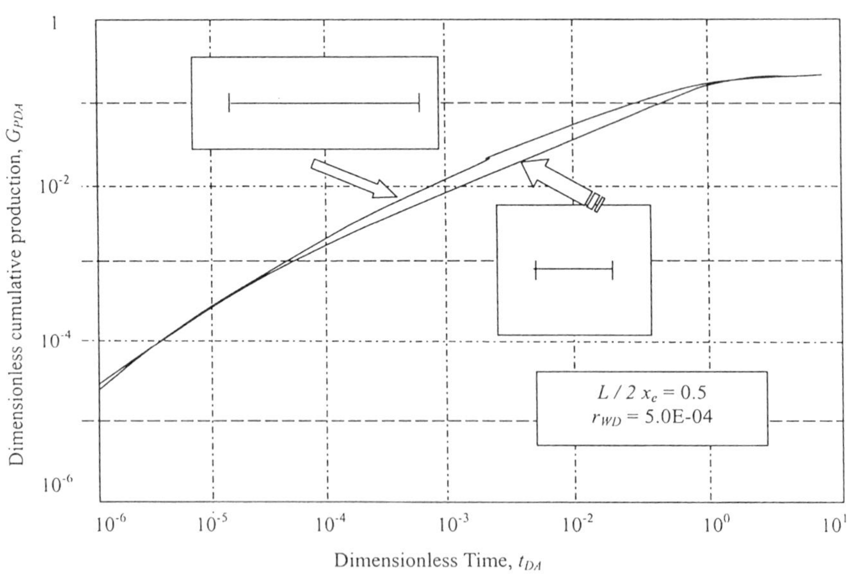

Figure 3–32. Production type curve for horizontal gas well in a rectangular reservoir.[18]

$$t_D = \frac{0.001055kt}{\phi \mu c_t L^2} \tag{3–118}$$

$$G_{pD} = \frac{9.009 G_P T}{h \phi \mu c_t L^2 \psi(\Delta p)} \tag{3–119}$$

where:

L = well length, ft; h = reservoir thickness, ft;
k = permeability, mD; t = time, hours;
k_v = permeability in vertical direction, mD;
k_h = permeability in x direction, mD;
ϕ = porosity, fraction;
μ = gas viscosity, cP;
μc_t = total compressibility, psi^{-1}
T = reservoir temperature, °R;
G_P = cumulative gas, mcf; and
$\psi(\Delta p)$ = pseudopressure, psia^2/cP

Rectangular Drainage Area

Aminian and Ameri[18] reported production type curves for a horizontal gas well in a rectangular reservoir. In these type curves various dimensionless terms are as follows.

Dimensionless time t_{DA} is defined as

$$t_{DA} = \left[\frac{0.001055 k_h}{\phi \mu c_t A}\right] t \tag{3–120}$$

Dimensionless cumulative production based on area is defined as:

$$G_{pDA} = \left[\frac{36T}{h \phi \mu c_t A \psi(\bar{p})}\right] G_P \tag{3–121}$$

where

k_h = horizontal permeability, mD;
c_t = total compressibility, psi^{-1};
A = drainage area, acres;
t = time, hours; ϕ = porosity, fraction;
μ = gas viscosity, cP;
$\bar{p}$ = reservoir pressure, psi,
G_P = cumulative production, mscf; and
T = reservoir temperature, °R

3.12 Summary

Chapter 3 summarizes a discussion of horizontal wells in gas reservoirs. The discussion presented indicates the advantages of horizontal wells in low- and high-permeability gas reservoirs. Horizontal wells enhance the drainage area in a given time period while, in high-permeability gas reservoirs, reducing near-wellbore turbulence and enhancing well deliverability. Horizontal wells have high potential in gas reservoirs.

References and Additional Reading

1. Smith, R. V., *Practical Natural Gas Engineering*, PennWell Publishing Co., Tulsa, OK, 1983.
2. Brown, K. E., *The Technology of Artificial Methods*, PennWell Publishing Co., Tulsa, OK, 1984.
3. Earlougher, R. C., Jr., *Advances in Well Test Analysis*, Monograph Vol. 5 of the Henry L. Doherty Series in Society of Petroleum Engineers of AIME, 1977.
4. Fetkovich, M. J., and Vienot, M. E., "Shape Factors, C_A, Expressed as a Skin, s_{CA}," *J. Petroleum Technol.* (Feb. 1985), 321–322.
5. Al-Hussainy, R., Ramey, H. J., Jr., and Crawford, P. B., "The Flow of Real Gases through Porous Media," *J. Petroleum Technol.* (1966) 624–636; *Trans. AIME* 237.
6. Goode, P. A., and Thambynayagam, R. K. M., "Pressure Drawdown and Buildup Analysis for Horizontal Wells in Anisotropic Media," *SPE Formation Evaluation* (Dec. 1987) 683–697.
7. Daviau, F., Mouronval, G., Bourdarot, G., and Curutchet, P., "Pressure Analysis for Horizontal Wells," SPE 14251, presented at the 1985 SPE Annual Technical Conference and Exhibition, Las Vegas, NV, Sept. 22–25.
8. Golan, M., and Whitson, C. H., *Well Performance*, International Human Resources Development Corporation, Boston, 1986.
9. Mutalik, P. N., Godbole, S. P., and Joshi, S. D., "Effect of Drainage Area Shapes on Horizontal Well Productivity," paper SPE 18301, presented at the SPE 63rd Annual Technical Conference, Houston, TX, Oct. 2–5, 1988.
10. Gringarten, A. C., "Reservoir Limit Testing for Fractured Wells," paper SPE 7452, presented at the SPE 53rd Annual Fall Technical Conference and Exhibition, Houston, TX, Oct. 1–3, 1978.
11. Gringarten, A. C., Ramey, H. J., Jr., and Raghavan, R., "Unsteady-State Pressure Distribution Created by a Well with a Single Infinite-Conductivity Vertical Fracture," *Soc. Petroleum Eng. J.* (Aug. 1974) 347–360.

12. Kuchuk, F. J., Goode, P. A., Wilkinson, D. J., and Thambynayagam, R. K. M., "Pressure Transient Behavior of Horizontal Wells with and without Gas Cap or Aquifer," paper SPE 17413, presented at the SPE California Regional Meeting, Long Beach, CA, March 23–25, 1988.
13. Odeh, A. S., and Babu, D. K., "Transient Flow Behavior of Horizontal Wells Pressure Drawdown and Buildup Analysis," *SPE Formation Eval.* (March 1990) 7–15.
14. Clonts, M. D., and Ramey, H. J., Jr., "Pressure Transient Analysis for Wells with Horizontal Drainageholes," paper SPE 15116, presented at the SPE California Regional Meeting, Oakland, CA, April 2–4, 1986.
15. Ozkan, E., Raghavan, R., and Joshi, S. D., "Horizontal Well Pressure Analysis," *SPE Formation Eval.* (Dec. 1989) 567–575.
16. Celier, G. C. M. R., Jouault, P., de Montigny, O. A. M. C., and Zuidwal, "A Gas Field Development with Horizontal Wells," paper SPE 19826, presented at the SPE 64th Annual Technical Conference and Exhibition of the Society of Petroleum Engineers, San Antonio, TX, Oct. 8–11, 1989.
17. Duda, J. R., "Type Curves for Predicting Production Performance from Horizontal Wells in Low Permeability Gas Reservoirs," paper SPE 18993, Richardson, TX.
18. Aminian, K, and Ameri, S., "Predicting Horizontal Well Production Performance Using Type Curves," paper SPE 19342 presented at the SPE Eastern Regional Meeting, Morgantown, WV, Oct. 1989.

Chapter 4

Deliverability Testing and Well Production Potential Analysis Methods

4.1 Introduction

This chapter discusses basic flow equations expressed in terms of the pseudopressure $\psi(p)$ and of approximations to the pseudopressure approach that are valid at high and low pressures. This is followed by deliverability tests of gas well flow-after-flow, isochronal, and modified isochronal deliverability tests including a simplified procedure for gas deliverability calculations using dimensionless IPR curves. The purpose of this chapter is to provide a complete reference work for various deliverability testing techniques. The mathematical determinations of the equations are avoided; this role is filled much better by other publications.[1–2] Field examples are included to provide a hands-on understanding of various deliverability testing techniques, their interpretations and their field applications.

4.2 Gas Flow in Infinite-Acting Reservoirs

References 1 and 3 have shown that gas flow in an infinite-acting reservoir can be expressed by an equation similar to that for flow of slightly compressible liquids if pseudopressure $\psi(p)$ is used instead of pressure. The equation in SI units is

$$\psi(p_{wf}) = \psi(p_i) + 3.733 \frac{p_{sc}}{T_{sc}} \frac{q_g T}{kh} \left[1.151 \log\left(\frac{125.3 \phi \mu_i C_{ti} r_w^2}{kt} \right) - s + D|q_g| \right]$$

In field units this equation becomes

$$\psi(p_{wf}) = \psi(p_i) + 50,300\frac{p_{sc}}{T_{sc}}\frac{q_g T}{kh} \times \left[1.151\log - In\left(\frac{1,688\phi\mu_{gi}C_{ti}r_w^2}{kt}\right) - s + D|q_g|\right] \quad (4\text{–}1)$$

where the pseudopressure is defined by the integral

$$\psi(p) = 2\int_{p_B}^{p} \frac{p}{\mu_g z}dp \quad (4\text{–}2)$$

where p_B is some arbitrary low base pressure. To evaluate $\psi(p_{wf})$ at some value of p, we can evaluate the integral in Eq. 4–2 numerically, using values for μ and z for the specific gas under consideration, evaluated at reservoir temperature. The term $D|q_g|$ gives a non-Darcy flow pressure drop, i.e., it takes into account the fact that, at high velocities near the producing well, Darcy's law does not predict correctly the relationship between flow rate and pressure drop. Therefore this additional pressure drop can be added to the Darcy's law pressure drop, just as pressure drop across the altered zone is, and D can be considered constant. The absolute value of q_g, $|q_g|$, is used so that the term $D|q_g|$, is positive for either production or injection.

4.3 Stabilized Flow Equations

For stabilized[3] ($r_i \leq r_e$), flow,

$$\psi(p_{wf}) = \psi(p_R) - 1.422 \times 10^6 \frac{q_{sc}T}{kh}\left[In\left(\frac{r_e}{r_w}\right) - 0.75 + s + D|q_{sc}|\right] \quad (4\text{–}3)$$

where p_R is any uniform drainage-area pressure. Equations 4–1 and 4–3 provide the basis for analysis of gas well tests. For $p > 3000$ psia, these equations assume a simple form (in terms of pressure, p); for $p < 2000$ psia, they assume another simple form in terms of p^2. Thus we can develop procedures for analyzing gas well tests with equations in terms of $\psi(p)$, p, and p^2. In most of this chapter, equations will be written in terms of $\psi(p)$ and p^2, not because p^2 is more generally applicable or more accurate (the equations in ψ best fit this role), but because the p^2 equation illustrate the general method and permit easier comparison with other methods of gas well test analysis. The stabalized flow equation in terms of pressure squared is

$$p_{wf}^2 = p_R^2 - 1.422 \times 10^6 \frac{q_{sc}\overline{\mu}_g \overline{z} T}{kh}\left[In\left(\frac{r_e}{r_w}\right) - 0.75 + s + D|q_{sc}|\right] \quad (4\text{–}4)$$

Equations 4–3 and 4–4 are complete deliverability equations. Given a value of flowing bottom-hole pressure, p_{wf}, corresponding to a given pipeline or backpressure, we can estimate the flow rate, q_{sc}, at which the well will deliver gas. However, certain parameters must be determined before the equations can be used in this way. The well flows at rate q_{sc} until $r_i \geq r_e$ (stabilized flow). In this case, Eq. 4–3 has the form

$$\psi(p_R) - \psi(p_{wf}) = Aq_{sc} + Bq_{sc}^2 \quad (4\text{–}5)$$

where

$$A = 1.422 \times 10^6 \frac{T}{kh}\left[In\left(\frac{r_e}{r_w}\right) - 0.75 + s\right] \quad (4\text{–}6)$$

and

$$B = 1.422 \times 10^6 \frac{T}{kh} D \quad (4\text{–}7)$$

Equation 4–4 has the form

$$p_R^2 - p_{wf}^2 = A' q_{sc} + B' q_{sc}^2 \quad (4\text{–}8)$$

where

$$A' = 1.422 \times 10^6 \frac{\overline{\mu}_g \overline{z} T}{kh}\left[In\left(\frac{r_e}{r_w}\right) - 0.75 + s\right] \quad (4\text{–}9)$$

$$B' = 1.422 \times 10^6 \frac{\overline{\mu}_g \overline{z} T}{kh} D \quad (4\text{–}10)$$

The constants A, B, A', and B' can be determined from flow tests for at least two rates in which q_{sc} and the corresponding value of p_{wf} are measured; p_R also must be known.

4.4 Application of Transient Flow Equations

When $r_i < r_e$, the flow conditions are said to be transient and for transient flow. In terms of pseudopressure:

$$\psi(p_R) - \psi(p_{wf}) = A_t q_{sc} + Bq_{sc}^2 \quad (4\text{–}11)$$

where B has the same meaning as for stabilized flow and where A_t, a function of time, is given by

$$A_t = \frac{1.637 \times 10^6 T}{kh}\left[\log\left(\frac{kt}{\phi \mu_{gi} C_i r_w^2}\right) - 3.23 + .869s\right] \tag{4–12}$$

In terms of pressure squared

$$p_R^2 - p_{wf}^2 = A_t' q_{sc} + B' q_{sc}^2 \tag{4–13}$$

$$A_t' = 1.422 \times 10^6 \frac{\overline{\mu}_g \overline{z} T}{kh}\left[\frac{1}{2} In\left(\frac{kt}{1{,}688 \phi \mu_g C_t r_w^2}\right) + s\right] \tag{4–14}$$

$$B' = 1.422 \times 10^6 \frac{\overline{\mu}_g \overline{z} T}{kh} D \tag{4–15}$$

Turbulence or Non-Darcy Effects on Completion Efficiency

Reference 3 can be applied to gas well testing to determine real or present time inflow performance relationships. No transient tests are required to evaluate the completion efficiency, if this method is applied. Reference 3 also suggested methods to estimate the improvement in inflow performance which would result from re-perforating a well to lengthen the completion interval and presents guidelines to determine whether the turbulent effects are excessive. Equation 4–13 can be divided by q_{sc} and written as

$$\frac{P_R^2 - P_{wf}^2}{q_{sc}} = A' + B' q_{sc} \tag{4–15a}$$

where A' and B' are the laminar and turbulent coefficients, respectively, and are defined in Eqs. 4–9 and 4–10. From Eq. 4–15a, it is apparent that a plot of $(P_R^2 - P_{wf}^2)/q_{sc}$ versus q_{sc} on Cartesian coordinates will yield a line that has a slope of B' and an intercept of $A' = \Delta(p^2)/q_{sc}$ as q_{sc} approaches zero. These plots apply to both linear and radial flow, but definition of A' and B' would depend on the type of flow. In order to have some qualitative measure of the importance of the turbulence contribution to the total drawdown. Reference 3 suggested comparison of the value of A' calculated at the AOF of the well (AA), to the stabilized value of A'. The value of AA can be calculated from

$$AA = A' + B'(AOF) \tag{4–15b}$$

where

$$AOF = \frac{-A' + \left[A'^2 + 4B'P_R^2\right]^{0.5}}{2B'} \quad (4\text{–}15c)$$

Reference 3 suggested that the ratio of *AA* to A' be replaced by the length of the completed zone, h_P, since most of the turbulent pressure drop occurs very near the wellbore. The effect of changing completion zone length on B' and therefore on inflow performance can be estimated from

$$B_2 = B_1\left(\frac{h_{P1}}{h_{P2}}\right)^2 \quad (4\text{–}15d)$$

where

B_2 = turbulence multiplier after recompletion
B_1 = turbulence multiplier before completion
h_{P1} = gas completion length, and
h_{P2} = new completion length

In term of real pseudopressure,

$$\frac{\psi(P_R) - \psi(P_{wf})}{q_{sc}} = A + Bq_{sc} \quad (4\text{–}15e)$$

where A and B are the laminar and turbulent coefficients, respectively, and are defined in Eqs. 4–6 and 4–7. From Eq. 4–15e it is apparent that a plot of $\psi(P_R) - \psi(P_{wf})/q_{sc}$ versus q_{sc} on Cartesian coordinates will yield a straight line that has a slope B and an intercept of $A = \psi(P)/q_{sc}$ as q_{sc} approaches zero. These plots apply to both linear and radial flow, but the definitions of A and B would depend on the type of flow. The value of A is calculated at the *AOF* of the well (*AAA*) to the stabilized value of A. The value of *AAA* can be calculated from

$$AAA = A + B(AOF) \quad (4\text{–}15f)$$

where

$$AOF = \frac{-A + [A^2 + 4B\psi(P_R)]^{0.5}}{4B} \quad (4\text{–}15g)$$

and

$$B_4 = B_3(h_{P1}/h_{P2})^2 \quad (4\text{–}15h)$$

where

B_4 = turbulence multiplier after recompletion
B_3 = turbulence multiplier before recompletion

The applications of these equations are illustrated in the following field examples.

Example 4–1 *Analyzing Completion Efficiency*

A four-point test is conducted on a gas well that has a perforated zone of 25 ft. Static reservoir pressure is 1660 psia. Determine the followings: (1) A and B, (2) AOF, (3) the ratio A'/A, and (4) new AOF if the perforated interval is increased to 35 ft. $P_R = 408{,}2$ psia, $\psi(P_R) = 772.56$ mmpsia^2/cP.

Four-Point Test Data

Test #	q_{sc} (mmscfd)	P_{wf} (psia)
1	4.288	403.1
2	9.265	394.0
3	15.552	378.5
4	20.177	362.6

Solution Equation 4–15a can be divided through by q_{sc} and written as

$$\frac{P_R^2 - P_{wf}^2}{q_{sc}} = A' + B' q_{sc}$$

where A' and B' are the laminar and turbulent coefficients, respectively, and are defined in Eqs. 4–9 and 4–10. It is apparent that a plot of $(P_R^2 - P_{wf}^2)/q_{sc}$ versus q_{sc} on Cartesian coordinates will yield a straight line that has a slope of B' and an intercept of $A' = \Delta(P^2)/q_{sc}$ as q_{sc} approaches zero.

Data from Table 4–1 are plotted for both empirical and theoretical analysis. Figure 4–1 is a plot of $(P_R^2 - P_{wf}^2)/q_{sc}$ versus q_{sc} on log-log paper and is almost linear, but there is sufficient curvature to cause a 15% error in calculated AOF. Therefore AOF is 51.8 mmscfd. Figure 4–2 is a plot of $(P_R^2 - P_{wf}^2)/q_{sc}$ versus q_{sc} on Cartesian paper and it is found that intercept $A' = 773$ psia^2/mscfd,

Table 4–1
Calculated Four-Point Test Data for Stabilized Flow Analysis

Test #	q_{sc} (mmscfd)	$(P_R^2 - P_{wf}^2)$ (psia^2)	$(P_R^2 - P_{wf}^2)/q_{sc}$ (psia^2/mscfd)
1	4.288	4,138	33.9
2	9.265	11,391	1,229
3	15.552	23,365	1,502
4	20.177	35,148	1,742

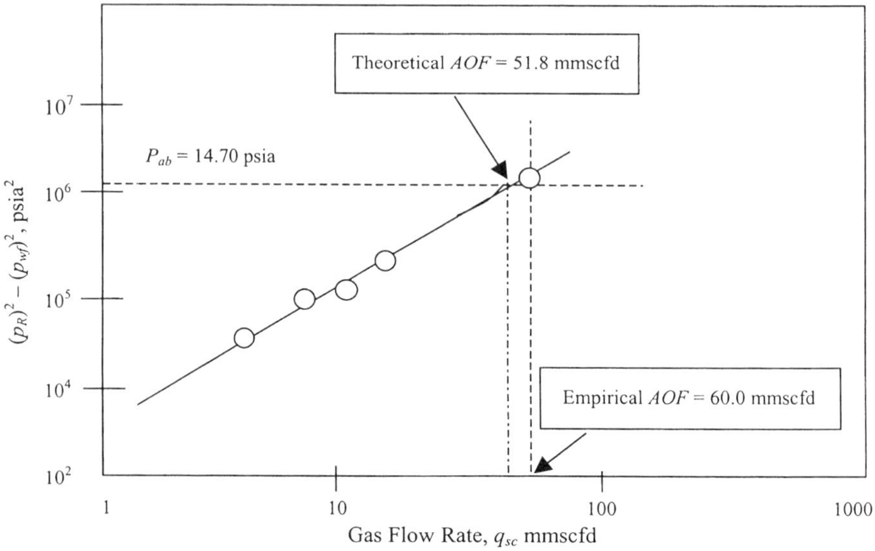

Figure 4–1. Plot of ΔP^2 versus q_{sc}.

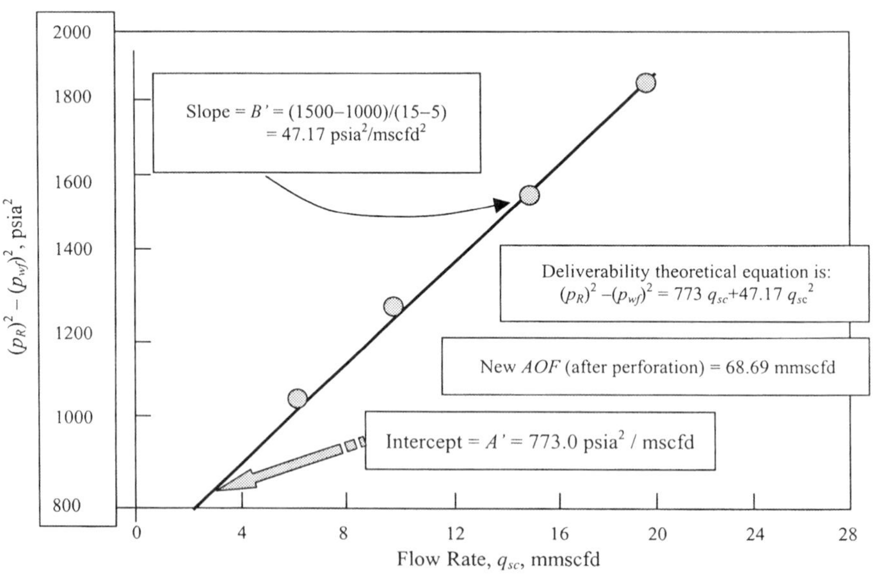

Figure 4–2. Stabilized deliverability test, showing theoretical flow equation and constants.

and slope $B' = 47.17$ psia^2/mscfd^2. Absolute open flow potential (*AOF*) is given by

$$AOF = \frac{-A' + \left[A'^2 + 4B'P_R^2\right]^{0.5}}{2B'} = 51.8 \text{ mmscfd}$$

Using Eq. 4–15b:

$$AA = A' + B'(AOF) = 773 + 47.17(51.8) = 245{,}113.60$$

$$AA/A' = 245{,}113.60/773 = 317.094$$

Using Eq. 4–15d:

$$B_2 = B_1(h_{P1}/h_{P2}) = 47.17(25/35)^2 = 24.063 \quad [\text{where } B_1 = B']$$

$$AOF_2 = \frac{-A' + \left[A'^2 + 4B_2P_R^2\right]^{0.5}}{2B_2}$$

$$= \frac{-773 + [773^2 + 4 \times 24.063 \times 408.2^2]^{0.5}}{2 \times 24.063}$$

$$= 68.69 \text{ mscfd}$$

The value of A' calculated in the previous example indicates a large degree of turbulence. The effect of increasing the perforated interval on the *AOF* is substantial.

4.5 Classifications, Limitations, and Use of Deliverability Tests

Figure 4–3 shows types, limitations, and uses of deliverability tests. In designing a deliverability test, collect and utilize all information, which may include logs, drill-stem tests, previous deliverability tests conducted on that well, production history, gas and liquid compositions, temperature, cores, and geological studies. Knowledge of the time required for stabilization is a very important factor in deciding the type of test to be used for determining the deliverability of a gas well. This may be known directly from previous tests, such as drill-stem or deliverability tests, conducted on the well or from the production characteristics of the well. If such information is not available, it may be assumed that the well will behave in a manner similar to neighboring wells in the same pool, for which data are available. When the approximate time to stabilization is not known, it may be estimated from

$$t_s \cong \frac{1000\phi\overline{\mu}_g r_e^2}{k\overline{p}_R} \qquad (4\text{–}16)$$

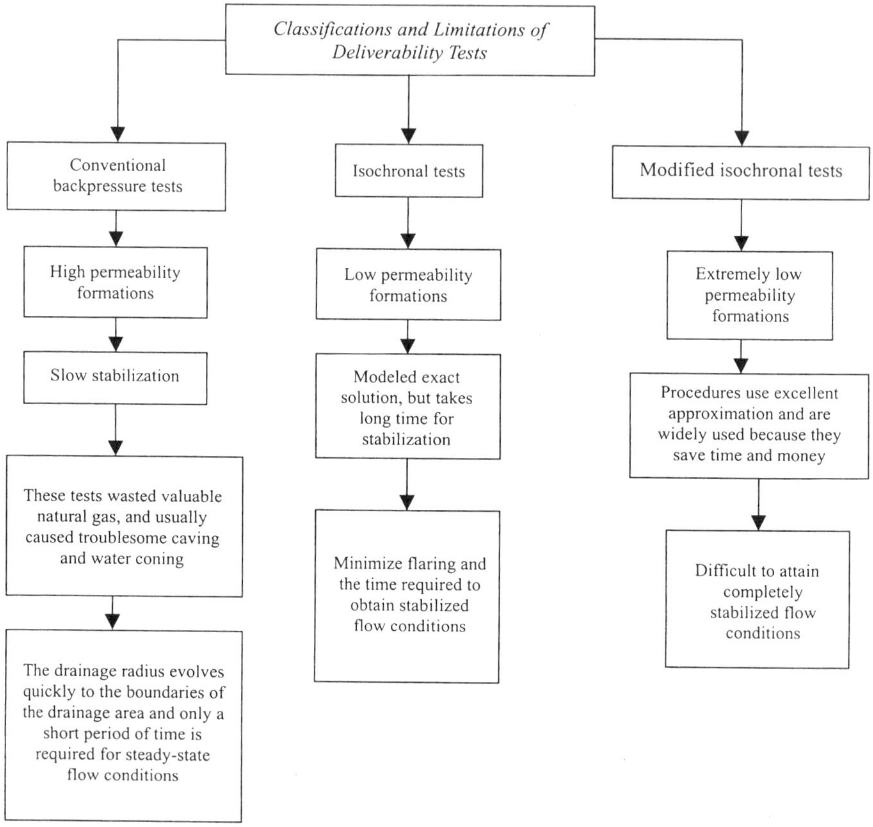

Figure 4–3. Types, limitations, and uses of deliverability tests.

where t_s is time of stabilization, and the radius of investigation can be found from

$$r_{inv} = 0.032\sqrt{\frac{k\overline{p}_R}{\phi\overline{\mu}_g}} \tag{4–17}$$

Applications of Eqs. 4–19 and 4–20 are as follows: if $r_{inv} = r_e \rightarrow$ pseudo-steady-state; $r_{inv} < r_e \rightarrow$ transient state; and $r_d = 0.472 \rightarrow$ effective drainage radius. If $t < t_S$, both C and n changes, and if $t > t_S$, both C and n will stay constant. If the time to stabilization is of the order of a few hours, a conventional backpressure may be conducted. Otherwise one of the isochronal tests is preferable. The isochronal test is more accurate than the modified isochronal test and should be used if the greater accuracy is required. Types, limitations, advantages and disadvantages of deliverability tests are indicated

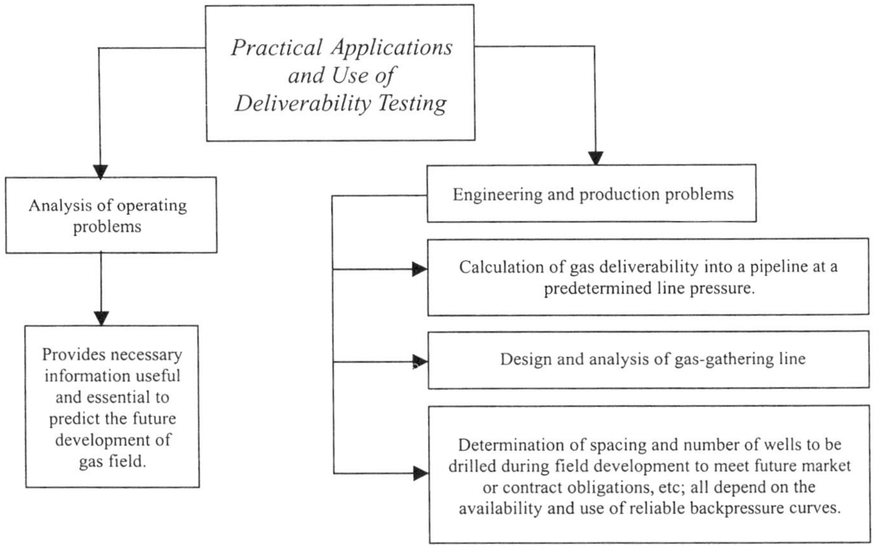

Figure 4–4. Practical applications and useful engineering practices.

in Figure 4–3, and practical applications and useful engineering practices are illustrated in Figure 4–4.

4.6 Flow-Rate, Pressure Behavior, and Deliverability Plots

In the past the behavior of gas wells was evaluated by open-flow tests. These tests wasted valuable natural gas, and usually caused troublesome caving and water coning. The need for better testing methods was first felt about 25 years ago. For many years, the U.S. Bureau of Mines[14] (Monograph 7) has served as a guide for evaluating the performance of gas wells by backpressure tests. Since Monograph 7, various methods of testing of gas wells have been published and put into practice. These methods,[13–15] also called flow-after-flow, isochronal, and modified isochronal performance methods, have all been based on experimental data and permit the determination of the exponent, n, and the performance coefficient C, from direct flow tests.

Conventional Backpressure Test

Figure 4–5 shows flow rate and pressure with time for q_{sc} increases in sequences. The method is based on the well-known Monograph 7 (Rawlins and Schellhardt, 1936),[14] which was the result of a large number of empirical

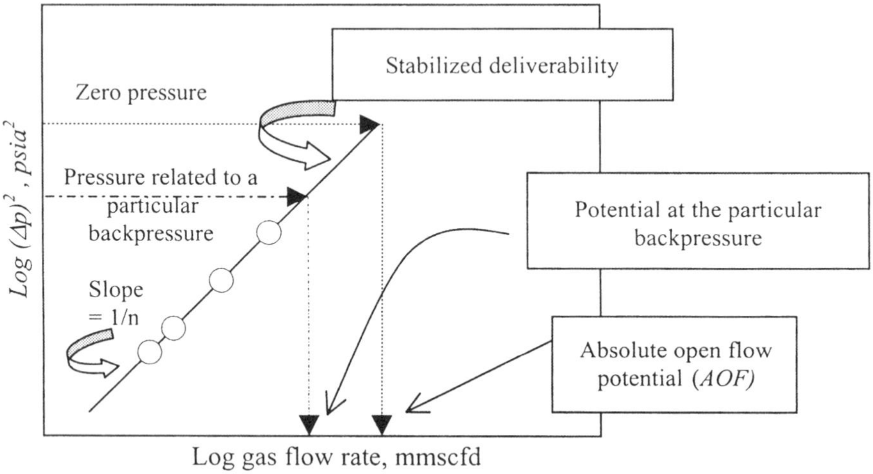

Figure 4–5. Conventional backpressure behavior curves.

observations. The relationship between the gas delivery rates and the bottom-hole pressure take, in general, the form

$$q_{sc} = C\left(\overline{p}_R^2 - p_{wf}^2\right)^n \tag{4–18}$$

$$C = \frac{q_{sc}}{\left(\overline{p}_R^2 - p_{wf}^2\right)^n} \tag{4–19}$$

where C is the performance coefficient, and n is the exponent corresponding to the slope of the straight-line relationship between q_{sc} and $(\overline{p}_R^2 - p_{wf}^2)$ plotted on logarithmic coordinates (see Figure 4–5). Exponents of $n < 0.5$ may be caused by liquid accumulation in the wellbore.

Exponents apparently greater than 1.0 may be caused by fluid removal during testing. If n is outside the range of 0.5 to 1.0, the test may be in error because of insufficient cleanup or liquid loading in the gas well. Performance coefficient C is considered as a variable with respect to time and as a constant only with respect to a specific time. Thus the backpressure curve represents the performance of the gas well at the end of a given time of interest. The value of C with respect to time does not obscure the true value of the slope.

Isochronal Testing

The isochronal test consists of alternately closing in the well until a stabilized or very nearly stabilized pressure $\overline{p}_R$ is reached and the well is flowed at different rates for a set period of time t, the flowing bottom-hole pressure p_{wf} at time t being recorded. One flow test is conducted for a time period long enough

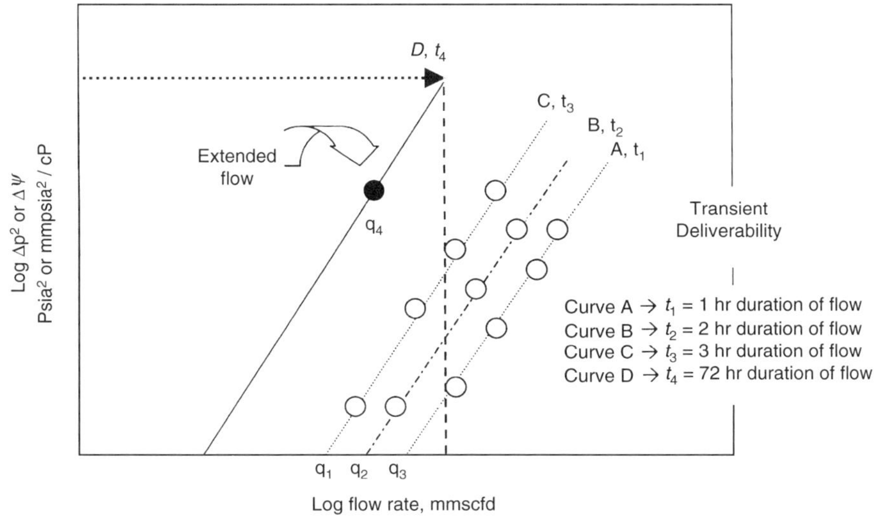

Figure 4–6. Isochronal performance curves.

to attain stabilized conditions and is usually referred to as the extended flow period. The behavior of the flow rate and pressure with various time periods is shown in Figure 4–6. The characteristic slope n, developed under short flow conditions, is applicable to long-time flow conditions. Also, the decline in the performance coefficient C is a variable with respect to time.

$$n = \frac{\log q_{sc1} - \log q_{sc2}}{\log \Delta(p)_1^2 - \log \Delta(p)_2^2} \tag{4–20}$$

where C is the performance coefficient, and n is the exponent corresponding to the slope of the straight-line relationship between q_{sc} and $(\overline{p}_R^2 - p_{wf}^2)$ plotted on logarithmic coordinates (see Figure 4–5). Exponents of $n < 0.5$ may be caused by liquid accumulation in the wellbore.

Modified Isochronal Testing

This type of testing is the same as the preceding isochronal method except that of $\overline{p}_R$. The preceding shut-in pressure is used in obtaining Δp^2 or $\Delta\psi$. The shut-in pressure to be used for the stabilized point is $\overline{p}_R$, the true stabilized shut-in pressure. The pressure and flow rate characteristic of the modified isochronal test is shown in Figure 4–7.

Transient deliverability equation:

$$\psi(\overline{p}_R) - \psi(p_{wf}) = A_t q_{sc} + B q_{sc}^2 \tag{4–21}$$

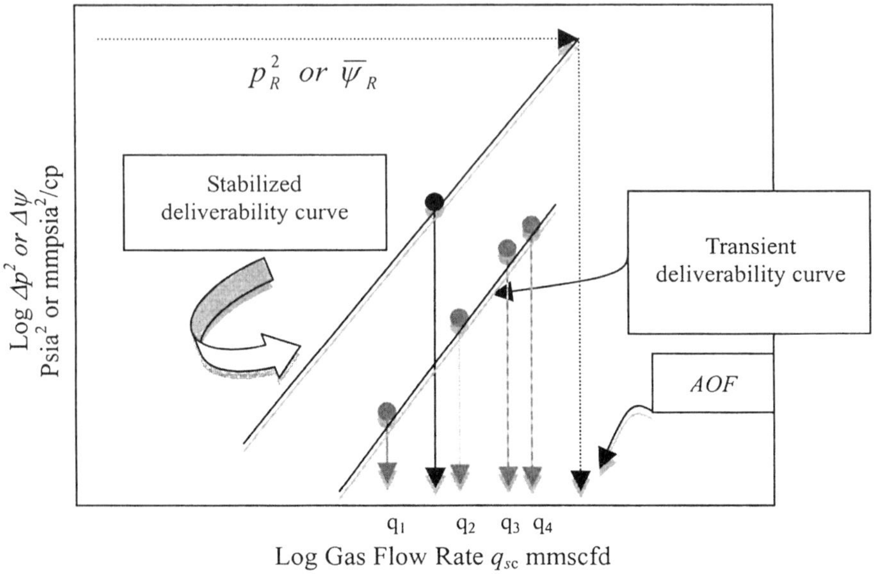

Figure 4–7. Modified isochronal test pressure–flow rate behavior.

Absolute Open Flow Potential $|_{\text{Transient}} = (AOF)_t$

$$= \frac{-A_t \pm \sqrt{A_t^2 + 4B[\psi(p_R)]}}{2B} \qquad (4\text{–}22)$$

Stabilized deliverability equation:

$$\psi(\overline{p}_R) - \psi(p_{wf}) = Aq_{sc} + Bq_{sc}^2 \qquad (4\text{–}23)$$

Absolute Open Flow Potential $|_{\text{Stabilized}} = AOF$

$$= \frac{-A \pm \sqrt{A^2 + 4B[\psi(p_R)]}}{2B} \qquad (4\text{–}24)$$

where

$$A_t = \frac{\sum \frac{\Delta\psi}{q_{sc}} \sum q_{sc}^2 - \sum q_{sc} \sum \Delta\psi}{N \sum q_{sc}^2 - \sum q_{sc} \sum q_{sc}} \qquad (4\text{–}25)$$

$$B = \frac{N \sum \Delta\psi \sum q_{sc}^2 - \sum q_{sc} \sum \Delta\psi}{N \sum q_{sc}^2 - \sum q_{sc} \sum q_{sc}} \qquad (4\text{–}26)$$

$$A = \frac{\Delta\psi - Bq_{sc}^2}{q_{sc}} \qquad (4\text{–}27)$$

4.7 Gas Well Deliverability Testing and Production Potential Analysis

Deliverability tests have been called "backpressure" tests. The purpose of these tests is to predict the manner in which the flow rate will decline with reservoir depletion. The stabilized flow capacity or deliverability of a gas well is required for planning the operation of any gas field. The flow capacity must be determined for different backpressures or flowing bottom-hole pressures at any time in the life of the reservoir and the change of flow capacity with average reservoir pressure change must be considered. The flow equations developed earlier are used in deliverability testing with some of the unknown parameters being evaluated empirically from well tests. The Absolute Open Flow (*AOF*) potential of a well is defined as the rate at which the well will produce against a zero backpressure. It cannot be measured directly but may be obtained from deliverability tests. Regulatory authorities often use it as a guide in setting maximum allowable producing rates.

Flow-after-Flow Tests

Gas well deliverability tests have been called backpressure tests because they test flow against particular pipeline backpressure greater than atmospheric pressure. The backpressure test is also referred to as a flow-after-flow test, or a multipoint test. In this testing method, a well flows at a selected constant rate until pressure stabilizes, i.e., pseudo-steady-state is reached. The stabilized rate and pressure are recorded; the rate is then changed and the well flows until the pressure stabilizes again at the new rate. The process is repeated for a total of three, four, or five rates. The behavior of flow rate and pressure with time is illustrated in Figure 4–8 for q_{sc} increasing in sequence. The tests may be

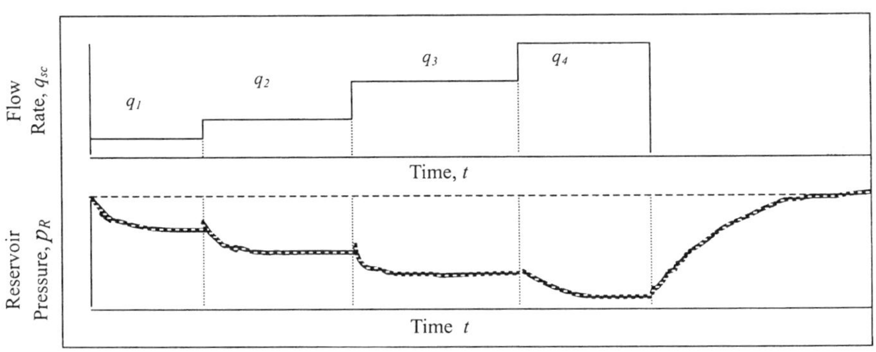

Figure 4–8. Conventional flow rate and pressure diagrams.

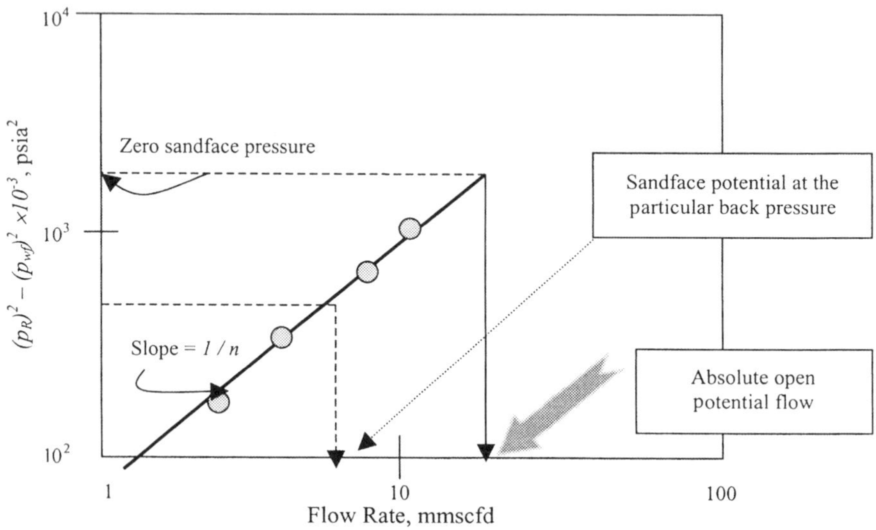

Figure 4–9. Deliverability test plot.

run in the reverse sequence. A plot of typical flow-after-flow data is shown in Figure 4–9.

Empirical Method

The method is based on the well-known Monograph 7 (Rawlins and Schellhardt, 1936),[14] which was the result of a large number of empirical observations. The relationship is commonly expressed in the form

$$q_{sc} = C\left(\overline{P}_R^2 - P_{wf}^2\right)^n = C(\Delta P^2)^n \tag{4–28}$$

Examination of Eq. 4–28 reveals that a plot of $\Delta(P^2) = \overline{P}_R^2 - P_{wf}^2$ versus q_{sc} on log-log scales should result in a straight line having a slope of $1/n$. At a value of $\Delta(P^2)$ equal to 1, $C = q_{sc}$. This is made evident by taking the log of both sides of Eq. 4–28:

$$\log\left(P_R^2 - P_{wf}^2\right) = \frac{1}{n}\log q_{sc} - \frac{1}{n}\log C \tag{4–29}$$

Once a value of n has been determined from the plot, the value C can be calculated by using data from one of the tests that falls on the line. That is,

$$C = \frac{q_{sc}}{\left(P_R^2 - P_{wf}^2\right)^n} \tag{4–30}$$

For wells in which turbulence is important, the value of n approaches 0.5, whereas for wells in which turbulence is negligible, n is obtained from well tests will fall between 0.5 and 1.0. If the values for the flow coefficient C and exponent n can be determined, the flow rate corresponding to any value of P_{wf} can be calculated and an inflow performance curve can be constructed. A parameter commonly used to characterize or compare gas wells is the flow rate that would occur if P_{wf} could be brought to zero. This is called the absolute open low potential, or *AOF*.

Theoretical Methods

The plot of $\Delta(P^2)$ versus $\log q_{sc}$ that we have discussed so far are based on empirical correlations of field data. Extrapolation of the deliverability curve much beyond the range of test data may be required to estimate *AOF*. An *AOF* determined from such a lengthy extrapolation may be incorrect. The apparent line of the deliverability curve should be slightly concave with unit slope at low flow rates and somewhat greater slope at high flow rates. The change of slope is because of increased turbulence near the wellbore and changes in the rate-dependent skin factor as the flow rate increases. Based on this analysis, a plot of $\Delta P/q_{sc}$, $\Delta P^2/q_{sxc}$, $\psi(\Delta P)/q_{sc}$ versus q_{sc} on Cartesian coordinate paper should be a straight line with slope b and intercept a. The *AOF* determined using this curve should be in less error. The deliverability equations[9] in this case are as follows:

Case 1: Using pressure solution technique:

$$\Delta P = \overline{P}_R - P_{wf} = a_1 q_{sc} + b_1 q_{sc}^2 \tag{4–31}$$

Case 2: Using pressure-squared technique:

$$\Delta P^2 = \overline{P}_R^2 - P_{wf}^2 = a_2 q_{sc} + b_2 q_{sc}^2 \tag{4–32}$$

Case 3: Using pseudopressure technique:

$$\psi(\Delta P) = \psi(P_R) - \psi(P_{wf}) = a_3 q_{sc} + b_3 q_{sc}^2 \tag{4–33}$$

Interpreting Flow Tests

More information, and greater accuracy, can result from the proper conducting and analysis of tests. It will be shown in a later section that the analysis of data from an isochronal type test using the laminar-inertial-turbulent (LIT) flow equation will yield considerable information concerning the reservoir in addition to providing reliable deliverability data. This may be achieved even without conducting the extended flow test, which is normally associated with the isochronal tests, thus saving still more time and a reduction in flared gas. For these reasons, the approach utilizing the LIT flow analysis is introduced and its use in determining deliverability is illustrated in the following section.

Fundamental Flow Equations

Case 1: For stabilized flow ($r_i \geq r_e$), using pressure-squared approach:

$$\overline{P}_R^2 - P_{wf}^2 = A'q_{sc} + B'q_{sc}^2 \tag{4–34}$$

where

$$A' = 1.422 \times 10^6 \frac{\overline{\mu}_g \overline{z} T}{kh}\left[In\left(\frac{r_e}{r_w}\right) - 0.75 + s\right] \tag{4–35}$$

and

$$B' = 1.422 \times 10^6 \frac{\overline{\mu}_g \overline{z} T}{kh} D \tag{4–36}$$

For stabilized flow ($r_i \geq r_e$), using pseudopressure approach:

$$\psi(p_R) - \psi(p_{wf}) = Aq_{sc} + Bq_{sc}^2 \tag{4–37}$$

where

$$A = 1.422 \times 10^6 \frac{T}{kh}\left[In\left(\frac{r_e}{r_w}\right) - 0.75 + s\right] \tag{4–38}$$

and

$$B = 1.422 \times 10^6 \frac{T}{kh} D \tag{4–39}$$

Case 2: For nonstabilized flow or transient flow ($r_i < r_e$):
Using pressure-squared approach:

$$\overline{p}_R^2 - p_{wf}^2 = A_t'q_{sc} + B'q_{sc}^2 \tag{4–40}$$

where B' has the same meaning as for stabilized flow and where A_t', a function of time, is given by

$$A_t' = 1.422 \times 10^6 \frac{\overline{\mu}_g \overline{z} T}{kh}\left[\frac{1}{2} In\left(\frac{kt}{1{,}688\phi c_t r_w^2} + s\right)\right] \tag{4–41}$$

Using pseudopressure approach:

$$\overline{\Psi}_R - \Psi_{wf} = A_t q_{sc} + Bq_{sc}^2 \tag{4–42}$$

where B has the same value for transient and stabilized flow as shown by Eqs. 4–40 and 4–42. A_t is obviously a function of the duration of flow.

For equal duration of flow, as in an isochronal test, t is a constant and therefore A_t is a constant:

$$A_t = \frac{1.632 \times 10^6 T}{kh}\left[\log\left(\frac{kt}{\phi \mu_{gi} c_i r_w^2}\right) - 3.23 + 0.869s\right] \quad (4\text{–}43)$$

Determination of Stabilized Flow Constants

Deliverability tests have to be conducted on wells to determine, among other things, the values of the stabilized constants. Several analysis techniques are available to evaluate C and n, of simplified analysis, and a, b of the LIT(ψ) flow analysis from deliverability tests. A deliverability test plot (Figure 4–10) may be used for simplified flow analysis to obtain the *AOF* and the well inflow performance without calculating values for C and n. The *AOF* is determined

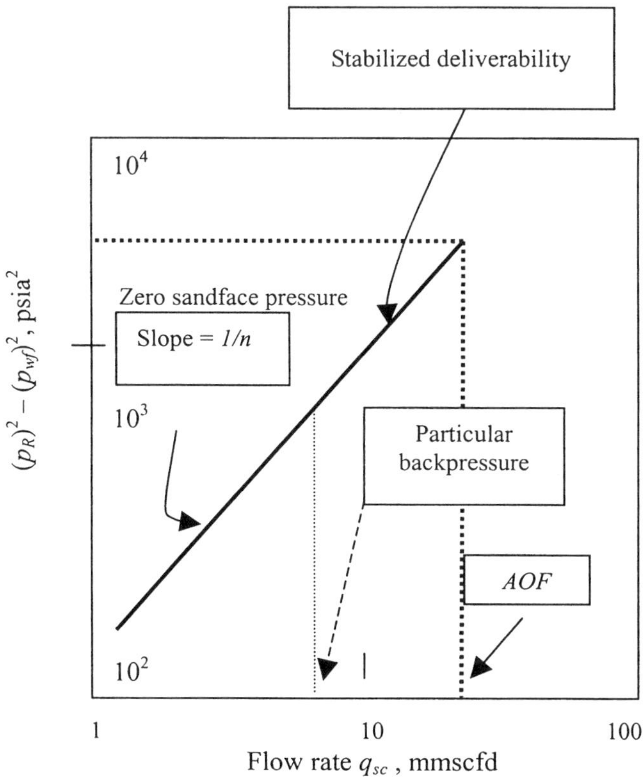

Figure 4–10. Deliverability test plot—simplified flow analysis.

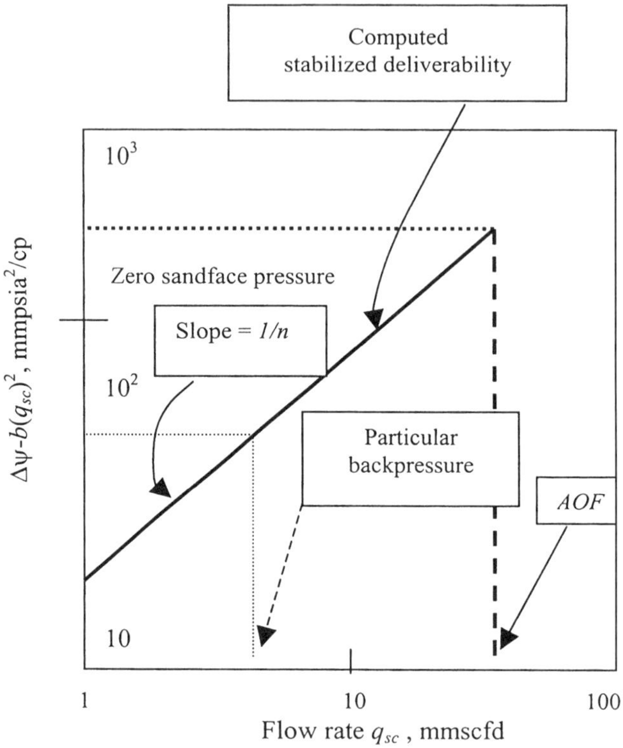

Figure 4–11. Deliverability test plot—LIT(ψ) analysis.

by entering the ordinate at $\overline{p}_R^2$ and reading the *AOF*. For LIT(ψ) flow analysis, a straight line may be obtained by plotting $(\Delta\psi - bq_{sc}^2)$ versus q_{sc} as shown in Figure 4–11. This particular method is chosen since the ordinate then represents the pseudopressure drop due to laminar flow effects, a concept that is consistent with the simplified analysis. To perform a conventional test, the stabilized shut-in reservoir pressure, $\overline{p}_R$, is determined. A flow rate, q_{sc}, is then selected and the well is flowed to stabilization. The stabilized flowing pressure, P_{wf}, is recorded. The flow rate is changed three or four times and every time the well is flowed to pressure stabilization. Figures 4–10 through 4–12 show the behavior of flow rate and pressure with time for simplified, LIT(ψ), and flow after-flow tests.

Case 1: Simplified Analysis

A plot of $(p_R^2 - p_{wf}^2) = \Delta p^2$ versus q_{sc} on a 3×3 log-log graph paper is constructed. This gives a straight line of slope $\frac{1}{n}$ or reciprocal slope n, known as the "backpressure line" or the deliverability relationship. The exponent n

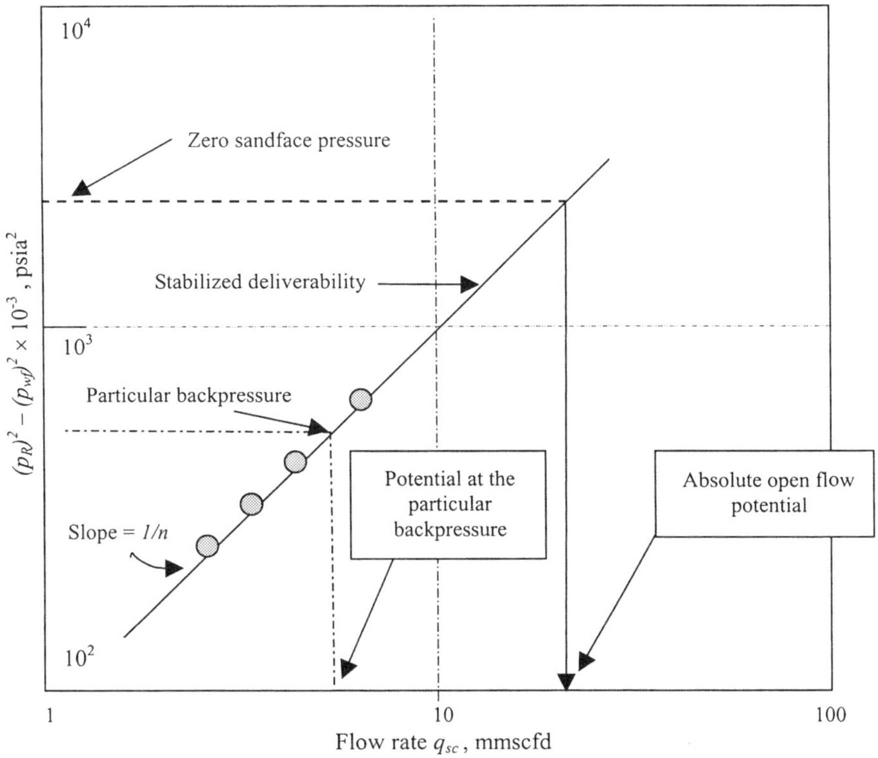

Figure 4–12. Flow-after-flow test data plot.

can be calculated by using

$$n = \log\left[\frac{(\overline{p}_R^2 - p_{wf}^2)q_{sc1}}{(p_R^2 - p_{wf}^2)q_{sc2}}\right] \tag{4–44}$$

or

$$1/n = \left[\log\left(P_R^2 - P_{wf}^2\right)_2 - \log\left(P_R^2 - P_{wf}^2\right)_1\right] / \left[\log q_{sc2} - \log q_{sc1}\right]$$

$(p_R^2 - p_{wf}^2)q_2$ should be read on the straight line corresponding to q_1 and q_2, respectively, exactly one log cycle apart. The value of n may also be obtained from the angle the straight line makes with the vertical, in which case $n = \frac{1}{\tan\theta}$. The value of performance coefficient C is then obtained from

$$C = \frac{q_{sc}}{\left(\overline{p}_R^2 - p_{wf}^2\right)^n} \tag{4–45}$$

The value of C can also be determined by extrapolating the straight line until the value of $(p_R^2 - p_{wf}^2)$ is equal to 1.0. The deliverability potential (AOF)

may be obtained from the straight line (or its extrapolation) at $\overline{p}_R^2$ if $p_{wf}^2 = 0$ psi, or at $(p_R^2 - p_{wf}^2)$ when p_{wf} is the atmospheric pressure. The following equation represents the straight-line deliverability curve:

$$q_{sc} = C\left(\overline{p}_R^2 - \overline{p}_{wf}^2\right)^n \tag{4–46}$$

The value of n ranges from 0.5 to 1.0. Exponents of $n < 0.5$ may be caused by liquid accumulation in the wellbore. Exponents apparently greater than 1.0 may be caused by fluid removal during testing. When a test is conducted using decreasing rate sequence in slow stabilizing reservoirs, an exponent greater than 1.0 may be experienced. If n is outside the range of 0.5 to 1.0, the test data may be in error because of insufficient cleanup or liquid loading in the gas well. Bottom-hole static and flowing pressures are determined by Amerada-type downhole pressure gauges or by converting the stabilized static and flowing tubing pressures (determined at the surface) to bottom-hole conditions using the Cullender and Smith method.[26]

Example 4–2 *Stabilized Flow Test Analysis*

A flow-after-flow test was performed on a gas well located in a low-pressure reservoir. Using the following test data, determine the values of n and C for the deliverability equation, *AOF*, and flow rate for $P_{wf} = 175$ psia.

Solution Flow-after-flow Test Data are shown in Table 4–2.

A plot of q_{sc} versus $(P_R^2 - P_{wf}^2)$ is shown in Figure 4–13. From the plot it is apparent that tests 1 and 4 lie on the straight line and can thus be used to determine n. From Eq. 4–44,

$$n = \frac{\log q_{sc1} - \log q_{sc4}}{\log(\Delta p^2)_1 - \log(\Delta P^2)_4} = \frac{\log(2730) - \log(5550)}{\log(1.985 \times 10^3) - \log(4.301 \times 10^3)} = 0.92$$

Table 4–2
Flow-after-Flow Test Data

Test	q_{sc} (mscfd)	P_{wf} (psia)	$(p_R)^2 - (p_{wf})^2$ ($\times 10^{-3}$ psia^2)
	0	201	40.4
1	2730	196	1.985
2	3970	195	2.376
3	4440	193	3.152
4	5550	190	4.301

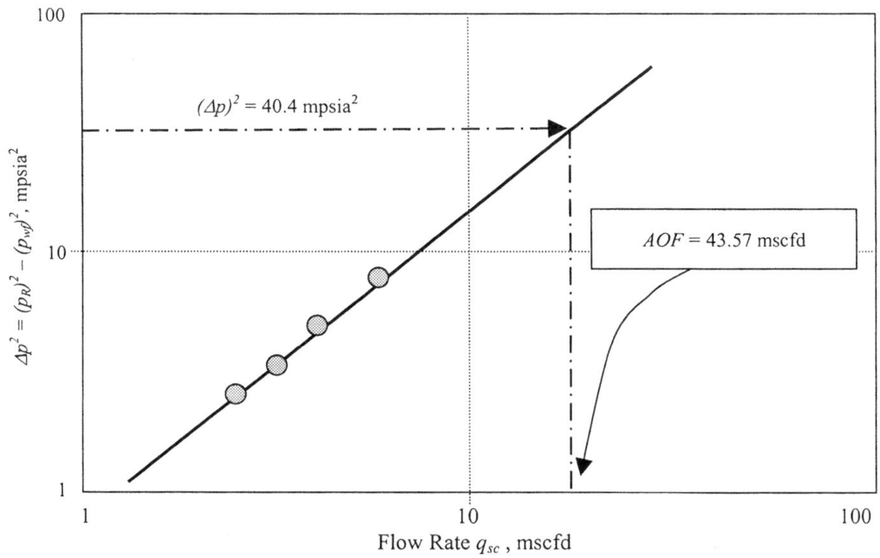

Figure 4–13. $\Delta p^2 = (p_R)^2 - (p_{wf})^2$, mpsia², versus flow rate, q_{sc}, mscfd.

From test 4, calculate C using Eq. 4–45:

$$C = \frac{q_{sc}}{\left(P_R^2 - P_{wf}^2\right)^n}$$

$$= \frac{5550}{(4.301 \times 10^3)^{0.92}}$$

$$= 2.52 \text{ mscfd/psia}$$

Therefore, the deliverability equation is

$$q_{sc} = 2.52\left(P_R^2 - P_{wf}^2\right)^{0.92}$$

$$P_{wf} = 0,$$

$$q_{sc}(AOF) = 2.52(201^2 - 0^2)^{0.92}$$

$$= 43579 \text{ mscfd}$$

$$P_{wf} = 175 \text{ psia}$$

$$q_{sc} = 2.52(201^2 - 175^2)^{0.92}$$

$$= 11812.691 \text{ mscfd}$$

Case 2: Theoretical Method of Backpressure Test Analysis

The theoretical deliverability equation is

$$\frac{\left(\overline{P}_R^2 - P_{wf}^2\right)}{q_{sc}} = a + bq_{sc} \tag{4–47}$$

A plot of $(P_R^2 - P_{wf}^2)/q_{sc}$ versus q_{sc} is made on Cartesian coordinates. The slope b may be determined either by using regression analysis or from the line drawn through the points with greatest pressure drawdown and, thus, least potential error. Two points are selected on this best straight line and slope is calculated using

$$\text{slope, } b = \frac{\frac{(P_R^2 - P_{wf}^2)_2}{q_{sc2}} - \frac{(P_R^2 - P_{wf}^2)_1}{q_{sc1}}}{q_{sc2} - q_{sc1}} \tag{4–48}$$

From the stabilized test, the intercept a may be found as

$$a = \frac{\left(\overline{P}_R^2 - P_{wf}^2\right)_{stabilized} - bq_{stabilized}}{q_{stabilized}} \tag{4–49}$$

Substituting these values in Eq. 4–47 gives a quadratic equation; this quadratic equation is then solved for *AOF* using

$$AOF = \frac{-a + \sqrt{a^2 + 4b\left(p_R^2\right)}}{2b} \tag{4–50}$$

Example 4–3 *Backpressure Test Analysis Using Theoretical Method*
Using the theoretical method of gas well test analysis, analyze the test data in Table 4–3.

Table 4–3
Isochronal Test Data Analysis

Flow Rate (mmscfd)	$\overline{P}_R^2 - P_{wf}^2$ (psia^2)	$(\overline{P}_R^2 - P_{wf}^2)/q_{sc}$ (psia^2/mmscfd)
2.397	2,925,039	1,220,292
5.214	7,105,644	1,362,800
6.144	9,033,036	1,470,221
7.148	10,319,104	1,443,635
(Stabilized) 6.148	10,707,471	1,741,619

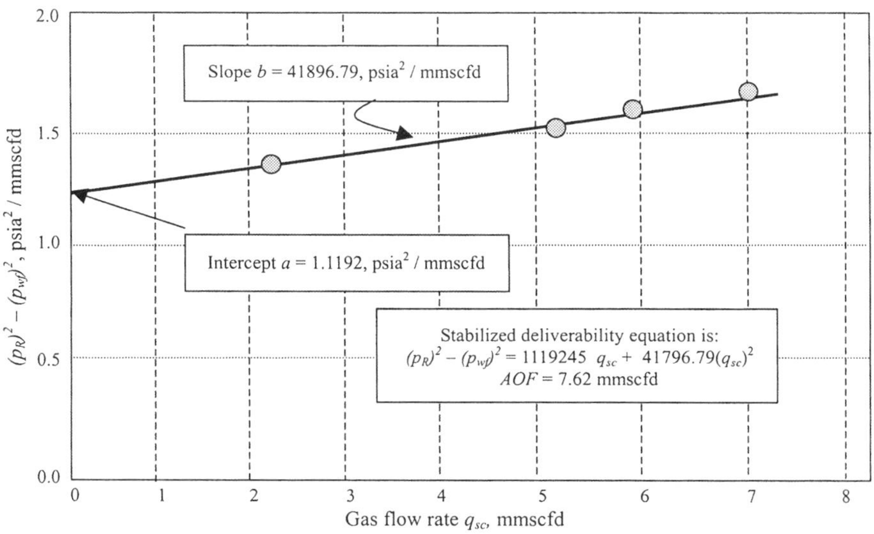

Figure 4–14. Data plot of $(p_R)^2 - (p_{wf})^2$ versus flow rate—Example 4–3.

Solution Figure 4–14 is a plot of $(P_R^2 - P_{wf}^2)/q_{sc}$ versus q_{sc} for the test data in Table 4–3. Two points on the best straight line through the data are (1,362,800, 5.214) and (1,443,635, 7.148). Substitutung these values in Eq. 4–48, the slope is given by

$$\text{Slope } b, \text{ is} = \frac{1{,}443{,}635 - 1{,}362{,}800}{7.148 - 5.214} = 41{,}796.79 \frac{\text{psia}^2}{\text{mmscfd}^2}.$$

From the stabilized test, $q_{sc} = 6.148$ mmscfd and $P_R^2 - P_{wf}^2 = 10{,}707{,}471$ psia^2; thus from Eq. 4–49,

$$a = \frac{\left(\overline{P}_R^2 - P_{wf}^2\right)_{stabilized} - bq_{stabilized}}{q_{stabilized}}$$

$$= \frac{10{,}707{,}471 - (41{,}796.79)(6.148)}{6.148} = \frac{10{,}707{,}471 - 1{,}579{,}831}{6.148}$$

$$= \frac{9{,}127{,}640}{6.148} = 1{,}484{,}651.92 \frac{\text{psia}^2}{\text{mmscfd}}$$

Thus, the stabilized deliverability curve is $\overline{P}_R^2 - P_{wf}^2 = 1{,}484{,}651.92 q_{sc} + 41{,}796.79 q_{sc}^2$. Solving AOF, we find that it is equal to $AOF = \frac{-a + \sqrt{a^2 + 4b(P_R^2)}}{2b}$.

Substituting the values of a and b in this equation, we have

$$AOF = \frac{-1{,}484{,}851.92 + \sqrt{(1{,}484{,}851.92)^2 + 4(41{,}796.79)(3700)^2}}{2 \times 41{,}796.79}$$

$$= \frac{634{,}953.97}{83{,}593.58} = 7.62\,\text{mmscfd.}$$

This value is quite close to the value established using the empirical method.

Case 3: LIT (ψ) Flow Analysis

The values of p_{wf} are converted to ψ_{wf} using $\psi - p$ curve. The LIT flow equation is given by

$$\Delta\psi = \overline{\psi}_R - \psi_{wf} = Aq_{sc} + Bq_{sc}^2 \qquad (4\text{–}51)$$

where

$\overline{\psi}_R$ = pseudopressure corresponding to $\overline{p}_R$
ψ_{wf} = pseudopressure corresponding to p_{wf}
Aq_{sc} = pseudopressure drop due to laminar flow and well conditions
Bq_{sc}^2 = pseudopressure drop due to inertial-turbulent flow effects

A plot of $(\Delta\psi - bq_{sc}^2)$ versus q_{sc}, on logarithmic coordinates, should give the stabilized deliverability line. The values of A and B may be obtained from the equations given below (Kulczycki, 1955),[29] which are derived by the curve fitting method of least squares.

$$A = \frac{\sum \frac{\Delta\psi}{q_{sc}} \sum q_{sc}^2 - \sum q_{sc} \sum \Delta\psi}{N \sum q_{sc}^2 - \sum q_{sc} \sum q_{sc}} \qquad (4\text{–}52)$$

$$B = \frac{N \sum \Delta\psi \sum q_{sc}^2 - \sum q_{sc} \sum \Delta\psi}{N \sum q_{sc}^2 - \sum q_{sc} \sum q_{sc}} \qquad (4\text{–}53)$$

where

N = number of data points

The deliverability potential of a gas well against any sandface pressure may be obtained by solving the quadratic equation for the particular value of $\Delta\psi$:

$$q_{sc} = \frac{-A \pm \sqrt{(A^2 + 4B\Delta\psi)}}{2B} \qquad (4\text{–}54)$$

The values of A and B in the simplified LIT(ψ) flow analysis depend on the same gas and reservoir properties as do C and n in the simplified analysis, except for viscosity and compressibility factor. These two variables have been taken into account in the conversion of p to ψ and consequently will not affect the deliverability relationship constants A and B. It follows, therefore, that the stabilized deliverability Equation 4–51 is more likely to be applicable throughout the life of a reservoir. In a reservoir of very high permeability, the time required to obtain stabilized flow rates and flowing pressures, as well as a stabilized shut-in formation pressure, is usually not excessive. In this type of reservoir a stabilized conventional deliverability test may be conducted in a reservoir period of time. On the other hand, in low-permeability reservoirs the time required to even approximate stabilized flow conditions may be very long. In this situation, it is not practical to conduct a completely stabilized test, and since the results of an unstabilized test can be misleading, other methods of testing should be used to predict well behavior. The application of these method of analysis to calculate C, n, a, b, and AOF is illustrated by field examples.

Example 4–4 *Stabilized Flow Test Analysis*

An isochronal test was conducted on a well located in a reservoir that had an average pressure of 1952 psia. The well was flowed on four choke sizes, and the flow rate and flowing bottom-hole pressure were measured at 3 hr and 6 hr for each choke size. An extended test was conducted for a period of 72 hr at a rate of 6.0 mmscfd, at which time p_{mf} was measured at 1151 psia. Using the data in Table 4–4, find the followings: (1) Stabilized deliverability equation; (2) *AOF*; (3) an inflow performance curve.

The slopes of both the 3-hr and 6-hr lines are apparently equal (see Figure 4–15). Use the first and last points on the 6-hr test to calculate n from Eq. 4–44, which gives

$$n = \frac{\log q_1 - \log q_4}{\log(\Delta P^2)_1 - \log(\Delta P^2)_4} = \frac{\log(2{,}600 - \log(6{,}300)}{\log(709) - \log(2{,}068)} = 0.83$$

Table 4–4
Isochronal Test Data

	$t = 3$ hr		$t = 6$ hr	
q_{sc} (mscfd)	p_{wf} (psia)	$\overline{p}_R^2 - p_{wf}^2$ $\times 10^3$ (psia^2)	p_{wf} (psia)	$p_R^2 - p_{wf}^2$ $\times 10^3$ (psia^2)
2600	1793	597	1761	709
3300	1757	724	1657	1064
5000	1623	1177	1510	1530
6300	1505	1545	1320	2068
6000	Extended flow $t = 72$ hr		1151	2485

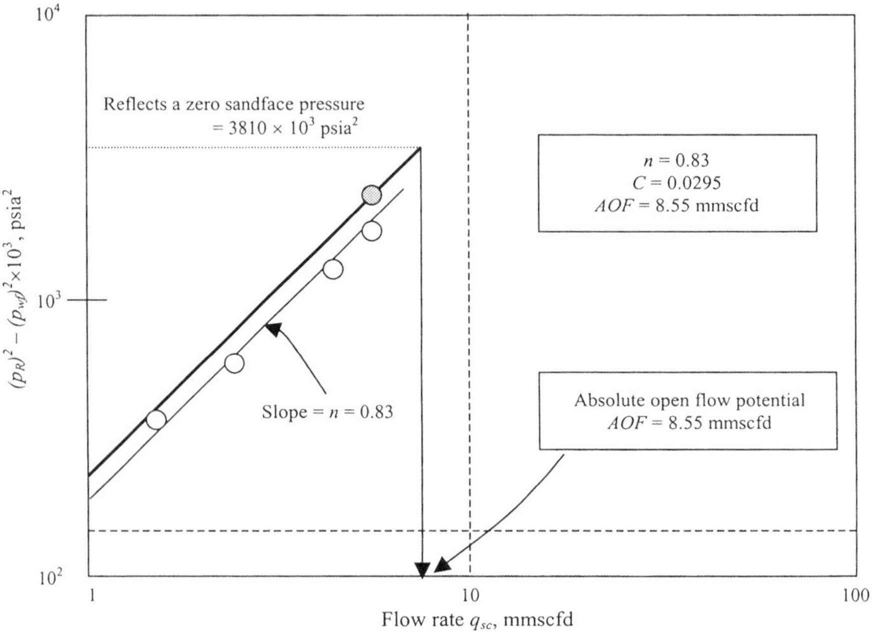

Figure 4–15. Deliverability data plot—Example 4–4.

Using the extended flow test to calculate C using Eq. 4–45:

$$C = \frac{q_{sc}}{\left(\overline{p}_R^2 - p_{wf}^2\right)^n} = \frac{6000}{(2485 \times 10^3)^{0.83}} = 0.0295$$

Solution

1. Given the data in Table 4–4, the deliverability equation for q_{sc} in mscfd is

$$q_{sc} = 0.0295\left(\overline{p}_R^2 - p_{wf}^2\right)^{0.83}$$

2. To calculate AOF, set $p_{wf} = 0$:

$$q_{sc} = 0.0295(1952^2 - 0)^{0.83} = 8551 \text{ mscfd}$$

3. In order to generate an inflow performance curve, pick several values of p_{wf} and calculate the corresponding q_{sc}.

Well inflow performance responses are shown in Table 4–5.

The inflow performance curve is plotted in Figure 4–16. If the log-log plot is used to determine the absolute open flow or the inflow performance, the line drawn through the stabilized test must be used.

Table 4–5
Well Inflow Performance Responses

p_{wf} (psia)	q_{sc} (mscfd)
1.952	0
1800	1768
1400	4695
1000	6642
600	7875
200	8477
0	8551

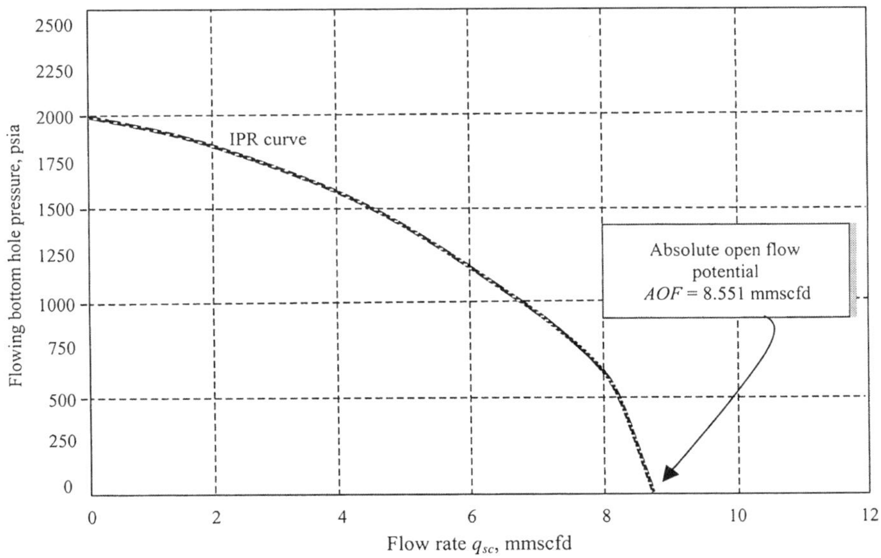

Figure 4–16. Well inflow performance response—Example 4–4.

Unstabilized Flow-after-Flow Test Data Analysis

The following equation provides a convenient and useful way for correcting unstabilized flow-after-flow test data into approximate isochronal data:

$$\frac{\left(P_i^2 - P_{wf}^2\right)_{Desires}}{\left(P_i^2 - P_{wf}^2\right)_{Actual}} = \frac{q_{sc}\left[\left(In\, t_D^* + 0.809\right)\right]}{\sum_{j-1}^{n}(\Delta q_{sc})[(In\, t_{Dj} + 0.809)]} \tag{4–55}$$

Table 4–6
Conventional Drawdown Test Data

t (hr)	T (minutes)	BHP (psia)	Δp^2 $\times 10^3$ (psia)	Rate, q_{sc} (mscfd)	$\Delta q_{sc,\text{ actual}}$ (mscfd)
0	0	3609	—	0	0
1	60	3131	3221.720	2397	0
2	120	2652	5991.777	5214	2817
3	180	2206	8158.445	6144	930
4	240	1903	9403.472	7148	1004

If the pressure drop due to turbulence and the skin factor are small relative to the total pressure drop, this equation will provide reasonable corrections. If enough pressure data are available for the first pressure drawdown, so that the reservoir properties could be estimated using conventional drawdown analysis techniques, then the following equation will provide better results without meeting previous assumptions:

$$\left(P_i^2 - P_{wf}^2\right)_{Desired} = \left(P_i^2 - P_{wf}^2\right)_{Actual} - \text{Correction term} \tag{4–56}$$

where the correction term is $(0.8718)(m/q_{sc}) \times \sum_{j=1}^{n}[(\Delta q_{sc} \log t_j) - q_{sc} \log t_D]$ and t_D^* is based on the isochronal producing time, and $t_D^* = (0.000264kt/\Phi\mu c_i r_w^2)$. The next example will clarify the application of this concept.

Example 4–5 *Unstabilized Flow-after-Flow Test Analysis*

A well is tested by flowing it at four different flow rates. The test data are given in Table 4–6. Calculate the approximately 10 hr isochronal test data. Other well/reservoir data are as follows:

$c_t = 0.00023$ psi^{-1}, $\mu_I = 0.0235$ cP, $r_w = 0.4271$ ft, $\phi = 0.1004$ fraction, $k = 8.21$ mD

Solution

$$t_D^* = \frac{0.000264kt}{\phi\mu c_t r_w^2} = \frac{0.000264 \times 8.21 \times (t_{\min})}{0.1004 \times 0.0235 \times 0.00023 \times 0.4271^2}$$

$$= 21.896 \times 10^3 t_{\min}$$

$$t_{60} = 1313.76 \times 10^3$$

$$t_{120} = 2627.52 \times 10^3$$

$$t_{180} = 3941.28 \times 10^3$$

$$t_{240} = 5255.04 \times 10^3$$

For $t = 60$ minutes:

$$(\Delta p^2)_{desired} = (\Delta p^2)_{actual} \frac{2397(In\, t_{60} + 0.809)}{2397(In\, t_{60} + 0.809)} = 3609^2 - 3131^2$$

$$= 3221.720 \times 10^3 \text{psia}^2$$

For $t = 120$ minutes:

$$(\Delta p^2)_{desired} = (5991.777) \frac{5214(In\, t_{60} + 0.809)}{2397(In\, t_{120} + 0.809) + 2817(In\, t_{120} + 0.809)}$$

$$= 5859.251 \times 10^3 \text{ psia}^2$$

For $t = 180$ minutes:

$$(\Delta p^2)_{desired} = (8158.445) \frac{6144(In\, t_{60} + 0.809)}{2397(In\, t_{120} + 0.809) + 930(In\, t_{120} + 0.809) + 2817(In\, t_{120} + 0.809)}$$

$$= 7860.758 \times 10^3 \text{ psia}^2$$

For $t = 240$ minutes:

$$(\Delta p^2)_{desired} = 9340.585 \times 10^3 \text{ psia}^2$$

Thus the stabilized deliverability curve on log-log graph paper will consist of the following points:

$p_R^2 - p_{wf}^2$ (psia$^2 \times 10^3$)	q_{sc} (mscf/d)
3221.720	2397
5859.251	5214
7860.758	6144
9340.585	7148

Isochronal Tests

The isochronal test consists of alternately closing in the well until a stabilized, or very nearly stabilized, pressure $\overline{p}_R$ is reached, and flowing the well at different rates for a set period of time t, the flowing bottom-hole pressure, p_{wf}, at time t being recorded. One flow test is conducted for a time period long enough to attain stabilized conditions and is usually referred to as the extended flow period. The behavior of the flow rate and pressure with time is illustrated in Figure 4–10 for increasing flow rates. The reverse order should also be used. Figures 4–18, 4–18a, and 4–19 show plots of isochronal test data for increasing flow rates. From the isochronal flow rates and the corresponding pseudopressures, A_t and B can be obtained from Eqs. 4–52 and 4–53; A_t refers to the value of A at the isochronal time t. A logarithmic plot of $(\Delta\psi - Bq_{sc}^2)$ versus q_{sc} is made and the isochronal data also plotted. This plot is used to identify erroneous data which must be rejected and A_t and B are recalculated, if necessary. The data obtained from the extended flow rate, $\Delta\psi$, and q_{sc} are used with the value of B already determined in Eq. 4–52 to obtain the stabilized value of A. This is given by:

$$A = \frac{\Delta\psi - Bq_{sc}^2}{q_{sc}} \tag{4–57}$$

A and B are now known and the stabilized deliverability relationship may be evaluated from Eq. 4–51. A sample calculation of stabilized deliverability from an isochronal test is shown in Example 4–6 and Figure 4–17. The LIT(ψ) flow analysis does give a more correct value and should be used instead of simplified analysis.

Example 4–6[27] *Isochronal Test Analysis*

The data in Table 4–7 were reported for an isochronal test in Reference 23. Estimate *AOF* of the well.

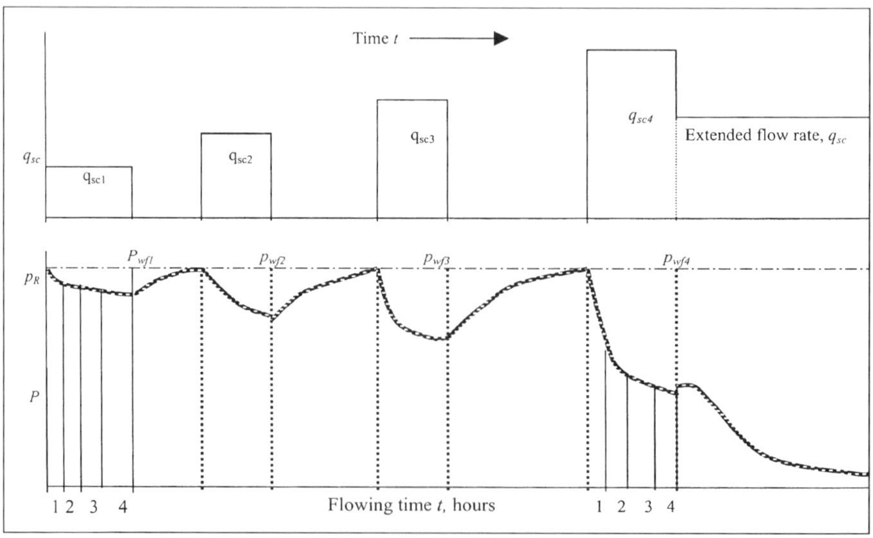

Figure 4–17. Isochronal test.

Table 4–7
Isochronal Test Data

Test	Duration (hr)	p_{ws} or p_{wf} (psi)	q_q (mmscfd)	$\overline{P}^2 - p_{wf}^2$ (psi² × 10³)
Initial shut-in	48	1952		
First flow	12	1761	2.6	709
First shut-in	15	1952		
Second flow	12	1694	3.3	941
Second shut-in	17	1952		
Third flow	12	1510	5.0	1530
Third shut-in	18	1952		
Fourth flow	12	1320	6.3	2070
Extended flow	72	1151	6.0	2486
Final shut-in	100	1952		

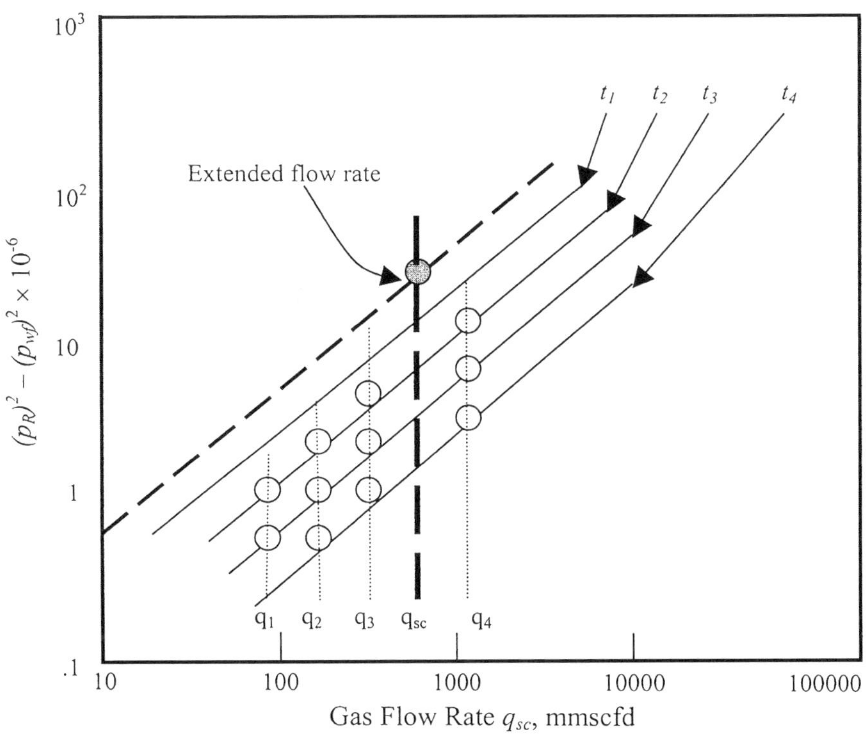

Figure 4–18. Isochronal test data.

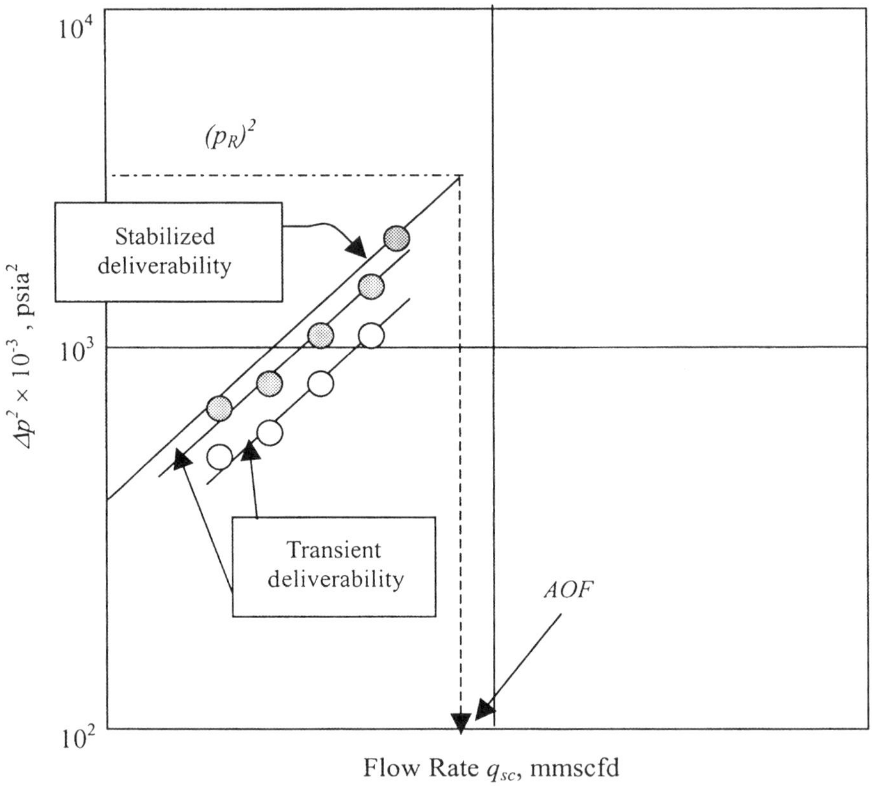

Figure 4–18a. Plot of Δp^2 versus q_{sc}—Isochronal test.

Solution Plot the data $(\overline{p}_R^2 - p_{wf}^2)$ versus q_g on log-log paper as shown in Figure 4–19. From the stabilized deliverability curve in Figure 4–19, the *AOF* is 8.4 mmscfd.

$$(\overline{p}_R^2 - p_{wf}^2) = 220, \quad q_1 = 1.0 \quad \text{and} \quad (\overline{p}_R^2 - p_{wf}^2) = 4600, \quad q_2 = 10.0$$

$$1/n = \log\left[\frac{(\overline{p}_R^2 - p_{wf}^2)_2}{(\overline{p}_R^2 - p_{wf}^2)_1}\right] = \log\left(\frac{4600}{220}\right) = 1320$$

$$n = 0.6$$

$$C = \frac{q_2}{(\overline{p}_R^2 - p_{wf}^2)_2^n} = \frac{10}{(4600)^{0.76}} = 0.016$$

Then the stabilized deliverability equation is given by

$$q_q = 0.016(\overline{p}_R^2 - p_{wf}^2)^{0.76}$$

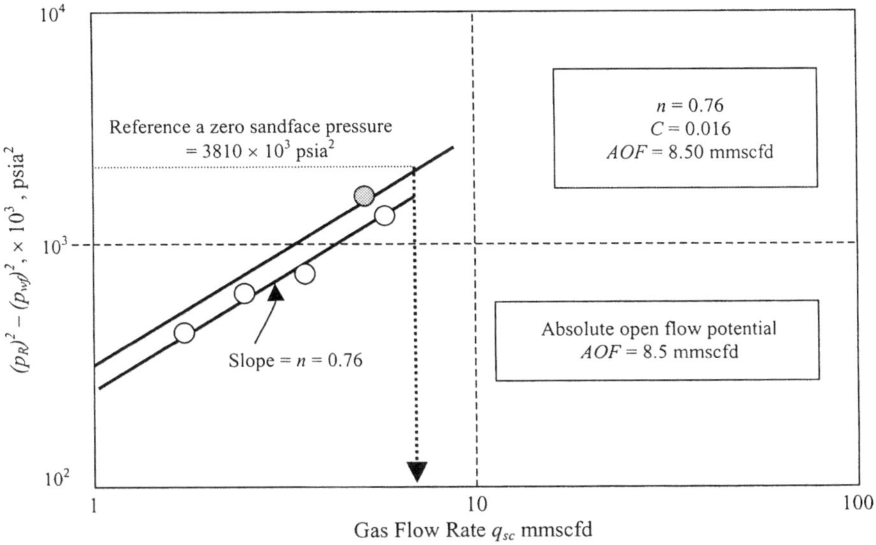

Figure 4–19. $(\overline{P}_R^2 - P_{wf}^2)$ versus q_{sc} data plot.

Modified Isochronal Tests

The objective of modified isochronal tests is to obtain the same data as in an isochronal test without using the sometimes lengthy shut-in periods required for pressure to stabilize completely before each flow test is run. As in the isochronal test, two lines are obtained, one for the isochronal data and one through the stabilized point. This latter line is the desired stabilized deliverability curve. This method, referred to as the modified isochronal test, does not yield a true isochronal curve but closely approximates the true curve. The pressure and flow rate sequence of the modified isochronal flow test are depicted in Figures 4–20 and 4–21.

The method of analysis of the modified isochronal test data is the same as that of the preceding isochronal method except that instead of $\overline{p}_R$, the preceding shut-in pressure is used in obtaining Δp^2 or $\Delta\psi$. The shut-in pressure to be used for the stabilized point is $\overline{p}_R$, the true stabilized shut-in pressure. Note that the modified isochronal procedure uses approximations. Isochronal tests are modeled exactly; modified isochronal tests are not. However, modified isochronal tests are used widely because they save time and money and because they have proved to be excellent approximations to true isochronal tests. A sample calculation of stabilized deliverability from a modified isochronal test is shown in Example 4–7.

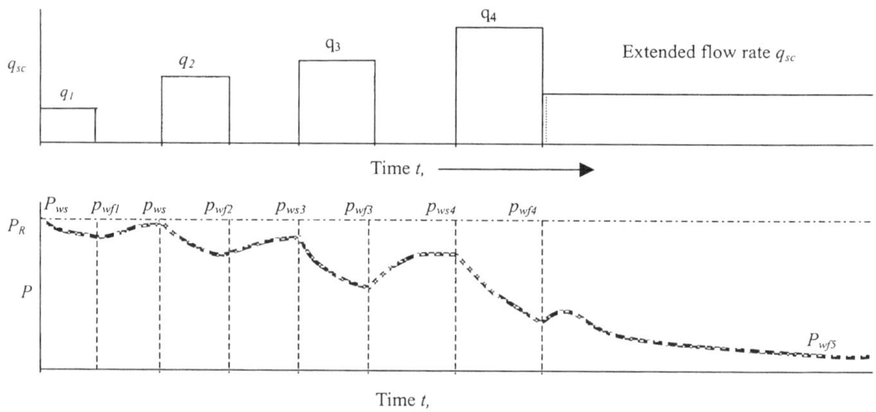

Figure 4–20. Modified isochronal test.

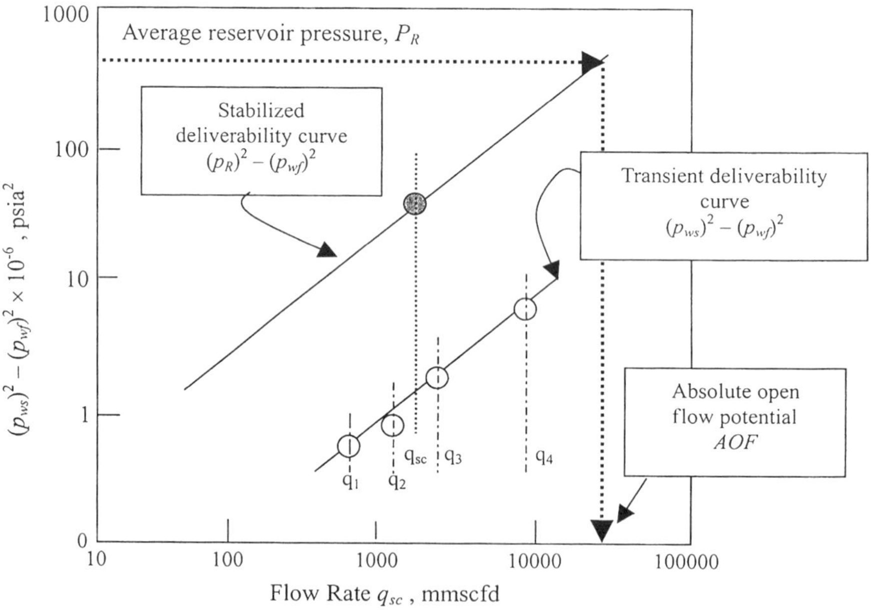

Figure 4–21. Modified isochronal test data.

Example 4–7[27] *Modified Isochronal Test Analysis*

A modified isochronal test was conducted on a gas well located in a reservoir that had average wellhead and reservoir pressures of 2388 psia and 3700 psia, respectively. The well was flowed on four choke sizes: 16, 24, 32, and 48 inches.

The flow rate, wellhead, and flowing bottom-hole pressures were measured at 6 hr for each choke size. An extended test was conducted for a period of 24 hr at a rate of 6.148 mmscfd at which time P_{wh} and P_{wf} were measured at 1015 and 1727 psia. Well test data are presented in Tables 4–8 through 4–15 and are given directly in the solution of this problem. The gas properties, pseudopressures, and numerical values of coefficients for predicting PVT properties are given below:

Compositional Gas Analysis Gas Properties and Pseudopressure

Component	mole %	Properties
N_2	0.11	MW = 21.20, T_c = 380.16°R
CO_2	7.82	s_g = 0.732, P_c = 645.08 psia
H_2S	0.0	P_{sc} = 14.65 psia, T_{sc} = 60°F
C_1	80.55	T_{wh} = 86⁰F, T_R = 710°R
C_2	5.10	
C_3	4.36	
iC_4	0.87	
nC_4	0.77	
iC_5	0.22	
nC_5	0.09	
C_6	0.11	
C_7	0.00	
Total	100.0	

Table 4–8
Calculated PVT Gas Properties and Pseudopressure

Pressure (psia)	Z —	μ_g (cP)	Real gas pseudopressure (mmpsia^2/cP)
4000	0.9647	0.024580	872.920
3600	0.9445	0.023151	739.560
3200	0.9282	0.021721	610.280
2800	0.9169	0.020329	486.770
2400	0.9113	0.019008	371.180
2000	0.9120	0.017784	266.410
1600	0.9189	0.016681	175.330
1200	0.9319	0.015723	100.830
800	0.9503	0.014932	45.510
400	0.9733	0.014337	11.470
14.65	0.9995	0.013978	0.517

1. Using the simplified analysis approach:
   (i) Find the values of stabilized flow constants n, C, and AOF at wellhead and bottom-hole conditions.
2. Using the LIT(ψ) analysis approach:
   (ii) Find the values of A_t, B, A, and AOF, and the equation of the stabilized deliverability curve and inflow performance response at wellhead conditions.
   (iii) Find the values of A_t, B, A, and AOF, and the equation of the stabilized deliverability curve including inflow performance response at bottom-hole pressure.

Solution Gas properties and necessary data were calculated from available literature and gas viscosity, and real gas pseudopressure versus pressures are shown in Figures 4–22 and 4–23. Empirical data equations were enveloped to predict PVT properties and are shown in Table 4–9.

1. Using Simplified Analysis Approach

Gas well deliverability calculations at wellhead conditions is shown in Table 4–10.

(i) Figure 4–24 shows the data plot for simplified analysis. This is a plot of $(p_R^2 - p_{wh}^2)$ versus q_{sc} on log-log paper and extrapolation of this plot to $p_R^2 - p_{wf}^2 = 5703$ (where $p_{wf} = 0$ psig or 14.65 psia, $AOF = 7.50$ mmscfd).

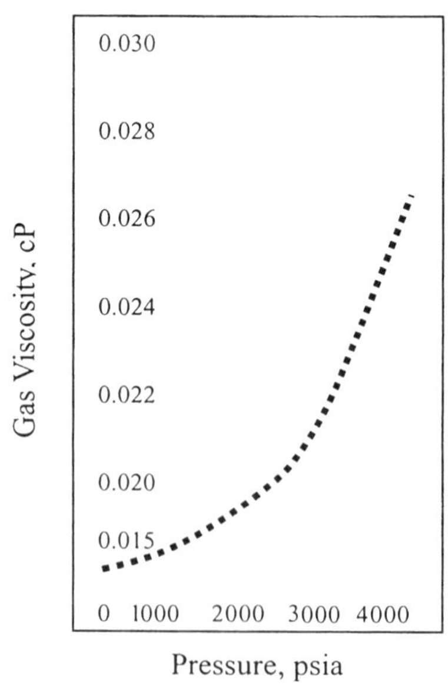

Figure 4–22. Gas viscosity versus pressure.

Table 4–9
Numerical Values of Coefficients for Predicting PVT Properties

Polynomial coefficient	Z-Factor —	Gas viscosity (cP)	Pseudopressure function (mmpsia^2/cP)
A	0.999513	0.0139689	39,453
B	−6.810505E-05	6.044023E-07	−222.976
C	4.707337E-09	8.323752E-10	72.0827
D	5.011202E-12	−1.145527E-17	5.287041E-04
E	−6.626846E-16	1.550466E-17	−1.993697E-06
F	1.094491E-20	−1.721434E-21	1.92384E-10

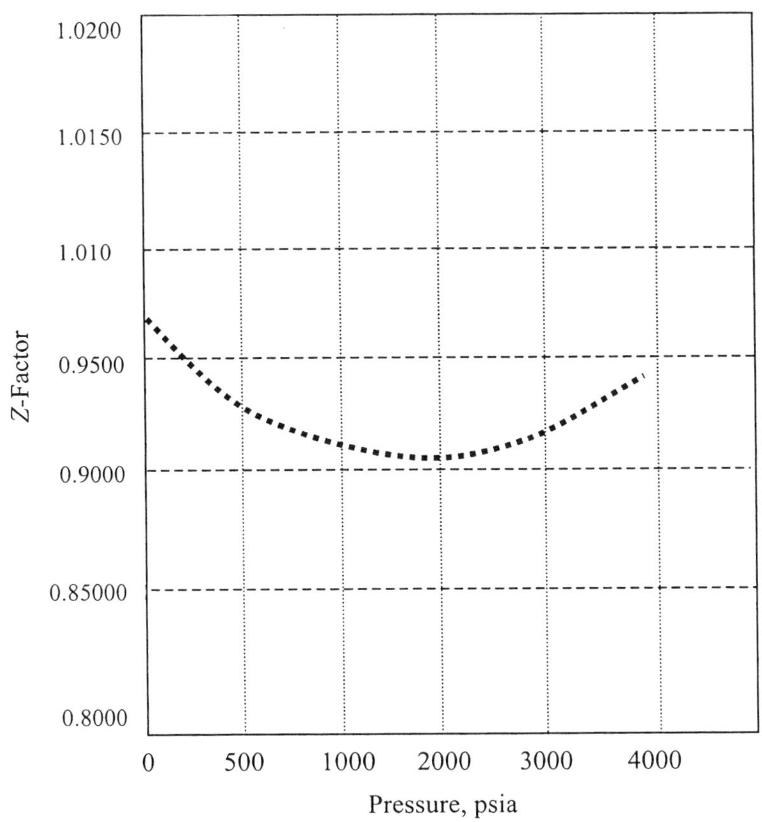

Figure 4–23. Z-factor versus pressure (Example 4–9).

Table 4–10
Gas Well Deliverability Calculations at Wellhead Conditions

	Duration (hr)	Surface pressure (psia)	Choke size (inches)	$p^2 \times 10^3$ (psia^2)	$\Delta p^2 \times 10^3$ (psia^2)	Flow rate (mmscfd)
Initial shut-in	147.2	2388		5703		
Flow 1	6	2015	16	4060	1642	2.397
Shut-in	6	2388		5703		
Flow 2	6	1640	24	2690	3013	5.214
Shut-in	6	2388		5703		
Flow 3	6	1365	32	1863	3744	6.144
Shut-in	6	2368		5607		
Flow 4	6	1015	48	1030	4673	7.186
Extended flow	24	1015	32	1030	4673	6.148
Final shut-in	22.75	2388		5703		

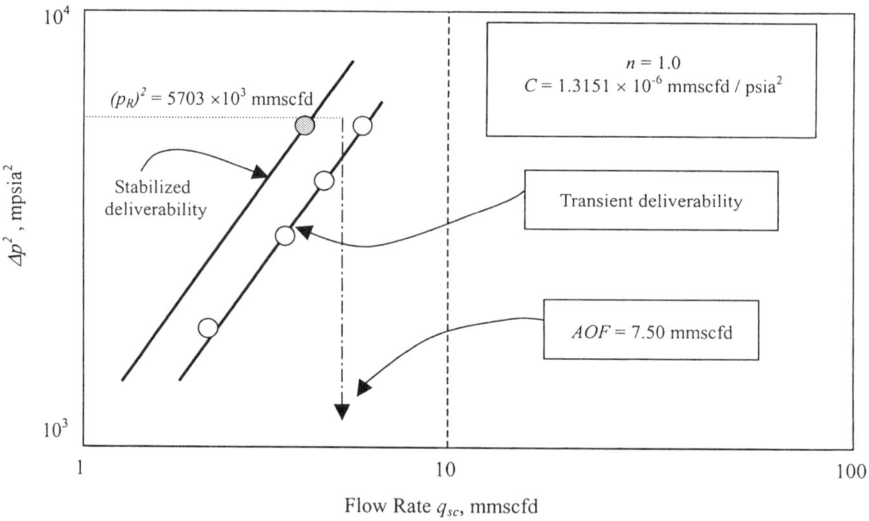

Figure 4–24. Wellhead deliverability plot using Eq. 4–44.

Using Eq. 4–44, the slope of the curve, $1/n$, is

$$1/n = \frac{\log\left(p_R^2 - p_{wf}^2\right)_2 - \log\left(p_R^2 - p_{wf}^2\right)_1}{\log q_{sc,2} - \log q_{sc,1}}$$

$$= \frac{\log(450/150)}{\log(6/3)} = \frac{0.47712}{0.47712} = 1.0$$

Table 4–11
Gas Well Deliverability Calculations at Bottom-Hole Pressure Conditions

	Duration (hr)	Bottom-hole pressure (psia)	Choke size (inches)	$p^2 \times 10^3$ (psia^2)	$\Delta p^2 \times 10^3$ (psia^2)	Flow rate (mmscfd)
Initial shut-in	147.12	3,700		13,690		
Flow 1	6	3,144	16	9,985	3,805	2.397
Shut-in	6	3,700		13,690		
Flow 2	6	2,566	24	6,584	7,106	5.214
Shut-in	6	3,700		13,690		
Flow 3	6	2,158	32	4,657	9,033	6.144
Shut-in	6	3,698		13,690		
Flow 4	6	1,836	48	3,371	10,352	7.186
Shut-in	22.75	3,690		13,616		
Extended	24	1,727	32	2,962	10,730	6.148

Thus, $n = 1.0$. Then using Eq. 4–45,

$$C = \frac{q_{sc}}{\left(p_R^2 - p_{wf}^2\right)^n} = 1.3151 \times 10^{-6} \text{ mmscfd/psia}^2$$

The stabilized deliverability equation is

$$q_{sc} = 1.3151 \times 10^{-6}\left(p_R^2 - p_{wf}^2\right)$$

To determine *AOF* (absolute open flow potential), we substitute in the above equation as follows:

$$q_{sc}(AOF) = 1.3151 \times 10^{-6}(2388^2 - 14.65^2) = 7.5 \text{ mmscfd}$$

Table 4–11 shows gas well deliverability calculations at bottom-hole pressure conditions.

(ii) Figure 4–25 shows the data plot for simplified analysis. This is a plot of $(p_R^2 - p_{wf}^2)$ versus q_{sc} on log-log paper and extrapolation of this plot to $(\overline{p}_R^2 - p_{wf}^2) = 13{,}690$ mpsia^2, where $p_{wf} = 0$ psig or 14.65 psia, $AOF = 8.21$ mmscfd. Using Equation 4–44, the slope of the curve, $1/n$ is

$$1/n = \frac{\log\left(\overline{p}_R^2 - p_{wf}^2\right)_2 - \log\left(\overline{p}_R^2 - p_{wf}^2\right)_1}{\log q_{sc,2} - \log q_{sc,1}} = \frac{\log\left(\frac{10{,}000}{2{,}000}\right)}{\log\left(\frac{6}{1}\right)}$$

$$= \frac{0.69897}{0.69897} = 1.000$$

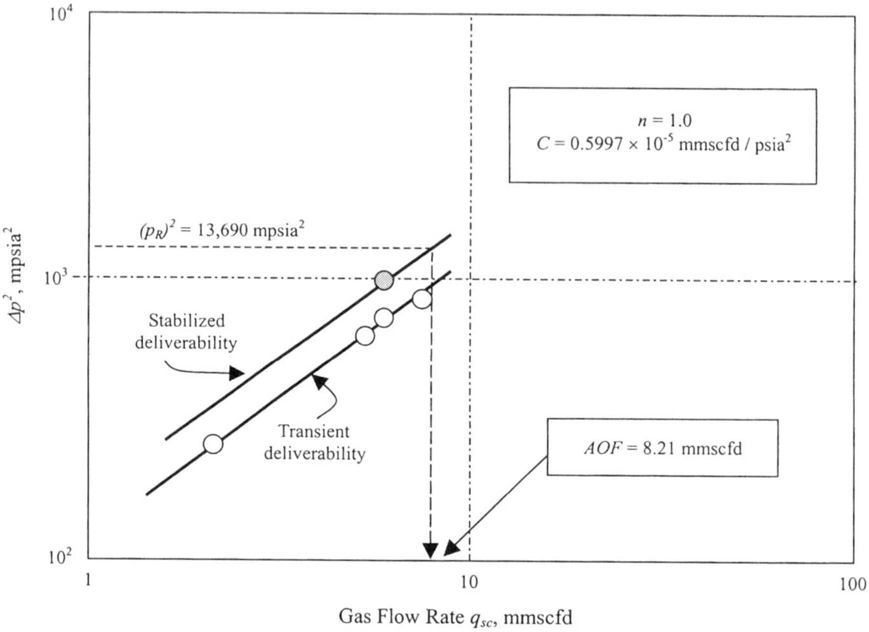

Figure 4–25. Bottom hole deliverability plot.

Thus, $n = 1.000$, then using Equation 4–45

$$C = \frac{q_{sc}}{\left(\overline{p}_R^2 - p_{wf}^2\right)^n} = \frac{6.148}{10{,}352} = 0.594 \times 10^{-4} \text{ mmscfd/psia}^2$$

Stabilized deliverability is given by: $q_{sc} = 0.594 \times 10^{-4}(\overline{p}_R^2 - p_{wf}^2)$

To determine AOF, we substitute in the above equation as follows:

Discussion: Pressure-Squared Approach

Flow rates q_{sc} and wellhead and bottom-hole pressure were calculated. A plot of $p^2(= \overline{p}_R^2 - p_{wf}^2)$ versus q_{sc} on logarithmic coordinates gives a straight line of slope $1/n$ as shown in Figures 4–24 and 4–25. Such plots are used to obtain the deliverability potential of this well against any sandface pressure, including the AOF, which is deliverability against a zero sandface pressure. The values of slope n, coefficient C, and AOF were found to be as follows:

	Wellhead conditions	**Bottom hole conditions**
n	1.00	1.00
C	1.3151×10^{-6} mmscfd/psia^2	0.594×10^{-4} mmscfd/psia^2
AOF	7.500 mmscfd	8.12 mmscfd

Table 4–12
Gas Well Deliverability Calculations at Wellhead Conditions

	Duration (hr)	Wellhead pressure (psia)	ψ (mmpsia^2/cP)	$\Delta\psi$ (mmpsia^2/cP)	Flow rate q_{sc} (mmscfd)	$\Delta\psi/q_{sc}$	q_{sc}^2	$\Delta\psi$-$bq^2{}_{sc}$
Initial shut-in	147.12	2388	452.51					
Flow 1	6	2015	336.61	115.91	2.397	48.36	5.746	103.59
Shut-in	6	2388	452.51					
Flow 2	6	1640	230.89	221.62	5.214	42.51	27.186	163.37
Shut-in	6	2388	452.51					
Flow 3	6	1365	162.99	289.52	6.144	47.12	37.749	208.63
Shut-in	6	2368	446.10					
Flow 4	6	2015	91.43	354.67	7.186	49.36	51.639	244.02
Total				981.72	20.941	187.34	122.319	
Extended flow	24	1015	91.43	361.08	6.148	58.73	37.798	280.09
Final shut-in	22.75	2388	452.51					

The performance coefficients, C were calculated using Equation 4–46:

$$q_{sc} = C\left(\overline{p}_R^2 - p_{wf}^2\right)^n$$

2. Using the LIT (ψ) Analysis Approach

Table 4–12 shows gas well deliverability calculations at wellhead conditions.

Discarded point: None

$$N = 5,\ \text{and}\ \psi(\overline{p}_R) = 452.51$$

Calculate the values of A_t, B, and A from Eqs. 4–52, 4–53, and 4–57:

$$A_t = \frac{\sum \frac{\Delta\Psi}{q_{sc}} \sum q_{sc}^2 - \sum q_{sc} \sum \Delta\Psi}{N \sum q_{sc}^2 - \sum q_{sc} \sum q_{sc}} = \frac{187.34 \times 122.319 - 20.941 \times 981.72}{5 \times 122.319 - 20.941 \times 20.941}$$

$$= 33.9572$$

$$B = \frac{N \sum \Delta\Psi - \sum q_{sc} \sum \frac{\Delta\Psi}{q_{sc}}}{N \sum q_{sc}^2 - \sum q_{sc} \sum q_{sc}} = \frac{5 \times 981.72 - 20.941 \times 187.34}{5 \times 122.319 - 20.941 \times 20.941}$$

$$= 2.1429$$

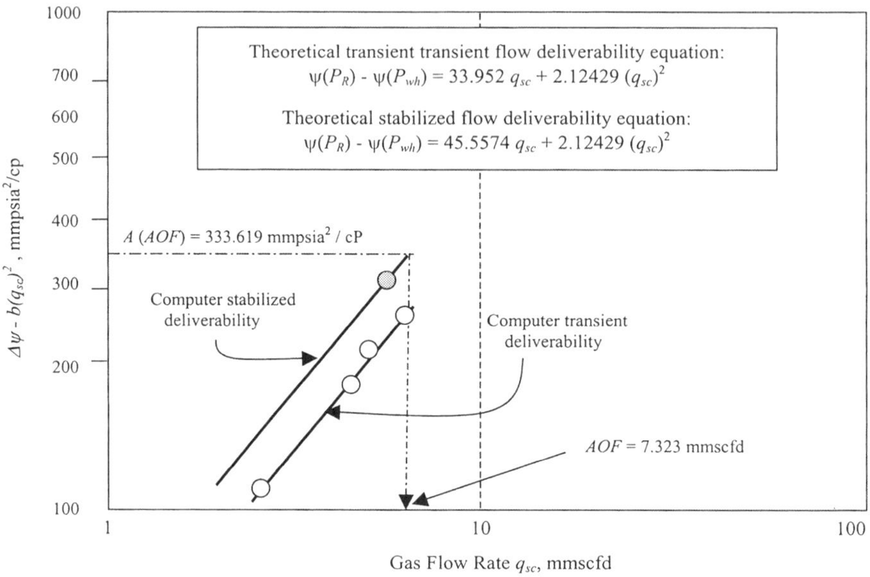

Figure 4–26. Plot of $(\Delta\psi - bq_{sc})^2$ versus flow rate q_{sc} using modified isochronal test—wellhead conditions.

For extended flow, $\Delta\Psi = 361.08$, $q_{sc} = 6.148$, $B = 2.1429$, and using Eq. 4–57:

$$A = \frac{\Delta\Psi - Bq_{sc}^2}{q_{sc}} = \frac{361.08 - 2.1429 \times 6.148^2}{6.148} = 45.5574$$

Results

The theoretical transient flow deliverability equation is

$$\text{For } r_i < r_e, \quad \Psi(p_R) - \Psi(p_{wh}) = 33.9572\, q_{sc} + 2.1429\, q_{sc}^2$$

Figure 4–26 shows the Plot of $(\Delta\Psi - bq_{sc})^2$ versus flow rate q_{sc} using a modified isochronal test in wellhead conditions.

The theoretical stabilized flow deliverability equation is

$$\text{For } r_i \geq r_e, \quad \Psi(p_R) - \Psi(p_{wh}) = 45.5574\, q_{sc} + 2.1429\, q_{sc}^2$$

Calculate deliverability from Equation 4–54 as follows:

$$q_{sc} = \frac{-A + \sqrt{(A^2 + 4B[\Psi(p_R) - \Psi(p_{wh})]}}{2B}$$

$$= \frac{-45.5574 + \sqrt{45.5574 + 4(2.1429)(452.51 - \Psi(p_{wh})}}{2 \times 2.1429}$$

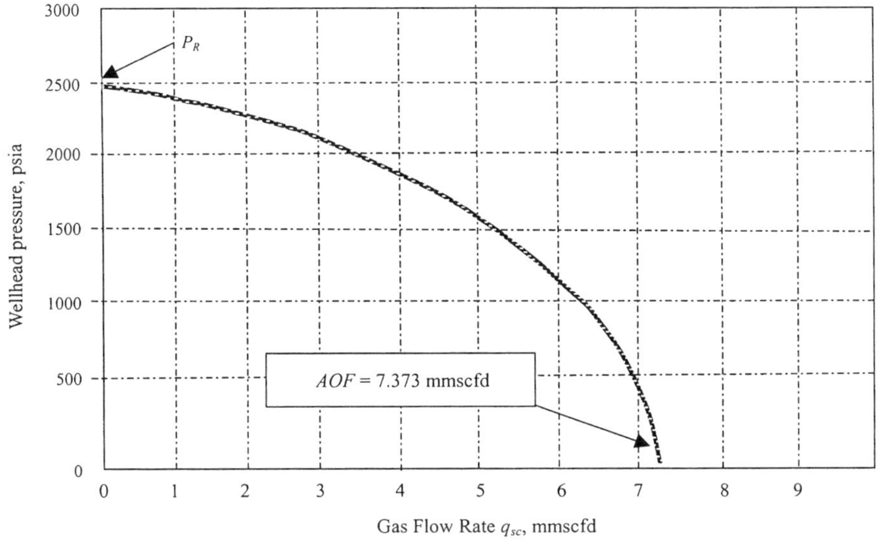

Figure 4–27. Inflow performance response using LIT(ψ) flow equation—wellhead conditions (Example 4–7).

For $\Psi(p_{wh}) = 0$, $q_{sc}(AOF) = 7.323$ mmscfd.

Well inflow performance response using the LIT(ψ) flow equation is shown in Table 4–13.

Figure 4–28 shows a data plot of $\Delta\psi - bq_{sc}^2$ versus q_{sc} (wellhead). Figure 4–29 shows the inflow performance curve (wellhead). Gas well deliverability calculations at bottom-hole pressure conditions are shown in Table 4–14.
Discarded point—None

$$N = 5, \text{ and } \Psi(P_R) = 772.56$$

Calculate the values of A_t, B, and A from Eqs. 4–52, 4–53, and 4–57:

$$A_t = \frac{\sum \frac{\Delta\psi}{q_{sc}} \sum q_{sc}^2 - \sum q_{sc} \sum \Delta\psi}{N \sum q_{sc}^2 - \sum q_{sc} \sum q_{sc}}$$

$$= \frac{294.97 \times 122.379 - 20.941 \times 1545.00}{5 \times 122.379 - 20.941 \times 20.941} = 72.7577$$

$$B = \frac{N \sum \Delta\Psi - \sum q_{sc} \sum \Delta\Psi/q_{sc}}{N \sum q_{sc}^2 - \sum q_{sc} \sum q_{sc}}$$

$$= \frac{5 \times 1545.00 - 20.941 \times 294.97}{5 \times 122.379 - 20.941 \times 20.941} = 0.1785$$

(text continued on page 186)

Table 4–13
Well Inflow Performance Response for Example 4–7 Using LIT(ψ) Flow Equation (Wellhead)

Wellhead pressure p_{wh} (psia)	$\psi(p_{wh})$ (mmpsia^2/cP)	Stabilized deliverability q_{sc} (mmscfd)
2388	452.51	0.000
2300	424.45	0.599
2200	393.04	1.234
2000	332.14	2.377
1800	274.37	3.375
1600	220.45	4.246
1400	171.10	5.001
1200	127.01	5.646
1000	88.78	6.185
800	56.96	6.621
600	31.97	6.956
400	14.10	7.191
200	3.49	7.329
100	0.90	7.363
14.65 (AOF)	0.10	7.373

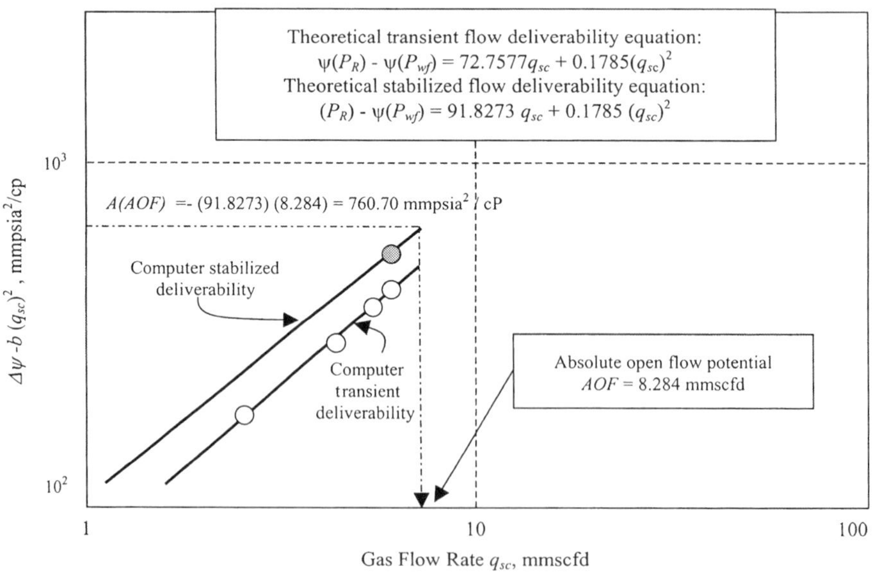

Figure 4–28. Plot of $\Delta\psi - B(q_{sc})^{22}$ versus q_{sc} using modified isochronal test—bottom-hole conditions.

Table 4–14
Gas Well Deliverability Calculations at Bottom-Hole Pressure Conditions

	Duration (hr)	Sandface pressure (psia)	$\psi(p)$ (mmpsia²/cP)	$\Delta\psi$ (mmpsia²/cP)	Flow rate q_{sc} (mmscfd)	ψ/q_{sc}	q_{sc}^2	$\Delta\psi - Bq_{sc}^2$
Initial shut-in	147.12	3700	772.56					
Flow 1	6	3144	592.59	179.98	2.397	75.08	5.746	178.95
Shut-in	6	3700	772.56					
Flow 2	6	2566	419.59	354.58	5.214	68.00	27.186	349.73
Shut-in	6	3700	772.56					
Flow 3	6	2158	306.32	466.24	6.144	75.89	37.749	459.50
Shut-in	6	3698	771.90					
Flow 4	6	1836	227.24	544.66	7.186	75.79	51.639	535.44
Total				1545.00	20.941	294.97	122.379	
Extended flow	24	1721	201.25	571.30	6.148	92.92	37.798	564.55
Final shut-in	22.5	3698	771.00					

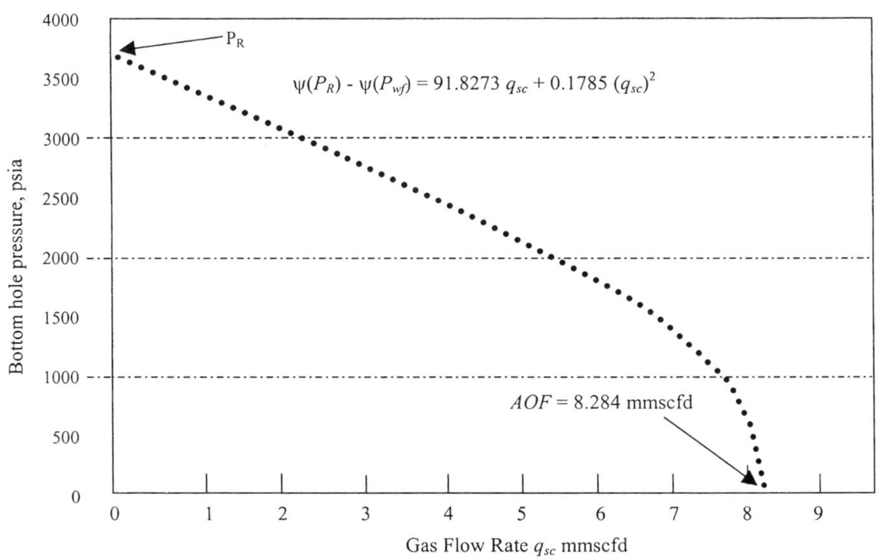

Figure 4–29. Inflow performance response using LIT(ψ) flow equation—bottom-hole conditions.

(text continued from page 183)

For extended flow,

$$\Delta\Psi = 571.30, \quad q_{sc} = 6.148 \text{ mmscfd}, \quad B = 0.1785$$

Using Eq. 4–57:

$$A = \frac{\Delta\Psi - Bq_{sc}^2}{q_{sc}} = \frac{571.30 - 0.1785 \times 6.148^2}{6.148} = 91.8273$$

Figure 4–28 shows a data plot of $\Delta\psi - bq_{sc}^2$ versus q_{sc} (bottom-hole conditions).

Results

The theoretical transient flow deliverability equation is

$$\text{For } r_i < r_e, \quad \Psi(\overline{p}_R) - \Psi(p_{wf}) = 72.7577\, q_{sc} + 0.1785\, q_{sc}^2$$

The theoretical stabilized flow deliverability equation is

$$\text{For } r_i \geq r_e, \quad \Psi(\overline{p}_R) - \Psi(p_{wf}) = 91.8273\, q_{sc} + 0.1785\, q_{sc}^2$$

Calculate deliverability from Eq. 4–54 as follows:

$$\begin{aligned} q_{sc} &= \frac{-A + \sqrt{A^2 + 4B[\psi(p_R) - \psi(p_{wf})]}}{2B} \\ &= \frac{-91.8273 + \sqrt{(91.8273)^2 + 4(0.1785)[772.56 - 0)]}}{2(0.1785)} \\ &= 8.284 \text{ mmscfd} \end{aligned}$$

Well inflow performance response using the LIT (ψ) flow equation is shown in Table 4–15. Figure 4–29 shows the inflow performance curve (bottom-hole pressure).

General Remarks

Pseudopressure Approach

A straight line is obtained by plotting $(\Delta\Psi - Bq_{sc}^2)$ versus q_{sc} on logarithmic coordinates as shown in Figure 4–28. This particular method is chosen since the ordinate then represents the pseudopressure drop due to laminar flow effects, a concept which is consistent with the simplified analysis. The deliverability potential of a well against any sandface pressure is obtained by solving the quadratic equation (Eq. 4–58) for the particular value of Ψ:

$$q_{sc} = \frac{-A + [A^2 + 4B(\Delta\Psi)]^{0.5}}{2B} \tag{4–58}$$

Table 4–15
Well Inflow Performance Response Using LIT(ψ) Flow Equation (Bottom-Hole Pressure)
$(\psi(p_R) - \psi(p_{wf}) = 91.8273\, q_{sc} + 0.1785\, q_{sc}^2)$

Bottom-hole pressure (psia)	$\psi(p_{wf})$ (mmpsia^2/cP)	Stabilized deliverability q_{sc} (mmscfd)
3700	772.56	0.000
3500	706.80	0.715
3000	547.65	2.438
2500	399.17	4.035
2000	266.41	5.454
1500	155.04	6.639
1250	109.14	7.126
1000	70.63	7.534
750	40.06	7.857
500	17.90	8.091
400	11.47	8.159
200	2.88	8.250
100	0.74	8.272
14.65 (*AOF*)	0.05	8.284

A and B in the LIT(ψ) flow analysis depend on the same gas and reservoir properties as do C and n in the simplified analysis, except for viscosity and compressibility factor. These two variables have been taken into account in the conversion of p to ψ and consequently will not affect the deliverability relationship constants A and B. It follows, therefore, that the stabilized deliverability equation or its graphical representation is more likely to be applicable throughout the life of a reservoir.

Least Square Method

A plot of $(\Delta\Psi - Bq_{sc}^2)$ versus q_{sc}, on logarithmic coordinates, should give the stabilized deliverability line. A_t and B may be obtained from Eqs. 4–52 and 4–53, which are derived by the curve fitting method of least squares.

$$A_t = \frac{\sum \frac{\Delta\Psi}{q_{sc}} \sum q_{sc}^2 - \sum q_{sc} \sum \Delta\Psi}{N \sum q_{sc}^2 - \sum q_{sc} \sum q_{sc}} \tag{4–59}$$

$$B = \frac{N \sum \Delta\Psi - \sum q_{sc} \sum \Delta\Psi / q_{sc}}{N \sum q_{sc}^2 - \sum q_{sc} \sum q_{sc}} \tag{4–60}$$

LIT (ψ) Flow Analysis

From the isochronal flow rates and the corresponding pseudopressure, A_t and B can be obtained from the foregoing equations. A logarithmic plot of $(\Delta\Psi - Bq_{sc}^2)$ versus q_{sc} is made and the isochronal data are also plotted as shown in Figure 4–28. This plot is used as before to identify erroneous data which must be rejected and A_t and B are recalculated, if necessary. The data obtained from the extended flow rate, $\Delta\Psi$, and q_{sc} are used with the value of B already determined in Eq. 4–53 to obtain the stabilized value of A. Equation 4–57 gives this:

$$A = \frac{\Delta\Psi - Bq_{sc}^2}{q_{sc}} \tag{4–61}$$

A and B are now known and the stabilized deliverability relationship has been evaluated by using the following equation:

$$\Delta\Psi = \Psi(\overline{p}_R) - \Psi(p_{wf}) = Aq_{sc} + Bq_{sc}^2 \tag{4–62}$$

Single-Point Test

If the value of slope n or the inertial-turbulent (IT) flow effect constant, b, is known, only a one-point test will provide the stabilized deliverability curve. This is done by selecting one flow rate and flowing the well at that rate for 1 to 3 days to stabilized conditions.

A sample calculation of stabilized deliverability from a single-point test is given in Example 4–8 ($n = 1.0$ and $B = 0.178$).

Example 4–8[27] *Calculating Deliverability for a Single-Point Test*
Calculate stabilized deliverability from a single-point test knowing $n = 1.0$, $B = 0.1785$, for the Ψ–p curve in Figure 4–23.

Solution Using simplified analysis, single rate test data and calculations for single rate test (as shown in Tables 4–16 and 4–17).

$$q_{sc} = C\left(\overline{p}_R^2 - p_{wf}^2\right)^n$$

$$C = \frac{q}{\left(\overline{p}_R^2 - p_{wf}^2\right)^n} = \frac{6.148}{(3700^2 - 1727^2)} = \frac{6.148}{10{,}707{,}471}$$

$$= 0.5742 \times 10^{-6} \text{ mmscfd/psia}^2$$

where slope $n = 1.0$, $\overline{p}_R = 3700$ psia.

Therefore, $q_{sc} = AOF = 0.5742 \times 10^{-6}(3700^2 - 14.65^2) = 7.86$ mmscfd. Figure 4–30 shows a plot of Δp^2 versus q_{sc}.

Table 4–16
Single-Rate Test Data

	Duration (hr)	Sandface pressure (psia)	$P^2 \times 10^3$ (psia^2)	$\Delta P^2 \times 10^3$ (psia^2)	Flow rate (mmscfd)
Extended flow	24	1,727	2,962	10,730	6.148
Final shut-in	147.12	3,700	13,690		

Table 4–17
Calculations for Single-Rate Test

	Duration (hr)	Sandface pressure (psia)	$\psi(p)$ (mmpsia^2/cP)	$\Delta\psi$ (mmpsia^2/cP)	Flow rate q_{sc} (mmscfd)	$\Delta\psi/q_{sc}$	q_{sc}^2	$\Delta\psi$-Bq_{sc}^2
Extended flow	24	1721	201.25	571.30	6.148	92.92	37.798	564.55
Final shut-in	22.5	3698	771.00					

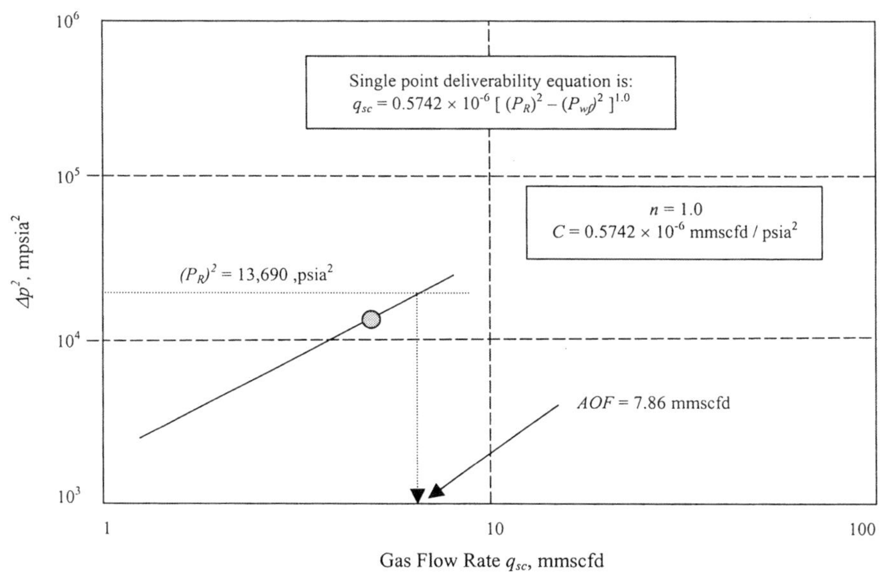

Figure 4–30. Plot of Δp^2 versus q_{sc} (single-point test).

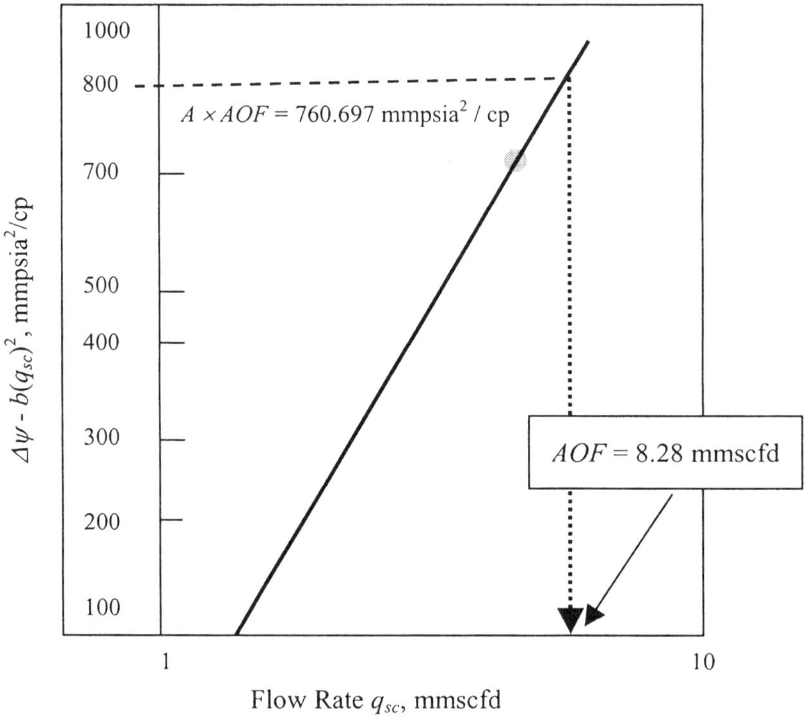

Figure 4–31. Plot of $\Delta\psi - b(q_{sc})^2$ versus q_{sc} (single-point test).

Using LIT(ψ) Analysis

Calculate the value of A for extended flow from Eq. 4–57 as follows:

$$A = \frac{\Delta\Psi - Bq_{sc}^2}{q_{sc}} = \frac{571.30 - 0.1785 \times 6.148^2}{6.148^2} = 91.8273$$

Calculate deliverability from Equation 4–54 as follows:

$$q_{sc} = \frac{-91.8273 + \sqrt{91.8273^2 + 4.0 \times 0.1785(772.56 - 0)}}{2 \times 0.1785}$$
$$= 8.284 \text{ mmscfd}$$

Figure 4–31 shows a plot of $\Delta\psi - bq_{sc}^2$ versus q_{sc}. For a single-point test, the deliverability equation is

$$q_{sc} = \frac{-A + \sqrt{A^2 + 4 \times B[(\psi(P_R) - \psi(P_{wf})]}}{2 \times B}$$

where

$A = 91.8273$
$B = 0.1705$
$q_{sc} = 8.284$ mmscfd

Wellhead Deliverability

In practice it is sometime more convenient to measure the pressures at the wellhead. These pressures may be converted to bottom-hole conditions by the calculation procedure suggested by Cullender and Smith.[26] However, in some instances, the wellhead pressures might be plotted versus flow rate in a manner similar to the bottom-hole curves of Figs. 4–25 or 4–28. The relationship thus obtained is known as the wellhead deliverability and is shown in Figures 4–24 and 4–27.

On logrithmatic coordinates the slope of the wellhead deliverability plot is not necessarily equal to that obtained using bottom-hole pressures. A wellhead deliverability plot is useful because it relates to a surface situation, for example, the gathering pipeline backpressure, which is more accessible than the reservoir. Because the wellhead deliverability relationship is not constant throughout the life of a well, different curves are needed to represent the different average reservoir pressures, as shown in Figures 4–32 and 4–33. A sample calculation is shown in Example 4–7.

Time to Stabilization

Stabilization is more properly defined in terms of a radius of investigation. By radius of investigation, r_{inv}, we mean the distance that a pressure transient has moved into a formation following a rate change in a well. As time increases, this radius moves outward into the formation until it reaches the outer boundary of the reservoir or the no-flow boundary between adjacent flowing wells. From then on, it stays constant, that is, $r_{inv} = r_e$, and stabilization is said to have been attained. This condition is also called the pseudo-steady-state. The pressure does not become constant but the rate of pressure decline does. The time to stabilization can be determined approximately by

$$t_s \cong 1000 \frac{\phi \overline{\mu}_g r_e^2}{k \overline{p}_R} \tag{4–63}$$

where

t_s = time of stabilization, hr
$\overline{\mu}_g$ = gas viscosity at $\overline{p}_R$, cP
ϕ = gas-filled porosity, fraction
k = effective permeability to gas, mD, and
r_e = outer radius of the drainage area, ft

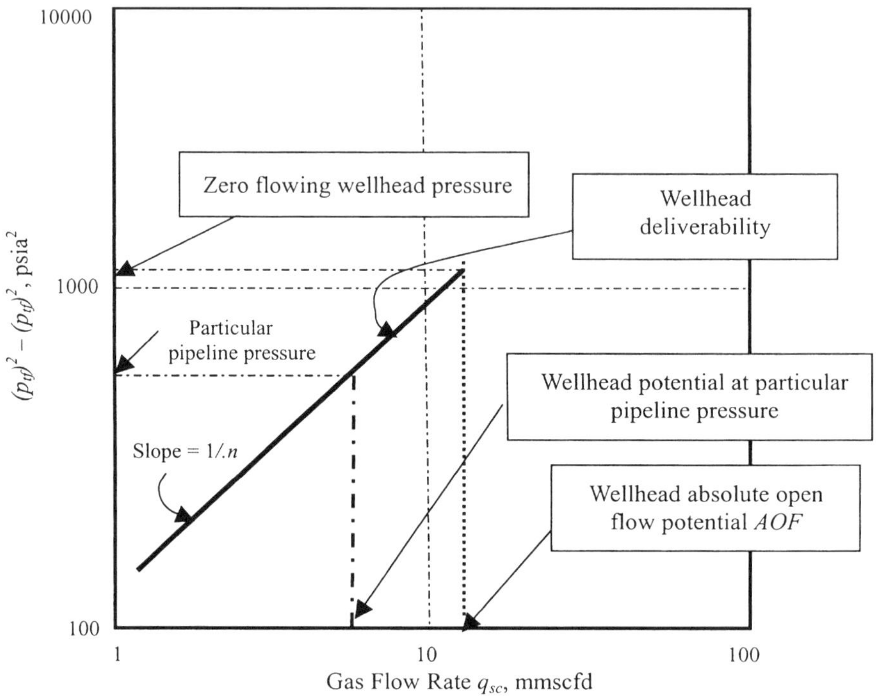

Figure 4–32. Wellhead deliverability plot.

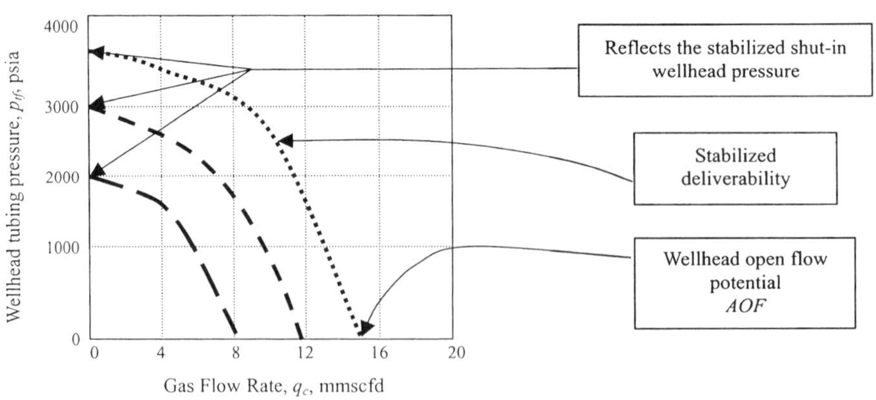

Figure 4–33. Wellhead deliverability versus flowing wellhead pressure at various stabilized shut-in pressure.

The rate of pressure decline at the well is

$$\frac{\partial p_{wf}}{\partial t} = -374 \frac{\bar{z} T q_{sc}}{\phi h r_e^2} \tag{4–64}$$

The radius of investigation, r_{inv}, after t hours of flow is

$$r_{inv} = 0.032 \sqrt{\left(\frac{k \bar{p}_R t}{\phi \bar{\mu}_g} \right)} \tag{4–65}$$

for $r_{inv} < r_e$.

As long as the radius of investigation is less than r_e, stabilization has not been reached and the flow is said to be transient. Gas well tests often involve interpretation of data obtained in the transient flow regime. Both C and A will change with time until stabilization is reached. From this time on, performance coefficient, C and A (see Eqs. 4–45 and 4–57) will stay constant. When the radius of investigation reaches the exterior boundary, r_e, of a closed reservoir, the effective drainage radius is given by

$$r_d = 0.472\, r_e \tag{4–66}$$

Example 4–9 *Calculating Radius of Investigation*

Given the following data, calculate the radius of investigation: $k = 6.282$ mD, $\bar{p}_R = 3700$ psia, $\phi = 0.1004$, $\mu_g = 0.02350$ cP, $r_e = 2200$ ft, $t = 147.2$ hr.

Solution Using Eq. 4–63, time to stabilization is

$$t_s \cong 1000 \frac{\phi \mu_g r_e^2}{k \bar{p}_R} = 1000 \times \frac{0.1004 \times 0.0235 \times 2200^2}{6.282 \times 3700} = 491 \text{ hr}$$

Using Eq. 4–65, the radius of investigation is

$$r_{inv} = 0.32 \sqrt{\frac{k \bar{p}_R t}{\phi \mu_g}} = \frac{0.032\sqrt{6.282 \times 3700 \times 147.2}}{0.1004 \times 0.0235} = 1219 \text{ ft}$$

Reservoir Parameter Estimation Techniques

Brigham related the empirical constants C and n in Eqs. 4–45 and 4–44 to the reservoir parameters in the following form of the Forchheimer equation:[24]

$$q_{sc} = \frac{1.987 \times 10^{-5} k h T_{sc} \left(p_R^2 - p_{wf}^2 \right)}{\bar{\mu}_g \bar{z} T P_{sc} \left[In(C_A \sqrt{A / r_w}) + s + F q_{sc} \right]} \tag{4–67}$$

Equation 4–67 can be written as

$$q_{sc} = \frac{a\left(p_R^2 - p_{wf}^2\right)}{b + Fq_{sc}} \tag{4–68}$$

where

$$a = \frac{1.987 \times 10^{-5} khT_{sc}}{\overline{\mu}_g \overline{z} TP_{sc}} \text{ (reservoir flow term)}$$

$$b = In(C_A\sqrt{A/r_w}) + s \text{ (Darcy geometric flow term)}$$

and non-Darcy term

$$Fq_{sc} = b\left(\frac{1-n}{2n-1}\right) = \left[In(C_A\sqrt{A/r_w + s\left(\frac{1-n}{2n-1}\right)}\right] \tag{4–69}$$

The geometric mean of the flow rates should be used to evaluate this equation because this is the midpoint on log-log paper. The constant C in Eq. 4–70 can also be related to the reservoir parameters as shown below:

$$C = q_{sc}^{1-n}\left(\frac{a}{b + F_b q_{sc}}\right)^n \tag{4–70}$$

$$\frac{C^{1/n}}{(q_{sc})^{(1-n)/n}} = \frac{1.987 \times 10^{-5} khT_{sc}}{\overline{\mu}_g \overline{z} P_{sc}\left[In(C_A\sqrt{A/r_w}) + s + Fq_{sc}\right]} \tag{4–71}$$

Example 4–10 *Reservoir Parameters Calculations Using Backpressure Equation*

A backpressure test was conducted on a gas well. Using the test data and the following reservoir data, calculate the reservoir parameter $kh/T\overline{\mu}\,\overline{z}$, given: $T_{sc} = 520°\text{R}$, $P_{sc} = 14.65$ psia, $C_A = 31.62$, $A = 360$ acres, $r_w = 0.29$ ft, $s = -1.5$.

Solution Using the methods discussed in the previous sections, the following information is obtained from the deliverability plot in Figure 4–34. Table 4–18 shows backpressure test data.

$$C = 0.00229 \text{ mmscfd/psia}^2, \quad n = 0.93, \quad AOF = 44.000 \text{ mmscfd}$$

Table 4–18
Backpressure Test Data

P_{ws} (psia)	$p_{ws}^2 (10^3\ psia^2)$	$\Delta p^2\ (psia^2)$	q_{sc} (mmscfd)
201	40.4	—	—
196	38.4	2.0	2.730
195	38.0	2.4	3.970
193	37.2	3.2	4.440
190	36.1	4.3	5.500

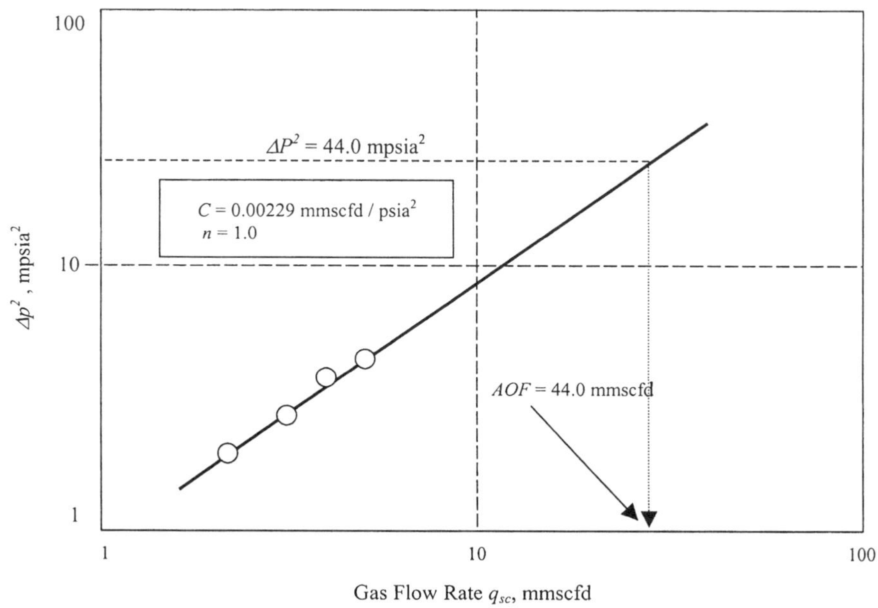

Figure 4–34. Deliverability plot for Example 4–10.

Calculate the Darcy geometric flow term:

$$b = In\left(C_A\sqrt{\frac{A}{r_w}} + s\right)$$

$$= In\left(31.62\sqrt{\frac{360 \times 43{,}560 \times 7}{22 \times .29}}\right) + (-1.5) = 6.83$$

Calculate the geometric mean flow rate by choosing the two flow rates of 8.0 and 25.0 mmscfd, which fall on a straight deliverability line:

$$q_{sc} = \sqrt{(8.0)(25.0)} = 14.142 \text{ mmscfd}$$

Using Eq. 4–69:

$$Fq_{sc} = b\left(\frac{1-n}{2n-1}\right) = 6.83\left(\frac{1-0.93}{2(0.93)-1}\right) = \frac{0.07}{0.86} = 0.556$$

The values of Fq_{sc}, C, and n are then substituted into Eq. 4–71, to evaluate the reservoir parameters:

$$\frac{kh}{T\overline{\mu}_g\overline{z}} = \frac{C^{1/n}}{(q_{sc})^{\frac{1-n}{n}}} \times \frac{P_{sc}\left[In\left(C_A\sqrt{A/r_w}\right) + s + Fq_{sc}\right]}{1.987 \times 10^{-5} T_{sc}}$$

$$= \frac{0.00229^{(1/0.93)}}{(14.142)(1-0.93)/0.93}$$

$$\times \frac{14.65\left[In\left(31.62\sqrt{\frac{17.50\times 43{,}560\times 7}{22\times 0.29}}\right) + (-1.5) + 0.556\right]}{1.987 \times 10^{-5} \times 520}$$

$$= \frac{1.00021}{14.142^{0.0753}} \times \frac{14.65[10.272 - 1.5 + 0.556]}{1.987 \times 10^{-5} \times 520}$$

$$= 0.8193 \times 13{,}276.187 = 10{,}836.215 \text{ mD-ft/cP-}^{\circ}\text{R}$$

4.8 Stabilized Deliverability Equation

The buildup and drawdown tests discussed in Chapters 5 and 6 result in knowledge of various reservoir parameters and flow characteristics of gas wells. However, these detailed tests are not always successful and in some cases may be uneconomical to conduct. It becomes necessary to get the maximum possible information from the limited data available through the use of limited or short flow tests or short-time data to estimate reservoir parameters.

This section discusses a few methods for utilizing limited data to estimate the reservoir parameters kh, s, and $\overline{p}_R$ and the stabilized deliverability equation for a gas well. Since these methods involve a substantial number of approximations, the added accuracy is not warranted. Accordingly, any of the three approaches, p, p^2, or ψ, is used, as and when convenient.

The stabilized deliverability and the LIT flow equations in terms of pressure-squared and pseudopressure, have been derived in the previous section and are given below:

$$\overline{p}_R^2 - p_{wf}^2 = A'q_{sc} + B'q_{sc}^2 \qquad (4\text{–}72)$$

The parameters A' and B' are defined, respectively, as

$$A' = \frac{1.422 \times 10^6 \overline{\mu} z T}{kh}\left[In\left(\frac{0.472r}{rw}\right) + s\right] \tag{4–73}$$

$$B' = \frac{1.422 \times 10^6 \mu z T}{kh} D \tag{4–74}$$

It is sometimes convenient to establish a relationship between the two parameters that indicates the degree of turbulence occurring in a gas reservoir. These parameters are the velocity coefficient, β, and the turbulence coefficient, D. It is frequently necessary to solve pressure or pressure drop for a known flow rate, q_{sc}:

$$\overline{p}_R^2 - p_{wf}^2 = \frac{1.422 \times 10^6 \mu z T q_{sc}}{kh}\left[In\left(\frac{0.472r_e}{r_w}\right) + s + Dq_{sc}\right] \tag{4–75}$$

This equation may be written as follows:

$$\overline{p}_R^2 - p_{wf}^2 = \frac{1.422 \times 10^6 \mu z T}{kh}\left[In\left(\frac{0.472r_e}{r_w}\right) + s\right] q_{sc} + Dq_{sc}^2 \tag{4–76}$$

The deliverability potential of a gas well may be obtained by solving the quadratic equation for the particular value of $\Delta\overline{p}_R^2$.

$$q_{sc} = \frac{-A' + \sqrt{A'^2 + 4B'(\overline{p}_R^2)}}{2B'} \tag{4–77}$$

In terms of pseudopressure, the equation is

$$\psi(\overline{p}_R) - \psi(p_{wf}) = \frac{1.422 \times 10^6 T}{kh}\left[In\left(\frac{0.472r_e}{r_w}\right) + s\right] q_{sc} + Dq_{sc}^2 \tag{4–78}$$

The parameters A and B are also defined as follows:

$$A = \frac{1.422 \times 10^6 T}{kh}\left[In\left(\frac{0.472r_e}{r_w}\right) + s\right] \tag{4–79}$$

$$B = \frac{1.422 \times 10^6 T}{kh} D \tag{4–80}$$

Equation 4–78 can be written as follows:

$$\psi(\overline{p}_R) - \psi(p_{wf}) = Aq_{sc+} + Bq_{sc}^2 \tag{4–81}$$

In quadratic form Eq. 4–81 is

$$q_{sc} = \frac{-A + \sqrt{A^2 + 4B\psi(\overline{p}_R)}}{2B} \tag{4–82}$$

Calculation of Inertial–Turbulent (IT) Flow Factor

The value of the IT flow factor D may be estimated by relating D to the turbulence factor, β:

$$D = \frac{2.715 \times 10^{-15} \beta k M P_{sc}}{h \overline{\mu} r_w T_{sc}} \tag{4–83}$$

The velocity coefficient or turbulence factor β is found to be related to absolute permeability by Ref. 10,

$$\beta = \frac{4.11 \times 10^{10}}{k^{1.3333}} \tag{4–84}$$

However, Fetkovich et al.[29] observed that β values calculated using Eq. 4–84 were about 100 times lower than calculated from the field data. Thus, Eq. 4–84 was modified to

$$\beta = \frac{4.11 \times 10^{12}}{k^{1.3333}} \tag{4–85}$$

Using values of $P_{sc} = 14.65$ psia, $T_{sc} = 520°\text{R}$, $M = 28.966 \times \gamma_g$, and evaluating viscosity at present average reservoir pressure, D is calculated as

$$D = \frac{9.106 \times 10^{-3} \gamma_g}{h k^{0.3333} \overline{\mu} r_w} \tag{4–86}$$

For convenience, the values of B and B' may be expressed in terms of β rather than D by substituting for D from Eq. 4–83 to give

$$B' = \frac{1.422 \times 10^6 \overline{\mu}\,\overline{z} T}{kh} \left[\frac{2.715 \times 10^{-15} \beta k M P_{sc}}{h \overline{\mu}\,\overline{r}_w T_{sc}}\right]$$

$$- \frac{1.422 \times 10^6 \mu z T}{kh} \left[\frac{2.715 \times 10^{-15} \beta k (28.966\gamma_g)(14.65)}{h \overline{\mu} r_w (520)}\right]$$

$$- \frac{3.1506 \times 10^{-9} \overline{z} T \beta \gamma_g}{h^2 r_w} \tag{4–87}$$

or

$$B' = \frac{3.1506 \times 10^{-9} T \beta}{h^2 r_w} \tag{4–88}$$

where

T = reservoir temperature, °R; h = net pay thickness, ft;
r_e = external radius of the reservoir, ft; r_w = wellbore radius, ft;
s = skin factor, dimensionless; D = IT flow resistance, mmscfd^{-1};
q_{sc} = flow rate, mmscfd; k = reservoir permeability, mD;
z = compressibility factor of the gas; M = molecular weight;
$\overline{\mu}$ = gas viscosity, cP; β = turbulent factor, ft^{-1}

Reservoir characteristics and gas properties may have to be evaluated in order to obtain the stabilized deliverability relationship. Accordingly the next section covers the methods for estimating k_h, s, and D that are used in subsequent sections to develop the LIT flow equation. Calculate A', A, B', and B using above equations and develop LIT flow equations as illustrated in Examples 4–11 through 4–13.

Example 4–11[27] *Estimating Reservoir Parameters and Flow Behavior from Limited Data Using Pressure-Squared Approach*

Given the following gas and well/reservoir properties:

$\overline{p}_R$ = 3965 psia; T = 710°R; γ_g = 0.733 (from recombined gas);
k = 2.85 mD (from type curve analysis);
r_w = 0.4271 ft (from field data);
r_e = 2200 ft (from well spacing for the area);
s = −0.22 (from type curve analysis); h = 41 ft (from well logs)

and using pressure-squared method, estimate

1. The turbulence factor, β
2. The IT flow factor, D
3. The values of A' and B'
4. The absolute open flow potential of the well

Solution

1. From Eq. 4–85:

$$\beta = \frac{4.11 \times 10^{12}}{k^{1.3333}} = \frac{4.11 \times 10^{12}}{2.85^{1.3333}} = 1.0172 \times 10^{12}\ \text{ft}^{-1}$$

2. From Eq. 4–86:

$$D = \frac{9.106 \times 10^{-3}\gamma_g}{hk^{0.3333}\overline{\mu}r_w} = \frac{9.106 \times 10^{-3} \times 0.733}{41 \times 2.85^{.3333} \times 0.0235 \times 0.4271}$$
$$= 0.011441\ \text{mmscfd}^{-1}$$

3. From Eq. 4–87,

$$B' = \frac{3.1506 \times 10^{-9} \beta \gamma_g z T}{h^2 r_w}$$

$$= \frac{3.1506 \times 10^{-9} \times 1.0172 \times 10^{12} \times 0.733 \times 0.81 \times 710}{41^2 \times 0.4271}$$

$$= 1882 \frac{\text{psia}^2}{\text{mmscfd}^2}$$

From Equation 4–73:

$$A' = \frac{1.422 \times 10^6 \mu z T}{kh} \left[In \left(\frac{0.472 r}{rw} \right) + s \right]$$

$$= \frac{1.422 \times 10^6 \times 0.0235 \times 0.81 \times 710}{2.85 \times 41}$$

$$\times \left[In \left(\frac{0.472 \times 2,200}{0.4271} \right) - 0.22 \right]$$

$$= 1,246,040 \frac{\text{psia}^2}{\text{mmscfd}}$$

$$\overline{p}_R^2 - p_{wf}^2 = 1,246,040\, q_{sc} + 1882\, q_{sc}^2$$

4. From Eq. 4–77:

$$q_{sc} = \frac{-A' + \sqrt{A'^2 + 4B'(\overline{p}_R^2)}}{2B'}$$

$$= \frac{-1,246,040 + \sqrt{1,246,040^2 + 4(1882)(3965)^2}}{2(1882)}$$

$$= 12.39 \text{ mmscfd}$$

The above equation may be plotted on logarithmic coordinates to give a stabilized deliverability line.

Example 4–12[27] *Estimating Reservoir Parameters and Flow Behavior from Limited Data Using $\Psi(p)$ Approach*

Rework Example 4–14, given: $\Psi(p_R) = 861.120 \times 10^6$ psia^2/cP, or $p_R =$ 3965 psia

Find the following: (1) the values of A and B and (2) the AOF of this well.

Solution From Example 4–11, $\beta = 1.0172 \times 10^{12}$ ft^{-1}, $D = 0.011441$ mmscfd^{-1}.

1. Using Eq. 4–79:

$$A = \frac{1.422 \times 10^6 T}{kh}\left[In\left(\frac{0.472 r_e}{r_w}\right) + s\right]$$

$$= \frac{1.422 \times 10^6 \times 710}{2.85 \times 41}\left[In\left(\frac{0.472(2{,}200)}{0.4271}\right) + (-0.22)\right]$$

$$= 8.6403 \times 10^6[7.7962 - 0.22] = 65.4605 \times 10^6 \frac{\text{psia}^2/\text{cP}}{\text{mmscfd}}$$

From Eq. 4–80:

$$B = \frac{1.422 \times 10^6 \times T}{kh} D = \frac{1.422 \times 10^6 \times 710}{2.85 \times 41}(0.011441)$$

$$= 0.098854 \times 10^6 \frac{\text{psia}^2/\text{cP}}{\text{mmscfd}^2}$$

From Eq. 4–81:

$$\psi(\overline{p}_R) - \psi(p_{wf}) = Aq_{sc+} + Bq_{sc}^2$$

$$\Psi(p_R) - \Psi(p_{wf}) = 65.4605 \times 10^6 q_{sc} + 0.098854 \times 10^6 q_{sc}^2$$

The above equation may be plotted on logarithmic coordinate to give a stabilized deliverability line.

2. From Equation 4–82,

$$q_{sc} = \frac{-A + \sqrt{A^2 + 4B\psi(\overline{p}_R)}}{2B}$$

$$= \frac{-65.4605 \times 10^6 + \sqrt{(65.4605 \times 10^6)^2 + 4 \times 0.098854 \times 10^6(861.12 \times 10^6)}}{2 \times 0.098854 \times 10^6}$$

$$= \frac{-65.4605 \times 10^6 + 68.012 \times 10^6}{0.19771 \times 10^6} = 12.91 \text{ mmscfd}$$

Example 4–13 *Estimating Stabilized Flow Equation from a Single Stabilized Flow Test*

Given the data of the previous example, except for the skin factor, s, estimate the stabilized deliverability equation for a gas well that gives a stabilized flowing pressure, p_{wf}, of 1640 psia at a flow rate of 5.214 mmscfd. The average reservoir pressure, $\overline{p}_R$, at time of the test is 2388 psia.

Solution From Example 4–12:

$$B' = 1{,}882 \frac{\text{psia}^2}{\text{mmscfd}^2}$$

Using Eq. 4–72:

$$\overline{p}_R^2 - p_{wf}^2 = A'q_{sc} + B'q_{sc}^2$$

$$(2388)^2 - (1640)^2 = A'(5.214) + 1882(5.214)^2$$

$$5{,}702{,}544 - 26{,}889{,}600 = A'(5.214) + 51{,}163.668$$

Therefore

$$A' = \frac{2{,}961{,}780.332}{5.214} = 568{,}043.79 \frac{\text{psia}^2}{\text{mmscfd}}$$

The stabilized deliverability is given by $\overline{p}_R^2 - p_{wf}^2 = 568{,}043.79q_{sc} + 1882q_{sc}^2$.

4.9 Stabilized Deliverability Relationship Using Graphical Method

Riley[28] developed the set of curves, assuming 640-acre spacing, a zero skin factor, $In(r_e/r_w) = 9$, and various test times of interest. The use of these figures is quite simple. The test duration (4, 8,12 or 72 hrs) and a shut-in pressure are taken from the available short-term test. The permeability may be estimated from core data or any other reliable source. The stabilization factor, SF is obtained corresponding to the permeability and shut-in pressure. This factor, when applied to the short-term flow rate, gives a reasonable approximation of the stabilized flow rate at the backpressure obtained in the flow test. The following equations can be used:

$$SF\ (\text{corrected}) = \frac{SF(\text{from curves}^{28}) \times 9 + s}{In(r_e/r_w) + s} \tag{4–89}$$

The fully laminar flow equation is

$$\overline{p}_R^2 - p_{wf}^2 = A'q_{sc} \tag{4–90}$$

or

$$A' = \frac{P_R^2 - P_{wf}^2}{q_{sc}} \tag{4–90a}$$

Using the value of SF, calculated above, to obtain a stabilized q_{sc} corresponding to

$$\Delta p^2 = \left(\overline{p}_R^2 - p_{wf}^2\right),$$

A' may be calculated as illustrated by Example 4–14.

Example 4–14 *Calculation of Stabilized Deliverability Relationship Assuming Negligible Turbulence Effects*

A short flow test (6 hr) is conducted on a gas well, given: $p_{wf} = 2015$ psia, $q_{sc} = 2.397$ mscfd, $r_e = 2200$ ft, $r_w = 0.4271$ ft, $s = -0.22$, $k = 10.0$ mD (core data), and $\overline{p}_R = 2388$ psia. Determine the stabilized deliverability relationship assuming no turbulence effects.

Solution From Ref. 28, test duration 4 hr:

$$SF = 0.69$$

From Ref. 28, test duration 8 hr:

$$SF = 0.74$$

$$SF = (0.69 + 0.74)/2 = 0.715$$

From Eq. 4–89:

$$SF(\text{corrected}) = \frac{SF \times 9 + s}{In(r_e/r_w) + s}$$

$$= \frac{0.715 \times 9 + (-0.22)}{In(2,200/0.4271) + (-0.22)} = \frac{6.215}{8.327} = 0.746$$

therefore equivalent stabilized flow rate $= 0.746 \times 2.397 = 1.788$ mmscfd.

From Eq. 4–90:

$$\overline{p}_R^2 - p_{wf}^2 = A'q_{sc} = 2388^2 - 2015^2 = A'(1.788)$$

$$A' = \frac{5{,}702{,}544 - 4{,}060{,}223}{1.788} = 918{,}522.93 \frac{\text{psia}^2}{\text{mmscfd}}$$

Hence the stabilized deliverability is given by

$$\overline{p}_R^2 - p_{wf}^2 = 918{,}522.93\, q_{sc}$$

Average Reservoir Pressure Prediction

If the middle-time data are not available, then the following methods may be used to predict average reservoir pressures from deliverability and short-flow tests described in the following sections.

Case 1: From Known Stabilized Deliverability Equation

If the variables in the parameters appearing in Eq. 4–72 with declining pressure or production life are small enough to be neglected, and the LIT flow equation may be considered to be valid over a lengthy period of time. The $\overline{p}_R$ can be calculated quite easily. The test simply involves the measurement of the stabilized flow rate and the associated bottom hole pressure. The $\overline{p}_R$ can be estimated by the following equations:

$$\overline{p}_R = \left[p_{wf}^2 + A'q_{sc} + B'q_{sc}^2\right]^{0.5} \tag{4–91}$$

$$\psi(\overline{p}_R) = \psi(p_{wf}) + Aq_{sc} + Bq_{sc}^2 \tag{4–92}$$

Case 2: Not Knowing Stabilized Deliverability Equation

When the stabilized deliverability equation is not known or when there is reason to believe that a previously obtained deliverability equation no longer applies to a partially depleted reservoir, the average reservoir pressure at any time during the producing life of the well may be obtained as follows.

If the value of B' is known, say from an isochronal deliverability test, a procedure similar to a two-rate test may be used. When a well is producing at a stabilized rate, measure q_{scq} and the corresponding flowing bottom hole pressure, p_{wf1}. Then Eq. 4–72 may be written as

$$\overline{p}_R^2 - p_{wf1}^2 = A'q_{sc1} + B'q_{sc1}^2 \tag{4–93}$$

Immediately change the flow rate to q_{sc2} and when the pressure has stabilized, determine the flowing bottom-hole pressure, p_{wf2}. Again, from Eq. 4–72:

$$\overline{p}_R^2 - p_{wf2}^2 = A'q_{sc2} + B'q_{sc2}^2 \tag{4–94}$$

Eliminating A' from Eqs. 4–93 and 4–94 gives

$$\overline{p}_R^2 = \frac{q_{sc2}p_{wf2}^2 - q_{sc1}p_{wf2}^2}{q_{sc2} - q_{sc1}} - B'q_{sc1}q_{sc2} \tag{4–70c}$$

In terms of pseudopressure

$$\psi(\overline{p}_R) = \frac{q_{sc2}\psi(p_{wf2}) - q_{sc1}\psi(p_{wf1})}{q_{sc2} - q_{sc1}} - Bq_{sc1}q_{sc2} \tag{4–95}$$

If n is calculated from an isochronal deliverability test, using exactly the same testing procedure as described above, the following relationship may be derived:

$$\overline{p}_R^2 = \frac{q_{sc2}^{1/n}\,p_{wf1} - q_{sc1}^{1/n}\,p_{wf2}^2}{q_{sc2}^{1/n} - q_{sc1}^{1/n}} \tag{4–96}$$

When the value of B' or n is not known, the foregoing analysis is simply extended to include a third flow rate. This yields a set of three simultaneous equations in three unknowns, which may be solved to obtain $\overline{p}_R$. Examples 4–15 and 4–16 illustrate calculation procedure.

Example 4–15 *Calculating Average Reservoir Pressure, Knowing Stabilized Deliverability Equation*

Calculate the average reservoir pressure if the flowing bottom-hole pressure, p_{wf}, is 3144 psia corresponding to a stabilized flow rate of 2.397 mmscfd, $A = 65.4605$, $B = 0.098854$, $P_{wf} = 3144$ psia or $\psi(p_{wf}) = 592.59$ mmpsia²/cP.

Solution From Eq. 4–92,

$$\begin{aligned}\psi(\overline{p}_R) &= \psi(p_{wf}) + Aq_{sc} + Bq_{sc}^2\\ &= [592.59 + 65.4605 \times (2.397) + 0.098854 \times (2.397)^2] \times 10^6\\ &= 748.75 \times 10^6 \text{ mmpsia}^2/\text{cP} \times 3{,}595 \text{ psia}\end{aligned}$$

From Eq. 4–91,

$$\begin{aligned}\overline{p}_R &= \left[p_{wf}^2 + A'q_{sc} + B'q_{sc}^2\right]^{0.5}\\ &= [3144^2 + 1{,}246{,}040(2.397) + 1882(2.397)^2]^{0.5}\\ &= \sqrt{12{,}876{,}005} = 3588 \text{ psia}\end{aligned}$$

Example 4–16 *Calculating Average Reservoir Pressure, Not Knowing Stabilized Deliverability Equation*

A gas well that had been producing for some time gave a stabilized flow rate of 2.397 mmscfd and a corresponding bottom-hole pressure of 3144 psia. When the rate was changed to 5.214 mmscfd and the pressure permitted to stabilize a flowing bottom-hole pressure of 2566 psia was obtained. A previously conducted isochronal test on the same well gave a value for $B = 0.1785$. Assuming that this value of B may still be considered valid for the well, calculate the average reservoir pressure at the time of the test.

Solution $q_{sc1} = 2.379$ mmscfd; $p_{wf1} = 3144$ psia $= 592.59$ mmpsia²/cP; $q_{sc2} = 5.214$ mmscfd; $p_{wf2} = 2566$ psia $= 417.99$ mmpsia²/cP.

From Eq. 4–96:

$$\begin{aligned}\psi(\overline{p}_R) &= \frac{5.214(592.59 - 2.397(417.99)}{5.214 - 2.397}\\ &\quad - 0.1785 \times 2.397 \times 5.214 - \frac{3089.764 - 1001.922}{2.817} - 2.231\\ &= 738.93 \text{ mmpsia}^2/\text{ cP} \longleftrightarrow 3530 \text{ psia}\end{aligned} \tag{4–97}$$

4.10 Estimation of Gas Well Deliverability from Short Flow Tests

To predict gas-well deliverability or inflow performance requires at least one test conducted for a period long enough to reach stabilization. The following equation can be used to calculate the approximate time to stabilization:

$$t_s = 1000\frac{\phi\overline{\mu}r_e^2}{kp_R} \tag{4–98}$$

This can be a very long period of time in low-permeability reservoirs, especially if a well is draining a large area. Several methods have been proposed for obtaining a deliverability equation without a stabilized test.[6] Essentially the only difference in these methods is the method used to obtain the coefficients *AA* and *BB* in Eq. 4–72. A method presented by Brar and Aziz[7] will be described in the following section.

Using Pseudo-Steady-State and Transient-Flow Deliverability Equations

The pseudo-steady-state equation for gas flow is

$$\Delta(p^2) = \overline{p}_R^2 - p_{wf}^2 = AAq_{sc} + BBq_{sc}^2 \tag{4–99}$$

For unsteady-state flow, *AA* varies with time and will be written as AA_t. The equation can be written in terms of common logs.

For pseudo-steady-state:

$$\Delta(p^2) = 2m\left[\log\left(\frac{0.472r_e}{r_w}\right) + \frac{s}{2.303}\right]q_{sc} + 0.869mDq_{sc}^2 \tag{4–100}$$

where

$$m = \frac{1{,}637\mu zT}{kh} \tag{4–101}$$

For transient flow:

$$\Delta(p^2) = m\left[\log\left(\frac{kt}{\phi\overline{\mu}\,\overline{c}r_w^2}\right) - 3.23 + 0.869s\right]q_{sc} + 0.869mDq_{sc}^2 \tag{4–102}$$

Comparing Eqs. 4–100 and 4–101 to Eq. 4–99 implies that

$$AA = 2m\left[\log\left(\frac{0.472r_e}{r_w}\right) + \frac{s}{2.303}\right] \tag{4–103}$$

$$AA_t = m\left[\log\left(\frac{kt}{\phi\bar{\mu}\,\bar{c}r_w^2}\right) - 3.23 + 0.869s\right] \tag{4–104}$$

and

$$BB = 0.869\,mD \tag{4–105}$$

The object of the analysis is to determine the values of AA and BB for stabilized flow, and then Eq. 4–99 can be used to calculate inflow performance. The skin factor s, the turbulence coefficient D, and the permeability k can also be determined. Equation 4–99 may be written as

$$\frac{\Delta(p^2)}{q_{sc}} = AA_t + BBq_{sc} \tag{4–106}$$

where AA_t and BB are defined in Eqs. 4–104 and 4–105, respectively. The value of AA_t will increase with time until stabilized flow is reached. A plot of $\Delta(p^2)/q_{sc}$ versus q_{sc} on Cartesian coordinates will result in a series of straight, parallel lines having slopes equal to BB and intercepts AA_t equal to $\Delta(p^2)/q_{sc}$ for each flow time. The slopes and intercepts can also be determined using least square analysis. Equation 4–104 can be expressed as

$$AA_t = m\left[\log\left(\frac{k}{\phi\mu c r_w^2}\right) - 3.23 + 0.869s\right] + m\log t \tag{4–107}$$

Therefore a plot of AA_t versus t on semilog paper will result in a straight line having a slope equal to m and an intercept at $t = 1$ hr ($\log 1 = 1$) equal to AA_{t1}. The procedure for analyzing short-term multirate flow tests is as follows:

1. Determine AA_t and BB from transient tests for several flow rates using plots of Eq. 4–99 or least squares.
2. Plot AA_t versus t on semilog scales to determine m and AA_{t1}.
3. Using the value of m, calculate k from Eq. 4–101:

$$k = \frac{1{,}637\bar{\mu}\,\bar{z}T}{mh} \tag{4–108}$$

4. Using values of AA_t, m, k at $t = 1$ hr, calculate s:

$$s = 1.151\left[\frac{AA_{t1}}{m} - \log\left(\frac{k(1)}{\phi\bar{\mu}\,\bar{c}r_w^2}\right) + 3.23\right] \tag{4–109}$$

5. Calculate a stabilized value for AA using Eq. 4–103.
6. Using the value of BB from step 1, calculate D using Eq. 4–105:

$$D = \frac{BB}{0.869\,m} \tag{4–110}$$

7. Calculate the stabilized well performance from Eq. 4–99 using the stabilized values for AA and BB. The method of least squares may be used to determine AA_t and BB from N transient flow tests:

$$AA_t = \frac{\frac{\sum \Delta(p^2)}{q_{sc} \sum q_{sc}^2} - \sum \Delta(p^2) \sum q_{sc}}{N \sum q_{sc}^2 - \sum q_{sc} \sum q_{sc}} \tag{4–111}$$

$$BB = \frac{N \sum \Delta(p^2) - \sum \frac{\Delta(p^2)}{q_{sc} \sum q_{sc}}}{N \sum q_{sc}^2 - \sum q_{sc} \sum q_{sc}} \tag{4–112}$$

Values for AA_t and BB will be obtained for each time at which p_{wf} was measured. The value of BB obtained from the longest flow test is the representative value. Equation 4–99 may be solved for q_{sc} to obtain

$$q_{sc} = \frac{-AA + \left[(AA^2) + 4BB\left(\overline{p}_R^2 - p_{wf}^2\right)\right]^{0.5}}{2BB} \tag{4–113}$$

If pseudopressure $\psi(p)$ is used instead of pressure squared, the pseudo-steady-state equation is written as

$$\Delta\psi = \psi(\overline{p}_R) - \psi(p_{wf}) = AA' q_{sc} + BB' q_{sc}^2 \tag{4–114}$$

where the coefficients AA' and BB' are given by

$$AA' = 1.637 \times 10^6 \frac{T}{kh}\left[\log\left(\frac{A}{r_w^2}\right) + \log\left(\frac{2.2458}{C_A}\right) + 0.869s\right] \tag{4–115}$$

$$BB' = 0.869\, m' D \tag{4–116}$$

and

$$m' = 1.637 \times 10^6 \frac{T}{kh} \tag{4–117}$$

For transient flow,

$$\Delta\psi = \psi(p'_R) - \psi(p_{wf}) = AA'_t q_{sc} + BB' q_{sc}^2 \tag{4–118}$$

where

$$AA'_t = m'\left[\log\left(\frac{t}{\phi \overline{\mu}\, \overline{c} r_w^2}\right) - 3.23 + 0.869s\right] \tag{4–119}$$

Equation 4–118 may be written as

$$\frac{\Delta\psi}{q_{sc}} = AA'_t + BB'q_{sc} \tag{4–120}$$

where AA'_t and BB' are defined in Eqs. 4–119 and 4–116; respectively. Equation 4–119 can be expressed as

$$AA'_t = m'\left[\log\left(\frac{k}{\phi\bar{\mu}\,\bar{c}r_w^2}\right) - 3.23 + 0.869s\right] + m'\log t \tag{4–121}$$

Therefore a plot of AA'_t versus t on semilog graph paper will result in a straight-line having a slope equal to m' and an intercept at $t = 1$ hr equal to AA'_t. The method of least squares may be used to determine AA'_t and BB' from N transient flow tests.

$$AA'_t = \frac{\sum \frac{\Delta\psi}{q_{sc}} \sum q_{sc}^2 - \sum q_{sc} \sum \Delta\psi}{N\sum q_{sc}^2 - \sum q_{sc}\sum q_{sc}} \tag{4–122}$$

$$BB' = \frac{N\sum \Delta\psi - \sum q_{sc} \sum \frac{\Delta\psi}{q_{sc}}}{N\sum q_{sc}^2 - \sum q_{sc}\sum q_{sc}} \tag{4–123}$$

Equation 4–114 may be solved for q_{sc} to obtain

$$q_{sc} = \frac{-AA' + [(AA')^2 + 4BB'(\psi(p_R) - \psi(p_{wf})]^{0.5}}{2BB'} \tag{4–124}$$

Calculation procedures are shown in the following Examples 4–17 through 4–20 for both unfractured and fractured gas reservoirs.

Example 4–17 *Calculating Deliverability Equations from Short Flow Tests Using Pressure-Squared Approach*

Short flow tests were conducted using four different flow rates, 0.4746, 0.8797, 1.2716, and 1.6589 mmscfd, respectively, and the following bottom-hole pressure was recorded at periods of 1, 2, 4, 6, and 8 hr. The test, reservoir, and well data are tabulated below.

$\bar{p}_R = 922.6$ psia; $\bar{\mu}_g = 0.0116$ cP; $\bar{z} = 0.9720$; $\bar{c} = 0.00109$ psi^{-1};
$\phi = 0.23$; $h = 12$ ft; $T = 582°$R; $r_e = 2000$ ft; $r_w = 0.23$ ft.

Shut-in Pressure, P_{ws}, psia			
922.6	921.9	919.9	917.6

Flowing Bottom-Hole Pressure, P_{wf}, psia

t, hrs	$q_{sc} = 0.4746$ mmscfd	$q_{sc} = 0.8797$ mmscfd	$q_{sc} = 1.2716$ mmscfd	$q_{sc} = 1.6589$ mmscfd
1	900.1	863.0	798.9	676.3
2	897.1	853.9	769.9	662.2
4	892.2	833.0	754.9	642.0
6	890.1	827.9	732.8	635.2
8	888.1	825.1	727.3	629.3

Determine the following:

1. Permeability k and skin factor s
2. Turbulence coefficient D
3. AOF for this well
4. Inflow performance response

Solution Table 4–19 shows the calculated values for AA_t and B. To further illustrate the procedure, some of the entries for $t = 4$ hr are calculated. For $q_{sc} = 0.8797$ mmscfd,

$$\Delta p^2 = p_{ws}^2 - p_{wf}^2 = (921.9)^2 - (833.0)^2 = 849{,}899.61 - 693{,}889$$
$$= 156{,}010.61 \text{ psia}^2 = 156.01 \text{ mpsia}^2$$
$$\frac{\Delta p^2}{q_{sc}} = \frac{156.01}{0.8797} = 177.345 \text{ mpsia}^2/\text{mmscfd}.$$

Using Eq. 4–111:

$$AA_t = \frac{\frac{\sum \Delta(p^2)}{q_{sc}\sum q_{sc}^2} - \sum \Delta(p^2)\sum q_{sc}}{N\sum q_{sc}^2 - \sum q_{sc}\sum q_{sc}}$$

$$AA_t = \frac{(770.012)(5.3680) - (917.35)(4.2848)}{4(5.3680) - (4.2048)(4.2848)}$$

$$= \frac{4{,}133.424 - 3{,}930.661}{21.52 - 18.3595} = \frac{202.763}{3.1605} = 64.16 \text{ mpsia}^2/\text{mmscfd}$$

Using Eq. 4–112,

$$BB = \frac{N\sum \Delta(p^2) - \sum \frac{\Delta(p^2)}{q_{sc}\sum q_{sc}}}{N\sum q_{sc}^2 - \sum q_{sc}\sum q_{sc}}$$

$$BB = \frac{4(917.35) - (770.012)(4.2848)}{4(5.3680) - (4.2848)(4.2848)}$$

$$= \frac{3669.400 - 3299.347}{3.1605} = \frac{370.053}{3.1605} = 117.09\ \frac{\text{psia}^2}{\text{mscfd}^2}$$

Table 4–19

Flow Rate #	q_{sc}	q_{sc}^2	$t = 1$		$t = 2$		$t = 4$		$t = 6$		$t = 8$	
			Δp^2	$\Delta p^2/q_{sc}$	$10\Delta p^2$	$\Delta p^2/q_{sc}$	Δp^2	$\Delta p^2/q_{sc}$	Δp^2	$\Delta p^2/q_{sc}$	Δp^2	p^2/q_{sc}
1	0.4746	0.2252	41.01	86.410	46.40	97.77	55.17	116.25	58.91	124.13	62.47	131.63
2	0.8797	0.7739	105.13	119.51	120.75	137.26	156.01	177.35	164.48	186.97	169.11	192.24
3	1.2716	1.6170	207.97	163.55	253.47	199.33	276.34	217.32	309.22	243.17	317.25	249.49
4	1.6589	2.7519	384.61	231.85	403.48	243.22	429.83	259.11	438.51	264.34	445.97	268.84
$\sum$	4.2848	5.3680	738.72	601.31	824.10	677.58	917.35	770.01	971.12	818.61	994.80	842.19
			$AA_t = 20.11$		$AA_t = 34.11$		$AA_t = 64.16$		$AA_t = 74.94$		$AA_t = 83.00$	
			$BB = 121.57$		$BB = 126.30$		$BB = 117.09$		$BB = 121.09$		$BB = 119.07$	

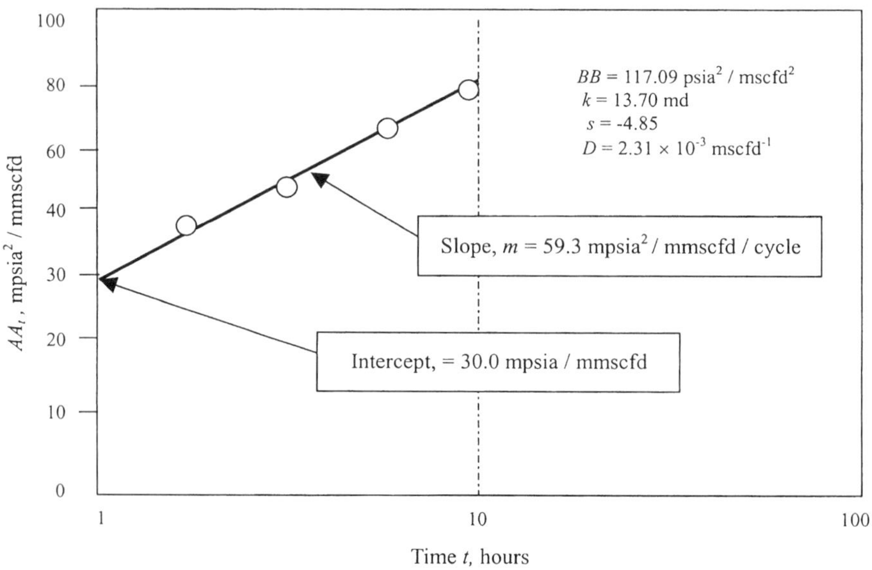

Figure 4–35. AA_t versus log t plot using pressure-squared approach.

The values calculated for AA_t are plotted versus t on semilog paper in Figure 4–35. The slope of the line is $m = 59.9$ mpsia^2/mmscfd/cycle and is obtained by drawing a straight line through the last three points. The intercept at $t = 1$ hr can be read from the graph as 30.0 mpsia^2/mmscfd.

1. From Eq. 4–101:

$$k = \frac{1637T\mu z}{mh} = \frac{1637 \times (528)(0.0116)(0.972)}{(59.9)(12)} = 13.70\,\text{mD}$$

For intercept, AA_t at 1 hr = 30.0 mpsia^2/mmscfd and using Eq. 4–109:

$$s = 1.151\left[\frac{AA_{t1}}{m} - \log\left(\frac{kt}{\phi\bar{\mu}\bar{c}r_w^2}\right) + 3.23\right]$$

$$= 1.151\left[\frac{30.0}{59.3} - \log\left(\frac{(13.70)(1)}{0.23 \times 0.0116 \times 0.00109 \times 0.23^2}\right) + 3.23\right]$$

$$= 1.151[0.5059 - 7.950 + 3.23]$$

$$= -4.85\,(\text{indicating well is stimulated})$$

2. To obtain the turbulence coefficient, solve Eq. 4–110:

$$D = \frac{BB}{0.869\,m} = \frac{119.07}{0.869(59.3)} = 2.311 \text{ mmscfd}^{-1}$$
$$= 2.31 \times 10^{-3} \text{ mscfd}^{-1}$$

3. To calculate *AOF* of this well, first find stabilized value of *AA*, using Eq. 4–103:

$$AA = 2m\left[\log\left(\frac{0.472r_e}{r_w}\right) + \frac{s}{2.303}\right]$$
$$= 2(59.3)\left[\log\frac{0.472(2000)}{0.23} + \frac{-4.85}{2.303}\right]$$
$$= 118.6[3.613 - 2.106] = 178.73 \text{ psia}^2/\text{mscfd}$$

The value chosen for *BB* is 117.09 mpsia^2/mmscfd^2 = 0.11709-psia/mcfd^2. The stabilized flow equation for determining inflow performance is then $\overline{p}_R^2 - p_{wf}^2 = 178.73q_{sc} + 0.11907q_{sc}^2$ for p in psia and q_{sc} in mscfd. Using Eq. 4–113, find q_{sc}:

$$q_{sc} = \frac{-AA + \left[(AA^2) + 4BB\left(\overline{p}_R^2 - p_{wf}^2\right)\right]^{0.5}}{2BB}$$
$$= \frac{-178.73 + [178.73^2 + 4 \times 0.11709(922.6^2 - 0)]}{2 \times 0.11709}$$
$$= \frac{-178.73 + [31{,}944.41 + 398{,}663.70]^{0.5}}{0.23418}$$
$$= -\frac{477.48}{0.23418} = 2{,}038.933 \text{ mscfd} = 2.039 \text{ mmscfd}$$

Bottom-hole pressure P_{wf} (psia)	**Stabilized deliverability q_{sc} (mmscfd)**
922.6	0
900.0	0.020
800	0.078
700	1.152
600	1.423
500	1.628
400	1.783
300	1.898
200	1.977
100	2.024
(AOF)	2.039

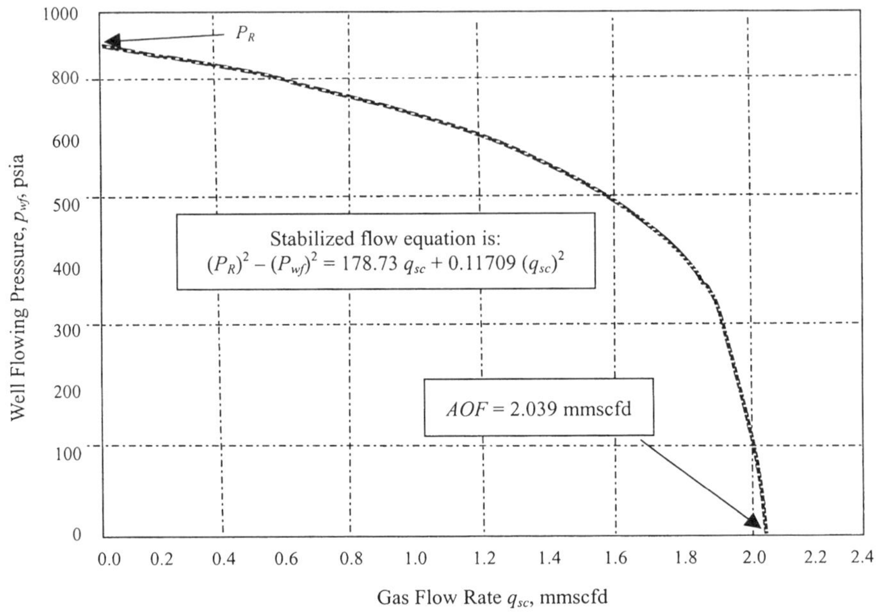

Figure 4–36. Inflow performance response using pressure-squared approach.

4. Figure 4–36 shows inflow performance response using the pressure-squared approach.

$$\bar{p}_R^2 - p_{wf}^2 = 178.73\, q_{sc} + 0.11709\, q_{sc}^2$$

Example 4–18[27] *Analyzing Short-Term Flow Test Using Pseudopressure Approach*

A modified isochronal test was conducted using four different flow rates, and the flowing bottom hole pressures were measured at periods of 1, 2, 3, 4, 5, and 6 hr. The test data are tabulated below.

$$\bar{p}_R = 3700 \text{ psia};\ \bar{\mu}_g = 0.0235\,\text{cP};\ \bar{z} = 0.9491;\ \bar{c} = 0.00023\,\text{psi}^{-1};$$
$$s_g = 0.733;\ \phi = 0.137;\ h = 41\,ft;\ T = 710^\circ R;\ r_e = 2200 \text{ ft};$$
$$r_w = 0.4271 \text{ ft},\ \phi_{HC} = 0.1004$$

Determine:

1. permeability k and skin factor s
2. turbulence coefficient D
3. *AOF* for this well using least square method,
4. inflow performance response
5. *AOF*

Table 4–20
Flowing Bottom-Hole Pressure, P_{wf}, (psia)

Shut-in pressure P_{ws} (psia)			
3700	3700	3700	3698
Shut-in pressure		**$\psi(p_R)$, (mmpsia^2/cP)**	
772.56	772.56	772.56	771.90

Flowing bottom-hole pressure, P_{wf} (psia)

t (hr)	$q_{sc} = 2.397$ mmscfd (psia	mmpsia^2/cP)	$q_{sc} = 5.214$ mmscfd (psia	mmpsia^2/cP)	$q_{sc} = 6.144$ mmscfd (psia	mmpsia^2/cP)	$q_{sc} = 7.186$ mmscfd (psia	mmpsia^2/cP)
1	3,130	588.23	2,652.15	442.94	2,206.25	318.91	1,903.25	243.01
2	3,127.75	587.48	2,602.25	428.43	2,189.55	314.53	1,888.65	239.56
3	3,133.85	589.39	2,590.15	424.91	2,180.25	312.11	1,875.65	236.49
4	3,136.65	590.27	2,579.85	421.97	2,172.15	310.00	1,870.35	235.25
5	3,139.85	591.28	2,572.85	419.95	2,164.45	308.00	1,855.00	234.01
6	3,144.00	592.45	2,566.45	418.12	2,157.55	306.21	1,836.00	227.24

Solution Table 4–20 shows flowing bottom-hole pressure and Table 4–21 shows the calculated values for AA'_t and BB'. To further illustrate the procedure, some of the entries for $t = 6$ hr are calculated. For $q_{sc} = 7.186$ mmscfd,

$$\Delta\psi = \psi(\overline{p}_R) - \psi(p_{wf}) = 771.90 - 227.24 = 544.66\,\text{mmpsia}^2/\text{cP}$$

$$\frac{\Delta\psi}{q_{sc}} = \frac{544.66}{7.186} = 75.79\,\frac{\text{mmpsia}^2/\text{cp}}{\text{mmscfd}}$$

Using Eq. 4–122,

$$\begin{aligned}
AA'_t &= \frac{\sum \frac{\Delta\psi}{q_{sc}} \sum q_{sc}^2 - \sum q_{sc} \sum \Delta\psi}{N\sum q_{sc}^2 - \sum q_{sc} \sum q_{sc}} \\
&= \frac{(294.81)(122.379) - (20.941)(1{,}545.56)}{4(122.379) - (20.941)(20.941)} \\
&= \frac{36{,}078.55 - 32{,}365.57}{489.52 - 438.53} \\
&= \frac{3{,}712.98}{50.99} = 72.82\,\frac{\text{mmpsia}^2/\text{cP}}{\text{mmscfd}}
\end{aligned}$$

Table 4–21

Flow rate #	q_{sc} (mmscfd)	q_{sc}^2 (mmscfd)	$t = 1$		$t = 2$		$t = 3$		$t = 4$		$t = 6$	
			$\Delta\psi$	$\frac{\Delta\psi}{q_{sc}}$	$\Delta\psi$	$\frac{\Delta\psi}{q_{sc}}$	$\Delta\psi$	$\frac{\Delta\psi}{q_{sc}}$	$\Delta\psi$	$\frac{\Delta\psi}{q_{sc}}$	$\Delta\psi$	$\frac{\Delta\psi}{q_{sc}}$
1	2.397	5.746	184.00	76.76	185.08	77.21	183.17	76.42	182.29	76.05	180.11	75.14
2	5.214	27.186	329.62	63.22	344.13	66.00	347.65	66.68	350.59	67.24	354.44	67.98
3	6.144	37.749	453.65	73.84	458.03	74.55	460.45	74.94	462.56	75.29	466.35	75.90
4	7.186	51.639	528.89	73.60	534.34	74.36	535.47	74.90	536.65	75.08	544.66	75.79
$\sum$	20.941	122.38	1,496.2	287.42	1,521.6	292.12	1,526.7	292.94	1,538.1	293.66	1,545.6	194.81
			$AA'_t = 72.73$		$AA'_t = 73.4$		$AA'_t = 74.01$		$AA'_t = 74.43$		$AA'_t = 72.82$	
			$BB' = .0127$		$BB' = .1234$		$BB' = .1148$		$BB' = .1239$		$BB' = .1690$	

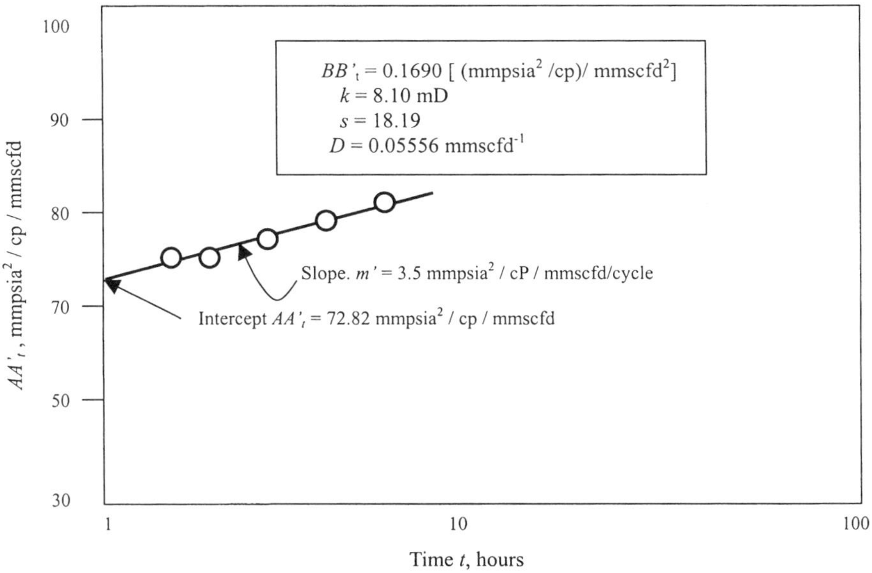

Figure 4–37. AA_t' versus log t using pseudopressure approach.

Using Eq. 4–123:

$$BB' = \frac{N\sum \Delta\psi - \sum q_{sc} \sum \frac{\Delta\psi}{q_{sc}}}{N\sum q_{sc}^2 - \sum q_{sc} \sum q_{sc}}$$

$$= \frac{4(1{,}545.56) - (20.941)(294.81)}{4(122.379) - (20.941)(20.941)} = \frac{6182.24 - 6173.62}{489.52 - 438.53}$$

$$= \frac{8.620}{50.99} = 0.1690\ \frac{\text{mmpsia}^2/\text{cP}}{\text{mmscfd}^2}$$

The values calculated for AA_t' are plotted versus t on semilog paper in Figure 4–37. The slope of the line is $m' = 3.5$ mmpsia^2/mmscfd/cycle obtained by drawing a straight line through the best points. The intercept at $t = 1$ hr can be read from the graph as 72.3 mmpsia^2/mmscfd.

1. From Eq. 4–108:

$$k = \frac{1.637 \times 10^6 \times T}{m'h} = \frac{1.637 \times 10^6 \times 710}{3.5 \times 10^6 \times 41} = 8.10\ \text{mD}$$

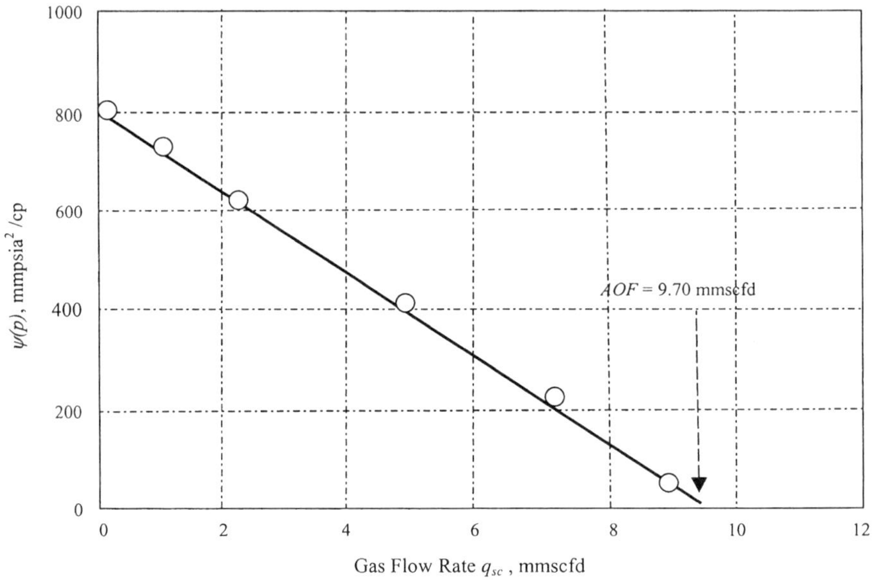

Figure 4–38. Inflow performance response.

From Eq. 4–109:

$$s = 1.151\left[\frac{AA'_{t1}}{m'} - \log\left(\frac{k(1)}{\phi\overline{\mu c}r_w^2}\right) + 3.23\right]$$

$$= 1.151\left[\frac{72.3\times10^6}{3.5\times10^6} - \log\left(\frac{8.10\times(1)}{.1004\times.0235\times.00023\times.4271^2}\right)\right.$$

$$\left.+3.23\right]$$

$$= 1.151[20.66 - 7.97 + 3.23] = 18.39 \text{ (indicating well damage)}$$

2. To obtain the turbulence coefficient using Eq. 4–110:

$$D = \frac{BB'}{0.869\,m'} = \frac{0.1690}{0.869\times3.5} = 0.05556\,\text{mmscfd}^{-1}$$

3. To determine *AOF* of this well, first find the stabilized value of AA' and BB', using Eqs. 4–115 and 4–116, respectively.

$$AA' = 1.637 \times 10^6 \frac{T}{kh}\left[\log\left(\frac{A}{r_w^2}\right) + \log\left(\frac{2.2458}{C_A}\right) + 0.869s\right]$$

$$= 1.637 \times 10^6 \frac{710}{(8.10)(41)}$$

$$\times\left[\log\left(\frac{22/7(2200)^2}{0.4271^2}\right) + \log\left(\frac{2.2458}{31.62}\right) + 0.869(18.39)\right]$$

$$= 3.499 \times 10^6\left[7.921 - 1.146 + 15.98\right] = 79.62\frac{\text{mmmpsia}^2/\text{cP}}{\text{mmscfd}}$$

and

$$BB' = 0.869\, m'D = 0.869(3.5)(0.05556) = 0.1690\ \frac{\text{mmpsia}^2/\text{cP}}{\text{mmscfd}^2}$$

4. The stabilized flow equation for determining inflow performance is

$$\psi(p_R) - \psi(p_{wf}) = 79.62\, q_{sc} + 0.1690\, q_{sc}^2$$

5. The preceding equation may be solved for q_{sc} to obtain

$$q_{sc} = \frac{-AA' + [(AA')^2 + 4\,BB'(\psi(p_R) - \psi(p_{wf})]^{0.5}}{2\,BB'}$$

For $AA' = 79.62$ and $BB' = 0.1690$:

$$q_{sc} = \frac{-79.62 + [79.62^2 + 4(0.1690)(772.56 - 0)]^{0.5}}{2(0.1690)}$$

$$= \frac{-79.62 + 82.88}{0.338} = 9.7 \text{ mmscfd}$$

Figure 4–38 shows inflow performance response.

4.11 Predicting Gas Well Deliverability Using Type Curves

To analyze wellbore storage controlled early-time data, the following procedure is described by Earlougher and Kerch[25] to estimate the permeability k and skin factor s.

Unfractured Gas Wells

1. Plot the observed test data as $(p_i - p_{wf})/t$ versus t on log-log graph paper of the same size as Earlougher and Kerch's curve.[25]
2. Estimate the wellbore storage coefficient using Eq. 4–125:

$$C_s = V_{ws} C_{ws} \tag{4–125}$$

3. Calculate the location of the horizontal asymptote on the data plot:

$$\left(\frac{\Delta p}{t}\right)_{1.0} = \frac{(q_{sc} \times 10^6)\beta_g}{24 C_s} \tag{4–126}$$

where

$$\beta_g = \frac{P_{sc}}{T_{sc}} \frac{T}{p} z, \quad scf/ft^3$$

4. From a convenient match point, read the following values from the data plot (see Fig. 4–39):

$$\left(\frac{\Delta p}{t}\right)_M, (t)_M, (C_{sd}e^{2s})_M, \left(\frac{\Delta p}{t} \frac{24C_s}{(q_{sc} \times 10^6)\beta_g}\right)_M, \left(\frac{kh}{\mu} 5.61 \frac{t}{C_s}\right)_M$$

where subscript M refers to a match point.

5. Recalculate the wellbore storage coefficient:

$$C_s = \frac{(q_{sc} \times 10^6)\beta_g \left(\frac{\Delta p}{t} \frac{24C_s}{(q_{sc} \times 10^6)\beta_g}\right)_M}{24\left(\frac{\Delta p}{t}\right)_M} \tag{4–127}$$

6. Estimate the permeability from

$$k = \frac{\mu}{h} \cdot \frac{C_S\left[\frac{kh}{\mu} 5.61 \frac{t}{C_S}\right]_M}{5.61\ (t)_M} \tag{4–128}$$

7. Estimate the skin factor from the value of $C_{sd}e^{2s}$ obtained in step 4:

$$s = 0.5\ In\left[\frac{1}{C_{sd}}(C_{sd}e^{2s})_M\right] \tag{4–129}$$

where C_{sd} is defined by Eq. 4–130:

$$C_{sd} = \frac{0.159\, C_s}{\phi\, h\, c\, r_w^2} \tag{4–130}$$

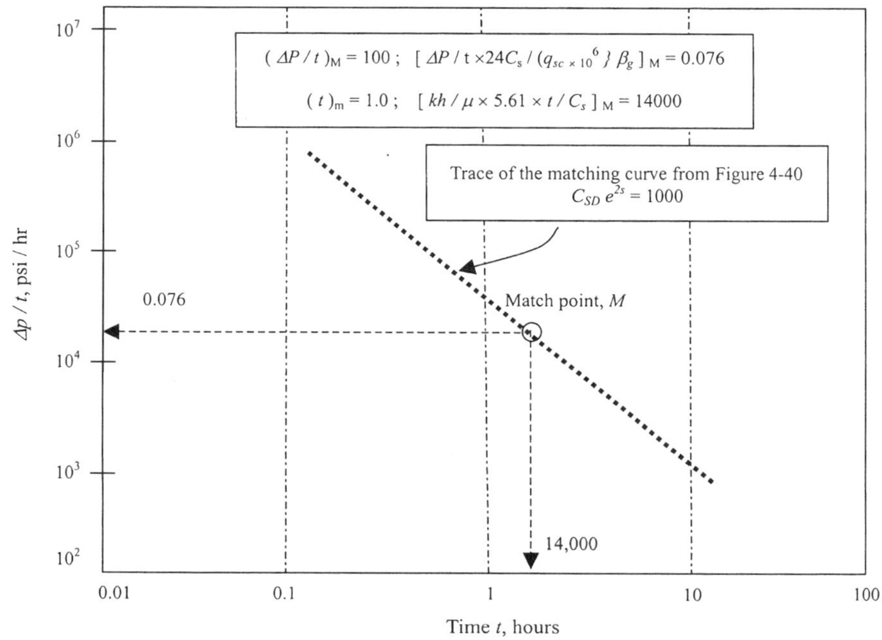

Figure 4–39. Data plot and type curve match for the short flow test—Example 4–19.

The values of k and s obtained by this type curve matching technique are not exact and should be compared with values obtained from other sources to improve their reliability.

Example 4–19 *Short Flow Tests Analysis Using Type Curve (Unfractured Well)*

A short-flow test was conducted on an unfractured well which was produced at a constant rate of 6.148 mmscfd. The pressure p_i in the reservoir prior to the test was 3965 psia. Reservoir and well data are given below. The early pressure–time data are also tabulated and are given directly in the solution to this problem. Well/reservoir data are as follows: $T = 710°R$; $h = 41$ ft; $r_w = 0.4271$ ft; $r_e = 2200$ ft; $s_g = 0.733$; well depth $= 12{,}860$ ft; $C_{ws} = 0.00027\ \text{psi}^{-1}$; $\bar{c} = 0.00023\,\text{psi}^{-1}$; $\bar{\mu} = 0.0235$ cP. Determine the permeability k of the reservoir and the skin factor s.

Solution From the available short-flow test data, make the tabulations shown in Tables 4–22 and 4–23.

Table 4–22
Calculations for Short-Flow Test Data

t (hr)	p_{wf} (psia)	$\frac{\Delta p}{t} = \left(\frac{p_i - p_{wf}}{t}\right) \times \frac{\Delta p}{(t \times q_{sc} \times 10^6)}$
0.07	1799	30,942.9
0.10	1786	21,790.0
0.25	1768	8,788.0
0.33	1765	6,666.7
0.50	1758	4,414.0
0.75	1755	2,946.7
1.00	1756	2,209.0
1.50	1758	1,471.3
2.00	1755	1,105.0
2.50	1754	884.4
3.00	1751	738.0
4.00	1747	554.5
5.00	1745	444.0
5.50	1742	404.2
6.00	1741	370.7
6.50	1739	342.7
7.00	1738	318.1
7.50	1736	296.7
8.00	1738	278.4
8.50	1737	262.1
9.00	1736	247.7
9.50	1735	234.7
10.00	1735	223.0

1. Plot $\Delta p/t$ versus t on log-log graph paper (of the same size as the type curve[6,20]) as shown in Figure 4–39.
2. From Eq. 4–125,

$$C_S = V_{WS} C_{ws} = \pi (0.4271)^2 (12{,}860)(0.00027) = 1.9906 \text{ ft}^3/\text{scf}$$

3. Calculate the formation volume factor:

$$\beta_g = \frac{P_{sc}}{T_{sc}} \frac{T}{p} z = \frac{14.65}{520} \times \frac{710}{1810} \times 0.9155 = 0.010117 \text{ ft}^3/\text{scf}$$

$$\left(\frac{\Delta p}{t}\right)_{1.0} = \frac{\left(q_{sc} \times 10^6\right) \beta_g}{24 C_S} = \frac{(6.148 \times 10^6)}{(24)(1.9906)} (0.010117) = 1298.4$$

Table 4–23
Short Flow Test Data

t (hr)	P_{wf} (psia)	$\psi(P_{wf})$ (mmpsia²/cP)	$\Delta\psi = \psi(P_i) - \psi(P_{wf})$ (mmpsia²/cP)
0.2	3670	848.90	24.10
0.3	3662	845.68	27.32
0.4	3652	841.98	31.02
0.6	3632	834.81	38.19
0.8	3621	830.68	42.32
1.0	3605	824.83	48.17
1.5	3581	815.77	57.23
2.0	3563	808.99	64.01
2.5	3547	802.89	70.11
3.0	3530	796.81	76.19
4.0	3505	787.46	85.54
5.07	3497	784.46	88.75
6.13	3492	782.70	90.30
7.00	3480	778.17	94.83
8.00	3466	772.79	100.21
10.13	3460	770.67	102.33
15.20	3433	760.61	112.39
20.00	3412	752.77	120.23
30.13	3398	747.63	125.37
40.00	3343	727.22	145.78
60.00	3324	720.28	152.72
80.00	3317	717.36	155.64
100.00	3307	713.79	159.21
120.00	3297	710.09	162.91
150.00	3277	702.55	170.45
200.00	3250	692.64	180.36

4. A match of the data plot (Figure 4–39) with the type curve $C_{SD}e^{2s} = 10^3$ of Ref. 24 is possible from a convenient match point. Match points are

$$\left(\frac{\Delta p}{t}\right)_M = 100, \quad \left(\frac{\Delta p}{t} \frac{24C_S}{(q_{sc}10^6)\,\beta_g}\right)_M = 0.076$$

and

$$(t)_M = 1.0, \quad \left(\frac{kh}{\mu_g} \frac{5.61\,t}{C_S}\right)_M = 14{,}000$$

5. From Eq. 4–128:

$$k = \frac{\mu}{h}\frac{C_s\left(\frac{kh}{\mu}5.61\frac{t}{C_s}\right)_M}{5.61(t)_M} = \frac{0.0235}{41}(1.9906)\frac{14{,}000}{5.61(1)} = 2.85\,\text{mD}$$

6. From Eqs. 4–129 and 4–130:

$$s = 0.5\,In\left[\frac{1}{C_{sd}}(C_{sd}e^{2s})_M\right] = 0.5\,In\left[\frac{\phi\,hcr_w^2}{0.159C_S}(C_{SD}e^{2s})_M\right]$$

$$= 0.5\,In\left[\frac{0.1004 \times 41 \times 0.00027 \times 0.4271^2}{0.159(2.9906)} \times 1000\right]$$

$$= 0.5(-0.606) = -0.22$$

Check

Recalculate the wellbore storage coefficient using Eq. 4–127:

$$C_S = \frac{(q_{sc} \times 10^6)\beta_g}{24}\frac{\left(\frac{\Delta p}{t}\frac{24C_S}{(q_{sc}\times 10^6)\beta_g}\right)_M}{\left(\frac{\Delta p}{t}\right)_M}$$

where

$$\beta_g = \frac{P_{sc}}{T_{sc}\frac{T}{p}z} = \frac{14.65}{520} \times \frac{710}{1810} \times .9155 = 0.010117\,\text{ft}^3/\text{scf}$$

Therefore,

$$C_S = \frac{(6.148 \times 10^6)(0.010117)}{24} \times \frac{0.076}{100} = 1.9696$$

This is close to the calculated value.

Fractured Gas Wells

Linear flow through fracture controls early-time data. Gringarten, Ramey, and Raghavan's type curves[6,20] may be used to analyze such data. The following procedure is described to estimate the flow capacity kh, the fracture half-length x_f, and the skin factor s:

1. Plot the test data as $\Delta\psi = \psi(p_i) - \psi(p_{wf})$ versus t on log-log graph paper of the same size used for Ramey's curves.
2. Slide the data plot over either Figure 4–40 or Figure 4–41, both horizontally and vertically, until the best match is obtained. The most likely curve is that for $x_e/x_f = \infty$, except where the fracture length and the duration of the test are usually large.

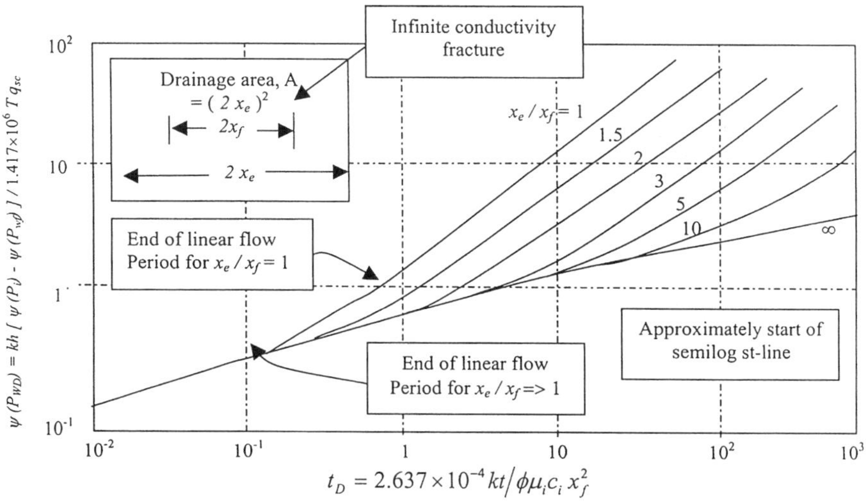

Figure 4–40. Type curves for an infinite conductivity vertical fracture. After Gringarten, Ramey, and Raghavan, © SPE, 1972, 1974.[6,20]

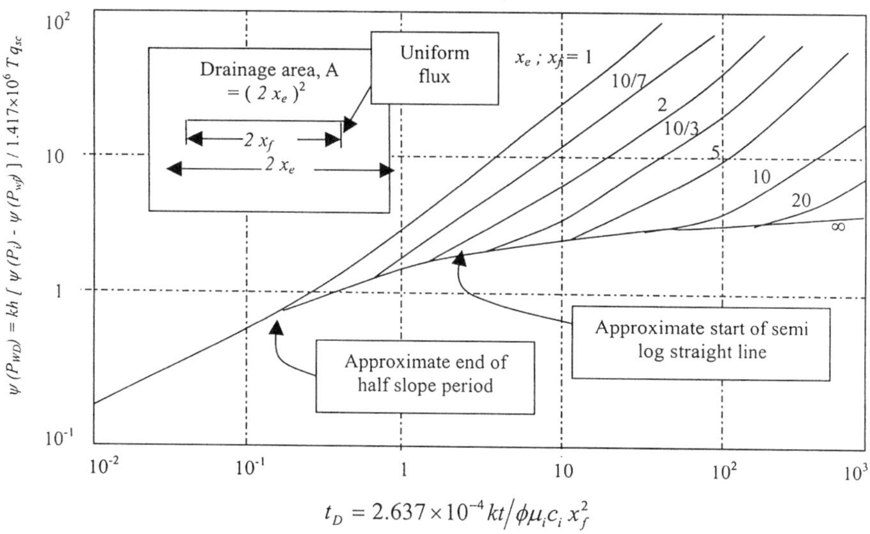

Figure 4–41. Type curves for uniform-flux fracture. After Gringarten, Ramey, and Raghavan © SPE, 1972, 1974.[6,20]

3. Sketch the match curve on to the data plot. Pick any convenient match point and read

$$(t)_M \quad \text{and} \quad \lfloor \Delta\psi = \psi(p_i) - \psi(p_{wf}) \rfloor_M$$

from the data plot.
4. Estimate flow capacity from

$$kh = \frac{50.300 \times 10^6 T P_{sc} q_{sc}}{T_{sc}} \frac{\left(\frac{kh\Delta\psi}{1.417\times10^6 T q_{sc}}\right)_M}{(\Delta\psi)_M} \tag{4–131}$$

5. Estimate the fracture half-length from

$$x_f = \left[\frac{2.637 \times 10^{-4} k}{\phi\mu_i c_i} \frac{(t)_M}{\left(\frac{2.637\times10^{-4} kt}{\phi\mu_i c_i x_f^2}\right)_M}\right]^{0.5} \tag{4–132}$$

6. Estimate the skin factor from

$$s = -In\left(\frac{0.5\, x_f}{r_w}\right) \tag{4–133}$$

7. Calculate IT turbulent factor D from

$$D = \frac{Fkh}{1.422 \times 10^6 T} \tag{4–134}$$

where

$$F = \frac{3.161 \times 10^{-12} T \gamma_g k\beta}{\mu h^2 r_w} \tag{4–135}$$

$$\beta = \frac{2.73 \times 10^{10}}{k^{1.1045}} \tag{4–136}$$

Example 4–20[27] *Analyzing Short Flow Tests Using Type Curve (Fractured Well)*

A short-flow test was conducted on a fractured well which was produced at a constant rate of 5.650 mmscfd. The pressure p_i in the reservoir prior to the test was 3732 psia. Reservoir and well data are given below. The early pressure-time data are also tabulated and are given directly in the solution to this problem. Well/reservoir data: $T = 673°R$; $h = 20$ ft; $r_w = 0.29$ ft; $r_e = 2640$ ft; $r_g = 0.680$; well depth $= 10{,}600$ ft; $C_{ws} = 0.00026$ psi^{-1}; $\bar{c} = 0.00022\, \psi^{-1}$; $\bar{\mu} = 0.0208$ cP. The real gas pseudopressure $\Psi(P)$ can be calculated from the following equation and vice versa:

$$\psi = -516.4 + 0.372P, \text{ mmpsia}^2/\text{cP}$$

$$P = \frac{\psi + 516.4}{0.372}, \text{ psia}$$

Determine the reservoir permeability k and the skin factor s.

Solution Make the following tabulations from the available short-flow test data.

Plot ψ versus t on log-log graph paper of the same size as the type curves of Figure 4–40 or 4–41. A match of the data plot (Figure 4–42) with the type curve is possible. From a convenient match point:

$$(t)_M = 3.5, (\Delta\psi)_M = 80 \times 10^6 \text{ psia}^2/\text{cP(data plot in Figure 4–42)}$$

$$(t_D)_M = 1, (\Delta\psi_D)_M = 1.68 \text{ (type curve in Figure 4–41)}$$

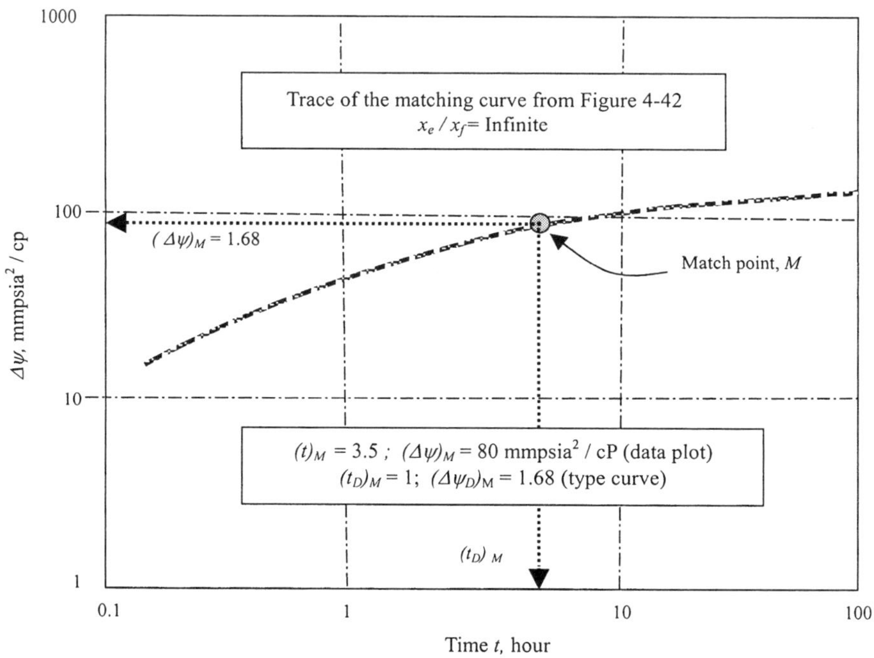

Figure 4–42. Data plot and type curve match for fractured gas well—Example 4–20.

From Eq. 4–131:

$$kh = \frac{50.300 \times 10^6 T P_{sc} q_{sc}}{T_{sc}} \frac{\left(\frac{kh\Delta\psi}{1.417\times10^6 T q_{sc}}\right)_M}{(\Delta\psi)_M}$$

$$k = \frac{50.300 \times 10^6 \times 673 \times 5.650 \times 14.65}{20 \times 520} \frac{1.68}{80 \times 10^6} = 5.66\,\text{mD}$$

From Eq. 4–132:

$$x_f = \left[\frac{2.637 \times 10^{-4} k}{\phi \mu_i c_i} \frac{(t)_M}{\left(\frac{2.637\times10^{-4} kt}{\phi\mu_i c_i x_f^2}\right)_M}\right]^{0.5}$$

$$x_f = \left[\frac{2.637 \times 10^{-4} \times 5.66}{0.1 \times 0.0208 \times 0.00022} \frac{(1)}{(3.5)}\right]^{0.5} = \sqrt{906.02} = 30.10\,\text{ft}$$

Calculate the skin factor using Equation 4–133:

$$s = -In\left(\frac{0.5x_f}{r_w}\right) = -In\left(\frac{0.5 \times 30.10}{0.29}\right) = -3.95$$

From Eq. 4–136:

$$\beta = \frac{2.73 \times 10^{10}}{k^{1.1045}} = \frac{2.73 \times 10^{10}}{5.66^{1.1045}} = 0.4024 \times 10^{10}\ \text{ft}^{-1}$$

From Eq. 4–135:

$$F = \frac{3.161 \times 10^{-12} T \gamma_g k \beta}{\mu h^2 r_w}$$

$$= \frac{3.161 \times 10^{-12} \times 673 \times 5.66 \times 0.4024 \times 10^{10}}{0.0208 \times 20 \times 0.29} = 2.2207$$

From Eq. 4–134:

$$D = \frac{Fkh}{1.422 \times 10^6 T} = \frac{2.2207 \times 5.66 \times 20}{1.422 \times 10^6 \times 673} = 2.61056 \times 10^{-5}\ \text{mmscfd}^{-1}$$

4.12 Estimation of Skin Factors from Well Completion Data

To improve the skin factor for a gas well, various types of stimulation treatment are often performed. Acidizing or fracturing techniques may be applied in low permeability wells. Table 4–24 lists the possible skin factors that may result under different well completion conditions.

Table 4–24

Type of stimulation	Skin factor s
Natural completion	0
Light acid	−0.5
Medium acid or light fracture	−1.0
Heavy acid or medium fracture	−2.0
Heavy fracture	−3.0
Heavy fracture in low permeability	−4.0
Very large fracture in low permeability	−5.0

The classification of acidizing or fracturing as light, medium, or heavy is purely qualitative. In general, light fracturing involves up to 1000 gallons of fracturing fluid per foot of net pay, whereas more than 4000 gal/ft is usually considered fairly heavy. Light aciding involves up to 300 gal/ft, whereas heavy acidizing usually exceeds 600 gal/ft. The effect of either acidizing or fracturing is much greater for tight sands than for fairly permeable sands. This too is a qualitative classification requiring some engineering judgement. Generally, formations of permeability less than 5 mD are considered tight, while those of permeability less than 1 mD are considered very tight. A formation permeability more than 25 mD is considered to be high.

4.13 Laminar–Inertial Turbulent Flow Analysis

Pseudo-Steady State (Laminar Flow)

Pressure-Squared Relationship

Equation 4–28 is the commonly used Rawlins and Schellhardt deliverability equation and is obtained empirically but may be related to a theoretically derived relationship, Eq. 4–32, also called the Δp^2 and LIT(p) flow equation. Combining Eqs. 2–93 and 2–99 and substituting for various dimensionless variables, for stabilized flow (pseudo-steady-state), and assuming laminar flow in the reservoir,

$$\overline{p}_R^2 - p_{wf}^2 = \frac{116.246 \times 10^6 q_{sc} \mu z T}{kh} \frac{P_{sc}}{T_{sc}} \log\left(0.472\frac{r_e}{r_w}\right)$$

or

$$\overline{p}_R^2 - p_{wf}^2 = \frac{50.474 \times 10^6 q_{sc} \mu z T}{kh} \frac{P_{sc}}{T_{sc}} In\left(0.472\frac{r_e}{r_w}\right) \tag{4–137}$$

The skin factor s and inertial-turbulent flow effects Dq_{sc} may be introduced to give

$$\overline{p}_R^2 - p_{wf}^2 = \frac{116.246 \times 10^6 \mu z T}{kh}\frac{P_{sc}}{T_{sc}}\left[\log\left(\frac{0.472 r_e}{r_w}\right) + \frac{s}{2.303}\right] q_{sc} + \frac{50.474 \times 10^6 \mu z T}{kh} D q_{sc}^2$$
$$= aa' q_{sc} + bb' q_{sc}^2 \tag{4–138}$$

Therefore,

$$aa' = \frac{116.246 \times 10^6 \mu z T}{kh}\frac{P_{sc}}{T_{sc}}\left[\log\left(\frac{0.472 r_e}{r_w}\right) + \frac{s}{2.303}\right] \tag{4–139}$$

$$bb' = \frac{50.474 \times 10^6 \mu z T}{kh}\frac{P_{sc}}{T_{sc}} D \tag{4–140}$$

The interrelationship of aa' and bb' to C and n of Eq. 4–28 has been given in various forms by Houpeurt (1959), Willis (1965), Carter *et al.* (1963), and Cornelson (1974), who gave similar relationships in graphical form for various ranges of flow rates. One form of the interrelationship, as expressed by Carter (1985), assumes the following:

1. Equation 4–28 is valid for $q_{min} \leq q_{sc} \leq q_{max}$.
2. $\overline{p}_R^2 - p_{wf}^2$ versus q_{sc} plot is a straight-line on a log-log plot.
3. Equation 4–32 is valid for $0 \leq q_{sc} \leq AOF$.
4. The function $\overline{p}_R^2 - p_{wf}^2$ from Eqs. 4–28 and 4–32 is equal to the range q_{min} to q_{max}.
5. The rate of change of the above functions is equal at the geometric mean of q_{min} and q_{max}, to give

$$aa' = \left(\frac{1}{C}\right)^{1/n} q_{sc}^{(1/n-1)}\left(2 - \frac{1}{n}\right) \tag{4–141}$$

$$bb' = \left(\frac{1}{C}\right)^{1/n} q_{sc}^{(1/n-2)}\left(\frac{1}{n} - 1\right) \tag{4–142}$$

and

$$C = \frac{q_{sc}}{\left(aa' q_{sc} + bb' q_{sc}^2\right)^{\frac{aa' + bb' q_{sc}}{aa' + 2bb' q_{sc}}}} \tag{4–143}$$

$$n = \frac{aa' + bb' q_{sc}}{aa' + 2bb' q_{sc}} \tag{4–144}$$

For very low flow rates;

$$aa'q_{sc} \gg bb'q_{sc}^2, \quad \Delta p^2 \cong aa'q_{sc}$$

and n of Eq. 4–32 reduces to Eq. 4–28 = 1.0 for $n = 1, aa' = (\frac{1}{C})$.

For high flow rates,

$$aa'q_{sc} \ll bb'q_{sc}^2, \; \Delta p^2 \cong \left(\frac{1}{C}\right)^2$$

and Eq. 4–32 reduces to Eq. 4–28.

Hence n may vary 1.0 for fully laminar flow to 0.5 for turbulent flow. An approximate AOF may be obtained from

$$AOF \cong \frac{kh\left(\overline{p}_R^2 - p_{wf=0}^2\right)}{116.246 \times 10^6 \mu z T \frac{P_{sc}}{T_{sc}}\left[\log\left(0.472\frac{r_e}{r_w}\right) + \frac{s}{2.303}\right]} \tag{4–145}$$

or

$$AOF \cong \frac{kh\left(\overline{p}_R^2 - p_{wf=0}^2\right)}{50.474 \times 10^6 \mu z T \frac{P_{sc}}{T_{sc}}\left[In\left(0.472\frac{r_e}{r_w}\right) + \frac{s}{2.303}\right]} \tag{4–146}$$

Pseudopressure Relationship

If pseudopressure $\psi(p)$ is used instead of pressure-squared, the pseudo-steady-state equation is written as

$$\psi(\overline{p}_R - \psi(p_{wf}) = \frac{116.246 \times 10^6 T}{kh}\frac{P_{sc}}{T_{sc}}\left[\log\left(\frac{0.472 r_e}{r_w}\right) + \frac{s}{2.303}\right] q_{sc}$$

$$+ \frac{50.474 \times 10^6 T}{kh} D q_{sc}^2 \tag{4–147}$$

$$= aaq_{sc} + bbq_{sc}^2$$

Therefore,

$$aa = \frac{116.246 \times 10^6 T}{kh}\frac{P_{sc}}{T_{sc}}\left[\log\left(\frac{0.472 r_e}{r_w}\right) + \frac{s}{2.303}\right] \tag{4–148}$$

$$bb = \frac{50.474 \times 10^6 T}{kh}\frac{P_{sc}}{T_{sc}} D \tag{4–149}$$

The interrelationship of aa and bb to C and n can be obtained by replacing aa' and bb' by aa and bb. An approximate idea of the absolute open flow potential of a gas well may be obtained from

$$AOF \cong \frac{kh\lfloor\psi(\overline{p}_R) - \psi(p_{wf=0})\rfloor}{1.637 \times 10^6 T\left[\log\left(0.472\frac{r_e}{r_w}\right) + \frac{s}{2.303}\right]} \tag{4–150}$$

or

$$AOF \cong \frac{kh\lfloor\psi(\overline{p}_R) - \psi(p_{wf=0})\rfloor}{1.422 \times 10^6 T\left[In\left(0.472\frac{r_e}{r_w}\right) + s\right]} \tag{4–151}$$

Equations 4–138 and 4–147 can be applied to stabilized conditions only; that is, $t > t_s$, the time of stabilization:

$$t_s \cong 1000\frac{\phi\overline{\mu}r_e^2}{k\overline{p}_R} \tag{4–152}$$

Pressure Relationship

At stabilization, the flow equation (excluding skin and IT flow effects) can be written in terms of pressure:

$$\overline{p}_R - p_{wf} = \frac{7.085 \times 10^5 \overline{z}\,\overline{\mu} T q_{sc}}{\overline{p}kh}\left[\frac{2(2.637 \times 10^{-4})kt}{\phi\overline{\mu}\,\overline{c}r_e^2}\right] + In\left(\frac{r_e}{r_w}\right) - 0.75 \tag{4–153}$$

The rate of pressure decline is obtained by

$$\begin{aligned}\frac{\partial p_{wf}}{\partial t} &= -\frac{2(7.085 \times 10^5)(2.637 \times 10^{-4})\overline{z}\,T q_{sc}}{\overline{p}\,\phi\, h\,\overline{c}\, r_e^2} \\ &= -374\frac{\overline{z}\,T q_{sc}}{\overline{p}\,\phi\, h\,\overline{c}\, r_e^2} \\ &= -374\frac{\overline{z}\,T q_{sc}}{\phi\, h\, r_e^2} \quad \text{For } (\overline{p} = \overline{c})\end{aligned} \tag{4–154}$$

Before stabilization is achieved, the radius of investigation, r_{inv} is given by

$$r_{inv} = 0.032\sqrt{\frac{k\overline{p}_R t}{\phi\overline{\mu}}} \tag{4–155}$$

Transient Relationship

The deliverability relationships represented by Equations 4–138 and 4–147 apply at stabilized conditions, that is, for $r_{inv} = r_e$. When $r_{inv} < r_e$, the flow conditions are said to be transient. The transient flow equations in terms of pressure-squared and pseudopressure have the following forms:

In terms of pressure squared:

$$\overline{p}_R^2 - p_{wf}^2 = \frac{1.637 \times 10^6 T \mu z}{kh}\left[\log\left(\frac{kt}{\phi\overline{\mu}\,\overline{c}r_w^2}\right) - 3.23 + 0.869s\right] q_{sc} + \frac{1.422 \times 10^6 T \mu z}{kh} D q_{sc}^2 \qquad (4\text{–}156)$$

$$= aa_t' q_{sc} + bb' q_{sc}^2 \qquad (4\text{–}157)$$

Therefore,

$$aa_t' = \frac{2.637 \times 10^6 T \mu z}{kh}\left[\log\left(\frac{kt}{\phi\overline{\mu}\overline{c}r_w^2}\right) - 3.23 + 0.869s\right] \qquad (4\text{–}158)$$

In terms of pseudopressure:

$$\psi(p_R) - \psi(p_{wf}) = \frac{3.275 \times 10^6 T}{kh}\left[0.5\left(\log\left(\frac{2.637 \times 10^{-4} kt}{\phi \mu_i c_i r_w^2}\right) + \frac{0.809}{2.303}\right)\frac{s}{2.303}\right] q_{sc} + \frac{1.422 \times 10^6 T}{kh} D q_{sc}^2 \qquad (4\text{–}160)$$

$$= aa_t q_{sc} + bb q_{sc}^2 \qquad (4\text{–}161)$$

Therefore,

$$aa_t = \frac{3.275 \times 10^6 T}{kh} 0.5\left(\log\left(\frac{2.637 \times 10^{-4} kt}{\phi \mu_i c_i r_w^2} + \frac{0.809}{2.303}\right) + \frac{s}{2.303}\right)$$

$$= \frac{1.637 \times 10^6 T}{kh}\left[\log\left(\frac{kt}{\phi \mu_i c_i r_w^2}\right) - 3.23 + 0.869s\right] \qquad (4\text{–}162)$$

and

$$bb = \frac{1.422 \times 10^6 T}{kh} D \qquad (4\text{–}163)$$

Estimation of Wellbore Storage Time

The time at which wellbore storage effects are significant is given by

$$t_{ws} = \left(\frac{\eta}{\lambda}60\right)\frac{\overline{\mu}V_{ws}C_{ws}}{kh} \tag{4–164}$$

where

$$\eta = 0.159, \text{ when } V_{ws} \text{ is in ft}^3$$
$$\lambda = 2.637 \times 10^{-4}$$
$$V_{ws} = \pi r_w^2 L \text{ ft}^3$$
$$C_{ws} = \text{compressibility of wellbore fluid, psi}^{-1}$$

Therefore,

$$t_{ws} = \frac{0.159 \times 60}{2.637 \times 10^{-4}}\left(\frac{\overline{\mu}V_{ws}C_w}{kh}\right) = 36{,}177.5\frac{\overline{\mu}V_{ws}C_{ws}}{kh} \tag{4–165}$$

Example 4–21 *Analyzing Wellbore Storage Effect*
Calculate the time t_{ws} required for wellbore storage effects to become negligible for a gas well with no bottom-hole packer, given the following characteristics: $L = 12{,}860$ ft; $r_w = 0.4271$ ft; $h = 41$ ft; $C_{ws} = 0.000552$ psi^{-1}; $k = 8.96$ mD; and $\mu = 0.01723$ cP.

Solution $V_{ws} = \pi r_w^2$ (depth of well) $= 22/7 \times 0.4271 \times 0.4271 \times 12{,}860 =$ 7273 ft^3. From Equation 3–164,

$$t_{ws} = 36{,}177.5\frac{\mu V_{ws}C_{ws}}{kh} = \frac{36{,}177.5 \times 0.01732 \times 7273 \times 0.000552}{8.96 \times 41}$$
$$= 6.81 \text{ hr}$$

After a time of 6.81 hr, wellbore storage effects become negligible and the analytical solution for transient flow can apply.

4.14 Summary

Chapter 4 deals with deliverability testing and commonly used techniques for predicting short-term and long-term behavior of unfractured and fractured gas wells. Deliverability tests can be analyzed to provide reliable values of kh, s, and D within the usual limits of engineering accuracy when costlier buildup tests are not warranted. Data derived from a backpressure test can be valuable in determining permeability distribution for subsequent use in engineering calculations and gas reservoir simulation study. In addition, the data obtained from the test can be used to analyze and predict gas well performance using a numerical model which accounts for effects of turbulence, skin, afterflow, partial penetration, pressure dependent k, and any degree of crossflow ranging from complete to none.

References and Additional Reading

1. Al-Hussainy, R., Ramey, H. J., Jr., and Crawford, P. B., "The Flow of Real Gases through Porous Media," *J. Petroleum Technol.* (May 1966) 624–636; *Trans. AIME* 237.
2. Cullender, M. H., "The Isochronal Performance Method of Determining the Flow Characteristics of Gas Wells," *Trans. AIME* (1955) 204, 137–142.
3. Jones, L. G., Blount, E. M., and Glaze, O. H., "Use of Short Term Multiple Rate Flow Tests to Predict Performance of Wells Having Turbulence," paper SPE 6133 presented at the SPE 51st Annual Meeting, New Orleans, Oct. 3–6, 1976.
4. *Theory and Practice of the Testing of Gas Wells,* 3rd ed., Alberta Energy Resources Conservation Board, 1975.
5. Al-Hussainy, R., and Ramey, H. J., Jr., "Application of Real Gas Flow Theory to Well Testing and Deliverability Forecasting," *J. Petroleum Technol.* (1996) 18, 637–642.
6. Gringarten, A. C., Ramey, Henry. J., Jr., and Raghavan, R., "Pressure Analysis for Fractured Wells." Paper SPE 4051 presented at the SPE. AIME 47th Annual Fall Meeting, San Antonio, TX, Oct. 8–11, 1972.
7. Brar, G. S., and Aziz, K., "The Analysis of Modified Isochronal Tests to Predict the Stabilized Deliverability of Gas Wells without Using Stabilized Flow Data," paper SPE 6134, presented at the SPE 51st Annual Meeting, New Orleans, Oct. 3–6, 1976.
8. Wattenbarger, R. A., and Ramey, H. J., Jr., "Gas Well Testing with Turbulence Damage, and Wellbore Storage," *J. Petroleum Technol.* (Aug. 1968) 877–887; *Trans. AIME* 243.
9. Al-Hussainy, R., and Ramey, H. J., "Application of Real Gas Flow Theory to Well Testing and Deliverability Forecasting," *J. Petroleum Technol.* (May 1966).
10. Katz, D. L., Cornell, D., Kobayashi, R., Poettmann, F. H., Vary, J. A., Elenbaas, J. R., and Weinaug, C. F., *Handbook of Natural Gas Engineering.* McGraw-Hill, New York, 1959.
11. Dake, L. P., *Fundamentals of Reservoir Engineering.* Elsevier Scientific, 1978.
12. Muskat, M., *The Flow of Homogeneous Fluids through Porous Media.* McGraw-Hill, New York, 1937.
13. *Manual of Back Pressure Testing of Gas Wells.* Kansas State Corporation Commission, 1959.

14. *Back Pressure Test for Natural Gas Wells,* revised edition. Railroad Commission of Texas, 1951.
15. *Manual of Back Pressure Testing of Gas Wells.* Interstate Oil Compact Commission, 1962.
16. *Engineering Data Book*; 9th ed. Gas Processors Suppliers Association, 1972, revised 1974.
17. Wattenbarger, R. A., "Effects of Turbulence, Wellbore Damage, Wellbore Storage and Vertical Fractures on Gas Well Testing," Ph.D. Thesis, Stanford University, Stanford, CA, 1967.
18. Ramey, H. J., Jr., "Short-Time Well Test Data Interpretation in the Presence of Skin Effect and Wellbore Storage," *J. Petroleum Technol.* (1970) 22, 97–104.
19. Slider, H. C., "A Simplified Method of Pressure Analysis for a Stabilized Well," *J. Petroleum Technol.* (1971) 23, 1155–1160.
20. Gringarten, A. C., Ramey, H. J., Jr., and Raghavan, R., "Unsteady-State Pressure Distribution Created by a Well with a Single Infinite-Conductivity Vertical Fracture," *Soc. Petroleum Eng. J.* (Aug. 1974) 347–360; *Trans. AIME* 257.
21. Ramey, H. J., Jr., "Non-Darcy Flow and Wellbore Storage Effects in Pressure Build-up and Drawdown of Gas Wells," *J. Petroleum Technol.* (1965) 7, 223–233.
22. Ramey, H. J., Jr., Kumar, A., and Gulati, M. S., *Gas Well Test Analysis under Water-Drive Conditions.* American Gas Association, VA, 1973.
23. Kazemi, H., "Determining Average Reservoir Pressure from Pressure Buildup Tests," *Soc. Petroleum Eng. J.* (1974) 14, 55–62.
24. Brigham, W. E., "Estimating Reservoir Parameters from the Gas Backpressure Equation," SPE Reservoir Engineers, May (1998), 649–650.
25. Earlougher, R. C., Jr., and Kerch, K. M., "Analysis of Short-Time Transient Test Data by Type-Curve Matching," *J. Petroleum Technol.* (1974) 26, 793–800.
26. Cullender, M. H., and Smith, R. V., "Practical Solution of Gas-Flow Equations for Well and Pipelines with Large Temperature Gradient," *Trans. AIME* (1956) 207, 281–287.
27. Amanat, U. C., "Pressure Transient Test Analysis User's Handbook" Twpsom Petroleum Software Series by Advanced TWPSOM Petroleum Systems Inc., Houston, TX, Vol. 8, Oct. 1995.
28. Riley, H. G., "A Short Cut to Stabilized Gas Well Productivity," *J. Petroleum Technol.* (1970) 22, 537–542.
29. Fetkovich, M. J. "Multipoint Testing of Gas Wells," Paper presented at the SPE-AIME Mid-Continent Section Continuing Education Course on Well Test Analysis (March 1975).

Chapter 5

Fundamentals of Drawdown Test Analysis Methods

5.1 Introduction

Important reservoir parameters can be determined by flowing a well at a constant rate and measuring flowing wellbore pressure as a function of time. This is called drawdown testing and it can utilize information obtained in both the transient and pseudo-steady-state flow regimes. If the flow extends to the pseudo-steady state, the test is referred to as a reservoir limit test and can be used to estimate in-place gas and shape of the reservoir. Both single-rate and two-rate tests are utilized depending on the information required. The purpose of the drawdown testing is to determine the reservoir characteristics that will affect flow performance. Some of the important characteristics are the flow capacity kh, skin factor s, and turbulence coefficient D.

5.2 Characteristics of Flow and Gas Well Transient Testing

Much of that information can be obtained from pressure transient tests. Pressure transient testing techniques, such as buildup, drawdown, interference, and pulse, are an important part of reservoir and production engineering. As the term is used in this book, pressure transient testing includes generating and measuring pressure variations with time in gas wells and subsequently, estimating rock, fluid, and well properties and predicting reservoir/well behavior. Practical information obtainable from transient testing includes wellbore volume, damage, and improvement; reservoir pressure; permeability; porosity; reserves; reservoir and fluid discontinuities; and other related data. All this information can be used to help analyze, improve, and forecast reservoir performance.

Pressure transient testing and analysis is an important diagnostic tool to define near-wellbore and interwell conditions as opposed to composite properties that would be indicated by steady-state productivity index data. In other cases, simpler approach is adequate, or a different or combined approach is needed to solve a problem.

Pressure interference or pulse testing could establish the possible existence and orientation of vertical fracture of a gas reservoir. However, other information (such as profile surveys, production logs, stimulation history, well production tests, packer tests, core descriptions, and other geological data about reservoir lithology and continuity) would be useful in distinguishing between directional permeability and fractures or estimating whether the fractures were induced or natural. It is generally good practice to run a base pressure transient test on a producing well shortly after completion or an injection well after a suitable period of injection. This can lead to early recognition and correction of many problems, of which insufficient stimulation is only the most obvious. Such tests also provide *in situ* data for reservoir simulation and a base for comparison with reservoir or well problems as they arise.

5.3 Pressure–Time History for Constant-Rate Drawdown Test

Figure 5–1 shows the flow history of an unfractured well and can be divided into three periods for analysis:

The transient or early flow period is usually used to analyze flow characteristics

The late transient period is more complete

The semisteady-state flow period is used in reservoir limit tests

As shown in Figure 5–1, radial flow is preceded by a period of linear flow when wells contain fractures. If the pay interval is partly penetrated or perforated, a spherical-flow-dominated period should be expected between the linear and radial flow times. Also, the first flow unloads the well while accepting a contribution from the reservoir. Thus, a group of curves must be constructed to analyze well tests properly. Flow tests may better represent well performance than buildup tests since particle movement, turbulence, and capillary constrictions are then included.

5.4 Characteristics of Various Flow Regimes

The different flow regimes are depicted in Figure 5–1. It is convenient to treat each one separately.

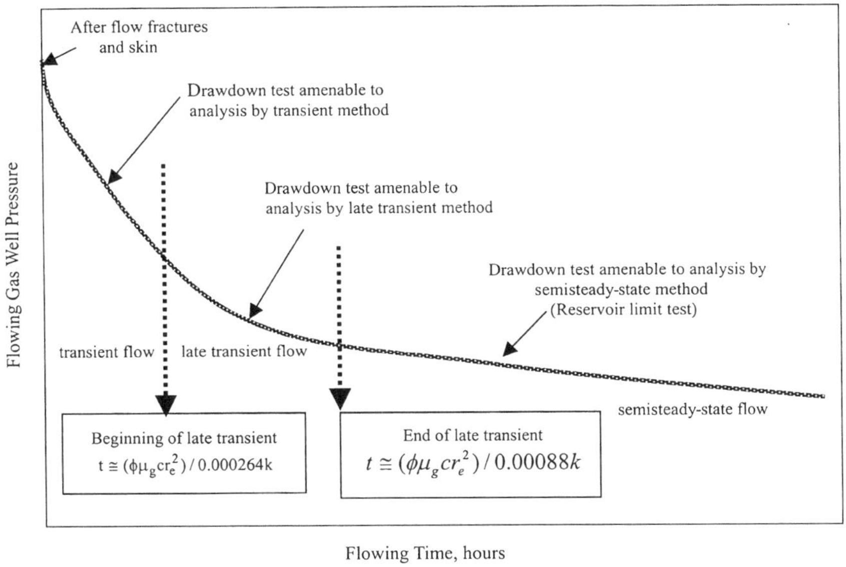

Figure 5–1. Schematic pressure–time histories for a constant-rate drawdown test (after Odeh and Nabor, *JPT*, Oct. 1966).[1]

Early-Time Flow Regime

Initially during early-time flow, wellbore storage and skin effects dominate the flow. When the well is opened at the surface for flow at a constant rate, the initial flow comes primarily from the wellbore itself, rather than from the formation. In fact, flow from the reservoir increases gradually from zero until the specified wellhead flow rate q is reached in a length of time, t_{wb}, given by

$$t_{wb} = \frac{const\, \bar{\mu}_g C_s}{kh} \tag{5–1}$$

where $const$ = a constant = 36,177 when V_{wb} is in ft^3, and is 203,413 when V_{wb} is in bbl, field units. kh is formation flow capacity, mD-ft. C_s is the wellbore constant, is defined as the rate unloading of, or storage in, the wellbore per unit pressure difference, and is given by

$$C_s = V_{wb}\, C_{wb} \tag{5–2}$$

where V_{wb} = volume of wellbore tubing (well with bottom-hole packer) or volume of wellbore annulus (well without bottom-hole packer); C_s = compressibility of the wellbore fluid evaluated at the mean wellbore pressure and temperature, and not at reservoir conditions as is usually the case.

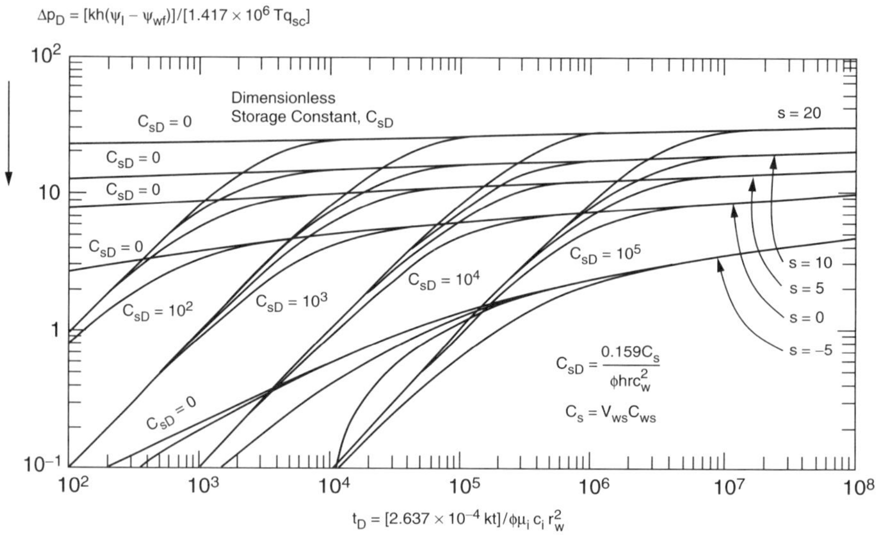

Figure 5–2. Dimensionless pressure p_D versus dimensionless time t_D, including wellbore storage and skin effects (after Agarwal, Al-Hussainy, and Ramey).[2]

Equation 5–1 applies to wells with zero skin effects. Agarwal, Al-Hussainy, and Ramey presented the combined effects of wellbore storage and skin in the form of the type curves of Figures 5–2 through 5–7. These type curves can be used quite effectively to define the time of start of transient flow and its use, as illustrated in next section. Although early-time data are not analyzed in this section, it is of interest to note that in the presence of wellbore storage effects, a plot of Δp_D versus t_D on logarithmic coordinates will give a straight line of slope 1.0 for the initial data.

Transient Flow Regime

In this flow regime the pressure is the same as that created by a line-source well with a constant skin. Since a plot of Δp_D versus t_D on semilogarithmic coordinates will yield a straight line, the analysis of transient flow data is often referred to as a semilog analysis. The semilog analysis of drawdown data yields consistent values of reservoir parameters. Only the permeability thickness kh the skin factor s, and the inertial-turbulence factor D may be determined from such an analysis. This semilog straight line continues as long as the reservoir is infinite-acting. If a fault is encountered in the reservoir, the slope of the

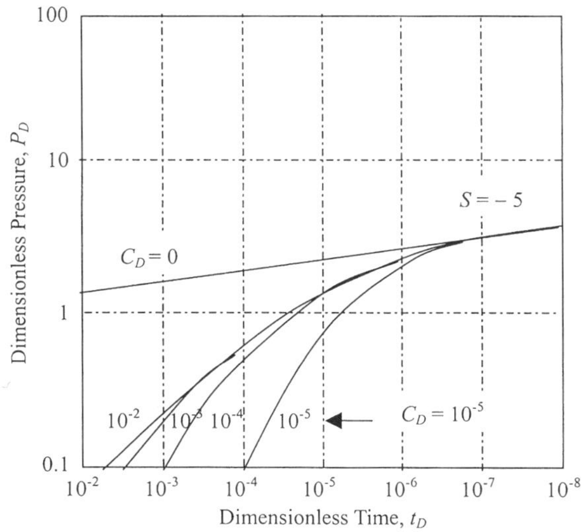

Figure 5–3. Type curves to determine end of wellbore storage distortion.[2]

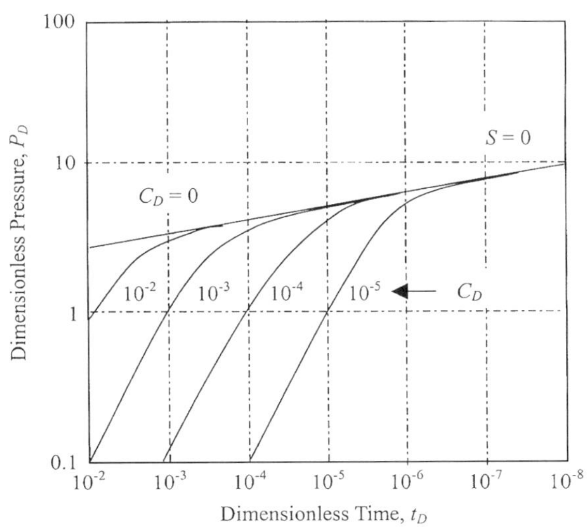

Figure 5–4. Type curves to determine end of wellbore storage distortion.[2]

line will double, and a new straight line will be established. The effects of a fault/barrier are discussed further in this chapter. When the reservoir boundary begins to have a significant effect on well drawdown, the transient region ends; the pseudo-steady-state or depletion phase directly follows the transient period.

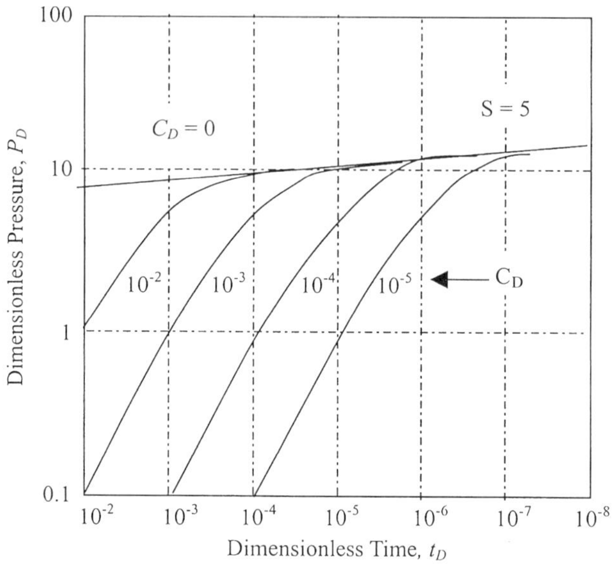

Figure 5–5. Type curves to determine end of wellbore storage distortion.[2]

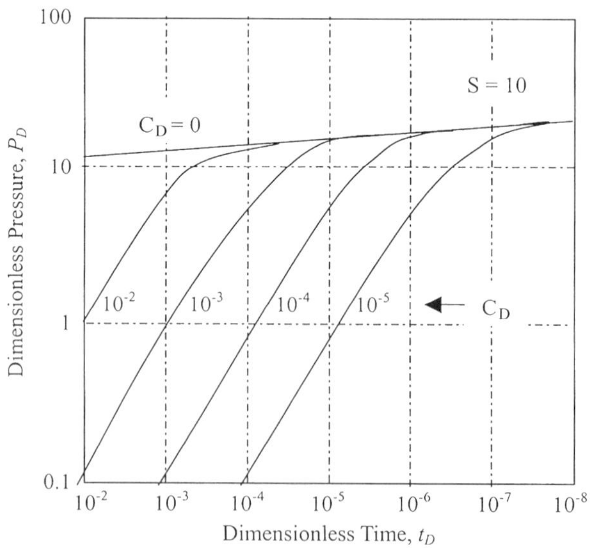

Figure 5–6. Type curves to determine end of wellbore storage distortion.[2]

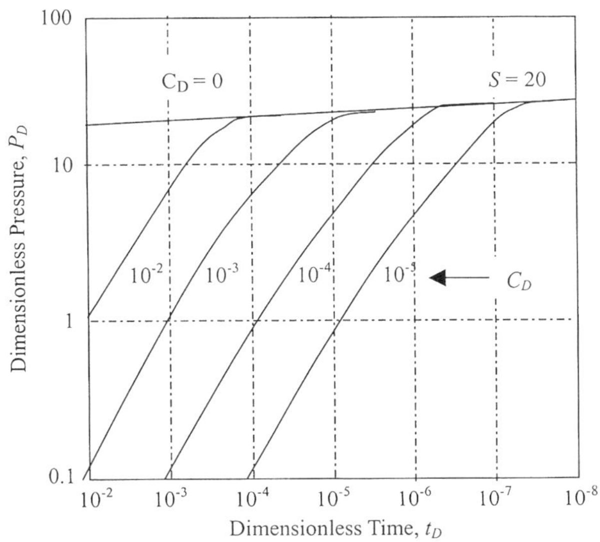

Figure 5–7. Type curves to determine end of wellbore storage distortion.[2]

Pseudo-Steady-State Flow Regime

When a constant-rate drawdown test is run for a long period of time, the boundary effects eventually dominate the pressure behavior at the well. The pressure starts declining at the same rate at all points in the reservoir; hence the name pseudo-steady-state. In effect, then, the total drainage area is being depleted at a constant rate. A plot of Δp_D versus t_D on arithmetic coordinates will yield a straight line from which the reservoir pore volume occupied by gas and the reservoir limits can be calculated. Tests utilizing this regime of the drawdown history are often known as reservoir limit tests.

Type Curve Applications to Drawdown Testing

From early-time to pseudo-steady-state can be combined and expressed graphically. Such graphic representations give type curves. The most useful type curves are Figures 5–2 through 5–7. A type curve analysis essentially consists of matching the test data to the appropriate type curve. When a match is obtained, the coordinates of the axes of the data plot and the type curve plot are said to correspond to each other, providing the scales of these axes also correspond. The use and merits of type curve analysis are discussed by Ramey[12] and by several of the authors mentioned in the previous sections. The use of type curves for determining the time of start of the transient flow period and reservoir parameters are discussed in next section.

5.5 Pressure–Time Behavior in Gas Wells with Horizontal and Vertical Fractures

Wattenbarger[3] studied the effects of a vertical fracture and confirmed that the net effect is equivalent to an effective wellbore radius equal to $x_f/2$, where x_f is the distance from the midpoint of the well to the tip of the fracture. In terms of a skin, this is equivalent to a negative skin factor s and given by

$$s = In\left(\frac{2r_w}{x_f}\right) \tag{5–3}$$

When flow into the fracture first starts, it is linear, and the pressure behavior is proportional to $\sqrt{t_D}$. This means that a plot of Δp_D versus t_D on logarithmic coordinates will give a straight line of slope 0.5 for early-time data. Such a plot would then deviate from these characteristics of the transient region. Gringarten, Ramey and Raghavan[4] presented the effects of a vertical hydraulic fracture in the form of the type curves of Figures 4–40 and 4–41. It is essentially a combination of the linear and radial flow equations. These curves can also be used to define the time of start of transient flow by noting from Figure 4–40 or 4–41 that transient flow starts at a time given by

$$t \cong \frac{15{,}168.75\phi\mu_i c_i x_f^2}{k} \tag{5–4}$$

The pressure–time behavior in infinite-acting reservoirs with horizontal or vertical fractures is shown in Figure 5–8. The dimensionless formation thickness, h_D, is defined as

$$h_D = \frac{h}{r_f}\left(\frac{k_r}{k_z}\right)^{0.5} \tag{5–5}$$

Figure 5–8 compares the behavior of horizontal and vertical fractures, and in the horizontal fracture case the transition zone involves a change in slope from 0.5 toward 1.0 for values of h_D less than 0.7. In practice, h_D is often less than 0.7, but it is quite difficult to recognize the characteristic of transition for horizontal fractures unless data are taken early enough during a flow test. Analysis of early-time data and short flow tests is described in the next section.

5.6 Uses of Pressure Drawdown Tests

Producing the well at a constant flow rate while continuously recording bottom-hole pressure runs the drawdown test. In this type of test, well-completion data details must be known so the effect and duration of wellbore storage may be estimated. While most reservoir information obtained from a

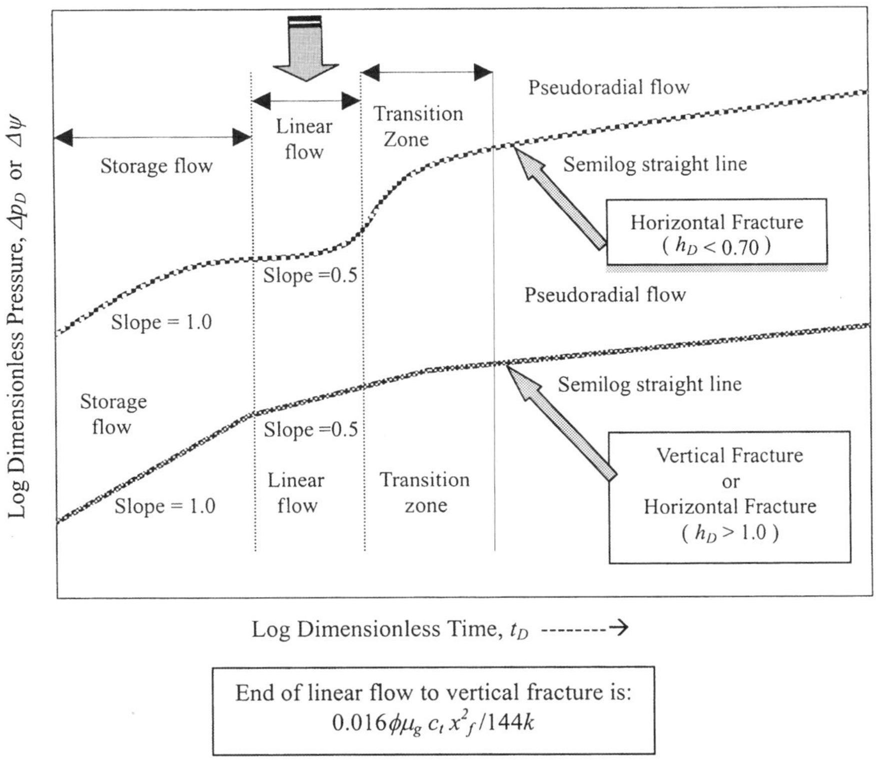

Figure 5–8. Pressure-time behaviors in infinite-acting reservoirs with horizontal and vertical fractures (after Gringarten, Ramey, and Raghavan).[4]

drawdown test also can be obtained from a pressure buildup test (Chapter 6), there is an economic advantage to drawdown testing since the well is produced during the test. Properly run drawdown tests may provide information about formation permeability k, skin factor s, and the reservoir volume communicating with the well. The main technical advantage of drawdown testing is the possibility for estimating reservoir volume. The major disadvantage is the difficulty of maintaining a constant production rate.

5.7 Analysis of Early-Time Flow Data

In this region a pressure transient is moving through the formation nearest the well bore. Early-time data may be used to determine the time of start of transient flow. In some instances, however, the available data are not amenable to a transient analysis, in which case it becomes necessary to analyze early-time data. Type-curve matching techniques are suited to this purpose. In unfractured wells, the early-time data are controlled by wellbore storage and skin effect.[5,12] Figures 5–2 through 5–7 are particularly useful for analyzing wellbore storage controlled early-time data. The theory of Ramey's type curves leads to the following procedure for using the curves for test analysis:

Step 1. Plot pressure change versus time on log-log paper the same size as Ramey's type curve on tracing paper. This plot is referred to as the data plot.

Step 2. If the test has a uniform slope region (45° line at earliest times), choose any point t (change in time) on the unit-slope line and calculate the wellbore storage constant C_s:

For $p < 3000$ psia:

$$C_s = \frac{q_{sc} \times 10^6 \times \beta_g}{24}\left(\frac{t}{p_i - p_{wf}}\right)_{unit\ slope\ line} \tag{5–6}$$

where

$$\beta_g = \frac{p_{sc}}{T_{sc}} \times \frac{T}{p} \times z\ \frac{ft^3}{scf}$$

For $p > 2000$ psia:

$$C_s = \frac{q_{sc} \times 10^6 \times \beta_g}{24}\left(\frac{t}{p_i^2 - p_{wf}^2}\right)_{unit\ slope\ line} \tag{5–7}$$

In terms of pseudopressure, as a high-order accuracy approximation,

$$C_s = \frac{q_{sc} \times 10^6 \times \beta_g}{24}\left(\frac{t}{[\Psi(p_i) - \Psi(p_{wf})]}\right)_{unit\ slope\ line} \tag{5–8}$$

Then calculate the dimensionless wellbore storage constant:

$$C_{SD} = \frac{0.159 C_s}{\phi c_t h r_w^2} \tag{5–9}$$

If a unit-slope line is not present, C_s and C_{SD} must be calculated from wellbore properties, and inaccuracy may result if these properties do not describe actual test behavior.

Step 3. Using type curves with C_{SD} as calculated in step 2, find the curve that most nearly fits all the plotted data. This curve will be characterized by some skin factor s; record its value. Interpolation between curves should improve the precision of the analysis, but may prove difficult.

Step 4. With the actual test data plot placed in the position of best fit, record corresponding values from any convenient match point. To show this, we note that dimensionless quantities are

$[(P_i - P_{wf}), P_D]$ and (t, t_D)
$[(P_i^2 - P_{wf}^2), P_D]$ and (t, t_D)
$[\psi(P_i) - \psi(P_{wf}), P_D]$, and (t, t_D)

$$t_D = \frac{0.000264kt}{\phi \mu_{gi} C_{gi} r_w^2} \tag{5–10}$$

$$p_D = \frac{kh(p_i - p_{wf})}{141.2\, q_{sc} \mu_{gi} \beta_{gi}} \tag{5–11}$$

$$p_D = \frac{khT_{sc}\left(p_i^2 - p_{wf}^2\right)}{50{,}300\, p_{sc} q_{sc} T} \tag{5–12}$$

$$\Psi_D = \frac{khT_{sc}[\Psi(P_i) - \Psi(P_{wf})]}{50{,}300\, p_{sc} q_{sc} T} \tag{5–13}$$

$$s' = s + D\,|q_{sc}| \tag{5–14}$$

$$\Psi(p) = 2\int_{p_b}^{p} \frac{p}{\mu(p)z(p)}\,dp \tag{5–15}$$

where

$$\beta_{gi} = 0.00504\,\frac{z_i T}{p_i}\ \text{rb/mmscf} \tag{5–16}$$

Step 5. Calculate k and ϕc_t from match point, MP, using the following equations:

$$k = 141.2 \frac{q_{sc} \mu_{gi} \beta_{gi}}{h} \left(\frac{p_D}{p_i - p_{wf}}\right)_{MP,} \tag{5–17}$$

Equation 5–17 may be written as

$$k = 50{,}300\,\frac{p_{sc} q_{sc} \mu_{gi} z_i T}{T_{sc} h} \left(\frac{p_D}{p_i^2 - p_{wf}^2}\right)_{MP} \tag{5–18}$$

$$k = \frac{50{,}300\, p_{sc} q_{sc} T}{hT_{sc}} \left(\frac{\Psi_D}{[\Psi(p_i) - \Psi(p_{wf})]}\right)_{MP} \tag{5–19}$$

$$\phi C_{ti} = \frac{0.000264k}{\mu_{gi} r_w^2} \left(\frac{t}{t_D} \right)_{MP} \tag{5–20}$$

Step 6. Compare those with values used to determine C_{SD} from C_S.

In summary, the procedure outlined in steps 1 through 6 provides estimation of k, s, and C_s in terms of pressure, pressure squared, and pseudopressure cases. Next example will clarify the use of early-time drawdown test data.

Example 5–1[21] *Drawdown Test Analysis Using Ramey's Type Curves*

Determine k, s, and C_s from the data below and in Table 5–1, which were obtained in a pressure drawdown test on a gas well: $P_I = 3000$, $s_{wi} = 0.22$, $V_W = 290$ cuft, $h = 12$ ft, $T = 210°$F, $r_w = 0.39$ ft, $q_g = 1000$ mcf/d, $P_{sc} = 14.65$ psia, $T_{sc} = 520°$R, $C_{ti} = 0.000333$ psi^{-1}, $\phi = 0.20$, $\mu_i = 0.01925$ cP, $\gamma_g = 0.655$, $D = 11{,}000$ ft, Drainage area = 640 acres (square); well is centered in drainage area.

Solution We must first prepare the data for plotting (Table 5–1). The data are plotted in Figure 5–9.

From Eq. 5–2, find

$$\begin{aligned} C_S &= V_{WS} C_{WS} = 11{,}000 \times 22/7 \times 0.39^2 \times 0.000333 \\ &= 1.751 \text{ psi}^{-1} \end{aligned}$$

Then, from Eq. 5–6,

$$C_{SD} = \frac{0.159\, C_S}{\phi h c r_w^2} = \frac{0.159 \times 1.751}{0.2 \times 12 \times 0.000333 \times 0.39^2} = 2290 \approx 10^3$$

For $C_{SD} = 10^3$, the best fitting type curve is for $s = 10$. A time match point is $t = 1$ hr when $t_D = 4.8 \times 10^3$. A pressure match point is $\psi(p_i) - \psi(p_{wf}) = 660 \times 10^6$, when $\psi(p_D) = 0.107$. From the match point, we also note that wellbore storage distortion ends at $t = 0.5$ hr (i.e., the type curve for $C_{SD} = 10^3$ becomes identical to the type curve for $C_{SD} = 0$). Using Eq. 5–16 find permeability, k from the pressure match point, and from the time match point, using Eq. 5–20, find ϕc. We also note that wellbore storage distortion ends at $t = 0.5$ hr (i.e., the type curve for $C_{SD} = 10^3$ becomes identical to the type curve for $C_{SD} = 0$).

Table 5–1
Single-Rate Drawdown Test Data for Ramey's Type Curve Analysis

Time t hr	Flowing pressure p_{wf} (psia)	$\psi(p_{wf})$ (mmpsia^2/cP)	$\Delta\psi = \psi(p_i) - \psi(p_{wf})$
0.02	1810.65	221.41	639.71
0.03	1807.45	220.68	640.44
0.07	1798.95	218.74	642.38
0.10	1786.35	215.87	645.24
0.17	1775.75	213.47	647.64
0.25	1768.05	211.74	649.38
0.33	1764.75	211.00	650.12
0.50	1757.45	209.36	651.76
0.75	1754.65	208.73	652.38
1.00	1755.45	208.91	652.20
1.50	1757.85	209.45	651.67
2.00	1754.65	208.73	652.38
2.50	1754.65	208.73	652.38
3.00	1751.35	208.00	653.12
3.50	1748.95	207.46	653.66
4.00	1747.35	207.10	654.01
5.00	1745.25	206.64	654.48
5.50	1742.05	205.92	655.19
6.00	1740.45	205.57	655.55
6.50	1739.25	205.30	655.82
7.00	1738.35	205.10	656.02
7.50	1738.35	205.10	656.02
8.00	1737.95	205.01	656.10
8.50	1737.55	204.92	656.19
9.00	1737.15	204.83	656.28
9.50	1737.15	204.83	656.28
10.00	1735.55	204.48	656.64
11.00	1735.55	204.48	656.64
12.00	1734.35	204.21	656.90
13.00	1733.55	204.04	657.08
14.00	1733.55	204.04	657.08
15.00	1732.75	203.86	657.26
16.00	1731.05	203.48	657.63
17.00	1730.65	203.39	657.72
18.00	1730.25	203.31	657.81
19.00	1729.45	203.13	657.99
20.00	1719.75	200.99	660.13
21.00	1724.55	202.05	659.07
22.00	1721.75	201.43	659.69
23.00	1720.95	201.25	659.86
24.00	1720.95	201.25	659.86

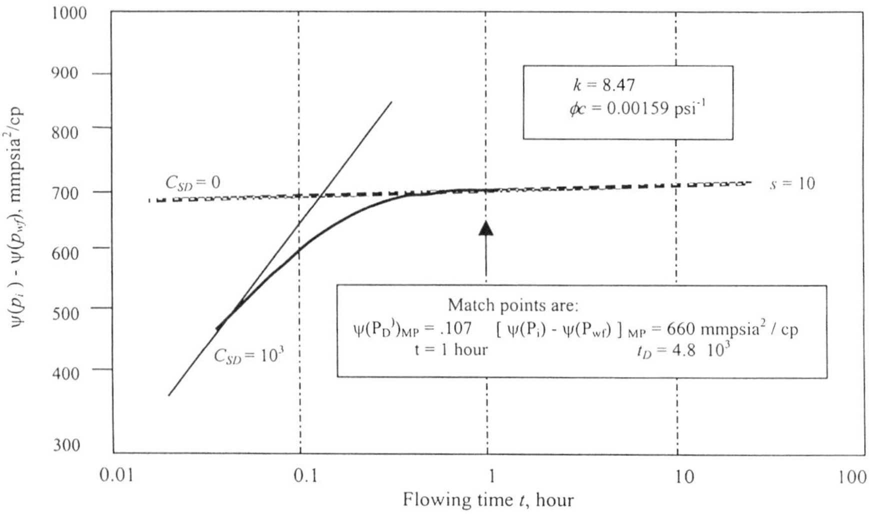

Figure 5–9. Drawdown test analysis with Ramey's type curve.

$$k = \frac{50,300\ P_{sc} q_{sc} T}{h\ T_{sc}} \left[\frac{\psi_D}{\psi(p_i) - \psi(p_{wf})} \right]_{MP}$$

$$= \frac{50,300 \times 14.65 \times 1.0 \times 10^3 \times 670}{12 \times 52} \left[\frac{0.107}{660 \times 10^6} \right] = 8.47 \text{ mD}$$

From the match point, using Eq. 5–20, find ϕc:

$$\phi c = \frac{.0002637\, k}{\mu r_w^2} \left[\frac{t}{t_D} \right]_{MP} = \frac{.0002637 \times 8.47}{.01925 \times .39^2} \left[\frac{1}{4.8 \times 10^3} \right]$$

$$= .000159 \text{ psi}^{-1}$$

Compare those values of ϕc with values used to determine C_{SD} and C_S, as shown in Figure 5–9:

$$\phi c = \frac{0.159\, C_S}{h r_w^2 C_{SD}} = \frac{0.159 \times 1.751}{12 \times 0.39^2 \times 10^3} = 0.000153 \text{ psi}^{-1}$$

Check

$$\phi c = 0.2 \times 0.000333 = 0.0001 \text{ psi}^{-1}$$

5.8 Estimating Formation Characteristics from Transient Flow Test Data

Early-time data may be used to determine when transient flow theory becomes applicable, with well-bore storage effects. Data should be in the transient flow regime since reservoir parameters calculated by transient flow analysis are far more reliable than those calculated by an early-time flow analysis. Various drawdown tests, utilizing transient flow data, that may be used to determine well and reservoir parameters are discussed in detailed along with examples illustrating the analysis procedures.

Single-Rate Drawdown Test

This test consists of flowing the well at a constant rate and continuously recording the flowing bottom-hole pressure p_{wf} as a function of time of flow, t. Flow starts from stabilized shut-in conditions. The data obtained from a single-rate test may be analyzed as described below to give values of kh and apparent skin factor, s'. s' is composed of two parts: s due to the well completion, and $D'q_{sc}$ due to turbulence effects. The values of s and D may be obtained separately from two-rate tests, discussed in next section. For analyzing pressure drawdown tests, we plotted $(p_i - p_{wf})$ versus log t on semilogarithmic coordinates, and $\log(p_i - p_{wf})$ versus log t using the log-log plot to identify the beginning of transient flow period. To analyze pressure drawdown tests in gas reservoirs, the ordinates of the plots mentioned above may be $p_i - p_{wf}$, $p_i^2 - p_{wf}^2$, or the pseudopressure $\Psi(p_i) - \Psi(p_{wf})$. Now the question arises as when to use which pressure. The rule of thumb is to use:

1. $p_i - p_{wf}$ if reservoir pressures are greater than 3000 psi
2. $p_i^2 - p_{wf}^2$ if reservoir pressures are less than 2000 psi
3. $\Psi(p_i) - \Psi(p_{wf})$ if 1 and 2 are not valid or may be used in all cases

Wattenbarger and Ramey[6] proposed plotting the product $\mu_g z$ versus pressure p_{wf} as shown in Figure 5–10.

1. If the variation in the $\mu_g z$ product with pressure is linear, use $p_i - p_{wf}$.
2. If the variation in the $\mu_g z$ product with pressure is small, use $p_i^2 - p_{wf}^2$.
3. If 1 and 2 are not applicable, then the $\Psi(p_i) - \Psi(p_{wf})$ approach should be used.

Pressure drawdown equations and solved examples are provided for each case to clarify the analysis techniques. All other rules suggested for identifying the early-time region, transient flow region, and pseudo-steady-state region are also applicable.

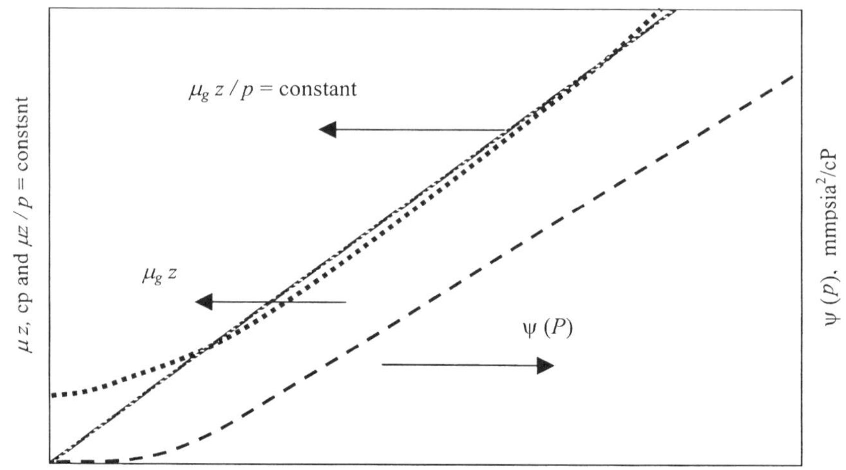

Figure 5–10. Variation of ψ and μz with pressure.[6]

Drawdown Test Analysis Using p_{wf} Approach

After the transient flow region is identified, the following equations are used when the use of p_{wf} is appropriate to analyze a gas well drawdown test. Equation 5–21 models transient flow at constant rate from an infinite-acting gas reservoir.

$$p_i - p_{wf} = \frac{162.6 q_{sc}\,\beta_{gi}\,\mu_{gi}}{kh}\left[\log\left(\frac{1{,}688\,\phi\,\mu_{gi}\,C_{ti}}{ht}\right)\right] - \frac{(s + D\,q_{sc})}{1.151} \tag{5–21}$$

$$kh = \frac{162.6 q_{sc}\,\beta_{gi}\,\mu_{gi}}{m} \tag{5–22}$$

$$s' = 1.151\left[\frac{p_i - p_{1hr}}{m} - \log\frac{k}{\phi\,\mu_{gi} C_{ti} r_w^2} + 3.23\right] \tag{5–23}$$

where q_{sc} is conveniently expressed in mmscfd, and the gas formation volume factor, β_{gi}, is then expressed in reservoir barrels per mmscf, so that the product $q_{sc}\beta_{gi}$ is in reservoir barrels per day (rb/d) as in the analogous equation for slightly compressible liquids. All gas properties (β_{gi}, μ_{gi}, and C_g) are evaluated at original reservoir pressure, p_i. In Eq. 5–21,

$$\beta_{gi} = \frac{0.1781\, z_i T p_{sc}}{p_i T_{sc}} \text{ (rb/mmscfd)} \tag{5–24}$$

$$\begin{aligned} C_{ti} &= C_{gi} S_g + C_w S_w + C_f \\ &\cong C_{gi} S_g \end{aligned} \tag{5–25}$$

The factor D is a measure of non-Darcy or turbulent pressure loss (i.e., a pressure drop in addition to that predicted by Darcy's law). It cannot be calculated separately from the skin factor from a single buildup or drawdown test; thus, the concept of apparent skin factor $s' = s + Dq_{sc}$, is sometimes convenient since it can be determined from a single test. A plot of $\Delta p = (p_i - p_{wf})$ versus t on semilogarithmic coordinates should give a straight line of slope m, from which formation permeability can be calculated. The apparent skin factor s' can then be calculated using Eq. 5–23, where the value of p_{1hr} must be obtained from the straight-line portion of the semilog plot (extrapolated, if necessary). The pressure drop due to skin effects may be obtained from

$$(\Delta p)_{skin} = 0.869\, ms' \tag{5–26}$$

Similarly, the pressure drop due to IT flow effects may be obtained from

$$(\Delta p)_{IT} = 0.869\, m D q_{sc} \tag{5–27}$$

The total pressure drop may then be obtained from

$$(\Delta p)_{s'} = 0.869\, ms' = 0.869\, m\,(s + Dq_{sc}) \tag{5–28}$$

The well flow efficiency, *FE*, is defined as the ratio of the drawdown at the well, without skin or IT flow effects, to the actual drawdown and may be calculated from

$$FE = \frac{(p_i - p_{wf}) - (\Delta p)_{s'}}{(p_i - p_{wf})} \tag{5–29}$$

Drawdown Test Analysis Using p_{wf}^2 Approach

The most useful solution for transient flow is the so-called line source solution. The solution is

$$\Delta p_D = 0.5\,(Int_D + 0.909) \tag{5–30}$$

Equation 5–30 may be written including formation damage and turbulence effects as

$$\Delta p_D = 0.5\,(Int_D + 0.809) + s' \tag{5–31}$$

where

$$s' = s + Dq_{sc} \tag{5–32}$$

In terms of real variables and common logs, Eq. 5–31 becomes

$$p_i^2 - p_{wf}^2 = \frac{57.920 \times 10^6 q_{sc} T \bar{\mu}_g \bar{z} p_{sc}}{khT_{sc}} \left[\log \frac{k}{\phi \bar{\mu}_g \bar{c}\, r_w^2} - 3.23 + 869s'\right] \tag{5–33}$$

A plot of $\Delta p^2 (= p_i^2 - p_{wf}^2)$ versus t on semilogarithmic coordinates should give a straight line of slope m, from which

$$m = \frac{57.920 \times 10^6 q_{sc} T \bar{\mu}_g \bar{z} p_{sc}}{khT_{sc}} \tag{5–34}$$

From this, kh can be calculated. To obtain s', let $t = 1$ hr $(\log 1 = 0)$. Then

$$p_i^2 - p_{1hr}^2 = m\left[\log \frac{k}{\phi \bar{\mu}_g \bar{c} r_w^2} + 3.23 + 0.869\, s'\right] \tag{5–35}$$

where p_{1hr} is obtained from an extrapolation of the linear segment of the plot. Solving for s' in Eq. 5–36 gives

$$s' = 1.151\left[\frac{p_i^2 - p_{1hr}^2}{m} - \log \frac{k}{\phi \bar{\mu}_g \bar{c} r_w^2} + 3.23\right] \tag{5–36}$$

Since s' is rate dependent, two single-rate drawdown tests may be conducted to determine s and D. The removable pressure drop due to actual damage can be calculated from

$$\Delta(p^2)_s = 0.869\, ms \tag{5–37}$$

and the rate dependent pressure drop from

$$\Delta(p^2)_D = 0.869\, m D q_{sc} \tag{5–38}$$

Drawdown Test Analysis Using Pseudopressure Approach

Transient flow at constant rate from an infinite-acting reservoir in terms of pseudopressure $\Psi(p_{wf})$ is modeled by Eq. 5–39:

$$\Psi(p_i) - \Psi(p_{wf}) = 57.920 \times 10^6 \frac{q_{sc} T p_{sc}}{khT_{sc}}\left[\log t + \log\left(\frac{k}{\phi \mu_{gi} c_i r_w^2}\right) - 3.23 + 0.869 s'\right] \tag{5–39}$$

A plot of $\Delta\Psi[= \Psi(p_i) - \Psi(p_{wf})]$ versus t on semilog coordinates should give a straight line of slope m, from which

$$kh = \frac{57.920 \times 10^6 q_{sc} T p_{sc}}{m T_{sc}} \tag{5–40}$$

The apparent skin factor s' can then be calculated using Eq. 5–41:

$$s' = 1.151 \left[\frac{\Delta\Psi_1}{m} - \log\left(\frac{k}{\phi \mu_{gi} c_i r_w^2} \right) + 3.23 \right] \tag{5–41}$$

where $\Delta\psi_1$ is the value of $\Delta\Psi$ at $t = 1$. This value must be obtained from the straight-line portion of the plot (extrapolated, if necessary). The pressure drop due to skin effects may be obtained from

$$(\Delta\psi)_s = 0.869\, ms' = 0.869\, m(s + Dq_{sc}) \tag{5–42}$$

Similarly, the pressure drop due to IT flow effects may be obtained from

$$(\Delta\Psi)_{IT} = 0.869\, m\, Dq_{sc} \tag{5–42a}$$

The total pressure drop directly attributed to skin and IT flow effects may then be obtained from

$$(\Delta\Psi)_{S'} = 0.869\, ms' = 0.869\, m\,(s + Dq_{sc}) \tag{5–43}$$

The well flow efficiency, *FE*, may be calculated from

$$FE = \frac{\lfloor \Psi(p_i) - \Psi(p_{wf}) \rfloor - (\Delta\Psi)_{s'}}{\lfloor \Psi(p_i) - \Psi(p_{wf}) \rfloor} \tag{5–44}$$

Sometimes it is convenient to express the drawdown in dimensionless forms. This is easily done as follows:

$$\Delta\Psi_D = \frac{\lfloor \Psi(p_i) - \Psi(p_{wf}) \rfloor}{0.869\, m} \tag{5–45}$$

The analysis of a single-rate test in terms of p_{wf}, p_{wf}^2, and $\Psi(p_{wf})$ is illustrated by the following Example 5–2.

Example 5–2[21] *Single-Rate Drawdown Test Analysis Using Pseudopressure Approach*

A gas well in an infinite-acting reservoir was produced at a constant rate of 6.148 mmscfd. The pressure, p_I, throughout the reservoir prior to the test was 3965 psia. General data pertinent to the test are given below. From the recombined gas analysis: $CO_2 = 7.84\%$, $H_2S = 0.0\%$, $N_2 = 0.11\%$,

$G = 0.732$, $P_c = 380.16$ psia, $T_c = 645.06°$R, $\mu_I = 0.02458$ cP, $C_i = 0.00023$ psi^{-1}. Well/reservoir data = 710°R, $h = 41$ ft, $r_e = 2200$ ft, $r_w = 0.4271$ ft, $s_w = 0.267$, $s_g = 0.733$, $c_w = 3.2 \times 10^{-6}$ psi^{-1}, $C_g = 0.000252$ psi^{-1}, $c_f = 3.9 \times 10^{-6}$ psi^{-1}, $\phi = 0.137$, $\phi_{HC} = 0.1004$. Calculate permeability, apparent skin factor s', and pressure drop across skin. Also calculate the flow efficiency of the well.

Solution From Figure 5–11, find the following: $P_i = 3965$ psia $\leftrightarrow \psi(p_i) = 861.12$ mmpsia^2/cP; $P_{wfo} = 1811$ psia $\leftrightarrow \psi(p_{wfo}) = 221.41$ mmpsia^2/cP. Using this value of $\psi(p_i)$ and Figure 5–12, tabulations may be made (Table 5–2).

To analyze this single-rate drawdown test follow these steps:

1. Plot $\Delta\psi$ versus t on 3 × 5 log-log graph paper (the same as for the type curves of Figure 5–2).
2. A match of the above drawdown plot with the type curve $s = 5$, $C_{SD} = 10^3$ of Figure 5–12 indicates that the time of start of the semilog straight-line data is approximately 9 hr.
3. Plot $\Delta\psi$ versus log t and draw the best straight line through the semilog straight-line data identified in step 2, as shown in Figure 5–13.

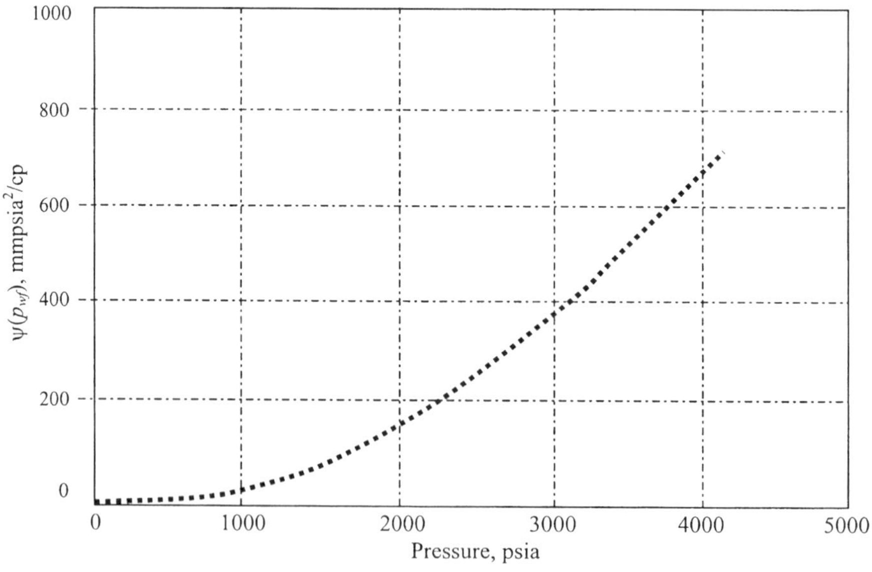

Figure 5–11. $\psi - p$ curve—Example 5–2.

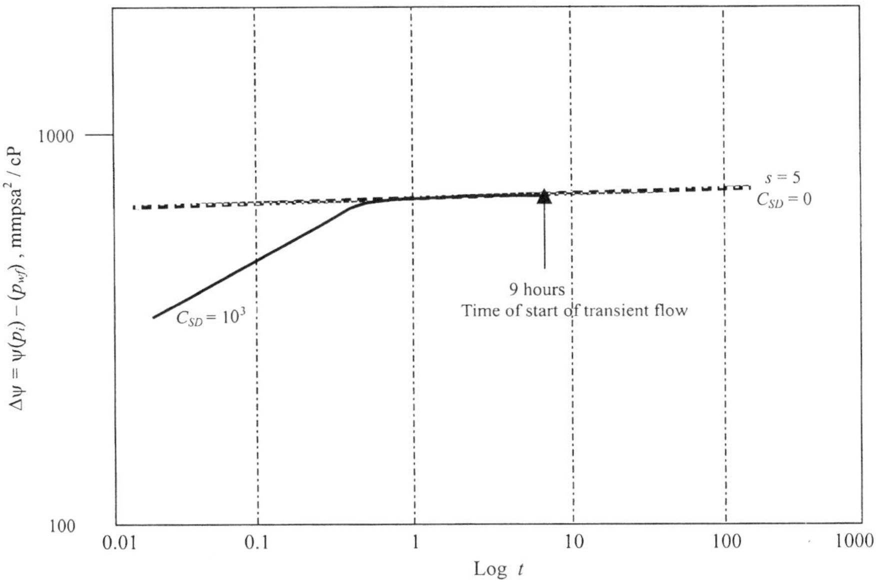

Figure 5–12. Single-rate drawdown test analysis using pseudopressure approach.

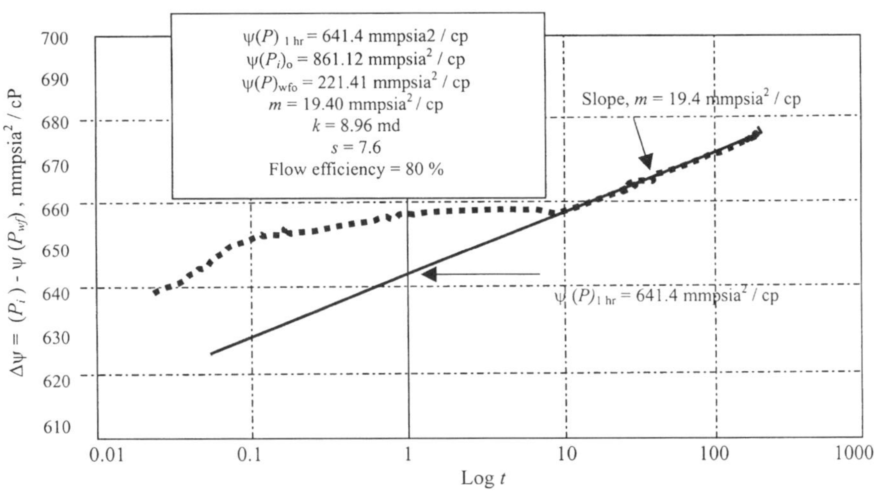

Figure 5–13. Single-rate drawdown test analysis using pseudopressure approach.

Table 5–2
Single-Rate Drawdown Test Data

Time t hr	Flowing pressure p_{wf} (psia)	$\psi(p_{wf})$ (mmpsia^2/cP)	$\Delta\psi = \psi(p_i) - \psi(p_{wf})$ (mmpsia^2/cP)
0.0	1810.65	221.41	639.71
0.03	1807.45	220.68	640.44
0.07	1798.95	218.74	642.38
0.10	1786.35	215.87	645.24
0.13	1781.45	214.76	646.35
0.17	1775.75	213.47	647.64
0.25	1768.05	211.74	649.38
0.33	1764.75	211.00	650.12
0.50	1757.45	209.36	651.76
0.75	1754.65	208.73	652.38
1.00	1755.45	208.91	652.20
1.50	1757.85	209.45	651.67
2.00	1754.65	208.73	652.38
2.50	1754.65	208.73	652.38
3.00	1751.35	208.00	653.12
3.50	1748.95	207.46	653.66
4.00	1747.35	207.10	654.01
5.00	1745.25	206.64	654.48
5.50	1742.05	205.92	655.19
6.00	1740.45	205.57	655.55
6.50	1739.25	205.30	655.82
7.00	1738.35	205.10	656.02
7.50	1738.35	205.10	656.02
8.00	1737.95	205.01	656.10
8.50	1737.55	204.92	656.19
9.00	1737.15	204.83	656.28
9.50	1737.15	204.83	656.28
10.0	1735.55	204.48	656.64
11.0	1735.55	204.48	656.64
12.0	1734.35	204.21	656.90
13.0	1733.55	204.04	657.08
14.0	1733.55	204.04	657.08
15.0	1732.75	203.86	657.26
16.0	1731.05	203.48	657.63
17.0	1730.65	203.39	657.72
18.0	1730.25	203.31	657.82
19.0	1729.45	203.13	657.99
20.0	1719.75	200.99	660.13
21.0	1724.55	202.05	659.07
22.0	1721.75	201.43	659.69

Table 5–2 (Continued)

Time t hr	Flowing pressure p_{wf} (psia)	$\psi(p_{wf})$ (mmpsia^2/cP)	$\Delta\psi = \psi(p_i) - \psi(p_{wf})$ (mmpsia^2/cP)
23.0	1720.95	201.25	659.86
24.0	1720.95	201.25	659.86
30.0	1720.51	199.12	662.00
40.0	1720.21	197.92	663.20
50.0	1720.15	196.42	664.70
60.0	1720.10	195.62	665.50
70.0	1720.00	195.12	666.00
80.0	1719.80	193.72	667.40
90.0	1719.30	192.62	668.50
100.0	1719.00	191.92	669.20
110.0	1718.80	191.62	669.50
120.0	1718.40	191.12	670.00
150.0	1718.10	189.62	671.50
180.0	1717.8	188.32	672.80
200.0	1717.5	188.12	673.00

4. Find the absolute value of the slope m of the MTR line and $\Delta\psi$ $(p_{wf\,at\,1\,hour})$:

$$m = \frac{(669.2 - 649.8) \times 10^6}{\log 100 - \log 10} = 19.40 \frac{\text{mmpsia}^2/\text{cP}}{\text{cycle}}$$

and

$$\Delta\psi\ (p_{wf\,at\,1\,hour}) = 641.4\ \text{mmpsia}^2/\text{cP}$$

5. Find formation permeability k from Eq. 5–40:

$$k = \frac{57.920 \times 10^6\,(6.148)(710)(14.65)}{19.40 \times 10^6(41)(520)} = 8.96\,\text{mD}$$

6. Estimate apparent skin factor s' from Eq. 5–41:

$$s' = 1.151\left[\frac{(861.12 - 641.4)10^6}{19.4 \times 10^6} - \log\left(\frac{8.96}{0.1004 \times 0.0235 \times 0.00023 \times 0.4271^2}\right) + 3.23\right]$$
$$= 7.60$$

7. Determine the pressure drop across the skin from Eq. 5–42:

$$(\Delta\psi)_{skin} = 0.869(19.40)(7.60) = 128.13 \text{ mmpsia}^2/\text{cP}$$

8. Calculate the well flow efficiency *FE* from Eq. 5–44:

$$FE = \frac{861.12 - 221.41 - 128.13}{861.12 - 221.41} \times 100 = 80\%$$

Two-Rate Drawdown Test

Two-rate test consists of flowing the well at a constant rate q_{sc1} for a period of time t, and then changing the flow rate to q_{sc2}. The first rate is usually the actual production rate of the well. Before the flow rate is changed, the flowing bottom-hole pressure is measured with a bottom-hole pressure gauge and flowing bottom-hole pressure after the rate change is recorded continuously. Such data may be analyzed by the methods of the single-rate test analysis to obtain kh and s'. It should be noted that the duration t of the first flow must be long enough to ensure that it is in the transient flow regime. Two single-rate tests are necessary to determine the IT flow factor using the following equations:

$$s_1' = s + Dq_{sc1} \tag{5–46}$$

$$s_2' = s + Dq_{sc2} \tag{5–47}$$

Solving the simultaneous equations:

$$D = \frac{s_1' - s_2'}{q_{sc1} - q_{sc2}} \tag{5–48}$$

$$s = s_1' - \frac{\left(s_1' - s_2'\right)}{\left(q_{sc1} - q_{sc2}\right)} q_{sc1} \tag{5–49}$$

where s may be positive (well damage) or negative (well improvement); D must always be positive. Zero replaces it, and s becomes the average of s_1' and s_2'. Example 5–3 will clarify the use of two sets of single-rate drawdown test data. When wellbore storage effects are significant, a two-rate test has a definite advantage: a two-rate test eliminates the problems caused by redistribution of the gas and liquid phases, and in fact it has become the standard test in some instances.[13] The analysis of such a test will give kh, s, and D if p_i is available. If p_i is not available, the analysis will yield kh, s, and p_i. Pressure response obtained by changing the flow rate from q_{sc1} to q_{sc2} may be analyzed by applying the principle of superposition in time. For the second flow period of a two-rate test, the pseudopressure drawdown is given by

$$\Psi(p_i) - \Psi(p_{wf}) = 57.920 \times 10^6 \frac{q_{sc1}Tp_{sc}}{khT_{sc}} \left[\log \frac{(t_1 + \Delta t)}{\Delta t} + \frac{q_{sc2}}{q_{sc1}} \log \Delta t\right]$$

$$+ 52.920 \times 10^6 \frac{q_{sc2}Tp_{sc}}{khT_{sc}} \left[\log\left(\frac{k}{\phi \mu_{gi} c_i r_w^2}\right) - 3.23 + 0.869 s_2'\right] \qquad (5\text{–}50)$$

A plot of $\Psi(p_{wf})$ versus $\log(\frac{t_1+\Delta t}{\Delta t}) + \frac{q_{sc2}}{q_{sc1}} \log \Delta t$ on arithmetic coordinates should give a straight line of slope m, from which

$$kh = \frac{57.920 \times 10^6 q_{sc1} Tp_{sc}}{mT_{sc}} \qquad (5\text{–}51)$$

The apparent skin factor, s_2', associated with the flow rate, q_{sc2}, may then be calculated from

$$q_{sc1}\, s_1' - q_{sc2}\, s_2' = \frac{\lfloor \Psi(p_{wf1}) - \Psi(p_{wfo}) \rfloor q_{sc1}}{0.869\, m}$$

$$- \frac{q_{sc1} - q_{sc2}}{0.869} \left[\log\left(\frac{k}{\phi \mu_{gi} c_i r_w^2}\right) - 3.23\right] \qquad (5\text{–}52)$$

where s_1' = apparent skin factor associated with the flow rate q_{sc1}; $\psi(p_{wf1})$ = flowing bottom hole pseudopressure at $\Delta t = 1$, obtained from the straight line (extrapolated, if necessary); and $\psi(p_{wfn})$ = flowing bottom-hole pseudopressure at the time of changing the flow rate from q_{sc1} to q_{sc2}. To utilize Eq. 5–52, we need some additional information. Two alternative approaches may be considered.

Case 1: $\Psi(P_I)$ Is Known

Since the single-rate analysis applies to the first flow period of a two-rate test, the apparent factor s_1' related to the flow rate q_{sc1} may be obtained from Eq. 5–53:

$$s_1' = 1.151 \left[\frac{\Psi(p_i) - \Psi(p_{wfo})}{m} - \log\left(\frac{kt_1}{\phi\, \mu_{gi} c_i\, r_w^2}\right) + 3.23\right] \qquad (5\text{–}53)$$

where t_1 = time of changing the flow rate from q_{sc1} to q_{sc2}, i.e., time corresponding to the pseudopressure $\psi(p_{wfn})$. The apparent skin factor s_2' related to the flow rate q_{sc2} may be calculated as

$$s_2' = \frac{q_{sc1}}{q_{sc2}} s_1' - \frac{\lfloor \Psi(p_{wf1}) - \Psi(p_{wfo}) \rfloor}{0.869\, m q_{sc2}} q_{sc1} + \frac{(q_{sc1} - q_{sc2})}{0.869\, m q_{sc2}} \left[\log\left(\frac{k}{\phi \mu_{gi} c_i r_w^2}\right) - 3.23 \right] \tag{5–53a}$$

The skin factor s and the IT flow factor D may then be calculated using the above values of s_1', s_2', and Eqs. 5–46 through 5–49.

Case 2: $\Psi(p_i)$ Is Not Known

In this case, the skin and IT flow effects cannot be separated. However, $\Psi(p_i)$ may be estimated by assuming s_1' and s_2' to be equal to an average s', calculated from Eq. 5–52, which may be written as

$$s' = 1.151 \left[\left(\frac{q_{sc1}}{q_{sc1} - q_{sc2}} \right) \left(\frac{\Psi(p_{wf1}) - \Psi(p_{wfo})}{m} \right) - \log\left(\frac{k}{\phi \mu_{gi} c_i r_w^2} + 3.23 \right) + 3.23 \right] \tag{5–54}$$

and using the calculated value of s' becomes the following equation, which is a form of Eq. 5–53:

$$\Psi(p_i) = \Psi(p_{wfo}) + m \left[\log\left(\frac{k}{\phi \mu_{gi} c_i r_w^2} \right) - 3.23 + 0.869 s' \right] \tag{5–55}$$

$\Psi(p_i)$ may then be converted back to p_i.

Using Pressure-Squared Approach

A plot of $(p_i^2 - p_{wf}^2)$ versus $\log(\frac{t_1 + \Delta t}{\Delta t}) + \frac{q_{sc2}}{q_{sc1}} \log \Delta t$ on Cartesian coordinates will yield a straight line of slope

$$m = \frac{57.920 \times 10^6 q_{sc1} T \bar{\mu}_g \bar{z} p_{sc}}{k h T_{sc}} \tag{5–56}$$

If the data from the first flow period are analyzed to obtain s_1', a value for s_2' can be calculated from

$$q_{sc1} s_1' - q_{sc2} s_2' = \frac{q_{sc1} \left[p_{wf1}^2 - p_{wfo}^2 \right]}{0.869\, m} - \frac{q_{sc1} - q_{sc2}}{0.869} \left[\log \frac{k}{\phi \bar{\mu}_g \bar{c} r_w^2} - 3.23 \right] \tag{5–57}$$

where

p_{wf1}^2 is read at $\Delta t = 1$

p_{wfo}^2 is read at $\Delta t = 0$

$$s_1' = 1.151\left[\frac{p_i^2 - p_{1hr}^2}{m} - \log\frac{k}{\phi\bar{\mu}_g\bar{c}r_w^2} + 3.23\right] \tag{5–58}$$

and

$$s_2' = \frac{q_{sc1}}{q_{sc2}}s_1' - \frac{\lfloor p_{wf1}^2 - p_{wfo}^2\rfloor q_{sc1}}{0.869\, m\, q_{sc2}} + \frac{q_{sc1} - q_{sc2}}{0.869\, m q_{sc2}} \times\left[\log\left(\frac{k}{\phi\,\mu_{gi}c_i\,r_w^2}\right) - 3.23\right] \tag{5–59}$$

s can be determined:

$$s_1' = s + Dq_{sc1} \tag{5–46}$$

$$s_2' = s + Dq_{sc2} \tag{5–47}$$

From Eqs. 5–46 and 5–47:

$$D = \frac{s_1' - s_2'}{q_{sc1} - q_{sc2}} \tag{5–48}$$

$$s = s_1' - Dq_{sc1} \tag{5–49}$$

or

$$s = s_1' - \frac{(s_1' - s_2')}{(q_{sc1} - q_{sc2})}q_{sc1} \tag{5–49a}$$

The pressure drop due to actual damage can be calculated from

$$\Delta(p^2)_{skin} = 0.869\, ms \tag{5–60}$$

and the rate-dependent pressure drop from

$$\Delta(p^2)_D = 0.869\, m D\frac{q_{sc2}}{q_{sc1}} \tag{5–61}$$

If initial pressure p_i is not known, the skin factor s, can be determined:

$$s = 1.151\left[\frac{q_{sc2}}{q_{sc1} - q_{sc2}}\left(\frac{p_{1hr}^2 - p_{wfo}^2}{m}\right) - \log\left(\frac{k}{\phi\,\mu_g c\, r_w^2}\right) + 3.23\right] \tag{5–62}$$

Having found values for k and s, one may now proceed to determine p^*:

$$\overset{*}{p}{}^2 = p_{wf}^2(t_1) + (-m)\left[\left(\log \frac{k\,t_1}{\phi\mu_g c r_w^2}\right) - 3.23 + 0.869s\right] \tag{5–63}$$

or

$$\overset{*}{p}{}^2 = p_{wf}^2(\text{at intercept}) - \frac{q_{sc2}}{q_{sc1} - q_{sc2}}\left[p_{wf}^2(\Delta t = 0) - p_{1hr}^2\right] \tag{5–64}$$

To estimate pressure drop across the skin at rates q_{sc1} and q_{sc2} respectively, at q_{sc1},

$$\Delta(p^2)_{skin} = 0.869(-m)(s) \tag{5–65}$$

and at q_{sc2},

$$\Delta(p^2)_{skin} = 0.869(q_{sc2}/q_{sc1})\,(-m)\,(s) \tag{5–66}$$

A sample calculation of two-rate drawdown test is shown in Example 5–3.

Example 5–3[21] *Two-Rate Drawdown Test Analysis When p_i Is Known*

A two-rate test was conducted on a well in gas reservoir, $p_i = 2925$ psia. The pressure–time data for the first flow rate, $q_{sc1} = 28.000$ mmscfd, was not recorded. The flow rate was changed at $t_0 = 6$ hr, at which the flowing bottom-hole pressure p_{wfo} was 2505 psia. The second flow rate, $q_{sc2} = 21.3$ mmscfd, was continued for 52 hr during which time the following bottom-hole pressures were recorded continuously. These pressure–time data are given in the tabulations for the solution to this problem. Gas properties and well/reservoir data are as follows: $p_c = 380.16$; $T_c = 645.08°R$; $G = 0.732$; $CO_2 = 7.84\%$; $N_2 = 0.11\%$; $H_2S = 0.00\%$; $\mu_I = 0.0186$ cP; $c_i = 0.000274$ psi^{-1}; $T = 710°R$; $h = 41$ ft; $r_w = 0.25$; $\phi = 0.137$ fraction; $\phi_{HC} = 0.1004$; $s_g = 0.733$.

Calculate the following:

1. The formation permeability k
2. Apparent skin factors s_1' and s_2'
3. Inertial-turbulent flow factor D
4. True skin factor s
5. Pressure drop due to actual skin
6. Rate-dependent pressure drop

Solution The $\psi - p$ curve, shown in Figure 5–11, is applicable to this problem.

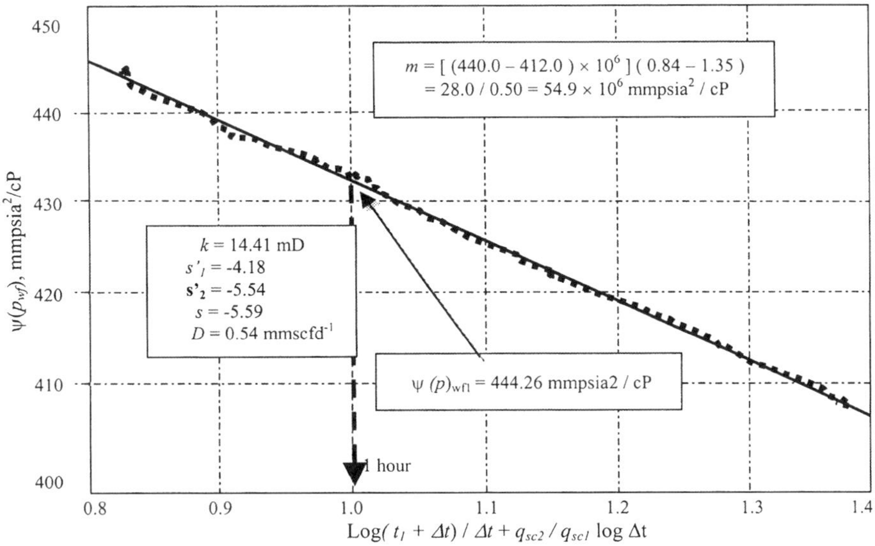

Figure 5–14. Data plot for two-rate drawdown test—Example 5–3.

Given

$$\psi(p_i) = 542.35 \text{ mmpsia}^2/\text{cP} \leftrightarrow 2{,}925 \text{ psia}$$
$$\psi(p_{wfo}) = 401.25 \text{ mmpsia}^2/\text{cP} \leftrightarrow 2{,}505 \text{ psia}$$

and $t_1 = 6$ hr, using the $\psi - p$ curve of Figure 5–11, the following tabulations may be made.

Plot $\psi(p_{wf})$ versus $\log(\frac{t_1+\Delta t}{\Delta t}) + \frac{q_{sc2}}{q_{sc1}} \log \Delta t$ on Cartesian coordinate paper and draw the best straight line as shown in Figure 5–14. From Figure 5–14, find the slope m:

$$m = \frac{(440.0 - 412.0) \times 10^6}{1.35 - 0.84} = 54.9 \times 10^6 \text{ mmpsia}^2/\text{cP}$$
$$\psi(p_{wf1})_{@\,\Delta t = 1hr} = 444.26 \text{ mmpsia}^2/\text{cP} \leftrightarrow 2683 \text{ psia}$$

1. From Eq. 5–51

$$kh = \frac{57.92 \times 10^6 q_{sc1} T P_{sc}}{m T_{sc}} = \frac{57.920 \times 28.0 \times 10^6 \times 710 \times 14.65}{54.9 \times 10^6 \times 520}$$
$$= 590.89 \text{ mD-ft}$$

$$\therefore\ k = 590.89/41 = 14.41 \text{ mD}$$

2. Since p_i is known, it is possible to separate the skin and IT effects.

Table 5–3
Calculations of Gas PVT Properties and Pseudopressure

Pressure (psia)	Compressibility factor z	Gas viscosity (cP)	$\psi(p)$ (mmpsia^2/cP)
4000	0.9647	0.024580	872.92
3600	0.9445	0.023151	739.56
3200	0.9292	0.021721	610.28
2800	0.9169	0.020329	486.72
2400	0.9113	0.019008	371.18
2000	0.9120	0.017784	266.41
1600	0.9189	0.016681	175.33
1200	0.9319	0.015723	100.83
800	0.9503	0.014932	45.51
400	0.9733	0.014337	11.47
14.65	0.9995	0.013978	5.17

Table 5–3 shows calculations of gas PVT properties and pseudopressure. Two-rate drawdown test data for wells are shown in Table 5–4. Using Eq. 5–58:

$$s_1' = 1.151\left[\frac{\psi(p_i) - \psi(p_{wfo})}{m} - \log\left(\frac{kt_1}{\phi\mu_i c_i r_w^2}\right) + 3.23\right]$$

$$= 1.151\left[\frac{(542.35 - 401.25) \times 10^6}{54.9 \times 10^6}\right.$$

$$\left. - \log\left(\frac{14.41 \times 6}{0.1004 \times 0.0186 \times 0.000274 \times 0.25^2}\right) + 3.23\right]$$

$$= 1.151[2.5701 - 9.4319 + 3.23] = -4.18$$

(indicating well improvement)

Using Eq. 5–59:

$$s_2' = \frac{q_{sc1}}{q_{sc2}} \cdot s_1' - \frac{[\psi(p_{wf1}) - \psi(p_{wfo})]q_{sc1}}{0.869mq_{sc2}} + \frac{q_{sc1} - q_{sc2}}{0.869mq_{sc2}}$$

$$\times \left[\log\left(\frac{k}{\phi\mu_i c_i r_w^2}\right) - 3.23\right]$$

Table 5–4
Two-Rate Drawdown Test Data for Gas Well

Time Δt (hrs) (1)	Drawdown pressure, P_{wf} (psig) (2)	Drawdown pressure, P_{wf} (psia) (3)	Function (4)	Pseudopressure $\psi(P_{wf})$ (mmscfd/cP) (5)
1.00	2668	2683	0.8445	444.26
1.25	2666	2681	0.8371	444.01
1.50	2664	2679	0.8330	443.89
2.00	2659	2672	0.8311	443.23
2.50	2644	2666	0.8342	442.10
3.00	2638	2660	0.8400	441.27
4.00	2640	2655	0.8559	440.31
5.00	2633	2648	0.8741	439.51
6.00	2629	2644	0.8929	438.71
7.00	2624	2640	0.9118	436.35
8.00	2620	2634	0.9300	436.61
9.00	2615	2630	0.9478	436.79
10.00	2612	2627	0.9648	435.69
11.00	2607	2622	0.9813	435.21
12.00	2606	2621	0.9971	433.29
14.00	2600	2615	1.0268	431.19
16.00	2598	2613	1.0543	427.27
18.00	2594	2609	1.0800	423.25
20.00	2588	2603	1.1036	425.21
24.00	2577	2592	1.1469	422.37
28.00	2580	2595	1.1852	418.21
32.00	2571	2586	1.2196	417.05
36.00	2564	2579	1.2509	416.79
40.00	2559	2574	1.2794	415.20
44.00	2552	2567	1.3057	414.01
48.00	2550	2565	1.3301	412.95
52.00	2547	2562	1.3528	410.12

$$= \frac{28.0}{21.3}(-418) - \left[\frac{(444.26 - 401.25) \times 10^6 \times 28.0}{0.869 \times 54.9 \times 10^6 \times 21.3} + \frac{28.0 - 21.3}{0.869 \times 21.3}\left(\log \frac{14.41}{0.1004 \times 0.0186 \times 0.000274 \times 0.25^2}\right) - 3.23\right]$$

$$= -5.49 - 1.1837 + 6.7/18.53[8.6538 - 3.23]$$

$$= -5.49 - 1.1837 + 1.9611 = -4.54$$

(indicating well improvement)

3. From Eq. 5–48,

$$D = \frac{s_1' - s_2'}{q_{sc1} - q_{sc2}} = \frac{-4.18 - (-4.54)}{28.0 - 21.3} = .054 \text{ mmscfd}^{-1}$$

4. From Eq. 5–49,

$$s = s_1' - Dq_{sc1} = -4.18 - 0.054(28.0) = -4.18 - 1.512 = -5.69$$

5. From Eq. 5–60, the pressure drop due to actual skin is

$$(\Delta\psi)_{skin} = 0.869\,ms = 0.869 \times 54.9 \times -5.69$$
$$= -271.46 \text{ mmpsia}^2/\text{cP}$$

6. From Eq. 5–61, rate-dependent pressure drop is

$$(\Delta\psi)_{skin} \text{ at rate } q_{sc2} = 0.869 \times s \times m \times \frac{q_{sc2}}{q_{sc1}}$$
$$= 0.869 \times (-5.69) \times 54.9 \times \frac{21.3}{28.0}$$
$$= -206.50 \text{ mmpsia}^2/\text{cP}$$

Example 5–4[21] *Rework Example 5–3, Assuming p_i Is Unknown. Estimate $\psi(p_i)$.*

Solution From Example 5–3 $q_{sc1} = 28.0$ mmscfd, $q_{sc2} = 21.3$ mmscfd, $\Psi(p_{wfo}) = 401.25$ mmpsia²/cP, $\psi(p_{wf1}) = 444.26$ mmpsia²/cP, $m = 54.9 \times 10^6$, and $k = 14.41$ mD. Calculate average value of apparent skin factor s' by using Eq. 5–41:

$$s' = 1.151\left[\left(\frac{q_{sc1}}{q_{sc1} - q_{sc2}}\right)\left(\frac{\psi(p_{wf1}) - \psi(p_{wfo})}{m}\right) - \log\left(\frac{k}{\phi\mu_i c_i r_w^2} + 3.23\right)\right]$$

$$= 1.151\left[\frac{28.0}{28.0 - 21.3}\left(\frac{(444.26 - 401.25) \times 10^6}{54.9 \times 10^6}\right) - \log\left(\frac{14.41}{0.1004 \times 0.0186 \times 0.000274 \times 0.25^2} + 3.23\right) + 3.23\right]$$

$$= 1.151\left[\frac{28.0}{6.7}(0.7834 - 8.5638 + 5.3338)\right]$$

$$= 1.151\,[3.2739 - 5.3338] = -2.37$$

Calculate $\psi(p_i)$ using Eq. 5–55:

$$\psi(p_i) = \psi(p_{wfo}) + m\left[\log \frac{k}{\phi\mu_i c_i r_w^2} - 3.23 + 0.869s'\right]$$

$$= 401.25 + 54.9\left[\log\left(\frac{14.41}{0.1004 \times 0.0186 \times 0.000274 \times 0.25^2}\right)\right.$$

$$\left. -3.23 + 0.869(-2.37)\right]$$

$$= 401.25 + 54.9[8.6522 - 3.23 - 2.0595]$$

$$= 401.25 + 184.61 = 585.66 \text{ mmpsia}^2/\text{cP}$$

$$= 585.66 \text{ mmpsia}^2/\text{cP} \leftrightarrow 2985 \text{ psia}$$

Example 5–5[21] *Two-Rate Drawdown Test Analysis*

A isochronal flow test is performed on a gas well at two different rates. Given the reservoir data and fluid properties below, determine the following using pressure-squared and real pseudopressure approaches:

1. Flow capacity kh
2. Apparent skin factors s'_1 and s'_2
3. Non-Darcy flow coefficient D for the well
4. True skin factor s

Solution This well is completed with a tubing-annulus packer. Reservoir and well data are as follows (and as shown in Table 5–5): $p_i = 2300$ psia; $r_w = 0.5$ ft; $r_e = 2980$ ft (640-acre spacing); $T = 590°$R; $z_i = 0.805$; $\mu_I = 0.0176$ cP; $\phi = 0.077$; and $c_i = 0.00041$ psi^{-1}.

Table 5–5
Calculations of PVT Gas Properties and Pseudopressure

P_{wf} (psia)	Z —	μ (cP)	$z(p/\mu z)$ —	Mean $z(p/\mu z)$	Δp (psia)	$(zp/\mu z) \times \Delta p$	$\psi(p_{wf})$ (mmpsia^2/cP)
400	0.95	0.0117	71.975	35.988	499	14.4×10^6	14.4
800	0.90	0.0125	142.222	107.099	400	42.9×10^6	57.3
1200	0.86	0.0132	211.416	176.819	400	70.7×10^6	128.0
1600	0.81	0.0146	270.590	241.003	400	96.5×10^6	224.5
2000	0.80	0.0163	306.748	288.669	400	115.5×10^6	340.0
2400	0.81	0.0180	329.218	319.000	400	127.6×10^6	467.6

Table 5–6
Two-Rate Drawdown Test Data

Flowing Time (hrs)	Flow 1 $q_{sc1} = 1.6$ mmscfd			Flow 2 $q_{sc2} = 3.2$ mmscfd		
	p_{wf} (psia)	p^2_{wf} (psia^2 $\times 10^6$)	$\psi(p_{wf})$ (mmpsia^2/cP)	p_{wf} (psia)	p^2_{wf} (psia^2 $\times 10^6$)	$\psi(p_{wf})$ (mmpsia^2/cP)
0.23	1855	3.44	298.21	1105	1.22	112.12
0.40	1836	3.37	294.69	1020	1.04	94.34
0.60	1814	3.29	290.17	954	0.91	85.10
0.80	1806	3.26	285.32	906	0.82	74.25
1.0	1797	3.23	279.95	860	0.74	68.41
2.0	1758	3.09	268.19	700	0.49	50.31
4.0	1723	2.97	260.04	539	0.29	26.22
6.0	1703	2.90	252.13	387	0.15	10.12

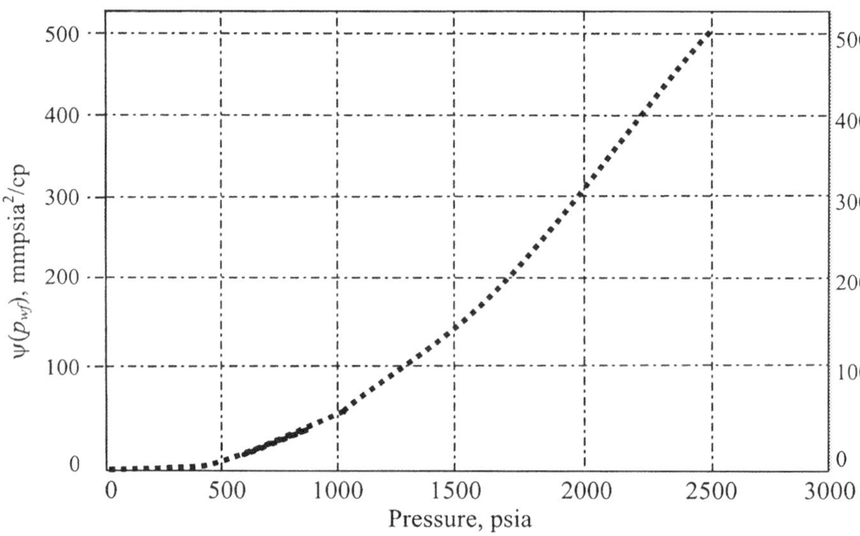

Figure 5–15. $\psi(p_{wf})$ versus pressure—Example 5–5.

Table 5–6 shows two-rate drawdown test data.

It can be assumed that wellbore storage effects are negligible, since the well is completed with a down-hole packer. The first step is to find $\psi(p_{wf})$ versus P for this gas. The gas properties are tabulated above; the quantities $2(p/\mu z)$ and $\psi(p_{wf})$ can be calculated and plotted versus pressure, as shown in Figure 5–15.

Using Pressure–Squared Approach

Figures 5–16 and 5–17 represent the drawdown data plotted for conventional manner in terms of p^2_{wf}.

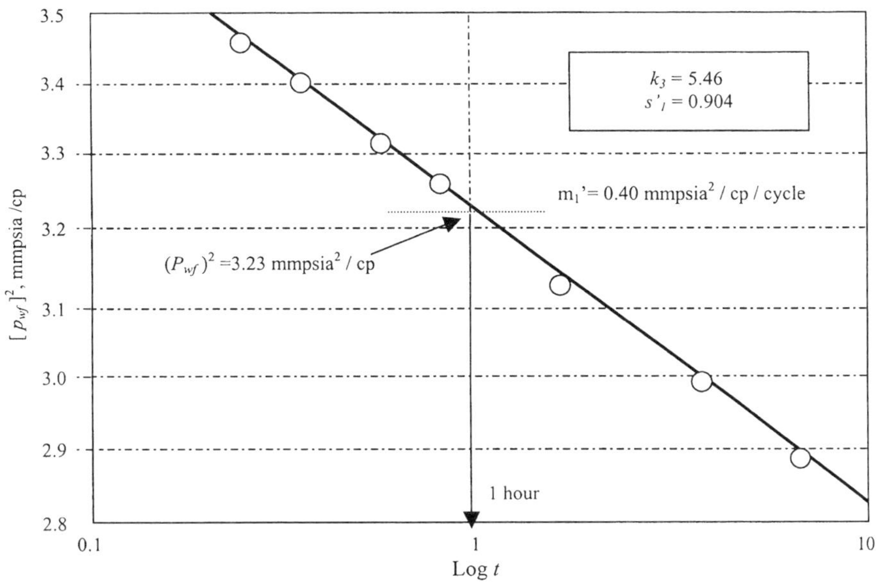

Figure 5–16. $(p_{wf})^2$ versus log time—flow rate 1.6 mmscfd (Example 5–5).

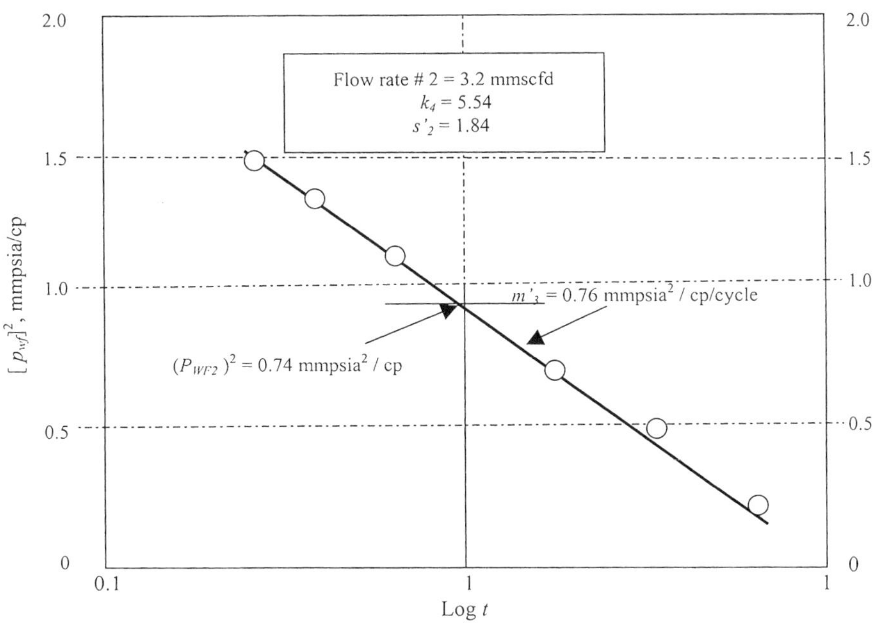

Figure 5–17. $(p_{wf})^2$ versus log time—flow rate 3.2 mmscfd (Example 5–5).

1. The flow capacity for flow no. 1 can be calculated from Eq. 5–34:

$$(kh)_{Flow1} = \frac{57.920 q_{sc1} T \mu_g z P_{sc}}{m_1^* T_{sc}}$$

$$= \frac{57.920 \times 10^6 \times 1.6 \times 590 \times 0.0176 \times 0.805 \times 14.65}{0.40 \times 10^6 \times 520}$$

$$= 54.56 \text{ mD-ft}$$

$$\therefore k_3 = 54.56/10 = 5.46 \text{ mD}$$

The flow capacity for flow no. 2 is

$$(kh)_{Flow2} = \frac{57.920 q_{sc2} T \mu_g z P_{sc}}{m_2^* T_{sc}}$$

$$= \frac{57.920 \times 10^6 \times 3.2 \times 590 \times 0.0176 \times 0.805 \times 14.65}{0.76 \times 10^6 \times 520}$$

$$= 57.43 \text{ mD-ft}$$

$$\therefore k_4 = 57.43/10 = 5.74 \text{ mD}$$

2. Apparent skin factor for flow no. 1 is

$$s_1^* = 1.151 \left[\frac{p_i^2 - p_{wf1}^2}{m_1^*} - \log \frac{k3}{\phi \mu_i c_i r_w^2} + 3.23 \right]$$

$$= 1.151 \left[\frac{(5.29 - 3.23) \times 10^6}{0.40 \times 10^6} - \log \left(\frac{5.46}{0.077 \times 0.0176 \times 0.00041 \times 0.5^2} \right) + 3.23 \right]$$

$$= 1.151[5.15 - 7.59 + 3.23] = 0.904$$

Apparent skin factor for flow no. 2 is

$$s_2^* = 1.151 \left[\frac{p_i^2 - p_{wf2}^2}{m_2^*} - \log \frac{k_4}{\phi \mu_i c_i r_w^2} + 3.23 \right]$$

$$= 1.151 \left[\frac{(5.29 - 0.74) \times 10^6}{0.76 \times 10^6} - \log \left(\frac{5.74}{0.077 \times 0.0176 \times 0.00041 \times 0.5^2} \right) + 3.23 \right]$$

$$= 1.151[5.987 - 7.616 + 3.23] = 1.84$$

3. Calculate the non-Darcy flow coefficient D^* using Eq. 5–48:

$$D^* = \frac{s_1^* - s_2^*}{q_{sc1} - q_{sc2}} = \frac{0.904 - 1.84}{1.6 - 3.2} = \frac{-1.036}{-1.6} = 0.6475 \text{ mmscfd}^{-1}$$

4. Calculate the true skin factor s using Eq. 5–49:

$$s = s_1^* - D^* q_{sc1} = 0.904 - 0.647 \times 1.6 = 0.904 - 1.036 = -0.232$$

Using Real Pseudopressure Approach

Figures 5–18 and 5–19 represent the drawdown data plotted in the conventional manner in terms of $\psi(p_{wf})$.

1. The flow capacity for flow no. 1 can be estimated from Eq. 5–51:

$$(kh)_{Flow\,1} = \frac{57.920 q_{sc1} T P_{sc}}{m_1 T_{sc}} = \frac{57.920 \times 10^6 \times 1.6 \times 590 \times 14.65}{32 \times 10^6 \times 520}$$
$$= 48.14 \text{ mD-ft}$$
$$\therefore k_1 = 48.14/10 = 4.81 \text{ mD}$$

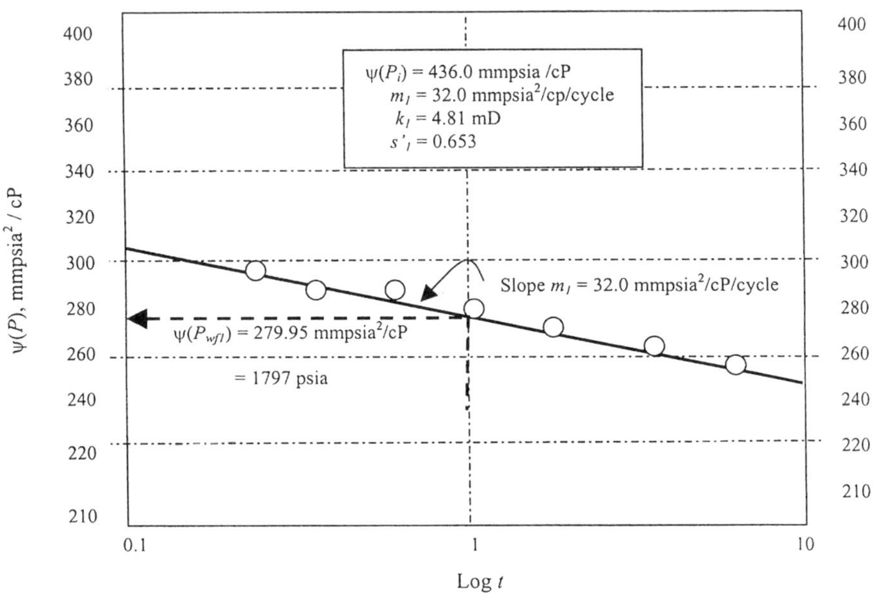

Figure 5–18. $\psi(P_{wff})$ versus log t—flow rate = 1.6 mmscfd for Example 5–5.

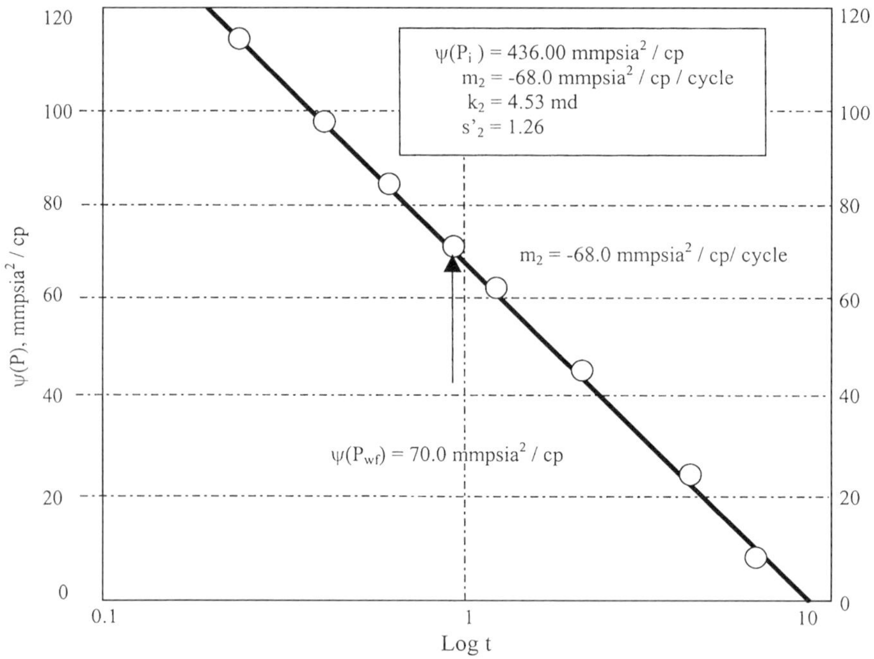

Figure 5–19. $\psi(P_{wf})$ versus log t—flow rate 4.2 mmscfd for Example 5–5.

The flow capacity for flow no. 2 is

$$(kh)_{Flow2} = \frac{57.920 q_{sc2} T P_{sc}}{m_2^* T_{sc}} = \frac{57.920 \times 10^6 \times 3.2 \times 590 \times 14.65}{68 \times 10^6 \times 520}$$

$$= 45.3 \text{ mD} - \text{ft}$$

$$k_2 = 45.3/10 = 4.53 \text{ mD}$$

2. Apparent skin factor for flow no. 1 is

$$s_1' = 1.151 \left[\frac{\psi(p_i) - \psi(p_{wf1})}{m_1} - \log \frac{k_1}{\phi \mu_i c_i r_w{}^2} + 3.23 \right]$$

$$= 1.151 \left[\frac{(436.0 - 279.95) \times 10^6}{32 \times 10^6} \right.$$

$$\left. - \log \left(\frac{4.81}{0.077 \times 0.0176 \times 0.00041 \times 0.5^2} \right) + 3.23 \right]$$

$$= 1.151[4.877 - 7.539 + 3.23] = 0.653$$

Apparent skin factor for flow no. 2 is

$$s_2' = 1.151\left[\frac{\psi(p_i) - \psi(p_{wf2})}{m_2} - \log\frac{k_4}{\phi\mu_i c_i r_w^2} + 3.23\right]$$

$$= 1.151\left[\frac{(436.0 - 70.0)\times 10^6}{68\times 10^6}\right.$$

$$\left. - \log\left(\frac{4.53}{0.077\times 0.0176\times 0.00041\times 0.5^2}\right) + 3.23\right]$$

$$= 1.151[5.382 - 7.513 + 3.23] = 1.26$$

3. Calculate the non-Darcy flow coefficient D^* using Eq. 5–48:

$$D^* = \frac{s_1' - s_2'}{q_{sc1} - q_{sc2}} = \frac{0.653 - (1.26)}{1.6 - 3.2} = \frac{-0.607}{-1.6}$$
$$= 0.37938 \text{ mmscfd}^{-1}$$

4. Calculate the true skin factor s using Eq. 5–49:

$$s = s_1' - D^* q_{sc1} = 0.653 - 0.37938\times 1.6 = 0.653 - 0.607 = 0.05$$

Example 5–6[21] *Analyzing Two-Rate Drawdown Test and Predicting Well Inflow Response (When P_R Is Not Known)*

A two-rate drawdown test was conducted on a gas well. The reservoir data and fluid properties were as follows, and as shown in Table 5–7:

$T = 622°R$; $r_w = 0.42$ ft; $r_e = 2640$ ft; $h = 56$ ft; $\phi = 0.05$ fraction; $\mu_{gi} = 0.022$ cP; $c_t = 0.000188$ psi^{-1}; initial pressure prior to test = 4327 psia ↔ 1176.92 mmpsia^2/cP; first flow rate $q_{sc1} = 27.76$ mmscfd; time t_1 at which first flow rate changed = 6 hr. At that time, flowing bottom-hole pressure = 3838 psia ↔ 980.84 mmpsia^2/cP; second flow rate $q_{sc2} = 20.16$ mmscfd.

Calculate the following:

1. Formation permeability k
2. Apparent skin factor s_1' related to flow rate q_{sc1}
3. Apparent skin factor s_2' related to flow rate q_{sc2}
4. Non-Darcy flow coefficient D
5. True skin factor s
6. Pressure drop across the skin related to flow rate q_{sc}
7. Pressure drop across the skin related to flow rate q_{sc}
8. Reservoir pressure p_R
9. The values of deliverability constants A and B
10. Absolute open flow potential AOF
11. Inflow performance response

Table 5–7
Two-Rate Drawdown Test Data for Gas Well

Time, Δt (hrs) (1)	Pressure, P_{wf} (psig) (2)	Pressure, P_{wf} (psia) (3)	Function X (4)	Pseudopressure $\psi(P_{wf})$ (mmscfd/cP) (5)
1.00	3956	3971	0.8449	1029.37
3.00	3958	3973	0.8235	1030.18
5.00	3956	3971	0.8499	1029.37
7.00	3950	3965	0.8824	1026.91
9.00	3945	3960	0.9147	1024.87
11.00	3941	3956	0.9452	1023.23
13.00	3936	3951	0.9736	1021.19
15.00	3929	3944	1.0001	1018.32
17.00	3926	3941	1.0247	1017.09
19.00	3919	3934	1.0477	1014.22
21.00	3916	3931	1.0692	1012.99
23.00	3913	3928	1.0894	1011.76
25.00	3910	3925	1.1084	1010.53
27.00	3904	3919	1.1264	1008.06
29.00	3898	3913	1.1435	1005.59
31.00	3895	3910	1.1597	1004.36
33.00	3892	3907	1.1751	1003.13
35.00	3889	3904	1.1898	1001.89
37.00	3886	3901	1.2039	1000.66

Where X is equal to $\log \frac{(t+\Delta t)}{\Delta t} + \frac{q_2}{q_1} \times \log \Delta t$

Solution The $\psi - p$ curve, shown in Figure 5–20 and is applicable to this problem. Plot $\psi(p_{wf})$ versus $\log(\frac{t_1+\Delta t}{\Delta t}) + \frac{q_{sc2}}{q_{sc1}} \log \Delta t$ on Cartesian coordinate paper and draw the best straight line as shown in Figure 5–21. From the straight line of Figure 5–21, find the slope m:

$$m = 86.67 \text{ mmpsia}^2/\text{cP} \quad \text{and} \quad \psi(p_{1hr}) = 1033.0 \text{ mmpsia}^2/\text{cP}$$

1. From Equation 5–51:

$$kh = \frac{57.92 \times 10^6 q_{sc1} T P_{sc}}{m T_{sc}}$$

$$= \frac{57.920 \times 10^6 \times 27.76 \times 622 \times 14.65}{86.67 \times 10^6 \times 520} = 325.36 \text{ mD-ft}$$

$$\therefore k = 325.36/56 = 5.81 \text{ mD}$$

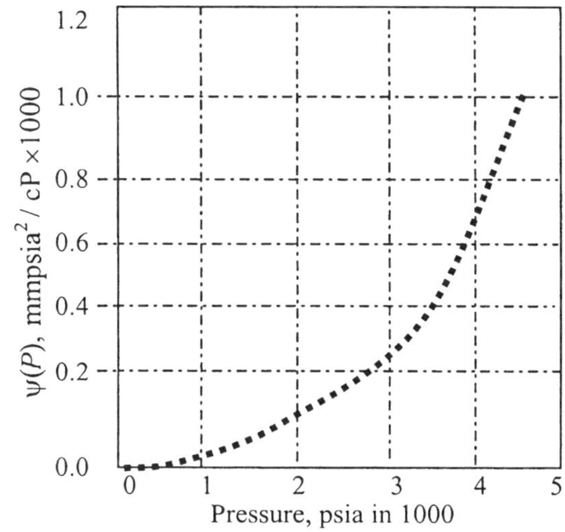

Figure 5–20. $\psi - P$ curve for two-rate test.

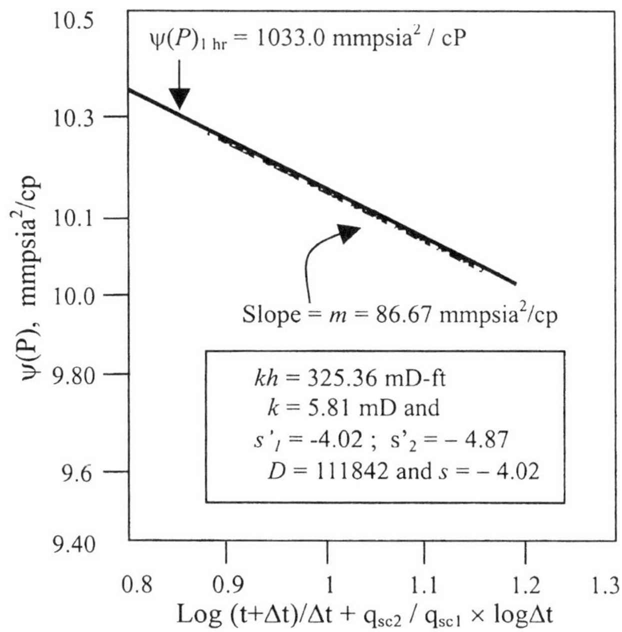

Figure 5–21. Semilog plot.

2. Since p_i is known, it is possible to separate the skin and IT effects. Calculate apparent skin factor s_1' using Eq. 5–41:

$$s_1' = 1.151\left[\frac{\psi(p_i) - \psi(p_{wfo})}{m} - \log\left(\frac{k\,t_1}{\phi\,\mu_i c_i r_w^2}\right) + 3.23\right]$$

$$= 1.151\left[\frac{(1176.92 - 980.84)\times 10^6}{86.67\times 10^6}\right.$$

$$\left. - \log\left(\frac{5.81\times 6}{0.05\times 0.022\times 0.000188\times 0.42^2}\right) + 3.23\right]$$

$$= 1.151\,[2.26 - 8.98 + 3.23]$$

$$= -4.02 \text{ (indicating well improvement)}$$

3. Calculate apparent skin factor s_2' using Eq. 5–53a:

$$s_2' = \frac{q_{sc1}}{q_{sc2}}\cdot s_1' - \frac{\left[\psi(p_{wf1}) - \psi(p_{wfo})\right] q_{sc1}}{0.869\,m\,q_{sc2}}$$

$$+ \frac{q_{sc1} - q_{sc2}}{0.869\,m q_{sc2}}\left[\log\left(\frac{k}{\phi\,\mu_i c_i r_w^2}\right) - 3.23\right]$$

$$= \frac{27.76}{20.16}(-4.02)$$

$$-\left[\frac{(1029.37 - 980.84)\times 10^6\times 20.16}{0.869\times 54.9\times 10^6\times 21.3} + \frac{27.76 - 20.16}{0.869\times 86.67(20.16)}\right.$$

$$\left.\times\left(\log\frac{5.81}{0.05\times 0.022\times 0.000188\times 0.42^2}\right) - 3.23\right]$$

$$= -5.54 - [0.644 + 0.024]$$

$$= -4.87 \text{ (indicating well improvement)}$$

4. Calculate non-Darcy flow coefficient D from Eq. 5–48:

$$D = \frac{s_1' - s_2'}{q_{sc1} - q_{sc2}} = \frac{-4.02 - (-4.87)}{27.76 - 20.16} = 0.111842 \text{ mmscfd}^{-1}$$

5. Calculate true skin factor from Eq. 5–49:

$$s = s_1' - Dq_{sc1} = -4.02 - 0.111842\,(27.76) = -4.022$$

6. Calculate pressure drop due to actual skin from Eq. 5–66:

$$(\Delta\psi)_{skin} = 0.869\,ms = 0.869 \times 86.67 \times (-4.022)$$
$$= -302.92 \text{ mmpsia}^2/\text{cP}$$

7. Calculate pressure drop related to flow rate q_{sc2} from Eq. 5–61:

$$(\Delta\psi)_{skin} = 0.869\,ms\frac{q_{sc2}}{q_{sc1}} = 0.869 \times 86.67 \times (-4.022) \times \frac{20.16}{27.76}$$
$$= -220.0 \text{ mmpsia}^2/\text{cP}$$

8. From Table 5–8, reservoir pressure P_R is 3999 psia $\leftrightarrow$ $\psi(p_R) = 1040.0$ mmpsia^2/cP.

Table 5–8
Average Reservoir Pressure Calculations from Various Assumed Values

Assumed average reservoir pressures P_R psia							
4300	**4250**	**4100**	**4050**	**4000**	**3999**	**3990**	**3980**
Assumed average reservoir pressures $\psi(P_R)$, mmpsia^2/cP							
1161.00	**1141.00**	**1082.00**	**1062.00**	**1041.00**	**1040.00**	**1037.00**	**1033.00**
Predicted flow rate, mmscfd							
11.078	11.203	4.942	2.586	0.053	0.000	−0.479	−1.019
20.255	18.632	13.349	11.424	9.402	9.360	8.984	8.562
26.727	25.280	20.635	18.972	17.244	17.208	16.890	16.533
32.610	31.292	27.101	25.616	24.084	24.052	23.771	23.456
37.982	36.764	32.912	31.558	30.166	30.137	29.882	29.598
42.883	41.743	38.155	36.900	35.613	35.587	35.352	35.090
47.333	46.255	42.876	41.699	40.495	40.470	40.251	40.005
51.342	50.315	47.103	45.988	44.848	44.825	44.618	44.386
54.914	53.929	50.853	49.787	48.700	48.678	48.479	48.259
58.056	57.104	54.138	53.112	52.066	52.045	51.855	51.642
60.773	59.849	56.972	55.978	54.966	54.945	54.761	54.556
63.080	62.177	59.372	58.403	57.417	57.398	57.218	57.018
64.995	64.110	61.361	60.412	59.448	59.428	59.253	59.057
66.548	65.677	62.972	62.039	61.091	61.072	60.899	60.707
67.781	66.920	64.249	63.329	62.393	62.374	62.204	62.015
68.752	67.899	65.254	64.343	63.417	63.399	63.230	63.043
69.536	68.690	66.066	65.162	64.244	64.225	64.058	63.872
70.230	69.390	66.784	65.886	64.975	64.956	64.791	64.606
70.956	70.121	67.534	66.643	65.739	65.721	65.556	65.373

9. Determine the values of deliverability constants A and B from Eqs. 4–79 and 4–80, respectively (see Table 5–7):

$$A = \frac{1.422 \times 10^6 T}{kh}\left[In\left(\frac{0.472\,r_e}{r_w}\right) + s\right]$$

$$= \frac{1.422 \times 10^6 (622)}{5.81 \times 56}\left[In\left(\frac{0.472 \times 2{,}640}{0.42}\right) + (-4.022)\right]$$

$$= 10.8005 \times 10^6\ \frac{\text{psia}^2/\text{cP}}{\text{mmscfd}}$$

$$B = \frac{1.422 \times 10^6 T}{kh} D = \frac{1.422 \times 10^6 (622)}{5.81(56)}(0.111842)$$

$$= 0.30404 \times 10^6\ \frac{\text{psia}^2/\text{cp}}{\text{mmscfd}^2}$$

10. Estimate the absolute open flow potential (AOF) of the gas well from Eq. 4–58:

$$q_{sc} = \frac{-A + \sqrt{A^2 + 4B\lfloor\psi(\bar{p}_R) - \psi(p_{wf})\rfloor}}{2B}$$

$$= \frac{-10.8005 + \sqrt{10.8005^2 + 4\,(0.30404)\ [1040.89 - 0]}}{4\,(0.30404)}$$

$$= \frac{26.38}{0.60808} = 43.52 \text{ mmscfd } (q_{\max})$$

Therefore, AOF of gas well is 43.52 mmscfd. The inflow performance (IPR) responses are shown in Table 5–9 and results are plotted in Figures 5–22 and 5–23, respectively.

Long-term gas deliverability equation s:

$$[\Psi(P_R) - \Psi(P_{WF})] = 10.8005 \times 10^6 q_g + 0.30404 \times 10^6 q_g^2$$

Multirate Drawdown Tests

The multirate test is similar to the conventional deliverability test described in Chapter 4, except that each of the flow periods is not continued to pressure stabilization. In fact, a multirate test is intended to investigate the transient flow regime only so that kh, s, and D may be determined by a semilog analysis

Table 5–9
Long-Term Gas Deliverability Calculations

Bottom-hole pressure, P_{wf} (psia) (1)	Pseudopressure $\psi(P_{wf})$ (mmpsia²/cP) (2)	Pressure ratio $\psi(P_{wf})/\psi(P_R)$ — (3)	Flow rate ratio q_1/q_{max} — (4)	Calculated flow rate, q_g (mmscfd) (5)
3999	1040.89	1.0000	0.0000	0.00
3800	958.97	0.92130	0.1475	6.42
3600	875.28	0.84090	0.2659	11.57
3400	791.13	0.76005	0.3667	15.91
3200	707.58	0.67978	0.4575	19.91
3000	625.76	0.60117	0.5340	23.24
2800	546.85	0.52536	0.6041	26.29
2600	472.00	0.45346	0.6664	29.00
2400	402.28	0.38648	0.7213	31.39
2200	338.64	0.32534	0.7693	33.48
2000	281.81	0.27074	0.8104	35.27
1800	232.29	0.22317	0.8451	36.78
1600	190.28	0.18281	0.8739	38.03
1400	155.61	0.14950	0.8973	39.05
1200	127.70	0.12268	0.9157	39.85
1000	105.49	0.10135	0.9302	40.48
800	87.40	0.08397	0.9419	40.99
600	71.27	0.06847	0.9522	41.44
$AOF \Rightarrow$ 14.73	5.59	0.0000	1.0000	43.52

approach. The analysis of multirate tests is not always reliable as far as the calculation of kh, s, and D is concerned. However, the application of the principle of superposition in time is described below. In a multirate test, flow starts from stabilized reservoir conditions. A constant flow rate q_{sc1} is maintained for a period of time t_1. The flow rate is then changed to q_{sc2} up to time t_2, after which it is changed to q_{sc3} up to time t_3, and so on. In general, the flow rate history may be summarized as

$$
\begin{aligned}
q_{sc1} &= q_1 \quad \text{for} \quad 0 < t < t_1 \\
q_{sc2} &= q_2, \quad t_1 < t < t_2 \\
&\cdot \\
&\cdot \\
q_{scn} &= q_n, \quad t_{n-1} < t
\end{aligned}
$$

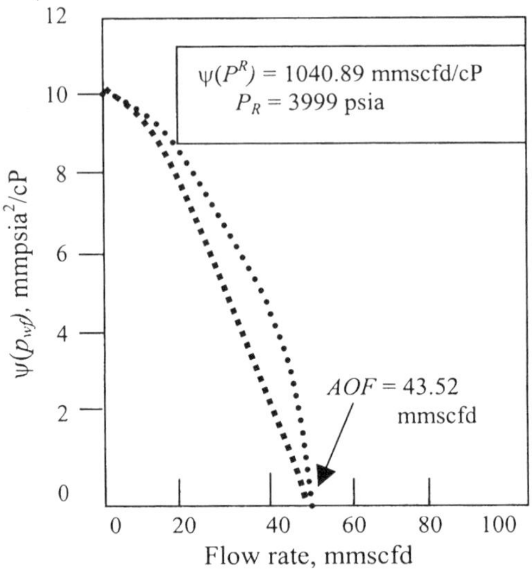

Figure 5–22. $\psi(p_{wf})$ versus flow rate.

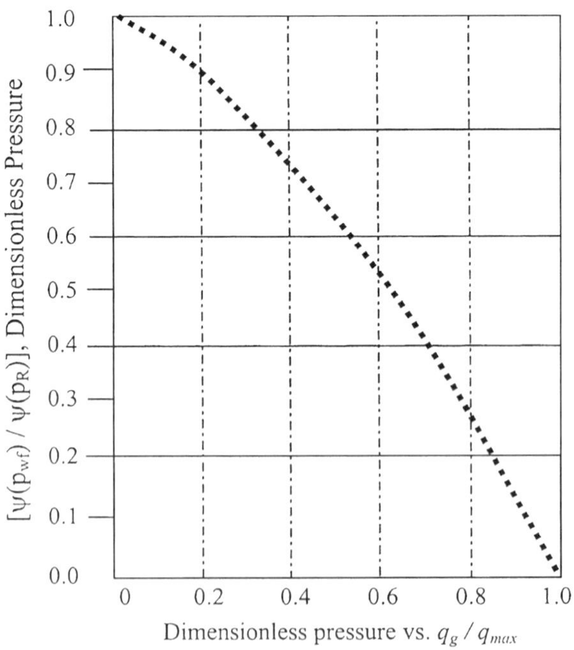

Figure 5–23. q_g/q_{max}, dimensionless flow rate.

During the nth flow period of a multirate test, the pressure drawdown is given by the equation below using the pseudopressure approach:

$$\frac{\Psi(p_i) - \Psi(p_{wf})}{q_n} = m \sum_{j-i}^{n} \left[\frac{\Delta q_j}{q_n} \log(t - t_{j-1}) \right] + m' \left[\log\left(\frac{k}{\phi \mu_{gi} c_i r_w^2} \right) - 3.23 + 0.869 s' \right] \quad (5\text{–}67)$$

where

$$m' = \frac{57.920 \times 10^6 T p_{sc}}{k h T_{sc}}$$

$$\Delta q_j = q_j - q_{j-i}$$

$$t_0 = q_0 = 0$$

A plot of

$$\lfloor \Psi(p_i) - \Psi(p_{wf}) \rfloor / q_n \text{ versus } \sum_{j-i}^{n} {}^{\Delta q_j}/_{q_n} \log(t - t_{j-1})$$

on arithmetic coordinates should give a straight line of slope m', from which

$$kh = \frac{57.920 \times 10^6 T p_{sc}}{m' T_{sc}} \quad (5\text{–}68)$$

The apparent skin factor s'_n associated with the flow rate q_n may then be calculated from

$$s'_n = 1.151 \left[\frac{\Delta \Psi_0}{m'} - \log\left(\frac{k}{\phi \mu_{gi} c_i r_w^2} \right) + 3.23 \right] \quad (5\text{–}69)$$

where

$$\Delta \Psi_0 = \text{intercept} = \text{value of } [\Psi(p_i) - \Psi(p_{wf})]/q_n$$

corresponding to a value of zero on the abscissa, obtained from the straight line (extrapolated, if necessary). Data from each of the preceding flow periods may also be analyzed by the method described for the nth flow period. Such an analysis would yield values of $s'_1, s'_2, s'_3, \ldots q_{n-1}$.

$$\begin{aligned} s'_1 &= s + Dq_1 \\ s'_2 &= s + Dq_2 \\ &\cdot \\ &\cdot \\ s'_n &= s + Dq_n \end{aligned} \quad (5\text{–}70)$$

This may then be solved by the method of least squares to give

$$s = \frac{\sum s' \sum q^2 - \sum s'q \sum q}{N \sum q^2 - \sum q \sum q} \tag{5–71}$$

$$D = \frac{N \sum s'q - \sum s' \sum q}{N \sum q^2 - \sum q \sum q} \tag{5–72}$$

where N = number of flow periods.

Alternatively, a plot of s' versus q_{sc} may be fitted with a best straight line which will give a slope equal to D and intercept on the $q_{sc} = 0$ axis equal to s. Next, Example 5–7 will clarify the use of the multirate test.

Example 5–7[21] *Analyzing Multirate Drawdown Test (Assuming Stabilized Flow Conditions)*

Multirate flow tests are performed on a gas well in Indonesia at four different rates. Given the reservoir data and fluid properties below, determine the non-Darcy flow coefficient for this well. The well is completed with a tubing annulus packer.

The reservoir data are as follows: $\bar{p}_R = 3700$ psia; $h = 41$ ft; $c_t = 0.00023$ psi^{-1}; $T = 710°$R; $r_e = 2{,}200$ ft; $r_w = 0.4271$ ft; $s_g = 0.733$; $P_{sc} = 14.65$ psia; $T_{sc} = 520°$R; $\bar{\mu} = 0.0235$ cP; $\bar{z} = 0.9491$; $\phi = 0.137$; $\phi_{HC} = 0.1004$. Gas properties are as follows: $T = 250°$F; gas gravity $= 0.732$; $T_c = 380.16°$R; $P_c = 645.08$ psia; $N_2 = 0.11$; $H_2S = 0.0$; $CO_2 = 7.84$; $\beta_g = 0.00513$ ft^3/scf; $= 0.000913$ bbl/scf $= 1095$ scf/bbl $= 1.0948$ mscf/bbl $= 0.9134$ bbl/mscf $= 194.732$ scf/ft^3. Determine the values of k, s, and D assuming flow rates from stabilized reservoir conditions (a) using the pressure-squared approach and (b) using the pseudopressure approach.

Solution Calculated PVT gas properties are given in Table 5–10 and results are plotted (real gas pseudopressure versus pressure) in Figure 5–24.

(a) Using Pressure-Squared Approach

Figure 5–25 is a plot of $(p_{wf})^2$ versus log time for each flow rate. From this plot the following results are obtained: Using Eq. 5–34, the values of k can be calculated at a different flow rate as follows:

$$k = \frac{1.637 \times q_{sc}\bar{\mu}\bar{z}T}{m'h}$$

Table 5–11 shows multirate drawdown test data. The necessary calculations to carry out the method proposed in this section are given in Table 5–12.

Table 5–10
Calculated PVT Gas Properties and Pseudopressure

Pressure (psia)	Gas compressibility z	Gas viscosity (cP)	Real gas pseudopressure (mmpsia^2/cP)
4000	0.9647	0.024580	872.72
3600	0.9445	0.023151	739.56
3200	0.9282	0.021721	610.28
2800	0.9169	0.020329	486.72
2400	0.9113	0.019008	371.18
2000	0.9120	0.017784	266.41
1600	0.9189	0.016681	175.33
1200	0.9319	0.015723	100.83
800	0.9503	0.014932	45.51
400	0.9733	0.014337	11.47
14.65	0.9995	0.013978	0.52

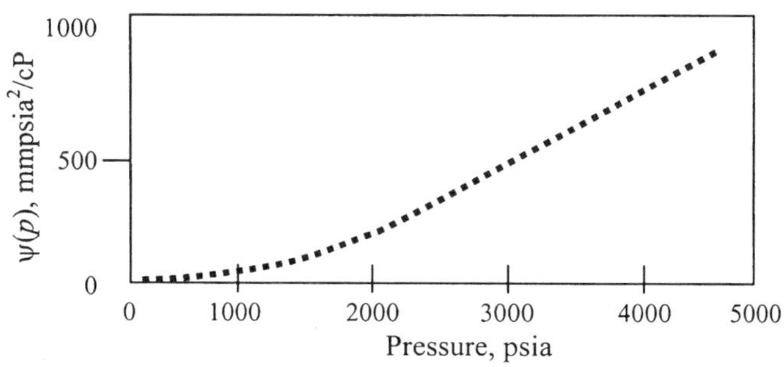

Figure 5–24. $\psi - p$ curve—Example 5-7.

$$\text{For flow 1,}\quad k = \frac{1.637 \times 10^6 \times 2.397 \times 0.0235 \times 0.9491 \times 710}{8.83 \times 41} = 4.33 \text{ mD}$$

$$\text{For flow 2,}\quad k = \frac{1.637 \times 10^6 \times 5.214 \times 0.0235 \times 0.9491 \times 710}{7.18 \times 41} = 8.68 \text{ mD}$$

$$\text{For flow 3,}\quad k = \frac{1.637 \times 10^6 \times 6.144 \times 0.0235 \times 0.9491 \times 710}{5.10 \times 41} = 9.75 \text{ mD}$$

(text continued on page 288)

Table 5–11
Multirate Drawdown Test Data

Flowing time (hrs)	Flow no. 1 $q_1 = 2.397$ (mmscfd) p_{wf}, (psia)	Flow no. 2 $q_2 = 5.214$ (mmscfd) p_{wf}, (psia)	Flow no. 3 $q_3 = 6.144$ (mmscfd) p_{wf}, (psia)	Flow no. 4 $q_4 = 7.148$ (mmscfd) p_{wf}, (psia)
0.02	3670	3577	3597	2315
0.03	3485	3455	3069	2375
0.07	3310	3300	2721	2236
0.10	3188	3183	2571	2153
0.13	3068	3040	2479	2090
0.17	3066	2956	2414	2048
0.25	3040	2826	2331	2023
0.33	3033	2757	2323	2004
0.50	3000	2710	2257	1937
0.75	2987	2714	2219	1911
1.00	2972	2652	2206	1903
1.50	2952	2611	2196	1893
2.00	2933	2602	2190	1889
2.50	2930	2595	2184	1881
3.00	2900	2591	2180	1876
4.00	2899	2580	2172	1870
5.00	2896	2573	2165	1855
6.00	2881	2567	2158	1836

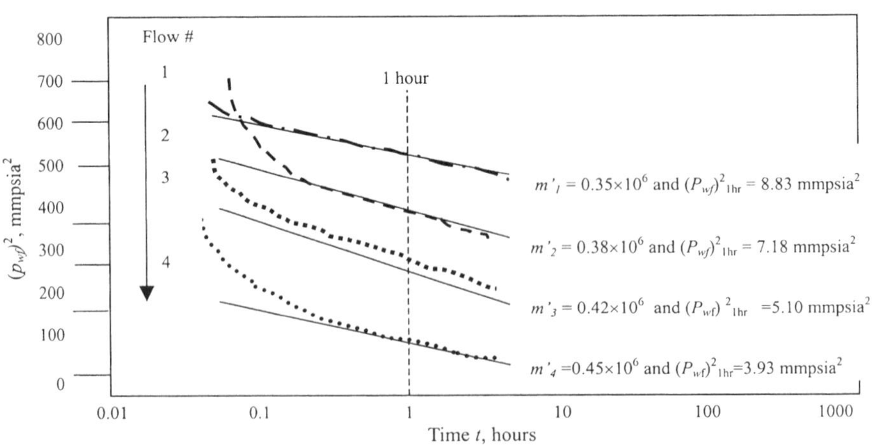

Figure 5–25. $(p_{wf})^2$ versus log time—Example 5–7.

Table 5–12
Calculations for Multirate Drawdown Test Data Analysis

Flowing	Flow no. 1 $q_1 = 2.397$ mmscfd		Flow no. 2 $q_2 = 5.214$ mmscfd		Flow no. 3 $q_3 = 6.144$ mmscfd		Flow no. 4 $q_4 = 7.148$ mmscfd	
time t (hrs)	P_{wf} (mmpsia^2)	$\psi(P_{wf})$ (mpsia^2/cP)	P_{wf} (mmpsia^2)	$\psi(P_{wf})$ (mpsia^2/cP)	P_{wf} (mmpsia^2)	$\psi(P_{wf})$ (mpsia^2/cP)	P_{wf} (mmpsia^2)	$\psi(P_{wf})$ (mpsia^2/cP)
0.02	13.469	642.50	12.795	732.02	12.938	738.62	5.359	710.33
0.03	12.145	621.30	11.937	692.04	9.419	569.16	5.641	360.25
0.07	10.956	600.37	10.890	642.12	7.404	463.22	5.000	400.75
0.10	10.164	582.01	10.131	604.88	6.610	419.29	4.635	304.94
0.13	9.412	579.23	9.242	560.03	6.145	393.28	4.368	288.81
0.17	9.400	575.03	8.738	534.07	5.827	374.99	4.194	278.18
0.25	9.241	564.12	7.986	494.52	5.434	352.25	4.093	272.01
0.33	9.201	560.43	7.601	473.82	5.396	350.13	4.016	267.38
0.50	9.000	550.16	7.344	459.77	5.094	332.35	3.752	251.07
0.75	8.925	542.07	7.366	461.07	4.924	322.12	3.652	244.85
1.00	8.830	535.15	7.033	442.94	4.866	318.91	3.621	243.01
1.50	8.715	530.07	6.817	431.01	4.822	316.13	3.583	240.52
2.00	8.603	522.17	6.770	428.43	4.796	314.53	3.568	239.56
2.50	8.587	520.11	6.734	426.41	4.770	313.18	3.538	237.83
3.00	8.501	513.37	6.713	424.91	4.752	312.11	3.519	236.49
4.00	8.407	510.12	6.656	421.97	4.718	310.00	3.497	235.25
5.00	8.385	500.00	6.620	419.95	4.687	308.00	3.441	234.10
6.00	8.301	492.45	6.589	418.12	4.657	306.21	3.371	233.01

(text continued from page 286)

$$\text{For flow 4,}\ k = \frac{1.637 \times 10^6 \times 7.148 \times 0.0235 \times 0.9491 \times 710}{3.95 \times 41}$$
$$= 10.04\ \text{mD}$$

Using Eq. 5–36, the values of apparent skin factor s' can be estimated at a different flow rate as follows:

$$s' = 1.151\left[\frac{p_i^2 - p_{wf\,@\,1hr}^2}{m'} - \log\left(\frac{k}{\phi\bar{\mu}_i\bar{c}_i r_w^2}\right) + 3.23\right]$$

$$\text{For flow 1,}\ s_1' = 1.151\left[\frac{(13.69 - 8.83) \times 10^6}{0.35 \times 10^6} - \log\left(\frac{4.33}{0.1004 \times 0.0235 \times 0.00023 \times 0.4271^2}\right) + 3.23\right]$$
$$= 9.48$$

$$\text{For flow 2,}\ s_2' = 1.151\left[\frac{(13.69 - 7.18) \times 10^6}{0.39 \times 10^6} - \log\left(\frac{8.68}{0.1004 \times 0.0235 \times 0.00023 \times 0.4271^2}\right) + 3.23\right]$$
$$= 12.42$$

$$\text{For flow 3,}\ s_3' = 1.151\left[\frac{(13.69 - 5.10) \times 10^6}{0.42 \times 10^6} - \log\left(\frac{9.25}{0.1004 \times 0.0235 \times 0.00023 \times 0.4271^2}\right) + 3.23\right]$$
$$= 15.23$$

$$\text{For flow 4,}\ s_4' = 1.151\left[\frac{(13.69 - 3.95) \times 10^6}{0.45 \times 10^6} - \log\left(\frac{10.04}{0.1004 \times 0.0235 \times 0.00023 \times 0.4271^2}\right) + 3.23\right]$$
$$= 16.85$$

Calculate s and D using the least squares method.

Calculations for true skin factor using the pressure-squared approach are shown in Table 5–13.

Table 5–13
Calculations for True Skin Factor for Example 5–7
(Using Pressure-Squared Approach)

N	q (mmscfd)	q^2 (mmscfd^2)	s'	$s'q$ (mmscfd)
1	2.397	5.746	9.48	22.7236
2	5.214	27.186	12.42	64.7579
3	6.144	37.749	15.23	93.5731
4	7.148	51.094	16.85	120.4438
$\sum$	20.885	121.775	53.98	301.498

Using Eq. 5–71,

$$s = \frac{\sum s' \sum q^2 - \sum s'q \sum q}{N \sum q^2 - \sum q \sum q}$$

$$= \frac{(53.98 \times 121.775) - (301.498)(20.885)}{4(121.775) - (20.885)(20.885)} = 5.43$$

Using Eq. 5–72,

$$D = \frac{N \sum s'q - \sum s' \sum q}{N \sum q^2 - \sum q \sum q}$$

$$= \frac{4(301.498) - (53.98)(20.885)}{4(121.775) - (20.885)(20.885)} = 1.54408 \text{ mmscfd}^{-1}$$

Alternatively, a plot of s' versus q is shown in Figure 5–26 with a best straight line. From this plot intercept, $s = 5.24$ and slope $D = 1.69418$ mmscfd^{-1}.

(b) Using Pseudopressure Approach

Figure 5–27 is a plot of $\psi(p_{wf})$ versus log time for each flow rate. From this plot the following results are obtained. Using Eq. 5–40, the values of k can be calculated at a different flow rate as follows:

$$k = \frac{5.792 \times 10^4 q_{sc} T P_{sc}}{m h T_{sc}}$$

$$\text{For flow 1, } k = \frac{5.792 \times 10^4 \times 2.397 \times 710 \times 14.65}{17.90 \times 10^6 \times 41} = 3.78 \text{ mD}$$

$$\text{For flow 2, } k = \frac{5.792 \times 10^4 \times 5.214 \times 710 \times 14.65}{17.20 \times 10^6 \times 41} = 8.57 \text{ mD}$$

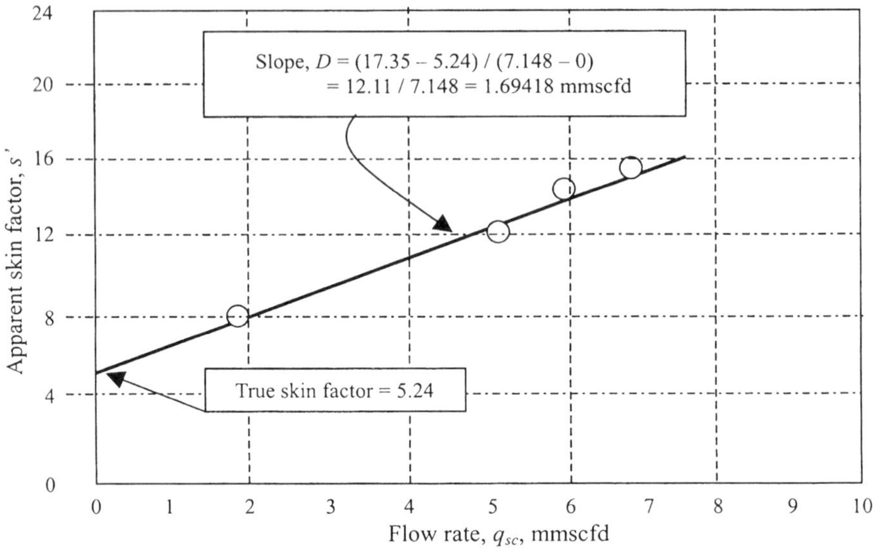

Figure 5–26. True skin factor determination using pressure squared approach—Example 5–7.

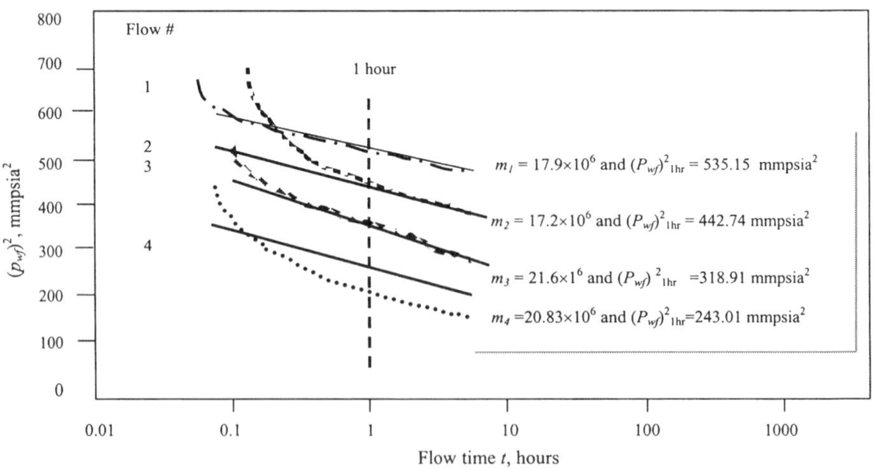

Figure 5–27. $\psi(p_{wf})$ versus log time—Example 5–7.

$$\text{For flow 3, } k = \frac{5.792 \times 10^4 \times 6.144 \times 710 \times 14.65}{21.6 \times 10^6 \times 41} = 8.04 \text{ mD}$$

$$\text{For flow 4, } k = \frac{5.792 \times 10^4 \times 7.148 \times 710 \times 14.65}{19.5 \times 10^6 \times 41} = 10.36 \text{ mD}$$

Using Eq. 5–41, the values of apparent skin factor s' can be estimated at a different flow rate as follows:

$$s = 1.151 \left[\frac{\psi(p_i) - \psi(p_{1hr})}{m} - \log\left(\frac{k}{\phi \bar{\mu}_i \bar{c}_i r_w^2} \right) + 3.23 \right]$$

$$\text{For flow 1, } s_1' = 1.151 \left[\frac{(772.56 - 535.35) \times 10^6}{17.90 \times 10^6} - \log\left(\frac{3.78}{0.1004 \times 0.0235 \times 0.00023 \times 0.4271^2} \right) + 3.23 \right] = 8.91$$

$$\text{For flow 2, } s_2' = 1.151 \left[\frac{(772.56 - 442.94) \times 10^6}{17.20 \times 10^6} - \log\left(\frac{8.57}{0.1004 \times 0.0235 \times 0.00023 \times 0.4271^2} \right) + 3.23 \right] = 14.46$$

$$\text{For flow 3, } s_3' = 1.151 \left[\frac{(772.56 - 318.91) \times 10^6}{21.60 \times 10^6} - \log\left(\frac{8.04}{0.1004 \times 0.0235 \times 0.00023 \times 0.4271^2} \right) + 3.23 \right] = 16.32$$

$$\text{For flow 4, } s_4' = 1.151 \left[\frac{(772.56 - 243.01) \times 10^6}{20.83 \times 10^6} - \log\left(\frac{10.36}{0.1004 \times 0.0235 \times 0.00023 \times 0.4271^2} \right) + 3.23 \right] = 20.63$$

Table 5–14
Calculations for True Skin Factor for Example 5–7
(Using Pseudopressure Approach)

N	q	q^2	s	sq
1	2.397	5.746	9.06	21.72
2	5.214	27.186	14.58	76.02
3	6.144	37.749	16.32	100.27
4	7.148	51.094	20.63	147.46
$\sum$	20.885	121.775	60.59	345.47

Calculate s and D using the least squares method. Table 5–14 shows calculations for true skin factor using the pseudopressure approach.
Using Eq. 5–71,

$$s = \frac{\sum s \sum q^2 - \sum sq \sum q}{N \sum q^2 - \sum q \sum q}$$

$$= \frac{(60.59 \times 121.775) - (345.47)(20.885)}{4(121.775) - (20.885)(20.885)} = 3.21$$

Using Eq. 5–72,

$$D = \frac{N \sum sq - \sum s \sum q}{N \sum q^2 - \sum q \sum q}$$

$$= \frac{4(345.47) - (60.59)(20.885)}{4(121.775) - (20.885)(20.885)} = 2.28725 \text{ mmscfd}^{-1}$$

Alternatively, a plot of s versus q is shown in Figure 5–28 with a best straight line. From this plot intercept, $s = 3.51$ and slope $D = 2.39508$ mmscfd^{-1}.

Example 5–8[21] *Analyzing Multirate Drawdown Test (Assuming Semisteady-State Conditions)*

A gas well is tested by producing at four different rates for periods of 6 hr, followed by a 147-hr pressure buildup. The rate, times, and observed pressures during the flow test are listed below. The reservoir temperature and

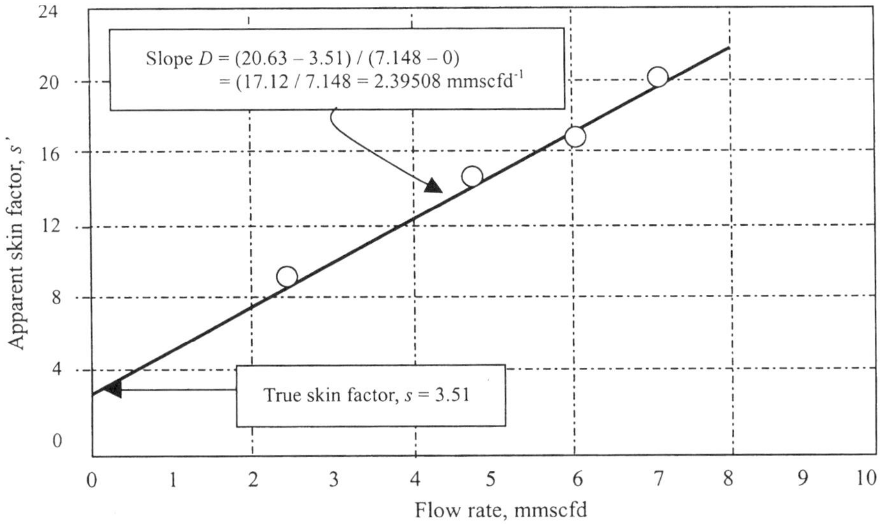

Figure 5–28. True skin factor determination using pseudopressure approach —Example 5–7.

gas properties are the same as detailed in Table 5–10. The reservoir and well data are as follows: $p_R = 3700$ psia; $s_g = 0.733$; $\mu = 0.0235$ cP; $z = 0.9491$; $c_t = 0.00023$ psi^{-1}; $\phi = 0.137$; $\phi_{HC} = 0.1004$; $T = 710°$R; $h = 41$ ft; $r_w = 0.4271$ ft. Calculate k, s, and F using transient flow conditions.

Cumulative time, (hr)	P_{wf} (psia)	$\psi(P_{wf})$ (mmpsia^2/cP)	q_{sc} (mmscfd)
0	3700	772.56	—
6	3144	592.45	2.397
12	2310	418.12	5.214
18	2158	306.21	6.144
24	1836	227.24	7.148

Solution The necessary calculation to carry out the method proposed in this section is given in Table 5–15.

Table 5–15
Calculations for Multirate Drawdown Test for Example 5–8 (Assuming Semisteady State)

t (hr)	p_{wf} (psia)	$\psi(p_{wf})$ (mmpsia²/cP)	$\Delta\psi$ (mmpsia²/cP)	q_{sc} (mmscfd)	Function X —	Function Y $F = 0.00$	$F = 0.04$	$F = 0.05$	$F = 0.07$
0	3700	772.56	0	—	0				
6	3144	592.45	180.11	2.397	0.7782	75.14	75.04	75.02	74.97
12	2310	378.74	393.82	5.214	0.9166	75.53	75.35	75.27	75.20
18	2158	306.21	466.35	6.144	1.1023	75.90	75.66	75.60	75.47
24	1836	227.24	575.32	7.148	1.2073	76.29	76.00	75.93	75.79

For $t = 24$ hr, $q_n = 7.148$ mmscfd, $t_n = 24$, the function X is

$$\frac{1}{q_n}\sum_{j=1}^{n}(q_j - q_{j-1})\log(t_n - t_{j-1})$$

$$= \frac{1}{7.148}\begin{bmatrix}(2.397)\log(24-0)\\ +(5.214-2.397)\log(24-6)\\ +6.144-5.214)\log(24-12)\\ +(7.148-6.144)\log(24-18)\end{bmatrix}$$

$$= \frac{1}{7.148}[3308.33 + 3536.18 + 1003.66 + 781.31]$$

$$= \frac{8629.48}{7.148} = 1.2073$$

A plot of

$$\left[\frac{\psi(p_R) - \psi(p_w) - Fq_n^2}{q_n}\right] \text{ versus } \frac{1}{q_n}\sum_{j=1}^{n}(q_j - q_{j-1})\log(t_n - t_{j-1})$$

is given in Figure 5–29. In this type of analysis the value of F must be obtained by trial and error until a straight line is achieved. The value of $F = 0.05$ gives an exact straight line. The analysis is shown in Table 5–15. From this plot, the following results are obtained: slope $m' = 4.739$ mmpsia²-cP/mmscfd = 4739 mmpsia²-cP/mscfd; intercept $b' = 73.05$. Thus, from Eq. 5–51,

$$k = \frac{57{,}920TP_{sc}}{m'hT_{sc}}$$

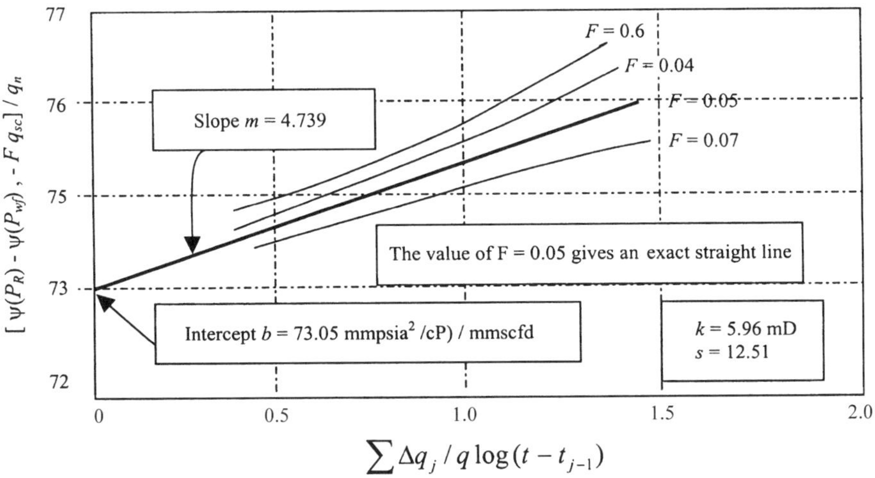

Figure 5–29. Multirate drawdown test under assumed semisteady-state conditions.

for standard conditions at $P_{sc} = 14.65$ psia and $T_{sc} = 520°$R.

$$k = \frac{57{,}920 \times 710 \times 14.65}{4739 \times 41 \times 520} = 5.96 \text{ mD}$$

From Eq. 5–41,

$$s = 1.151 \left[\frac{\Delta\psi_0}{m'} - \log \frac{k}{\phi\, \mu_i c_i\, r_w^2} + 3.23 \right]$$

$$= 1.151 \left[\frac{73.05}{4.739} - \log \left(\frac{5.96}{0.1004 \times 0.0235 \times 0.00023 \times 0.4271^2} \right) + 3.23 \right]$$

$$= 1.151\,[15.415 - 7.78 + 3.23] = 12.51$$

Example 5–9[21] *Analyzing Multirate Drawdown Tests (Assuming Steady-State Conditions)*

Producing it at four different rates over a total period of 48 hr tests a well in a gas reservoir. The rate–time sequence and pressures recorded at the end of each separate flow period are listed in Table 5–16. The reservoir and well data are as follows: $P_I = 4290$ psia; $h = 40$ ft; $r_w = 0.3$ ft; $(\mu c)_i = 3.6 \times 10^6$ cP/psi;

Table 5–16
Gas Well Test Analysis Assuming Steady-State Conditions

Time, t (hr) (1)	Flow rate mmscfd (2)	Assumed $P_i = 4200$ (psia) (3)	Assumed $P_i = 4280$ (psia) (4)	Assumed $P_i = 4290$ (psia) (5)	Assumed $P_i = 4300$ (psia) (6)
		Function Y	Function Y	Function Y	Function Y
12.00	10.00	625	3393	3738	4083
24.00	20.00	2666	4050	4222	4395
36.00	30.00	3681	4604	4719	4834
48.00	40.00	4428	5120	5206	5292

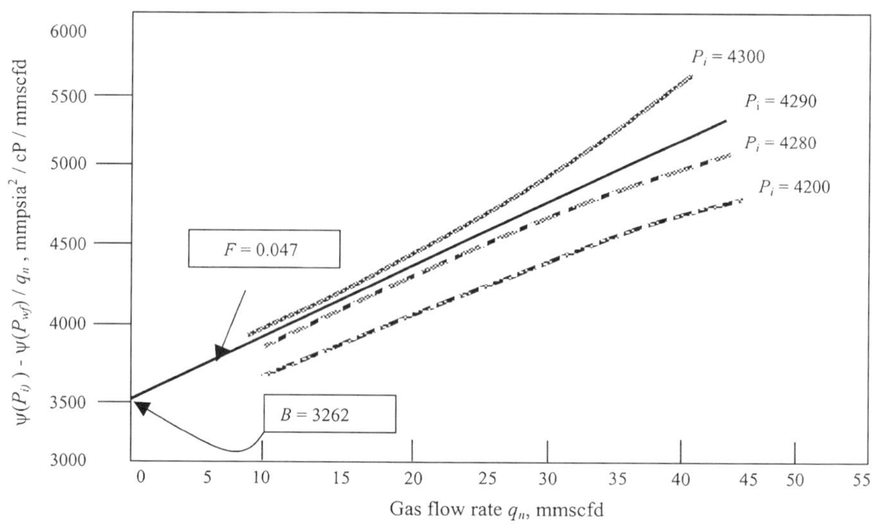

Figure 5–30. Gas well test analysis assuming semilog steady state.

$T = 200°\text{F}$; $\phi = 0.20$ fraction. Find the following: (1) the values of B and F; (2) the long-term deliverability equation and inflow performance response (IPR); and (3) the pore volume drained. The solution is shown in Figure 5–30, which is a plot of

$$\left[\frac{\Psi(P_i) - \Psi(P_{WF})}{q_n}\right]$$

Table 5–17
Evaluation of Non-Darcy Flow Coefficient

Cum. time (hrs) (7)	Flow time (hrs) (8)	Gas rate mmscfd (9)	Sum (hrs) (10)	Function (11)	Assume $F = 0.047$ (12)	Assume $F = 0.040$ (13)	Assume $F = 0.035$ (14)	Assume $F = 0.030$ (15)
12.00	12.00	10.00	12.00	37	3268	3338	3388	3438
24.00	12.00	20.00	18.00	84	3282	3422	3522	3622
36.00	30.00	30.00	24.00	142	3309	3519	3669	3819
48.00	40.00	40.00	30.00	208	3326	3606	3806	4006

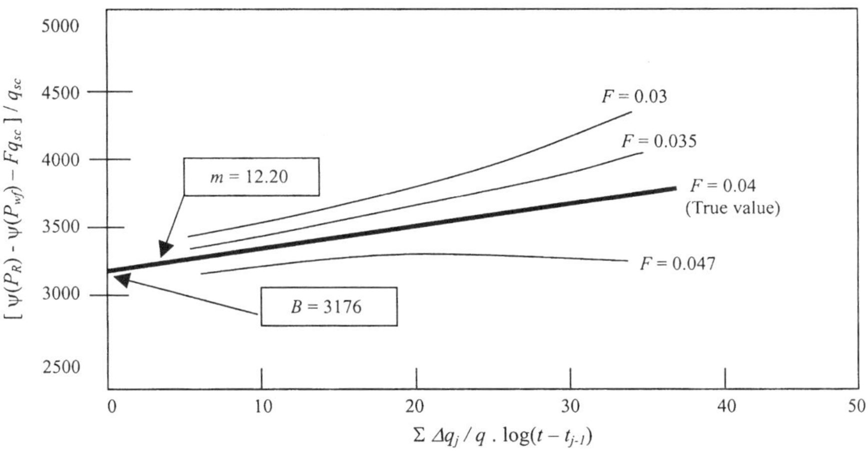

Figure 5–31. Gas well test analysis assuming semisteady-state conditions—Example 5–9.

versus flow rate, on coordinate paper. The intercept $B = 3262$ psia^2 /cP/mscfd and the slope $F = 0.047$ psia^2 /cP/mscfd^2.

$$\psi(\bar{p}) - \psi(p_{wf}) = 3262\, q_{sc} + 0.047\, q_{sc}^2$$

Table 5–17 shows evaluation of non-Darcy flow.

1. Plot columns (12, 13, 14, and 15) versus column 10 on coordinate paper. If the plot is not linearized, then choose other values of F.
2. Find the slope, which is 12.20 psia^2/cP/mscf/d/hr, and the intercept, $B = 3176$ psia^2/cP/mscfd (see Figure 5–31).
3. The stabilized long-term gas deliverability equation (see Table 5–18) is

$$\psi(\bar{p}) - \psi(p_{wf}) = 3176 q_{sc} + 0.04 q_{sc}^2$$

Table 5–18
Predicted Long-Term Gas Deliverability

Bottom-hole pressure P_{wf} (psia) (1)	Pseudopressure $\psi(P_{wf})$ psia^2/cP (2)	Pressure ratio — (3)	Flow rate ratio — (4)	Predicted flow rate q_g (mmscfd) (5)
4273	1062.80E+06	1.00000	0.00000	0
4200	1037.53E+06	0.97623	0.14292	1.314
4000	9677.70E+05	0.91058	0.28960	2.663
3800	8976.12E+05	0.84457	0.38615	3.551
3600	8275.12E+05	0.77861	0.46351	4.263
3400	7578.78E+05	0.71309	0.52957	4.871
3200	6890.79E+05	0.64836	0.58776	5.406
3000	6214.61E+05	0.58474	0.63993	5.885
2800	5553.47E+05	0.52253	0.68719	6.320
2600	4910.55E+05	0.46204	0.73028	6.716
2400	4288.98E+05	0.40355	0.76969	7.079
2200	3692.00E+05	0.34738	0.80576	7.411
2000	3132.00E+05	0.29385	0.83872	7.714
1800	2585.65E+05	0.24329	0.86872	7.990
1600	2083.94E+05	0.19608	0.89583	8.239
1400	1622.32E+05	0.15265	0.92009	8.462
1200	1205.75E+05	0.11345	0.94145	8.659
1000	8398.06E+04	0.07902	0.95983	8.828
800	5307.66E+04	0.04994	0.97508	8.968
600	2856.97E+04	0.02688	0.98701	9.078
500	1895.99E+04	0.01784	0.99165	9.120
AOF 14.65	1549.71E+03	0.00000	1.00000	9.188

$$\text{Pore volume (drained)} = \frac{2.359T}{(\mu c)_i \phi\, F} = \frac{2.359\,(660)}{3.6 \times 10^{-6} \times 0.2 \times 0.04}$$

$$= 35.5 \times 10^6\ \text{ft}^3$$

$$\phi h A = 35.5 \times 10^6$$

or

$$\text{Area(acres)} = \frac{35.5 \times 10^6}{0.2\,(40)\,(43560)} = 101\ \text{acres}$$

Variable-Rate DrawdownTests

In this section a theory that handles the drawdown analysis for variable rates is presented for gas wells. We assume that the outer boundary of the reservoir never affects the pressure behavior at the well. This allows us to set correction terms from Ref. 8. An approximate equation relating time and drainage radius is

$$t = -\mu\phi r_d^2 \, In\, C \tag{5–73}$$

where t = time (sec), r_d = drainage radius (cm), and C = correlation term = 0.410. For infinite reservoir behavior, from Refs. 8 and 9, considering q positive for a flowing well, we obtain

$$p_{(@\,t=0)}^2 - p_{wf}^2 = \frac{W\mu z_{av} RT}{kh\pi M} In\left(\frac{r_d}{r_e}\right) - \frac{\beta W^2 z_{av} RT}{2\pi^2 r_w h^2 M} \tag{5–74}$$

From Equation 5–73,

$$In\left(\frac{r_d}{r_w}\right) = 0.5\, In(t) - 0.5\, In\left(\frac{-\phi\mu r_w^2\, InC}{2kp_{(@\,t=0)}}\right) \tag{5–75}$$

Combining Eqs. 5–74 and 5–75, we have

$$p_{(@\,t=0)}^2 - p_{wf}^2 = 0.5\frac{W\mu z_{av} RT}{kh\pi M}\left[(In\, t) - \left(In\frac{-\phi r_w^2 InC}{2kp_{(@\,t=0)}}\right)\right] - \frac{\beta W^2 z_{av} RT}{2\pi^2 r_w h^2 M} \tag{5–76}$$

The pressure drawdown for a radial gas flow at constant rate in an infinite reservoir is described by[7]

$$p_i^2 - p_{wf}^2 = 0.5\frac{W\mu z_{av} RT}{kh\pi M}\left[(Int) - \left(In\frac{-\phi r_w^2 InC}{2kp_{(@\,t=0)}}\right)\right] - \frac{\beta W^2 z_{av} RT}{2\pi^2 r_w h^2 M} \tag{5–77}$$

where W = mass flow rate in g/sec, C = a correlation constant $\cong$ 0.41, and all other symbols denote the usual terms in cgs units. This solution is analogous to the point source solution for liquid flow. After converting terms in Eq. 5–77 to engineering units, we have

$$p_i^2 - p_{wf}^2 = \frac{712.4q\mu z_{av} T}{kh}\left[In\frac{ktp(@t=0)}{\phi\mu r_w^2} - 7.432\right] - 0.11 \times 10^{-12}\frac{Mq^2\beta z_{av} T}{r_w h^2} \tag{5–78}$$

where q = gas flow rate, mscfd; z_{av} = average gas compressibility factor; T = reservoir temperature, R°; t = flow duration, hr; ϕ = gas-filled porosity; M = gas molecular weight, lb/lb mole; and β = quadratic flow constant, ft^{-1} (must be negative).

Equation 5–78 describes pressure drawdown of a gas well where the formation around it is neither damaged nor improved. If a condition of permeability damage or improvement exists, an addition term must be added to the equation to amount for these effects. Thus Eq. 5–78 becomes

$$p_i^2 - p_{wf}^2 = m\left[q\,In\,t + q\left(In\frac{kp_{@(t=0)}}{\phi\mu r_w^2}\right) - 7.432 + 2s\right] + D'q^2 \quad (5\text{–}79)$$

where s is the skin factor and is dimensionless, and

$$m = \frac{712.4\mu z_{av}T}{kh}$$

$$D' = -0.11 \times 10^{-12}\frac{M\beta z_{av}T}{r_w h^2} \quad (5\text{–}80)$$

D' is related to the non-Darcy flow constant D of Ramey[9] by D':

$$D' = 2mD \quad (5\text{–}81)$$

When considering a variable-rate drawdown, we apply superposition. We consider the variable flow rate as a series of constant flow rates $q_1, q_2, q_3 \ldots q_n$. Although the total pressure-squared drop is given by three quantities; namely,

$$mq\left(In\frac{k\,p_{@(\Delta t=0)}}{\phi\mu r_w} - 7.432 + 2\,s\right),\ In\,t,\ \text{and}\ D'q^2$$

superposition affects only the first. The other two are dependent only on the instantaneous rate q_n. We define q_n as the sum of a series of delta rates Δq_i, where $\Delta q_0 = q_1 - q_0$, $\Delta q_1 = q_2 - q_1$, $\Delta q_2 = q_3 - q_2$, etc., and $q_0 = 0$. For a series of rates the pressure-squared drop due to the term $mqIn\,t$ will then be

$$m\sum_{i=0}^{n=1}\Delta q_i In\,(t_n - t_i)$$

and total pressure-squared drop is given by

$$p_{(@\,t=0)}^2 - p_{wf}^2 = m\left[\sum_{i=0}^{n-1}\Delta q_i In(t_n - t_0) + q_n\left(In\frac{kp_{(@\,t=0)}}{\phi\mu r_w^2} - 7.432 + 2s\right)\right] + D'q_n^2 \quad (5\text{–}82)$$

where t_n is the total flow time, t_i is the time when change in rate was initiated, and $t_0 = 0$. In many cases the quadratic term contributes only a small amount to the total drop and we can neglect it. Then Eq. 5–82 will become

$$p^2_{(@\,t=0)} - p^2_{wf} = m\left[\sum_{i=0}^{n-1} \Delta q_i In(t_n - t_i) + q_n\left(In\frac{kp_{(@\,t=0)}}{\phi\mu r_w^2} - 7.432 + 2s\right)\right] \tag{5–83}$$

If we plot

$$\left(p^2_{@\,t=0} - p^2_{wf}\right)/q_n \text{ versus } \left[\frac{1}{q_n}\sum_{i=0}^{n-1} \Delta q_i In\,(t_n - t_i)\right]$$

we should have a straight line with slope of m and intercept B equal to $m(In\frac{kp_{(@\,t=0)}}{\phi\,\mu\,r_w^2} - 7.43 + 2s)$, provided that the quadratic flow term is negligible. Then

$$kh = \frac{712.4\mu z_{av} T}{m} \tag{5–84}$$

When the gas rate of flow per unit thickness is high, the quadratic term can no longer be neglected. The quadratic effect puts a bow in the plot of

$$\left(p^2_{(@\,t=0)} - p^2_{wf}\right)/q_n \text{ versus } \frac{1}{q_n}\sum_{i=0}^{n-1} \Delta q_i In(t_n - t_i)$$

making it impossible to determine slope m or intercept B. This problem can be overcome quite easily, however, by a trial-and-error correction of the $p^2_{(@\,t=0)} - p^2_{wf}$ values. If the quadratic effect were not present, the plot would give a straight line. Therefore, we can correct the pressure data by using different D' values. The D' value that, when multiplied by q_n^2, results in correction factors that straighten out the plot is the proper value. The correct plot is then used to determine m and B as previous explained. The skin factor s is given by

$$s = 0.5\left(\frac{B}{m} + 7.432 - In\frac{kp^2_{(@\,t=0)}}{\phi\mu r_w^2}\right) \tag{5–85}$$

and the pressure-squared drop due to skin is

$$(\Delta p^2)_{skin} = 2(s)(m)(q_n) \tag{5–86}$$

If log is used in place of *In*, Eqs. 5–79, 5–81, 5–85, and 5–86 become

$$p^2_{(@\,t=0)} - p^2_{wf} = \frac{1640 q \mu z_{av} T}{kh}\left[\log\frac{ktp_{(@\,t=0)}}{\phi\mu r_w^2} - 3.23 + 0.87s\right] - 0.11\times 10^{-12}\frac{Mq^2\beta z_{av}T}{r_w h^2} \tag{5–87}$$

$$D' = 0.869\, m\, D \tag{5–88}$$

$$s = 1.151\left[\frac{B}{m} + 3.23 - \log\frac{kp_{(@\,t=0)}}{\phi\mu r_w^2}\right] \tag{5–89}$$

and

$$(\Delta p^2)_{skin} = 0.869\, m\, q_n \tag{5–90}$$

where m is the slope of the straight line when $(p^2_{(@\,t=0)} - p^2_{wf})/q_n$ versus $\frac{1}{q_n}\sum_{i=0}^{n-1}\Delta q_i \log(t_n - t_i)$ is plotted, and kh is given by

$$kh = \frac{1640\,\mu\, z_{av} T}{m} \tag{5–91}$$

The following example will illustrate the applications of these equations.

Example 5–10[21] *Analyzing Variable-Rate Drawdown Test*

A backpressure test is used to illustrate the method of analysis. The data used to simulate the gas well are presented in Table 5–19 and as follows: r_w = 0.23 ft; ϕ = 0.16; s_w = 0.20; h = 40 ft; μ_g = 0.017 cP; c_t = 6.89×10^{-4} psi^{-1}; $\beta_g = 8.28 \times 10^{-3}$ cu ft/cu ft; gas gravity = 0.70; T_R = 560°R; T_{sc} = 520°R; P_{sc} = 14.7 psia.

Solution Figure 5–32 is a plot of Δp^2 versus $\frac{1}{q_n}\sum_{j=0}^{n}\Delta q_i \log(t_n - t_j)$ from the backpressure test. From this Figure 5–32, we find the slope m = 370.7 and

Table 5–19
Backpressure Test Data

Point #	Duration t (hr)	P_{WF} (psia)	Gas rate (mscfd)
—	0	1691	0
1	1.25	1682	1048
2	2.25	1667	2101
3	3.25	1637	4167
4	4.25	1609	5116

Table 5–20
Pressure and Production Rate Data and Calculation of Ordinate

n	t (hr)	P_{wf} (psig)	q_n mscfd	$\Delta p^2/q_n$ (sia^2/mscfd)	$\frac{1}{q_n}\sum_{i=1}^{n-1}\frac{q_i - q_{i-1}}{q_n}\log(t_n - t_{i-1})$
	0	1676	0	—	—
1	1.25	1667	1048	0.07729	0.09691
2	2.25	1652	2101	0.27416	0.17568
3	3.25	1622	4167	0.69978	0.20481
4	4.25	1594	5116	1.31431	0.34849

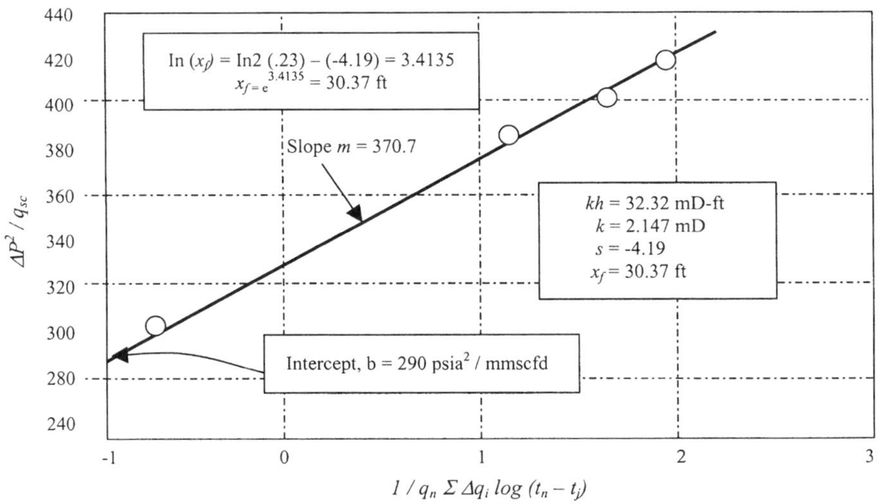

Figure 5–32. Calculation of *kh*, *k*, and *s* from backpressure test.

intercept $B = 290$ psia^2 /mmscfd and calculate k and s from Eqs. 5–91 and 5–89, respectively (see Table 5–20):

$$k = \frac{1{,}640\,\mu\, z_{av} T}{mh} = (1640 \times 0.013 \times 1.0 \times 560)/(370.7 \times 15)$$
$$= 2.147 \text{ mD}$$

and

$$s = 1.151\left[\frac{290}{370.7} + 3.23 - \log\left(\frac{2.147}{0.1 \times 0.013 \times 6.89 \times 10^{-4} \times 0.23^2}\right)\right]$$
$$= -4.19$$

For $n = 3$, $q_n = 4167$ mscfd, $t_n = 3.25$ hr, $q_0 = 0$, $t_0 = 0$, the ordinate is

$$\begin{aligned}\text{ordinate} &= \frac{1}{4167}[(q_1 - q_0)\log(t_n - t_0) + (q_2 - q_1)\log(t_n - t_1) \\ &\quad + (q_3 - q_2)\log(t_n - t_2)] \\ &= \frac{1}{4167}[(1048 - 0)\log(3.25 - 0) + (2101 - 1048) \\ &\quad \times \log(3.25 - 1.25) + (4167 - 2.101)\log(3.25 - 2.25)] \\ &= \frac{1}{4167}[(1048)(0.5119) + (1053)(0.3010) + (2066)(0)] \\ &= \frac{1}{4167}[536.47 + 316.95 + 0] \\ &= 0.2048\end{aligned}$$

From the plot (see Figure 5–33) we have $m = 0.02904$, $b = 0.00625$. Thus, from Eq. 5–91, find

$$kh = \frac{28{,}958\mu_g\beta_g}{m} = \frac{(28{,}958)(0.017)(8.28 \times 10^{-3})}{0.02904} = 140 \text{ mD-ft}$$

$$k = 140/40 = 3.5\text{mD}$$

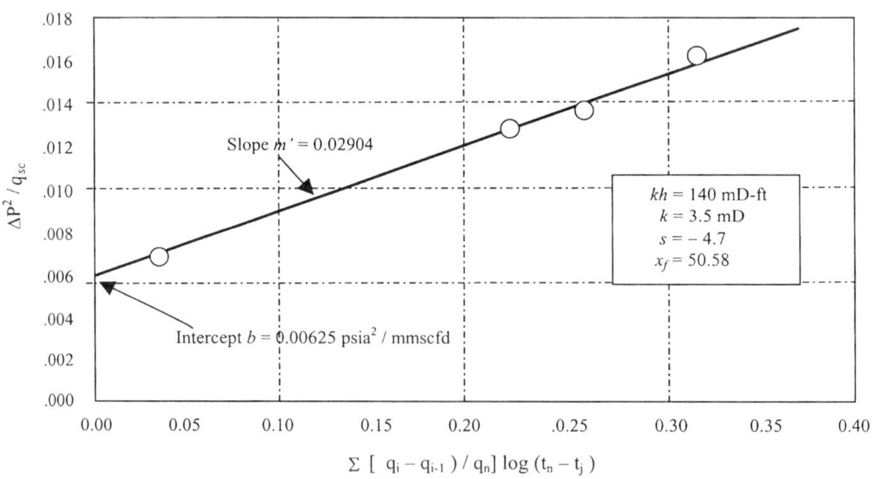

Figure 5–33. Calculation of *kh*, *k*, and *s* from variable-rate test—Example 5–10.

From Eq. 5–89:

$$s = 1.151\left[\frac{b}{m} - \log\frac{k}{\phi\mu_g c_t r_w^2}\right]$$

$$= 1.151\left[\frac{0.00625}{0.02904} - \log\frac{3.5}{(0.16)(0.017)(6.89\times 10^{-4})(0.052)} + 3.23\right]$$

$$= -4.7$$

Rate Normalization of Drawdown Pressure by Using After-Flow Data

Equation 5–83 can be expressed in terms of the real gas pseudopressure, $\Psi(p)$, as

$$\psi(p_i) - \psi(p_{wfn}) - D'q_{gn}^2 = \frac{711\,T_r}{kh}\,[In(t_D + 0.809) + 2s] \tag{5–92}$$

where

$$D' = \frac{711T_r}{kh}(2D) \tag{5–93}$$

A superposition equation for any variation of rate was given by Odeh and Jones[8] with p_D being approximately by the logarithm of time as

$$\frac{p_i^2 - p_{wfn} - D'q_{gn}^2}{q_{gn}} = \frac{711(\mu_g z)T_r}{kh}\left\{\frac{1}{q_{gn}}\left[\sum_{i=1}^{n}(q_{gi} - q_{gi-1})In(t_n - t_{i-1})\right] + In\frac{k}{\phi\mu_g c_t r_w^2} - 7.432 + 2s\right\} \tag{5–94}$$

A plot of

$$\frac{p_i^2 - p_{wfn}}{q_{gn}} \text{ versus } \frac{1}{q_{gn}}\sum_{i=1}^{n}(q_{gi} - q_{gi-1})In(t_n - t_{i-1})$$

should be linear if $D' = 0$, with slope m' and intercept b yielding kh and s, respectively. Flow capacity is evaluated from the slope m' by

$$kh = \frac{711(\mu_g z)T_r}{m'} \tag{5–95}$$

and the skin from intercept b by

$$s = 0.5\left(\frac{b}{m'} - In\frac{k}{\phi\mu_g c_t r_w^2} + 7.432\right) \tag{5–96}$$

Odeh and Jones[8] further recommended that if the plot bows, the data should be corrected for the quadratic effect such as $D'q^2$ until the plot is made linear. If this approach is not applicable, the logarithm of time approximation to p_D is used by normalization. Equation 5–97 can be expressed in terms of the real gas pseudopressure as

$$\frac{\Psi(p_i) - \psi(p_{wfn}) - D'q_{gn}^2}{q_{gn}} = \frac{1422T_r}{kh}\left\{\frac{1}{q_{gn}}\left[\sum_{i=1}^{n}(q_{gi} - q_{gi-1})p_D(t_n - t_{i-1})_D\right] + s\right\} \tag{5–97}$$

which, if $D' = 0$, also should result in a straight line when plotted as

$$\frac{\psi(p_i) - \psi(p_{wfn})}{q_{gn}} \text{ versus } \frac{1}{q_{gn}}\sum_{i=1}^{n}(q_{gi} - q_{gi-1})p_D(t_n - t_{i-1})_D$$

with slope m' given by

$$m' = \frac{1422T_r}{kh} \tag{5–98}$$

and intercept b equal to

$$b = \frac{1422T_r}{kh}(s)$$

or

$$s = b\frac{kh}{1422T_r} \tag{5–99}$$

and

$$x_f = e^{In(x_f)} \tag{5–99a}$$

where

$$In(x_f) = In(2r_w) - s \tag{5–99b}$$

Table 5–21
Calculations for Drawdown Rate Normalization

Time (hr)	p_{wf} (psia)	Gas rage (mscfd)	$\psi(p_{wf})$ (mmpsia²/cp)	$\Delta\psi(p)$ (mmpsia²/cp)	$\Delta\psi(p)/q$ (mpsia²/cp/mscfd)	$\sum$ —	p_D–t_D —
0	4185	0	1202.50	0	0	—	—
0.25	4079	1757	1153.95	48.55	27.80	−1.300	0.145
0.75	4025	1468	1129.32	73.18	49.85	−0.200	0.289
1.00	4000	1482	1117.95	84.55	57.05	0.120	0.295
7.00	3926	1494	1084.38	118.12	79.06	0.950	0.405
3.00	3888	1443	1067.20	135.30	93.76	1.180	0.491
6.00	3794	1443	1024.91	177.60	123.07	1.902	0.625
24.00	3650	1141	960.73	241.80	211.83	3.254	1.085
48.00	3562	1054	921.90	280.60	266.22	3.952	1.381
72.00	3478	1019	885.14	317.36	311.44	4.252	1.524

Example 5–11[21] *Normalizing Drawdown Rate*

Table 5-21 shows both the drawdown pressure and rate data declining as a function of time. Other reservoir and well data are as follows: $p_I = 4185$ psia; $T = 635°\text{R}$; $z = 0.7487$; $r_w = 0.29$ ft; $\mu = 0.0125$ cP; $c_t = 0.0000105$ psi^{-1}; $\phi = 0.11$; $h = 106$ ft.

Calculate k, s, and fracture half-length.

Solution Figure 5–34 is an Odeh-Jones superposition plot of the drawdown data where the logarithm of time approximation p_D–t_D is used. From this plot, slope m' and intercept b are 111,200 psi²/cP/mscfd and 182,400 psi²/cP/mscfd, respectively.

Flow capacity is obtained from Eq. 5–95:

$$kh = \frac{711\mu_g z T_r}{m'} = \frac{711 \times 0.0125 \times 0.7487 \times 635}{111{,}200}$$
$$= 4.028 \text{ mD-ft}$$
$$k = \frac{4.028}{106} = 0.038 \text{ mD}$$

and the skin factor s is obtained from Eq. 5–96:

$$s = 0.5\left(\frac{b}{m'} - In\frac{k}{\phi\mu_g c_t r_w^2} + 7.432\right)$$
$$= 0.5\left(\frac{182{,}400}{111{,}200} - In\frac{0.038}{0.11 \times 0.0125 \times 0.0000105 \times 0.29^2} + 7.432\right)$$
$$= -4.11$$

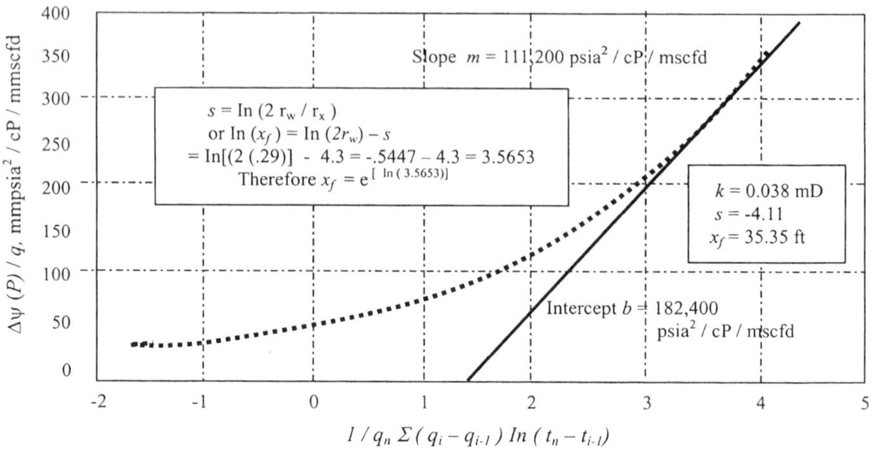

Figure 5–34. Superposition based on logarithm of time approximation to p_D—Example 5–11.

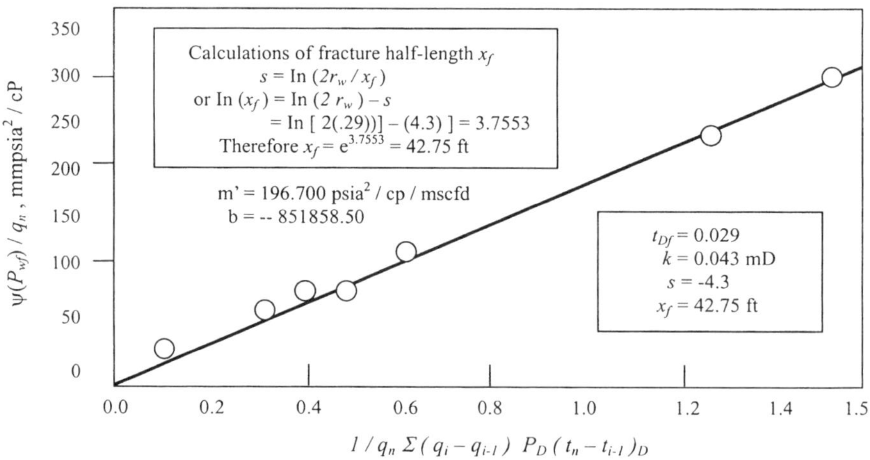

Figure 5–35. Superposition based on $p_D - t_D$ model—Example 5–11.

Figure 5–35 represents a superposition analysis based on values of p_D and t_D. They are obtained from the infinite-conductivity vertical fracture solution. All points fall on a straight line. From Eq. 5–98:

$$m' = 196{,}700 \text{ psi}^2/\text{cP/mscfd}$$

$$\therefore k = \frac{1422T_r}{m'h} = \frac{1422 \times 635}{196{,}700 \times 106} = 0.043 \text{ mD}$$

and the skin factor s is obtained from Eq. 5–99

$$s = b\frac{kh}{1422T_r} = -851{,}858.50\frac{0.043 \times 10^6}{1422 \times 635} = -4.3$$

The calculated P_D–t_D superposition value of permeability, $k = 0.043$ mD, is the same as that calculated from the normalized type curve analysis, whereas the fracture length x_f is 35.35 ft, an compared with 42.75 ft, an essentially identical result.

5.9 Analysis of Pseudo-Steady-State Flow Data

If the drawdown were continued, the pseudo-steady-state would be indicated by a deviation from the semilog straight line. As a matter of interest, the approximate time to stabilization is calculated from Eq. 2–143. Various tests utilizing pseudo-steady-state flow data are described. In particular, a useful variation of a reservoir limits test, called an economic limits test, which utilizes transient flow data instead of pseudo-steady-state flow data is discussed.

Reservoir Limits Test

If a single-rate drawdown test is allowed to flow until the reservoir boundary is felt (pseudo-steady-state), the pressure behavior is governed by equation given below for a circular reservoir.[14]

$$\Delta p_D = \frac{2t_D}{r_{eD}^2} + In\, r_D - \frac{3}{4} + s' \tag{5–100}$$

By rearranging Eq. 5–100, with appropriate substitution for the dimensionless terms, real variables and common log may be written as

$$\Psi(p_i) - \Psi(p_{wf}) = \frac{83{,}342\, q_{sc} T t p_{sc}}{\pi\, \phi\, \mu_{gi} c_i\, r_e^2 h T_{sc}} + 50.30 \times 10^6 \frac{q_{sc} T p_{sc}}{k\, h\, T_{sc}} \times \left[In\left(\frac{r_e}{r_w}\right) - \frac{3}{4} + s'\right] \tag{5–101}$$

In Eq. 5–101, $\pi\phi r_e^2 h$ represents the gas filled pore volume, v_P, of the reservoir. Equation 5–101 may also be written as

$$\Psi(p_i) - \Psi(p_{wf}) = \frac{83{,}342\, q_{sc} T t p_{sc}}{\pi\phi\mu_{gi} c_i\, r_e^2 h T_{sc}} + 115.820 \times 10^6 \frac{q_{sc} T p_{sc}}{k h T_{sc}} \times \left[\log\left(\frac{0.472 r_e}{r_w}\right) + \frac{s'}{2.303}\right] \tag{5–102}$$

A plot of $\Delta\Psi\lfloor= \Psi(p_i) - \Psi(p_{wf})\rfloor$ versus t on arithmetic coordinates will give a straight line of slope m''. Therefore,

$$v_P = \frac{83{,}342\, q_{sc} T p_{sc}}{\mu_{gi}\, c_i m''\, T_{sc}} \tag{5–103}$$

Also, v_P is equal to $\pi\phi h r_e^2$ and reservoir limits

$$r_e = \sqrt{\frac{v_P \times 10^6}{\pi\phi h}} \tag{5–104}$$

Defining a Minimum In-Place Gas Volume

The radius of investigation r_{inv} is, for $r_{inv} \leq r_e$,

$$r_{inv} = 2\sqrt{\left[\frac{2.637 \times 10^{-4} kt}{\phi\mu_{gi} c_i}\right]} \tag{5–105}$$

Define a minimum in-place gas volume, V_{pm}(in mmscf), as

$$V_{pm} \times 10^6 = \pi\phi h r_{inv}^2 \tag{5–106}$$

Equation 5–105 may be substituted in Eq. 5–104 to give

$$t = \frac{301.8 \times 10^6 \mu_{gi} c_i V_{mp}}{kh} \tag{5–107}$$

where t = duration of flow period required to conduct an economic limits test. To illustrate the technique just outlined, Example 5–12 will clarify the use of reservoir limits test data.

Example 5–12[21] *Estimating Reservoir Limit with Single-Rate Drawdown Test*

Estimate the pore volume of the reservoir with the single-rate drawdown test reported in Example 5–2. Given: p_i = 3965 psia; z_i = 0.9647; μ_I = 0.02458 cP; c_i = 0.000252 psi^{-1}; q_{sc} = 6.148 mmscfd.

Solution The first step is to plot p_{wf} versus time on ordinary Cartesian graph paper (Figure 5–36 and Table 5–22).

1. Plot column (5) versus column (2) on coordinate paper. (See Figure 5–36.)

Table 5–22
Reservoir Limit Test Data

Time t (min) (1)	Time t (hrs) (2)	Pressure P_{wf} (psia) (3)	$\psi(P_{wf})$ (psia²/cP) (4)	$\psi(P_i - P_{wf})$ (psia²/cP) (5)
1.2	0.02	1810.65	224.78E+06	636.34E+06
1.8	0.03	1807.45	224.04E+06	637.07E+06
4.2	0.07	1798.95	222.09E+06	639.03E+06
6.0	0.10	1786.35	219.21E+06	641.91E+06
7.8	0.13	1781.45	218.09E+06	643.02E+06
10.2	0.17	1775.75	216.90E+06	644.32E+06
15.0	0.25	1768.05	215.05E+06	646.07E+06
19.8	0.33	1764.75	214.30E+06	646.81E+06
30.0	0.50	1757.45	212.66E+06	648.46E+06
45.0	0.75	1754.65	212.03E+06	649.09E+06
60.0	1.00	1755.45	212.21E+06	648.91E+06
90.0	1.50	1757.85	212.75E+06	648.37E+06
120.0	2.00	1754.65	212.03E+06	649.09E+06
150.0	2.50	1754.65	212.03E+06	649.09E+06
180.0	3.00	1751.35	211.28E+06	649.83E+06
210.0	3.50	1748.95	210.74E+06	650.37E+06
240.0	4.00	1747.35	210.38E+06	650.73E+06
300.0	5.00	1745.25	209.91E+06	651.20E+06
330.0	5.50	1742.05	209.20E+06	651.92E+06
360.0	6.00	1740.45	208.84E+06	652.28E+06
390.0	6.50	1739.25	208.57E+06	652.55E+06
420.0	7.00	1738.35	208.37E+06	652.75E+06
450.0	7.50	1738.35	208.37E+06	652.75E+06
480.0	8.00	1737.95	208.28E+06	652.84E+06
510.0	8.50	1737.55	208.19E+06	652.93E+06
540.0	9.00	1737.15	208.10E+06	653.01E+06
570.0	9.50	1737.15	208.10E+06	653.01E+06
600.0	10.00	1735.55	207.74E+06	653.37E+06
660.0	11.00	1735.55	207.74E+06	653.37E+06
720.0	12.00	1734.35	207.48E+06	653.64E+06
780.0	13.00	1733.55	207.30E+06	653.82E+06
840.0	14.00	1733.55	207.30E+06	653.82E+06
900.0	15.00	1732.75	207.12E+06	654.00E+06
960.0	16.00	1731.05	206.74E+06	654.38E+06
1020.0	17.00	1730.65	206.65E+06	654.46E+06
1080.0	18.00	1730.25	206.56E+06	654.55E+06
1140.0	19.00	1729.45	206.38E+06	654.73E+06
1200.0	20.00	1719.75	204.23E+06	656.89E+06

Table 5–22 Continued

Time t (min) (1)	Time t (hrs) (2)	Pressure P_{wf} (psia) (3)	$\psi(P_{wf})$ (psia^2/cP) (4)	$\psi(P_i - P_{wf})$ (psia^2/cP) (5)
1260.0	21.00	1724.55	205.29E+06	655.82E+06
1320.0	22.00	1721.75	204.67E+06	656.44E+06
1380.0	23.00	1720.95	204.50E+06	656.62E+06
1440.0	24.00	1720.95	204.50E+06	656.62E+06

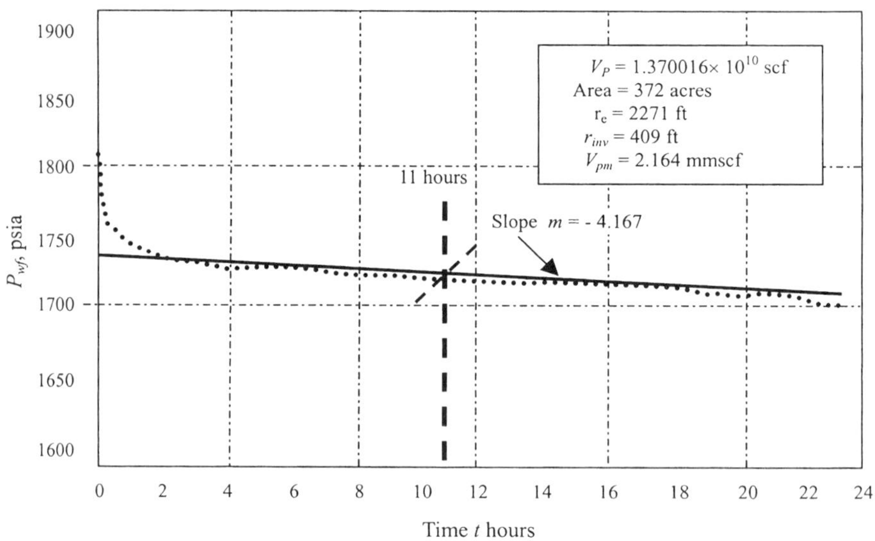

Figure 5–36. Reservoir limit test—Example 5–12.

2. Find slope in units of mmpsia^2/cP/hr.

$$\text{Slope} = m = \frac{1732 - 1734.5}{18 - 12} = -0.4167 \text{ psia/hr}$$

Thus from Eq. 5–103:

$$V_P = \frac{-0.234 q_{sc}}{c_i \frac{\partial p_{wf}}{\partial t}} = \frac{-0.234 \times 6.148 \times 10^6}{0.000252(-0.4167)} = 1.370016 \times 10^{10} \text{scf}$$

$$\beta_g = \frac{T_{sc}}{p_{sc}} \times \frac{p_i}{T z_i} = \frac{520}{14.65} \times \frac{3965}{710 \times 0.9647} = 205.475 \text{ scf/ft}^3$$

Therefore

$$V_P = \frac{1.370016 \times 10^{10}}{205.475} \frac{(\text{scf})}{(\text{scf/ft}^3)} = 66.675557 \times 10^6\ \text{ft}^3 = \phi h A$$

or

$$A = \frac{66.675557 \times 10^6}{0.1004 \times 41} = 16{,}197{,}541\ \text{ft}^2$$

From Eq. 5–104, acres = 16,197,541/43,560 = 372.

$$r_e = \sqrt{\frac{372 \times 43{,}560 \times 7}{22}} = 2271\ \text{ft}$$

The slope of this curve is constant for $t > 11$ hr; therefore the radius of investigation r_{inv} is calculated from Eq. 5–105:

$$r_{inv} = 2\left[\frac{2.637 \times 10^{-4} kt}{\phi \mu_i c_i}\right]^{0.5} = 2\left[\frac{2.637 \times 10^{-4} \times 8.96 \times 11}{0.1004 \times 0.02458 \times 0.000252}\right]^{0.5}$$
$$= 409\ \text{ft}$$

From Eq. 5–106, minimum in-place gas volume is given by

$$V_{pm} \times 10^6 = \pi \phi h r_{inv}^2 = 22/7 \times 0.1004 \times 41 \times 409^2$$
$$= 2{,}164{,}152.31\ \text{scf}$$
$$V_{pm} = \frac{2{,}164{,}157.31}{10^6} = 2.164\ \text{mmscf}$$

Example 5–13 *Estimating of Reservoir Pore Volume with Multirate Drawdown Tests*

Estimate the pore volume of the reservoir with the multirate drawdown tests reported in Example 5–9.

Solution The first step is to plot

$$\left[\frac{\psi(p_R) - \psi(p_{wf}) - F q_n^2}{q_n}\right] \text{ versus } \frac{1}{q_n}\sum_{j=1}^{n} \Delta q_j \log(t_n - t_{j-1})$$

as shown in Figure 5–37. The slope of this plot can be used and value of non-Darcy flow coefficient, F, decreased gradually until the plot becomes linear. The slope of the latter plot will then yield the value of the pore volume.

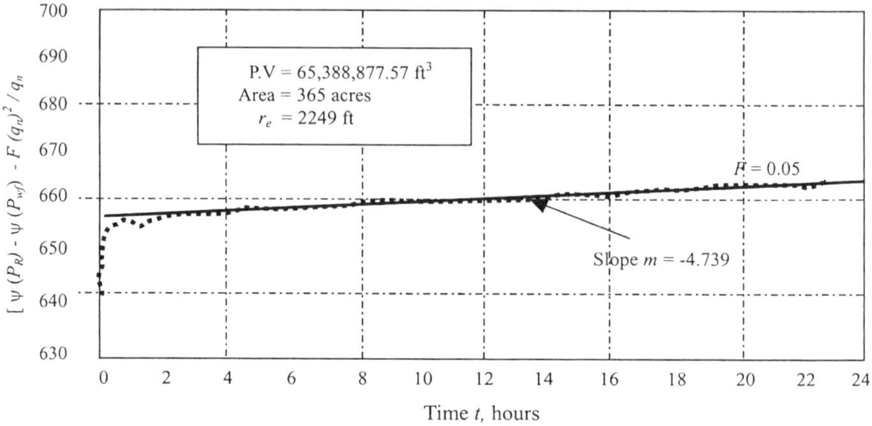

Figure 5–37. Estimation of pore volume with a multirate drawdown test—Example 5–13.

From Figure 5–37, we find $F = 0.05$, $m = -4.739$ and using Eq. 5–103:

$$\therefore A = \frac{65{,}388{,}877.57}{0.1004 \times 41} = 15{,}884{,}966.86\,\text{ft}^3 = 15{,}884{,}966.86/43{,}560$$

$$= 365 \text{ acres}$$

$$r_e = \sqrt{\frac{365 \times 43{,}560 \times 7}{22}} = 2249\,\text{ft}$$

Effect of Reservoir and Well Geometry

The pseudo-steady-state will start at a time given by the value of t_{DA} corresponding to the particular well and reservoir geometry:

$$t_{DA} = \frac{0.000264kt}{\phi \mu_{gi} c_i A} \tag{5–108}$$

or

$$t_{DA} = t_D \frac{r_w^2}{A} \tag{5–109}$$

$$\Delta p_D = p_t + s' \tag{5–110}$$

where p_t is dimensionless pressure drop at the well excluding skin and inertial-turbulent flow effects and is defined by Ramey and Cobb:[10]

$$p_t = \frac{1}{2} In \left(\frac{4A}{1.781 r_w^2 c A} \right) + 2\pi t_{DA} \tag{5–111}$$

Equation 5–110, with appropriate substitution for t_{DA} and dimensionless terms, may be written as

$$\Psi(p_i) - \Psi(p_{wf}) = 57.920 \times 10^6 \frac{q_{sc}Tp_{sc}}{khT_{sc}} \left[\frac{4.58 \times 10^{-4}kt}{\phi\mu_{gi}c_i A} \right] + 57.920 \times 10^6 \frac{q_{sc}Tp_{sc}}{khT_{sc}} \left[\log\left(\frac{2.246A}{r_w^2 cA} \right) + 0.869s' \right] \tag{5–112}$$

Equation 5–111 may be simplified further to give

$$\Psi(p_i) - \Psi(p_{wf}) = \frac{26{,}514.68q_{sc}Tp_{sc}t}{\phi\mu_{gi}c_i AhT_{sc}} + 57.920 \times 10^6 \frac{q_{sc}Tp_{sc}}{khT_{sc}} \times \left[\log\left(\frac{2.246A}{r_w^2 cA} \right) + 0.869s' \right] \tag{5–113}$$

5.10 Application of Stabilized Deliverability Equation

The most common reason for conducting drawdown tests, in addition to the calculation of the well reservoir parameter, is the determination of long-term deliverability. If flow tests are extended into the pseudo-steady-state region, the calculation of deliverability is relatively simple. The stabilized flow constants can be calculated by solving the simultaneous equations resulting from two single-rate tests at different flow rates or a two-rate test. The stabilized deliverability equations in terms of pressure squared is

$$\bar{p}_R^2 - p_{wf}^2 = A'q_{sc} + B'q_{sc}^2 \tag{5–114}$$

where

$$A' = \frac{50{,}300\, T\bar{\mu}\bar{z}p_{sc}}{khT_{sc}} \left(In\left(\frac{0.472r_e}{r_w} \right) + s \right)$$

and

$$B' = \frac{50{,}300\, T\bar{\mu}\bar{z}p_{sc}}{khT_{sc}} D$$

In terms of pseudopressure:

$$\Psi(p_i) - \Psi(p_{wf}) = Aq_{sc} + Bq_{sc}^2 \tag{5–115}$$

where

$$A = \frac{115.820 \times 10^6 T p_{sc}}{khT_{sc}} \left[\log\left(\frac{0.472 r_e}{r_w}\right) + \frac{s}{2.303} \right]$$

and

$$B = \frac{50.30 \times 10^6 T p_{sc}}{khT_{sc}} D$$

The single-rate and multirate tests described earlier yield values of kh, s, and D. Hence stabilized flow constants A', B', A, and B can be evaluated and substituted in Eqs. 5–114 and 5–115, respectively, to give the stabilized deliverability equations.

5.11 Alternative Form of the Deliverability Equation

If, for any reason, it is not possible to conduct these tests and evaluate s and D separately, a simplification may be made. An alternative form of the deliverability equations in terms of pressure squared and pseudopressure approaches is

$$\bar{p}_R^2 - p_{wf}^2 = \frac{50{,}300 T \overline{\mu c} p_{sc}}{khT_{sc}} \left[\log\left(\frac{0.472 r_e}{r_w}\right) + s' \right] \tag{5–116}$$

$$\Psi(p_i) - \Psi(p_{wf}) = \frac{115.820 \times 10^6 T q_{sc} p_{sc}}{khT_{sc}} \left[\log\left(\frac{0.472 r_e}{r_w}\right) + \frac{s'}{2.303} \right] \tag{5–117}$$

Equations 5–116 and 5–117 may be evaluated using the results of a single-rate test. However, these equations are then valid only for predicting the deliverability at flow rates near that used in the single-rate test.

5.12 Summary

A properly run drawdown test yields considerable information about the reservoir. However, the test may be hard to control because it is a flowing test. If a constant rate cannot be maintained, a multirate testing technique should be used. Those techniques also should be used if the well was not shut-in long enough to reach static reservoir pressure before the drawdown test. To ensure the best possible multiple-rate test, the engineer must have an idea of a well's flow characteristics. The rate change imposed must be large enough to give significant change in a pressure transient behavior of the well. Normally, rate is changed by a factor of two or three.

References and Additional Reading

1. Odeh, A. S., and Nabor, G. W., "The Effect of Production History on Determination of Formation Characteristics From Flow Tests," *J. Petroleum Technol.* (Oct. 1966) 1343–1350.
2. Agarwal, R. G., Al-Hussainy, R., and Ramey, H. J., Jr., "An Investigation of Wellbore Storage and Skin Effect in Unsteady Liquid Flow: I. Analytical Treatment," *Soc. Petroleum Eng. J.* (Sept. 1970) 279–290; *Trans. AIME* 240.
3. Wattenbarger, R. A., "Effects of Turbulence, Wellbore Damage, Wellbore Storage and Vertical Fractures on Gas Well Testing," Ph.D. Thesis, Stanford University, Stanford, CA, 1967.
4. Gringarten, A. C., Ramey, H. J., Jr., and Raghavan, R., "Applied Pressure Analysis for Fractured Wells," *J. Petroleum Technol.* (1975) 17, 887–892.
5. Earlougher, R. C., Jr., and Kerch, K. M., "Analysis of Short-Time Transient Test Data by Type-Curve Matching," *J. Petroleum Technol.* (1974) 26, 793–800.
6. Wattenbarger, R. A., and Ramey, H. J., Jr., "Gas Well Testing with Turbulence Damage, and Wellbore Storage," *J. Petroleum Technol.* (Aug. 1968) 877–887; *Trans. AIME* 243.
7. Muskat, M., *The Flow of Homogeneous Fluids through Porous Media,* McGraw-Hill, New York, 1937.
8. Odeh, A. S., and Jones, L. G., "Pressure Drawdown Analysis Variable-Rate Case," *J. Petroleum Technol.* (1965) 17, 960–964.
9. Ramey, H. J., Jr., "Non-Darcy Flow and Wellbore Storage Effects in Pressure Build-up and Drawdown of Gas Wells," *J. Petroleum Technol.* 7, 223–233.
10. Ramey. H. J., Jr., and Cobb, W. M., "A General Pressure Buildup Theory for a Well in a Closed Drainage Area," *J. Petroleum Technol.* (1971) 23, 1493–1505.
11. Jones, L. G., Blount, E. M., and Glaze, O. H., "Use of Short Term Multiple Rate Flow Tests to Predict Performance of Wells Having Turbulence," paper SPE 6133 presented at the SPE 51st Annual Meeting, New Orleans, Oct. 3–6, 1976.
12. Ramey, H. J., Jr., "Short-Time Well Test Data Interpretation in the Presence of Skin Effect and Wellbore Storage," *J. Petroleum Technol.* (1970) 22, 97–104.
13. Russell, D. G., "Determination of Formation Characteristics from Two-Rate Flow Tests," *J. Petroleum Technol.* (1963) 1317–1355.
14. Jones, P., "Reservoir Limits Test on Gas Wells," *J. Petroleum Technol.* 14, 613–619.
15. Dake, L. P., *Fundamentals of Reservoir Engineering*. Elsevier Scientific, 1978.

16. Raghavan, R., Cady, G. V., and Ramey, H. J., Jr., "Well Test Analysis for Vertically Fractured Wells," *J. Petroleum Technol.* (1972) 24, 1014–1020.
17. Van Poollen, H. K., "Radius-of-Drainage and Stabilization-Time Equations," *Oil Gas J.* (1964) 62, 138–146.
18. Katz, D. L., Cornell, D., Kobayashi, R., Poettmann, F. H., Vary, J. A., Elenbaas, J. R., and Weinaug, C. F., *Handbook of Natural Gas Engineering.* McGraw-Hill, New York, 1954.
19. Earlougher, R. C., Jr., "Estimating Drainage Shapes from Reservoir Limit Tests," *J. Petroleum Technol.* 23, 1266–1268.
20. Russell, D. G., Goodrich, J. H., Perry, G. E., and Bruskotter, J. F., "Methods for Predicting Gas Well Performance," *J. Petroleum Technol.* (Jan. 1966) 99–108.
21. Amanat U. C., *Pressure Transient Test Analysis, User's Handbook,* Vol. 8. Advanced TWPSOM Petroleum Systems Inc., Houston, TX, 1995.
22. Russell, D. G., and Prats, M., "The Practical Aspects of Interlayer Crossflow," *J. Petroleum Technol.* (June 1962) 589–594.

Chapter 6

Fundamentals of Pressure Buildup Analysis Methods

6.1 Introduction

This chapter discusses the most frequently used pressure buildup test. It is the simplest test that can be run on a gas well. If the effects of wellbore storage can be determined, much useful information can be obtained. This information includes permeability k, apparent skin factor s', and average reservoir pressure $\overline{p}_R$. The test consists of flowing the well at a constant rate q_{sc} for a period of time t_p, shutting the well in (at $\Delta t = 0$), and measuring wellbore pressure increase with shut-in time Δt. Horner developed the test, and this method of analysis is generally considered the best. Other conventional methods of analysis include the Horner plot, the Miller–Dyes–Hutchinson[2] plot (often abbreviated as the MDH plot), and the Muskat plot. Horner[1] showed that a plot of the shut-in pressure p_{ws} versus log $(t_P + \Delta t)/\Delta t$ should result in a straight line for an infinite-acting reservoir. In the buildup tests, t refers to the drawdown period prior to a buildup and Δt refers to the shut-in or buildup time. Matthews, Brons, and Hazebroek,[3] abbreviated as MBH, extended the application of the Horner plot to finite reservoirs.

A buildup test is always preceded by a drawdown, and the buildup data are directly affected by this drawdown. Usually, the drawdown starts from a stabilized reservoir condition represented by the stabilized reservoir pressure, p_I. At a time t, the well is shut in and the buildup is continued for a time Δt. Under these conditions, the behavior of the static sand face pressure, p_{ws}, is depicted in Figure 6–1.

6.2 Pressure Buildup Behavior Curves

Figure 6–2 shows pressure buildup curve shapes. As also shown in that figure, the fracture cases and large negative skin cases approach the semilog straight line from above when wellbore storage is small. That figure also shows

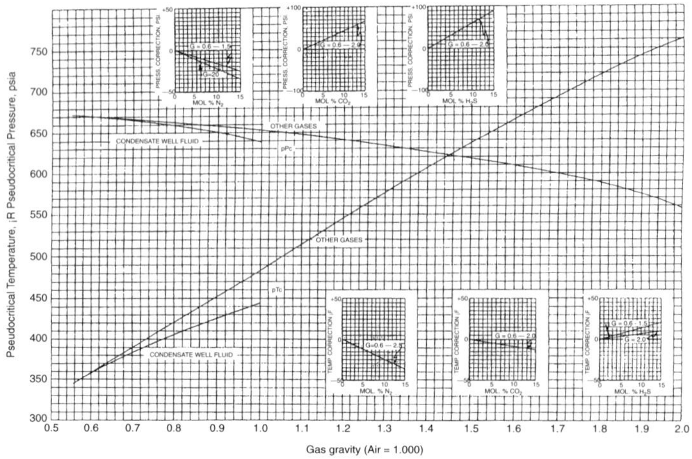

Figure 6–1. Behavior of the static sand face pressure upon shut-in a gas well.

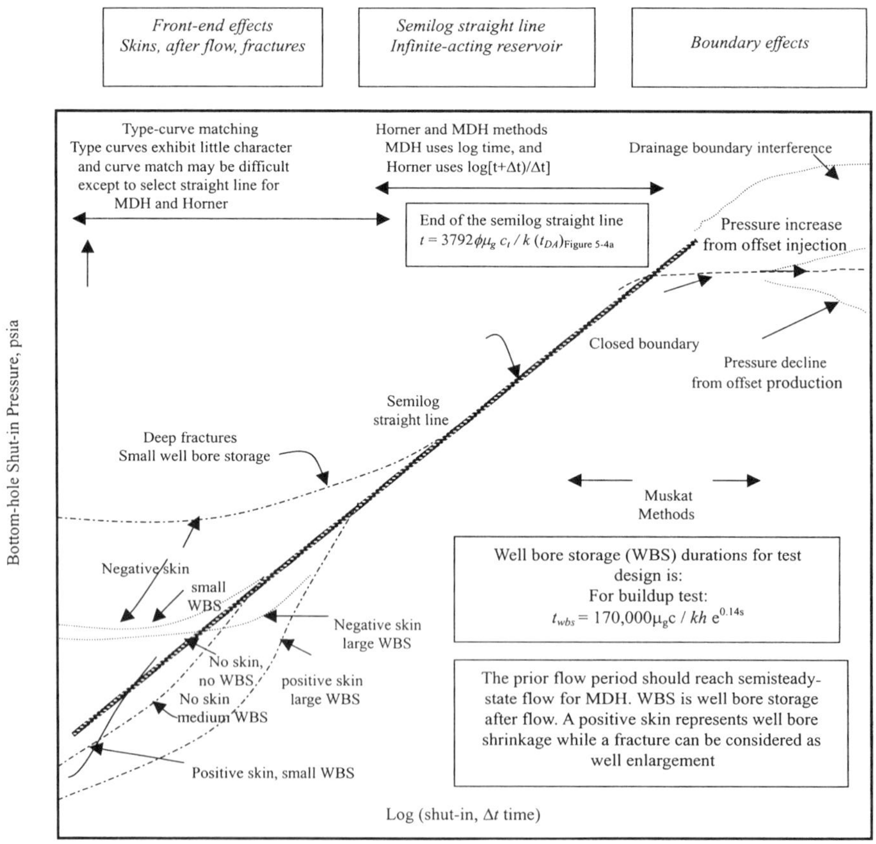

Figure 6–2. Idealized pressure buildup characteristics for a gas well (after Miller *et al.*).[2]

that behavior can be hidden by large wellbore storage effects, so the buildup curve may have the characteristic shape associated with wellbore storage or with a positive skin. For the figure shown Miller, Dyes, and Hutchinson scales[2] are used. The Horner plot is equally useful and is often preferred. Details discussion can be found in the next sections.

6.3 Uses and Practical Applications of Pressure Buildup Tests

Much information can be obtained from pressure transient tests. Pressure transient testing techniques, such as buildup, drawdown, interference, and pulse, are an important part of reservoir and production engineering. As the term is used in this book, pressure transient testing includes generating and measuring pressure variations with time on gas wells and, subsequently, estimating rock, fluid, and well properties and predicting reservoir/well behavior. Practical information obtainable from transient testing includes wellbore volume, damage, and improvements; reservoir pressure; formation permeability; porosity; reserves; reservoir and fluid discontinuities; and other related data. All this information can be used to help analyze, improve, and forecast reservoir performance.

It is generally good practice to run a base pressure transient test on a producing well shortly after completion or an injection well after a suitable period of injection. This can lead to early recognition and correction of many problems, of which insufficient stimulation is only the most obvious. Such tests also provide *in situ* data for reservoir simulation and a base for comparison with reservoir or well problem as they arise. Figure 6–3 shows types, uses, and practical applications of buildup tests.

6.4 Type Curves and Desuperposition

Type curves are also applicable to the analysis of buildup tests, though type curve matching techniques are not discussed fully in this section. Type curve matching provides a simple method for determining the time of start of transient flow during drawdown tests. A similar approach may be used to determine the time of start of semilog straight-line data in buildup tests with one additional step. Since a buildup is always preceded by a drawdown, the buildup data must be "desuperposed" before attempting a type curve match. It may be noted that desuperposition can also be performed on the second rate of a two-rate test. Early-time data obtained during a buildup may be desuperposed by assuming $\psi(P_{wfp})$ to be constant at a value $\psi(P_{wfo})$.

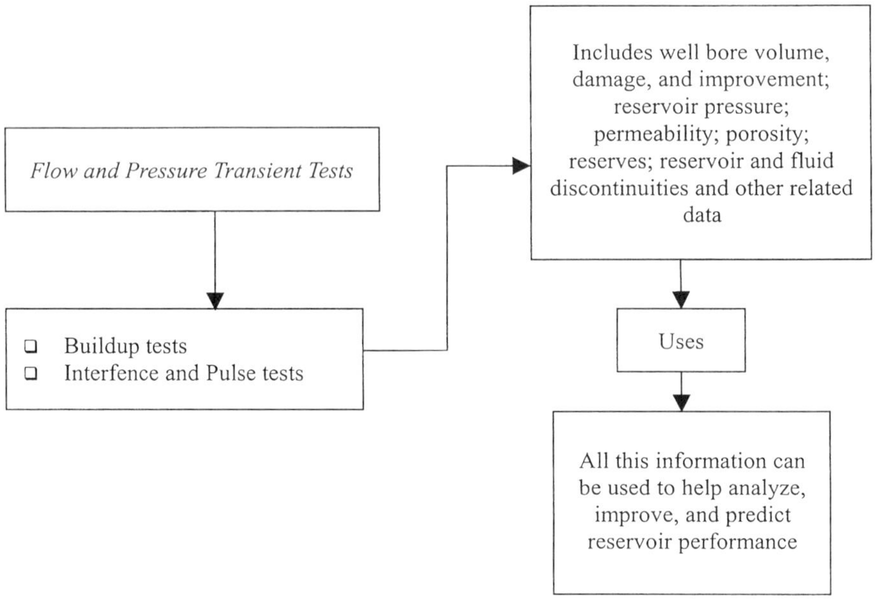

Figure 6–3. Uses and practical applications of buildup tests.

6.5 Tests Utilizing Early-Time Data

Analysis of early-time data may yield adequate approximations of kh. Such an analysis may be necessary only when middle-time data are not available. As mentioned in the previous section, a desuperposition of buildup data can give the equivalent of a drawdown plot and may be analyzed as such. Consequently, the discussions related to the early-time flow required in Chapter 5 generally apply to buildup tests as they do to drawdown tests.

6.6 Tests Utilizing Middle-Time and Late-Time Data

In Chapter 5 it has been shown that early-time data may be used to determine the time of start of transient flow data. A similar analysis applies to the early-time portion of a buildup. Data should be obtained in the transient flow regime since reservoir parameters calculated by an analysis of middle-time data are much more reliable than those calculated from early-time data.

Data obtained from a properly conducted buildup test that follows either a single-rate or a two-rate drawdown test and also a variable-rate drawdown test may be analyzed to yield values of kh, s', and $\overline{p}_R$. The pressure buildup behavior during the middle-time period is analogous to that during the transient flow period in a drawdown test. In other words, the reservoir is infinite acting and boundaries do not affect the pressure–time data. The analysis of middle-time data yields a semilog straight line, which should not be confused with the semilog straight line for a drawdown test. This straight line, when extrapolated, yields values of a false reservoir pseudopressure, Ψ^*, corresponding to p^*, which is used to calculate the average reservoir pseudopressure, $\overline{\psi}_R$, corresponding to the average reservoir pressure, $\overline{p}_R$.

6.7 Pressure–Time Behavior of Infinite-Acting Reservoirs

The most useful solution for transient flow is the so-called line source solution. The solution including formation damage and turbulence effects is

$$\Delta p_D = 0.5(\text{Int}_D + 0.809) + s' \tag{6–1}$$

where

$s' = s + Dq_{sc}$; s = actual well damage or improvement (may be positive or negative)

D = turbulence coefficient, which will always be positive

t_D = dimensionless time

Δp_D = dimensionless pressure drop

In terms of real variables and common log, Eq. 6–1 in terms of the pressure squared case becomes

$$p_i^2 - p_{wf}^2 = \frac{57.920 \times 10^6 q_{sc} T p_{sc} \overline{\mu}_g \overline{z}}{khT_{sc}} \times \left[\log t_P + \log \frac{k}{\phi \overline{\mu}_g \overline{c} r_w^2} - 3.23 + 0.869s'\right] \tag{6–2}$$

where k is in millidarcys. If the well is shut in at time t_P and allowed to build up for a time Δt, the effect of the shut-in may be obtained by the superposition of two effects. During the shut-in period, the static bottom hole pressure is given by the sum of the continuing effect of the drawdown rate q_{sc} and the superposed effect of the change in rate $(0 - q_{sc})$ and is represented by

$$p_i^2 - p_{wf}^2 = \frac{57.920 \times 10^6 q_{sc} T p_{sc} \overline{\mu}_g \overline{z}}{khT_{sc}} \left[\log(t_P + \Delta t) + \log\left(\frac{k}{\phi \overline{\mu}_g \overline{c} r_w^2}\right) - 3.23\right] + \frac{57.920 \times 10^6 q_{sc} T p_{sc} \overline{\mu}_g \overline{z}}{khT_{sc}} \times \left[\log \Delta t + \log\left(\frac{k}{\phi \overline{\mu}_g \overline{c} r_w^2} - 3.23\right)\right] \tag{6–3}$$

Note that the apparent skin, s', should not be superposed in time since it is a function only of the existing flow rate. The first term on the right-hand side of Eq. 6–3 represents the effect due to the drawdown at a rate q_{sc} for a time $(t_P + \Delta t)$. The second term is the effect of the change in rate from q_{sc} to 0 for a time Δt. Combining these terms and simplifying Eq. 6–3 gives

$$p_i^2 - p_{wf}^2 = \frac{57.920 \times 10^6 q_{sc} T p_{sc} \overline{\mu}_g \overline{z}}{khT_{sc}} \log\left(\frac{t_P + \Delta t}{\Delta t}\right) \tag{6–4}$$

This relationship represents the commonly used Horner plot. It is obtained from this equation that plot of $\Delta(p^2)$ versus $\log(t_P + \Delta t)/\Delta t$ on semilogarithmic coordinates will give a straight line of slope m, from which

$$kh = \frac{57.920 \times 10^6 q_{sc} T p_{sc} \overline{\mu}_g \overline{z}}{mT_{sc}} \tag{6–5}$$

It must be noted that in all semilog plots representing drawdown or buildup tests, only the magnitude and not the sign of the slope is considered. Defining p_{wfo} as the pressure just before shut-in, Eq. 6–2 may be written as

$$p_i^2 - p_{wfo}^2 = \frac{57.920 \times 10^6 q_{sc} T p_{sc} \overline{\mu}_g \overline{z}}{khT_{sc}} \times \left[\log t_P + \log\left(\frac{k}{\phi \overline{\mu}_g \overline{c} r_w^2}\right) - 3.23 + 0.869s'\right] \tag{6–6}$$

Subtracting Eq. 6–4 from Eq. 6–6 gives

$$p_{ws}^2 - p_{wfo}^2 = m\left[\log\frac{t\Delta t}{t + \Delta t} + \log\left(\frac{k}{\phi \overline{\mu}_g \overline{c} r_w^2}\right) - 3.23 + 0.869s'\right] \tag{6–7}$$

Defining p_{ihr} as the pressure at $\Delta t = 1$, and assuming $\frac{t_P}{t_P+1} \cong 1$, Eq. 6–7 may be simplified to give

$$s' = 1.151\left[\frac{p_{1hr}^2 - p_{wfo}^2}{m} - \log\left(\frac{k}{\phi \overline{\mu}_g \overline{c} r_w^2}\right) + 3.23\right] \tag{6–8}$$

where p_{1hr}^2 is obtained from an extrapolation of the linear segment of the plot at $\Delta t = 1$ hr ($\log 1 = 0$). Equation 6–8 may be used to calculate s'. There is no way to separate s' into its components s and Dq_{sc} from a single buildup test. However, if another is conducted following a different single-rate drawdown, a different value of s' is obtained. The two different values of s' may be used with Eqs. 5–46 and 5–47 to calculate s and D separately. Pressure buildup in term of the pseudopressure $\Psi(p_{ws})$ approach is

$$\Psi(p_i) - \Psi(p_{ws}) = \frac{57.920 \times 10^6 q_{sc} T p_{sc}}{kh T_{sc}} \log\left(\frac{t_P + \Delta t}{\Delta t}\right) \quad (6\text{–}9)$$

From this form of equation it can be seen that a plot of $\Psi(p_{ws})$ versus $\log(t_P + \Delta t)/\Delta t$ gives a straight line of slope m, from which flow capacity kh may be calculated by using Eq. 6–10:

$$kh = \frac{57.920 \times 10^6 q_{sc} T p_{sc}}{m T_{sc}} \quad (6\text{–}10)$$

$$s' = 1.151\left[\frac{\Psi(p_{1hr}) - \Psi(p_{wfo})}{m} - \log\left(\frac{k}{\phi \mu_{gi} c_i r_w^2}\right) + 3.23\right] \quad (6\text{–}11)$$

Noting that $\Psi(p_{1hr})$ should be obtained from the straight-line portion (extrapolated, if necessary) of the Horner plot, Eq. 6–11 may be used to calculate s', where $s' = s + Dq_{sc}$. s and Dq_{sc} can be found from two-rate buildup tests. The two different values of s' may be used with Eqs. 5–46 and 5–47 to calculate s and D separately.

$$\Psi(\Delta p)_{skin} = 0.869 m s' \quad (6\text{–}12)$$

$$FE = \frac{\Psi(\overset{*}{p}) - \Psi(p_{wfo}) - \Psi(\Delta p)_{skin}}{\Psi(\overset{*}{p}) - \Psi(p_{wfo})} \quad (6\text{–}13)$$

Equations 6–12 and 6–13 may be used to calculate pressure drop due to apparent skin and flow efficiency. The reservoir properties are evaluated at true p_i, average pressure $\overline{p}$ (if available) or $\sqrt{[(\overset{*}{p}^2 + p_{wf}^2)/2]}$. If $\Psi(\overline{p})$ is available, $\Psi(\overset{*}{p})$ should be replaced by $\Psi(\overline{p})$.

A buildup semilog plot for an infinite reservoir is shown in Figures 6–4 and 6–5. Commonly used alternative buildup plots are shown in Figures 6–6 and 6–7 (after Matthews and Russell, p. 123)[4] in which the time axis increases from left to the right. The use of a Horner plot to calculate reservoir parameter is illustrated in Example 6–1.

Example 6–1[26] *Analyzing Gas Well Single-Rate Buildup Test*

The well was shut in at a constant rate of 6.148 mmscfd for 147.12 hr, during which time the pressure buildup was monitored continuously. The pressure just prior to shut-in was 1735 psia. General data pertinent to the test are given below. The pressure–time data are also tabulated in Table 6–3, columns 1 and 3, and are given directly in the solution to this problem. From a recombined gas analysis: N_2 = 0.11%, CO_2 = 7.84%, H_2S = 0.0%, C_1 = 80.55%, C_2 = 5.10%,

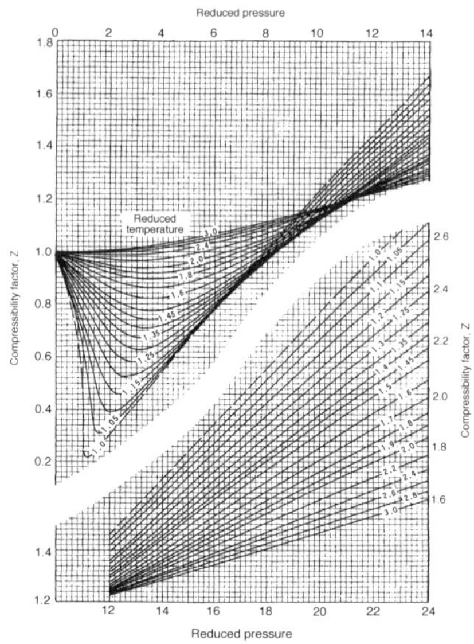

Figure 6–4. Buildup semilog plots.

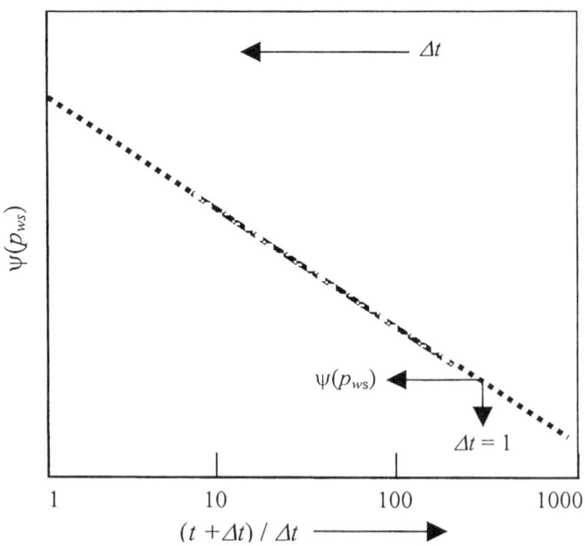

Figure 6–5. Buildup semilog plots.

C_3 = 4.36%, iC_4 = 0.87%, nC_4 = 0.77%, iC_5 = 0.22% , nC_5 = 0.09%, C_6 = 0.11%, $C7^+$ = 0.00.

Well/reservoir data are as follows: well depth = 12,550 ft; T = 250°F; h = 54 ft; ϕ = 0.179; c_g = 0.00025 psi^{-1}; c_w = 3.20×10^{-6} psi^{-1};

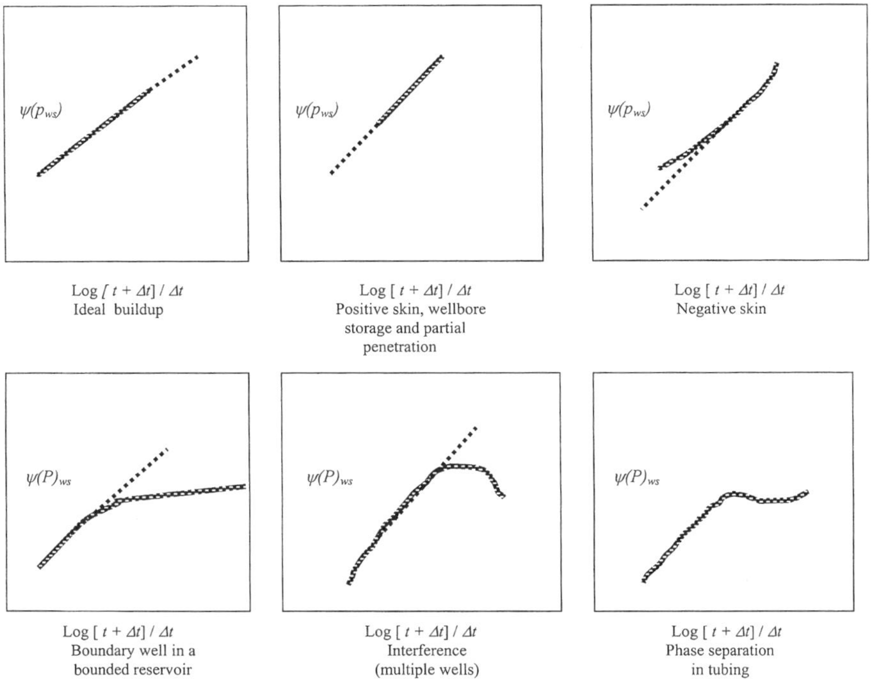

Figure 6–6. Horner buildup plots (after Matthews and Russell, p. 123).[4]

$c_0 = 3.30 \times 10^{-6}$ psi^{-1}; $c_f = 3.90 \times 10^{-6}$ psi^{-1}; $\bar{z} = 0.9467$; $\bar{\mu}_g = 0.02345$ cP, $s_w = 0.335$; $s_g = 0.665$; $c_t = 0.00023$ psi^{-1}; $p_R = 3700$ psia; production rate at shut-in time = 6.148 mmscfd; cumulative production prior to test = 11.382 mmscf; and

$$\begin{aligned} \beta_g &= 0.00513 \text{ ft}^3\text{/scf} = 0.000913 \text{ bbl/scf} \\ &= 194.9752 \text{ scf/ft}^3 = 1095 \text{ scf/bbl} \\ &= 1.0948 \text{ mscf/bbl} = 0.9134 \text{ bbl/mscf} \end{aligned}$$

Using the Horner method, determine the following:

1. Permeability k
2. Skin factor s and pressure drop due to skin, Δp_{skin}
3. Flow efficiency using p^*
4. Effective wellbore radius

Solution From gas compositional analysis calculate gas properties: MW = 21.20, G = 0.732, $P_c = 645.08$ psia, $T_c = 380.16°$R, $H_2S = 0.00\%$, $CO_2 = 7.84\%$, $N_2 = 0.11\%$, $P_i = 3965$ psia, $\mu_I = 0.02440$ cP, $c_i = 0.0002295$.

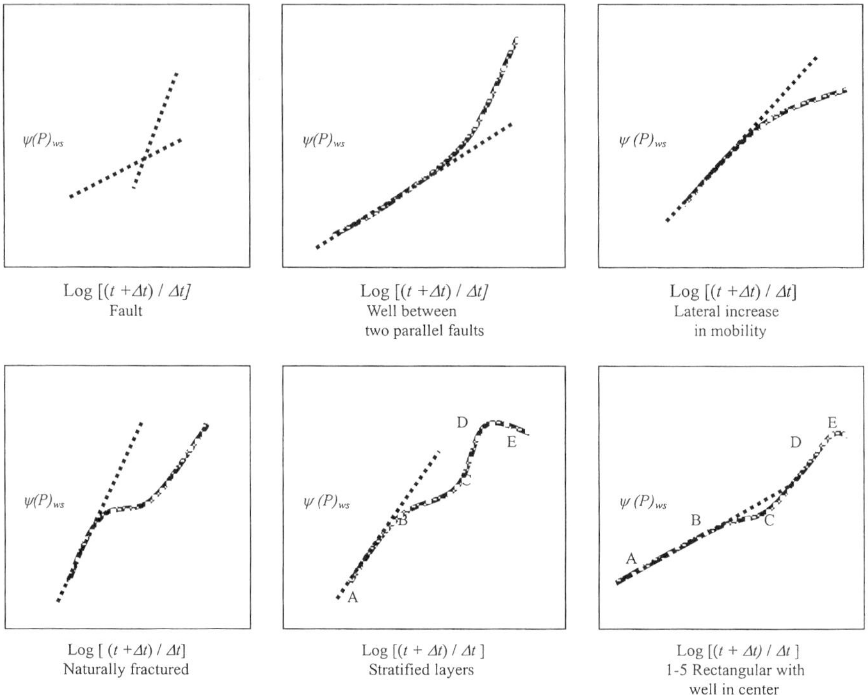

Figure 6–7. Horner buildup plots (after Matthews and Russell, p. 123).[4]

From known PVT correlations calculate z and μ_g, and find the real gas pseudopressure function

$$\Psi = \int_0^P \frac{P}{\mu z} dp$$

Table 6–1 shows PVT gas properties and pseudopressure calculations.

Figure 6–8 is a plot of compressibility factor versus pressure; Figure 6–9 is a plot of gas viscosity versus pressure; and Figure 6–10 is a plot of real gas pseudopressure versus pressure. Using Figures 6–8 through 6–10 develop six-degree polynomial coefficients. See Table 6–2 for numerical values of coefficients for predicting PVT properties. Table 6–3 shows the test data.

Calculate pseudoproducing time:

$$t_P = \frac{24 \times 11.382}{6.148} = 44.44 \text{ hr}$$

Table 6–1
PVT Gas Properties and Pseudopressure Calculations

Pressure (psia)	Compressibility factor (z)	Gas viscosity μ_g (cP)	Real gas pseudopressure (psia^2/cP)
4000	0.9647	0.024580	872.92E+06
3600	0.9445	0.023151	739.56E+06
3200	0.9282	0.217210	610.28E+06
2800	0.9169	0.020329	486.72E+06
2400	0.9113	0.019008	371.18E+06
2000	0.9120	0.017784	266.41E+06
1600	0.9189	0.016681	175.33E+06
1200	0.9319	0.015723	100.83E+06
800	0.9503	0.014932	45.51E+06
400	0.9733	0.014337	11.47E+06
14.65	0.9995	0.013978	5.17E+05

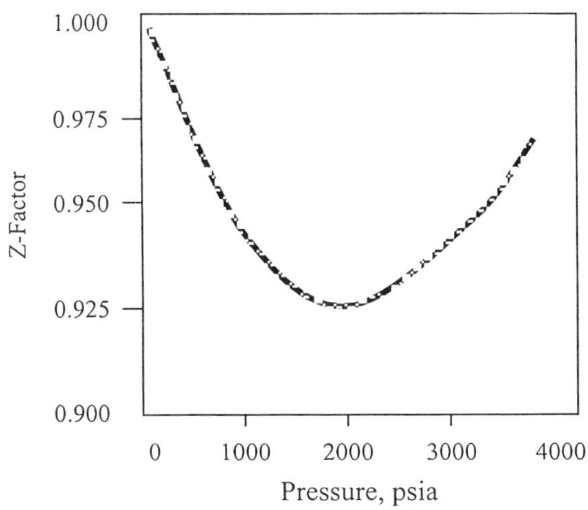

Figure 6–8. Z-factor versus pressure, psia.

Figure 6–11 shows a log-log plot. The beginning of the straight line by using the one and one-half log cycle rule is 3.75 hr. Figures 6–12 and 6–13 show Horner buildup data plot. From Figure 6–12 or 6–13, the following information is obtained: beginning of straight line = 5 hr; end of straight line = 9 hr; $m = 21$ mmpsia^2/cP/cycle.

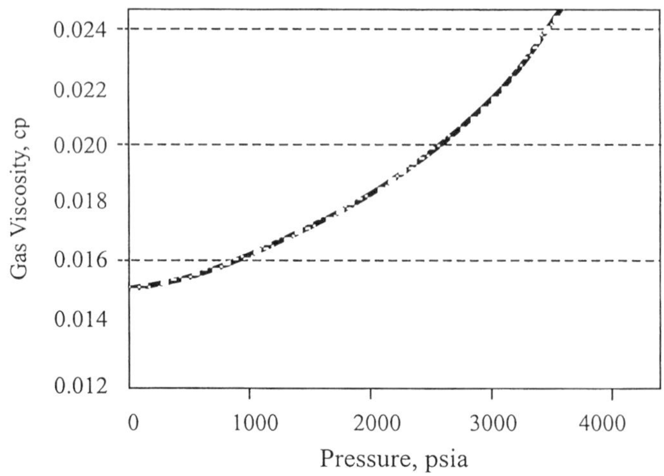

Figure 6–9. Gas viscosity versus pressure, psia.

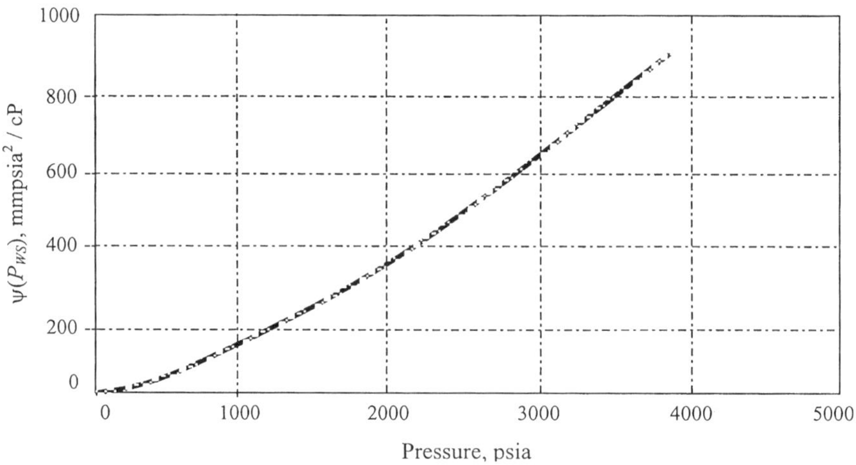

Figure 6–10. ψ–p curve for Example 6–1.

$$
\begin{aligned}
p_{1hr} &= 650.00\ \text{mmpsia}^2/\text{cP} \leftrightarrow 3330\ \text{psia} \\
p^* &= 861.12\ \text{mmpsia}^2/\text{cp} \leftrightarrow 3965\ \text{psia} \\
p_{wf(\Delta t=0)} &= 201.25\ \text{mmpsia}^2/\text{cP} \leftrightarrow 1720\ \text{psia} \\
c_t &= c_g s_g + c_w s_w + c_f \\
&= 2.5 \times 10^{-4} \times 0.665 + 3.2 \times 10^{-6} \times 0.335 + 3.9 \times 10^{-6} \\
&= 2.30 \times 10^{-4}\ \text{psi}^{-1}
\end{aligned}
$$

Table 6–2
Numerical Values of Coefficients for Predicting PVT Properties

Polynomial coefficients	Z-factor —	Gas viscosity (cP)	Real pseudopressure function (psia^2/cP)
A	0.999513	0.0139689	39453
B	−6.810505E-05	6.044023E-07	−222.976
C	4.707337E-09	8.323752E-10	72.0827
D	5.011202E-12	−1.145527E-17	5.287041E-04
E	−6.626846E-016	1.550466E-17	−1.993697E-06
F	1.094491E-20	−1.721434E-21	1.92384E-10

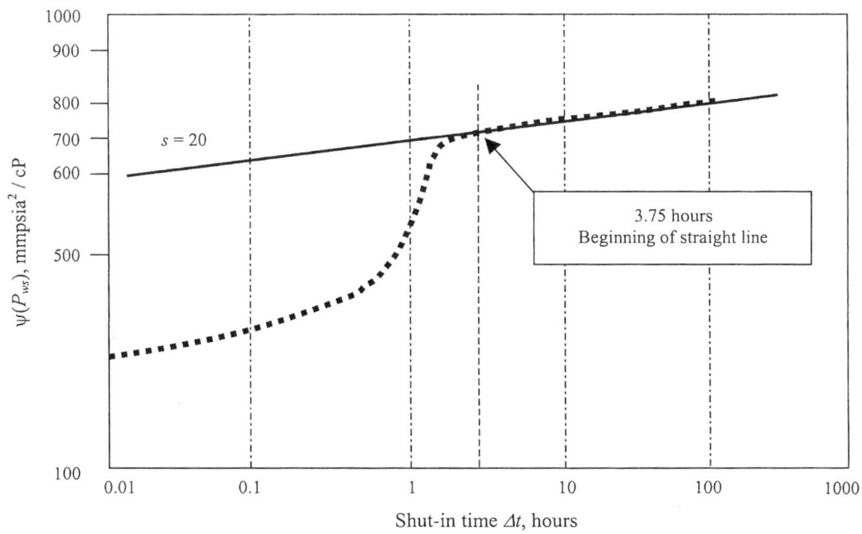

Figure 6–11. Type curve match for the desuperposed buildup data (log-log plot for Example 6–1).

1. Using Equation 6–10, permeability is

$$kh = \frac{57.92 \times 10^6 \times 6.148 \times 710 \times 14.65}{21 \times 520} = 339.07 \text{ mD/ft}$$

$$k = 339.07/54 = 8.27 \text{ mD}$$

(text continued on page 337)

Table 6–3
Pressure Buildup Test Data

Time, Δt hr (1)	$\frac{t_p+\Delta t}{\Delta t}$ — (2)	P_{ws} (psig) (3)	P_{ws} (psia) (4)	ΔP_{ws} psia (5)	$\psi(\Delta P_{ws})$ mmpsia^2/cP (6)	Drainage radius r_i (ft) (7)
0.00	—	1720	1735	204.35	0.00	0
0.02	2666.92	1723	1738	204.96	0.00	16
0.03	1333.96	1733	1747	207.12	0.059	22
0.07	667.48	1773	1788	216.16	14.89	31
0.10	445.32	1803	1818	223.10	21.83	38
0.13	334.24	1854	1869	234.89	33.62	44
0.17	267.59	1911	1925	248.28	47.01	49
0.25	178.73	2014	2028	273.44	72.17	60
0.33	134.30	2120	2135	300.34	99.07	69
0.50	89.86	2297	2312	347.09	145.82	85
0.75	60.24	2601	2615	432.19	230.92	104
1.00	45.43	2805	2819	492.58	291.31	120
1.50	32.62	3132	3146	593.35	392.07	147
2.00	23.22	3295	3310	645.38	444.11	170
2.50	18.77	3335	3350	658.17	456.89	190
3.00	15.81	3352	3366	663.42	462.14	208
3.50	13.69	3368	3382	668.65	467.37	225
4.00	12.11	3370	3385	669.42	468.14	240
4.83	10.19	3377	3391	671.52	470.24	264
5.00	9.89	3382	3397	673.23	471.96	269
5.50	9.08	3388	3403	675.21	473.93	282
6.00	8.41	3393	3407	676.66	475.38	294
6.50	7.84	3397	3411	677.96	476.68	306
7.00	7.35	3400	3415	679.15	477.88	318
7.50	6.92	3404	3418	680.32	479.04	329
8.00	6.55	3406	3421	681.13	479.85	340
8.50	6.23	3410	3425	682.30	481.02	350
9.00	5.94	3413	3428	683.37	482.09	360
9.50	5.68	3417	3432	684.66	483.39	370
10.0	5.44	3421	3436	685.99	484.72	380
10.50	5.23	3425	3440	687.32	486.05	389
11.00	5.04	3429	3443	688.36	487.08	398
11.50	4.86	3432	3447	689.43	488.16	407
12.00	4.70	3434	3448	689.95	488.68	416
12.50	4.55	3436	3451	690.86	489.59	425
13.00	4.42	3438	3453	691.55	490.27	433
13.50	4.29	3441	3456	692.46	491.18	441
14.00	4.17	3444	3459	693.40	492.12	449

Table 6–3 (Continued)

Time, Δt hr (1)	$\frac{t_p+\Delta t}{\Delta t}$ (2)	P_{ws} (psig) (3)	P_{ws} (psia) (4)	ΔP_{ws} psia (5)	$\psi(\Delta P_{ws})$ mmpsia^2/cP (6)	Drainage radius r_i (ft) (7)
14.50	4.06	3447	3461	694.18	492.90	457
15.00	3.96	3449	3464	694.96	493.68	465
15.50	3.87	3452	3466	695.88	494.62	473
16.00	3.78	3454	3468	696.53	495.27	480
16.50	3.69	3456	3471	697.31	496.06	488
17.00	3.61	3458	3473	697.86	496.61	495
17.50	3.54	3461	3475	698.77	497.52	502
18.00	3.47	3462	3477	699.29	498.04	509
18.50	3.40	3465	3479	700.11	498.85	516
19.00	3.34	3467	3481	700.76	499.51	523
19.50	3.28	3469	3483	701.41	500.16	530
20.00	3.22	3471	3486	702.22	500.97	537
20.50	3.17	3473	3488	702.75	501.49	544
21.00	3.12	3474	3489	703.14	501.88	550
21.50	3.07	3477	3491	703.92	502.67	557
22.00	3.02	3478	3493	704.47	503.22	563
22.50	2.97	3480	3494	704.99	503.74	570
23.00	2.93	3481	3496	705.52	504.26	576
23.50	2.89	3482	3497	705.78	504.53	582
24.00	2.85	3485	3499	706.59	505.34	588
24.50	2.81	3486	3501	706.98	505.73	594
25.00	2.78	3487	3502	707.38	506.12	600
26.00	2.71	3491	3505	708.58	507.33	612
27.00	2.65	3494	3509	709.76	508.51	624
28.00	2.59	3497	3512	710.71	509.45	635
29.00	2.53	3500	3515	711.62	510.37	647
30.00	2.48	3503	3518	712.53	511.28	658
31.00	2.43	3506	3521	713.61	512.36	669
32.00	2.39	3509	3524	714.66	513.41	679
33.00	2.35	3511	3526	715.21	513.96	690
34.00	2.31	3514	3529	716.26	515.01	760
35.00	2.27	3517	3532	717.21	515.96	710
36.00	2.23	3519	3533	717.73	516.48	720
37.00	2.20	3522	3536	718.65	517.40	730
38.00	2.17	3523	3538	719.17	517.92	740
39.00	2.14	3526	3541	720.12	518.87	750
40.00	2.22	3529	3544	721.17	519.92	759
41.00	2.08	3531	3546	721.72	520.47	769
42.00	2.06	3534	3548	722.64	521.39	778
43.00	2.03	3536	3550	723.30	522.04	787
44.00	2.01	3539	3553	724.25	522.99	797

Table 6–3 (Continued)

Time, Δt hr (1)	$\frac{t_p+\Delta t}{\Delta t}$ — (2)	P_{ws} (psig) (3)	P_{ws} (psia) (4)	ΔP_{ws} psia (5)	$\psi(\Delta P_{ws})$ mmpsia^2/cP (6)	Drainage radius r_i (ft) (7)
45.00	1.99	3540	3555	724.77	523.52	806
46.00	1.97	3543	3557	725.56	524.31	814
47.00	1.95	3445	3560	726.38	525.12	823
48.00	1.93	3547	3561	726.90	525.65	832
49.00	1.91	3549	3564	727.69	526.44	841
50.00	1.89	3551	3566	728.38	527.12	849
51.00	1.87	3552	3567	728.64	527.39	858
52.00	1.85	3555	3570	729.56	528.31	866
53.00	1.84	3559	3571	730.08	528.83	874
54.00	1.82	3560	3574	730.90	529.65	882
55.00	1.81	3663	3575	731.30	530.04	891
56.00	1.79	3565	3578	732.22	530.96	899
57.00	1.78	3567	3580	732.91	531.65	907
58.00	1.77	3568	3581	733.43	532.18	915
59.00	1.75	3570	3583	733.96	532.70	922
60.00	1.74	3572	3585	734.48	533.23	930
61.00	1.73	3574	3587	735.17	533.92	938
62.00	1.72	3576	3589	735.96	534.71	946
63.00	1.71	3578	3590	736.35	535.10	953
64.00	1.69	3579	3592	737.04	535.79	961
65.00	1.68	3580	3594	737.57	536.32	968
66.00	1.67	3583	3595	737.83	536.58	976
67.00	1.66	3584	3597	738.62	537.37	983
68.00	1.65	3586	3599	739.18	537.93	990
69.00	1.64	3588	3601	739.84	538.59	997
70.00	1.63	3589	3602	740.36	539.11	1005
71.00	1.63	3592	3604	740.89	539.64	1012
72.00	1.62	3592	3606	741.58	540.33	1019
73.00	1.61	3594	3607	741.84	540.59	1026
74.00	1.60	3596	3609	742.50	541.25	1033
75.00	1.59	3598	3611	743.16	541.91	1040
76.00	1.59	3599	3613	743.64	542.38	1047
77.00	1.58	3600	3614	744.12	542.86	1054
78.00	1.57	3604	3615	744.51	543.26	1061
79.00	1.56	3604	3618	745.60	544.34	1067
80.00	1.56	3606	3619	745.73	544.48	1074
81.00	1.55	3607	3621	746.39	545.13	1081
82.00	1.54	3609	3622	746.78	545.53	1087
83.00	1.54	3611	3624	747.31	546.06	1094
84.00	1.53	3612	3625	747.87	546.62	1101
85.00	1.52	3613	3626	748.26	547.01	1107
86.00	1.52	3615	3628	748.79	547.54	1114

Table 6–3 (Continued)

Time, Δt hr (1)	$\frac{t_p+\Delta t}{\Delta t}$ — (2)	P_{ws} (psig) (3)	P_{ws} (psia) (4)	ΔP_{ws} psia (5)	$\psi(\Delta P_{ws})$ mmpsia^2/cP (6)	Drainage radius r_i (ft) (7)
87.00	1.51	3616	3630	749.32	548.07	1120
88.00	1.50	3618	3631	749.71	548.46	1126
89.00	1.50	3619	3632	750.14	548.98	1133
90.00	1.49	3621	3634	750.67	549.42	1139
91.00	1.49	3624	3635	751.20	549.95	1145
92.00	1.48	3624	3638	752.15	550.90	1152
93.00	1.48	3625	3639	752.42	551.17	1158
94.00	1.47	3628	3640	752.68	551.43	1164
95.00	1.47	3628	3642	753.47	552.22	1170
96.00	1.46	3692	3643	753.74	552.49	1177
97.00	1.46	3631	3644	753.94	552.81	1183
98.00	1.45	3633	3645	754.63	553.18	1189
99.00	1.45	3635	3647	755.09	553.84	1195
100.00	1.44	3637	3650	755.88	554.63	1201
101.00	1.44	3640	3651	756.44	555.19	1207
102.00	1.44	3642	3654	757.50	556.25	1213
103.00	1.43	3644	3656	758.16	556.91	1219
104.00	1.43	3644	3658	758.72	557.47	1225
105.33	1.42	3644	3659	758.99	557.73	1232
106.00	1.42	3641	3659	758.85	557.60	1236
107.00	1.42	3644	3656	758.03	556.78	1242
109.00	1.41	3644	3658	758.72	557.47	1254
111.00	1.40	3644	3659	758.99	557.73	1265
113.00	1.39	3648	3661	759.65	558.39	1276
115.00	1.39	3651	3663	760.27	559.02	1288
117.00	1.38	3654	3666	761.26	560.01	1299
119.00	1.37	3656	3668	762.06	560.81	1310
121.00	1.37	3659	3671	762.98	561.73	1321
123.00	1.36	3662	3674	763.81	562.56	1332
125.00	1.36	3666	3676	764.74	563.48	1343
127.00	1.35	3666	3681	766.22	564.97	1353
129.00	1.34	3669	3683	766.89	565.63	1364
131.00	1.34	3669	3684	767.15	565.90	1374
133.00	1.33	3673	3684	767.15	565.90	1385
135.00	1.33	3677	3688	768.51	567.25	1395
137.00	1.32	3678	3691	769.73	568.48	1406
139.00	1.32	3680	3693	770.13	568.88	1416
141.00	1.32	3682	3695	770.92	569.67	1426
143.00	1.31	3685	3696	771.32	570.07	1436
145.00	1.31	3686	3700	772.41	571.16	1446
147.12	1.30	3686	3700	772.68	571.43	1446

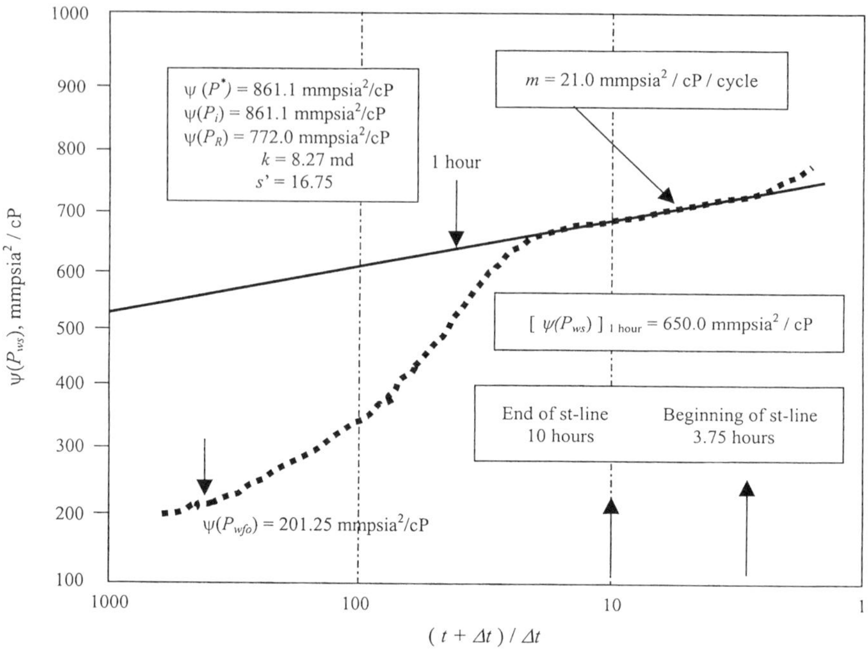

Figure 6–12. Horner buildup data plot for Example 6–1.

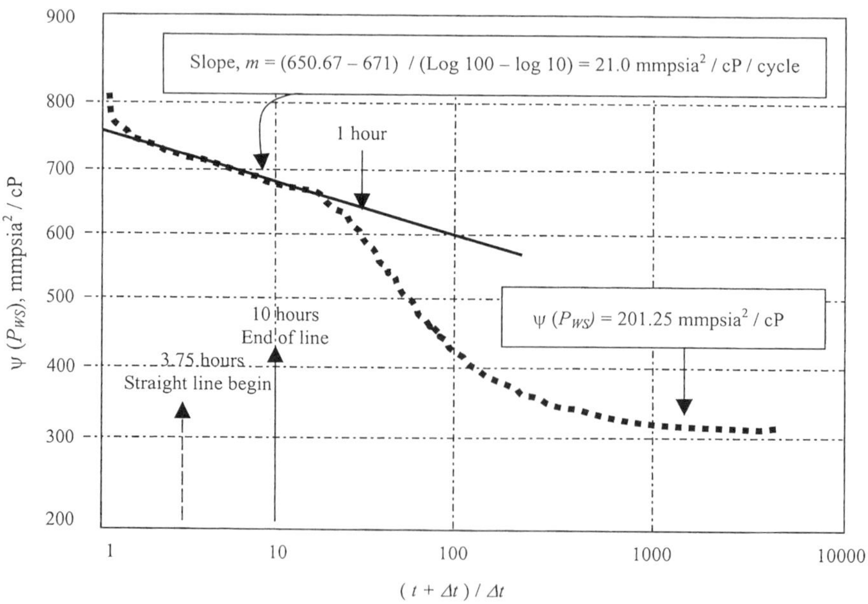

Figure 6–13. Horner buildup data plot for Example 6–1 (semilog plot).

(text continued from page 331)

2. Using Eq. 6–11, the apparent skin factor is

$$s' = 1.151\left[\frac{\psi(p_{1hr}) - \psi(p_{wfo})}{m} - \log\left(\frac{k}{\phi_{HC}\mu_g c_t r_w^2}\right) + 3.23\right]$$

$$= 1.151\left[\frac{650 - 201.25}{21}\right.$$

$$\left. - \log\left(\frac{8.27}{0.119 \times 0.02345 \times 2.3 \times 10^{-4} \times 0.4271^2}\right) + 3.23\right]$$

$$= 16.75$$

The well is damaged; the completion would probably benefit from stimulation. Using Eq. 6–12, the pressure drop due to skin is:

$$\psi(\Delta P)_{skin} = 0.869\,\mathrm{ms}' = 0.869(21)(16.75)$$

$$= 305.65\,\mathrm{mmpsia^2/cP} \leftrightarrow 2215\,\mathrm{psia}$$

3. Using Eq. 6–13, the flow efficiency is

$$FE = \frac{\psi(p^*) - \psi(p_{wfo}) - \psi(\Delta p)_{skin}}{\psi(p^*) - \psi(p_{wfo})}$$

$$= \frac{861.12 - 201.25 - 305.68}{861.12 - 201.25} = 0.54$$

This means that the well is producing about 54% as much gas with the given drawdown as an undamaged well in a completed perforated interval would produce.

4. The effective wellbore radius is

$$r_{wa} = r_w e^{-s} = 0.4271e^{-16.75} = 2.27 \times 10^{-8}\,\mathrm{ft}$$

The physical interpretation of this result is that the tested well is producing 6.148 mmscfd gas with the same pressure drawdown as would a well with a wellbore radius of 2.27×10^{-8} ft and permeability unaltered up to the sandface. Thus, of the total drawdown of approximately (861.12 − 201.25) = 659.87 mmpsia^2/cP ↔ 3340 psia, about 2125 is caused by damage.

6.8 Finite Reservoir Behavior

Equation 2–66 may represent flow from a finite reservoir, in the absence of skin and turbulent effects. Equation 2–66 may be written in terms of pseudopressure with substitutions for dimensionless quantities, and including an apparent skin factor s', as

$$\Psi(p_i) - \Psi(p_{wf}) = 57.920 \times 10^6 \frac{q_{sc} T p_{sc}}{k h T_{sc}} \left[\log t_P + \log \left(\frac{k}{\phi \mu_{gi} c_i r_w^2} \right) - 3.23 + \frac{4\pi t_{DA}}{2.303} - \frac{F}{2.303} + 0.869 s' \right] \quad (6\text{–}14)$$

Superposition of a buildup on the drawdown then gives

$$\Psi(p_i) - \Psi(p_{ws}) = 57.920 \times 10^6 \frac{q_{sc} T p_{sc}}{k h T_{sc}} \left[\log \left(\frac{t_P + \Delta t}{\Delta t} \right) + \frac{4\pi t_{DA}}{2.303} - \frac{1}{2.303} (F|_{tP+\Delta t} - F|_{\Delta t}) \right] \quad (6\text{–}15)$$

For $\Delta t \ll t_P$;

$$F|_{\Delta t} \cong 0 \; F|_{t_P+\Delta t} \cong F|_{\Delta t}$$

Equation 6–4, for $\Delta t \ll t_P$, then simplifies as

$$\Psi(p_i) - \Psi(p_{ws}) = 57.920 \times 10^6 \frac{q_{sc} T p_{sc}}{k h T_{sc}} \times \left[\log \left(\frac{t_P + \Delta t}{\Delta t} \right) + \frac{4\pi t_{DA}}{2.303} - \frac{F}{2.303} \right] \quad (6\text{–}16)$$

A plot of $\Psi(p_{ws})$ versus $(t_P + \Delta t)/\Delta t$ gives, initially, a straight line of slope m. Extrapolation of the line to an infinite shut-in time Δt, or $(\frac{t_P+\Delta t}{\Delta t} = 1)$, does not result in a value for $\Psi(p_i)$; the extrapolated value is called $\Psi(\overset{*}{p})$ and can be used to obtain $\overline{p}_R$, as described later. Figure 6–14 illustrates a typical buildup plot for a finite reservoir. Note that t_P is a pseudoproduction time in hours and is calculated by

$$t_P = \frac{24 G_P}{q_{sc}} \quad (6\text{–}17)$$

where G_p = cumulative production since well completion, mmscfd, q_{sc} = constant rate in mmscfd just before shut-in, and t_{DA} is a dimensionless time given by

$$t_{DA} = \frac{0.0002637 k t_P}{\phi \mu_{gi} c_i A} \quad (6\text{–}18)$$

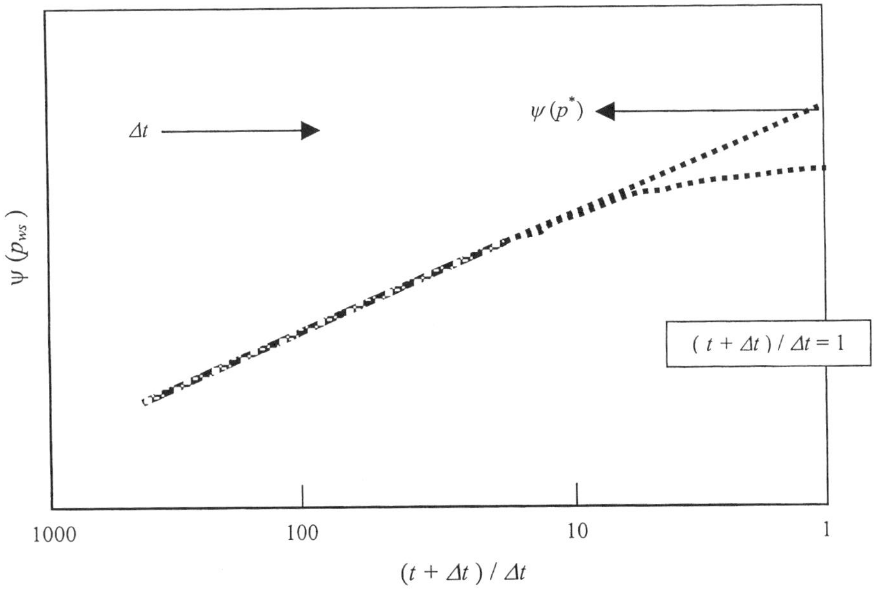

Figure 6–14. Buildup semilog plot—finite reservoir.

If real gas potential $\Psi(p)$ is replaced by bottom-hole pressure squared, p^2, Eq. 6–14 becomes

$$p_i^2 - p_{ws}^2 = 57.920 \times 10^6 \frac{q_{sc} T p_{sc} \mu_g z}{k h T_{sc}} \times \left[\log t_P + \log\left(\frac{k}{\phi \mu_g c r_w^2}\right) - 3.24 + \frac{4\pi t_{DA}}{2.303} - \frac{F}{2.303} + 0.869 s'\right] \quad (6\text{–}19)$$

Equation 6–4 becomes

$$p_i^2 - p_{ws}^2 = 57.920 \times 10^6 \frac{q_{sc} T p_{sc} \mu_g z}{k h T_{sc}} \times \left[\log\left(\frac{t_P + \Delta t}{\Delta t}\right) + \frac{4\pi t_{DA}}{2.303} - \frac{1}{2.303}\left(F|_{t_P+\Delta t} - F|_{\Delta t}\right)\right] \quad (6\text{–}20)$$

Equation 6–16 becomes

$$p_i^2 - p_{ws}^2 = 57.920 \times 10^6 \frac{q_{sc} T p_{sc} \mu_g z}{k h T_{sc}} \times \left[\log\left(\frac{t_P + \Delta t}{\Delta t}\right) + \frac{4\pi t_{DA}}{2.303} - \frac{F}{2.303}\right] \quad (6\text{–}21)$$

A plot of p_{ws}^2 versus $\frac{t_P+\Delta t}{\Delta t}$ gives a straight line of slope m. Extrapolation of the line to an infinite shut-in time Δt, or $\frac{t_P+\Delta t}{\Delta t} = 1$, does not result in a value for p_i^2. The extrapolated value is called $\overset{*}{p}{}^2$ and can be used to obtain $\overline{p}_R^2$.

6.9 Average Reservoir Pressure Estimating Techniques

Average reservoir pressures are used for characterizing a reservoir, computing its oil/gas in place, and predicting future behavior. In addition to these uses, the average reservoir pressure is required to find a quantitative use in volumetric-balance calculations of oil/gas in place in a reservoir. In this section we will present various methods to calculate average reservoir pressure in a gas reservoir.

Horner–MBH Method

The average reservoir pressure for a finite or bounded reservoir may be estimated as shown below using the values of m and $\Psi(\overset{*}{p})$ obtained from the Horner plot and the MBH curves.[3] From Equation 6–21 for $\frac{t_P+\Delta t}{\Delta t} = 1.0$,

$$\Psi(p_i) - \Psi(\overset{*}{p}) = \frac{57.920 \times 10^6 q_{sc} T p_{sc}}{kh T_{sc}} \left[\frac{4\pi t_{DA}}{2.303} - \frac{F}{2.303} \right] \qquad (6\text{–}22)$$

where Eq. 6–22 is the defining equation $\Psi(\overset{*}{p})$. The material balance equation may be written in terms of pseudopressure with substitution for dimensionless quantities as

$$t_{DA} = \frac{0.0002637 k t_P}{\phi \mu_{gi} c_i A} \qquad (6\text{–}23)$$

$$\Psi(p_i) - \Psi(\overline{p}_R) = 57.920 \times 10^6 \frac{q_{sc} T p_{sc}}{kh T_{sc}} \left(\frac{4\pi t_{DA}}{2.303} \right) \qquad (6\text{–}24)$$

Subtracting Eq. 6–22 from Eq. 6–24 gives

$$\Psi(\overset{*}{p}) - \Psi(\overline{p}_R) = m \frac{F}{2.303} \qquad (6\text{–}25)$$

or

$$\Psi(\overline{p}_R) = \Psi(\overset{*}{p}) - m \frac{F}{2.303} \qquad (6\text{–}26)$$

m is the absolute value of the slope of the straight-line section of the Horner plot:

$$m = 57.920 \times 10^6 \frac{q_{sc} T p_{sc}}{k h T_{sc}} \tag{6–27}$$

F is the MBH dimensionless pressure at t_{DA}, and the t_{DA} is the dimensionless time:

$$t_{DA} = \frac{0.0002637 k t_P}{\phi \mu_{gi} c_i A} \tag{6–28}$$

t_P is a pseudoproduction time in hours and is calculated using Eq. 6–17; $\psi(p^*)$ is the value of $\psi(p_{ws})$ corresponding to $\frac{t_P+\Delta t}{\Delta t} = 1$, from the extrapolated semilog straight line. F may be obtained from Table B–1 or Figures B–1 through B–5 corresponding to the appropriate well reservoir configuration and reservoir shape. Values of t_{DA} may be calculated from Eq. 6–28. If the MBH Figures B–1 through B–5 do not provide a particular configuration, F may be calculated from Eq. 6–29 or 6–30, whichever is appropriate:

$$F = In(C_A t_{DA}) \tag{6–29}$$

where C_A is a shape factor and is obtained from Table B–2. For small t_{DA}, that is, the transient region of flow, the well is infinite-acting and value of F is

$$F = 4\pi t_{DA} \tag{6–30}$$

The second term on the right-hand side of Eq. 6–15 is a correction term for finite or bounded reservoirs that is based on material balance. Thus, for an infinite reservoir,

$$\Psi(\overline{p}_R) = \Psi(p_i) = \Psi(\overset{*}{p}) \tag{6–31}$$

where p_i is the initial reservoir pressure for a bounded or finite reservoir, the procedure described above is applicable if t_{DA} can be calculated from a knowledge of k, ϕ, μ_{gi}, c_i, and A. If, however, all of these parameters are not known, the following method may be used to calculate $\Psi(\overline{p}_R)$.

Odeh and Al-Hussainy Method

This method[5] requires knowledge of $\Psi(p_i)$. A brief description of this method is given below.

Equation 6–24 may be written as

$$t_{DA} = \frac{2.303}{4\pi m}[\Psi(p_i) - \Psi(\overline{p}_R)] \tag{6–32}$$

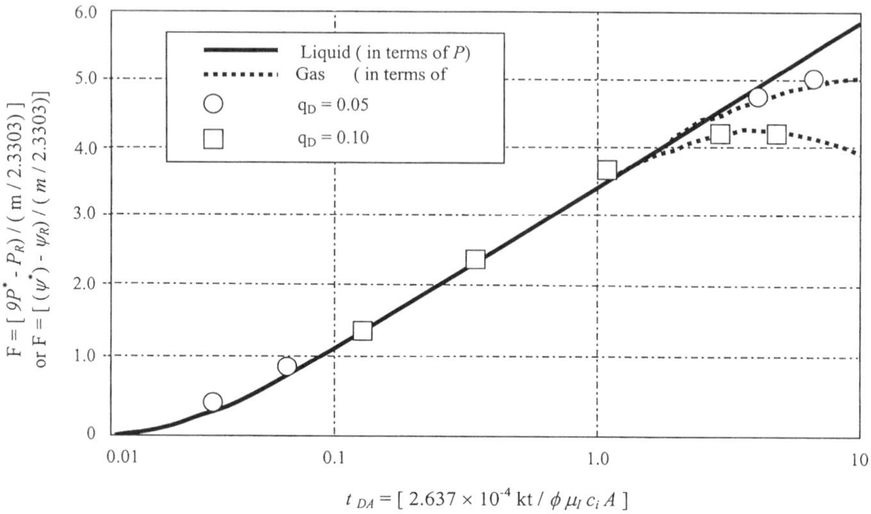

Figure 6–15. MBH dimensionless pressure function for one well in the center of a circle (after Odeh and Al-Hussainy, p. 61).[5]

Substituting Eq. 6–32 in Eq. 6–22 and rearranging gives

$$\frac{\Psi(p_i) - \Psi(\overset{*}{p})}{m} = \frac{\Psi(p_i) - \Psi(\overline{p}_R)}{m} - \frac{F}{2.303} \tag{6–33}$$

Figure 6–15 shows dimensionless pressure function F. Equation 6–33 may also be obtained by arranging Eq. 6–26. The following steps are followed to calculate $\Psi(\overline{p}_R)$:

1. Assume a number of values for $[\Psi(p_i) - \Psi(\overline{p}_R)]/m$.
2. Calculate corresponding values of t_{DA} from Eq. 6–32.
3. Choose the appropriate MBH curve, Figures B–1 through B–5 or Table B–1.
4. Obtain values for F for the calculated values of t_{DA}.
5. Calculate $[\Psi(p_i) - \Psi(\overset{*}{p})]/m$ using Eq. 6–33.
6. Plot $[\Psi(p_i) - \Psi(\overline{p}_R)]/m$ versus $[\Psi(p_i) - \Psi(\overset{*}{p})]/m$.
7. Since $\Psi(p_i)$ is known, $\Psi(\overset{*}{p})$ and m can be obtained from the Horner plot. $[\Psi(p_i) - \Psi(\overline{p}_R)]/m$ is easily obtained from the above plot. Hence $\Psi(\overline{p}_R)$ can be calculated.

Ramey and Cobb Method

Ramey and Cobb[6] have described a method for directly calculating $\Psi(\overline{p}_R)$ from a Horner plot. The solution yields

$$\left[\frac{t_p+\Delta t}{\Delta t}\right]_{\psi(P_{ws})=\psi(\overline{P}_R)} = e^F \tag{6–34}$$

F may be calculated either from Eq. 6–29 or from Eq. 6–30, whichever is appropriate. Thus $\Psi(\overline{p}_R)$ may be read directly from the extrapolation of the Horner straight line to a value of $(t_P + \Delta t)/\Delta t$ determined from Eq. 6–34.

6.10 Other Methods for Analyzing Pressure Buildup Test Data

Several other methods for analyzing pressure buildup data are available and are reviewed in this section.

MDH Method

This method[3] also uses the MDH plot, but the method is limited to estimating average reservoir pressure in a closed circular or square drainage region and to wells operating at pseudo-steady-state before the buildup test. The following equations and Figure 6–16 are used. Equation 6–9 can be written for large producing times, that is, for $t_P \gg \Delta t$, to become

$$\Psi(p_i) - \Psi(p_{ws}) = 57.920 \times 10^{10}\frac{q_{sc}Tp_{sc}}{khT_{sc}}[\log t_P - \log \Delta t] \tag{6–35}$$

Hence a plot of $\Psi(p_{ws})$ versus Δt on semilogarithmic coordinates should give a straight line of slope m, from which kh may be calculated. The apparent skin factor s' may be obtained from Eq. 6–11. The average reservoir pseudopressure $\Psi(\overline{p}_R)$ is obtained from Figure 5–12 for various shapes. The value of Δt_{De} is calculated at any chosen shut-in time Δt from

$$\Delta t_{De} = \frac{0.0002637k\Delta t}{\phi\mu_{gi}c_i r_e^2} \tag{6–36}$$

where $r_e^2 = \frac{A}{\pi}$ for noncircular geometry.

The corresponding value of Δp_D is obtained from Figure B–5. The average reservoir pressure is then calculated from

$$\Psi(\overline{p}_R) = \Psi(p_{ws}) + \frac{m}{1.151}\Delta\ p_D \tag{6–37}$$

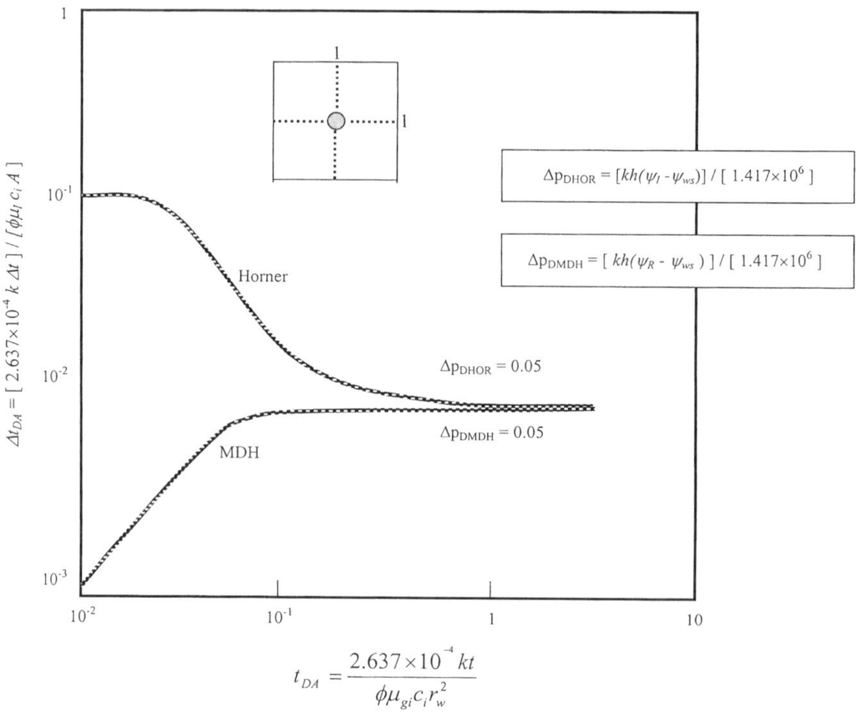

Figure 6–16. Comparison of times required to reach the end of Horner and MDH straight line (after Cobb and Smith).[7]

where m = slope of the MDH semilog straight line and $\psi(p_{ws})$ = value of the pseudopressure corresponding to the chosen shut-in time Δt from the straight line.

Extended Muskat Method

This method was developed by Muskat[8,9] and essentially gives a straight line for data that occur in the late-time region. The method requires plotting of $\log[\Psi(\overline{p}_R) - \Psi(p_{ws})]$ versus Δt. If the plot is not a straight line, another value of $\Psi(\overline{p}_R)$ is assumed, and the process is repeated. The method also requires that Δt and its corresponding $\Psi(p_{ws})$ value be chosen in the range given by the following relationship:

$$\frac{250\phi\mu_g cA}{k} \leq \Delta t \leq \frac{750\phi\mu_g cA}{k} \tag{6–38}$$

Equation 6–38 may be used to estimate the beginning and end of the Muskat straight line for closed square. Equation 6–38 may also be written as

$$0.066 \leq t_{DA} \leq 0.2 \tag{6–39}$$

where

$$t_{DA} = \frac{0.0002637 k \Delta t}{\phi \mu_{gi} c_i r_e^2} \tag{6–40}$$

The following equations are used to analyze late buildup data in terms of p_{ws}^2 and $\Psi(p_{ws})$.

Using bottom-hole pressure squared, p_{ws}^2:

$$k = \frac{50.300 \times 10^6 q_{sc} \overline{\mu}_g \overline{z} p_{sc}}{h T_{sc}} \left[\frac{p_{\text{Dint}}^2 (t_{PDA})}{(p_i^2 - p_{ws}^2)_{\Delta t=0}} \right] \tag{6–41}$$

$$\phi \overline{c} = -0.0559 \frac{k}{\overline{\mu}_g A} (\text{slope, } \log_{10} \text{cycle/hr}) \tag{6–42}$$

$$A = \frac{-0.00233k}{\phi \overline{\mu}_g \overline{c} m_M} \text{ (for closed square)} \tag{6–43}$$

$$= \frac{-0.00233k}{\phi \overline{\mu}_g \overline{c} m_M} \text{ (for a square with constant pressure boundary)} \tag{6–44}$$

Using the pseudopressure approach, $\Psi(p_{ws})$:

$$k = \frac{50.30 \times 10^6 q_{sc} T p_{sc}}{h T_{sc}} \left[\frac{\Psi_{\text{Dint}}(t_{PDA})}{(\Psi(\overline{p}_R) - \Psi(p_{ws}))_{\Delta t=0}} \right] \tag{6–45}$$

$$\phi c = -0.0559 \frac{k}{\mu_g A} (\text{slope, } \log_{10} \text{ cycle/hr}) \tag{6–46}$$

$$A = \frac{-0.00471k}{\phi \mu_g c\, m_M} \tag{6–47}$$

where $p_{\text{Dint}}^2(t_{PDA})^*$ or $\Psi_{\text{Dint}}(t_{PDA})^{**}$ can be found from Table 6–4, and $(p_i^2 - p_{ws}^2)_{\Delta t=0}$ or $[\Psi(\overline{p}_R) - \Psi(p_w)]_{\Delta t=0}$ is the intercept from the Muskat plot. The slope may also be estimated from the Muskat data plot:

$$\text{Slope} = \frac{\log\left(\overline{p}_R^2 - p_{ws}^2\right)_2 - \log\left(p_R^2 - p_{ws}^2\right)_1}{\Delta t_1 - \Delta t_2} \text{ cycle/hr}$$

or

$$\text{Slope} = \frac{\log\left[\Psi\left(\overline{p}_R^2 - p_{ws}^2\right)\right]_2 - \log[\Psi(\overline{p}_R) - \Psi(p_w)]_1}{\Delta t_1 - \Delta t_2} \text{ cycle/hr}$$

Table 6–4
Muskat Dimensionless Intercept Pressure[8]

t_{PDA}	Constant pressure boundary system* $p^2_{Dint}(t_{PDA})$	Closed square system** $\Psi_{Dint}(t_{PDA})$
0.001	0.0265	0.0265
0.002	0.0520	0.0520
0.003	0.0760	0.0760
0.004	0.100	0.100
0.005	0.126	0.126
0.006	0.140	0.140
0.007	0.168	0.168
0.008	0.195	0.195
0.009	0.220	0.220
0.01	0.240	0.240
0.02	0.440	0.440
0.03	0.600	0.500
0.04	0.725	0.550
0.05	0.850	0.583
0.06	0.900	0.615
0.07	0.968	0.630
0.08	1.035	0.645
0.09	1.098	0.658
0.10	1.160	0.670
0.20	1.340	0.670
0.30	1.340	0.670
0.40	1.340	0.670
0.50	1.340	0.670
0.60	1.340	0.670
0.70	1.340	0.670
0.80	1.340	0.670
0.90	1.340	0.670
1.00	1.340	0.670

* Ramey, Kumar, and Gulati (1973)[10]
** Ramey and Cobb (1971)[6]

$$t_{PDA} = \frac{0.0002637 k t_P}{\phi \mu_g c A}$$

$$A = \pi r_e^2$$

$$\text{Area} = \text{acre} \times 43{,}560 \text{ sq ft}$$

Table 6–4 shows the data for both systems. For the closed-square system:

$$p^2_{Dint}\,(t_{PDA} < 0.1) = 0.67 \quad \text{or} \quad \Psi_{Dint}(t_{PDA} < 0.1) = 0.67 \tag{6–47a}$$

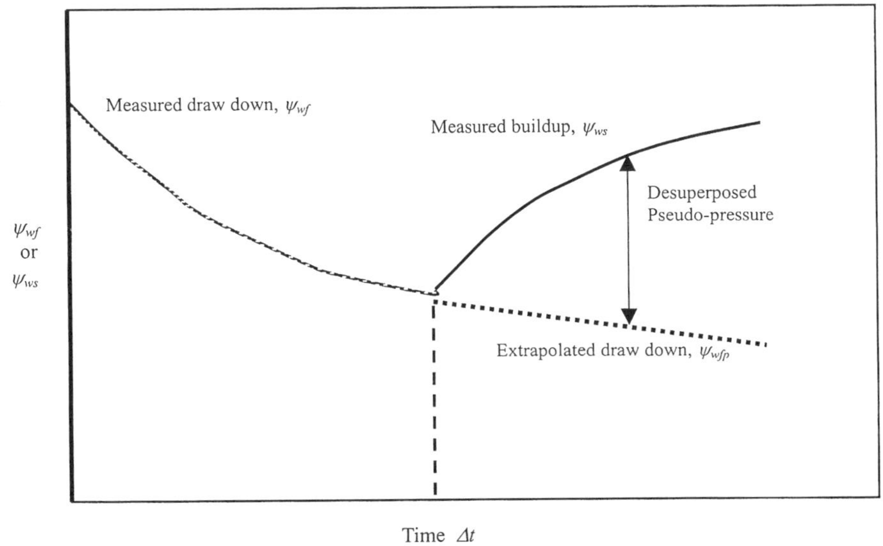

Figure 6–17. Desuperposition for the slider plot (after Slider).[11]

If producing time exceeds the time to pseudo-steady-state, then for the constant-pressure boundary system, Eq. 6–47a is applicable:

$$\begin{aligned} p_{D\text{int}}(t_{PDA} > 0.20) &= 1.34 \\ \Psi_{D\text{int}}(t_{PDA} > 0.20) &= 1.34 \end{aligned} \qquad (6.47b)$$

When producing time exceeds the time required to reach steady state, Eq. 6–47b is applicable.

Slider Method

The middle-time buildup data may be analyzed by using the approach proposed by Slider[11,12] and is illustrated by Figure 6–17. The drawdown behavior can be extrapolated as shown by the dotted line. This may be done either by type curve matching or by the use of appropriate equations. Equation 6–35 applies prior to the pseudo-steady-state ($t < t_p$); Eq. 6–36 applies thereafter.

$$\Psi(p_i) - \Psi(p_{wf}) = \frac{57.900 \times 10^6 q_{sc} T p_{sc}}{kh T_{sc}} \times \left[\log t + \log\left(\frac{k}{\phi \mu_{gi} c_i r_w^2}\right) - 3.23 + 0.869 s'\right] \qquad (6\text{–}48)$$

$$\Psi(p_i) - \Psi(p_{wf}) = \frac{2348 q_{sc} T t}{\pi \phi \mu_{gi} c_i r_e^2 h} + \frac{3.263 \times 10^6 q_{sc} T}{kh} \times \left[\log\left(\frac{0.472 r_e}{r_w}\right) + \frac{s'}{2.303}\right] \quad (6\text{–}49)$$

The desuperposed pseudopressure $[\Psi(p_{ws}) - \Psi(p_{wfp})]$ at any shut-in time Δt is obtained by subtracting the extrapolated drawdown, $\Psi(p_{wfp})$, from the measured buildup pseudopressure, $\Psi(p_{ws})$. This desuperposed pressure, if plotted against log Δt, should give a straight line of slope m given by Eq. 6–50:

$$m = \frac{57.920 \times 10^6 q_{sc} T p_{sc} \overline{\mu}_g \overline{z}}{kh T_{sc}} \quad (6\text{–}50)$$

or

$$kh = \frac{57.920 \times 10^6 q_{sc} T p_{sc} \overline{\mu}_g \overline{z}}{m T_{sc}} \quad (6\text{–}51)$$

The intercept is given by Eq. 6–51:

$$\Psi(p_{1hr}) = m\left[\log\left(\frac{k}{\phi \overline{\mu}_g \overline{c} r_w^2}\right) - 3.2275 + 0.8685 s\right] \quad (6\text{–}52)$$

The skin factor is estimated from

$$s = 1.151\left[\frac{\Psi(p_{1hr})}{m} - \log\left(\frac{k}{\phi \overline{\mu}_g \overline{c} r_w^2}\right) + 3.23\right] \quad (6\text{–}53)$$

Dietz Method

Dietz[13] suggested extrapolating the straight-line portion of an MDH plot (P_{ws} versus log Δt) directly to $\overline{P}$. The Dietz approach assumes that the well has been produced at a constant rate long enough to reach pseudo-steady-state before shut-in, and that a semilog straight line of appropriate slope will develop. Dietz determined the time when $\overline{P}$ may be read directly from the extrapolated semilog straight line:

$$\overline{P}_{circle} = \frac{\phi \mu_g c_t A}{0.0002637 C_A k} \quad (6\text{–}54)$$

For a well centrally located in a closed square drainage area:

$C_A = 30.8828$,

so

$$(\Delta t)_{\overline{P}_{square}} = 122.8\frac{\phi\mu_g c_t A}{k} \quad (6\text{–}55)$$

The use of various methods to analyze pressure buildup data is illustrated in Example 6–2.

Example 6–2[26] *Average Reservoir Pressure Computation Using Various Methods*

A pressure-buildup test was performed on a gas well in a finite reservoir. Data obtained were as follows (see Table 6–5): $q_{sc} = 10$ mmscfd, cumulative gas produced, $q_{sc} = 10^5$ scfd; $T = 605^0$ R; $z = 0.850$; $m_{gi} = 0.12$ cP; $c_i = 0.000436$ psi^{-1}; $h = 54$ ft; $\phi = 18\%$; $r_w = 0.3333$ ft; well spacing = 640 acres.

Assuming the well is in the center of a circle, compute the reservoir pressure in the drainage area of the well assuming finite boundary conditions:

1. Using the Horner or MBH method
2. Using the MDH method
3. Using the Ramey and Cobb method
4. Using the Dietz method

Table 6–5
Pressure Buildup Test Data

Shut-in time Δt (hr)	Well pressure p_{ws} (psia)	$\frac{t_p+\Delta t}{\Delta t}$	p_{ws}^2 (mmpsia^2)
0	1742	—	3.035
1	1865	2401	3.478
3	1979	801	3.917
6	2023	401	4.093
10	2054	241	4.219
15	2079	161	4.322
22	2102	110	4.418
34	2128	71.6	4.528
45	2145	54.3	4.601
65	2170	37.9	4.709
126	2190	20.0	4.796

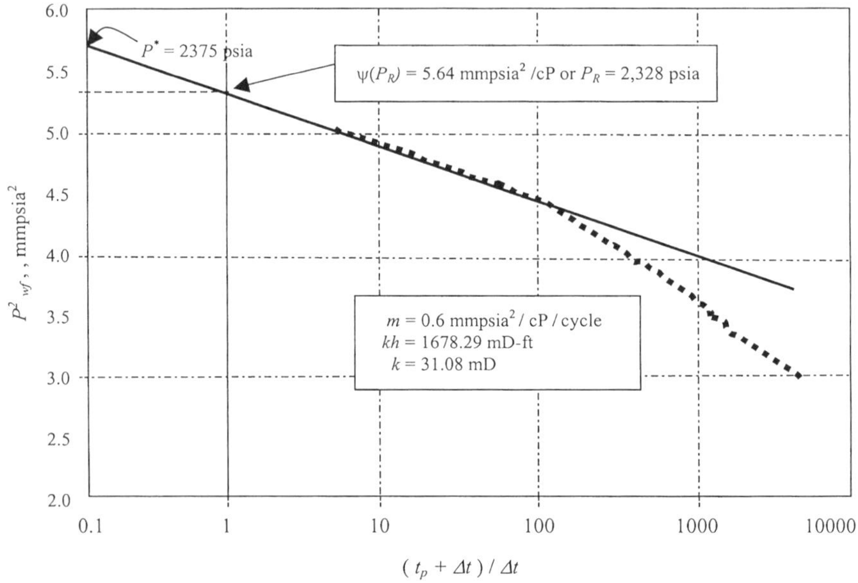

Figure 6–18. Pressure buildup curve, Horner's plot—Example 6–2.

Solution Calculate pseudoproducing time using Eq. 6–17:

$$t_P = \frac{24Q_P}{q_{sc}} = \frac{24 \times 10^5}{1000} = 2400 \text{ hr}$$

1. Solution Procedure for Horner and MBH Method
   From Figure 6–18

$$m = \frac{(5.05 - 4.45)10^6}{\log 100 - \log 10} = 6.0 \times 10^5 \text{psia}^2/\text{cycle}$$

$$p^{2*} = 5.64 \times 10^6 \text{psia}^2 \leftrightarrow p^* = 2375 \text{ psia}$$

Calculate k using Eq. 6–10:

$$kh = \frac{57.920 \times 10^3 \times q_{sc} T P_{sc} \mu z}{m T_{sc}}$$

For $q_{sc} = 10$ mscfd $= 0.01$ mmscfd, $T = 605^0$R, $p_{sc} = 14.65$, $T_{sc} = 520^0$R, $\mu = 0.12$ cP, $z = 0.85$.

$$kh = \frac{57.920 \times 10^6 \times 10 \times 605 \times 14.65 \times 0.12 \times 0.85}{6.0 \times 10^6 \times 520}$$

$$= 1678.290 \text{ mD–ft}$$
$$k = 1678.29/54 = 31.08 \text{ mD}$$

Also,

$$r_e = \sqrt{\frac{640 \times 43,560 \times 7}{22}} = 2978 \text{ ft}$$

Calculate dimensionless time using Eq. 6–40:

$$t_{DA} = \frac{0.0002637 \times kt_P}{\phi \mu_{gi} c_i A}$$
$$= \frac{0.0002637 \times 31.08 \times 2400}{0.18 \times 0.12 \times 0.000436 \times 640 \times 43{,}560} = 0.075$$

From Figure B–8 using 0.075, the ordinate (MBH dimensionless pressure function) reads as 0.85, and thus

$$\overline{p}_R = \left[\overset{*}{p}^2 - \frac{0.85 \times m}{2.3}\right]^{0.5}$$
$$= \left[5.64 \times 10^6 - \frac{0.85 \times 6.0 \times 10^5}{2.3}\right]^{0.5} = 2328 \text{ psia}$$

2. MDH Method
Figure 6–19 presents a Miller–Dyes–Hutchinson (MDH) type buildup graph for data of Example 6–2. The appearance of the graph is similar to that of the Horner graph, i.e., MBH (see Figure 6–18). The slope m of the straight-line portion is 5.9×10^5 psia^2/cycle (from MDH graph); the p_{1hr} is exactly the same, i.e., 1903 psia.

$$kh = \frac{57.920 \times 10^6 \times 10 \times 605 \times 14.65 \times 0.12 \times 0.85}{5.9 \times 10^5 \times 520}$$
$$= 1736.16 \text{ mD} - \text{ft}$$
$$k = 31.61 \text{ mD}$$

The difference in permeability between the two methods is only 2%.

$$s = 1.151\left[\frac{p_{1hr}^2 - p_{wfo}^2}{m} - \log\frac{k}{\phi \mu c r_w^2} + 3.23\right]$$
$$= 1.151\left[\frac{(3.62 - 3.03)10^6}{5.9 \times 10^5} - \log\left(\frac{31.61}{0.18 \times 0.12 \times 0.000436 \times 0.332}\right) + 3.23\right]$$
$$= -3.75 \text{ (well improvement)}$$

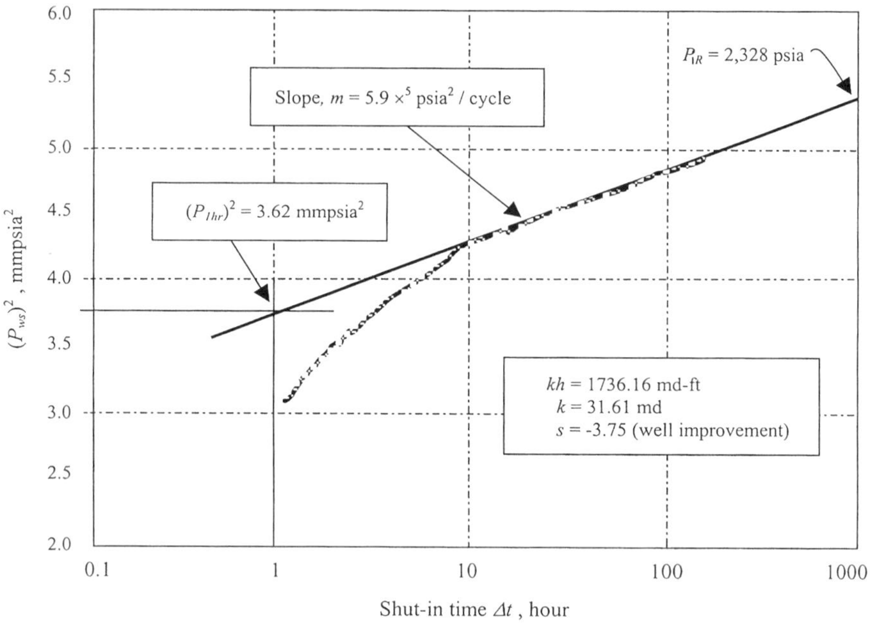

Figure 6–19. MDH graph—Example 6–2.

3. Ramey and Cobb Method
   Calculate the value of F from Eq. 6–29:

$$F = In(C_A t_{DA})$$

where C_A is shape factor = 31.62 (from Table B–2)

$$t_{DA} = \frac{0.0002637 k t_P}{\phi \mu c A}$$

$$= \frac{0.0002637 \times 31.61 \times 2400}{0.18 \times 0.12 \times 0.000436 \times 640 \times 43,560} = 0.076$$

$$\therefore F = In(31.62 \times 0.076) = 0.877$$

Using Equation 6–34:

$$\left(\frac{t_P + \Delta t}{\Delta t}\right)_{p_{ws}=p_R} = e^F = e^{0.897} = 2.40$$

Reading directly from the extrapolation of the Horner straight line to a value of 2.40 will give

$$p_{ws}^2 = \overline{p}_R^2 = 5.4 \times 10^6 \text{ psia}^2$$
$$p_R = 2324 \text{ psia.}$$

4. Dietz Method
   Calculate time when $\overline{p}$ may be read directly from the extrapolated semilog straight line from the MDH plot using Eq. 6–40:

$$(\Delta)_{\overline{p}} = \frac{\phi \mu_g c_t A}{0.0002637 C_A k}$$

$$= \frac{0.18 \times 0.12 \times 0.000436 \times 640 \times 43{,}560}{0.0002637 \times 31.62 \times 31.61} = 996.12\,\text{hr}$$

   Reading directly from the extrapolated of the MDH straight line to a value of 996.12 hr will give

$$P_{ws}^2 = \overline{P_R^2} = 5.39 \times 10^6\,\text{psia (see Figure 6–19)}$$
$$p_R = 2322\,\text{psia}$$

6.11 Pressure Behavior Analysis and Estimating Formation Characteristics

Buildup Following a Two-Rate Drawdown Test

The analysis of a buildup that follows a two-rate drawdown can yield values of kh that provide a check on the results of a drawdown analysis. The flow rate and time sequences used to develop multirate drawdown analysis still apply, but with a slight modification. A well is flowed at a rate q_{sc1} up to time t_1, at a rate q_{sc2} up to time, t, and then shut in. The shut-in time is represented, as before, by Δt. Using this notation, Eq. 4–50 may be modified to represent the buildup period as

$$\begin{aligned}\Psi(p_i) - \Psi(p_{ws}) &= \frac{57.920 \times 10^6 q_{sc1} T p_{sc}}{khT_{sc}} \\ &\times \left[\log(t + \Delta t) + \log\left(\frac{k}{\phi \mu_{gi} c_i r_w^2}\right) - 3.23\right] \\ &+ \frac{57.920 \times 10^6 (q_{sc2} - q_{sc1}) T p_{sc}}{khT_{sc}} \\ &\times \left[\log(t + \Delta t + t_1) + \log\left(\frac{k}{\phi \mu_{gi} c_i r_w^2}\right) - 3.23\right] \\ &+ \frac{57.920 \times 10^6 (0 - q_{sc2}) T p_{sc}}{khT_{sc}} \\ &\times \left[\log \Delta t + \log\left(\frac{k}{\phi \mu_{gi} c_i r_w^2}\right) - 3.23\right] \end{aligned} \quad (6\text{–}56)$$

Combining these terms and simplifying Eq. 6–56 gives

$$\Psi(p_i) - \Psi(p_{ws}) = \frac{57.920 \times 10^6 q_{sc1} T p_{sc}}{khT_{sc}} \left[\log\left(\frac{t+\Delta t}{t+\Delta t+t_1}\right) + \frac{q_{sc2}}{q_{sc1}} \log\left(\frac{t+\Delta t - t_1}{\Delta t}\right) \right] \quad (6\text{–}57)$$

Hence a plot of

$$\Psi(p_{ws}) \text{ versus } \log\left(\frac{t+\Delta t}{t+\Delta t - t_1}\right) + \frac{q_{sc2}}{q_{sc1}} \log\left(\frac{t+\Delta t - t_1}{\Delta t}\right)$$

on arithmetic coordinates should give a straight line of slope m from which permeability is calculated as

$$k = \frac{57.920 \times 10^6 q_{sc1} T p_{sc}}{mT_{sc}h} \quad (6\text{–}58)$$

Calculate the skin factors from

$$s = 1.151 \left[\frac{q_{sc1}}{(q_{sc1} - q_{sc2})} \frac{\Psi(p_{1hr}) - \Psi(p_{ws1})}{m} \right] - \log\left(\frac{k}{\phi \mu_{gi} c_i}\right) + 3.23 \Big] \quad (6\text{–}59)$$

The following equations can be used to estimate the pressure drop across the skin at rates q_{sc1} and q_{sc2}, respectively. Thus, at q_{sc1},

$$\Psi(\Delta p)_{skin} = 0.869(-m)(s) \quad (6\text{–}60)$$

and at q_{sc2},

$$\Psi(\Delta p)_{skin} = 0.869\left(\frac{q_{sc2}}{q_{sc1}}\right)(-m)(s) \quad (6\text{–}61)$$

Having found values for k and s, one may now proceed to determine p^*. The flowing bottom-hole pressure in a well at a time t_1 can be expressed as

$$\Psi(p_{wf}) = \Psi(p_i) - \frac{57.920 \times 10^6 q_{sc1} T p_{sc}}{khT_{sc}} \times \left[\log\left(\frac{kt_1}{\phi \mu_{gi} c_i r_w^2}\right) - 3.23 + 0.869s \right] \quad (6\text{–}62)$$

By setting $\Psi(p_i) = \Psi(\overset{*}{p})$ and rearranging Eq. 6–63, we have

$$\Psi(P^*) = \Psi(P_{wf}) + (-m)\left[\log\left(\frac{kt_1}{\phi \mu_{gi} c_i}\right) - 3.23 + 0.869s \right] \quad (6\text{–}63)$$

For a bounded reservoir, $\Psi(\overset{*}{P})$ may be used as described before to calculate $\Psi(P_R)$.

Example 6–3[26] *Buildup Following a Two-Rate Drawdown Test Analysis for Gas Well*

The gas well is tested for 6 hr at a rate of 2.397 mmscfd and finally by producing for a further 6 hr at a rate of 5.214 mmscfd. The pressures recorded during the flowing and closed-in periods are listed in Table 6–6. The well/reservoir data are as follows:

$$\begin{aligned}
P_I &= 3965 \text{ psia} \leftrightarrow \psi(p_i) = 861.12 \text{ mmpsia}^2/\text{cP};\\
P_R &= 3700 \text{ psia} \leftrightarrow \psi(P_R) = 772.00 \text{ mmpsia}^2/\text{cP};\\
h &= 41 \text{ ft};\ r_e = 2200 \text{ ft};\ r_w = 0.4271 \text{ ft};\ T = 710^0 R;\\
\phi_{HC} &= 0.119 \text{ fraction};\ \overline{\mu} = 0.02345 \text{ cP},\ \overline{c} = 0.00027 \text{ psi}^{-1};\\
T_{sc} &= 520^0 R;\\
P_{sc} &= 14.65 \text{ psia; and cumulative production} = 11.382 \text{ mmscf.}
\end{aligned}$$

1. From the pressure buildup determine k and s_1'.
2. From the flow tests determine k, s_1', s_2', and hence D, true skin factor s, and deliverability constants A and B. Develop an inflow (IPR) curve for this gas well. Use the same reservoir and well data given in Example 6–1.

Solution Table 6–7 shows pressure drawdown test data.

Since the fluid properties are the same as in Example 6–1, all other data presented in Example 6–1 can be used in the current example.

$$\text{Pseudoproducing time } t_p = \frac{11.382 \times 24}{6.148} = 44.432 \text{ hr}$$

1. *Buildup Analysis*

   For a flowing time of 147.12 hr, the data necessary to draw the Horner buildup plot are listed in Table 6–7. The corresponding build-up plot is shown in Figure 6–20, from which the slope has been determined as $m = 21.0$ psia^2/cP/cycle and using Eq. 6–10, permeability k as

$$k = \frac{57.92 \times 10^6 \times 6.148 \times 710 \times 14.65}{21 \times 10^6 \times 520 \times 41} = 8.27 \text{ mD}$$

(text continued on page 361)

Table 6–6
Pressure Buildup Test Data

Time Δt (hr) (1)	$\frac{t_p+\Delta t}{\Delta t}$ — (2)	P_{ws} (psig) (3)	P_{ws} (psia) (4)	$\psi(P_{ws})$ (mmpsia^2/cP) (5)	$\psi(\Delta P_{ws})$ (mmpsia^2/cP) (6)	Radius r, (ft) (7)
0.00	—	1720	1735	204.35	0.00	0
0.02	2666.92	1723	1738	204.96	0.61	16
0.03	1333.96	1733	1747	207.12	2.77	22
0.07	667.48	1773	1788	216.16	11.81	31
0.10	445.32	1803	1818	223.10	21.83	38
0.13	334.24	1854	1869	234.89	33.62	44
0.17	267.59	1911	1925	248.28	47.01	49
0.25	178.73	2014	2028	273.44	72.17	60
0.33	134.30	2120	2135	300.34	99.07	69
0.50	89.86	2297	2312	347.09	145.82	85
0.75	60.24	2601	2615	432.19	230.92	104
1.00	45.43	2805	2819	492.58	291.31	120
1.50	32.62	3132	3146	593.35	392.07	147
2.00	23.22	3295	3310	645.38	444.11	170
2.50	18.77	3335	3350	658.17	456.89	190
3.00	15.81	3352	3366	663.42	462.14	208
3.50	13.69	3368	3382	668.65	467.37	225
4.00	12.11	3370	3385	669.42	468.14	240
4.83	12.11	3370	3385	669.42	468.14	240
4.83	10.19	3377	3391	671.52	470.24	264
5.00	9.89	3382	3397	673.23	471.96	269
5.50	9.08	3388	3403	675.21	473.93	282
6.00	8.41	3393	3407	676.66	475.38	294
6.50	7.84	3397	3411	677.96	476.68	306
7.00	7.35	3400	3415	679.15	477.88	318
7.50	6.92	3404	3418	680.32	479.04	329
8.00	6.55	3406	3421	681.13	479.85	340
8.50	6.23	3410	3425	682.30	481.02	350
9.00	5.94	3413	3428	683.37	482.09	360
9.50	5.68	3417	3432	684.66	483.39	370
10.00	5.44	3421	3436	685.99	484.72	380
10.50	5.23	3425	3440	687.32	486.05	389
11.00	5.04	3429	3443	688.36	487.08	398
11.50	4.86	3432	3447	689.43	488.16	407
12.00	4.70	3434	3448	689.95	488.68	416
12.50	4.55	3436	451	690.86	489.59	425
13.00	4.42	3438	3453	691.55	490.27	433
13.50	4.29	3441	3456	692.46	491.18	441
14.00	4.17	3444	3459	693.40	492.12	449
14.50	4.06	3447	3461	694.18	492.90	457

Table 6–6 (Continued)

Time Δt (hr) (1)	$\frac{t_p+\Delta t}{\Delta t}$ — (2)	P_{ws} (psig) (3)	P_{ws} (psia) (4)	$\psi(P_{ws})$ (mmpsia^2/cP) (5)	$\psi(\Delta P_{ws})$ (mmpsia^2/cP) (6)	Radius r, (ft) (7)
15.00	3.96	3449	3464	694.96	493.68	465
15.50	3.87	3452	3466	695.88	494.62	473
16.00	3.78	3454	3468	696.53	495.27	480
16.50	3.69	3456	3471	697.31	496.06	488
17.00	3.61	3458	3473	697.86	496.61	495
17.50	3.54	3461	3475	698.77	497.52	502
18.00	3.47	3462	3477	699.29	498.04	509
18.50	3.40	3465	3479	700.11	498.85	516
19.00	3.34	3467	3481	700.76	499.51	523
19.50	3.28	3469	3483	701.41	500.16	530
20.00	3.22	3471	3486	702.22	500.97	537
20.50	3.17	3473	3488	702.75	501.49	544
21.00	3.12	3474	3489	703.14	501.88	550
21.50	3.07	3477	3491	703.92	502.67	557
22.00	3.02	3478	3493	704.47	503.22	563
22.50	2.97	3480	3494	704.99	503.74	570
23.00	2.93	3481	3496	705.52	504.26	576
23.50	2.89	3482	3497	705.78	504.53	582
24.00	2.85	3485	3499	706.59	505.34	588
24.50	2.81	3486	3501	706.98	505.73	594
25.00	2.78	3487	3502	707.38	506.12	600
26.00	2.71	3491	3505	708.58	507.33	612
27.00	2.65	3494	3509	709.76	508.51	624
28.00	2.59	3497	3512	710.71	509.45	635
29.00	2.53	3500	3515	711.62	510.37	647
30.00	2.48	3503	3518	712.53	511.28	658
31.00	2.43	3506	3521	713.61	512.26	669
32.00	2.39	3509	3524	714.66	513.41	679
33.00	2.35	3511	3526	715.21	513.96	690
34.00	2.31	3514	3529	716.26	515.01	760
35.00	2.27	3517	3532	717.21	515.96	710
36.00	2.23	3519	3533	717.73	516.48	720
37.00	2.20	3522	3536	718.65	517.40	730
38.00	2.17	3523	3538	719.17	517.92	740
39.00	2.14	3526	3541	720.12	518.87	750
40.00	2.22	3529	3544	721.17	519.92	759
41.00	2.08	3531	3546	721.72	520.47	769
42.00	2.06	3534	3548	722.64	521.39	778
43.00	2.03	3536	3550	723.30	522.04	787
44.00	2.01	3539	3553	724.25	522.99	797
45.00	1.99	3540	3555	724.77	523.52	806

Table 6–6 (Continued)

Time Δt (hr) (1)	$\frac{t_p+\Delta t}{\Delta t}$ — (2)	P_{ws} (psig) (3)	P_{ws} (psia) (4)	$\psi(P_{ws})$ (mmpsia^2/cP) (5)	$\psi(\Delta P_{ws})$ (mmpsia^2/cP) (6)	Radius r, (ft) (7)
46.00	1.97	3543	3557	725.56	524.31	814
47.00	1.95	3445	3560	726.38	525.12	823
48.00	1.93	3547	3561	726.90	525.65	832
49.00	1.91	3549	3564	727.69	526.44	841
50.00	1.89	3551	3566	728.38	527.12	849
51.00	1.87	3552	3567	728.64	527.39	858
52.00	1.85	3555	3570	729.56	528.31	866
53.00	1.84	3559	3571	730.08	528.83	874
54.00	1.82	3560	3574	730.90	529.65	882
55.00	1.81	3663	3575	731.30	530.04	891
56.00	1.79	3565	3578	732.22	530.96	899
57.00	1.78	3567	3580	732.91	531.65	907
58.00	1.77	3568	3581	733.43	532.18	915
59.00	1.75	3570	3583	733.96	532.70	922
60.00	1.74	3572	3585	734.48	533.23	930
61.00	1.73	3574	3587	735.17	533.92	938
62.00	1.72	3576	3589	735.96	534.71	946
63.00	1.71	3578	3590	736.35	535.10	953
64.00	1.69	3579	3592	737.04	535.79	961
65.00	1.68	3580	3594	737.57	536.32	968
66.00	1.67	3583	3595	737.83	536.58	976
67.00	1.66	3584	3597	738.62	537.37	983
68.00	1.65	3586	3599	739.18	537.93	990
69.00	1.64	3588	3601	739.84	538.59	997
70.00	1.63	3589	3602	740.36	539.11	1005
71.00	1.63	3592	3604	740.89	539.64	1012
72.00	1.62	3592	3606	741.58	540.33	1019
73.00	1.61	3594	3607	741.84	540.59	1026
74.00	1.60	3596	3609	742.50	541.25	1033
75.00	1.59	3598	3611	743.16	541.91	1040
76.00	1.58	3599	3613	743.72	542.47	1047
77.00	1.58	3600	3614	744.12	542.86	1054
78.00	1.57	3604	3615	744.51	543.26	1061
79.00	1.56	3604	3618	745.60	544.34	1067
80.00	1.56	3606	3619	745.73	544.48	1074
81.00	1.55	3607	3621	746.39	545.13	1081
82.00	1.54	3609	3622	746.78	545.53	1087
83.00	1.54	3611	3624	747.31	546.06	1094
84.00	1.53	3612	3625	747.87	546.62	1101
85.00	1.52	3613	3626	748.26	547.01	1107
86.00	1.52	3615	3628	748.79	547.54	1114

Table 6–6 (Continued)

Time Δt (hr) (1)	$\frac{t_p+\Delta t}{\Delta t}$ — (2)	P_{ws} (psig) (3)	P_{ws} (psia) (4)	$\psi(P_{ws})$ (mmpsia²/cP) (5)	$\psi(\Delta P_{ws})$ (mmpsia²/cP) (6)	Radius r, (ft) (7)
87.00	1.51	3616	3630	749.32	548.07	1120
88.00	1.50	3618	3631	749.71	548.46	1126
89.00	1.50	3619	3632	750.14	548.98	1133
90.00	1.49	3621	3634	750.67	549.42	1139
91.00	1.49	3624	3635	751.20	549.95	1145
92.00	1.48	3624	3638	752.15	550.90	1152
93.00	1.48	3625	3639	752.42	551.17	1158
94.00	1.47	3628	3640	752.68	551.43	1164
95.00	1.47	3628	3642	753.47	552.22	1170
96.00	1.46	3692	3643	753.74	552.49	1177
97.00	1.46	3631	3644	753.94	552.81	1183
98.00	1.45	3633	3645	754.63	553.18	1189
99.00	1.45	3635	3647	755.09	553.84	1195
100.00	1.44	3637	3650	755.88	554.63	1201
101.00	1.44	3640	3651	756.44	555.19	1207
102.00	1.44	3642	3654	757.50	556.25	1213
103.00	1.43	3644	3656	758.16	556.91	1219
104.00	1.43	3644	3658	758.72	557.47	1225
105.33	1.42	3644	3659	758.99	557.73	1232
106.00	1.42	3641	3659	758.85	557.60	1236
107.00	1.42	3644	3656	758.03	556.78	1242
109.00	1.41	3644	3658	758.72	557.47	1254
111.00	1.40	3644	3659	758.99	557.73	1265
113.00	1.39	3648	3661	759.65	558.39	1276
115.00	1.39	3651	3663	760.27	559.02	1288
117.00	1.38	3654	3666	761.26	560.01	1299
119.00	1.37	3656	3668	762.06	560.81	1310
121.00	1.37	3659	3671	762.98	561.73	1321
123.00	1.36	3662	3674	763.81	562.56	1332
125.00	1.36	3666	3676	764.74	563.48	1343
127.00	1.35	3666	3681	766.22	564.97	1353
129.00	1.34	3669	3683	766.89	565.63	1364
131.00	1.34	3669	3684	767.15	565.90	1374
133.00	1.33	3673	3684	767.15	565.90	1385
135.00	1.33	3677	3688	768.51	567.25	1395
137.00	1.32	3678	3691	769.73	568.48	1406
139.00	1.32	3680	3693	770.13	568.88	1416
141.00	1.32	3682	3695	770.92	569.67	1426
143.00	1.31	3685	3696	771.32	570.07	1436
145.00	1.31	3686	3700	772.41	571.16	1446
147.12	1.30	3686	3700	772.68	571.43	1446

Table 6–7
Pressure Drawdown Test Data

Flowing time t (hr)	First flow rate q_{sc} = 2.397 mmscfd P_{wf} (psia)	$\psi(p_{wf})$ (mmpsia²/cP)	Second flow rate q_{sc2} = 5.214 mmscfd P_{wf} (psia)	$\psi(p_{wf})$ (mmpsia²/cP)
0.02	3609	742.50	3577	732.08
0.03	3544	721.30	3455	692.04
0.07	3480	700.37	3300	642.12
0.10	3440	687.30	3183	604.88
0.13	3385	669.52	3040	560.03
0.17	3347	657.11	2956	534.07
0.25	3270	632.62	2826	494.52
0.33	3224	618.00	2757	473.82
0.50	3173	601.65	2710	459.77
0.75	3142	591.82	2714	461.07
1.00	3130	588.23	2652	442.94
1.50	3145	592.95	2611	431.01
2.00	3128	587.48	2602	428.43
2.50	3130	588.10	2595	426.41
3.00	3134	589.39	2590	424.91
4.00	3137	590.27	2580	421.97
5.00	3140	591.28	2573	419.95
6.00	3144	592.45	2567	418.12

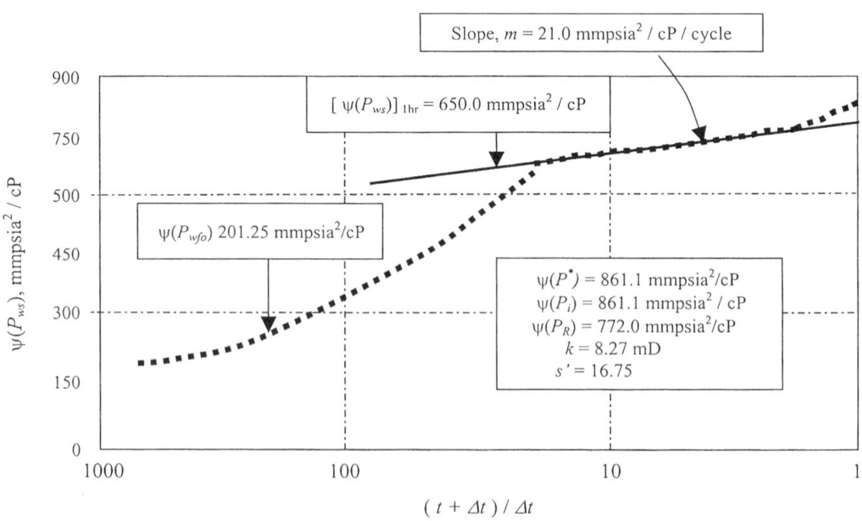

Figure 6–20. Horner buildup data plot—Example 6–3.

(text continued from page 355)

From Eq. 6–11:

$$s' = 1.151\left[\frac{650.0 - 201.25}{21} - \log\left(\frac{8.27}{0.119 \times 0.02345 \times 0.00023 \times 0.4271^2}\right) + 3.23\right]$$
$$= 16.75$$

2. *Flow Analysis*
   Plots of $\psi(p_{wf})$ versus log t for the data listed in Table 6–7 are shown as Figures 6–21 and 6–22, from which the following information is obtained. From Figure 6–21, first flow rate = 2.397 mmscfd; m_1 = 17.9 mmpsia^2/cP/cycle; and $\psi(p_{1hr})$ = 535.15 mmpsia^2/cP. From Eq. 6–10;

$$k_1 = \frac{1.632 \times 10^6 \times 2.397 \times 710}{17.9 \times 10^6 \times 41} = 3.78\,\text{mD}$$

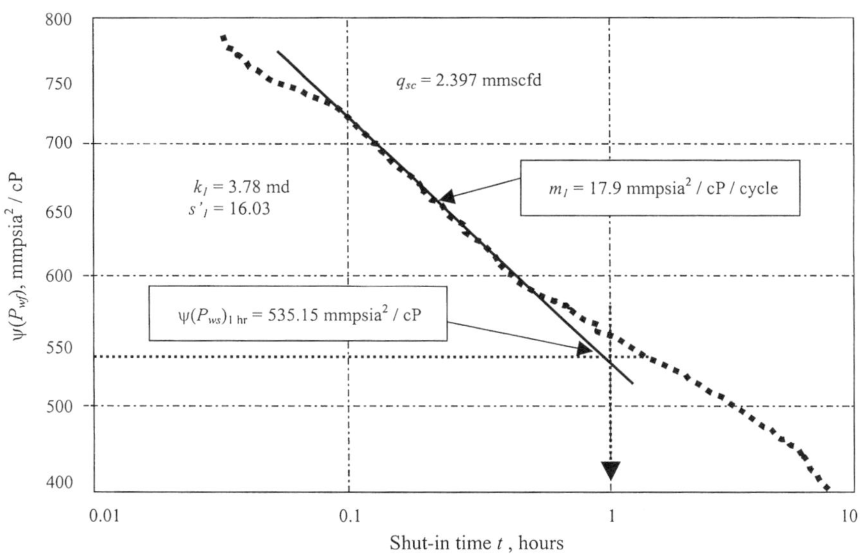

Figure 6–21. Transient flow analysis of first flow period (2.397 mmscfd)—Example 6–3.

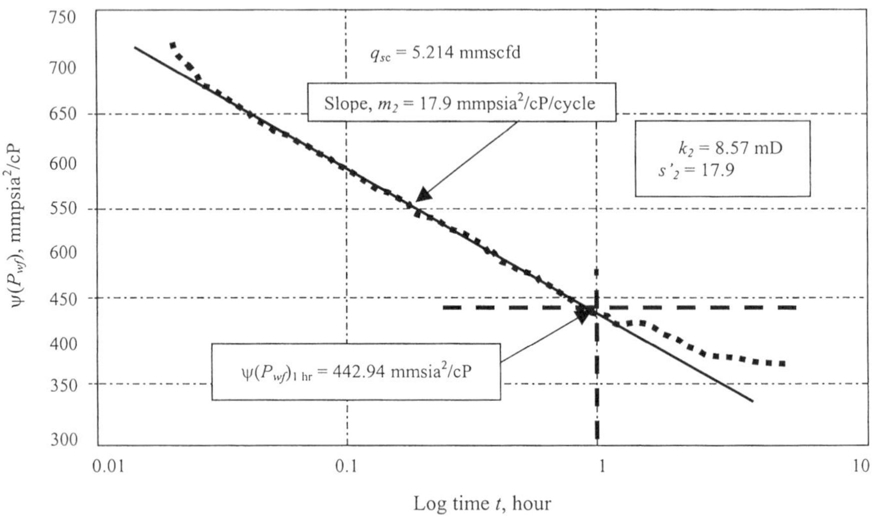

Figure 6–22. Transient flow analysis of second flow period (5.214 mmscfd)—Example 6–3.

From Eq. 6–11,

$$s'_1 = 1.151\left[\frac{(861.12 - 535.15) \times 10^6}{17.9 \times 10^6} - \log\left(\frac{3.78}{0.119 \times 0.02345 \times 0.00023 \times 0.4271^2}\right) + 3.23\right] = 16.03$$

From Figure 6–22, second flow rate = 5.214 mmscfd; m_2 = 17.2 mmpsia^2/cP/cycle, and $\psi(p_{1hr})$ = 442.94 mmpsia^2/cP. From Eq. 6–10;

$$k_2 = \frac{1.632 \times 10^6 \times 5.214 \times 710}{17.2 \times 10^6 \times 41} = 8.57\,\text{mD}$$

From Eq. 6–11,

$$s'_2 = 1.151\left[\frac{(861.12 - 442.94) \times 10^6}{17.2 \times 10^6} - \log\left(\frac{8.57}{0.119 \times 0.02345 \times 0.00023 \times 0.4271^2}\right) + 3.23\right] = 22.65$$

Finally, s and D can be calculated by solving the equations

$$s'_1 = s + Dq_{sc1}$$
$$s'_2 = s + Dq_{sc2}$$

or

$$16.03 = s + D\,(2.397)$$
$$22.65 = s + D\,(5.214)$$

From Eq. 5–48,

$$D = \frac{s_1' - s_2'}{q_{sc1} - q_{sc2}} = \frac{16.03 - 22.65}{2.397 - 5.214} = 2.3500 \times 10^{-6}\ \text{mmscfd}^{-1}$$

From Eq. 5–49,

$$s = s_1' - Dq_{sc1} = 16.03 - 2.35 \times 10^{-6} \times 2.397 \times 10^6 = 10.40$$

indicating the well is damaged.

From Eq. 5–115,

$$A = \frac{115.82 \times 10^6 TP_{sc}}{khT_{sc}}\left[\log\left(\frac{0.472r_e}{r_w}\right) + \frac{s}{2.303}\right]$$
$$= \frac{115.82 \times 10^6(710)(14.65)}{8.27(41)(520)}\left[\log\left(\frac{0.472(2200)}{0.4271}\right) + \frac{10.40}{2.303}\right]$$
$$= 54.187485 \times 10^6\,\frac{\text{psia}^2/\text{cP}}{\text{mmscfd}}$$

and

$$B = \frac{50.30 \times 10^6 TP_{sc}}{khT_{sc}}D$$
$$= \frac{50.30 \times 10^6(710)(14.65)}{8.27(41)(520)}(2.35)$$
$$= 6.997435 \times 10^6\,\frac{\text{psia}^2/\text{cP}}{\text{mmscfd}^2}$$

Hence the deliverability equation is

$$\psi(\overline{p}_R) - \psi(p_{wf}) = 54.187485 \times 10^6 q_{sc} + 6.997435 \times 10^6 q_{sc}^2$$

Solving the quadratic equation, the value of AOF is calculated as

$$AOF = \frac{-A + \sqrt{A^2 + 4B(\psi(\overline{p}_R))}}{2B}$$
$$= \frac{-54.187485 \times 10^6 + \sqrt{(54.187485)^2 + 4(6.997435 \times 10^6)(772.00 \times 10^6)}}{2(6.997435 \times 10^6)}$$
$$= \frac{102.478921}{13.994870} = 7.32\ \text{mmscfd}$$

Table 6–8
Predicted Long-Term Gas Deliverability Calculations

Bottom-hole pressure (psia)	$\psi(p_{wf})$ (mmpsia^2/cP)	Stabilized deliverability, q_{sc} (mmscfd)
3700	772.00	0.00
3500	706.80	0.63
3000	547.65	2.15
2500	399.17	3.57
2000	266.41	4.82
1500	155.04	5.87
1250	109.14	6.30
1000	70.63	6.66
750	40.06	6.94
500	17.90	7.15
400	11.47	7.21
200	2.88	7.26
100	0.74	7.31
14.65	0.00	7.32

The predicted long-term gas deliverability equation is

$$\psi(\overline{p}_R) - \psi(p_{wf}) = 54.187485 \times 10^6 q_{sc} + 6.997435 \times 10^6 q_{sc}^2$$

Table 6–8 shows calculated values and Figure 6–23 illustrates IPR gas well performance.

Example 6–4[26] *Analyzing Two-Rate Buildup Test*

The data are as follows: reservoir depth = 6550 ft; estimated initial reservoir pressure = 4290 psia; $T = 200^0$F; $h = 50$ ft; $\phi = 0.15$; $\mu_g = 0.0275$; $s_g = 58\%$; $s_o = 20\ \%$; $s_w = 22\%$; $c_g = 0.00022$ psi^{-1}; $c_o = 0.0003$ psi^{-1}; $c_w = 0.0000035$ psi^{-1}; $c_f = 0.0000039$ psi^{-1}; $c_t = 0.0003329$ psi^{-1}; $r_w = 0.30$ ft; $P_{sc} = 14.65$ psia; $T_{sc} = 520^0$R; drainage area = 100 acres; drainage radius = 1177.3 ft; $\psi(p_{wfo})_1 = 585.28$ mmpsia^2/cP; $\psi\ (p_{wfo})_2 = 222.27$ mmpsia/cP; and cumulative production = 5.0 mmscf.

First stabilized rate before rate change = 40 mmscfd

Time to change the first gas rate to second rate = 8 hours

Second stabilized rate after rate change = 60 mmscfd

Pressure at time of rate change = 3570 psia

Determine k, s, $\psi(\Delta p)_{skin}$, and deliverability constants A and B, and develop the inflow performance curve.

Table 6–9
First Flow Period Test Data

Flowing time, Δt, hr (1)	Log time Δt (2)	Flowing pressure P_{wf} (psia) (3)	Flowing pressure P_{wf} (psia^2/cP) (4)
0.75	—	3602	829.46E+06
1.00	0.00000	3596	827.39E+06
1.25	0.09689	3591	825.67E+06
1.50	0.17606	3587	824.29E+06
1.75	0.24299	3583	822.91E+06
2.00	0.30098	3580	821.88E+06
2.25	0.35212	3577	820.84E+06
2.50	0.39787	3575	820.15E+06
3.00	0.47704	3570	818.43E+06

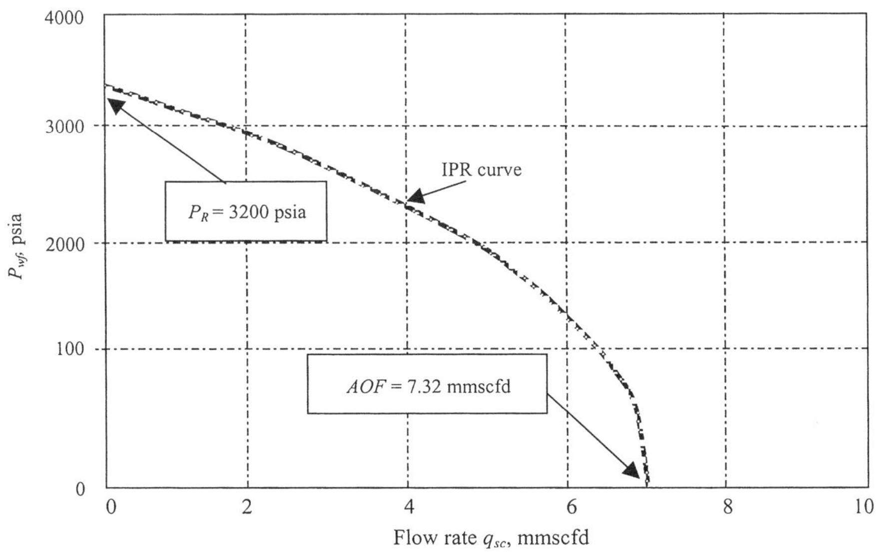

Figure 6–23. Inflow performance curve for Example 6–3.

Solution The pseudopressure function is shown in Figure 6–24. See Table 6–9 for first flow period test data.

From Figure 6–25, find the following:

Slope of the straight line $m_1 = 16.64$ mmpsia^2/cP/cycle

Pseudopressure at 1 hr, $\psi(p_{1hr}) = 827.39$ mmpsia^2/cP

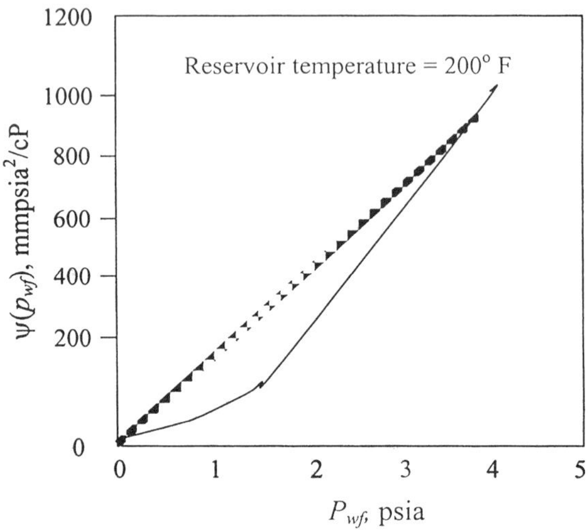

Figure 6–24. Real gas pseudopressures.

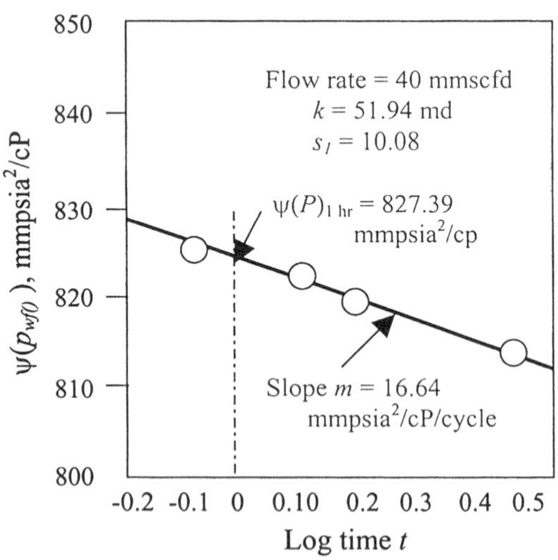

Figure 6–25. Transient flow analysis—First plot flow period.

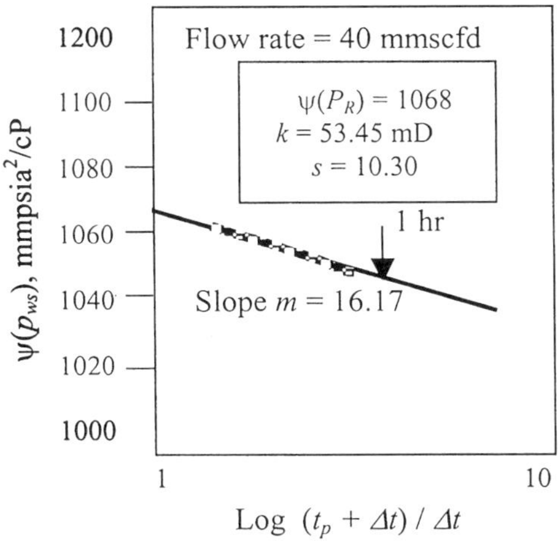

Figure 6–26. Pressure buildup test analysis.

From Eq. 5–40:

$$k = \frac{57.920 \times 10^6 \times 40 \times 660 \times 14.65}{16.64 \times 10^6 \times 520 \times 50} = 51.94 \text{ mD}$$

From Eq. 5–41:

$$s_1 = 1.151\left[\frac{(827.39 - 585.28) \times 10^6}{16.64 \times 10^6} - \log\left(\frac{51.94}{0.15(0.0275)(0.0001329)0.3^2}\right) + 3.23\right] = 10.08$$

From Fig. 6–26, find the following:

Slope of buildup's straight line $m_2 = 16.17$ mmpsia^2/cP/cycle
Pressure at 1 hr, $p_{1hr} = 4255$ psia
Pseudopressure at 1 hr $\psi(p_{1hr}) = 1057.0$ mmpsia^2/cP

Table 6–10 shows pressure buildup test data.
From Eq. 6–10:

$$k = \frac{57.92 \times 10^6 \times 40 \times 660 \times 14.65}{16.17 \times 520 \times 50} = 53.45 \text{ mD}$$

Table 6–10
Pressure Buildup Test Data

Time Δt (hr) (1)	Shut-in pressure P_{wf} (psia) (2)	Pseudopressure $\psi(P_{wf})$ (psia^2/cP) (3)	$\frac{t_p+\Delta t}{\Delta t}$ (4)	$\log \frac{t_p+\Delta t}{\Delta t}$ (5)
0.50	4085	100.23E+07	7.00	0.8449
1.00	4240	105.70E+07	4.00	0.6020
1.50	4248	105.98E+07	3.00	0.4770
2.00	4252	106.13E+07	2.50	0.3979
2.50	4254	106.20E+07	2.20	0.3424
3.00	4256	106.27E+07	2.00	0.3010
3.50	4257	106.30E+07	1.86	0.2688
4.00	4259	106.38E+07	1.75	0.2430
5.00	4261	106.45E+07	1.60	0.2041
6.00	4262	106.48E+07	1.50	0.1761
7.00	4263	106.52E+07	1.43	0.1549
8.00	4264	106.55E+07	1.38	0.1383

Table 6–11
Second Flow Period Test Data

Flowing time, t (hr) (1)	Log t (hr) (2)	Flowing pressure, P_{wf} (psia) (3)	Flowing pressure $\psi(P_{wf})$ (psia^2/cP) (4)
0.75	—	3076	649.20E+06
1.00	0.00000	3066	645.80E+06
1.25	0.09689	3059	643.43E+06
1.50	0.17606	3053	641.40E+06
1.75	0.24299	3048	639.70E+06
2.00	0.30098	3043	638.01E+06
2.25	0.35212	3038	636.31E+06
2.50	0.39787	3036	635.63E+06
3.00	0.47704	3029	633.26E+06

From Eq. 6–11:

$$s = 1.151\left[\frac{(1057.0 - 818.50) \times 10^6}{16.17 \times 10^6} - \log\left(\frac{53.45}{0.15(0.0275)(0.0001329)(0.3)^2}\right)\right] = 10.30$$

Table 6–11 shows second flow period test data.

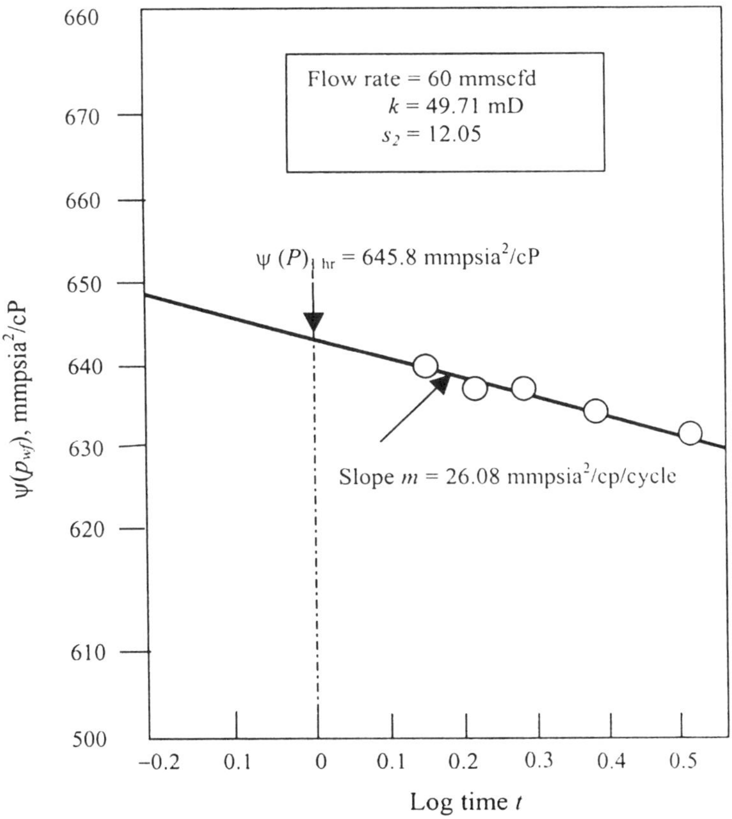

Figure 6–27. Transient flow analysis—second flow period.

From Figure 6–27, find the following:

Slope of the straight line $m_2 = 26.08$ mmpsia²/cP/cycle
Pseudopressure at 1 hr $\psi(p_{1hr}) = 645.80$ mmpsia²/cP

From Eq. 5–40,

$$k = \frac{57.920 \times 10^6 \times 60 \times 660 \times 14.65}{26.08 \times 10^6 \times 520 \times 50} = 49.71\,\text{mD}$$

From Eq. 5–41,

$$s_2 = 1.151\left[\frac{(645.80 - 222.28) \times 10^6}{26.08 \times 10^6} - \log\left(\frac{49.71}{0.15(0.0275)(0.0001329)0.3^2}\right) + 3.23\right] = 12.05$$

Using Eq. 5–48, find the rate-dependent skin factor D:

$$D = \frac{10.08 - 12.05}{40 - 60} = 0.09850 \, \text{mmscfd}^{-1}$$

Using Eq. 5–49, true skin factor

$$s = 10.08 - 0.0985 \times 40 = 6.13$$

Pressure drop due to skin is

$$\psi(\Delta p)_{skin} = 0.869(26.08)(6.13) = 138.93 \, \text{mmpsia}^2/\text{cP}$$

For a square centered well, the Dietz shape factor is 30.8828. From Eq. 6–28, the dimensionless time t_{DA} is

$$t_{DA} = \frac{0.0002637(53.45)(8)}{(0.15)(0.0275)(0.0001329)(22/7 \times 1177.3^2)} = 0.0472$$

The value of t_{DA} is small, indicating a transient region of flow; the well is infinite acting and Eq. 6–30 can be used to calculate the MBH dimensionless pressure function F:

$$F = 4\pi t_{DA} = 4(22/7)(0.0472) = 0.5934$$

From Eq. 6–26 the reservoir pressure is

$$\psi(\overline{p}_R) = \psi(p^*) - \frac{m}{2.303}(F) = 1070 - \frac{26.08}{2.303}(0.5934)$$
$$= 1063 \, \text{mmpsia/cP}$$

Calculate deliverability constants A and B using Eqs. 4–57 and 4–58, respectively, as (see Table 6–12):

$$A = 1.422 \times 10^6 \frac{T}{kh}\left[In\left(\frac{0.472 r_e}{r_w}\right) + s\right]$$

$$= 1.422 \times 10^6 \frac{660}{53.45 \times 50}\left[In\left(\frac{0.472 \times 1177.3}{.3}\right) + 6.13\right]$$

$$= 47768.23\text{E} + 02$$

$$B = 1.422 \times 10^6 \frac{T}{kh} D = 1.422 \times 10^6 \frac{660}{53.45 \times 50}(0.0985)$$

$$= 34559.66\text{E} - 03$$

Table 6–12
Predicted Long-Term Gas Deliverability Calculations

Bottom hole pressure, (P_{wf}) (psia) (1)	Pseudopressure $\psi(P_{wf})$ (psia^2/cP) (2)	Pressure ratio, $\frac{q_{gi}}{q_{gn}}$ — (3)	Flow rate ratio, $\frac{\psi(P_{wf})}{\psi(P_R)}$ — (4)	Predicted flow rate, q_{sc} (mmscfd) (5)
4273	106.34E+07	1.0000	0.00046	0.101
4200	103.75E+07	0.97521	0.02472	5.491
4000	967.32E+06	0.90925	0.09096	20.204
3800	897.86E+06	0.84395	0.15597	34.643
3600	828.77E+06	0.77901	0.22101	49.089
3400	759.94E+06	0.71431	0.28608	63.542
3200	691.43E+06	0.64991	0.35069	77.894
3000	623.45E+06	0.58602	0.41413	91.986
2800	556.38E+06	0.52297	0.47797	106.164
2600	490.66E+06	0.46120	0.53978	119.894
2400	426.84E+06	0.40122	0.59997	133.262
2200	365.53E+06	0.34358	0.65725	145.987
2000	307.34E+06	0.28888	0.71207	158.161
1800	252.89E+06	0.23771	0.76320	169.518
1600	202.79E+06	0.19061	0.81023	179.964
1400	157.57E+06	0.14811	0.85312	189.491
1200	117.72E+06	0.11065	0.89025	197.738
1000	835.73E+05	0.07855	0.92200	204.791
800	553.71E+05	0.05205	0.94884	210.752
600	331.80E+05	0.03119	0.96987	215.425
400	168.81E+05	0.1587	0.98515	218.817
AOF 14.73	671.30E+03	0.0000	1.00000	222.116

- If value of gas rate (calculated) at P_{AV} = zero, then P_{AV} (calc.) is **correct**.
- If value of gas rate (calculated) at P_{AV} < zero, then decrease the value of P_{AV}.
- If value of gas rate (calculated) at P_{AV} > zero, then increase the value of P_{AV}.

Average reservoir pressure $\psi(P_R) = 1063.407$ mmpsia^2/cP or average reservoir pressure $P_R = 4273$ psia. Data are plotted in Figures 6–28 and 6–29.

The long-term deliverability equation is

$$[\Psi(P_R) - \Psi(P_{WF})] = 47{,}768.23E + 02 \times q_g + 34{,}559.66E - 03 \times q_g^2$$

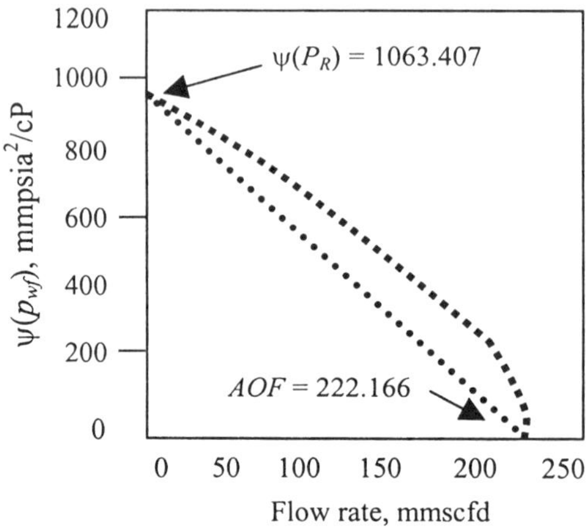

Figure 6–28. Long-term deliverability curve.

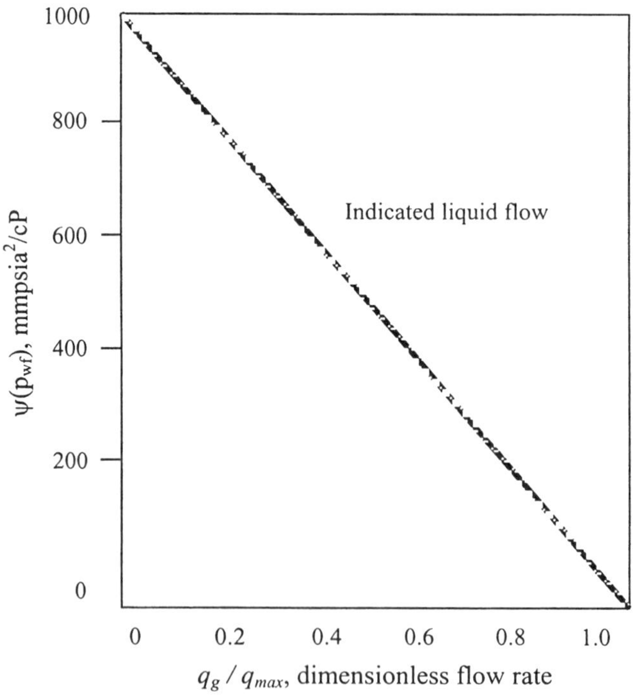

Figure 6–29. Inflow performance relationship using dimensionless IPR curves.

Buildup Test Preceded by Two Different Flow Rates

These types of analyses are applicable when a rate has been changed a very short time before the well is shut in for buildup. In this case there is not sufficient time for Horner's approximation to be valid. A Cartesian coordinate graph (Figure 6–30) is used to plot the following function:

$$\Psi(P_{ws}) \text{ versus } \left[\frac{q_{sc1}}{q_{sc2}}\log\left(\frac{t_{P1}+t_{P2}+\Delta t}{t_{P2}+\Delta t}\right)+\log\left(\frac{t_{P2}+\Delta t}{\Delta t}\right)\right] \quad (6\text{–}64)$$

We frequently can consider all production before time t_{P1} to have been at rate q_{sc1} and production just before the test to have been at rate q_{sc2} for time t_{p2}, and Δt is the shut-in time. The slope m' of this plot is related to formation permeability by

$$m' = \frac{57.920 \times 10^6 q_{sc} T P_{sc}}{k h T_{sc}} \quad (6\text{–}65)$$

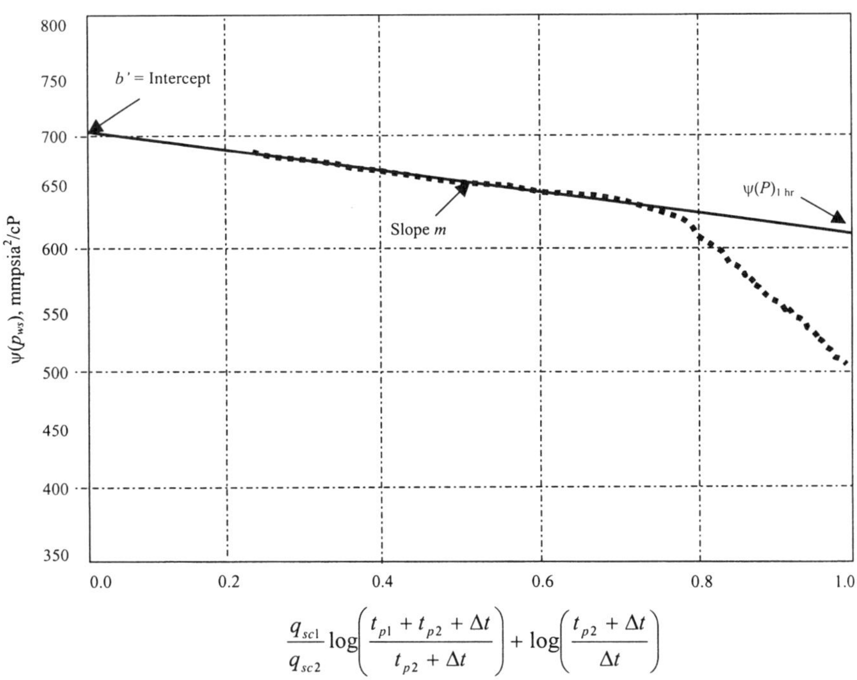

Figure 6–30. Cartesian coordinate graph plot—For buildup test preceded by two different rates.

or

$$k = \frac{57.920 \times 10^6 \times T \times P_{sc}}{m'hT_{sc}}$$

Calculate the skin factor s from

$$s = 1.151\left[\frac{\Psi(P_{1hr}) - \Psi(P_{wf})_t}{m'} - \log\left(\frac{k}{\phi\mu_i c_i r_w^2}\right) + 3.23\right] \quad (6\text{–}66)$$

where $\Psi(P_{1hr})$ from the plot at time 1 hr and $\Psi(P_{wf})$ at time $(t_{P1} + t_{P2})$ is calculated from

$$\Psi(P_i) - \Psi(P_{wf}) = m\frac{q_{sc1}}{q_{sc2}}\left[\log\left(\frac{t_{P1} + t_{P2}}{t_{P2}}\right) + \log(t_{P2}) + s\right] \quad (6\text{–}67)$$

The pressure drop across the skin at rate q_{sc1} is

$$\Psi(\Delta P)_{skin} = 0.869(-m')(s) \quad (6\text{–}68)$$

and at q_{sc2}

$$\Psi(\Delta P)_{skin} = 0.869\left(\frac{q_{sc2}}{q_{sc1}}\right)(-m')(s) \quad (6\text{–}69)$$

Estimate $\Psi(\overset{*}{P})$ by

$$\Psi(\overset{*}{P}) = \Psi(P_{mf}) + (-m')\left[\log\left(\frac{kt_{P2}}{\phi\mu_i c_i r_w^2}\right) - 3.23 + 0.869s\right] \quad (6\text{–}70)$$

For a bounded reservoir the intercept of the straight line on the ordinate of the plot should yield $\Psi(\overset{*}{P})$, which may then be used as described before to calculate $\Psi(P_R)$. A Cartesian coordinate graph plot for buildup test preceded by two different flow rates is shown in Figure 6–30.

Example 6–5 *Analyzing Pressure Buildup Preceded by Two Different Rates* The well is flowed at a rate 2.397 mmscfd up to time 6 hr, at a rate 5.214 mmscfd up to time 6 hr and then shut in. The shut-in time and pressures are recorded by Δt and p_{ws}, respectively, in Table 6–13. The reservoir/well data are the same as in Example 6–3. Determine formation permeability k, true skin factor s, pressure drop due to skin at rate q_{sc1}, and false reservoir pressure $\psi(p^*)$.

Solution $q_{sc1} = 2.397$ mmscfd; $t_{P1} = 6$ hr; $q_{sc2} = 5.214$ mmscfd; $t_{P2} = 6$ hr; $t = 6 + 6 = 12$ hr; $\psi(p_i) = 861.12$ mmpsia^2/cP

Table 6–13
Pressure Buildup Data Preceded by Two Different Rates

Shut-in time Δt, (hr)	Shut-in Pressure p_{ws}, (psia)	$\psi(p_{ws})$ (mmpsia²/cP)	Plotting function, X $\frac{q_{sc1}}{q_{sc2}}\log\left(\frac{t_{P1}+t_{P2}+\Delta t}{t_{P2}+\Delta t}\right)+\log\left(\frac{t_{P2}+\Delta t}{\Delta t}\right)$
0	2567	418.12	—
1.0	2819	492.56	0.9687
1.5	3146	593.33	0.8413
2.0	3310	645.36	0.7138
2.5	3350	658.14	0.6465
3.0	3366	663.39	0.5791
3.5	3382	668.62	0.5354
4.0	3385	669.40	0.4917
5.0	3397	673.21	0.4369
6.0	3407	676.64	0.3820
7.0	3415	697.13	0.3481
8.0	3421	681.10	0.3142
9.0	3428	683.34	0.2910
10.0	3436	685.92	0.2677

At $\Delta t = 4$ hr, the plotting function is

$$X = \frac{2.397}{5214}\log\left(\frac{6+6+4}{6+4}\right)+\log\left(\frac{6+4}{4}\right)$$
$$= 0.4597(0.2041) + 0.3979 = 0.4917$$

Buildup Analysis

For a shut-in of 10 hr, the data necessary to draw $\Delta(p_{ws})$ versus function X are listed in Table 6–13. The corresponding buildup plot is shown in Figure 6–31, from which

$$\text{Slope } m' = \frac{700-660}{0-0.7} = 57.14\,\text{mmpsia}^2/\text{cP}$$

$$\text{Intercept } b' = 705\,\text{mmpsia}^2/\text{cP and } \psi(p_{1\text{hr}}) = 635\,\text{mmpsia}^2/\text{cP}$$

From Eq. 6–65, formation permeability k is

$$k = \frac{57.920\times10^6\times5.214\times710\times14.65}{57.14\times41\times520} = 2.58\,\text{mD}$$

From Eq. 6–66, true skin s is

$$s = 1.151\left[\frac{\psi(p_{1hr})-\psi(p_{wf})_{t=0}}{m} - \log\left(\frac{k}{\phi\mu_i c_i r_w^2}\right)+3.23\right]$$

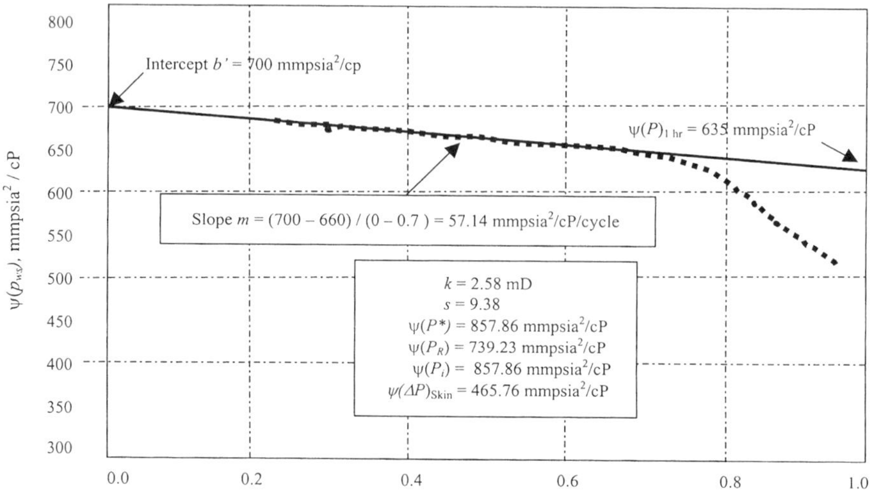

$$\frac{q_{sc1}}{q_{sc2}}\log\left(\frac{t_{p1}+t_{p2}+\Delta t}{t_{p2}+\Delta t}\right)+\log\left(\frac{t_{p2}+\Delta t}{\Delta t}\right)$$

Figure 6–31. Plot of $\psi(P_{ws})$ versus plotting function.

$$= 1.151\left[\frac{635 - 418.12}{57.14} - \log\left(\frac{2.58}{0.1004 \times 0.0235 \times 0.00023 \times 0.4271^2}\right) + 3.23\right] = 9.38$$

From Eq. 6–68, pressure drop due to skin at rate q_{sc1} is

$$\begin{aligned}\psi(\Delta p)_{skin}\text{ at }q_{sc1} &= 0.869(m)(s) = 0.869 \times -57.14 \times 9.38\\ &= 465.76\,\text{mmpsia}^2/\text{cP}\end{aligned}$$

From Eq. 6–70, $\psi(p^*)$ is

$$\begin{aligned}\psi(p^*) &= \psi(p_{wf0}) + (-m)\left[\frac{kt_{p2}}{\phi\mu_i c_i r_w^2} - 3.23 + 0.869s\right]\\ &= 201.21 + (-57.14)\\ &\quad\times\left[\frac{2.58 \times 6}{0.1114 \times 0.00023 \times 0.4271^2} - 3.23 + 0.869 \times 9.38\right]\\ &= 201.21 + (-57.14)[6.57 - 3.23 + 8.15]\\ &= 201.21 + 656.61 = 857.86\,\text{mmpsia}^2/\text{cP}\end{aligned}$$

Calculate $\psi(\overline{p}_R)$ from the MDH method:

$$e^F = 4\pi t_{DA} = (4)(22/7)\frac{0.000264(5.214)(44.44)}{(0.1004)(0.0027(0.42)\pi^2} = 0.1257$$

$$F = 1.1339$$

From Horner plot of Figure 6–20, $\psi(\overline{p}_R) = 772.00$ mmpsia^2/cP.

Buildup Following a Variable-Rate Drawdown Test

The methods of Odeh and Jones[14] may be used for analyzing a buildup following a variable-rate drawdown. The flow sequence may be summarized as

Flow rate q_1 up to time t_1
Flow rate q_2 up to time t_2
Flow rate q_n up to time t_n

The total drawdown time is again represented by t, that is, $t = t_n$. The shut-in following rate q_n extends over the time period Δt. Using these notations, Eq. 4–67 may be extended to include the shut-in period to give

$$\frac{\Psi(p_i) - \Psi(p_{ws})}{q_n} = m' \sum_{j=1}^{n+1} \left[\frac{\Delta q_j}{q_n} \log(t + \Delta t - t_{j-1})\right] + m'\left[\log\left(\frac{k}{\phi \mu_{gi} c_i r_w^2}\right) - 3.23 + 0.869s\right] \quad (6–71)$$

Defining $\Psi(P_{wfo})$ as the pseudopressure just before shut-in, Eq. 5–67 may be written as

$$\frac{\Psi(P_i) - \Psi(P_{wfo})}{q_n} = m' \sum_{j=1}^{n} \frac{\Delta q_i}{q_n} \log(t - t_{j-1}) + m'\left[\log\left(\frac{k}{\phi \mu_i c_i r_w^2}\right) - 3.23 + 0.869s'\right] \quad (6–72)$$

Subtracting Eq. 6–71 from Eq. 6–72 gives

$$\frac{\Psi(P_{ws}) - \Psi(P_{wfo})}{q_n} = m' \sum_{j=1}^{n} \frac{\Delta q_i}{q_n} \log\left(\frac{t + \Delta t - t_{j-1}}{t - t_{j-1}}\right) + m'\frac{(0 - q_n)}{q_n} - \log \Delta t + m'\left[\log\left(\frac{k}{\phi \mu_i c_i r_w^2}\right) - 3.23 + 0.869s'\right] \quad (6–73)$$

where

$$m' = \frac{57.920 \times 10^6 Tp_{sc}}{khT_{sc}}$$
$$\Delta q_j = q_j - q_{j-1}$$
$$q_{n+1} = 0$$
$$t_0 = q_0 = 0$$
$$t = t_n$$

A plot of

$$[\Psi(p_i) - \Psi(p_{ws})]/q_n \text{ versus } \sum_{j=1}^{n+1}\left[\frac{\Delta q_j}{q_n}\log(t + \Delta t - t_{j-1})\right]$$

on arithmetic coordinates should give a straight line from which kh may be obtained from

$$kh = \frac{57.920 \times 10^6 q_{sc} Tp_{sc}}{m' T_{sc}} \tag{6–74}$$

$$s' = 1.151\left[\frac{\Psi(p_{ws1}) - \Psi(p_{wfo})}{m' q_n} - \log\left(\frac{k}{\phi\mu_{gi}c_i r_w^2}\right) + 3.23\right] \tag{6–75}$$

where $\psi(p_{ws1}$ = the pseudopressure at $\Delta t = 1$; $\psi(p_{wfo})$ = the pseudopressure just before shut-in; and q_n = the rate just before shut-in. Equation 6–75 is valid when IT flow effects are negligible or when the assumptions $s' = s'_1 = s'_2 = \cdots\cdots = s'_n$ can be made; when IT flow effects cannot be neglected, the foregoing equations may be modified to include

$$s' = s + Dq_{sc}$$

Equation 6–71 may be written as

$$\frac{\Psi(p_i) - \Psi(p_{ws})}{q_n} = m' \sum_{j+1}^{n+1}\left[\frac{\Delta q_j}{q_n}\log(t + \Delta t - t_{j-1}) + 0.869 Dq_n\right]$$
$$+ m'\left[\log\left(\frac{k}{\phi\mu_{gi}c_i r_w^2}\right) - 3.23 + 0.869s\right] \tag{6–76}$$

A plot of

$$\frac{\psi(p_i)) - \psi(p_{ws})}{q_n} \text{ versus } \sum\left[\frac{\Delta q_j}{q_n}\log(t + \Delta t - t_{j-1})\right] + 0.869 Dq_n$$

on arithmetic coordinates should give a straight line from which good approximations of kh and s may be calculated. This method of analysis involves a graphical trial-and-error procedure in which values for D have to be guessed until a straight line is obtained. The correct plot is then used to determine slope m' and intercept B' equal to

$$B' = m'\left[In\left(\frac{k\Psi(p_{wfo})}{\phi\mu_{gi}c_i r_w^2}\right) - 7.432 + 2s\right] \tag{6–77}$$

If log is used in place of *In*, Eq. 6–77 becomes

$$B' = m'\left[\left(\log\frac{k\Psi(p_{wfo})}{\phi\mu_{gi}c_i r_w^2}\right) - 3.23 + 0.869s\right] \tag{6–78}$$

The values of kh and s may be determined by

$$kh = \frac{57.920 \times 10^6 q_{sc} T p_{sc}}{m' T_{sc}} \tag{6–79}$$

and

$$s = 1.151\left[\frac{B'}{m'} - \log\left(\frac{k\Psi(p_{wfo})}{\phi\mu_{gi}c_i r_w^2}\right) + 3.23\right] \tag{6–80}$$

Total pressure drop due to the skin effect then will be

$$\Psi(\Delta p)_{skin} = 0.869m'\ s\ q_n \tag{6–81}$$

Buildup Test Analysis When Rate Varies before Testing

Horner and MDH plotting techniques apply only for a constant production rate preceding the buildup test. However, as indicated by the equation $t_P = \frac{24V_p}{q_{sc}}$, variable-rate conditions may be handled approximately in many circumstances. Nevertheless, in buildup tests with relatively short flow periods or with widely varying rate before shut-in, it is important to include the effects of rate variation on test analysis for infinite-acting system and unfractured wells. The following equation may be used:

$$\Psi(P_{ws}) = \Psi(P_i) - m\sum_{j=1}^{n}\frac{q_j}{q_n}\log\left[\frac{t_n - t_{j-1} + \Delta t}{t_n - t_j + \Delta t}\right] \tag{6–82}$$

Figure 6–32 identifies the nomenclature for the variable-rate period. Equation 6–82 indicates that a plot of $\Psi(P_{ws})$ versus the summation term on the right-hand side should yield a straight-line portion with slope $-m$ given by equation

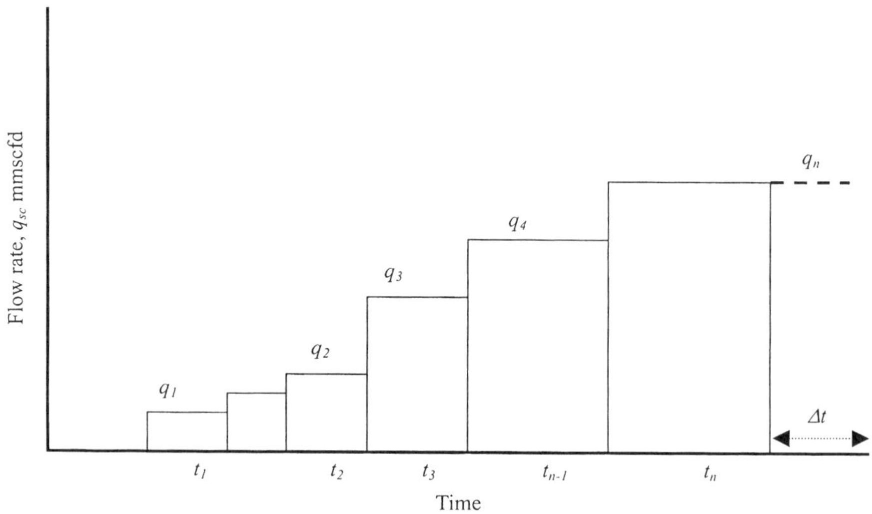

Figure 6–32. Schematic of rate variation preceding a pressure buildup test and skin factor from.

(with the final rate, q_n, used in place of q_{sc}) and intercept $\Psi(P_i)$. Permeability is estimated from.

$$k = \frac{57.920 \times 10^6 q_n T P_{sc}}{m T_{sc} h} \quad (6\text{–}83)$$

if $(t_n - t_{n-1}) \gg 1$ hr.

$$s = 1.151 \left[\frac{\Psi(P_{1hr}) - \Psi(P_{wf(\Delta t=0)})}{m} - \log\left(\frac{k}{\phi \mu c_t r_w^2}\right) + 3.23 \right] \quad (6\text{–}84)$$

Example 6–6[26] *Analyzing Pressure Buildup Preceded by Varying Flow Rates*

The gas well is tested with four varying flow rates. Flow rates sequences, shut-in time, and pressures along with calculated data are given in Tables 6–14 through 6–20. The reservoir/well data are as follows: $\overline{p}_R = 1660$ psia; $\psi(\overline{p}_R)$ or 189.00 mmpsia^2/cP; $\mu = 0.01639$ cP; $T = 686^0$R; $T_S = 89^0$; $T_c = 370.01^0$ R; $P_c = 650.59$ psia; $P_{sc} = 14.65$ psia; $T_{sc} = 60^0$F; $r_w = 0.4271$ ft; $r_e = 2106$ ft; $h = 69$ ft; $s_g = 0.603$; $s_w = 0.397$; $c_g = 0.00064$ psi^{-1}; $c_w = 3.01 \times 10^{-6}$ psi^{-1}; $c_f = 4.1 \times 10^{-6}$ psi^{-1}; $c_t = 0.000255$ psi^{-1}; $z = 0.9148$; $\beta_g = 93.9944$ scf/ft^3; $\phi = 14.6$; $\phi_{HC} = 0.088$.

Table 6–14
Calculated PVT Properties and Gas Pseudopressure, $\psi(p)$

Pressure (psia)	Compressibility factor (z)	Gas viscosity (cP)	Real gas pseudopressure $\psi(p)$, (mmpsia^2/cP)
4000	0.9470	0.023689	903.57
3700	0.9598	0.022859	816.26
3500	0.9354	0.022018	730.52
3250	0.9256	0.021176	646.66
3000	0.9177	0.020345	565.11
2750	0.9119	0.019533	486.41
2500	0.9085	0.018748	411.18
2250	0.9074	0.017997	340.12
2000	0.9189	0.017285	273.93
1750	0.9128	0.016618	213.36
1500	0.9192	0.016002	159.12
1250	0.9279	0.015441	111.91
1000	0.9389	0.014940	72.35
750	0.9518	0.014507	41.00
500	0.9665	0.014147	18.31
250	0.9825	0.013868	4.60
14.65	0.9985	0.013687	0.53

1. Determine formation permeability k and skin factor s', using the Horner and MDH plotting techniques (use long shut-in-time data).
2. Estimate k and s using Eq. 6–82 (summation term).

Solution Table 6–14 tabulates the calculated gas PVT properties and pseudopressure function.

Using Horner and MDH Plotting Techniques

$$\psi(\Delta p)_{skin} = 0.869 \times 6.6 \times 10.78 = 61.83 \text{ mmpsia}^2/\text{cP} \leftrightarrow 525 \text{ psia}$$

Pressure buildup data in Table 6–15 are shown in Figures 6–33 and 6–34. The log-log plot of the build-up data in Table 6–15 is used to check the significance of wellbore storage. Since there is no unit-slope line, we conclude that dominant wellbore storage has ended by 1.2 hr. However, the rapid pressure increase shown in Figure 6–33 does indicate that wellbore storage or skin effects are significant until about 0.50 hr. The data obtained about 0.50 hr can be analyzed. The following information can be obtained from Figure 6–33:

$$\text{Slope } m_1 = 6.6 \text{ mmpsia}^2/\text{cP/cycle and } \psi(p_{1hr}) = 189.40 \text{ mmpsia}^2/\text{cP}$$

Table 6–15
Pressure Buildup Data (t_{P1} = 6 hours, flow rate, q_{sc1} = 2.802 mmscfd, P_{wfo} = 1164.55 psia or $\psi(p)$ = 97.51 mmpsia^2/cP)

Δt (hr)	$\frac{t_{p1}+\Delta t}{\Delta t}$ —	Pressure, P_{ws} (psia)	$\psi(p_{ws})$ (mmpsia^2/cP)	$\Delta\psi = \Psi(p_{ws}) - \Psi(p_{wfo})$ (mmpsia^2/cp)
0.03	181	1216	106.00	8.50
0.07	91	1284	117.85	20.34
0.10	61	1330	126.18	28.68
0.13	46	1382	135.97	38.46
0.17	37	1432	145.63	48.12
0.25	25	1513	161.75	64.25
0.33	19	1581	175.94	78.43
0.50	13	1634	187.32	89.82
0.75	9	1644	189.60	92.09
1.00	7	1645	189.88	92.37
1.50	5	1648	190.43	92.92
2.00	4	1650	190.98	93.47
2.50	3.4	1653	191.53	94.02
3.00	3	1654	191.88	94.37
4.00	2.5	1657	192.43	94.93
5.00	2.2	1659	192.87	95.37
6.00	2.0	1660	193.12	95.61

Equation 6–10 is used to estimate permeability k from

$$k_1 = \frac{57.920 \times 10^6 q_{sc} T P_{sc}}{mhT_{sc}}$$
$$= \frac{57.920 \times 10^6 \times 2.802 \times 686 \times 14.65}{6.6 \times 10^6 \times 520 \times 69} = 6.89\,\text{mD}$$

Skin factor is estimated from Eq. 6–11 using $\psi(p_{1hr}) = 189.40$ mmpsia^2/cP from Eq. 6–12:

$$s_1' = 1.151\left[\frac{189.40 - 97.51}{6.6} - \log\left(\frac{6.89}{(.146)(.01639)(.000255)(.4271)^2}\right)\right]$$
$$= 10.78$$

We can estimate pressure drop across the skin from Eq. 6–12:

$$\psi(\Delta p)_{skin} = 0.869 \times 6.6 \times 10.78 = 61.83\,\text{mmpsia}^2/\text{cP} \leftrightarrow 525\,\text{psia}$$

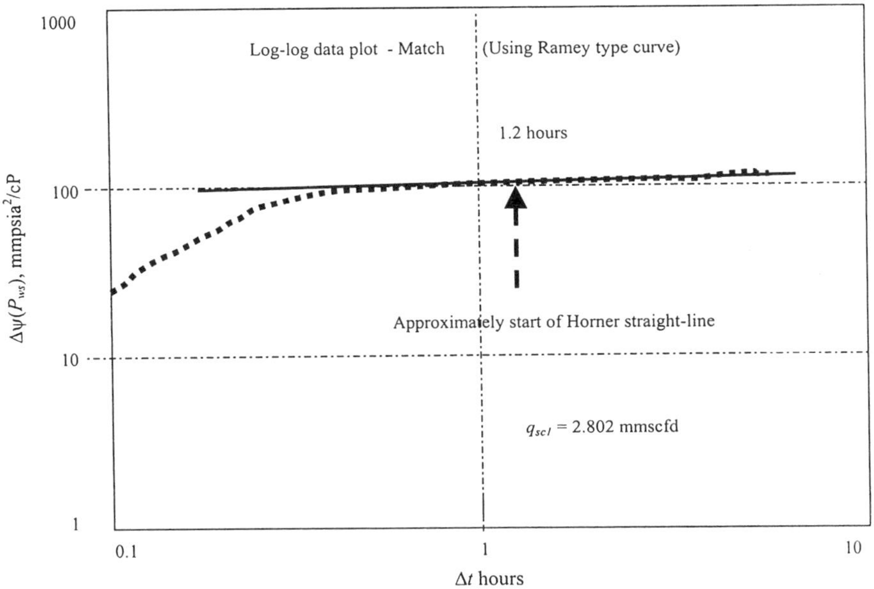

Figure 6–33. $\Delta\psi(p)$ versus Δt for buildup test # 1.

$$s_1' = 1.151\left[\frac{189.40 - 97.51}{6.6} - \log\left(\frac{6.89}{(.146)(.01639)(.000255)(.4271)^2}\right)\right]$$
$$= 10.78$$

Using Horner and MDH Plotting Techniques

Pressure buildup data in Table 6–16 are shown in Figures 6–35 and 6–36. The log-log plot of the buildup data in Table 6–16 is used to check the significance of wellbore storage. Since there is no unit-slope line, we conclude that dominant wellbore storage has ended by 1.5 hr. However, the rapid pressure increase shown in Figure 6–36 does indicate that wellbore storage or skin effects are significant until about 0.50 hr. The data obtained about 0.50 hr can be analyzed. The following information can be obtained from Figure 6–36:

$$\text{Slope}\, m_2 = 5.8\,\text{mmpsia}^2\text{/cP/cycle} \quad \text{and} \quad \psi(p_{1hr}) = 186.0\,\text{mmpsia}^2\text{/cP}$$

Equation 6–10 is used to estimate permeability k:

$$k_2 = \frac{57.920 \times 10^6 q_{sc} T P_{sc}}{mhT_{sc}}$$
$$= \frac{57.920 \times 10^6 \times 3.302 \times 686 \times 14.65}{5.8 \times 10^6 \times 520 \times 69} = 9.24\,\text{mD}$$

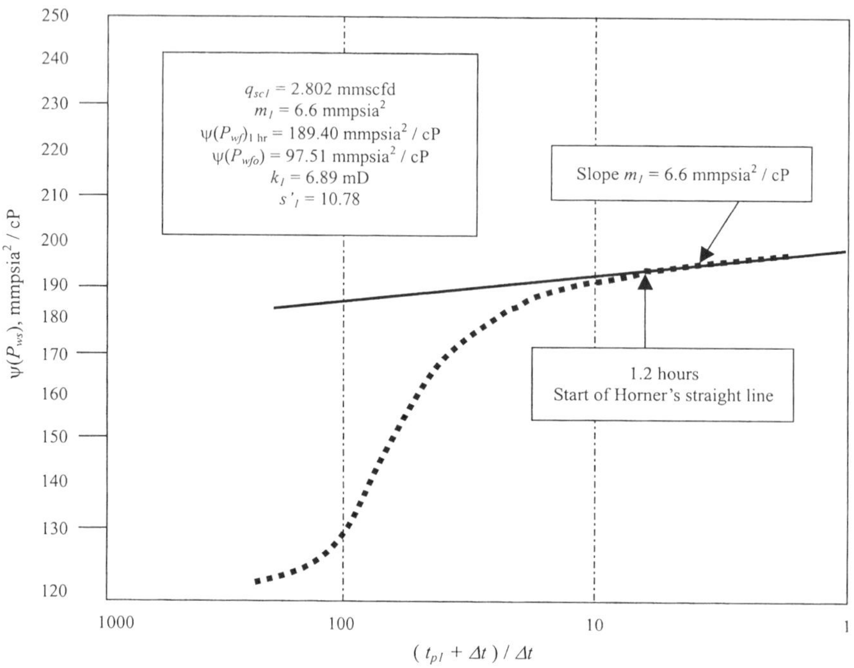

Figure 6–34. Horner's plot for buildup (rate # 1 = 2.802 mmscfd—Example 6–6).

Skin factor is estimated from Eq. 6–11 using $\psi(p_{1hr}) = 186.0$ mmpsia²/cP:

$$s_2' = 1.151\left[\frac{186.0 - 110.15}{5.8} - \log\left(\frac{9.24}{(0.146)(0.01639)(0.000255)(0.4271)^2}\right) + 3.23\right] = 8.39$$

Using Horner and MDH Plotting Techniques

Pressure buildup data in Table 6–17 are shown in Figures 6–37 and 6–38. The log-log plot of the buildup data in Table 6–17 is used to check the significance of wellbore storage. Since there is no unit-slope line, we conclude that dominant wellbore storage has ended by 1.3 hr. However, the rapid pressure increase shown in Figure 6–38 does indicate that wellbore storage or skin effects are significant until about 0.50 hr. The data obtained about 0.50 hr can be analyzed. The following information can be obtained from Figure 6–38:

Slope $m_3 = 5.9$ mmpsia²/cP/cycle and $\psi(p_{1hr}) = 184.30$ mmpsia²/cP

Table 6–16
Pressure Buildup Data (t_{P2} = 11.09146 hr; flow rate, q_{sc1} = 3.302 mmscfd; and P_{wfo} = 1215 psia or $\psi(p)$ = 110.15 mmpsia²/cP)

Δt (hrs)	$\frac{t_{p1}+\Delta t}{\Delta t}$	Pressure P_{ws} (psia)	$\psi(p_{ws})$ (mmpsia²/cP)	$\Delta\psi = \Psi(p_{ws}) - \Psi(p_{wfo})$ (mmpsia²/cP)
0.07	167.37	1255	112.98	24.01
0.1	111.91	1321	114.61	35.65
0.13	84.19	1368	113.32	44.36
0.17	67.55	1416	142.47	53.51
0.25	45.37	1506	160.32	71.36
0.33	34.27	1576	174.84	85.87
0.50	23.18	1622	184.76	95.79
0.75	15.79	1631	186.74	97.77
1.00	12.09	1633	187.13	98.16
1.50	8.39	1634	187.32	98.36
2.00	6.55	1636	187.87	98.90
2.50	5.44	1637	188.15	99.19
3.00	4.70	1640	188.61	99.65
4.00	3.77	1642	189.09	100.13
5.00	3.22	1644	189.49	100.52
6.00	2.85	1644	189.68	100.72

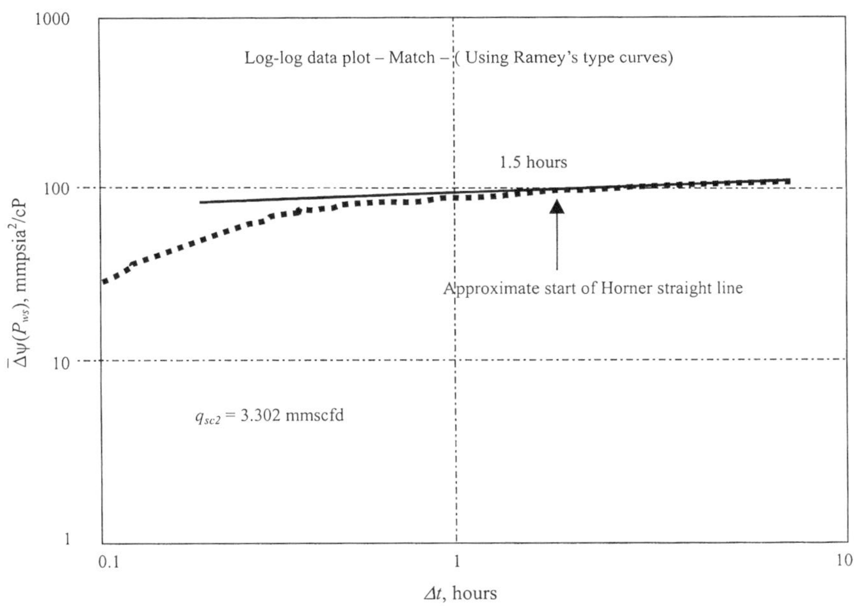

Figure 6–35. $\Delta\psi$(p) versus Δt for buildup test # 2—Example 6–6.

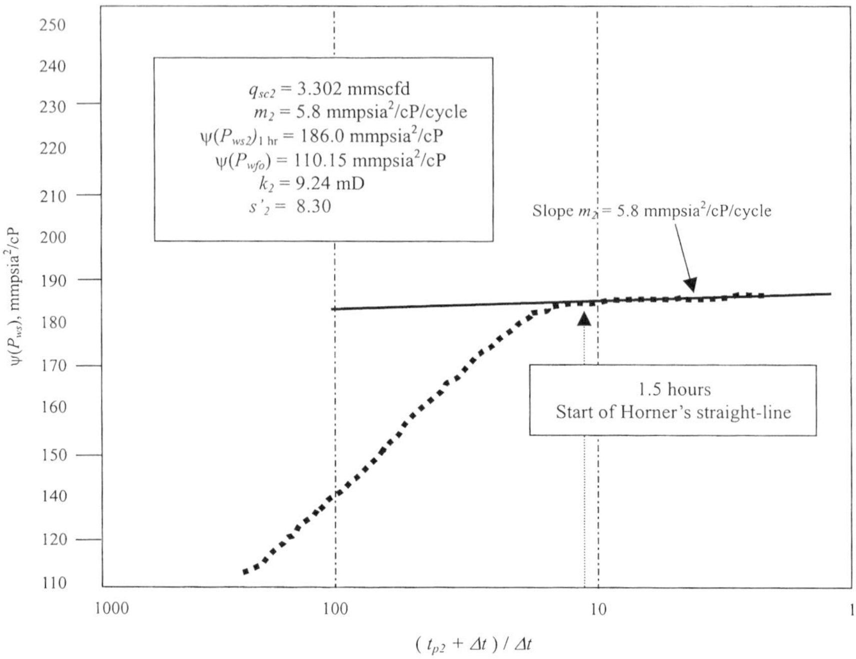

Figure 6–36. Horner's plot for buildup test (rate # 2 = 3.302 mmscfd)—Example 6–6.

Equation 6–10 is used to estimate permeability k:

$$k_3 = \frac{57.920 \times 10^6 q_{sc} T P_{sc}}{m h T_{sc}}$$

$$= \frac{57.920 \times 10^6 \times 3.524 \times 686 \times 14.65}{5.9 \times 10^6 \times 520 \times 69} = 9.69\,\text{mD}$$

Skin factor is estimated from Eq. 6–11 using $\psi(p_{1hr}) = 184.30$ mmpsia^2/cP from Eq. 6–11:

$$s'_3 = 1.151\left[\frac{184.3 - 84.09}{5.9} - \log\left(\frac{9.69}{(0.146)(0.01639)(0.000255)(0.4271)^2} + 3.23\right)\right] = 12.29$$

Using Horner and MDH Plotting Techniques

Pressure buildup data in Table 6–18 are shown in Figures 6–39 and 6–40. The log-log plot of the buildup data in Table 6–18 is used to check the

Table 6–17
Pressure Buildup Data (t_{P3} = 16.39274 hr; flow rate q_{sc1} = 3.524 mmscfd; and P_{wfo} =1079.55 psia or $\psi(p)$ = 84.09 mmpsia^2/cP)

Δt (hrs)	$\frac{t_{p1}+\Delta t}{\Delta t}$	Pressure P_{ws} (psia)	$\psi(p_{ws})$ (mmpsia^2/cP)	$\Delta\psi = \Psi(p_{ws}) - \Psi(p_{wfo})$ (mmpsia^2/cP)
0.02	984.56	1137	93.02	8.93
0.03	492.78	1189	101.53	17.45
0.07	246.89	1269	115.19	31.11
0.10	164.93	1330	126.22	42.14
0.13	123.95	1389	137.34	53.26
0.17	99.36	1436	146.35	62.27
0.25	66.57	1523	163.85	79.77
0.33	50.18	1578	175.45	91.37
0.50	33.79	1614	183.05	98.97
0.75	22.86	1621	184.61	100.53
1.00	17.39	1623	184.95	100.87
1.50	11.93	1626	185.65	101.57
2.00	9.20	1628	186.15	102.07
2.50	7.56	1630	186.52	102.44
3.00	6.46	1632	187.02	102.94
4.00	5.01	1635	187.61	103.53
5.00	4.28	1637	188.00	103.92
6.00	3.73	1639	188.42	104.33

significance of wellbore storage. Since there is no unit-slope line, we conclude that dominant wellbore storage has ended by 2.5 hr. However, the rapid pressure increase shown in Figure 6-40 does indicate that wellbore storage or skin effects are significant until about 0.75 hr. The data obtained about 0.75 hr can be analyzed. The following information can be obtained from Figure 6–40:

$$\text{Slope } m_4 = 5.7\,\text{mmpsia}^2/\text{cP/cycle} \quad \text{and} \quad \psi(p_{1hr}) = 178.6\,\text{mmpsia}^2/\text{cP}$$

Equation 6–10 is used to estimate permeability k:

$$k_4 = \frac{57.920 \times 10^6 q_{sc} T P_{sc}}{m h T_{sc}}$$

$$= \frac{57.920 \times 10^6 \times 3.543 \times 686 \times 14.65}{5.7 \times 10^6 \times 520 \times 69} = 10.08\,\text{mD}$$

(text continued on page 391)

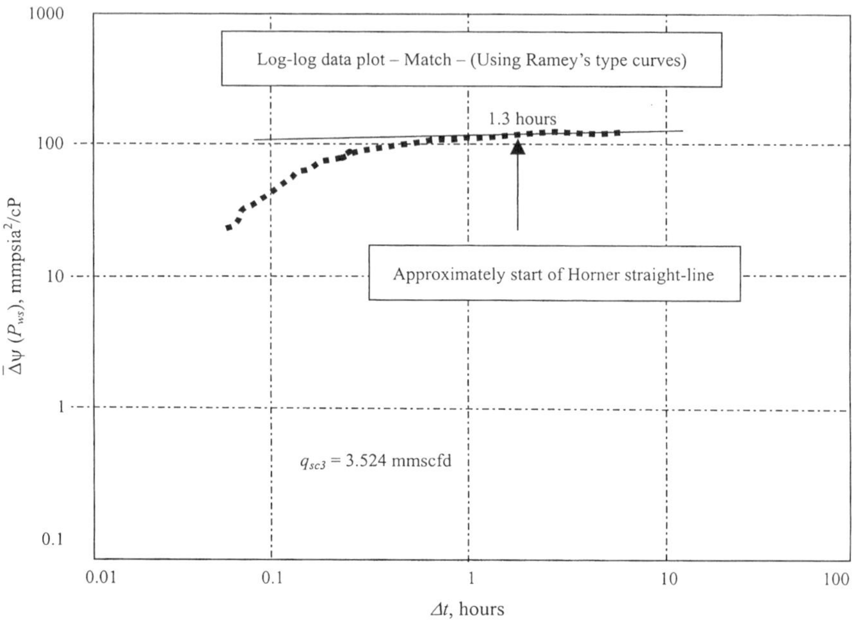

Figure 6–37. $\Delta\psi(P)$ versus Δt for buildup test # 3—Example 6–6.

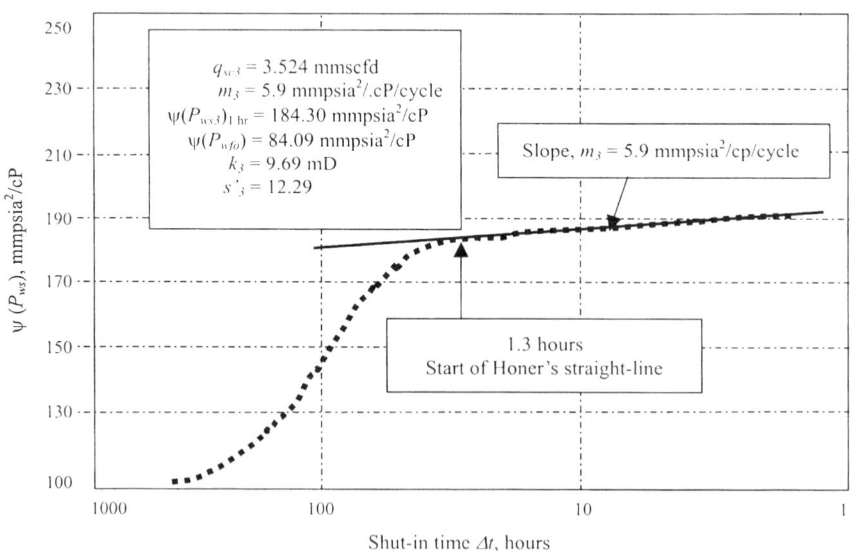

Figure 6–38. Horner's plot for buildup test # 3 (Rate # 3 = 3.524 mmscfd).

Table 6–18
Pressure Buildup Data (t_{P3} = 40.30483 hr; flow rate q_{sc4} = 3.543 mmscfd; P_{wfo} = 1232 psia or $\Psi(p)$ =113.45 mmpsia^2/cP)

Δt (hr)	$\frac{t_{p1}+\Delta t}{\Delta t}$	Pressure P_{ws} (psia)	$\psi(p_{ws})$ (mmpsia^2/cP)	$\Delta\psi = \Psi(p_{ws}) - \Psi(p_{wfo})$ (mmpsia^2/cP)
0.07	605.57	1250	113.95	22.90
0.10	404.05	1264	114.33	32.20
0.13	303.29	1318	123.99	41.87
0.17	242.83	1355	130.94	48.82
0.25	162.22	1449	148.89	66.77
0.33	121.91	1522	163.56	81.44
0.50	81.61	1576	174.94	92.82
0.75	54.74	1589	177.70	95.58
1.00	41.30	1593	178.62	96.50
1.50	27.87	1597	179.46	97.33
2.00	21.15	1599	179.84	97.72
2.50	17.12	1601	180.31	98.19
3.00	14.43	1603	180.68	98.56
3.50	12.52	1605	181.11	98.95
4.00	11.08	1607	181.45	99.33
5.00	9.96	1608	181.78	99.65
5.50	9.06	1609	182.03	100.15
6.00	8.33	1610	182.27	100.39
6.50	7.72	1611	182.51	100.45
7.00	7.20	1612	182.57	100.82
7.50	6.67	1613	182.94	100.88
8.00	6.37	1614	183.01	101.12
8.50	6.04	1615	183.24	101.23
9.00	5.74	1615	183.35	101.36
9.83	5.48	1616	183.48	101.57
10.00	5.10	1617	183.70	101.70
10.50	5.03	1618	183.83	101.85
11.00	4.84	1618	183.98	101.94
11.50	4.66	1619	184.06	102.05
12.00	4.50	1619	184.17	102.11
12.50	4.36	1620	184.24	102.29
13.00	4.22	1621	184.41	102.44
13.50	4.10	1622	184.56	102.59
14.00	3.99	1622	184.71	102.70
14.50	3.88	1622	184.82	102.74
15.00	3.78	1622	184.87	102.83
16.00	3.69	1623	184.95	103.07
17.00	3.52	1624	185.19	103.22
18.00	3.37	1625	185.34	103.44
19.00	3.24	1626	185.56	103.57

Table 6–18 (Continued)

Δt (hr)	$\frac{t_{p1}+\Delta t}{\Delta t}$	Pressure P_{ws} (psia)	$\psi(p_{ws})$ (mmpsia^2/cP)	$\Delta\psi = \Psi(p_{ws}) - \Psi(p_{wfo})$ (mmpsia^2/cP)
20.00	3.12	1626	185.69	103.81
21.00	3.02	1627	185.93	103.92
22.00	2.92	1628	186.04	104.00
23.00	2.83	1629	186.32	104.20
24.00	2.75	1630	186.43	104.31
25.00	2.68	1630	186.54	104.42
26.00	2.61	1631	186.74	104.61
27.00	2.55	1632	186.87	104.74
28.00	2.49	1632	186.97	104.85
29.00	2.44	1633	187.17	105.05
30.00	2.39	1633	187.21	105.09
31.00	2.34	1633	187.28	105.16
32.00	2.30	1634	187.48	105.35
33.00	2.26	1635	187.61	105.48
34.00	2.22	1635	187.67	105.55
35.00	2.19	1636	187.76	105.64
36.00	2.15	1636	187.91	105.79
37.00	2.12	1637	188.00	105.88
38.00	2.09	1637	188.11	105.99
39.00	2.06	1637	188.15	106.03
40.00	2.03	1638	188.26	106.14
41.00	2.00	1638	188.35	106.23
42.00	1.98	1639	188.42	106.29
43.00	1.96	1639	188.59	106.47
44.00	1.94	1640	188.66	106.53
45.00	1.92	1640	188.74	106.62
46.00	1.90	1640	188.81	106.69
47.00	1.88	1641	188.90	106.97
48.00	1.86	1641	189.01	106.88
49.00	1.84	1642	189.09	106.97
50.00	1.82	1642	189.16	107.06
51.00	1.81	1643	189.29	107.17
52.00	1.79	1643	189.36	107.23
53.00	1.78	1643	189.44	107.32
54.00	1.76	1644	189.53	107.41
55.00	1.75	1644	189.55	107.43
56.00	1.73	1652	191.33	109.26
57.00	1.72	1657	192.48	110.35
58.00	1.71	1660	193.01	110.88
59.00	1.69	1660	193.07	110.95
60.00	1.68	1659	192.81	110.68
61.00	1.67	1658	192.76	110.64
61.75	1.66	1660	193.01	110.88

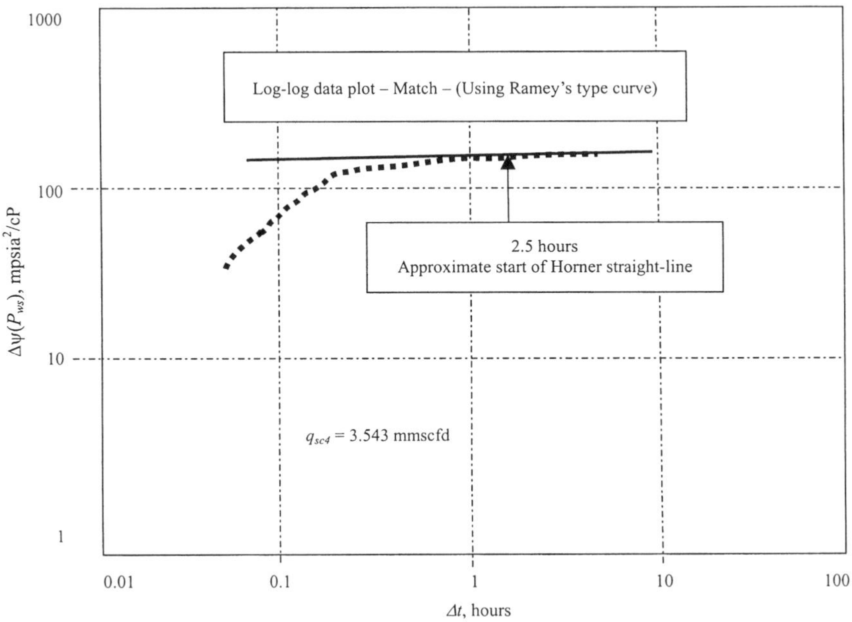

Figure 6–39. $\Delta\psi(P)$ versus Δt for buildup test # 4—Example 6–6.

(text continued from page 387)

Skin factor is estimated from Eq. 6–11 using $\psi(p_{1hr}) = 178.60$ mmpsia²/cP:

$$s_4' = 1.151\left[\frac{178.60 - 82.12}{6.6} - \log\left(\frac{10.08}{(0.146)(0.01639)(0.000255)(0.4271)^2}\right) + 3.23\right] = 12.20$$

Using Eq. 6–82

Figure 6–41 is a plot of the data in Tables 6–19 and 6–20, using Eq. 6–82, the summation term in that equation is written as follows for value of $n = 20$:

$$\sum_{j=1}^{n} \frac{q_j}{q_n} \log\left(\frac{t_n - t_{j-1} + \Delta t}{t_n - t_j + \Delta t}\right)$$

$$= \frac{2.802}{3.543} \log\left(\frac{18 - 0 + \Delta t}{18 - 6 + \Delta t}\right) + \frac{3.302}{3.543} \log\left(\frac{18 - 6 + \Delta t}{18 - 12 + \Delta t}\right) + \frac{3.543}{3.543} \log\left(\frac{18 - 12 + \Delta t}{\Delta t}\right)$$

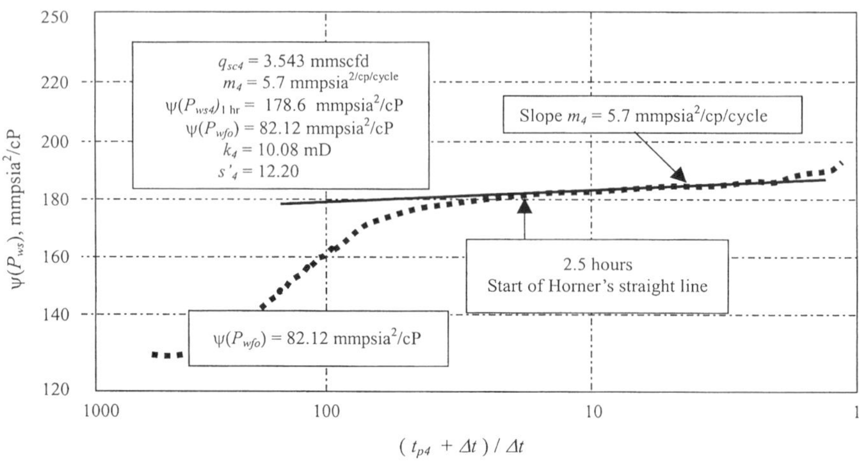

Figure 6–40. Horner's plot for buildup (rate # 4 = 3.543 mmscfd)—Example 6–6.

$$= 0.7909 \log\left(\frac{18+\Delta t}{12+\Delta t}\right) + 0.9320 \log\left(\frac{12+\Delta t}{6+\Delta t}\right) + \log\left(\frac{6+\Delta t}{\Delta t}\right)$$

$$= 0.7909 \log\left(\frac{18+20}{12+20}\right) + 0.9320 \log\left(\frac{12+20}{20}\right) + \log\left(\frac{6+20}{20}\right)$$

$$= 0.0590 + 0.0840 + 0.1139$$

$$= 0.2569$$

Table 6–20 summarizes the calculations. The slope in Figure 6–41 gives m = 10.71 mmpsia^2/cP. From Eq. 6–83;

$$k = \frac{57.920 \times 10^6 \times q_n \times T \times P_{sc}}{mhT_{sc}}$$

$$= \frac{57.920 \times 10^6 \times 3.543 \times 686 \times 14.65}{10.71 \times 10^6 \times 69 \times 520} = 5.37\,\text{mD}$$

From Eq. 6–84;

$$s = 1.151\left[\frac{(\psi(p_{1hr}) - \psi(p_{wf,\Delta t=0}))}{m} - \log\frac{k}{\phi\mu c_t r_w^2} + 3.23\right]$$

$$= 1.151\left[\frac{(175.5 - 113.45)10^6}{10.71 \times 10^6} - \log\frac{5.37}{0.088 \times 0.01639 \times 0.000255 \times 0.4271^2} + 3.23\right] = 1.12$$

Table 6–19
Pretest Rate and Pressure Data

j	t_j (hr)	q_{sc} (mmscfd)
0	0	0
1	6	2.802
2	12	3.302
3	18	3.543

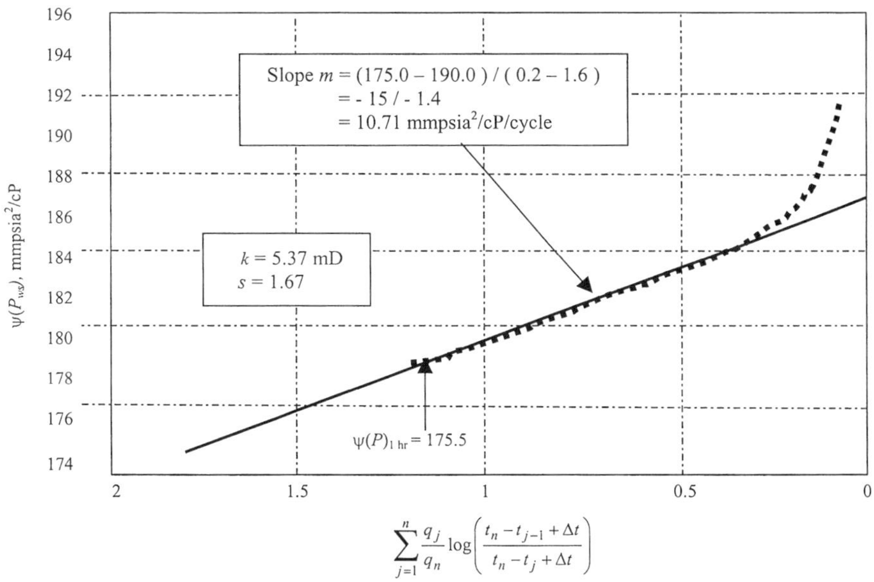

Figure 6–41. Buildup test analysis when rate varies before testing—Example 6–6.

6.12 Concept of Drainage Radius

The radius of investigation is also known as radius of drainage. Any equation for radius of investigation from a buildup test is very approximate. In a reservoir that is known to be infinite acting the radius of investigation is simply obtained by

$$r_{inv} = 0.03248\sqrt{\left(\frac{k\,\overline{p}_R(t_P + \Delta t)}{\phi\overline{\mu}_g}\right)} \tag{6–85}$$

Table 6–20
Buildup Data and Computations

Δt (hr)	$\frac{2.802}{3.543}\log\left(\frac{18+\Delta t}{12+\Delta t}\right)$ $0.7909\log\left(\frac{18+\Delta t}{12+\Delta t}\right)$	$\frac{3.302}{3.543}\log\left(\frac{12+\Delta t}{6+\Delta t}\right)$ $0.9320\log\left(\frac{12+\Delta t}{6+\Delta t}\right)$	$\frac{3.543}{3.543}\log\left(\frac{6+\Delta t}{\Delta t}\right)$ $\log\left(\frac{6+\Delta t}{\Delta t}\right)$	$\sum$ Term	P_{ws} (psia)	$\psi(P_{ws})$ (mmpsia^2/cP)
0.07	0.1386	0.2782	1.9381	2.3549	1210	105.02
0.10	0.1383	0.2772	1.7853	2.2008	1264	114.33
0.50	0.1347	0.2647	1.1139	1.5133	1576	174.94
1.00	0.1303	0.2506	0.8451	1.2260	1593	178.62
1.50	0.1263	0.2379	0.6990	1.0632	1597	179.46
2.00	0.1225	0.2265	0.6021	0.9511	1599	179.84
2.50	0.1189	0.2162	0.5315	0.8666	1601	180.31
3.00	0.1156	0.2068	0.4771	0.7995	1603	180.68
3.50	0.1124	0.1982	0.4337	0.7443	1605	181.11
4.00	0.1094	0.1902	0.3979	0.6975	1607	181.45
4.50	0.1065	0.1829	0.3680	0.6574	1608	181.78
5.00	0.1038	0.1762	0.3424	0.6224	1609	182.03
5.50	0.1013	0.1699	0.3203	0.5915	1610	182.27
6.00	0.0988	0.1641	0.3010	0.5639	1611	182.51
6.50	0.0965	0.1589	0.2840	0.5394	1612	182.57
7.00	0.0943	0.1536	0.2688	0.5167	1613	182.94
7.50	0.0921	0.1488	0.2553	0.4962	1614	183.01
8.00	0.0901	0.1444	0.2430	0.4748	1615	183.24
8.50	0.0882	0.1403	0.2324	0.4596	1615	183.35
9.00	0.0863	0.1362	0.2218	0.4443	1616	183.48
9.50	0.0846	0.1326	0.2130	0.4301	1617	183.70
10.00	0.0828	0.1289	0.2041	0.4158	1618	183.83
11.00	0.0797	0.1227	0.1901	0.3849	1619	184.06
12.00	0.0766	0.1164	0.1761	0.3691	1620	184.24
14.00	0.0717	0.1070	0.1572	0.3359	1622	184.82
16.00	0.0667	0.0976	0.1383	0.3026	1624	185.19
18.00	0.0629	0.0908	0.1261	0.2798	1626	185.56
20.00	0.0590	0.0840	0.1139	0.2569	1627	185.93
22.00	0.0558	0.0786	0.1047	0.2391	1628	186.13
24.00	0.0531	0.0741	0.0975	0.2247	1630	186.43
26.00	0.0504	0.0696	0.0902	0.2102	1631	186.74
28.00	0.0482	0.0660	0.0847	0.1989	1632	186.97
30.00	0.0459	0.0624	0.0792	0.1875	1633	187.21
32.00	0.0439	0.0593	0.0746	0.1778	1634	187.48
34.00	0.0422	0.0567	0.0708	0.1696	1635	187.67
36.00	0.0405	0.0540	0.0669	0.1614	1636	187.91
40.00	0.0375	0.0496	0.0607	0.1478	1638	188.26
44.00	0.0350	0.0459	0.0555	0.1364	1640	188.66
48.00	0.0329	0.0429	0.0515	0.1272	1641	189.01
52.00	0.0307	0.0398	0.0475	0.1180	1643	189.36
56.00	0.0291	0.0375	0.0445	0.1111	1652	191.33
60.00	0.0275	0.0352	0.0414	0.1041	1659	192.81
61.75	0.0269	0.0343	0.0403	0.1015	1660	193.11

This equation is valid only for $r_{inv} < r_e$ or $(t + \Delta t) < t_s$ where t_s is time of stabilization and is found from

$$t_s \cong 1000 \frac{\phi \overline{\mu}_g r_e^2}{k\, \overline{p}_R} \tag{6–86}$$

Example 6–7 *Calculating Radius of Investigation*

We wish to run a flow test on an exploratory well for sufficiently long to ensure that the well will drain a cylinder of more than 1000 ft radius. Preliminary well and fluid data analysis suggests that $k = 8.21$ mD, $\phi = 0.1004$, $c_t = 0.00023$ psi^{-1}, and $\mu_g = 0.02345$ cP. What length flow test appears advisable?

Solution The minimum length flow test would propagate a pressure transient approximately 2000 ft from the well (twice the minimum radius of investigation for safety).

$$t = \frac{984 \phi \mu c_t r_i^2}{k} = \frac{984 \times 0.1004 \times 0.02345 \times 0.00023 \times 2000^2}{8.21}$$
$$= 259.6\,\text{hr}$$

In practice, we require a flow rate large enough that pressure change with time can be recorded with sufficient precision to be useful for analysis; also, it depends on the particular pressure gauge used in the test.

6.13 Analysis of Responses in Composite Reservoirs

Several deviations that may include reservoir heterogeneity, phase redistribution, wellbore storage, and interference effects will affect the data collected during buildup tests. Some of these deviations may be recognized from a Horner plot. A few of the important deviations from the idealized reservoir model are shown in Figure 6–2.

6.14 Summary

In conclusion, practical considerations may prevent us from conducting pressure tests and the complexity may limit the information we can get. Thus, it is important for an engineer to determine the objectives of the well test and make some preliminary calculations about various flow periods and their duration. Based upon test duration, an engineer can then estimate the shut-in time required to obtain various flow periods, information obtained from these flow periods, and costs for conducting these tests. Then one can establish a reasonable basis to decide on conducting a well test. In some instances, it may be difficult to estimate a range of various reservoir parameters. The buildup test can then be undertaken based upon evaluation of the initial drawdown testing.

References and Additional Reading

1. Horner, D. R., "Pressure Buildup in Wells," *Third World Petroleum Congress Proceedings*, E. J. Brill, Leiden, 1951.
2. Miller, Dyes, and Hutchinson, "Estimation of Permeability and Reservoir Pressure from Bottom Hole Pressure Buildup Characteristics." *Trans. AIME* (1950), 189.
3. Matthews, L., Brons, G. S. F., and Hazebroek, P., "A Method for Determination of Average Pressure in a Bounded Reservoir," *Trans. AIME* (1954) 201, 182–191.
4. Matthews, C. S., and Russell, D. G., "Pressure Buildup and Flow Tests in Wells," SPE, AIME, Monograph 1 (1967).
5. Odeh, A. S., and Al-Hussainy, R., "A Method for Determining the Static Pressure of a Well from Buildup Data," *J. Petroleum Technol.* (1971) 23, 621–624.
6. Ramey, H. J., and Cobb, W. M., "A General Pressure Buildup Theory for a Well in a Closed Drainage Area," *J. Petroleum Technol.* (1971) 23, 1493–1505.
7. Cobb, W. M., and Smith, J. T., "An Investigation of Pressure Buildup Tests in Bounded Reservoirs," Paper SPE 5133, 49th Fall Meeting of AIME, Houston, TX, 1974.
8. Muskat, M., "Use of Data on the Build-Up of Bottom-Hole Pressures," (1936) paper presented Fort Worth Meeting, Oct. 1966.
9. Muskat, M., *The Flow of Homogeneous Fluids through Porous Media*. McGraw-Hill, New York, 1937.
10. Ramey, H. J., Kumar, A., and Gulati, M. S., "Gas Well Test Analysis Under Water-Drive Conditions," Amer. Gas Association, VA, 1973.
11. Slider, H. C., "Application of Pseudo-Steady-State Flow to Pressure-Buildup Analysis, Paper SPE 1403, 41st Fall Meeting of AIME, Amarillo, TX, 1966.
12. Slider, H. C., "A Simplified Method of Pressure Analysis for a Stabilized Well," *J. Petroleum Technol.* (1971) 23, 1155–1160.
13. Dietz, D. N., "Determination of Average Reservoir Pressure from Build-Up Surveys," (1965) *Trans. AIME.*
14. Odeh, A. S., and Jones, L. G., "Pressure Drawdown Analysis, Variable-Rate Case," *J. Petroleum Technol.* (1965) 17, 960–964.
15. Al-Hussainy, R., Ramey, H. J., Jr., and Crawford, P. B., "The Flow of Real Gases Through Porous Media," *J Petroleum Technol.* (May 1966) 624–636; *Trans. AIME*, 237.

16. Dake, L. P., *Fundamentals of Reservoir Engineering*, Elsevier Scientific, 1978.
17. Ramey, H. J., Jr. "Short-Time Well Test Data Interpretation in the Presence of Skin Effect and Wellbore Storage," *J. Petroleum Technol.* (1970) 22, 97–104.
18. Ramey, H. J., Jr. "Non-Darcy Flow and Wellbore Storage Effects in Pressure Build-up and Drawdown of Gas Wells," *J. Petroleum Technol.* (1965) 7, 223–233.
19. Earlougher, R. C., Jr., Ramey, H. J., Jr. Miller, F. G., and Mueller, T. D., "Pressure Distributions in Rectangular Reservoirs," *J. Petroleum Technol.* (1968) 20, 199–208.
20. Larson, V. C., "Understanding the Muskat Method of Analyzing Pressure Build-Up Curves," *J. Can. Petroleum Technol.* (1963) 2(3), 136–141.
21. McMahon, J. J., "Determination of Gas Well Stabilization Factors from Surface Flow Tests and Build-Up Tests," Paper SPE 114, 36th Fall Meeting of AIME, Dallas, TX, 1961.
22. Odeh, A. S., and Selig, F., "Pressure Build-Up Analysis, Variable-Rate Case," *J. Petroleum Technol.* (1963) 15, 790–794.
23. Martin, J. C., "Simplified Equations of Flow in Gas Drive Reservoirs and the Theoretical Foundation of Multiphase Pressure Buildup Analyses," *Trans. AIME* (1959) 216, 309–311.
24. Russell, D. G., "Extensions of Pressure Build-Up Analysis Methods," *J. Petroleum Technol.* (1966) 18, 1624–1636.
25. Russell, D. G., "Extensions of Pressure Build-Up Analysis Methods," *J. Petroleum Technol.* 18, 1624–2636.
26. Amanat U. C., *Pressure Transient Test Analysis User's Handbook*, Vol. 8, Twpsom Petroleum Software Series by Advanced TWPSOM Petroleum Systems Inc. Houston, TX, 1995.

Chapter 7

Predicting Future Deliverability Using Empirical Relationships

7.1 Introduction

Deliverability testing is a commonly used technique for predicting short-term and long-term behavior of gas wells. Typically, a well is flowed at different rates, and the pressure–rate–time response is recorded. From analysis of these data, information is obtained regarding the deliverability of the well, i.e., its ability to produce against a given backpressure at a given stage of reservoir depletion. Such forecasting is often required input for designing production facilities, planning field development, estimating payout time, setting allowable rates, etc. Deliverability testing has been done using multipoint flow tests. Empirical equations[5,7] to predict current and future gas well deliverability are presented. Deliverability calculations for both unfractured and fractured gas wells are also briefly discussed in this chapter.

7.2 Empirical Treatment

The basic assumptions are:

1. Homogeneous, isotropic, unfractured reservoir with a closed outer boundary
2. Single, fully penetrating well
3. Stabilized conditions prevail, i.e., pseudo-steady-state equations can be used to describe gas flow in the reservoir
4. Turbulent factor D and a rate-dependent skin Dq_{sc}.

Under these assumptions, the drawdown equation, in gas-field units, is

$$\Psi(p_i) - \Psi(p_{mf}) = \frac{50.300 \times 10^6 q_{sc} T p_{sc}}{khT_{sc}} \times \left[\frac{1}{2} In\left(\frac{2.2458}{C_A}\right) + 2\pi t_{DA} + s + Dq_{sc}\right] \quad (7\text{–}1)$$

After rearranging, Eq. 7–1 becomes

$$\left[\frac{\Psi(p_i) - \Psi(p_{wf})}{50.30 \times 10^6 \times \frac{q_{sc}Tp_{sc}}{khT_{sc}}}\right] = \frac{1}{2} In\left(\frac{A}{r_w^2}\right) + \frac{1}{2} In\left(\frac{2.2458}{C_A}\right) + 2\pi t_{DA} + s + Dq_{sc} \quad (7\text{–}2)$$

For a closed drainage volume, material balance gives

$$\left[\frac{\Psi(p_i) - \Psi(\overline{p}_R)}{50.300 \times 10^6 \frac{q_{sc}Tp_{sc}}{khT_{sc}}}\right] = 2\pi t_{DA} \quad (7\text{–}3)$$

Combining Eqs. 7–2 and 7–3, we have

$$\left[\frac{\Psi(p_i) - \Psi(\overline{p}_R)}{50.300 \times 10^6 \times \frac{q_{sc}Tp_{sc}}{khT_{sc}}}\right] = \frac{1}{2} In\left(\frac{A}{r_w^2}\right) + \frac{1}{2} In\left(\frac{2.2458}{C_A}\right) + s + Dq_{sc} \quad (7\text{–}4)$$

This equation can now be rewritten as the familiar quadratic deliverability equation:

$$\Psi(\overline{p}_R) - \Psi(p_{mf}) = aq_{sc} + bq_{sc}^2 \quad (7\text{–}5)$$

where the coefficients a and b are given by

$$a = 57.900 \times 10^6 \frac{T}{kh}\left[\log\left(\frac{A}{r_w^2}\right) + \log\left(\frac{2.2458}{C_A}\right) + 0.869s\right] \quad (7\text{–}6)$$

$$b = 50.300 \times 10^6 \frac{Tp_{sc}}{khT_{sc}} \quad (7\text{–}7)$$

Solving Eq. 7–5 and taking the positive root to be q_{sc},

$$q_{sc} = \frac{-a + \sqrt{a^2 + 4b[\Psi(\overline{p}_R) - \Psi(p_{mf})]}}{2b} \quad (7\text{–}8)$$

and corresponding to a zero sandface pressure, the absolute open flow potential (*AOFP*) is the theoretical maximum rate $(AOF)_{current}$:

$$(AOF)_{current}(\Psi(p_{mf}=0)) = \frac{-a+\sqrt{a^2+4b[\Psi(\overline{p}_R)]}}{2b} \tag{7–9}$$

Dividing Eq. 7–8 by Eq. 7–9 we get

$$\left[\frac{q_{sc}}{(AOF)_{current}}\right] = \frac{-a+\sqrt{a^2+4b[\Psi(\overline{p}_R)-\Psi(p_{mf})]}}{-a+\sqrt{a^2+4b[\Psi(\overline{p}_R)]}} \tag{7–10}$$

which can also be expressed in a form such as

$$\left[\frac{q_{sc}}{(AOF)_{current}}\right] = F1\left[\frac{\Psi(p_{mf})}{\Psi(\overline{p}_R)}\right] \tag{7–11}$$

where $F1$ is some functional form. The dimensionless groups $[\Psi(p_{mf})/\Psi(\overline{p}_R)]$ and $[q_{sc}/(AOF)_{current}]$ can be generated for a variety of cases and develop an empirical correlation of the form of Eq. 7–11. This will then be the IPR for current deliverability. To calculate deliverability at future conditions, Eq. 7–9 can be rewritten as

$$(AOF)_{future} = \frac{-a+\sqrt{a^2+4b[\Psi(\overline{p}_R)_f]}}{2b} \tag{7–12}$$

Now dividing Eq. 7–12 by Eq. 7–11, we get

$$\frac{(AOF)_{future}}{(AOF)_{current}} = \frac{-a+\sqrt{a^2+4b[\Psi(\overline{p}_R)_f]}}{-a+a^2+4b[\Psi(\overline{p}_R)]} \tag{7–13}$$

This equation can be restated as

$$\left[\frac{(AOF)_{future}}{(AOF)_{current}}\right] = F2\left[\frac{\Psi(\overline{p}_R)_f}{\Psi(\overline{p}_R)}\right] \tag{7–14}$$

where $F2$ is some other functional form. The other dimensionless groups $[\Psi(\overline{p}_R)_f/\Psi(\overline{p}_R)]$ and $[(AOF)_{future}/(AOF)_{current}]$ can be generated for a variety of cases and develop a second empirical relation, of the form of Eq. 7–14. This will then be the IPR for future deliverability.

Current Deliverability Calculations

Follow these steps:

1. Knowing $\Psi(\overline{P}_R)$ and $\Psi(P_{mf})$, calculate $\Psi(P_{mf})/\Psi(\overline{P}_R)$.
2. From the dimensionless IPR for current conditions (test fit curve), estimate $[q_{sc}/(AOF)_{current}]$.

3. Knowing q_{sc}, calculate AOFP as

$$AOFP = \frac{q_{sc}}{[q_{sc}/(AOF)_{current}]}$$

4. At any other sandface pressure $\Psi(P'_{wf})$, to find the flow rate q'_{sc}, first calculate $[\Psi(P'_{wf})/\Psi(\overline{P}_R)]$. Then, from the dimensionless IPR for current conditions (test fit curve), estimate $\lfloor q'_{scf}/(AOF)_{current}\rfloor$. Deliverability is next calculated as $q'_{sc} = (AOF)_{current} \times [q'_{sc}/(AOF)]$.

Future Deliverability Calculations

Follow these steps:

1. Given a future average reservoir pressure $\Psi(\overline{P}_R)_f$, and knowing the current average reservoir pressure $\Psi(\overline{P}_R)$, calculate $[\Psi(\overline{P}_R)_f/\Psi(\overline{P}_R)]$.
2. Using the dimensionless IPR for future conditions (best fit curve), estimate $\lfloor(AOF)_{future}/(AOF)_{current}\rfloor$.
3. Knowing current $(AOF)_{current}$, calculate future $(AOF)_{future}$ as

$$(AOF)_{future} = (AOF)_{current} \times \lfloor(AOF)_{future}/(AOF)_{current}\rfloor$$

4. At any future sandface pressure $\Psi(P_{wf})_f$, if the flow rate q_{scf} is desired, first calculate $\lfloor\Psi(P_{mf})_f/\Psi(\bar{P}_R)_f\rfloor$. Then, from the dimensionless IPR for current conditions (best fit curve), estimate $\lfloor q_{scf}/(AOF)_{future}\rfloor$. Deliverability can finally be calculated as $q_{scf} = (AOF)_{future} \times [q_{scf}/(AOF)_{future}]$.

A sample calculation is shown in Example 7–1.

Example 7–1 *Future Deliverability Calculations from Current Flow Test Data*

A gas well was flowed at a rate of 5.214 mmscfd. The stabilized sandface pressure at the end of the flow test was 2566 psia, and the current average reservoir pressure was estimated to be 3700 psia. For a gas gravity of 0.732 and a bottom-hole temperature of 710°R, the P–$\Psi(P)$ table was calculated and tabulated in Table 7–1. Figure 7–1 shows a plot of the ψ–p curve. The objective is to simplify the method suggested in this section by calculating the following parameters.

(a) $(AOF)_{current}$ at current conditions ($\overline{P}_R = 3700$ psia)
(b) Deliverability at a flowing bottom-hole pressure $P_{wf} = 1000$ psia
(c) $(AOF)_{future}$ at a future average pressure $(P_R)_{future} = 3000$ psia
(d) Deliverability at a future bottom-hole pressure $P_{wf} = 2566$ psia

Solution From Example 4–8, the deliverability coefficients are $a = 91.8273$ psia²/cP-mmscfd, and $b = 0.1785$ psia²/cP-mmscfd². Calculate current deliverability, $(AOF)_{current}$, using Eq. 7–9. Find the dimensionless parameters $\Psi(p_{wf})/\Psi(P_R)_{current}$ and $q_{sc}/(AOF)_{current}$. The results are reported in

Table 7–1
PVT Gas Property and Pseudopressure for Example 7–1

P (psia)	Z (—)	μ (cP)	$\Psi(P)$ (mmpsia^2/cP)
14.65	0.9995	0.013978	0.05
400	0.9733	0.014337	11.47
800	0.9503	0.014932	45.51
1200	0.9319	0.015723	100.83
1600	0.9189	0.016681	175.33
2000	0.9120	0.017784	266.41
2400	0.9113	0.019008	371.18
2800	0.9169	0.020329	486.72
3200	0.9282	0.021721	610.28
3600	0.9445	0.023151	739.56
4000	0.9647	0.024580	872.92

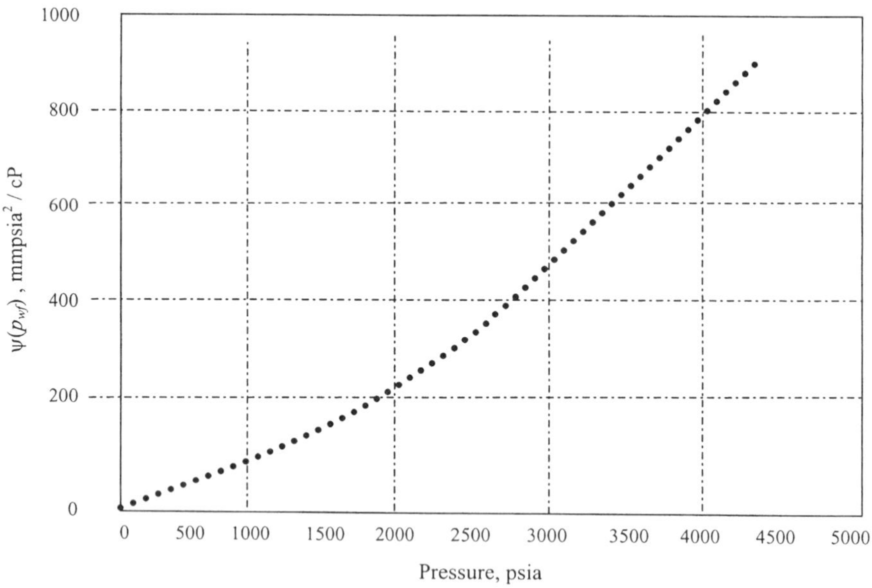

Figure 7–1. $\psi-P$ curve data plot for Example 7–1.

Table 7–2 and the data are plotted in Figure 7–2. Calculate future deliverability, $(AOF)_{future}$ using Eq. 7–12. Find the dimensionless parameters $\Psi(P_R)_f/\Psi(P_R)_{current}$ and $(AOF)_{future}/(AOF)_{current}$. The results are reported in Table 7–3 and data are plotted in Figure 7–3.

Table 7–2
Gas Well Deliverability at Current Conditions

Bottom-hole pressure P_{wf} (psia)	$\Psi(P_{wf})$ (mmpsia^2/cP)	Stabilized deliverability q (mmscfd)	Dimensionless parameters: Ratio = $\Psi(P_{wf})/\Psi(P_R)$ —	Ratio = $q/AOFC$ —
P_R 3700	772.56	0.0	1.0000	0.0000
3500	706.80	0.715	0.9149	0.0863
3000	547.65	2.438	0.7089	0.2943
2500	399.17	4.035	0.5167	0.4871
2000	266.41	5.454	0.3448	0.6584
1500	155.04	6.639	0.2007	0.8014
1250	109.14	7.126	0.1413	0.8602
1000	70.63	7.534	0.0914	0.9095
750	40.06	7.857	0.0519	0.9485
500	17.90	8.091	0.0232	0.9767
400	11.47	8.159	0.0148	0.9849
200	2.88	8.250	0.0037	0.9959
100	0.74	8.272	0.0010	0.9986
AOFC 14.65	0.05	8.284	0.0000	1.0000

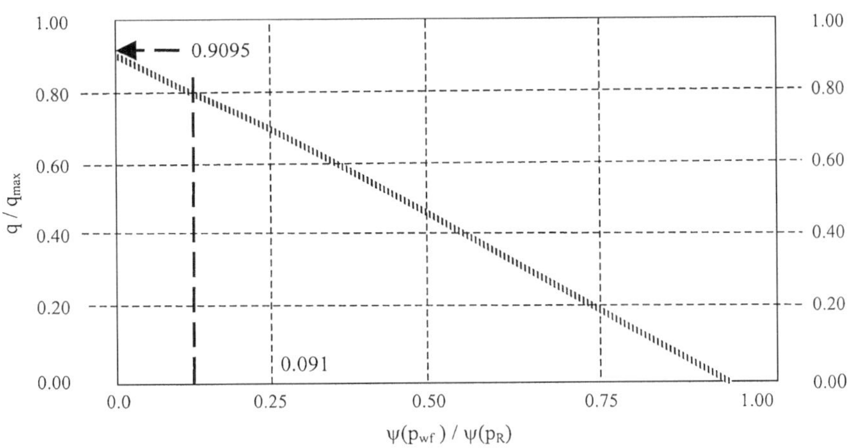

Figure 7–2. Dimensionless IPR for current conditions—Example 7–1.

Table 7–3
Gas Well Deliverability at Future Conditions

Future reservoir pressure (psia)	$\Psi(P_R)_{future}$ (mmpsia^2/cP)	Ratio = $\psi(p_R)_{future}/\psi(p_R)_{current}$ —	Ratio = $(AOF)_{future}/(AOF)_{current}$ —
3700	772.56	1.0000	1.0000
3500	706.80	0.9149	0.9156
3000	547.65	0.7089	0.7122
2500	399.17	0.5167	0.5203
2000	266.41	0.3448	0.3489
1500	155.04	0.2031	0.2007
1000	70.63	0.0914	0.0927

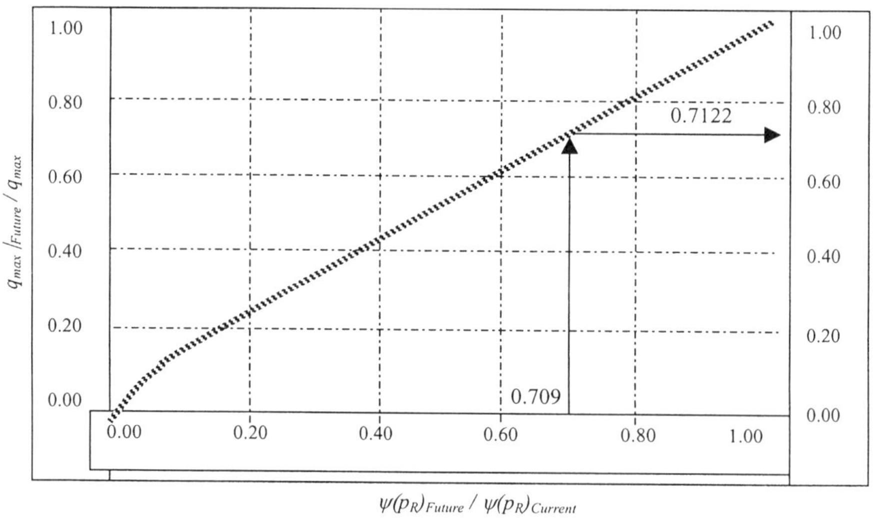

Figure 7–3. Dimensionless IPR for future conditions—Example 7–1.

Current Deliverability

(a) For $P_{wf} = 2566$ psia $\longleftrightarrow$ $\Psi(P_{wf}) = 417.59$ mmpsia^2/cP:

$$\overline{P}_R = 3700 \text{ psia} \longleftrightarrow \Psi(P_R) = 772.56 \text{ mmpsia}^2/\text{cP}$$

(b) Now $P_{wf} = 1000$ psia $\longleftrightarrow$ $\psi(p_{wf}) = 70.63$ mmpsia^2/cP:

$$\frac{\psi(P'_{wf})}{\psi(P_R)_{current}} = \frac{70.63}{772.56} = 0.091$$

$$\frac{q'}{(AOF)_{current0}} = 0.9095 \text{ (from Figure 7–2)}$$

$$q' = (AOF)_{current} \times \frac{q'}{(AOF)_{current}}$$

$$= (10.72)(0.9095) = 9.75 \text{ mmscfd}$$

Future Deliverability Calculations

(c) For $(P_R)_{future} = 3000$ psia $\longleftrightarrow$ $\psi(P_R)_{future} = 547.65$ mmpsia^2/cP:

$$\frac{\psi(P_R)_f}{\psi(P_R)_{current}} = \frac{547.65}{772.56} = 0.709$$

$$\frac{(AOF)_{future}}{(AOF)_{current}} = 0.7122 \text{ (from Figure 7–3)}$$

$$(AOF)_{future} = (AOF)_{current} \times \frac{(AOF)_{future}}{(AOF)_{current}} = (10.72)(0.7122)$$

$$= 7.63 \text{ mmpsia}^2\text{/cP}$$

(d) For $(P_{wf})_f = 2566$ psia $\longleftrightarrow$ $\psi(P_{wf})_f = 417.59$ mmpsia^2/cP:

$$\frac{\psi(P_R)_f}{\psi(P_R)_{current}} = \frac{547.65}{772.56} = 0.709$$

$$\frac{(AOF)_{future}}{(AOF)_{current}} = 0.7122 \text{ (from Figure 7–3)}$$

$$(AOF)_{future} = (AOF)_{current} \times \frac{(AOF)_{future}}{(AOF)_{current}}$$

$$= (7.63)(0.5185) = 3.96 \text{ mmscfd}$$

Sample calculations at $P_R = 3000 \longleftrightarrow 547.65$ mmpsia^2/cP:

$$q_{\max,f} = \frac{-a + \sqrt{a^2 + 4b}[\psi(\overline{P}_R)_f}{2b}$$

$$= \frac{-91.8273 + \sqrt{(91.8273)^2 + 4 \times 0.1785 \times 547.65}}{2 \times 0.1785}$$

$$= 5.90 \text{ mmscfd}$$

Therefore,

$$\frac{\psi(\overline{P}_R)_{future}}{\psi(\overline{P}_R)_{current}} = \frac{547.65}{772.56} = 0.7089 \quad \text{and} \quad \frac{(q_{max})_{future}}{(q_{max})_{current}} = \frac{5.90}{8.284} = 0.7122$$

General Remarks

The utility of any empirical correlation is essentially restricted to the conditions from which it was developed. It should only be used for gas wells draining from an unfractured reservoir under stabilized conditions. While the proposed method is certainly not universal in its application, we believe that it is a simple alternative to conventional deliverability testing methods for typical field situations.

7.3 Fractured Gas Well Deliverability Estimation Techniques

Under Darcy's Conditions

Hadinoto *et al.*[10] have used the following backpressure equation for fractured gas wells:

$$\frac{khT_{sc}}{0.50321 \times 10^5 qTp_{sc}}[\psi(\overline{p}_R - \psi(p_{wf})] = P_D(x_e/x_f, t_{DA}) - 2\pi t_{DA} \quad (7\text{–}15)$$

Equation 7–15 applies to short and long terms and can generate plots of $\log[\psi(P_R) - \psi(P_{wf})]$ versus $\log q$ provided that fracture length, fracture penetration ratio, and flow capacity are known. The previous unknown can be found from buildup and drawdown tests discussed previously in this chapter. Equation 7–15 can be rearranged to calculate *AOF* (absolute open flow potential) and stabilized deliverability:

$$q = \frac{khT_{sc}\lfloor\psi(\overline{p}_R) - \psi(p_{wf} = 14.65)\rfloor}{0.50327 \times 10^5 Tp_{sc}[p_D(x_e/x_f, t_{DA}) - 2\pi t_{DA}]} \quad (7\text{–}16)$$

where

$$t_{DA} = \frac{0.000264kt}{\phi\mu_g cA} \text{ and}$$

$$P_D = 2\pi t_{DA} + \frac{1}{2}In(x_e/x_f)^2 + \frac{1}{2}In(2.2458/c_f) + \frac{1}{2}In(16) \quad (7\text{–}17)$$

c_f = shape factor for a fractured vertical well

All the dimensionless variables and parameters considered important are taken into account in Figures 4–42 and 4–43, which are log-log plots of

$$p_D \text{ versus } \frac{2.637 \times 10^4 kt}{\phi_i \mu_i c_i x_f^2}.$$

In these parameters, x_f is the fracture half-length and x_e is the distance from the well to the side of the square drainage area in which it is assumed to be constant. Hadinoto *et al.*[10] found that the conventional methods of determining deliverability, namely isochronal testing and pressure drawdown variable-rate analysis, applied to fractured reservoirs provided that the duration of the flow period for each flow was well beyond the linear flow period. Reverse sequence flow-after-flow tests were discouraged. An important finding was that the slope (n) of a backpressure curve obtained from a short-time isochronal test on a fractured well was in general equal to 1. Consequently, deliverability could be calculated in principle, from long-time drawdown tests assuming a flow exponent (n) equal to 1.0. Flow of gas within a fracture reservoir, which accounts for turbulence around the wellbore, can be modeled by

$$\Delta p^2 = A_t' q + B_t' q^2 \tag{7–18}$$

$$\frac{\Delta p^2}{q} = A_t' + B_t' q \tag{7–19}$$

where

$$A_t' = \frac{1637 \mu_g T z}{kh} \left[\log \frac{1.42 \times 10^{-2} \times k_g t}{\phi \mu c r_w^2} \right] \tag{7–20}$$

$$B_t' = \frac{1422 \mu_g T z}{k_g h} D \tag{7–21}$$

μ_g = gas viscosity, cP; T = reservoir temperature; °R; z = gas deviation factor; h = net thickness, ft; k_g = effective permeability, mD; t = flowing time, days; r_w = wellbore radius, ft; c_g = gas compressibility, psi^{-1}; and D = non-Darcy flow constant.

In term of pseudopressure, the equations are

$$\psi(p) = A_t q + B_t q^2 \tag{7–22}$$

$$\frac{\psi(p)}{q} = A_t + B_t q^2 \tag{7–23}$$

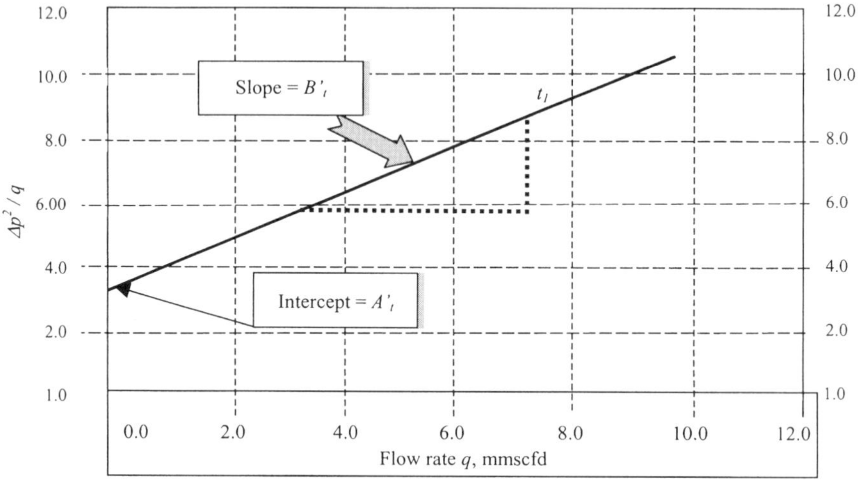

Figure 7–4. Diagram of $\Delta P^2/q$ versus q for one fixed time.

where

$$A_t = \frac{1637T}{kh}\left[\log\left(\frac{1.42 \times 10^{-2}kt}{\phi\mu c r_w^2}\right)\right] \tag{7–24}$$

$$B_t = \frac{1422T}{kh}D \tag{7–25}$$

Figure 7–4 is a diagram of $\Delta p^2/q$ versus q_{sc} for one fixed time. Equation 7–19 or 7–23 indicates that a plot of $\Delta p^2/q$ or $\psi(p)$ versus q should result in a straight line with slope equal to B_t and the intercept at zero flow equal to A_t. If a similar set of plots is prepared for fixed times during each flow period, the same slope should be obtained if Darcy's flow present in the formation. If turbulent flow is present in the formation, the resulting lines would have different slopes. In such a case, the remaining procedure could not be applied. Figure 7–5 is a plot of $\Delta p^2/q$ or $\psi(p)$ versus q at various fixed times. Under Darcy's conditions in the reservoir, a plot of AAA_t versus log time should result in a straight line as indicated in Figure 7–6. The stabilized value of AAA_t can be obtained by entering Figure 7–6 with the stabilized time (t_{stab}), which is given by the equation by Van Poollen:[14]

$$t_{stab} = \frac{\phi\mu_g c_g r_e^2}{0.25k_g} \tag{7–26}$$

The parameter AAA_t at stabilized conditions allows us to write a general equation that can be used to forecast future behavior of the reservoir. Such an

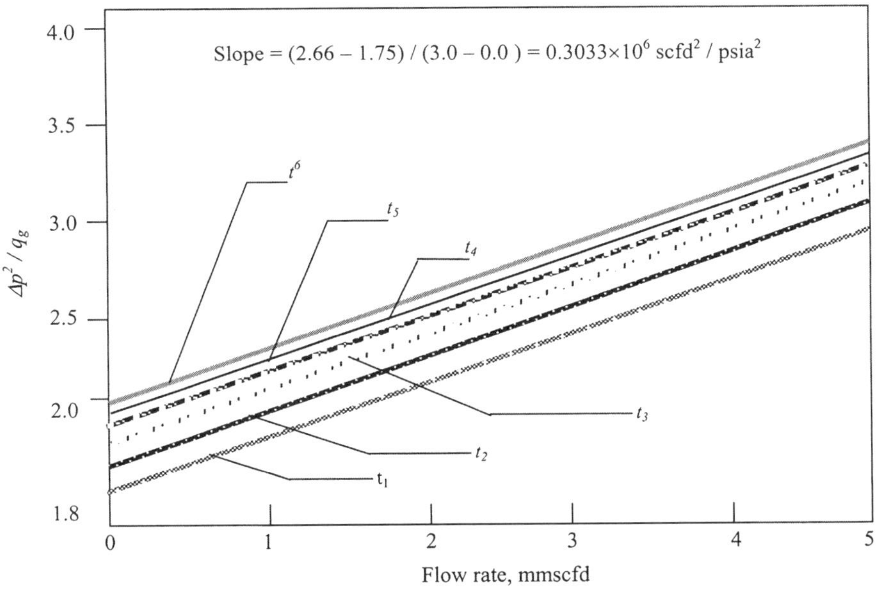

Figure 7–5. $\Delta p^2/q_g$ versus q_g for various fixed times—Example 7–2.

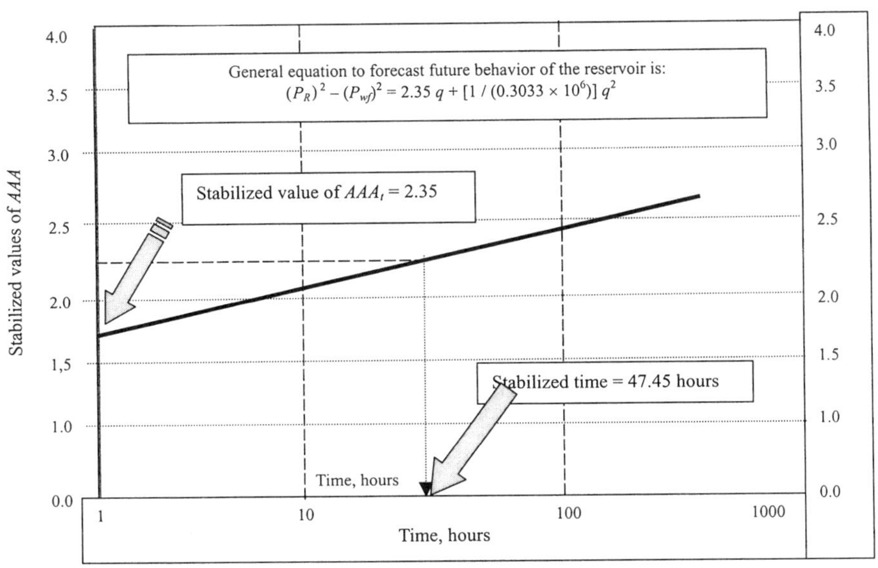

Figure 7–6. Plot of stabilized values of AAA_t versus log time—Example 7–2.

equation is

$$p_i^2 - p_{wf}^2 = \Delta p^2 = AAA_{t(stabilized)}q + BBB_t q^2 \tag{7–27}$$

Example 7–2 *Calculation of Future Deliverability of Vertical Fractured Gas Well under Darcy's Conditions*

An isochronal test was carried out in a gas well to four different rates as follows: $p_i = 3965$ psia; $k = 6.282$ mD; $r_e = 2200$ ft; $c_t = 0.00023$ psi^{-1}; $\phi = 0.119$ fraction; $\mu_g = 0.02345$ cP

Flow rate (mmscfd) **Flowing time (hr)**	$q_1 = \mathbf{2.379}$ p_{wf} **(psia)**	$q_2 = \mathbf{3.200}$ p_{wf} **(psia)**	$q_3 = \mathbf{3.850}$ p_{wf} **(psia)**	$q_4 = \mathbf{4.115}$ p_{wf} **(psia)**
1	3130	2652	2206	1903
2	3100	2602	2190	1889
3	3085	2509	2180	1876
4	3060	2580	2172	1870
5	3049	2573	2165	1855
6	3035	2567	2158	1836

Develop a general equation, which can be used to forecast future behavior of the reservoir.

Solution Values of $(\Delta p^2)/q$ were calculated as $(p_R^2 - p_{wf}^2)/q$ for each flow rate with the following results:

Flowing time (hr)	$(\Delta p^2)/q_1$ **(psia²/mmscfd)**	$(\Delta p^2)/q_2$ **(psia²/mmscfd)**	$(\Delta p^2)/q_3$ **(psia²/mmscfd)**	$(\Delta p^2)/q_4$ **(psia²/mmscfd)**
1	2.471	2.715	2.819	2.940
2	2.549	2.797	2.838	2.953
3	2.588	2.817	2.849	2.965
4	2.652	2.833	2.858	2.971
5	2.680	2.844	2.866	2.984
6	2.716	2.854	2.871	3.001

Figure 7–5 shows a plot of $\Delta p^2/q$ versus q. From this plot, the slope *BBB* was determined to be 0.3033×10^6 scfd^2/psia^2, and the intercepts at zero flow rate were found to be

1. $AAA_{t1} = 1.752$ at $t_1 = 1$ hr
2. $AAA_{t2} = 1.863$ at $t_2 = 2$ hr

3. $AAA_{t3} = 1.905$ at $t_3 = 3$ hr
4. $AAA_{t4} = 1.952$ at $t_4 = 4$ hr
5. $AAA_{t5} = 1.978$ at $t_5 = 5$ hr
6. $AAA_{t6} = 2.012$ at $t_6 = 6$ hr

Figure 7–6 shows a plot of AAA_t versus log of time. A straight line is obtained as predicted by theory. Calculation of stabilized is carried out with the use of Eq. 7–16:

$$t_{stabilized} = \frac{\phi \mu_g c_g r_e^2}{0.25 k_g} = \frac{0.119 \times 0.02345 \times 0.00023 \times 2200^2}{25 \times 6.282}$$

$$= 1.978 \text{ days} = 47.45 \text{ hours}$$

The stabilized value of AAA_t is 2.35 psia^2/mmscf from Figure 7–6. Consequently, the equation that controls the deliverability of this well at stabilized conditions is

$$(p_R^2 - p_{wf}) = 2.35q + \frac{1}{.3033 \times 10^6} q^2$$

This is a general equation, which can be used to forecast future behavior of the reservoir.

Under Turbulent Flow Conditions

Forchheimer[15] modified Darcy's law to account for turbulence effects as

$$p_R^2 - p_{wf}^2 = aaq + bbq^2 \tag{7–28}$$

or

$$\psi(p_R) - \psi(p_{wf}) = aa'q + bb'q^2 \tag{7–29}$$

where

$$aa = \frac{1422T\overline{\mu}\,\overline{z}}{kh}[In(r_e/r_w) - 0.75 + s + s_{CA,f} - 1.386] \tag{7–30}$$

$$bb = \frac{1422T\overline{\mu}\,\overline{z}}{kh}D \tag{7–31}$$

$$aa' = \frac{1422T}{kh}[In(r_e/r_w) - 0.75 + s + s_{CA,f} - 1.386] \tag{7–32}$$

$$bb' = \frac{1422T}{kh}D \tag{7–33}$$

Table 7–4
Shape Factors, c_f, for Fractured Vertical Wells in a Squared Drainage Area[11]

x_f/x_e	Shape factor c_f
0.1	2.6541
0.2	2.0348
0.3	1.9986
0.5	1.6620
0.7	1.3127
1.0	0.7887

and

$$D'' = \frac{2.222 \times 10^{-15} \times Gkh\beta'}{\mu_{pwf} r_w h_P^2} \quad (7\text{–}34)$$

$$\beta' = 2.33 \times 10^{10} k^{-1.201} \quad (7\text{–}35)$$

or

$$\beta' = 2.73 \times 10^{10} k^{-1.1045} \quad (7\text{–}36)$$

$$s_{CA,f} = In\sqrt{30.88/c_f} \quad (7\text{–}37)$$

The shape factor c_f is obtained from Table 7–4, where x_f/x_e = fracture penetration ratio; x_f = fracture half length, ft; x_e = half the side of the square drainage area, ft.

The effective wellbore radius of an infinite conductivity fully penetrating vertical fracture is $r'_w = x_f/2$ for $x_f/x_e < 0.4$. Also, for a square drainage area, area is $A = (2x_e)^2 = 4x_e^2$ for $x_f/x_e < 0.1$; c_f is calculated as

$$c_f = 0.25c'_f \quad (7\text{–}38)$$

where c'_f is obtained from Figure 7–7.

Example 7–3 *Estimation of Deliverability of Vertical Fracture Well under Both Darcy's and Turbulent Flow Conditions*

Given: $k = 8.27$ mD; $h = 41$ ft; $T = 710°$ R; $\overline{\mu} = 0.0235$ cP; $\overline{z} = 0.9491$; $p_i = 3965$ psia; $\psi(p_i) = 861.10$ mmpsia^2/cP; $p_R = 3700$; $\psi(p_R) = 772.25$ mmpsia^2/cP; $G = 0.732$; $p_{sc} = 14.65$ psia; $T_{sc} = 520°$R; $r_w = 0.4271$; $r_e = 2200$ ft; $\phi = 0.137$; $\phi_{HC} = 0.1004$; $s_g = 0.732$; perforated length

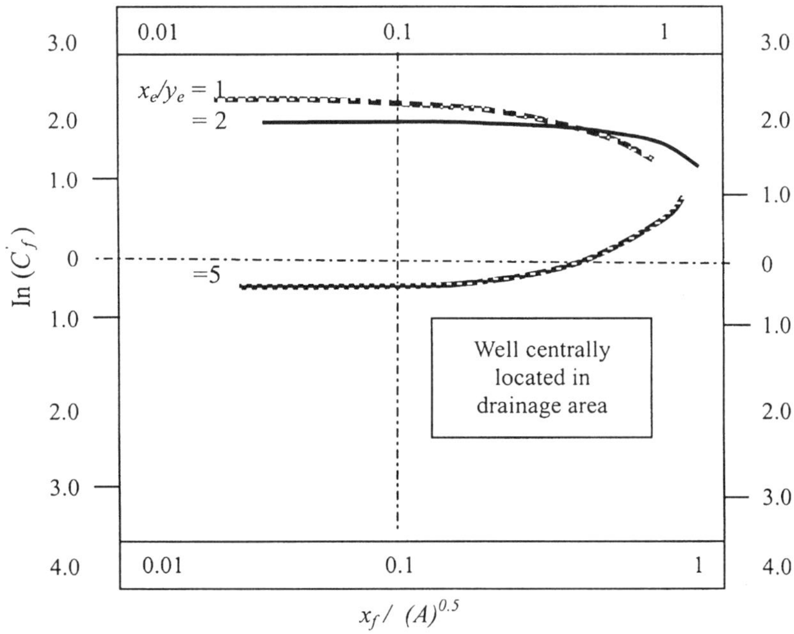

Figure 7–7. Shape factors for fractured vertical wells for different fracture formations.[12,13]

$h_P = 30$ ft; and $x_f = 86.5$ ft. Calculate the followings using the $\psi(p)$ equation:

(a) *AOF* and flow rate, assuming Darcy's flow
(b) *AOF* and flow rate, assuming influence of turbulence

Solution Calculation of skin factor s:

$$r'_w = x_f/2 = \frac{86.5}{2} = 43.25 \text{ ft} \quad \text{and} \quad s = -In(r'_w/r_w)$$
$$= -In(43.25/0.4271) = -4.62$$

$$\text{Acres} = \pi r_e^2/43{,}500 = 22/7 \times 2200 \times 2200 \div 43{,}560 = 350$$

Calculation of $s_{CA,f}$: Assuming a square drainage area with each side being $2x_e$ for 350 acres, we have

$$A = (2x_e)^2,\ 2x_e = (350 \times 43{,}560)^{0.5} = 3905 \text{ ft or } x_e = 1953 \text{ ft}$$
$$\text{Penetration ratio } x_f/x_e = 86.5/1953 = 0.00443$$

Since $x_f/x_e < 0.1$, Figure 7–7 is used to calculate c'_f. For $x_f/\sqrt{A} = 86.5/(12 \times 1953) = 0.022$, from Eq. 7–38, $c_f = 0.25 \times c'_f = 0.25 \times 7.389 = 1.847$, and using Eq. 7–37,

$$s_{CA,f} = In\left(\sqrt{\frac{30.88}{c_f}}\right) = In\left(\sqrt{\frac{30.88}{1.847}}\right) = 1.4082$$

(a) Substituting these parameters in Eq. 7–32 we have

$$\begin{aligned} aa' &= \frac{1422T}{kh}[In(r_e/r_w) - 0.75 + s + s_{CA,f} - 1.386] \\ &= \frac{1422 \times 710}{8.27 \times 41}[In(2200/0.4271) - 0.75 + (-4.62) + 1.4082 - 1.386] \\ &= 2977.62[8.5470 - 0.75 - 4.62 + 1.4082 - 1.386] = 9525.852 \end{aligned}$$

The value of bb' can be calculated using Eq. 7–33 as

$$bb' = \frac{1422T}{kh}D'$$

To calculate bb', first estimate D' and β' using Eqs. 7–34 and 7–36:

$$\begin{aligned} D' &= \frac{2.222 \times 10^{-15} \times G \times kh\beta'}{\mu r_w h_P^2} 16.162 \times 10^{-5} \\ &= \frac{2.222 \times 10^{-15} \times 0.732 \times 8.27 \times 41 \times 0.264715 \times 10^{10}}{0.0235 \times 0.4271 \times 30^2} \\ &= 16.162 \times 10^{-5} \end{aligned}$$

and

$$\begin{aligned} \beta' &= 2.73 \times 10^{10} k^{-1.1045} = 2.73 \times 10^{10}(8.27)^{-1.1045} \\ &= 0.264715 \times 10^{10}\ \text{ft}^{-1} \end{aligned}$$

Substituting these values into Eq. 7–33, bb' is calculated as

$$bb' = \frac{1422 \times 710 \times 16.162 \times 10^{-5}}{8.27 \times 41} = 48{,}122.890 \times 10^{-5}$$

Substituting these values of aa' and bb' into Eq. 7–29;

$$\psi(p_R) - \psi(p_{wf}) = 9525.852q + 48{,}122.890 \times 10^{-5} q^2$$

The above quadratic equation is rearranged as

$$48{,}122.890 \times 10^{-5} q^2 + 9525.852q - [\psi(p_R) - \psi(p_{wf})] = 0$$

Table 7–5
Calculated Flow Rates of Vertical Fractured Well (Example 7–3)

p_{wf} (psia)	$\psi(p_{wf})$ (mmpsia^2/cP)	$\psi(p_R)-\psi(p_{wf})$ (mmpsia^2/cP)	Darcy's flow, $D = 0$ q (mmscfd) $\psi(p)/a'$	Turbulence flow q (mmscfd)
3700	772.25	0	0	0
3500	706.80	65.45	6.87	5.40
3000	547.65	224.60	23.58	13.87
2500	399.17	373.08	39.17	19.65
2000	266.41	505.84	53.10	24.00
1500	155.04	617.21	64.79	27.26
1000	70.63	701.62	73.65	29.55
750	40.06	732.19	76.86	30.35
500	17.90	754.35	79.19	30.91
400	11.47	760.78	79.87	31.08
200	2.88	769.37	80.77	31.22
100	0.74	771.51	80.99	31.35
(*AOF*) 14.65	0	772.25	81.07	31.37

Solving the quadratic equation, the value of q is calculated as

$$q = \frac{-9525.852 + \sqrt{(9525.852)^2 + 4 \times 48{,}122.890 \times 10^{-5}[\psi(p_R) - \psi(p_{wf})]}}{2 \times 48{,}122.890 \times 10^{-5}}$$

Calculated values of q both with and without turbulence for various values of p_{wf} are summarized in Table 7–5. This table shows that turbulence has no effect on horizontal wells; horizontal wells minimize turbulence-related pressure drops. Thus, in high-permeability gas reservoirs, horizontal wells provide a method for minimizing wellbore turbulence.

7.4 Summary

Chapter 7 presents simplified procedures for gas deliverability calculations using dimensionless IPR curves and includes a discussion of a new method for deliverability calculations of fractured and unfractured gas wells, which eliminates the need for conventional multipoint tests.

References and Additional Reading

1. Cullender, M. H., "The Isochronal Performance Method of Determining the Flow Characteristics of Gas Wells," *Trans. AIME* (1955) 204, 137–142.

2. Katz, D. L., Cornell, D., Kobayashi, R., Poettmann, F. H., Vary, J. A., Elenbaas, J. R., and Weinaug, F., *Handbook of Natural Gas Engineering*. McGraw-Hill, New York, 1959.
3. Fetkovich, M. J., "Multi-point Testing of Gas Wells," Continuing Education Course, SPE Mid-Continent Section, March 17, 1975.
4. Lee, A. L., Gonzalez, M. H., and Eakin, B. E., "The Viscosity of Natural Gases," *Trans. AIME* (1966) 237, 997–1000.
5. Vogel, J. L., "Inflow Performance Relationships for Solution-Gas Drive Wells," *J. Petroleum Technol.* (Jan. 1968) 83–92.
6. Rawlins, E. L., and Schellhardt, M. A., "Back-Pressure Data on Natural Gas Wells and Their Application to Production Practices," Monograph 7, USBM, 1936.
7. Mishra, S., "Deliverability Testing of Gas Wells Using Dimensionless IPR Curves," M.S. Thesis, The University of Texas at Austin, May 1983.
8. *Theory and Practice of the Testing of Gas Wells*, 3rd ed., Alberta Energy Resources Conservation Board, 1979.
9. Swift, G. W., and Kiel, O. G., "The Prediction of Gas Well Performance Including the Effect of Non-Darcy Flow," *Trans. AIME* (1962) 225, 791–798.
10. Hadinoto, N., Raghavan, R., and Thomas, G. W., "Determination of Gas Well Deliverability of Vertically Fractured Wells," Paper SPE 6136 presented at the 51st Annual Fall Technical Conference and Exhibition of SPE of AIME, New Orleans (October 3–6, 1976).
11. Earlougher, R. C., Jr., "Advances in Well Test Analysis," Monograph Vol. 5 of the Henry L. Doherty Series in Society of Petroleum Engineers of AIME, 1977.
12. Al-Hussainy, R. Ramey, H. J., Jr., and Crawford, P. B., "The Flow of Real Gases through Porous Media," *J. Petroleum Technol.* (1966) 624–636; *Trans. AIME* V 237.
13. Gringarten, A. C., Ramey, H. J., Jr., and Raghavan, R., "Unsteady State Pressure Distribution Created by a Well with a Single Infinity-Conductivity Vertical Fracture," *SPE* (Aug. 1974) 347–360.
14. Van Poollen, H. K., "Radius-of-Drainage and Stabilization Time Equations," *Oil Gas J.* (1964) 62, 138–146.
15. Forchheimer, Ph., "Wasserbewegung durch Boden," *Z. Ver. Deutsch. Ing.* (1901) 45, 1781–1788.

Chapter 8

Application of Type Curve Matching Techniques

8.1 Introduction

Type curve matching provides methods[2,6,8] for analyzing transient gas well tests with known dimensionless pressure P_D and time t_D. Type curve matching techniques may be used for drawdown, buildup, interference, and constant pressure testing. For single well testing, type curve matching should be used only when conventional analysis techniques such as those illustrated in Chapter 5 cannot be used. In such cases, type curve analysis can provide approximate results. This chapter discusses the quantitative use and applications of type curves in gas well test analysis. The most generally useful type curves have been selected and are included herein. Fundamentals of type curve use are presented and will allow the reader to understand and to apply newer type curves as they appear in the literature.

8.2 Fundamentals of Type Curve Matching

A log-log plot of the line-source solution—the Theis solution—shown in Figure 8–1 may be used to analyze such tests. Plots such as the one shown in Figure 8–1 are known as type curves. The basis for the type curve approach can be understood if we consider the definition of $P_D(r_D, t_D)$ and t_D/r_D^2. Taking logarithms of these definitions, we obtain

$$\log[\mathrm{P_D}(r_D, t_D)] = \log \frac{kh}{1.417 \times 10^6 T q_{sc}} + \log[\psi(P_i) - \psi(P_{wf}(r, t))] \tag{8–1}$$

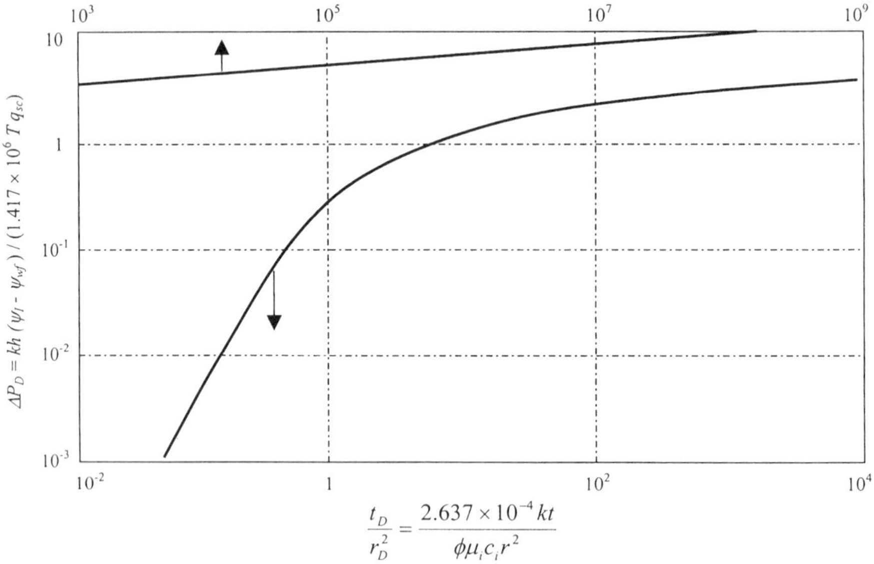

Figure 8–1. Line source solution.

and

$$\log\left(\frac{t_D}{r_D^2}\right) = \log\left(\frac{2.637 \times 10^{-4}k}{P_i c_i r^2}\right) + \log t \qquad (8\text{–}2)$$

The first terms on the right-hand sides of Eqs. 8–1 and 8–2 contain the characteristic constants that govern the pressure distribution in the reservoir. Inspection of Eqs. 8–1 and 8–2 suggests that a plot of the logarithm of the pressure difference versus the logarithm of time must look exactly like the logarithm of $P_D(r_D, t_D)$ versus the logarithm of t_D/r_D^2 curve. Thus, the basis for the type curve approach is as follows. If actual measurements are plotted as the logarithm of the change in pressure subsequent to the change in rate versus the logarithm of time since the rate changed, then a plot of the measurements should be similar to a log-log plot of $P_D(r_D, t_D)$ versus t_D/r_D^2. If a proper match or alignment is obtained, then kh/μ_g can be calculated from the vertical displacement of the horizontal axes and $kh/\phi c_t \mu_g$ can be calculated from the horizontal displacement of the vertical axes. Once the two plots are aligned, a point commonly referred to as the match point is chosen and the coordinates of the common point are noted. We then obtain the formation flow capacity, *kh*, by substituting the coordinates of p_D and $\psi(\Delta p)$ of the match point in the definition of dimensionless pressure, and the product, $\phi c_t h$, by substituting the coordinate of t and t_D/r_D^2 of the match point in the definition of t_D/r_D^2. It is preferable to take this approach because the effective system compressibility

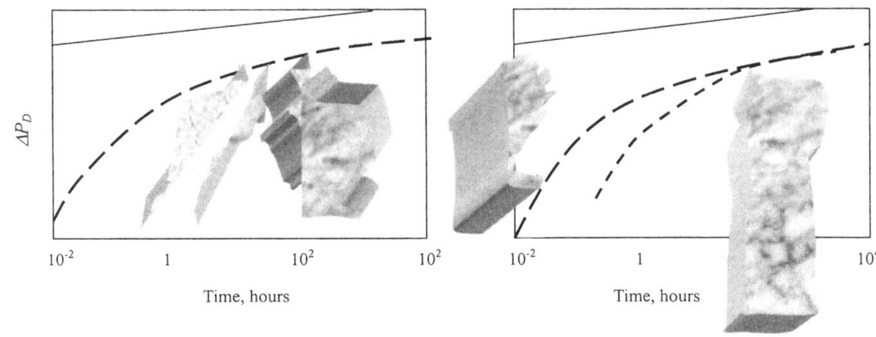

Figure 8–2. The type curve matching procedure.[3]

may not be known precisely. If the product $\phi c_t h$ is known, then it is possible to determine the formation flow capacity from the time match point.

8.3 Mechanics of Type Curve Matching

The basic steps involved in the type curve matching procedure may be explained as follows:

1. Plot the actual pressure changes versus time in any convenient units on log-log tracing paper. Use the same scale as the type curve.
2. Place the points plotted on the tracing paper over the type curve. Keeping the two coordinate axes parallel, shift the field curve to a position on the type curve that presents the best fit of the measurements.
3. To evaluate reservoir constants, select a match point anywhere on the overlapping portion of the curves, and record the coordinates of the convenient point on both sheets of paper. Once the match is obtained, use the coordinates of the match point to compute reservoir parameter (kh and $\phi c_t h$).

Figure 8–2 is a presentation of this procedure as presented by Earlougher.[3] In the petroleum engineering literature Elkins[4] appears to have been the first to analyze interference tests by type curve matching.

8.4 Type Curves for Constant Production Rate, Infinite-Acting Reservoirs

Figure 8–1 represents the line source solution at any radius for constant-rate production from an infinite-acting reservoir. It is very useful for analyzing interference effects. Type curve matching can be performed as follows:

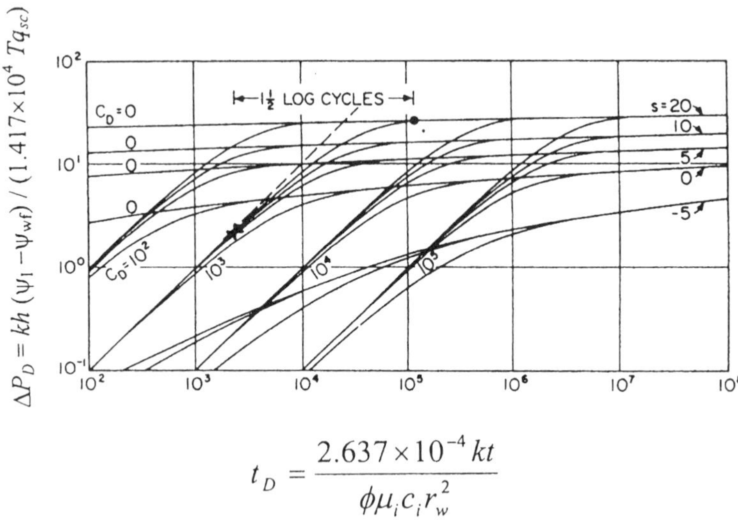

Figure 8–3. Dimensionless pressures for a single well in an infinite system; wellbore storage and skin included (copyright © 1970, SPE, from *Trans. AIME,* 1970).[7]

1. Plot $(\psi_i - \psi_{wf})$ drawdown tests or $(\psi_{ws} - \psi_{wf})$ buildup tests versus time on tracing paper that has the same scale as that of the type curve of Figure 8–3.
2. Align the $\Delta\psi$ axis of the field curve with the Δp_D axis of the type curve and match the field response with the type curve by horizontal movement of the field plots. Once the match is obtained, choose a match point and record the values of t and t_D/r_D^2.

$$(t)_M, \left(\frac{t_D}{r_D^2}\right)_M$$

3. Using the match determined in step 2, align the time axis (horizontal axes) of the field curve and the type curve, and then match the pressure response ($\Delta\psi$ versus t) with the conventional log-log type curve by vertical movement of the field curve. Once this match is obtained, then kh and ϕch can be obtained as follows. From the pressure match point:

$$kh = \left(\frac{50{,}300\, q_{sc} P_{sc}}{T_{sc}}\right)\left[\frac{\psi(\Delta p_D)}{\psi_i - \psi_{mf}}\right]_{match\ point}, mD - ft$$

$$k = \frac{kh}{h}, mD - ft$$

From the time match point, calculate

$$\phi c = \frac{0.0002637k}{\mu_i r^2}\left[\frac{t}{t_D/r_D^2}\right]_{match\ point}$$

Figure 8–3 can be used to depict the effects of skin and wellbore storage on constant-rate drawdown. Since surface valves control most gas well tests, wellbore storage may be of significance in tests of short duration. Figure 8–3 is a log-log plot of ψ_D versus t_D with parameter s and C_{SD}:

$$\psi_D = \psi(\Delta p_D) = \frac{khT_{sc}}{50{,}300\,Tq_{sc}P_{sc}}\Delta\psi \tag{8–3}$$

Tables 8–1 through 8–5 present values of dimensionless pressure Δp_D versus t_D with dimensionless storage constant C_D as a parameters and skin effects of zero, +5, +10, +20, and −5, respectively. Portions of the solutions are shown in Figure 8–3. Figure 8–3 is a log-log plot such as would be required if

Table 8–1
Value of $\Delta p_D(s, C_D, t_D)$ versus t_D Including Wellbore Storage and Skin Effects[3] ($S = 0$)

Dimensionless time	Dimensionless storage constant			
t_D	C_D 10^2	C_D 10^3	C_D 10^4	C_D 10^5
100	0.7975	0.09763	0.00998	0.00100
200	1.3724	0.1919	0.01992	0.00200
500	2.4357	0.4585	0.04956	0.00500
1,000	3.2681	0.8585	0.0984	0.00999
2,000	3.9274	1.5298	0.1944	0.01995
5,000	4.5585	2.8832	0.4697	0.0497
10,000	4.9567	4.0328	0.8925	0.0989
20,000	5.3288	4.9350	1.6275	0.1958
50,000	5.8027	5.6762	3.2109	0.4765
100,000	6.1548	6.0940	4.6773	0.9141
200,000	6.5043	6.4736	5.8871	1.6931
500,000	6.9643	6.9515	6.7895	3.4571
1,000,000	7.3116	7.3049	7.2309	5.2164
2,000,000	7.6585	7.6550	7.6185	6.7731
5,000,000	8.1168	8.1154	8.1004	7.8983
10,000,000	8.4635	8.4627	8.4550	8.3701
20,000,000	8.8101	8.8097	8.8057	8.7663
50,000,000	9.2683	9.2681	9.2664	9.2523
100,000,000	9.6149	9.6148	9.6139	9.6082

Table 8–2
Values of $\Delta p_D(s, C_D, t_D)$ versus t_D Including Wellbore Storage and Skin Effects[7] (S = +5)

Dimensionless time t_D	Dimensionless storage constant C_D 10^2	C_D 10^3	C_D 10^4	C_D 10^5
100	0.9319	0.09929	0.009993	0.00100
200	1.7512	0.1973	0.01997	0.00200
500	3.6982	0.4843	0.04984	0.00500
1,000	5.7984	0.9410	0.0994	0.00999
2,000	7.8403	1.7820	0.1977	0.01998
5,000	9.3823	3.8349	0.4863	0.0499
10,000	9.8913	6.1533	0.9480	0.0995
20,000	10.300	8.5524	1.8062	0.1979
50,000	10.792	10.436	3.9463	0.4878
100,000	11.150	11.025	6.4558	0.9536
200,000	11.693	11.445	9.1982	1.8256
500,000	12.311	11.941	11.488	4.0388
1,000,000	12.311	12.300	12.156	6.7163
2,000,000	12.658	12.652	12.859	9.7845
5,000,000	13.117	13.114	13.090	12.517
10,000,000	13.463	13.462	13.450	13.286
20,000,000	13.810	13.809	13.803	13.734
50,000,000	14.268	14.268	14.265	14.239
100,000,000	14.613	14.615	14.613	14.601

"type-curve" matching of field performance were to be used.

$$t_D = \frac{0.0002637\,kt}{\phi \mu_i c_{gi} r_w^2} \tag{8–4}$$

where

$$\Delta\psi = \psi_i - \psi_{mf}, \quad \text{(drawdown test)}$$
$$\Delta\psi = \psi_{ws} - \psi_{mf}, \quad \text{(buildup test)}$$

$$C_{SD} = \frac{0.159\,C_S}{\phi\, h\, c r_w^2} \tag{8–5}$$

where C_S is the wellbore storage constant and can be determined from the following relation:

$$C_S = \frac{q_{sc} \times 10^6 \beta_{gi}}{24} \times \left[\frac{t}{\Delta\psi}\right]_{unit\ slope\ line} \quad \text{(from type curve match)} \tag{8–6}$$

Table 8–3
Values of $\Delta p_D(s, C_D, t_D)$ versus t_D Including Wellbore Storage and Skin Effects[7] ($S = +10$)

Dimensionless time t_D	Dimensionless storage constant C_D 10^2	C_D 10^3	C_D 10^4	C_D 10^5
100	0.9594	0.09958	0.01000	0.001000
200	1.8463	0.1984	0.01998	0.002000
500	4.1401	0.4904	0.04990	0.005000
1,000	7.0124	0.9629	0.0996	0.0100
2,000	10.487	1.8587	0.1985	0.0200
5,000	13.852	4.2027	0.4911	0.0499
10,000	14.797	7.2010	0.9658	0.0997
20,000	15.269	10.995	1.8693	0.1986
50,000	15.781	14.811	4.2568	0.4918
100,000	16.144	15.917	7.3677	0.9683
200,000	16.499	16.413	11.382	1.8785
500,000	16.962	16.930	15.737	4.3043
1,000,000	17.311	17.295	17.031	7.5162
2,000,000	17.658	17.650	17.556	11.773
5,000,000	18.117	18.113	18.079	16.631
10,000,000	18.463	18.462	18.445	18.138
20,000,000	18.810	18.809	18.801	18.699
50,000,000	19.268	19.268	19.264	19.227
100,000,000	19.615	19.165	19.613	19.595

and

$$\beta_{gi} = \frac{P_{sc} T z}{T_{sc} p}, \text{ ft}^3/\text{scf}$$

and for a wellbore filled with a single-phase gas:

$$C_S = C_{wb} V_{wb} (\text{from field data}) \quad (8\text{–}7)$$

where C_{wb} is compressibility of the wellbore fluid at mean wellbore pressure and temperature, psi^{-1}; V_{wb} is volume of wellbore tubing, ft^3, which is equal to $\pi r_w^2 \times$(well depth).

The type curves of Figure 8–3 involve the following steps:

1. Plot $\Delta\psi$ versus time on log-log paper the same size as the type curve; if the test has a uniform slope region (45° line at earliest times), choose any point $[t, \Delta\psi]$ on the unit slope line and calculate the wellbore storage

Table 8–4
Values of $\Delta p_D(s, C_D, t_D)$ versus t_D Including Wellbore Storage and Skin Effects[7] ($S = +20$)

Dimensionless time	Dimensionless storage Constant			
t_D	C_D 10^2	C_D 10^3	C_D 10^4	C_D 10^5
100	0.9776	0.09977	0.01000	0.00100
200	1.9130	0.1991	0.02000	0.00200
500	4.4896	0.4946	0.0499	0.00500
1,000	8.1212	0.9787	0.0998	0.0100
2,000	13.478	1.9172	0.1992	0.0200
5,000	21.101	4.5125	0.4948	0.0500
10,000	24.241	8.1986	0.9797	0.0998
20,000	25.186	13.709	1.9209	0.1993
50,000	25.758	21.786	4.5333	0.4953
100,000	26.134	25.271	8.2698	0.9810
200,000	26.494	26.324	13.925	1.9252
500,000	26.960	26.907	22.443	4.5545
1,000,000	27.310	27.284	26.268	8.3394
2,000,000	27.657	27.645	27.460	14.133
5,000,000	28.116	28.112	28.055	23.085
10,000,000	28.463	28.461	28.434	27.297
20,000,000	28.810	28.809	28.795	28.606
50,000,000	29.268	29.268	29.262	29.216
100,000,000	29.615	29.615	29.612	29.596

constant C_S:

$$C_S = \frac{q_{sc} \times 10^6 \beta_{gi}}{24} \left[\frac{t}{\Delta\psi} \right]_{MP} \tag{8–8}$$

2. Calculate the dimensionless wellbore constant:

$$C_{SD} = \frac{0.159 C_S}{\phi h c_g r_w^2} \tag{8–5}$$

3. Calculate the time when wellbore storage effects become negligible:

$$t_{WS} = 36{,}177 \frac{\mu C_S}{kh}, \text{ hr}$$

4. If a unit-slope line is not present, C_S and C_{SD} must be calculated from wellbore properties.

Table 8–5
Values of $\Delta p_D(s, C_D, t_D)$ versus t_D Including Wellbore Storage and Skin Effects[7] ($S = -5$)

Dimensionless time	Dimensionless storage constant			
t_D	C_D 10^2	C_D 10^3	C_D 10^4	C_D 10^5
100	0.0697	0.0447	0.00896	0.00099
200	0.0992	0.0715	0.0172	0.00197
500	0.1557	0.1263	0.0394	0.00487
1,000	0.2164	0.1872	0.0718	0.00963
2,000	0.2977	0.2697	0.1267	0.01896
5,000	0.4446	0.4199	0.2518	0.0458
10,000	0.5913	0.5701	0.3990	0.0879
20,000	0.7722	0.7548	0.5972	0.1655
50,000	1.0646	1.0523	0.9313	0.3622
100,000	1.3232	1.3145	1.2254	0.6219
200,000	1.6086	1.6028	1.5422	0.9926
500,000	2.0170	2.0139	1.9806	1.6088
1,000,000	2.3420	2.3401	2.3201	2.0895
2,000,000	2.6757	2.6747	2.6630	2.5324
5,000,000	3.1248	3.1243	3.1197	3.0598
10,000,000	3.4677	3.4675	3.4644	3.4323
20,000,000	3.8124	3.8123	3.8107	3.7932
50,000,000	4.2693	4.2693	4.2685	4.2608
100,000,000	4.6154	4.6154	4.6150	4.6108

5. Using type curves with C_{SD} as calculated in steps 2 and 3, find the curve that best fits all the plotted data. Note the values of the match points $(\Delta p_D)_M$, $(\Delta\psi)_M$, (t_D), $(t \text{ or } \Delta t)_M$, and the value of s.

6. Calculate permeability k from the pressure match point:

$$k = \frac{50{,}300 T q_{sc} P_{sc}}{h T_{sc}} \left[\frac{\Delta p_D}{\Delta\psi}\right]_M \tag{8–9}$$

7. Calculate ϕc from the time match point:

$$\phi c = \frac{0.0002637k}{\mu r_w^2} \left[\frac{t \text{ or } \Delta t}{t_D}\right]_M \tag{8–10}$$

8. Compare the value of ϕc with values used to determine C_{SD} from C_S:

$$\phi c = \frac{0.159 C_S}{h r_w^2 C_{SD}} \tag{8–11}$$

Table 8–6
Constant-Rate Drawdown Test Data

Time t (hr)	Flowing pressure p_{wf} (psia)	Time t (hr)	Flowing pressure p_{wf} (psia)
0.05	3207	3.00	2995
0.10	3199	5.00	2965
0.15	3191	10.00	2925
0.30	3168	20.00	2915
0.50	3145	30.00	2907
0.80	3110	50.00	2901
1.00	3095	80.00	2896
1.50	3055	100.00	2893
2.00	3025	200.00	2886

Table 8–7
Drawdown Data Tabulated for Plotting

Time t (hr)	$\psi(p_i) - \psi(p_{wf})$ (mmpsia²/cP)	Time t (hr)	$\psi(p_i) - \psi(p_{wf})$ (mmpsia²/cP)
0.05	8.00	3.00	220.00
0.10	16.00	5.00	250.00
0.15	24.00	10.00	290.00
0.30	47.00	20.00	300.00
0.50	70.00	30.00	308.00
0.80	105.00	50.00	314.00
1.00	120.00	80.00	319.00
1.50	180.00	100.00	322.00
2.00	190.00	200.00	329.00

Example 8–1[15] *Analyzing Drawdown Test Using Ramey's Type Curves*
Determine the wellbore storage coefficients C_S, C_{SD}, s and formation permeability k from the data below and in Table 8–6, which were obtained in a pressure drawdown test on a gas well. Other reservoir and well data are as follows: $q_{sc} = 6.148$ mmscfd; $h = 41$ ft; $r_w = 0.4271$ ft; $\phi = 0.1004$; $\mu_i = 0.02441$ cP; $c_{ti} = 0.0002295$ psi^{-1}; $p_i = 3700$ psia $\leftrightarrow$ $\psi(p_i) = 861.12$ mmpsia^2/cP; $T_{SC} = 520°$R; $P_{SC} = 14.65$ psia; well depth $= 12{,}550$ ft; $C_{WS} = 0.000292$ psi^{-1} at $P_{WS} = 3420$ psi.

Solution First prepare the data for plotting (Table 8–7). The data are plotted in Figure 8–4. From the unit-slope line on which the data lie for $t \leq 5.05$ hr

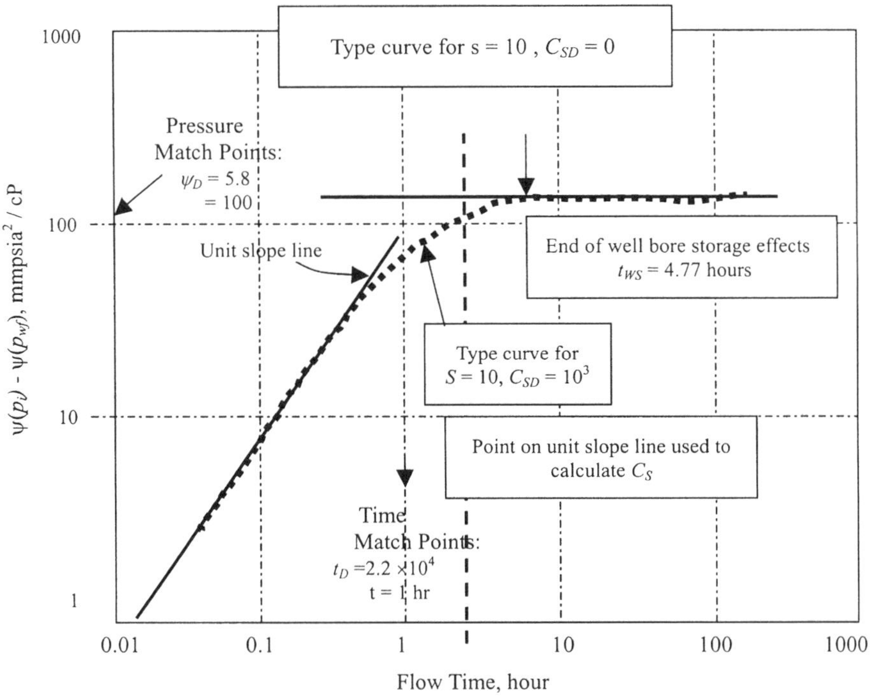

Figure 8–4. Drawdown test analysis with type curve—Example 8–1.

because field data are given, therefore calculate C_S from Eq. 8–7:

$$C_S = V_{WS}\, C_{WS} = \pi r_w^2 (\text{well depth}) \times C_{WS}$$

$$= \left[\frac{22}{7} \times 0.4271^2 \times 12{,}550 \times 0.000292\right] = 2.1038 \text{ ft}^3/\text{psi}$$

Then, from Eq. 8–5:

$$C_{SD} = \frac{0.159\, C_S}{\phi h c r_w^2} = \frac{0.159 \times 2.1038}{0.1004 \times 41 \times 0.0002295 \times 0.4271^2} = 1940$$

$$\approx 10^3$$

Calculate the time, t_{WS}, when wellborn storage effects become negligible from Eq. 2–141:

$$t_{WS} = 36{,}177 \frac{\mu C_S}{kh} = 36{,}177 \frac{0.0225 \times 2.1038}{8.75 \times 41} = 4.77\,\text{hr}$$

For $C_{SD} = 10^3$, the best fitting type curve is for $s = 10$. A time match point is $t = 1$ hr when t_D is 2.2×10^4; a pressure match point is $[\psi(p_i) - \psi(p_{wf})] =$ 100 mmpsia²/cP, when $\psi_D = 5.8$. From the match, we also note that wellbore storage distortion ends at $t = 4.77$ hr (i.e., the type curve for $C_{SD} = 0$). From the pressure match point, using Eq. 8–9:

$$k = \frac{50{,}300 T q_{sc} P_{sc}}{h T_{sc}} \left[\frac{\Delta p_D}{\Delta \psi}\right]_M$$

$$= \frac{50{,}300 \times 710 \times 6.148 \times 10^3 \times 14.65}{41 \times 520} \left[\frac{5.8}{100 \times 10^6}\right] = 8.75\,\text{mD}$$

From the time match point using Eq. 8–10, calculate ϕc_t as

$$\phi c_t = \frac{0.0002637k}{\mu r_w^2} \left[\frac{t \text{ or } \Delta t}{t_D}\right]_M$$

$$= \frac{0.0002637 \times 8.75}{0.02440 \times 0.4271^2} \left[\frac{1}{2.2 \times 10^4}\right] = 0.000024\ \text{psi}^{-1}$$

Compare those with values used to determine C_{SD} from C_S :

$$\phi c_t = (0.1004)(0.0002295) = 0.000023\ \text{psi}^{-1}$$

$$= 0.000023 \rightarrow \text{Values in}$$

$$= 0.000024 \rightarrow \text{Values out.}$$

Semilog Analysis of Constant-Rate Drawdown Test

From Figure 8–4, after a time of 4.77 hr, wellbore storage effects become negligible and the analytical solution for transient flow applies. Hence from Figure 8–5, we find the following:

$$\text{Slope } m = \frac{(320 - 299)}{(\log 100 - \log 10)} = 21.0\ \text{mmpsia}^2\text{/cP/cycle}$$

and

$$\Delta\psi(p_i)_{at\ 1hr} = 270\ \text{mmpsia}^2\text{/cP}$$

Using Eq. 5–40, formation permeability is

$$k = \frac{57.92 q_{sc} T P_{sc}}{m T_{sc} h} = \frac{57.92 \times 6.148 \times 710 \times 14.65}{21.0 \times 520 \times 41} = 8.27\ \text{mD}$$

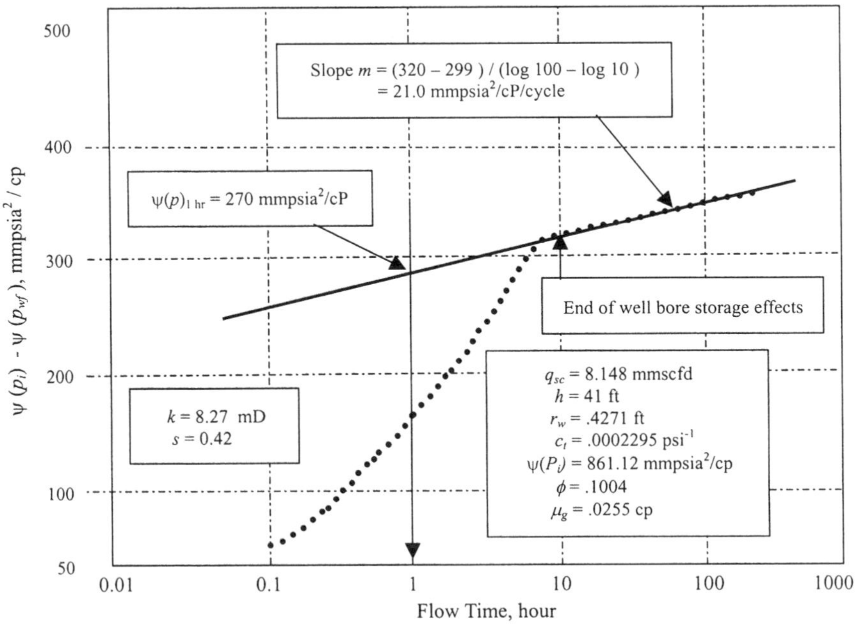

Figure 8–5. Semilog plot for single-rate drawdown test—Example 8–1.

and using Eq. 5–41, apparent skin factor is

$$s' = 1.151\left[\frac{\Delta\psi_i}{m} - \log\frac{k}{\phi\mu c r_w^2} + 3.23\right]$$

$$= 1.151\left[\frac{270 \times 10^6}{21.0 \times 10^6}\right.$$

$$\left. - \log\frac{8.27}{0.1004 \times 0.02441 \times 0.0002295 \times 0.4271^2} + 3.23\right]$$

$$= 1.151[12.86 - 7.91 + 3.23] = 9.42$$

Discussion

The use of a type curve match to identify the approximate time-of-start of the semilog straight line results in more reliable values of formation permeability k and apparent skin factor s'. As described in Chapter 5, type curves find similar applications to buildup testing in that the time for wellbore loading and other early-time effects can be identified. However, the application is approximate since a desuperposition of buildup data is involved. The separation of the skin and IT flow components of the apparent skin factor simply involves the analysis of another single-rate test by the methods described in this example with the only additional step being the simultaneous solution of Eqs. 5–46 and 5–47.

8.5 Storage and Skin Type Curve Matching Techniques

In unfractured gas wells, the early-time data are controlled by wellbore storage and skin effects.[2,6] Figure 4–40, developed by Earlougher and Kerch,[6] is particularly useful for analyzing wellbore storage controlled early-time data. The application of this type curve is to estimate kh and s, after it has been ascertained that an analysis of early-time data is desirable. The steps involved using this type curve are given in Chapter 4 (section 4.11). The values of kh and s, obtained by this type curve matching technique, are not exact and should be compared with values obtained from alternative sources to improve their reliability. In the absence of any other information, they serve merely as an indication of these parameters and should be treated as such. However, this curve-matching method will give a much more accurate value of kh, if C_S and s are known from different sources. The use of this type curve (Figure 4–39) to analyze wellbore storage controlled early-time data is illustrated in Example 4–19.

8.6 Fracture Type Curve Matching Techniques[5]

Infinite Conductivity Vertical Fractured Wells

Figure 4–40 represents the condition of constant-rate production for a vertically fractured well. It is a combination of the linear and radial flow equations. Its usefulness is readily seen since the majority of wells receive a hydraulic fracture upon completion. Such fractures are generally vertical and have an infinite conductivity. Figure 4–40 is a log-log plot of ψ_D versus t_D with various values of parameter x_e/x_f. Figure 4–40 (see $x_e/x_f = \infty$) reveals that the early portion of the type curve has a straight-line slope of one-half on a log-log graph, indicating linear flow. The deviation from the half-slope line represents elliptical flow or the transient from linear to radial flow response at the well. These type curves are applicable to drawdown, buildup, and fall-off tests in fractured wells and have been used for many years. Although the ordinate and abscissa of Figures 4–40 and 4–41 are given in terms of ψ, they may be modified quite easily for use with either the pressure or pressure square treatment methods by using appropriate definitions of Δp_D and t_D:

$$\psi_D = \frac{khT_{sc}}{50{,}300\, Tq_{sc}\, P_{sc}} \Delta\psi \tag{8–12}$$

and

$$t_D = \frac{0.0002637kt}{\phi \mu_i c_i x_f^2} \tag{8–13}$$

where

$\Delta\psi = \psi_i - \psi_{wf} \rightarrow$ drawdown test and $\Delta\psi = \psi_{ws} - \psi_{wf}, \rightarrow$ buildup test
x_f = fracture half-length, ft

After the desirability of an early-time data analysis has been established, the following procedure may be used to estimate the permeability thickness kh, the fracture half-length x_f, and the fracture skin factor s_f.

Steps in use of Figure 4–40 or Figure 4–41 as a type curve for well test analysis include the following:

1. Plot $(\psi_i - \psi_{wf})$, drawdown test, or $(\psi_{ws} - \psi_{wf})$, buildup test, on 3×5 cycle log-log paper.
2. Select the best match by sliding the actual test data plot both horizontally and vertically. The most likely curve is that for x_e/x_f = infinity, except where the fracture length and the duration of the test are unusually large.
3. Sketch the matched curve onto the data plot, and keep the data plot transfixed on the type curve (Figure 4–40 or Figure 4–41). Pick any convenient match point and read the following values:

   $(\Delta\psi)_M$ and $(t)_M \rightarrow$ from data plot
   $(\psi_D)_M$ and $(t_D)_M \rightarrow$ from Figure 4–40 or Figure 4–41

   where subscript M refers to a match point.
4. Estimate formation permeability from the pressure match point:

$$k = \frac{50.300 \times 10^6 T q_{sc} P_{sc}}{h T_{sc}} \times \frac{(\psi_D)_M}{(\psi_i - \psi_{wf})_M} \tag{8–14}$$

5. Estimate fracture half-length from the time match point:

$$x_f = \sqrt{\frac{0.0002637\, k\, (t)_M}{\phi \mu_i c_i\, (t_D)_M}} \tag{8–15}$$

6. Estimate the fracture skin factor from

$$s_f = -In\left(\frac{x_f}{2r_w}\right) \tag{8–15a}$$

Figure 4–41 is more applicable to a vertical fracture (natural fracture). Steps in the use of Figure 4–41 as a type curve for well test analysis are the same as for

Figure 4–40. Furthermore, if data are plotted as shown in Figure 4–40, instead of obtaining a single curve, a family of curves will be obtained that depends on fracture flow capacity, formation permeability, and fracture length. However, for infinite fracture flow capacity and a given formation permeability, a unique curve exists for each value of x_e/x_f.. Therefore, data from wells with finite flow-capacity fractures generally cannot be analyzed using the type curves developed for infinite flow-capacity fractures.

Numerical results, in the form of dimensionless pressure drop versus dimensionless time, were given for six different fracture penetration ratios, $x_f/x_e = 0.1, 0.2, 0.3, 0.5, 0.7$, and 1, respectively, and are shown in Table 8–8.

$$t_D > 0.7\ t_D + 1.4127\ t_D + 1.0560\ t_D + 0.8036\ t_D + 0.5371\ t_D \\ +0.3980\ t_D + 0.3331$$

Example 8–2[15] *Analyzing Pressure Drawdown Test Data for Vertical Fractured Gas Well Using Type Curve Matching Technique*

The reservoir and drawdown data of this gas well with a vertical fracture are presented in Table 8–9. Use type curve matching method to calculate permeability k and fracture length x_f. Compare your results with the values calculated using semilog analysis and the $\Delta\psi$ versus $\sqrt{t}$ plot.

Solution To establish the desirability of an early-time data analysis. Prepare the following plots:

1. A log-log plot of $\psi(p_i) - \psi(p_{wf})$ versus time was prepared on a sheet of tracing paper (data curve). The initial data formed a straight line with unit slope as shown in Figure 8–6. This gave an indication of linear flow and consequently the possible of a fracture. This linear flow period lasted less than 0.6 hr and there were no wellbore storage effects.
2. Prepare a Cartesian coordinate plot of $\Delta\psi$ versus t as shown in Figure 8–7.
3. Prepare a Cartesian plot of $\Delta\psi$ versus $\sqrt{t}$ as shown in Figure 8–8. This plot shows a straight line, which has slope $m_{vf} = 52.50$ mmpsia²/cP/hr$^{1/2}$, and at the end of the linear flow period, $(\Delta\psi)_{ef} = 105$ mmpsia²/cP. Substituting in the following equation, we get the beginning of the semilog straight line: $(\Delta\psi)_{bsl} = 2 \times (\Delta\psi)_{ef} = 2 \times 105 = 210$ mmpsia²/cP.
4. Prepare a plot (on tracing paper) for type curve matching. In this example, the data curve was placed over the type curve for vertical fractures (Figure 4–40) and was displaced until a match was obtained as shown in Figure 8–9. The match obtained was perfect over the entire 4-hr drawdown. In fact, most fracture below 3000 ft tends to be nearly vertical.

(text continued on page 440)

Table 8–8
Dimensionless Pressure Drop Function[5] ψ_D

t_D	$x_f/x_e = 0.1$	$x_f/x_e = 0.2$	$x_f/x_e = 0.3$	$x_f/x_e = 0.5$	$x_f/x_e = 0.7$	$x_f/x_e = 1$
0.000001	0.01128	0.005642	0.003761	0.002257	0.001612	0.001128
0.000002	0.01596	0.007979	0.005319	0.003192	0.002280	0.001596
0.000003	0.01954	0.009772	0.006515	0.003909	0.002792	0.001954
0.000004	0.02257	0.01128	0.007523	0.004514	0.003224	0.002257
0.000005	0.02523	0.01262	0.008410	0.005046	0.003604	0.002523
0.000006	0.02764	0.01382	0.009213	0.005528	0.003949	0.002764
0.000007	0.02985	0.01493	0.009951	0.005971	0.004265	0.002985
0.000008	0.03192	0.01596	0.01064	0.006383	0.004559	0.003192
0.000009	0.03385	0.01693	0.01128	0.006770	0.004836	0.003385
0.000010	0.03568	0.01784	0.01189	0.007137	0.005098	0.003568
0.00002	0.05046	0.02523	0.01682	0.01009	0.007209	0.005046
0.00003	0.06180	0.03090	0.02060	0.01236	0.008829	0.006180
0.00004	0.07136	0.03568	0.02379	0.01427	0.01019	0.007136
0.00005	0.07979	0.03989	0.02660	0.01596	0.01140	0.007979
0.00006	0.08740	0.04370	0.02913	0.01748	0.01249	0.008740
0.00007	0.09441	0.04720	0.03147	0.01888	0.01349	0.009441
0.00008	0.1009	0.05046	0.03364	0.02019	0.01442	0.01009
0.00009	0.1070	0.05352	0.03568	0.02141	0.01529	0.01070
0.00010	0.1128	0.05917	0.03761	0.02257	0.01612	0.01128
0.0002	0.1596	0.07979	0.05319	0.03192	0.02280	0.01596
0.0003	0.1954	0.09772	0.06515	0.03909	0.02792	0.01954
0.0004	0.2257	0.1128	0.07523	0.04514	0.03224	0.02257
0.0005	0.2523	0.1262	0.08410	0.05046	0.03604	0.02523
0.0006	0.2704	0.1382	0.09213	0.05528	0.03949	0.02764
0.0007	0.2875	0.1493	0.09951	0.05971	0.04265	0.02985
0.0008	0.3002	0.1576	0.1064	0.06383	0.04559	0.03192
0.0009	0.3145	0.1632	0.1128	0.06770	0.04836	0.03385
0.0010	0.3264	0.1714	0.1188	0.07129	0.05093	0.03568
0.0015	0.3768	0.2027	0.1391	0.08539	0.06161	0.04370
0.0020	0.4071	0.2276	0.1568	0.09722	0.07046	0.05046
0.0025	0.4360	0.2498	0.1724	0.10760	0.07817	0.05642
0.0030	0.4628	0.2698	0.1865	0.11690	0.08509	0.06180
0.0035	0.4874	0.2880	0.1994	0.12540	0.09142	0.06676
0.0040	0.5101	0.3047	0.2112	0.13320	0.09729	0.07136
0.0045	0.5312	0.3202	0.2223	0.14060	0.10280	0.07569
0.0050	0.5509	0.3347	0.2327	0.14750	0.10800	0.07979
0.0055	0.5693	0.3484	0.2425	0.15400	0.11290	0.08368

Table 8–8 (continued)

t_D	$x_f/x_e = 0.1$	$x_f/x_e = 0.2$	$x_f/x_e = 0.3$	$x_f/x_e = 0.5$	$x_f/x_e = 0.7$	$x_f/x_e = 1$
0.0065	0.6031	0.3735	0.2519	0.16620	0.12200	0.09097
0.0070	0.6186	0.3851	0.2607	0.17190	0.12630	0.09441
0.0075	0.6334	0.3962	0.2692	0.17740	0.13040	0.09772
0.0080	0.6475	0.4068	0.2852	0.18270	0.13440	0.10090
0.0085	0.6610	0.4170	0.2927	0.18780	0.13820	0.10400
0.0090	0.6738	0.4269	0.2999	0.19270	0.14190	0.10700
0.0095	0.6862	0.4363	0.3070	0.19750	0.14550	0.11000
0.010	0.6981	0.4454	0.3138	0.2021	0.1490	0.1128
0.011	0.7206	0.4628	0.3267	0.2110	0.1558	0.1183
0.012	0.7415	0.4791	0.3390	0.2194	0.1621	0.1236
0.013	0.7612	0.4944	0.3507	0.2274	0.1682	0.1287
0.014	0.7796	0.5090	0.3617	0.2351	0.1741	0.1335
0.015	0.7970	0.5228	0.3723	0.2424	0.1797	0.1382
0.016	0.8135	0.5360	0.3825	0.2495	0.1851	0.1427
0.017	0.8292	0.5486	0.3922	0.2563	0.1903	0.1471
0.018	0.8441	0.5607	0.4016	0.2629	0.1953	0.1514
0.019	0.8584	0.5723	0.4106	0.2693	0.002	0.1555
0.020	0.8720	0.5834	0.4193	0.2754	0.2050	0.1596
0.021	0.8851	0.5942	0.4278	0.2814	0.2096	0.1635
0.022	0.8976	0.6045	0.4359	0.2872	0.2140	0.1674
0.023	0.9097	0.6145	0.4438	0.2928	0.2184	0.1711
0.024	0.9213	0.6242	0.4515	0.2983	0.2227	0.1748
0.025	0.9325	0.6336	0.4590	0.3037	0.2268	0.1784
0.026	0.9433	0.6427	0.4663	0.3089	0.2309	0.1819
0.027	0.9538	0.6515	0.4733	0.3140	0.2348	0.1854
0.028	0.9639	0.6601	0.4802	0.3190	0.2387	0.1888
0.029	0.9737	0.6684	0.4869	0.3238	0.2425	0.1922
0.030	0.9833	0.6765	0.4935	0.3286	0.2462	0.1954
0.031	0.9925	0.6844	0.4999	0.3333	0.2499	0.1987
0.032	1.0015	0.6921	0.5062	0.3378	0.2535	0.2019
0.033	1.0102	0.6996	0.5123	0.3423	0.2570	0.2050
0.034	1.0187	0.7069	0.5183	0.3467	0.2604	0.2081
0.035	1.0270	0.7141	0.5242	0.3510	0.2638	0.2111
0.036	1.0351	0.7211	0.5299	0.3553	0.2672	0.2141
0.037	1.0429	0.7279	0.5356	0.3594	0.2704	0.2170
0.038	1.0506	0.7346	0.5411	0.3635	0.2737	0.2200
0.039	1.0581	0.7411	0.5465	0.3675	0.2769	0.2228
0.040	1.0654	0.7475	0.5518	0.3715	0.2800	0.2257

Table 8–8 (continued)

t_D	$x_f/$ $x_e = 0.1$	$x_f/$ $x_e = 0.2$	$x_f/x_e =$ 0.3	$x_f/x_e =$ 0.5	$x_f/x_e =$ 0.7	$x_f/x_e =$ 1
0.042	1.0796	0.7600	0.5622	0.3792	0.2861	0.2312
0.044	1.0931	0.7719	0.5722	0.3867	0.2921	0.2367
0.046	1.1061	0.7834	0.5818	0.3939	0.2979	0.2420
0.048	1.1186	0.7945	0.5912	0.4010	0.3036	0.2472
0.050	1.1306	0.8052	0.6002	0.4078	0.3091	0.2523
0.052	1.1422	0.8156	0.6090	0.4145	0.3145	0.2573
0.054	1.1534	0.8256	0.6176	0.4210	0.3198	0.2622
0.056	1.1642	0.8353	0.6259	0.3273	0.3249	0.2670
0.058	1.1746	0.8448	0.6339	0.4335	0.3300	0.2717
0.060	1.1847	0.8539	0.6418	0.4396	0.3349	0.2764
0.062	1.1945	0.8628	0.6495	0.4455	0.3398	0.2810
0.064	1.2040	0.8714	0.6569	0.4513	0.3446	0.2855
0.066	1.2133	0.8798	0.6642	0.4570	0.3493	0.2899
0.068	1.2222	0.8880	0.6713	0.4625	0.3539	0.2942
0.070	1.2309	0.8960	0.6783	0.4680	0.3584	0.2985
0.072	1.2394	0.9038	0.6851	0.4733	0.3629	0.3028
0.074	1.2477	0.9114	0.6917	0.4785	0.3672	0.3070
0.076	1.2557	0.9188	0.6982	0.4837	0.3716	0.3111
0.078	1.2636	0.9261	0.7046	0.4887	0.3758	0.3151
0.080	1.2713	0.9332	0.7109	0.4937	0.3800	0.3192
0.082	1.2788	0.9401	0.7170	0.4985	0.3841	0.3231
0.084	1.2861	0.9449	0.7230	0.5033	0.3882	0.3270
0.086	1.2932	0.9535	0.7289	0.5080	0.3922	0.3309
0.088	1.3002	0.9600	0.7346	0.5127	0.3962	0.3347
0.090	1.3071	0.9664	0.7403	0.5172	0.4001	0.3385
0.092	1.3138	0.9726	0.7459	0.5217	0.4040	0.3423
0.094	1.3203	0.9788	0.7513	0.5262	0.4078	0.3460
0.096	1.3268	0.9848	0.7567	0.5305	0.4116	0.3496
0.098	1.3330	0.9907	0.7620	0.5348	0.4153	0.3532
0.100	1.3392	0.9965	0.7672	0.5391	0.4190	0.3568
0.11	1.3685	1.0239	0.7919	0.5594	0.4396	0.3742
0.12	1.3952	1.0493	0.8149	0.5786	0.4539	0.3909
0.13	1.4200	1.0727	0.8364	0.5967	0.4702	0.4069
0.14	1.4430	1.0946	0.8565	0.6139	0.4858	0.4222
0.15	1.4644	1.1151	0.8755	0.6303	0.5009	0.4371
0.16	1.4846	1.1345	0.8935	0.6460	0.5154	0.4515
0.17	1.5036	1.1528	0.9106	0.6611	0.5295	0.4654
0.18	1.5217	1.1703	0.9270	0.6757	0.5432	0.4790

Table 8–8 (continued)

t_D	$x_f/x_e = 0.1$	$x_f/x_e = 0.2$	$x_f/x_e = 0.3$	$x_f/x_e = 0.5$	$x_f/x_e = 0.7$	$x_f/x_e = 1$
0.19	1.5397	1.1869	0.9427	0.6898	0.5565	0.4922
0.20	1.5553	1.2028	0.9578	0.7035	0.5696	0.5052
0.21	1.5711	1.2181	0.9723	0.7168	0.5823	0.5178
0.22	1.5862	1.2329	0.9864	0.7298	0.5947	0.5302
0.23	1.6008	1.2472	1.0001	0.7425	0.6070	0.5424
0.24	1.6150	1.2610	1.0134	0.7549	0.6190	0.5544
0.25	1.6287	1.2744	1.0264	0.7670	0.6308	0.5661
0.26	1.6420	1.2875	1.0391	0.7790	0.6424	0.5778
0.27	1.6551	1.3003	1.0515	0.7908	0.6539	0.5892
0.28	1.6678	1.3128	1.0636	0.8024	0.6653	0.6006
0.29	1.6802	1.3251	1.0756	0.8138	0.6765	0.6118
0.30	1.6924	1.3371	1.0874	0.8251	0.6876	0.6228
0.31	1.7044	1.3490	1.0990	0.8363	0.6986	0.6338
0.32	1.7162	1.3606	1.1104	0.8473	0.7095	0.6447
0.33	1.7278	1.3721	1.1217	0.8583	0.7203	0.6555
0.34	1.7393	1.3835	1.1329	0.8692	0.7311	0.6663
0.35	1.7506	1.3947	1.1439	0.8799	0.7418	0.6769
0.36	1.7618	1.4058	1.1549	0.8907	0.7524	0.6875
0.37	1.7728	1.4168	1.1657	0.9013	0.7629	0.6981
0.38	1.7838	1.4277	1.1765	0.9119	0.7734	0.7086
0.39	1.7947	1.4386	1.1872	0.9224	0.7839	0.7190
0.40	1.8055	1.4493	1.1979	0.9329	0.7943	0.7294
0.41	1.8162	1.4599	1.2084	0.9433	0.8046	0.7398
0.42	1.8269	1.4705	1.2190	0.9537	0.8150	0.7501
0.43	1.8374	1.4811	1.2294	0.9640	0.8253	0.7604
0.44	1.8480	1.4916	1.2399	0.9743	0.8356	0.7707
0.45	1.8585	1.5020	1.2503	0.9846	0.8458	0.7809
0.46	1.8689	1.5124	1.2606	0.9949	0.8560	0.7912
0.47	1.8793	1.5228	1.2709	1.0051	0.8662	0.8014
0.48	1.8896	1.5331	1.2812	1.0153	0.8764	0.8116
0.49	1.9000	1.5434	1.2915	1.0255	0.8866	0.8217
0.50	1.9103	1.5537	1.3017	1.0357	0.8968	0.8319
0.51	1.9206	1.5639	1.3119	1.0459	0.9069	0.8420
0.52	1.9308	1.5741	1.3221	1.0560	0.9170	0.8521
0.53	1.9410	1.5843	1.3323	1.0661	0.9271	0.8622
0.54	1.9512	1.5945	1.3424	1.0762	0.9372	0.8723

Table 8–8 (continued)

t_D	$x_f/x_e = 0.1$	$x_f/x_e = 0.2$	$x_f/x_e = 0.3$	$x_f/x_e = 0.5$	$x_f/x_e = 0.7$	$x_f/x_e = 1$
0.55	1.9614	1.6047	1.3526	1.0863	0.9473	0.8824
0.56	1.9715	1.6148	1.3627	1.0964	0.9574	0.8925
0.57	1.9817	1.6250	1.3728	1.1065	0.9675	0.9026
0.58	1.9918	1.6351	1.3829	1.1166	0.9776	0.9127
0.59	2.0019	1.6452	1.3930	1.1267	0.9876	0.9227
0.60	2.0120	1.6553	1.4031	1.1367	0.9977	0.9328
0.61	2.0221	1.6654	1.4132	1.1468	1.0077	0.9428
0.62	2.0322	1.6755	1.4233	1.1569	1.0178	0.9528
0.63	2.0423	1.6855	1.4333	1.1669	1.0278	0.9629
0.64	2.0524	1.6956	1.4434	1.1769	1.0379	0.9730
0.65	2.0625	1.7057	1.4535	1.1870	1.0475	0.9830
0.66	2.0726	1.7157	1.4635	1.1970	1.0579	0.9930
0.67	2.0827	1.7258	1.4736	1.2071	1.0680	1.0031
0.68	2.0927	1.7358	1.4836	1.2171	1.780	1.0131
0.69	2.1027	1.7459	1.4936	1.2271	1.0880	1.0231
0.70	2.1127	1.7560	1.5036	1.2371	1.0980	1.0331
$t_D > 0.7$	$t_D + 1.4127$	$t_D + 1.0560$	$t_D + 0.8036$	$t_D + 0.5371$	$t_D + 0.3980$	$t_D + 0.3331$

Table 8–9
Vertical Fracture Drawdown Test (T = 710°R; r_w = 0.39 ft; h = 22 ft; P_{sc} = 14.65; T_{sc} = 520°R; q_{sc} = 5.125 mmscfd; μ_i = .02134 cP; ϕ = 0.185; c_{tl} = 0.00027 psi^{-1}; well depth = 2950 ft)

Time t (hr)	Time $\sqrt{t}$ (hr)$^{1/2}$	$\psi(P_i)-\psi(P_{wf})$ (mmpsia^2/cP)	Time t (hr)	Time $\sqrt{t}$ (hr)$^{1/2}$	$\psi(P_i)-\psi(P_{wf})$ (mmpsia^2/cP)
0.00	0.0000	0	4.00	2.0000	104.87
0.15	0.3872	18.00	5.00	2.2361	114.21
0.20	0.4472	22.12	6.00	2.4495	121.29
0.30	0.5477	25.75	8.00	2.8284	134.39
0.40	0.6325	31.23	10.00	3.1623	144.52
0.50	0.7071	35.65	15.00	3.8730	163.66
0.60	0.7746	40.45	20.00	4.4721	177.72
0.80	0.8944	43.50	30.00	5.4772	197.35
1.00	1.0000	50.14	40.00	6.3246	211.32
1.50	1.2247	68.24	50.00	7.1711	222.45
2.00	1.4142	78.35	60.00	7.7460	232.21
3.00	1.7321	93.24	80.00	8.9443	245.07
			100.00	10.0000	256.07

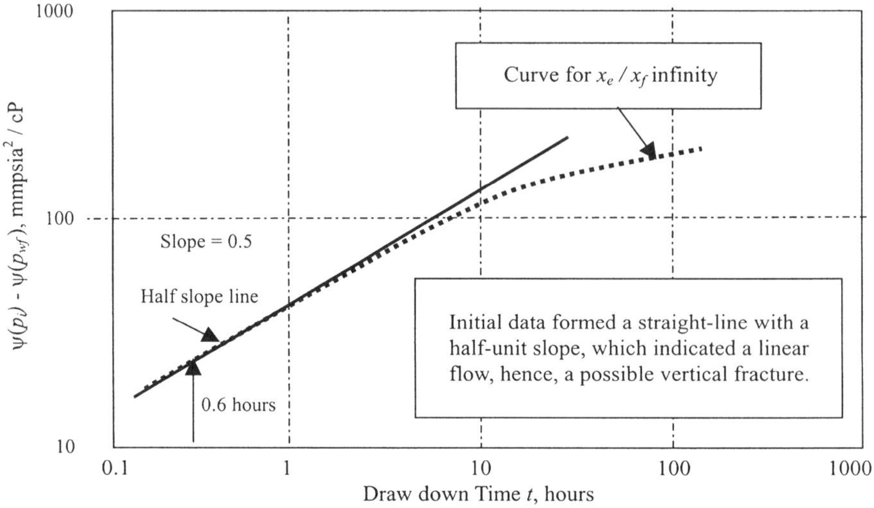

Figure 8–6. Log-log plot of data in Table 8–3—Example 8–2.

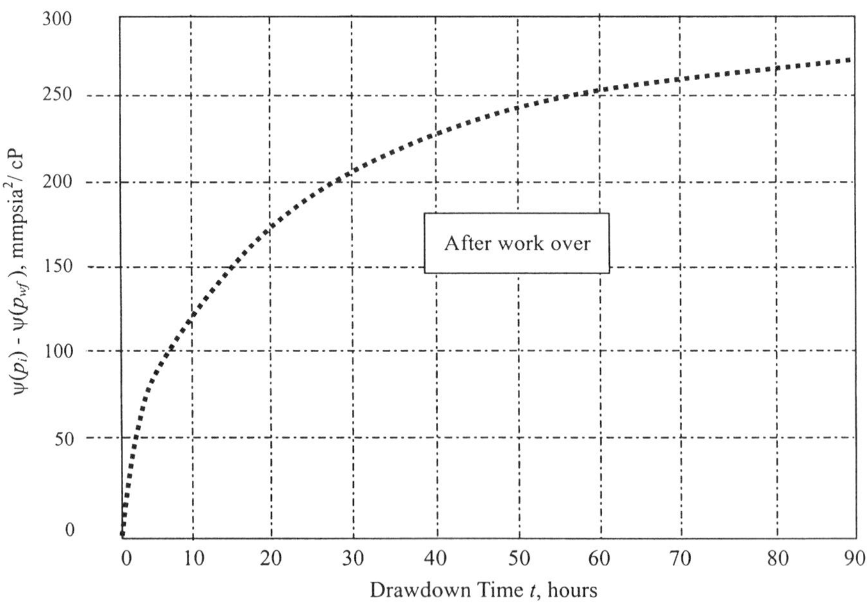

Figure 8–7. Cartesian coordinate plot—Drawdown test, fractured gas well—Example 8–2.

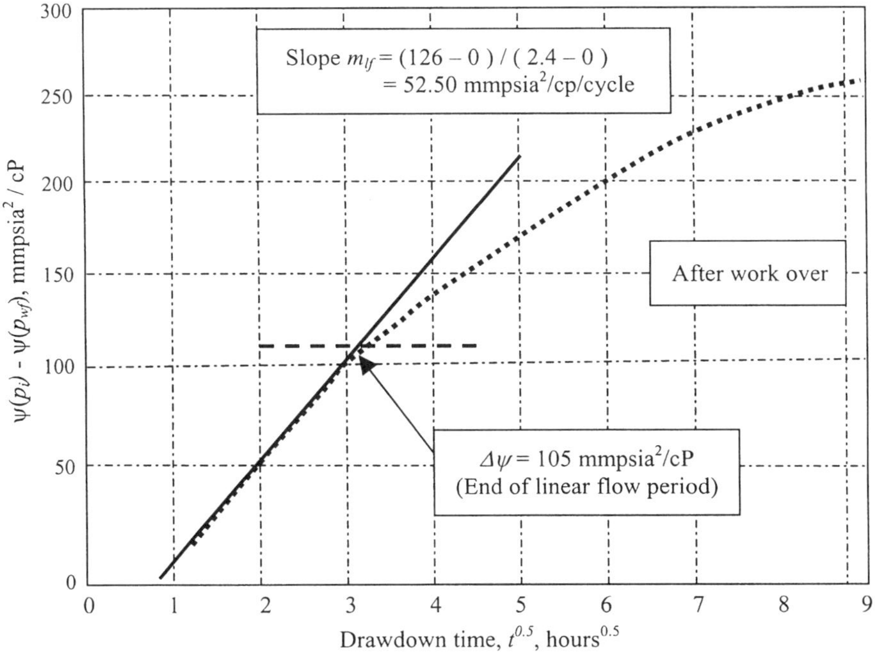

Figure 8–8. Plot of $\Delta\psi$ versus $t^{0.5}$ for data in Table 8–9—Example 8–2.

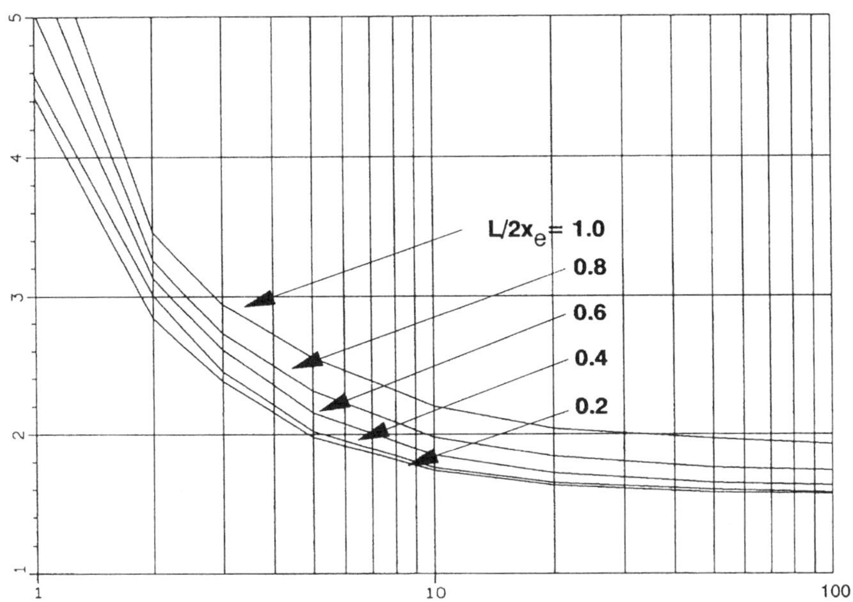

Figure 8–9. Type curve matching for vertical fractured well—Infinite fracture conductivity.

(text continued from page 432)

Estimates of effective reservoir permeability and fracture half-length can be obtained as follows:

1. Match points are obtained from Figure 8–9:

$$\Delta\psi_D = 1.55 \text{ at } \Delta\psi = 70 \text{ mmpsia}^2/\text{cP} \quad \text{and} \quad t_D = 200 \text{ at } t = 70 \text{ min}$$

2. Calculate effective reservoir permeability from Eq. 8–14 as follows:

$$k = \frac{50.300 \times 10^6 T q_{sc} P_{sc}}{h T_{sc}} \times \frac{(\psi_D)_M}{(\psi_i - \psi_{wf})_M}$$

$$= \frac{50.300 \times 10^6 \times 710 \times 5.125 \times 14.65}{22 \times 520} \cdot \frac{1.55}{200 \times 10^6} = 1.82 \text{ mD}$$

3. Calculate the fracture half-length (for distance that the vertical fracture extends from the center of the well) from Eq. 8–15 as

$$x_f = \sqrt{\frac{.0002637\, k}{\phi \mu_i c_i} \cdot \frac{(t)_M}{(t_D)_M}}$$

$$= \sqrt{\frac{0.0002637 \times 1.82}{0.185 \times 0.02134 \times 0.00027} \frac{70}{6.5}} = 69.63 \text{ ft}$$

Notice that if the match had gone as far as reaching any particular value of x_e/x_f, the distance to the outer reservoir boundary could have been calculated.

4. Calculate the fracture skin factor from Eq. 8–19 (given later) as follows:

$$s_f = -In\left[\frac{2r_w}{x_f}\right] = -In\left[\frac{2 \times .39}{69.63}\right] = -4.49$$

Apparent wellbore radius $r_{wa} = x_f/2 = 69.63/2 = 34.82$ ft

We can also calculate fracture skin factor by

$$s_f = -In\left(\frac{r_{wa}}{r_w}\right) = -In\left(\frac{34.82}{0.39}\right) = -4.49$$

Semilog Analysis

Now we are able to examine Figure 8–10, which is a semilog plot of $\Delta\psi$ versus t, and fit the semilog straight line as shown, beginning at $(\Delta\psi)_{bsl} \geq$

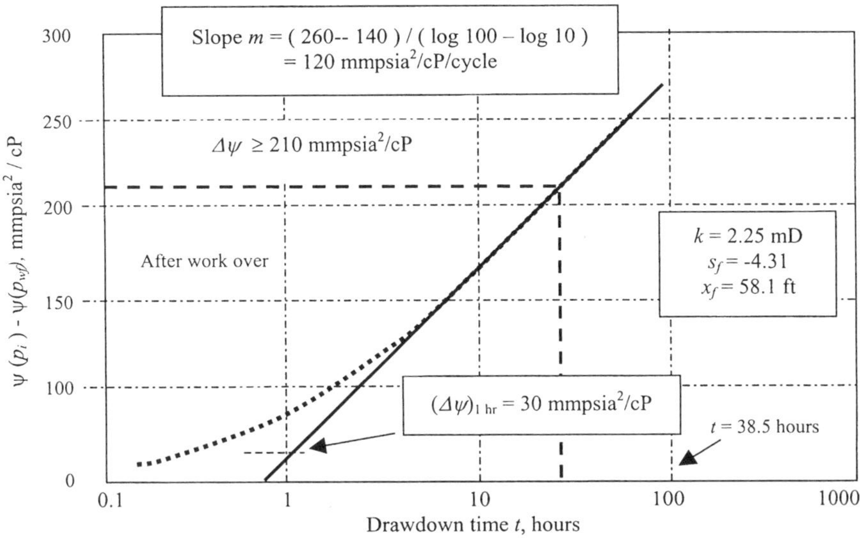

Figure 8–10. Semilog plot, drawdown test—Fractured gas well—Example 8–2.

210 mmpsia^2/cP. The slope of the semilog straight line is found to be 120 mmpsia^2/cP/cycle and $(\Delta\psi)_{1hr} = 20$ mmpsia^2/cP. Thus, from Eq. 5–40,

$$k = \frac{57.92 \times 10^6 \times 5.125 \times 14.65 \times 710}{120 \times 10^6 \times 22 \times 520} = 2.25 \text{ mD}$$

From Eq. 5–41, calculate fracture skin factor, s_f:

$$s_f = 1.151 \left[\frac{20 \times 10^6}{120 \times 10^6} - \log \frac{2.25}{0.185 \times 0.02134 \times 0.00027 \times 0.39^2} + 3.23 \right]$$

$$= 1.151\,[0.167 - 7.14 + 3.23] = -4.31$$

(indicating heavy fracture in a low-permeability gas reservoir)

Calculate fracture length x_f from

$$r'_w = r_w\, e^{-S_f}$$

$$= 0.39\, e^{-(-4.31)} = 29.03$$

Hence, $x_f = 2 \times r'_w = 2 \times 29.03 = 58.06$ ft.

A comparison between the log-log and semilog calculated values of parameters is shown in Table 8–10. Note, however, that the estimation of x_f is

Table 8–10
Comparison between Log-Log and Semilog Values of Several Parameters

Reservoir parameters	Log-log solution	Semilog solution
Effective permeability k (mD)	1.82	2.25
Half-fracture length x_f (ft)	69.63	58.06
Skin factor s	−4.49	−4.31

very sensitive to the reservoir parameters. For example, if ϕ is taken equal to 0.1004 instead of 0.185, x_f would be equal to 292 ft instead of 94.52 ft. Thus, one has to keep in mind that x_f is correct only if all the reservoir and fluid parameters are correct. Having calculated k and x_f, we substitute in Eq. 1–141 to calculate t, the time of the beginning of the semilog straight line, which is 38.5 hr. Therefore, the beginning of the semilog straight line must meet the following criteria:

$$\Delta\psi \geq 120 \text{ mmpsia}^2/\text{cP}$$
$$t = 38.5 \text{ hr}$$

Discussion

All data obtained from a few minutes to 100 hr have matched the appropriate type curve solution. The type curve solution presented in this example application is useful and appears to represent a large portion of field data. A combination of older semilog analytical methods with the log-log type curve permits a second-generation well test analysis with extraordinary confidence levels concerning the results. There are many instances, particularly in tight gas wells, in which the linear flow period lasts for several hundred hours. Under these conditions, neither the type curve nor the conventional approach is uniquely applicable. However, the last point on the half-slope line may be used to estimate an upper limit of the permeability–thickness product.

Uniform Flux and Infinite Conductivity Vertical Fracture Type Curves[9,10]

Alagoa *et al.*[10] presented the pressure derivative approach to analyze transient tests in both the uniform-flux and the infinite-conductivity vertical fracture models for a homogeneous infinite reservoir. The new type curves are shown in Figure 8–11. The two types of vertical fractures have been discussed in the section on buildup tests in vertically fractured reservoirs. At late time the derivative form of the infinite-acting radial-flow equation for both the uniform-flux and

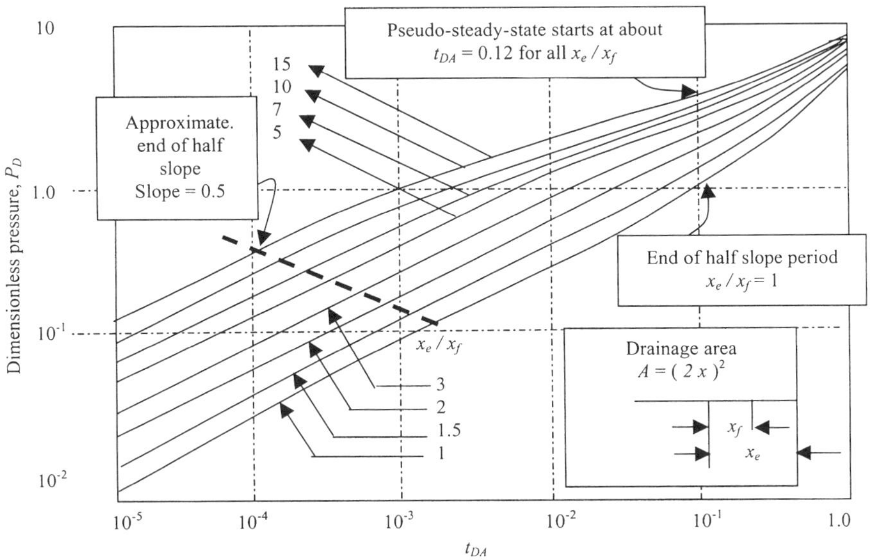

Figure 8–11. Dimensionless pressure for vertically fractured well in the center of a closed system. (Gringarten, Ramey).[16,17]

infinite-conductivity fractures is given by $t_{Df}p'_D = 0.5$. This is characterized by a 0.5 horizontal line asymptote. Within the half-unit slope straight line and the 0.5 line asymptote, the fracture model type considered defines pressure derivative response. The type curve analysis involves the following steps:

1. Plot $\psi(\Delta p)$ and derivative of pressure versus Δt on log-log graph paper with the same grid size as in Figure 8–11.
2. Match the constant derivative portion of the data plot with the 0.5 horizontal straight lines. This will fix the pressure match.
3. Move the data plot along the 0.5 horizontal line until early time data are matched on both derivative and pressure type curves. This gives the time match and the C_{Df} value.
4. Note the values of the match points

$$\left[\frac{p_D}{\Delta t}\right]_M, \left[\frac{t_{Df}}{\Delta t}\right]_M$$

5. Calculate kh from the pressure match, fracture half-length x_f from the time match, and wellbore storage constant C from the C_{Df} value. The following equations are used:

$$k = \frac{50{,}300\, q_g T P_{sc}}{h T_{sc}} \left[\frac{\psi(p_D)}{\Delta t}\right]_M \tag{8–16}$$

$$x_f = \sqrt{0.0002637 \frac{k}{\phi \mu_g c_t} \left[\frac{t_{Df}}{\Delta t} \right]_M} \tag{8–17}$$

$$C = \frac{\phi c_t x_f^2 C_{Df}}{0.8936} \tag{8–18}$$

No wellbore storage, infinite conductivity fracture (after Gringarten, Ramey, and Raghavan):[9,10]

$$s = In \left(\frac{2r_w}{x_f} \right) \tag{8–19}$$

Comparing Figures 8–3 and 8–11 shows a striking difference in the slopes at early dimensionless times. The curves for a well with skin and storage effect (Figure 8–3) start with a unit slope, while the curves for the vertical fracture (Figure 8–11) start with a slope of one-half. The one-half slope is caused by linear flow at very short times. A dashed line in Figure 8–11 indicates the end of the half-slope period.

8.7 Summary

Based on the material presented in this chapter, the following remarks are pertinent:

- A new technique is presented to analyze data in the bilinear flow period. It is shown that, during this flow period, a graph of $\psi(p_{wf})$ versus $t^{1/4}$ yields a straight line when slope is inversely proportional to $h_f(k_f b_f)^{1/2}$.
- New type curves are now available for pressure analysis of fractured gas wells, and the problem in the analysis is reduced considerably with the use of these type curves.
- Prefracture information about the reservoir is necessary to estimate fracture parameters.
- The type curve analysis method must be used simultaneously with the specific analysis methods $\psi(p_{wf})$ versus $t^{1/4}$, $\psi(p_{wf})$ versus $t^{1/2}$, and $\psi(p_{wf})$ versus log t to produce reliable results.

References and Additional Reading

1. Ramey, H. J., Jr., "Application of the Line Source Solution to Flow in Porous Media—A Review," *Prod. Monthly* (May 1967) 4–7, 25–27.
2. Ramey, H. J., Jr., "Short-Time Well Test Data Interpretation in the Presence of Skin Effect and Wellbore Storage," *J. Petroleum Technol.* (1970) 22, 97–104.

3. Earlougher, R. C., Jr., "Advances in Well Test Analysis," SPE Monograph, Volume 5, Society of Petroleum Engineers, Dallas, TX, 1977.
4. Elkins, L. F., "Reservoir Performance and Well Spacing—Silica Arbuckle Pool," *Drilling and Production Practice, API* (1946) 109–127.
5. Raghavan, R., and Hadingto, N., "Analysis of Pressure Data for Fractured Wells: The Constant-Pressure Outer Boundary," SPE-AIME 51th Annual Fall Technical Conference and Exhibition, New Orleans, Oct. 3–6, 1976.
6. Earlougher, R. C., Jr., and Kerch, K. M., "Analysis of Short-Time Transient Test Data by Type-Curve Matching," *J. Petroleum Technol.* (1974) 26, 793–800.
7. Agarwal, R. G., Al-Hussainy, R., and Ramey, H. J., Jr., "An Investigation of Wellbore Storage and Skin Effect in Unsteady Liquid Flow: I. Analytical Treatment," *Soc. Petroleum Eng. J.* (1970) 10, 279–290.
8. Fetkovich, M. J., "Decline Curve Analysis Using Type Curves," Paper SPE 4629, 48th Fall Meeting of AIME, Las Vegas, 1973.
9. Raghavan, R., Cady, G. V., and Ramey, H. J., Jr., "Well Test Analysis for Vertically Fractured Wells," *J. Petroleum Technol.* (1972) 24, 1014–1020.
10. Alagoa, A., Bourdet, D., and Ayoub, J. A., "How to Simplify the Analysis of Fractured Well Tests," *World Oil* (Oct. 1985).
11. Russell, D. G., and Truitt, N. E., "Transient Pressure Behavior in Vertically Fractured Reservoirs," *Soc. Petroleum Eng.* (Aug. 1964).
12. Aguilera, R., "Well Test Analysis of Naturally Fractured Reservoirs," *APEEJ* (Sept. 1987) 239–252.
13. Wattenbarger, R. A., and Ramey, H. J., Jr., "Well Test Interpretations of Vertically Fractured Gas Wells," *J. Petroleum Technol.* (May 1969) 625–632; *Trans. AIME* 246.
14. *Pressure Transient Testing Methods*, SPE Reprint Series No. 1. Society of Petroleum Engineers, Dallas, TX 1980.
15. Amanat, U. C., *Pressure Transient Test Analysis User's Handbook*, Vol. 8. Advanced TWPSOM Systems Inc., Houston, TX, Oct. 1995.
16. Gringarten, A. C., Ramey, H. J., Jr., and Raghavan, R., "Pressure Analysis for Fractured Wells," paper SPE 4051 presented at the SPE-AIME 47th Annual Fall Meeting, San Antonio, Tex; Oct. 8–11, 1972.
17. Gringarten, A. C., Ramey, H. J., Jr., and Raghavan, R., "Unsteady-State Pressure Distribution Created by a Well With a Single Infinite-Conductivity Vertical Fracture," Soc. Petroleum Engineering Journal (Aug. 1974) 347–360; Trans; AIME, 257.

Chapter 9

Pressure Derivative Method of Analysis

9.1 Introduction

The pressure derivative application to gas well test analysis involves the combined use of existing type curves in both the conventional dimensionless pressure form (P_D) and the new dimensionless pressure derivative grouping ($P'_D \times t_D/C_D$). Thus this new approach has combined the most powerful aspects of the two previously distinct methods into a single-stage interpretive plot. Use of the pressure derivative with pressure-behavior type curves reduces the uniqueness problem in type curve matching and gives greater confidence in the results. Features that are hardly visible on the Horner plot or are hard to distinguish because of similarities between are reservoir system and another are easier to recognize on the pressure-derivative plot.

9.2 Calculation of Pressure Derivative Functions

Figure 9–1 illustrates the calculation of the derivative at a given point A given by the following relationship:

$$\left\langle \frac{dy}{dx} \right\rangle_A = \frac{\left(\frac{Y_1}{X_1} X_2 + \frac{Y_2}{X_2} X_1 \right)}{(X_1 + X_2)} \tag{9–1}$$

9.3 Log-Log Diagonostic Plots of Pressure Change and Its Derivative

Derivative response patterns for the flow regimes are shown in Figures 9–2 through 9–6. For each flow there is a specified plot of the portion of the data exhibiting the characteristic derivative response pattern. On the specialized

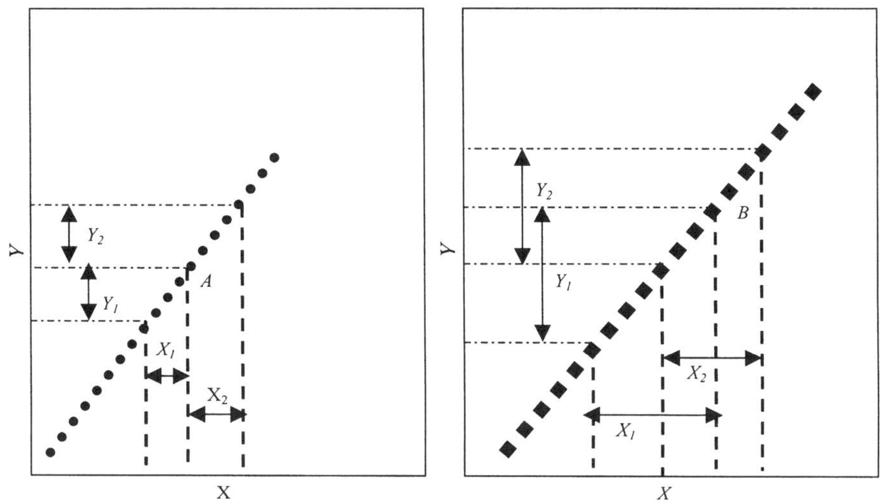

Figure 9–1. Illustration of differentiation algorithm.

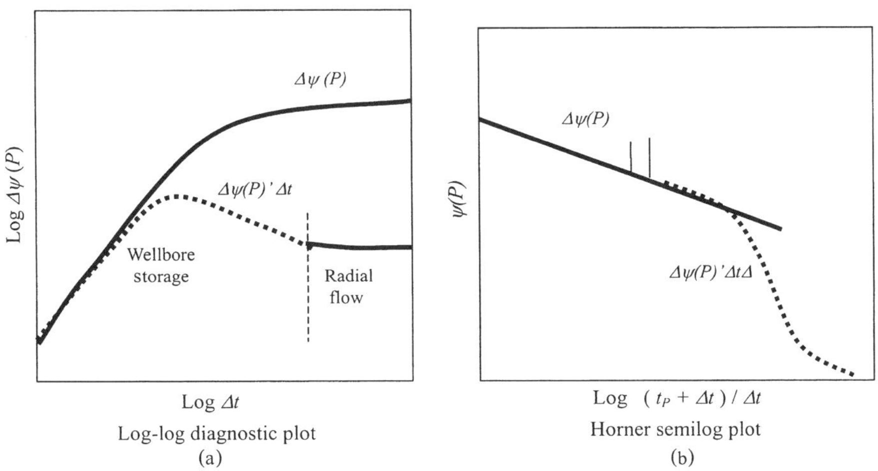

Figure 9–2. Well with wellbore storage and skin in a homogeneous reservoir.

plot, the data identified with the characteristic derivative response pattern lie on a straight line, and the slope and the intercepts of the line are used to compute well and/or reservoir parameters. Figures 9–2 through 9–6 show the specialized plot associated with respective trends identified on the log-log diagnostic plot of pressure change and its derivative.

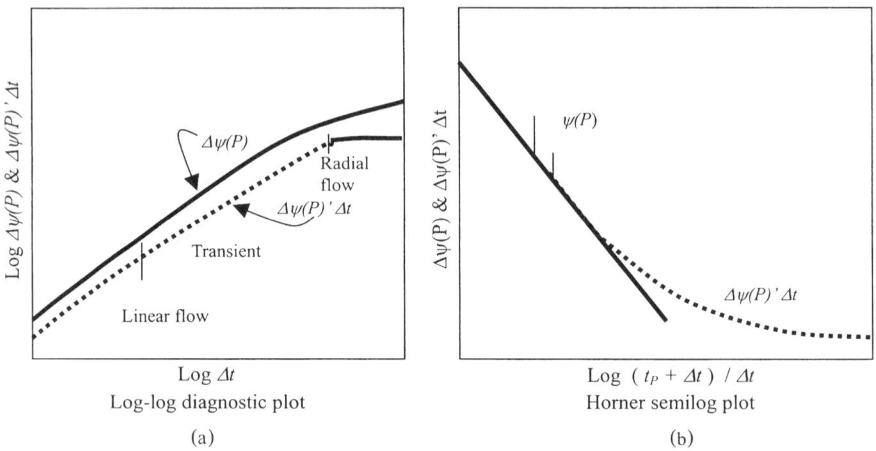

Figure 9–3. Well with infinite conductivity vertical fracture in a homogeneous reservoir.

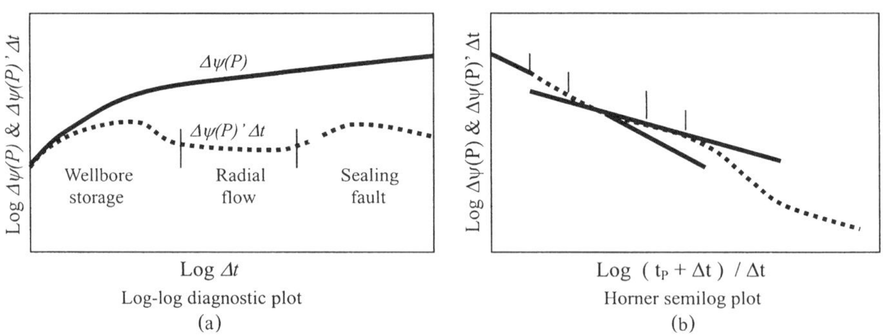

Figure 9–4. Well with wellbore storage and skin in a homogeneous reservoir with one sealing fault.

9.4 Pressure Derivative Trends for Other Common Flow Regimes

Figures 9–7 and 9–7a show pressure derivative trends for common flow regimes.

9.5 Homogenous Reservoir Systems

Bourdet *et al.*[17] introduced type curves in terms of the pressure derivative. Their type curves are largely based on the solutions obtained by Agarwal *et al.*[7]

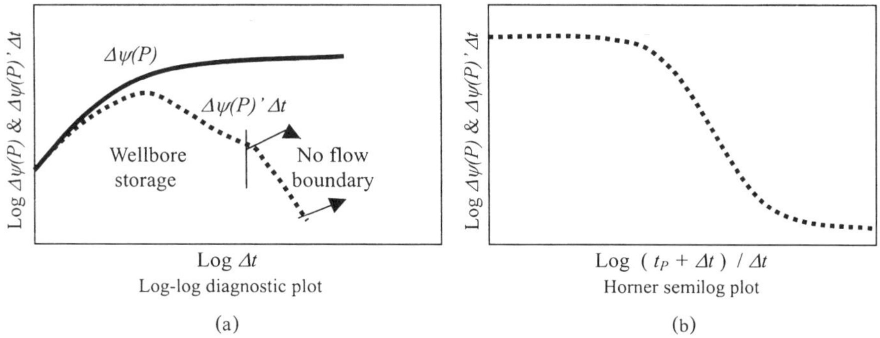

Figure 9–5. Well with wellbore storage and skin in a homogeneous reservoir with closed outer boundary.

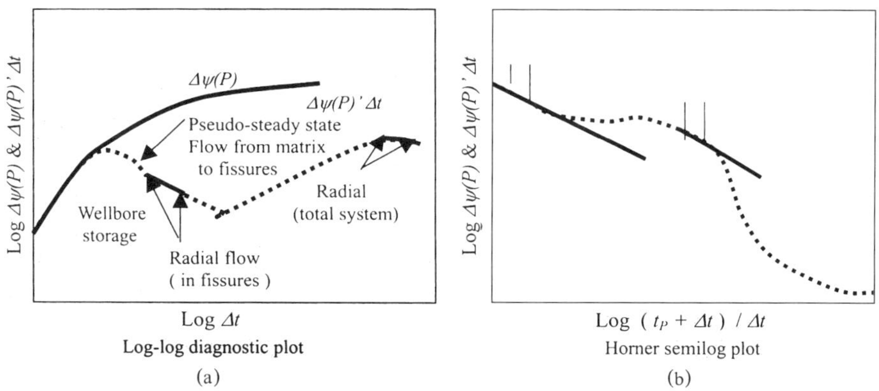

Figure 9–6. Well with wellbore storage and skin in a dual porosity system with pseudo-steady-state flow from matrix to fractures.

During early time, Agarwal *et al.* show that

$$p_{WD} = \frac{t_D}{C_D} \tag{9–2}$$

Thus,

$$dp_{WD}/d(t_D/C_D) = 1$$

or

$$p'_{WD} = 1 \tag{9–3}$$

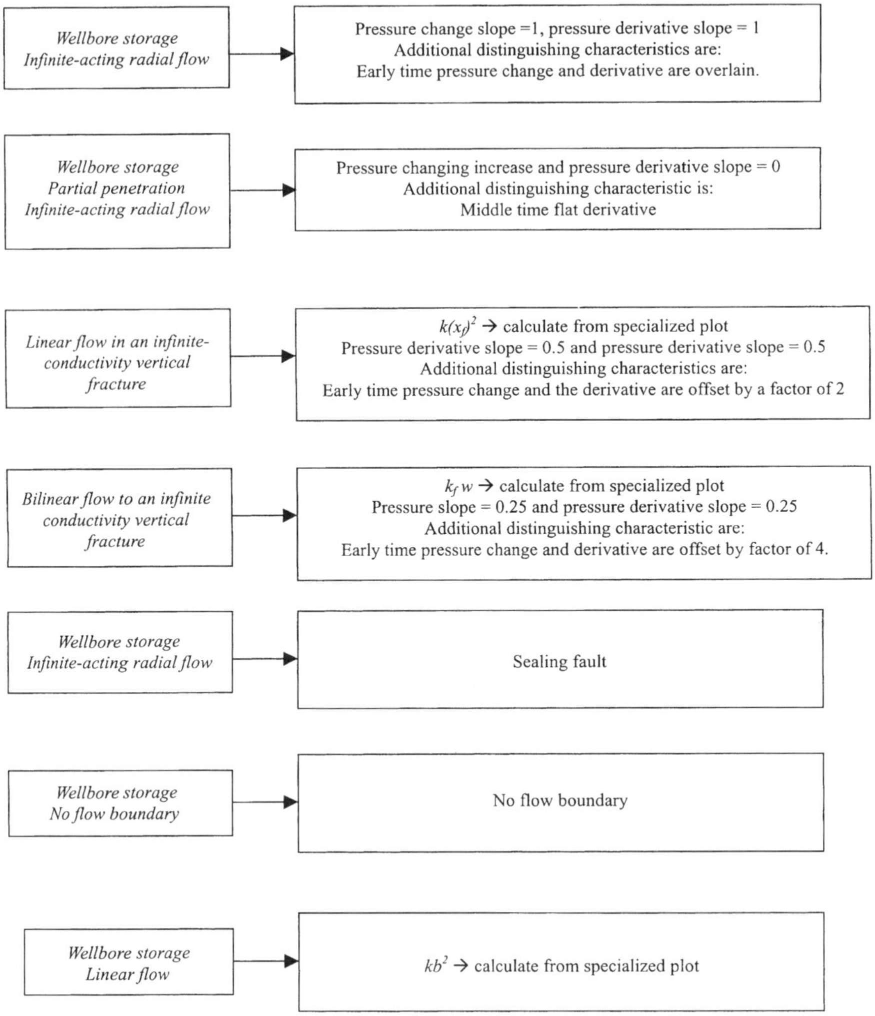

Figure 9–7. Pressure derivative trends for common flow regimes.

Equation 9–3 shows that during early time a plot of p'_{WD} versus (t_D/C_D) is a horizontal line and the intercept is equal to 1. In a reservoir that is homogeneous with respect to both rock and fluid properties and where the well is fully penetrating the pay section, the flow must become radial after wellbore storage effects have subsided. Therefore, during late time, Eq. 9–3 is applicable,

Wellbore storage dual porosity matrix to fissure flow	Semilog straight lines with slope = 1.151. Parallel straight-line responses are characteristics of naturally fractured reservoirs.
Dual porosity with pseodo-steady-state interporosity flow	Pressure change slope → increasing, leveling off increasing Pressure derivative slope = 0, valley = 0 Additional distinguishing characteristic is middle time valley trend during is more than I log cycle.
Dual porosity with transient interporosity flow	Pressure change slope $\bar{\Delta}\psi(P)$ → steepening Pressure derivative slope $\Delta\psi(P')\ \Delta t = 0$, upward trend = 0 Additional distinguishing characteristic → middle time slope doubles
Pseudo-steady-state	Pressure change slope, $\Delta\psi(P)$ → 1 for drawdown and zero for buildup Pressure derivative slope $\Delta\psi(P')\Delta t$ → 1 for drawdown and steeply descending for buildup Additional distinguishing characteristic → late time drawdown pressure change and derivative are overlain; slope of 1 occurs much earlier in the derivative.
Constant pressure boundary (steady-state)	Pressure change slope → 0 Pressure derivative slope → steeply descending Additional distinguishing characteristic → cannot be distinguished from pseudo-steady-state in pressure buildup test.
Single sealing fault (pseudoradial flow)	Pressure change slope → steeping Pressure derivative slope → 0, upward trend → 0 Additional distinguishing characteristic → late-time slope doubles
Elongated reservoir Linear flow	Pressure change slope → 0.5 Pressure derivative slope → 0.5 Additional distinguishing characteristic → late time pressure change and derivative are offset by factor of 2; slope of 0.5 occurs much earlier in the derivative

Figure 9–7a. Pressure derivative trends for common flow regimes.

assuming, of course, that the flow remains in the transient (infinite-acting) state. Thus,

$$\psi(p_i) - \psi(p_{wf}) = \frac{5.7920 \times 10^4 q_g T P_{sc}}{khT_{sc}} (\log\ t + \bar{s})$$

$$= \frac{5.7920 \times 10^4 q_g T\, P_{sc}}{khT_{sc}}$$

$$\times \left[\log t + \log \frac{k}{\phi \mu_g c_t r_w^2} - 3.23 + 0.869s\right]$$

$$= \frac{5.7920 \times 10^4}{2.303} \frac{q_g T P_{sc}}{khT_{sc}} \left[In \frac{5.922 \times 10^{-4} kt}{\phi \mu_g c_t r_w^2} + 2s\right] \tag{9–4}$$

Equation 9–4 may be written as

$$\frac{kh \lfloor \psi(p_i) - \psi(p_{wf}) \rfloor T_{sc}}{5.030 \times 10^4 TP_{sc}} = 0.5[In\, 2.2458\, t_D/C_D + InC_D + In\, e^{2S}]$$

or

$$\psi(p_{WD}) = 0.5\lfloor In(t_D/C_D) + 0.80907 + In(C_D e^{2S})\rfloor \tag{9–5}$$

Thus,

$$dp_{WD}/d(t_D/C_D) = 0.5/(t_D/C_D)$$

or

$$p'_{WD} = 0.5/(t_D/C_D) \tag{9–6}$$

and

$$\log p'_{WD} = -\log(t_D/C_D) + \log 0.5 \tag{9–7}$$

Equation 9–7 shows that during the transient state and after the wellbore storage effects have subsided, a plot of p'_{WD} versus (t_D/C_D) on log-log graph paper will be a straight line of slope –1. Equations 9–3 and 9–7 show that during early time and during the transient state p'_{WD} is independent of $C_D e^{2S}$. At intermediate times p'_{WD} is dependent on $C_D e^{2S}$ and could be evaluated from the solution of Agarwal *et al.* (see Figure 9–8). The resulting type curves as given by Bourdet *et al.* are shown in Figure 9–9, where p'_D is equivalent to p'_{WD}.

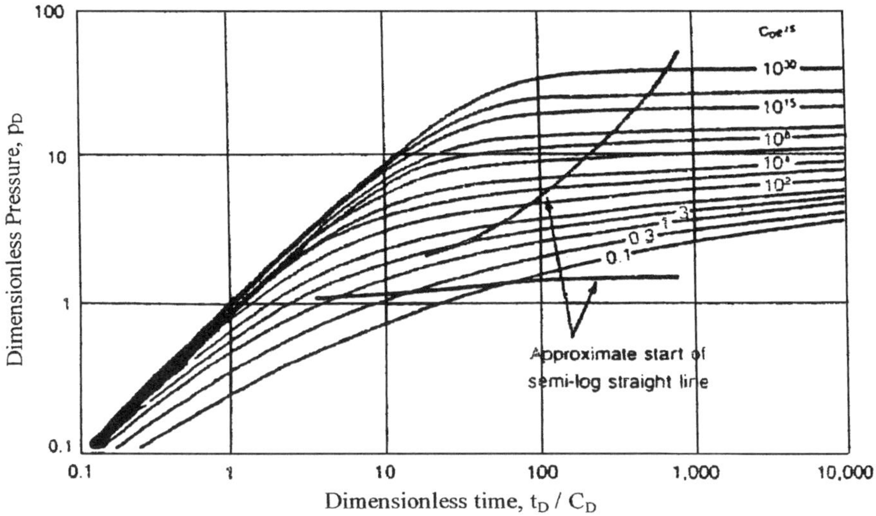

Figure 9–8. Type curves for a well with wellbore storage and skin in a reservoir with homogeneous behavior. Copyright © 1983 World Oil. Bourdet *et al.*, May 1983.[17]

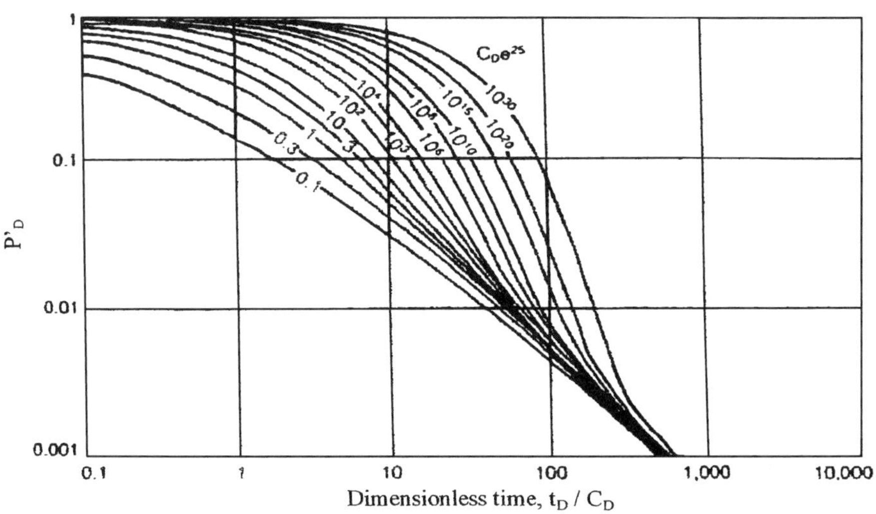

Figure 9–9. Pressure derivative type curve. Copyright © 1983 *World Oil*. Bourdet *et al.*, May 1983.[17]

Bourdet *et al.*[18] then defined the derivative of the dimensionless pressure as follows:

$$p'_D(t_D/C_D) = \Delta t \Delta p' \frac{khT_{sc}}{5.030 \times 10^{-4} q_g T} \frac{2}{P_{pc}} \int_{p_o}^{p_o+\Delta p} \frac{p}{\mu(p)z(p)} dp \tag{9–8}$$

where Δt = producing time, hr:

$$\psi(\Delta p') = d \lfloor \psi(p_i) - \psi(p_{wf}) \rfloor / dt$$

With the definition of Eq. 9–8, Bourdet *et al.* redrew their type curves and presented them as shown in Figure 9–10. To use the type curve of Figure 9–10, the field drawdown data must be plotted on log-log graph paper as $\Delta t \psi(\Delta p')$ versus Δt. At early time the data will fall on the straight line of unit slope and at late time during the transient state the data will fall as the horizontal line $\psi(p')(t_D/C_D) = 0.5$. In the case of buildup test, Δt represents the shut-in time, and the buildup data are plotted as $\Delta t \psi(\Delta p')(t_p + \Delta t)/t_p$ versus Δt. Note that since $\psi(\Delta p_{ws}) = \psi(p_{ws}) - \psi(p_{ws}$ at $\Delta t = 0)$ are functions of Δt

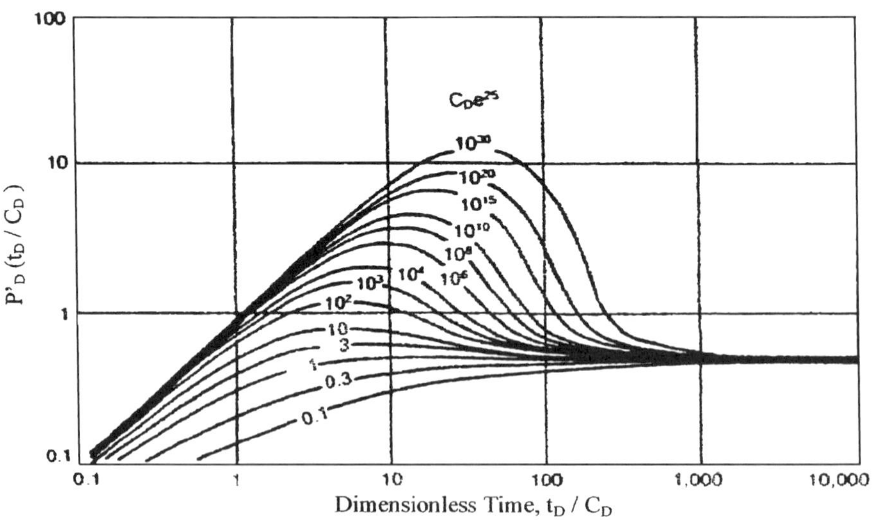

Figure 9–10. Pressure derivative type curve in terms of P'_D (t_D/C_D). Copyright © *World Oil*. Bourdet *et al.*, May 1983.[17]

and $(t_p + \Delta t / \Delta t)$, respectively, then

$$\frac{d\psi(\Delta p_{wf})}{d(ln\Delta t)} = \frac{d\psi(\Delta p_{wfs})}{d\Delta t} \times \frac{d\Delta t}{d(ln\Delta t)}$$

$$= \psi(\Delta p'_{wfs}) \times \frac{1}{1/(\Delta t)} = \psi(\Delta p_{wfs})\Delta t$$

and

$$\frac{d\psi(p_{ws})}{d[ln\{t_p + \Delta t/\Delta t\}]} = \psi(\Delta p'_{ws})\frac{d\Delta t}{d[In\{(t_p + \Delta t)/\Delta t\}]}$$

$$= -\psi(\Delta p'_{ws})\Delta t\frac{(t_p + \Delta t)}{t_p}$$

These results show that a plot of $\psi(\Delta p'_{wf})\Delta t$ versus Δt in the case of a drawdown is equivalent to a plot of $d\psi(\Delta p_{wf})/d(In\,\Delta t)$ versus Δt, where Δt here is producing time. Likewise, in the case of a buildup, a plot of $\lfloor \Delta t\psi(\Delta p'_{ws})(t_p + \Delta t)/t_p \rfloor$ versus Δt is equivalent to a plot of $\psi(\Delta p_{ws})/d[ln\{(t_p + \Delta t)/\Delta t\}]$ versus Δt, where Δt is the shut-in time. In the usual way, the pressure match gives kh, the time match C, and the curve match s. The procedure to be followed for matching with the type curve in Figure 9–8 may be summarized as follows:

1. Compute $d\psi(\Delta p)/d\,In\,t$, and plot $\psi(\Delta p)$ and $d\psi(\Delta p')/d\,In\,t$ versus time on tracing paper that has the same scale as the type curve. Align late-time derivative responses with the reference line in Figure 9–11, and slide the field curve horizontally until responses in the storage-dominated period are matched with the unit slope lines of Figure 9–11.
2. Obtain a best fit of field responses using both the pressure response and the derivative response, the derivative plot being the primary basis for obtaining a best fit.
3. Note the values of the match points

$$(p_D)_M, \psi(\Delta p)_M, \left(\frac{\Delta t}{t_D/C_D}\right)_M, \text{ and } (C_De^{2S})_M$$

4. Compute k, C, C_D and s by the following relations, respectively:
   Reservoir permeability k:

$$k = \frac{50,300 T q_g P_{sc}}{h T_{sc}} \cdot \frac{\psi[p_{WD}]_M}{[\psi(\Delta P)]_M} \quad (9\text{–}9)$$

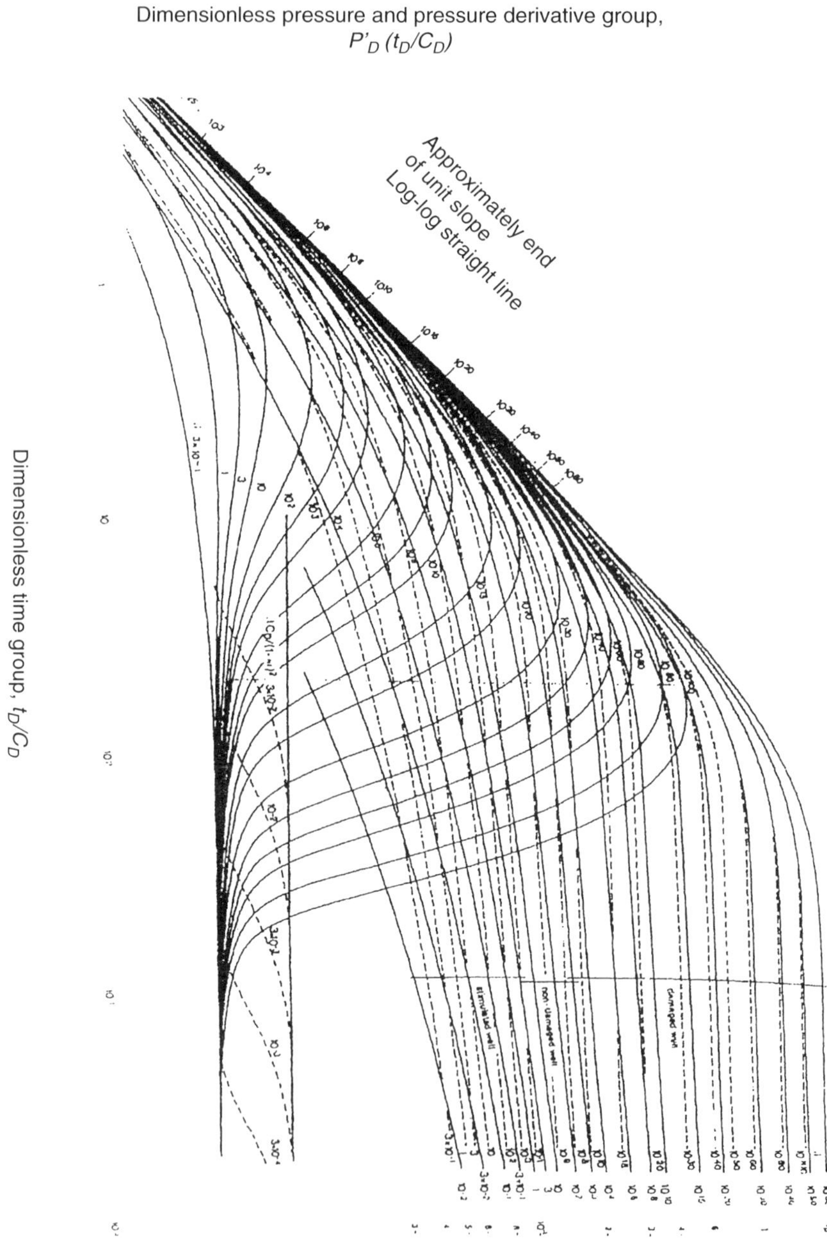

Figure 9–11. Combined and pressure type curves for a well with wellbore storage and skin in infinite-acting homogeneous reservoir (courtesy of Flopetrol Johnson Schlumberger).[14]

Wellbore storage constant C:

$$C = 0.000295\frac{kh}{\mu_g} \cdot \frac{(t)_M}{(t_D/C_D)_M} \tag{9–10}$$

Dimensionless wellbore constant C_D:

$$C_D = 0.8936\frac{C}{\phi h c_t r_w^2} \tag{9–11}$$

Skin factor s:

$$s = 0.5\, In\left[\frac{(C_D e^{2S})_M}{C_D}\right] \tag{9–12}$$

Some typical matched data are shown later in Figures 9–14 and 9–15. The following example illustrates the analysis of the pressure buildup test using Bourdet *et al.* type curves for homogeneous reservoirs.

Example 9–1 *Analyzing Pressure Buildup Test Using Bourdet* et al. *Type Curves for Homogeneous Reservoirs*

The test is a build-up preceded by a long production period at a single constant flow rate. The pressure–time data along with calculated values are shown in Table 9–1. The well/reservoir data are as follows: $T = 710°R$; $P_{SC} = 14.65$ psia; $T_{SC} = 520°R$; $h = 54$ ft; $r_w = 0.4271$ ft; $q_g = 6.148$ mmscfd; $\phi = 0.119$; $\mu_g = 0.02345$ cP; $c_t = 0.00023$ psi^{-1}; and $t_P = 44.432$ hr. Determine reservoir parameters using both type curves (Figures 9–9 and 9–10) and semilog analyses.

Solution In view of the knowledge about the reservoir indicating that it is not fractured in the zone near the well, the interpretation is normally made by comparison with the "homogeneous reservoir." First prepare the data for plotting (see Table 9–1). The data are plotted in Figures 9–12 and 9–13 respectively. Figure 9–12 is a log-log plot of $\Delta\psi$ versus Δt and Figure 9–13 is a log-log plot of $[\Delta t \Delta\psi'(t_P + \Delta t)/t_P]$ versus Δt.

Type Curve Analysis

Figure 9–13 shows a match of the pressure–time data with the type curve of Figure 9–8. From the match point, the following parameters may be obtained:

Type curve is $C_D e^{2S} = 10^{15}$

$\Delta t = 1$ hr, $\Delta\psi = 100$ mmpsia^2/cP, $t_D/C_D = 14.5$,

and $P_D = 5.5$

(text continued on page 460)

Table 9–1
Buildup Data for Plotting

Δt (hr) (1)	Shut-in pressure P_{WS} (psia) (2)	$\frac{t_p+\Delta t}{\Delta t}$ (3)	$\psi(P_{ws})$ (mmpsia^2/cP) (4)	$\Delta\psi$ (mmpsia^2/cP) (5)	Slope (mmpsia^2/cP/hr) (6)	$\Delta\psi'$ (mmpsia^2/cP/hr) (7)	$\Delta t\Delta\psi'$ $(t_p + \Delta t/t_p)$ (Pressure derivative function) (8)
0.00	1721	—	201.25	0.0	—	—	—
0.02	1735	2666.92	204.39	3.14	157.00	—	—
0.03	1747	1333.96	207.10	5.85	271.00	214.00	4.28
0.04	1758	1111.80	209.15	7.65	180.00	225.50	10.00
0.05	1763	889.64	212.19	10.94	329.00	254.50	14.04
0.06	1777	741.53	214.85	13.60	266.00	297.50	16.05
0.07	1788	667.48	216.14	14.89	226.00	248.5	17.43
0.10	1818	445.32	223.08	21.83	231.33	228.67	22.92
0.13	1819	334.24	234.87	33.62	393.00	312.17	40.70
0.17	1625	267.59	248.26	47.01	334.75	363.88	62.04
0.25	2028	178.73	273.42	72.17	314.50	324.63	81.62
0.33	2135	134.30	300.32	99.07	336.25	325.38	108.18
0.50	2312	89.86	347.07	145.82	275.00	305.63	154.54
0.75	2615	60.24	432.17	230.92	340.40	307.70	234.68
1.00	2819	45.43	492.56	291.31	241.56	290.98	297.53
1.50	3146	32.62	593.33	392.07	201.52	221.54	343.57
*2.00	3310	23.22	645.36	444.11	104.08*	152.80*	319.36*
2.50	3350	18.77	658.14	456.89	25.56	64.82	171.17
3.00	3366	15.81	663.39	462.14	12.00	18.78	60.14
3.50	3382	13.69	668.62	467.37	10.46	11.30	42.67
4.00	3385	12.11	669.40	468.14	1.54	6.00	26.16
4.83	3391	10.19	671.49	470.24	2.53	2.04	10.92
5.00	3397	9.89	673.21	471.96	10.12	6.33	35.21
5.50	3403	9.08	675.18	473.93	3.94	7.03	43.46
6.00	3407	8.41	676.64	475.38	2.90	3.42	23.29
6.50	3411	7.84	677.93	476.68	2.60	2.75	20.40
7.00	3415	7.35	679.13	477.87	2.40	2.50	20.26
7.50	3418	6.92	680.29	479.04	2.32	2.36	20.69
8.00	3421	6.55	681.10	479.85	1.62	1.97	18.60
8.50	3425	6.23	682.27	481.02	2.34	1.98	20.05
9.00	3428	5.94	683.34	482.09	2.14	2.24	24.24
9.50	3432	5.68	684.64	483.39	2.60	2.37	27.33
10.00	3436	5.44	685.97	484.72	2.66	2.63	32.22
10.50	3440	5.23	687.3	486.05	2.66	2.66	34.53
11.00	3443	5.04	688.34	487.08	2.06	2.36	32.39
11.50	3447	4.86	689.41	488.16	2.16	2.11	30.55
12.00	3448	4.70	689.93	488.68	1.04	1.60	24.39
12.50	3451	4.55	690.84	489.59	1.82	1.43	22.91
13.00	3453	4.42	691.52	490.27	1.36	1.59	26.72
13.50	3456	4.29	692.43	491.18	1.82	1.58	27.81

*Column (6) → (444.11 − 392.07)/(2.0 − 1.5) = 104.08
Column (7) → (201.52 + 104.08)/2 = 152.80
Column (8) → 2.00 (152.80) [(44.432 + 2.0)/44.432)] = 319.36

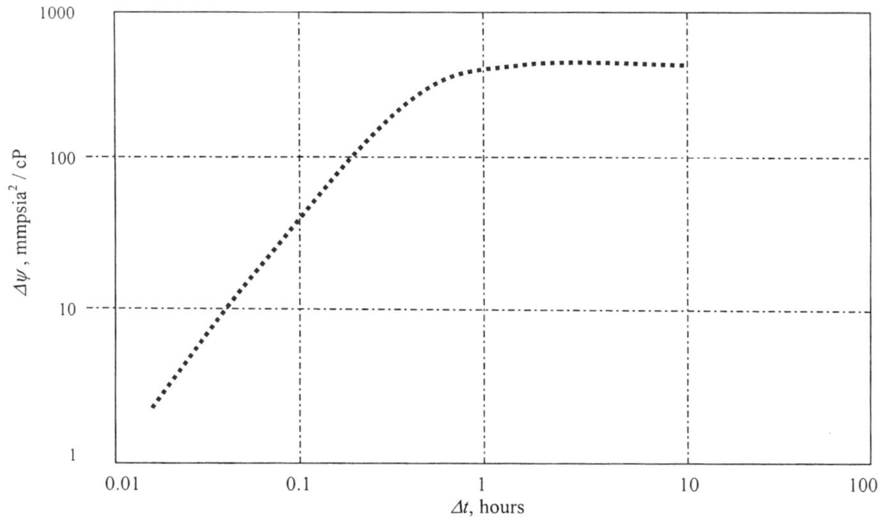

Figure 9–12. Log-log plot—Data from Table 9–1.

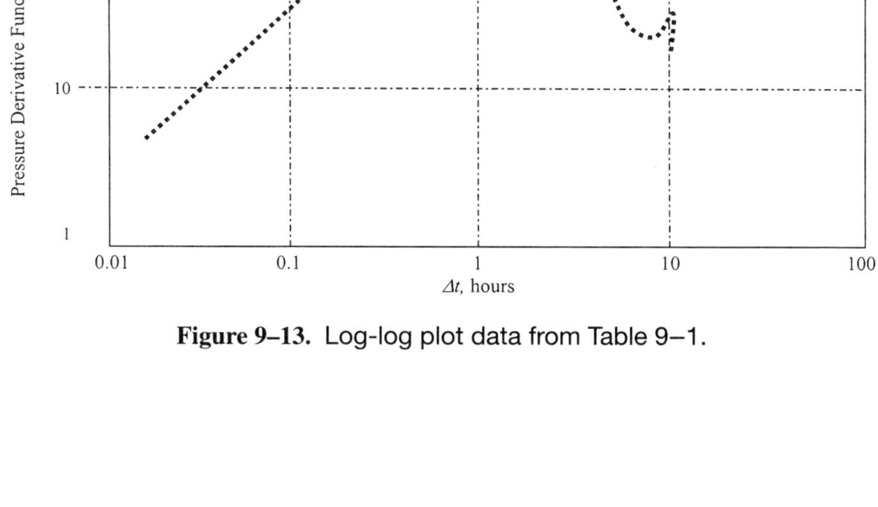

Figure 9–13. Log-log plot data from Table 9–1.

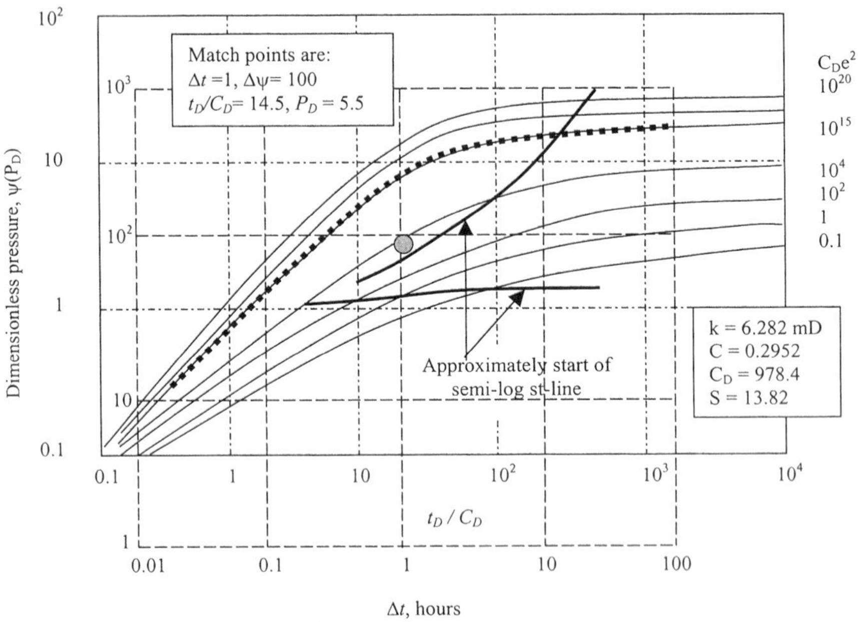

Figure 9–14. Buildup data plotted on log-log graph paper and matched to type curve of Figure 9–4—Example 9–1.

(text continued from page 457)

Using Eqs. 9–9, 9–10, 9–11, and 9–12 respectively:

Effective permeability:

$$k = \frac{50{,}300 \times 6.148 \times 10^3 \times 710 \times 14.65}{54 \times 520} \cdot \frac{5.5}{100 \times 10^6} = 6.30\,\text{mD}$$

Wellbore storage constant:

$$C = 0.000295 \frac{6.30 \times 54}{0.02345} \cdot \frac{1}{14.5} = 0.2952$$

Dimensionless wellbore constant:

$$C_D = 0.8936 \frac{0.2952}{0.119 \times 54 \times 0.00023 \times 0.4271^2} = 978.4$$

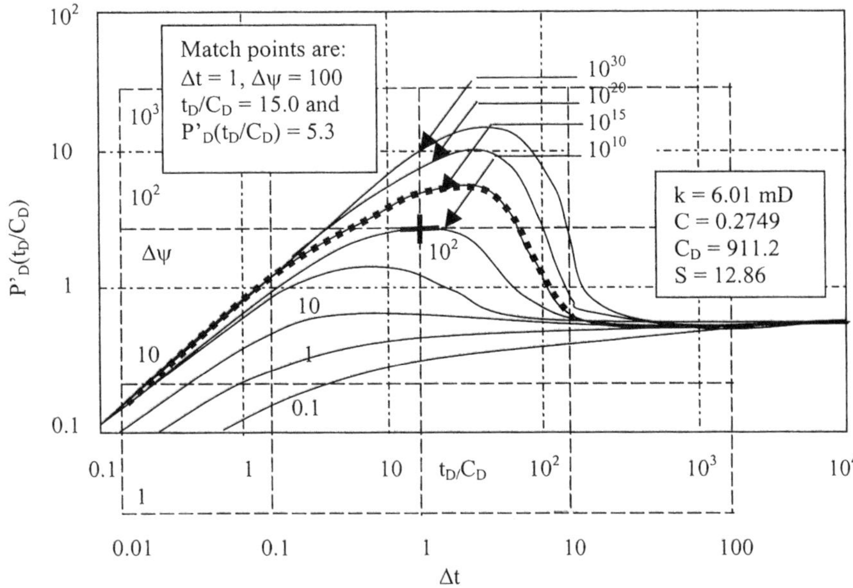

Figure 9–15. Buildup data plotted on log-log graph paper and matched to pressure derivative type curve of Figure 9–10—Example 9–1.

Skin factor:

$$s = 0.5\, ln\left[\frac{10^{15}}{978.4}\right] = 13.62$$

Figure 9–15 shows a match of the pressure derivative function versus time data with the type curve of Figure 9–10. From the match point, the following parameters may be obtained:

Type curve $C_D e^{2S} = 10^{15}$

$\Delta t = 1$ hr, $\Delta\psi = 100$ mmpsia^2/cP, $\frac{t_D}{C_D} = 15.0$, and $P_D'(t_D/C_D) = 5.3$

Using Eqs. 9–9 through 9–12, respectively,

Effective permeability:

$$k = \frac{50{,}300 \times 6.148 \times 10^3 \times 710 \times 14.65}{54 \times 520} \cdot \frac{5.3}{100 \times 10^6} = 6.07\,\text{mD}$$

Wellbore storage constant:

$$C = 0.000295 \frac{6.07 \times 54}{0.02345} \cdot \frac{1}{15.0} = 0.2749$$

Dimensionless wellbore constant:

$$C_D = 0.8936 \frac{0.2749}{0.119 \times 54 \times 0.00023 \times 0.4271^2} = 911.2$$

Skin factor:

$$s = 0.5\, In \left[\frac{10^{15}}{911.2} \right] = 13.86$$

Semilog Analysis

Now examining Figure 9–16, which is a semilog plot of $\Delta\psi$ versus Δt, fit the semilog straight line as shown, beginning at $\Delta\psi \geq 470$ mmpsia²/cP. The slope m of the semilog straight line is 22.0 mmpsia²/cP/cycle and $(\Delta\psi)_{1hr} =$ 448.0 mmpsia²/cP. Thus from Eq. 5–40, the effective permeability is

$$k = \frac{57.92 \times 10^6 \times 6.148 \times 14.65 \times 710}{22.0 \times 10^6 \times 54 \times 520} = 5.80\,\text{mD}$$

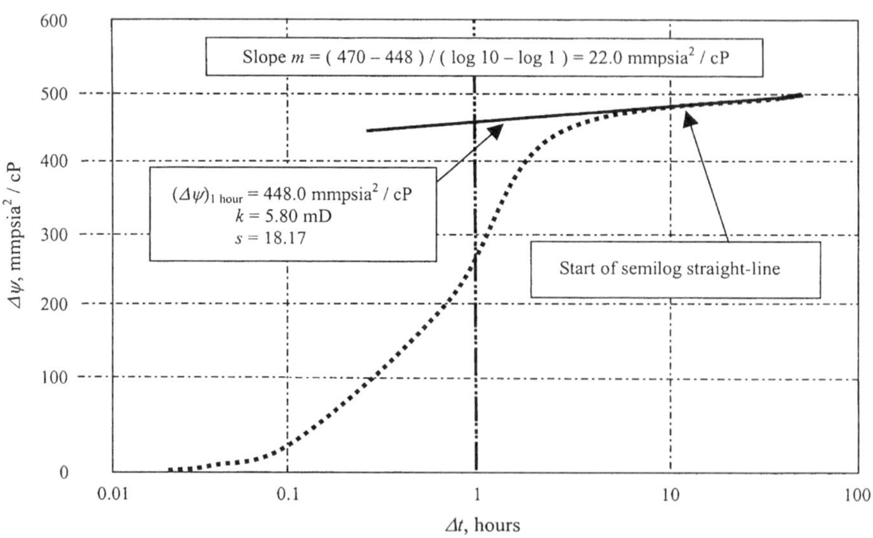

Figure 9–16. Semilog graph of $\Delta\psi$ versus log Δt—Data from Table 9–1.

Table 9–2
Comparison between Pressure, Pressure Derivative, and Semilog Values of Several Parameters

Reservoir parameters	Pressure solution	Pressure derivative solution	Semilog solution
Effective permeability k	6.282	6.07	5.80
Wellbore storage constant	0.2952	0.2749	—
Dimensionless wellbore storage constant	978.4	911.2	—
Skin factor s	13.62	13.86	18.17

From Eq. 5–41, the skin factor is

$$s = 1.151\left[\frac{448 \times 10^6}{22.0 \times 10^6} - \log\frac{5.80}{0.119 \times 0.02345 \times 0.00023 \times 0.4271^2} + 3.23\right]$$

$$\doteq 1.151[20.26 - 7.70 + 3.23] = 18.17$$

Table 9–2 shows a comparison between pressure deriative and semilog values of several parameters.

Discussion

All data obtained from a few minutes to 13 hr have matched the appropriate type curve solutuion. The type-curve solutions presented in this example application are useful and appear to represent large portions of the field data. The well is damaged and completion would probably benefit from stimulation.

9.6 Fractured Reservoir Systems with Double Porosity Behavior

New type curves suitable for practical applications based on the model by Warren and Root were introduced by Bourdet *et al.*[17,18] These curves are primarily used for diagnosing dual-porosity behavior and for ensuring that an optimum, conclusive test is obtained. The idea behind these curves is that the log-log plot consists of three typical flow regimes as follows:

1. The first flow regime represents radial flow in a homogeneous reservoir with wellbore storage, skin, permeability k_f, and reservoir storage s_t.
2. The second flow represents a transient period.
3. The third flow represents radial flow in a homogeneous system with wellbore storage, skin, permeability k_f, and reservoir storage $(s_f + s_m)$.

Pseudo-Steady-State Interporosity Flow

Bourdet *et al.* type curves as shown in Figure 9–17 can be used for the analysis of fractured reservoirs with pseudo-steady-state interporosity flow. Only two parameters (ω and λ) characterize the reservoir heterogeneity. The parameter ω is defined as the ratio of fracture storage to total storage. The interporosity flow parameter λ is proportional to the ratio of matrix permeability to fracture permeability. Thus,

$$\omega = \frac{s_f}{s_f + s_m} \tag{9–13}$$

where

s_f = fracture storage $= \phi_f c_f h_f$

s_m = matrix storage $= \phi_m c_m h_m$

and

$$\lambda \propto k_m / k_f$$

or

$$\lambda = \alpha r_w^2 k_m / k_f \tag{9–14}$$

where

α = geometric parameter, $1/L^2$
L = length of matrix, ft

In Figure 9–17, the dimensionless pressure (p_D versus t_D/C_D) curves show two families of component curves as the $C_D e^{2S}$ that correspond to homogeneous behavior and the λe^{2S} curves that show pressure behavior during transition. The pressure derivative curve response follows this sequence:

1. Initially, because of wellbore storage effects, the derivative curve follows $(C_D e^{2S})_f = 1$ type curve.
2. When the infinite-acting radial flow occurs in the fissured system, the pressure derivative group will follow the 0.5 horizontal straight line.
3. During the transition period when pressure stabilizes, the derivative with respect to natural logarithm of time drops and follows the $(\lambda C_D)/\omega(1 - \omega)$ type curve until it reaches a minimum and then bounces back up along the $(\lambda C_D/(1 - \omega)^2$ curve before returning to the 0.5 straight line. The 0.5 horizontal lines corresponding to the infinite-acting radial flow in the total system $(C_D e^{2S})_{f+m}$.

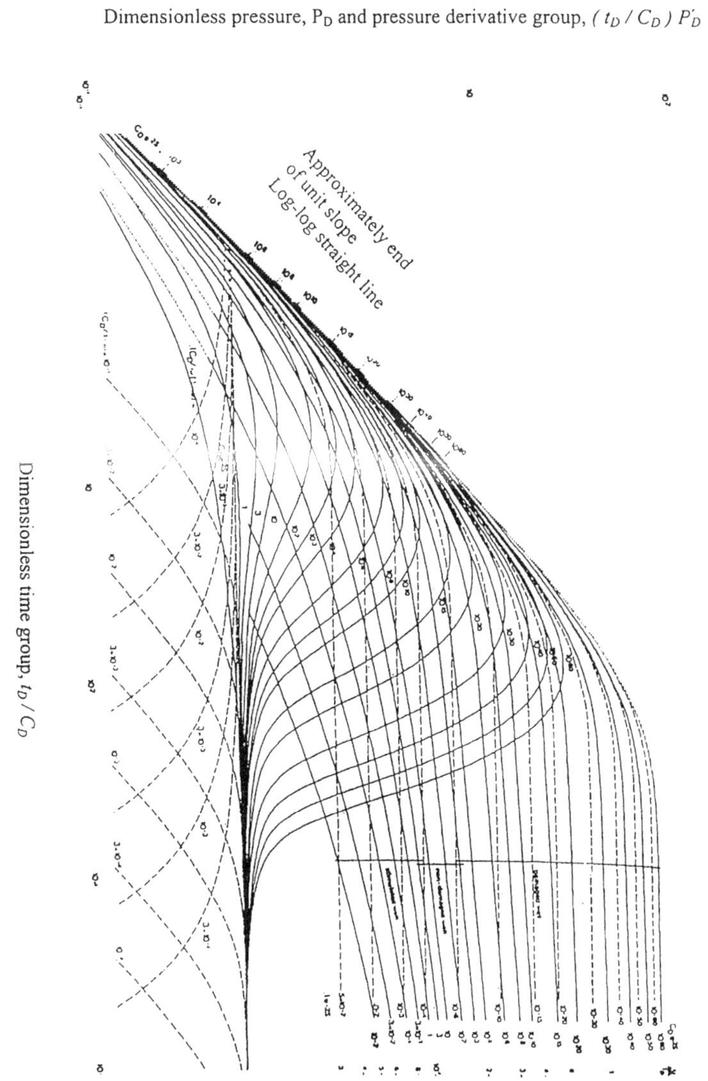

Figure 9–17. Type curve showing both behavior of pressure and its derivative (after Bourdet *et al.*[18]). It can be used to analyze test data from fractured reservoirs. The behavior of the pressure derivative is reasonably resolutive for the identification of the transition period.

To use the type curve of Figure 9–17, one has to match the early data with one of the type curves labeled $C_D e^{2S}$. The label of the matched curve is now referred to as $(C_D e^{2S})_f$. The permeability k_f is calculated from the pressure match, and C is calculated from the time match. The matching procedure is as follows:

1. Plot pressure derivative versus Δt on log-log graph paper.
2. Plot Δp_{ws} versus Δt on log-log paper.
3. Copy the curve (step 1) on curve (step 2).
4. Match the derivative curve of step 3 with one of the derivative type curves of Figure 9–17.
5. Choose any point and read its coordinates on both figures. Thus, $\psi(\Delta p)_M$, $(\Delta t)_M$, $(p_D)_M$, and $(t_D/C_D)_M$ would become known. Also read the matched derivative curve label $\lambda C_D/(1 - \omega)^2$, where C_D here is $C_{D_{f+m}}$.
6. Now, with the match still maintained, change your focus from the derivative curve to the data curve in step 3. Read the values of the curves labeled $C_D e^{2S}$ which match the initial and final segments of the data curve, $(C_D e^{2S})_f$ and $(C_D e^{2S})_{f+m}$, respectively.
7. Calculate the different parameters as follows:

$$\omega = \frac{(C_D e^{2S})_{f+m}}{(C_D e^{2S})_f} \tag{9–15}$$

$$k_f = \frac{50,300 q_g T P_{sc}}{h T_{sc}} \cdot \frac{(p_D)_M}{\psi(\Delta p)_M} \text{ mD-ft} \tag{9–16}$$

$$C = \frac{0.000295 kh}{\mu_g} \cdot \frac{(\Delta t)_M}{\left(\frac{t_D}{C_D}\right)_M} \tag{9–17}$$

$$C_{D_{f+m}} = \frac{0.8936C}{\phi c_t h r_w^2} \tag{9–18}$$

Assuming that the total reservoir storage $(s_f + s_m)$ is known from well logs,

$$s = 0.5\, In[(C_D e^{2S})_{f+m}/C_{D_{f+m}}] \tag{9–19}$$

and λ can be calculated from the label of the matched derivative curve $\lambda C_{D_{f+m}}/(1 - \omega)^2$.

$$\lambda = \frac{(1-\omega)^2}{C_{D_{f+m}}} \cdot \left[\lambda C_{D_{f+m}}/(1-\omega)^2\right]_M \tag{9–20}$$

For a gas well drawdown test, the derivative is calculated as follows:

$$\frac{d\psi(\Delta p)}{d\, In\Delta t} = \frac{d\psi(\Delta p)}{d\Delta p} \cdot \frac{d\Delta p}{d\Delta t}\frac{d\Delta t}{d\, In\Delta t}$$
$$= \frac{2\Delta p}{\mu(p_{wf})\, z(p_{wf})}\Delta p'\Delta t$$

where Δt is the producing time and $\psi(\Delta p) = \psi(p_i) - \psi(p_{wf})$. For a buildup test, the differentiation is carried out with respect to $In\frac{t_p+\Delta t}{\Delta t}$ and the result would be

$$\frac{d\,\psi(\Delta p)}{d\, In\frac{t_p+\Delta t}{\Delta t}} = \frac{2\Delta p_{ws}}{\mu(p_{ws})\, z(p_{ws})} \cdot \Delta p'_{ws}\frac{t_p+\Delta t}{\Delta t}\Delta t$$

where Δt is the shut-in time and $\psi(\Delta p) = \psi(p_{ws}) - \psi(p_{wf,\,\Delta t=0})$.

Transient Interporosity Flow

Bourdet *et al.* type curves as shown in Figure 9–18 can be used for the analysis of fractured reservoir with transient interporosity flow.[17,18] The transient period is described by a family of β' curves, which are identical to homogeneous C_De^{2S} curves except that two divides pressure and time. In the transition interporosity flow period, the double porosity responses do not flatten out but tend to develop a semilog straight line, the slope of which is half of the true radial-flow slope. The dimensionless interporosity transient flow parameter β' is defined by

$$\beta' = \delta' \frac{(C_De^{2S})_{f+m}}{\lambda e^{-2S}} \qquad (9\text{–}21)$$

where

δ' = matrix block shape factor
= 1.8914 for slab matrix block
= 1.0508 for spherical matrix block

(The choice of matrix geometry for interpretation has to be supported by geological models.)

Figure 9–18 shows the following characteristics:

1. β': At early time, the fissured flow $(C_De^{2S})_f$ is masked by wellbore storage, and the pressure response starts on the transition curve.
2. $(C_De^{2S})_{f+m}$: At the late time, the homogeneous behaviors corresponding to the total system parameters are reached.

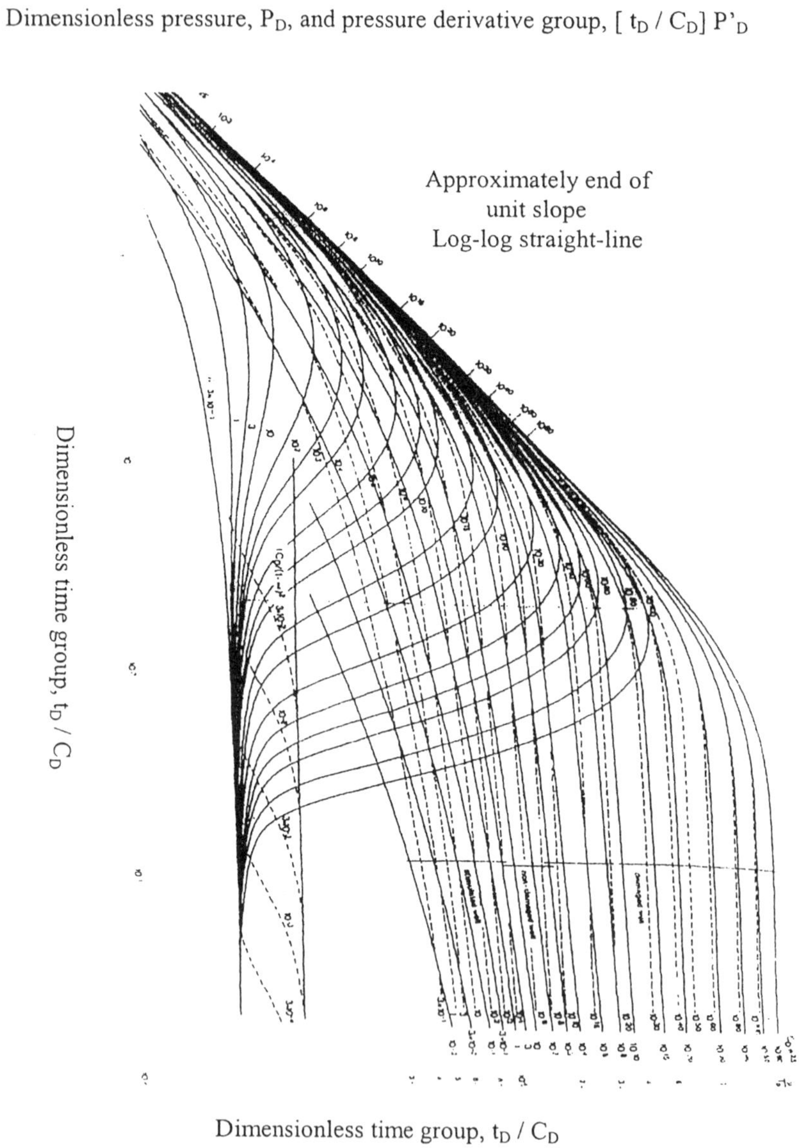

Figure 9–18. Type curve matching both the behavior of the pressure and its derivative (after Bourdet *et al.*).[18] Transient interporosity fluid flow behavior.

The derivative of pressure curves is three component curves:

1. β': At early time, the response first follows an early transition derivative curve.
2. $\gamma(C_D)_{f+m}/(1-\omega)^2$: A late transition curved is reached.
3. 0.5: At late time, the homogeneous behavior corresponding to $(C_D e^{2S})_{f+m}$ is, in general reached on the 0.5 line.

As presented in the previous sections, k_f and C are calculated from pressure and time match, respectively. The skin factor s is obtained from the curve match and Equation 9–18. The parameter λ is calculated from the transition curve match and the following equation:

$$\lambda = \delta' \frac{(C_D e^{2S})_{f+m}}{\beta'} \cdot \frac{1}{e^{2S}} \tag{9–22}$$

Choice between Pseudo-Steady-State or Transient Interporosity Flow

The choice between pseudo-steady-state or transient interporosity flow for interpreting doubleporosity data has been subject to debate for many years. As a general observation, for the same values of ω, the derivative of the pseudo-steady-state model drops well below the 0.25 straight lines. Field data that produce a drop of more than one log cycle when differentiated are observed, thus justifying the applicability of Warren and Root model.[1] Selection of an appropriate model may sometimes be difficult, especially when pressure recording quality is low. If the value of ω is large (for example, 0.2), the derivative of the pseudo-steady-state model will not drop below the 0.25 line, and in some cases the transient model can produced a very similar shape but with a much smaller value of $\omega(10^{-3})$. In such cases the physical description of the formation under consideration must be carefully studied.

9.7 Summary

Based on the material presented in this chapter, the following remarks are pertinent:

- A new technique is presented to analyze data in the bilinear flow period. It is shown that, during this flow period, a graph of $\psi(p_{wf})$ versus $t^{1/4}$ yields a straight line when the slope is inversely proportional to $h_f(k_f b_f)^{1/2}$.
- New type curves are now available for pressure analysis of fractured gas wells, and the problem in the analysis is reduced considerably with the use of these type curves.

- Prefracture information about the reservoir is necessary to estimate fracture parameters.
- The type curve analysis method must be used simultaneously with the specific analysis methods $\psi(p_{wf})$ versus $t^{1/4}$, $\psi(p_{wf})$ versus $t^{1/2}$, and $\psi(p_{wf})$ versus $\log t$ to produce reliable results.

References and Additional Reading

1. Warren, J. E., and Root, P. J., "The Behavior of Naturally Fractured Reservoirs," *Soc. Petroleum Eng. J.* (1963) 3, 245–255.
2. Ramey, H. J., Jr., "Short-Time Well Test Data Interpretation in the Presence of Skin Effect and Wellbore Storage," *J. Petroleum Technol.* (1970) 22, 97–104.
3. Raghavan, R., Cady, G. V., and Ramey, H. J., Jr., "Well Test Analysis for Vertically Fractured Wells," *J. Petroleum Technol.* (1972) 24, 1014–1020.
4. McKinly, R. M. "Wellbore Transmissibility from after Flow-Dominated Pressure Buildup Data," Paper SPE 2416, 45th Fall Meeting of AIME, Houston, TX.
5. Gringarten, A. C., Ramey, H. J., Jr., and Raghavan, R., "Applied Pressure Analysis for Fractured Wells," *J. Petroleum Technol.* (1975) 17, 887–892.
6. Earlougher, R. C., Jr., and Kerch, K. M., "Analysis of Short-Time Transient Test Data by Type-Curve Matching," *J. Petroleum Technol.* (1974) 26, 793–800.
7. Agarwal, R. G., Al-Hussainy, R., and Ramey, H. J., Jr., "An Investigation of Wellbore Storage and Skin Effect in Unsteady Liquid Flow: I. Analytical Treatment," *Soc. Petroleum Eng. J.* (1970) 10, 279–290.
8. Fetkovich, M. J., "Decline Curve Analysis Using Type Curves, Paper SPE 4629, 48th Fall Meeting of AIME, Las Vegas, 1973.
9. Maer, N. K., Jr., "Type Curves for Analysis of after Flow-Dominated Gas Well Build-up Data," Paper SPE 5134, 49th Fall Meeting of AIME, Houston, TX, 1974.
10. Uldrich, D. O., and Ershaghi, I., "A Method for Estimating the Interporosity Flow Parameter in Naturally Fractured Reservoirs," *Soc. Petroleum Eng. J.* (Oct. 1979) 324–332.
11. Wattenbarger, R. A., and Ramey, H. J., Jr., "Well Test Interpretations of Vertically Fractured Gas Wells," *J. Petroleum Technol.* (May 1969) 625–632; *Trans. AIME*, 246.
12. Agarwall, R. G., Carter, R. D., and Pollock, C. B., "Evaluation and Prediction of Performance of Low-Permeability Gas Wells Stimulated by Massive Hydraulic Fracturing," *J. Petroleum* (March 1979) 362–372; *Trans. AIME* 267.
13. Cinco, H., and Samaniego, F., "Effect of Wellbore Storage and Damage on the Transient Pressure Behavior for a Well with a Finite-Conductivity Vertical Fracture," *Soc. Petroleum Eng. J.* (Aug. 1978) 253–264.

14. Flopetrol Johnston Schlumberger, *Course Manual*, 100 Macco Boulevard, Sugarland, TX 77478.
15. Economides, C. E., "Use of the Pressure Derivative for Diagnosing Pressure-Transient Behavior," *J. Petroleum Technol.* (Oct. 1988) 1280–1282.
16. Bourdet, D., Whittle, T. M., Douglas, A. A., and Pirard, Y. M., "A New Set of Type Curves Simplifies Well Test Analysis," *World Oil* (May 1983).
17. Bourdet, D., Ayoub, J. A., Whittle, T. M., Pirard, Y. M., and Kniazeff, Y., "Interpreting Well Tests in Fractured Reservoirs," *World Oil* (Oct. 1983).
18. Bourdet, D., Alagoa, A., Ayoub, J. A., and Pirard, Y. M., "New Type Curves Aid Analysis of Fissured Zone Well Tests," *World Oil* (April 1984).
19. Alagoa, A., Bourdet, D., and Ayoub, J. A., "How to Simplify the Analysis of Fractured Well Tests," *World Oil* (Oct. 1985).

Chapter 10

Massive Hydraulic Fractured Gas Well Behavior Analysis

10.1 Introduction

References 1 and 2 presented a new set of type curves. These type curves were specifically needed for massive hydraulic fractured (MHF) wells to handle production under constant pressure and constant rate. A fracture is said to have an infinite flow capacity when there is little or no pressure drop along the axis of the fracture. The fracture is said to have a finite flow capacity when there is a significant pressure drop along its axis. Since the distinction between the definitions of fracture flow capacity and formation flow capacity is often confusing, it may be worthwhile to restate the definition of the formation flow capacity:

$$\text{Formation flow capacity} = kh, \text{mD-ft} \tag{10–1}$$

$$\text{Fracture flow capacity} = k_f w, \text{mD-ft} \tag{10–2}$$

10.2 Methods of Evaluating MHF Fractured Gas Wells

Figure 10–1 illustrates method of analyzing massive hydraulic fractured gas wells.

10.3 Evaluation of Fracturing Treatments

Constant Wellbore Pressure and Finite Flow Capacity Fracture

Figure 10–2 presents constant wellbore pressure type curves for finite flow-capacity fractures. These type curves are especially useful when analyzing performance data (production rate versus time) for MHF gas wells that

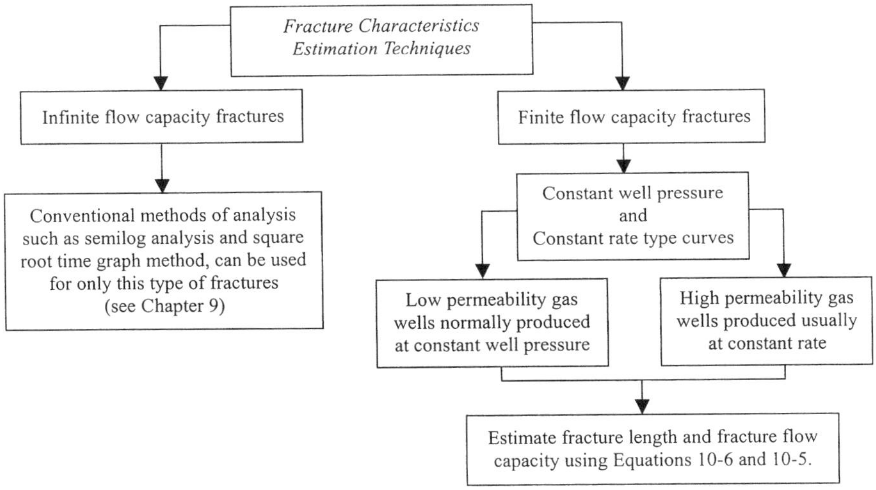

Figure 10–1. Methods of evaluating MHF gas wells.

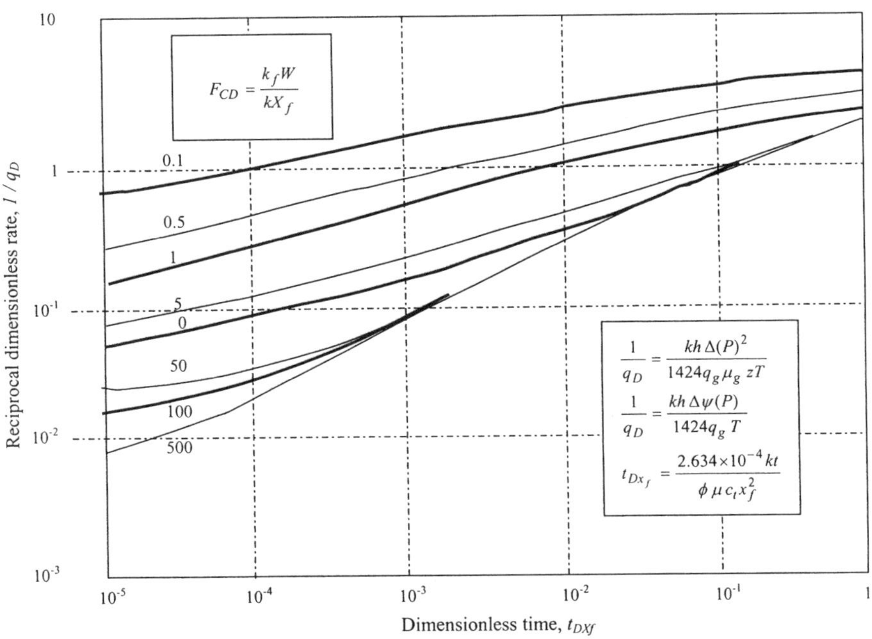

Figure 10–2. Log-log type curves for finite capacity vertical fractures—Constant wellbore pressure (after Agarwal, Carter and Pollock).[1]

generally are produced at a constant wellbore pressure, rather than a constant rate. The reciprocal of the dimensionless rate, $1/q_D$, was plotted as a function of dimensionless time, t_{Dxf}, on log-log paper with dimensionless fracture flow capacity, F_{CD}, as a parameter. Definitions of $1/q_D$, t_{Dxf}, and F_{CD} are as follows:

Dimensionless rate, $1/q_D$:

$$1/q_D = \frac{kh\lfloor\psi(p_i) - \psi(p_{wf})\rfloor T_{sc}}{50{,}300\, q_g T P_{sc}} \tag{10–3}$$

Dimensionless time, t_{Dxf}:

$$t_{Dxf} = \frac{0.0002637kt}{\phi\mu_i c_i x_f^2} \tag{10–4}$$

Dimensionless flow capacity, F_{CD}:

$$F_{CD} = \frac{k_f w}{k x_f} \tag{10–5}$$

Fracture half length, x_f:

$$x_f = \sqrt{\frac{0.0002637k(t)_M}{\phi\mu_i c_i (t_{DXF})_M}} \tag{10–6}$$

Fracture skin, s_f:

$$s_f = \left[s_f + In\left(\frac{x_f}{r_w}\right)\right]_{Table10-1} - In\left(\frac{x_f}{r_w}\right) \tag{10–6a}$$

The application of Agarwal's type curves for the constant wellbore pressure case to analyze massive hydraulic fractured gas well is illustrated in Example 10–1.

Example 10–1[13] *Analyzing Pressure Drawdown Test for Massive Hydraulic Fractured Gas Well—Constant Wellbore Pressure Case*

A buildup test was conducted on a gas well producing from a low-permeability reservoir. A prefractured buildup test yielded a permeability value k of 0.015 mD. Following a massive hydraulic fracturing (MHF) treatment, the well was produced at a constant pressure, i.e., $\lfloor\psi(p_i) - \psi(p_{wf})\rfloor$ remained constant. The rate time data are shown in Table 10–2. Additional reservoir parameters are given as follows: $\lfloor\psi(p_i) - \psi(p_{wf})\rfloor = 401$ mmpsia²/cP; $T_{sc} = 520°$R; $P_{sc} = 14.7$ psia; $h = 35$ ft; $q_g = 1050$ mscfd; $\mu_i = 0.0175$; $c_i = 2.35 \times 10^{-4}$ psi^{-1}; $\phi = 0.11$; $T = 640°$R.

Table 10–1
Pseudoskin Factor for a Well with a Finite Conductivity Vertical Fracture (After Cinco-Ley and Samaniego)[7]

$k_{fD}W_{fD} = k_fW/kx_f$	$S_f + In(x_f/r_w)$	$k_{fD}W_{fD} = k_fW/kx_f$	$S_f + In(x_f/r_w)$
0.1	3	10	0.82
0.2	2.10	20	0.80
0.3	2.40	30	0.79
0.4	2.20	40	0.77
0.5	2.00	50	0.785
0.6	1.90	60	0.778
0.7	1.85	70	0.777
0.8	1.75	80	0.776
0.9	1.73	90	0.775
1	1.60	100	0.774
2	1.20	200	0.772
3	1.10	300	0.772
4	1.00	400	0.772
5	0.94	500	0.772
6	0.90	600	0.772
7	0.85	700	0.772
8	0.88	800	0.772
9	0.84	900	0.772
10	0.82	1000	0.772

Table 10–2
Drawdown Test of MHF Gas Well

t (days)	q (mscfd)	$1/q(\times 10^{-3})$ d/mscf
25	610	1.639
35	475	2.105
65	410	2.439
100	314	3.185
150	252	3.968
250	210	4.762
300	195	5.128

Use Agarwal's type curves to compute the following:

1. Fracture permeability k_f
2. Fracture length x_f
3. Fracture flow capacity $k_f w$
4. Match with past performance and prediction of future performance of this MHF gas well

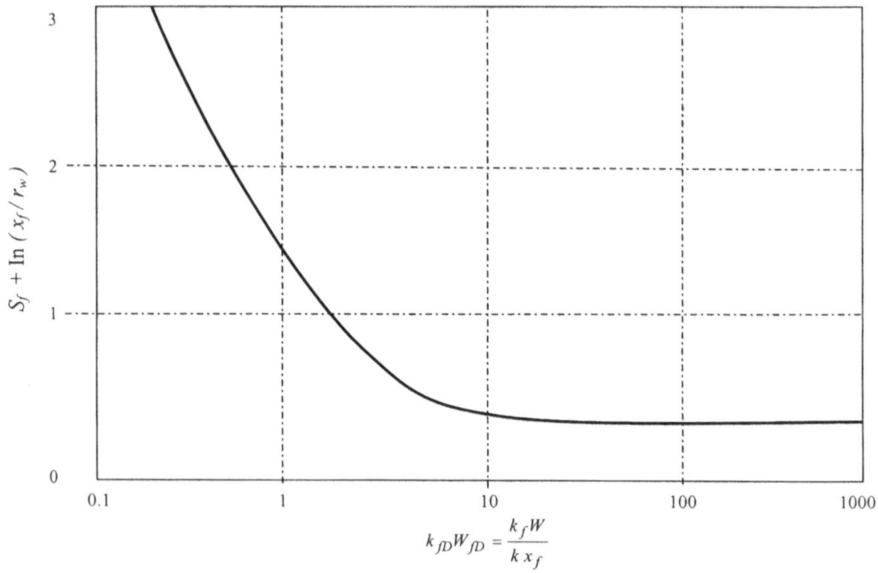

Figure 10–3. Pseudoskin factor for a well with a finite conductivity vertical fracture (after Cinco-Ley and Samaniego).[2]

Solution Before we match, we must know k and $\psi(p_i) - \psi(p_{wf})$. The first steps in analyzing this type of problem are as follows:

Step 1. Prepare a plot production rate q versus time as shown in Figure 10–4.

Step 2. $1/q$ versus time data (Table 10–2) are plotted on tracing paper using the log-log scale of the type curves. Main x and y axes also are drawn on the tracing paper. Such a plot is shown in Figure 10–5.

Step 3. Since formation flow capacity, kh, is known from the prefracturing test,

$$1/q_D = \frac{kh\lfloor\psi(p_i) - \psi(p_{wf})\rfloor T_{sc}}{50{,}300\, q_g\, P_{sc}}$$

$$= \frac{(0.015)(35)(401 \times 10^6)(520)}{50{,}300(1050)(1000)(14.7)} = 0.141$$

Also,

$$1/q_g = \frac{1}{1050} = 0.952 \times 10^{-3}, \text{ day/mscf}$$

Place the data plot on Figure 10–6 such that the axes labeled 0.952×10^{-3} and 0.141 coincide as shown in Figure 10–5. Then move the data plot

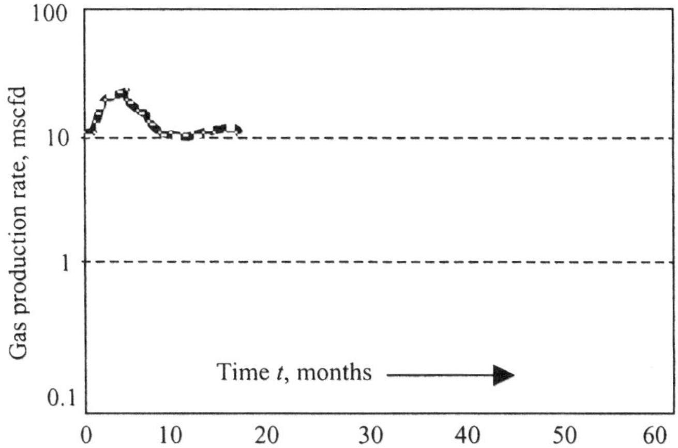

Figure 10–4. Actual performance (gas rate versus time) data.

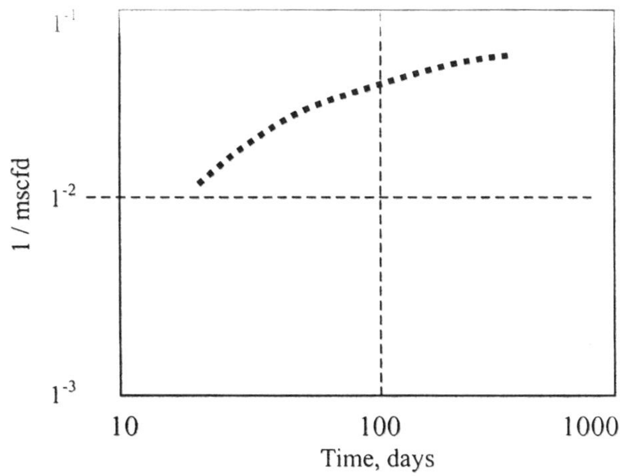

Figure 10–5. Reciprocal smooth rate versus time for MHF—Example 10–1.

horizontally until it matches one of the curves of Figure 10–2. From the match point given in Figure 10–4, read the values of t_M, t_{DXF}, and F_{CD}. The calculation of x_f is made by using

$$x_f = \sqrt{\frac{0.0002637k(t)_M}{\phi\mu_i c_i (t_{Dxf})_M}}$$

$$= \sqrt{\frac{0.0002637(0.015)(2{,}400)}{(0.11)(0.0175)(2.35 \times 10^{-4})(215 \times 10^{-2})}} = 988 \text{ ft}$$

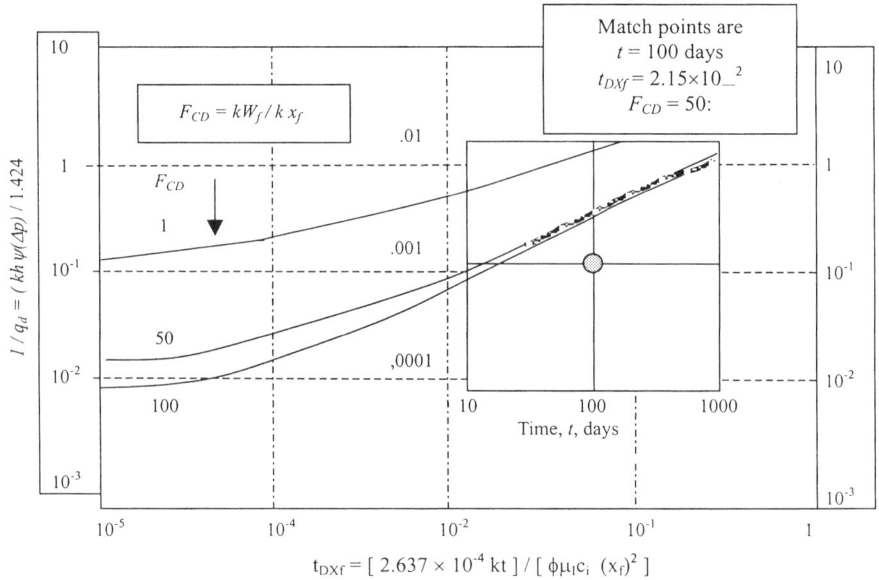

Figure 10–6. Type curve matching for MHF gas well—Example 10–1.

Thus, the total fracture length $2x_f = 1976$ ft.

The matched curve is labeled $[F_{CD}]_M = 50$. Thus, from Eq. 10–5, we get

$$k_f w = [F_{CD}]_M \, k x_f = (50)(0.015)(988) = 741 \text{ mD-ft}$$

Discussion

If the prefracturing buildup test data were not available, matching would require shifting the tracing paper along both the x and y axes, which could be difficult. This emphasizes the need for determining kh from a prefracturing test.

Match of Past Performance Data

Figure 10–7 shows a match of past production data.

Long-Term Production Forecasting

To predict the future performance of a massive hydraulic fractured gas well, the fracture characteristics determined from the type curve analysis, together with reservoir and fluid properties and drainage area geometry, are entered in the MHF simulator. The summary of the results is given in Table 10–3. Figure 10–7 shows the match of the past performance and prediction for the future performance of the MHF gas well.

Table 10–3
Summary of Results of Past and Future Performance of MHF Gas Well

Time t (months)	Past performance q (msc/day)	Time t (months)	Future performance q (msc/day)
1	110.11	0	1000.01
2	175.34	6	175.20
3	195.42	12	115.19
4	199.66	18	100.26
5	202.13	24	95.78
6	147.43	30	90.23
7	117.00	36	85.12
8	123.58	42	80.67
9	107.94	48	75.44
10	100.00	54	70.22
11	98.00	60	68.55
12	95.00	66	66.34
13	93.00	72	64.12
16	91.00		
18	90.00		

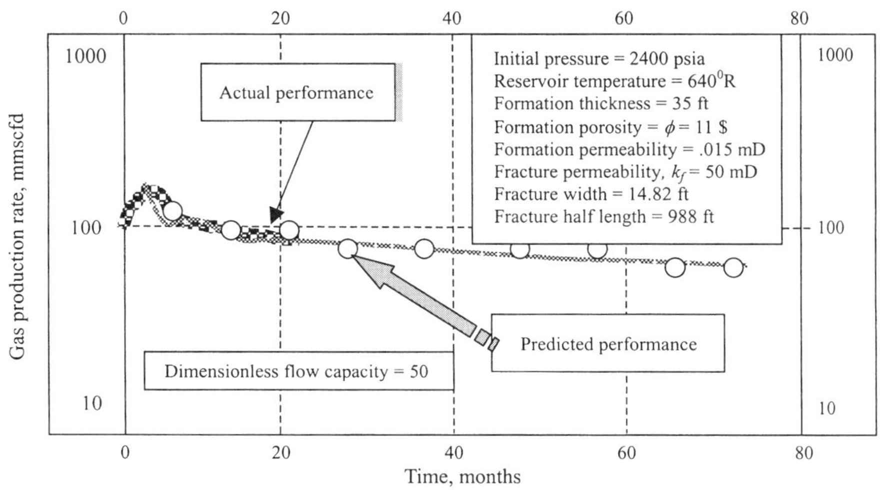

Figure 10–7. Match and performance prediction for MHF gas well—Example 10–1.

Discussion of Results

Fracture characteristics determined by type curve analysis, together with reservoir and fluid properties, when entered in the MHF simulator gave reasonably good matches of past performance and provided confidence in future predictions. However, type curve analysis does not provide unique values of fracture length and capacity. There are many factors such as the following:

1. Wellbore storage
2. Fracture face damage due to liquid invasion
3. Relative permeability effects
4. Gas compressibility
5. Confining pressure
6. Turbulence
7. Varying fracture capacity
8. Lateral and vertical reservoir heterogeneities

These factors may complicate the analysis. As for future performance predictions, it should be noted that because of the slow response of pressure transients in tight gas formations, long-term performance data will be needed before the lateral extent of the drainage area of the well can be estimated reasonably.

Constant Rate and Finite Flow Capacity Fracture

Figure 10–8 presents constant-rate type curves for finite flow-capacity fractures. Dimensionless pressure drop P_{WD} has been plotted as a function of dimensionless time t_{Dxf} on log-log paper with the dimensionless fracture flow capacity, F_{CD}, as a parameter. Dimensionless variables shown in Figure 10–8 are defined as follows:

Dimensionless pressure p_{WD}:

$$\psi(p_{WD}) = \frac{kh\lfloor\psi(p_i) - \psi(p_{wf})\rfloor T_{sc}}{50{,}300\, q_g T P_{sc}} \tag{10–7}$$

Dimensionless time t_{Dxf}:

$$t_{Dxf} = \frac{0.0002637kt}{\phi\mu_i c_i x_f^2} \tag{10–4}$$

Dimensionless flow capacity F_{CD}:

$$F_{CD} = \frac{k_f w}{k x_f} \tag{10–5}$$

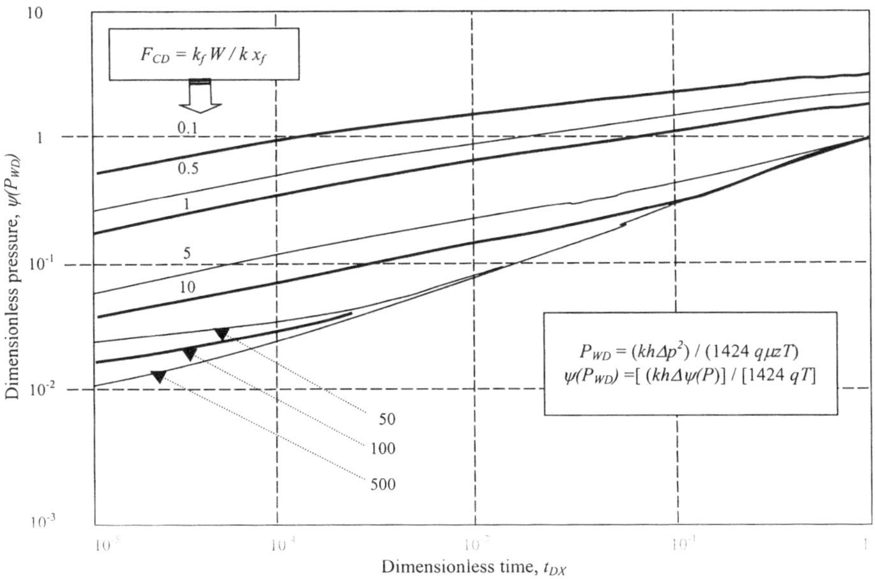

Figure 10–8. Log-log type curve for finite capacity vertical fractures (constant well rate) (after Cinco-Ley and Samaniego).[6]

This definition of the dimensionless fracture flow capacity is slightly different from that used by earlier investigators, but appears more convenient. The definition of Prats *et al.*[5] in our terms is

$$\alpha = \frac{\pi}{2F_{CD}} \tag{10–5a}$$

The definition of Cinco-Ley and Samaniego[6] is

$$C_r = \frac{F_{CD}}{\pi} \tag{10–5b}$$

Fracture conductivity:

$$k_f W = F_{CD}(kx_f) \tag{10–5c}$$

Fracture half-length x_f:

$$x_f = \sqrt{\frac{0.0002637k(t)_M}{\phi\mu_i c_i (t_{Dxf})_M}} \tag{10–6}$$

Fracture skin s_f:

$$s_f = In\left(\frac{x_f}{r_w}\right) \quad (10\text{–}5d)$$

The value of fracture skin s_f may be calculated by using Eq. 10–6a or from Table 10–1 or Figure 10–3.

$$s_f = \left[s_f + In\left(\frac{x_f}{r_w}\right)\right]_{Table10-1} - In\left(\frac{x_f}{r_w}\right) \quad (10\text{–}6a)$$

In Figure 10–8, dimensionless fracture flow capacity F_{CD} ranges from 0.2 to 100. Note that the higher values of F_{CD} normally correspond to higher fracture flow capacity. However, higher values of F_{CD} also may be caused by lower formation permeability or short fracture length. The infinite flow-capacity fracture solution is shown in Figure 10–8 by the dotted line. A curve for F_{CD} values of 100 or greater should approximately represent an infinite flow-capacity fracture. This accounts for the utility of the infinite flow capacity type curves of Gringarten *et al.* For the analysis of wells stimulated with conventional fractures. For greater values of t_{Dxf}, Cinco-Ley and Samaniego type curves may be used. For t_{Dxf} values smaller than 10^{-5}, type curves are influenced by porosity and compressibility in the fracture.

Example 10–2[13] *Analyzing Drawdown Test Using Agarwal et al. Constant–Rate Type Curves*

The reservoir and drawdown data are presented in Table 10–4. Estimate the following:

1. Formation permeability k
2. Fracture half length x_f
3. Fracture conductivity $k_f w$
4. Skin in the fracture s_f
5. Compare the results with the semilog analysis technique.

Solution A log-log cross plot of $\psi(p_i) - \psi(p_{wf})$ versus time was prepared on a sheet of tracing paper (Figure 10–9). Notice that at early times the slope is smaller than 0.5. The data curve was placed over the type curve for a constant-rate finite conductivity vertical fracture (Figure 10–9) and was displaced until a match was obtained (Figure 10–10). This match was obtained for

$$F_{CD} = \frac{k_f w}{k x_f} = k_{fD} w_{fD} = 2\pi$$

Table 10–4
Pressure Drawdown Data for a Gas Well Crossed by a Constant-Rate Finite Conductivity Vertical Fracture (T = 710°R; h = 24 ft; r_w = 0.4271 ft; c_t = 0.0002155 psi^{-1}; q_{sc} = 12.255 mmscfd; ϕ = 0.1004 fraction; μ = 0.01925 cP)

Time t (hr)	$\psi(p_i) - \psi(p_{wf})$ (mmscfd)	Time t (hr)	$\psi(p_i) - \psi(p_{wf})$ (mmscfd)
0.25	72.10	40.00	276.22
0.50	83.08	50.00	295.14
1.00	94.75	60.00	313.89
2.50	121.85	70.00	326.78
5.00	149.35	80.00	336.87
10.00	182.60	90.00	349.32
20.00	225.12	100.00	358.45
30.00	253.78	150.00	399.01

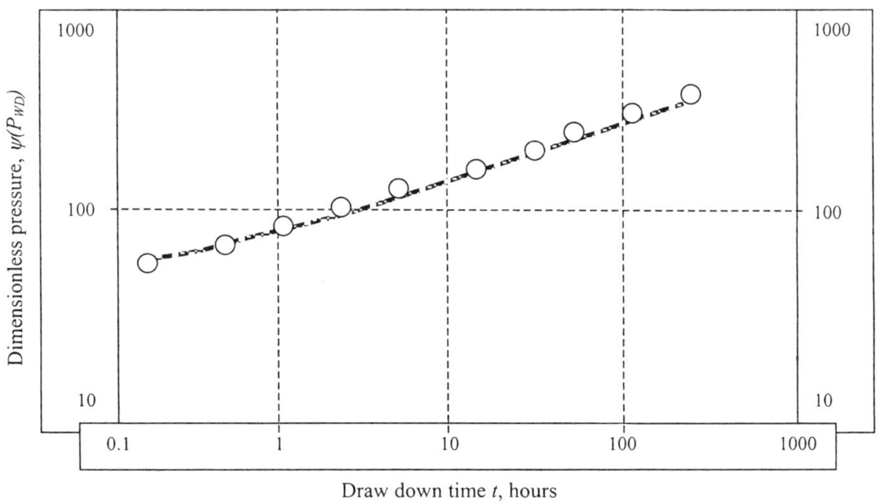

Figure 10–9. Data curve for Example 10–2.

and indicates that the test was not run long enough to reach the semilog straight line. Match points from Figure 10–10:

$$t = 10 \text{ hr}, \quad \Delta\psi = \psi(p_i) - \psi(p_{wf}) = 100 \text{ mmpsia}^2/\text{cP},$$
$$t_D = 0.65, \quad \psi(p_{WD}) = 0.45, \quad F_{CD} = 5$$

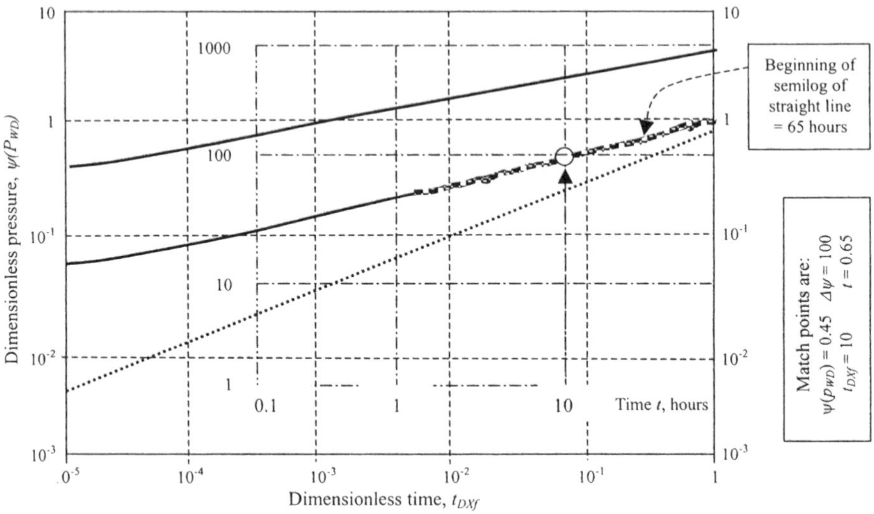

Figure 10–10. Application of type curves matching techniques—Example 10–2.

Calculate formation permeability k from Eq. 10–7:

$$k = \frac{50{,}300\, q_{sc} T P_{sc}}{h\psi(p_i) - \psi(p_{wf}) T_{sc}} [\psi(p_{WD})]_M$$

$$= \frac{50{,}300 \times 12.255 \times 10^3 \times 710 \times 14.65 \times 0.45}{24 \times 100 \times 10^6 \times 520} = 2.32 \text{ mD}$$

Calculate the half fracture length from Eq. 10–6:

$$x_f = \sqrt{\frac{.0002637k}{\phi\mu_i c_i}} \left[\frac{t}{t_{Dxf}}\right]_M$$

$$= \sqrt{\frac{0.0002637(2.32)(10)}{0.1004(0.01925)(0.0002155)(0.65)}} = 150.3 \text{ ft}$$

Calculate fracture conductivity $k_f w$ from Eq. 10–5a:

$$k_f w = 2\pi(x_f)(k) = 2(22/7)(2.32)(204.9) = 1.494 \times 10^3 \text{ mD-ft}$$

Calculate fracture skin factor with the use of Eq. 10–5d:

$$s_f = In\left[\frac{2r_w}{x_f}\right] = In\left[\frac{2 \times 0.4271}{150.3}\right] = -5.17$$

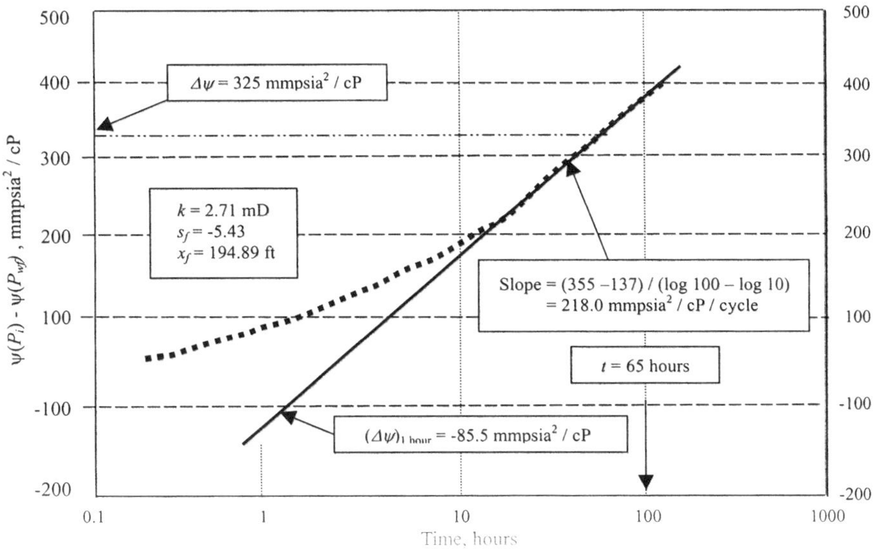

Figure 10–11. Semilog cross plot—Example 10–2.

Compare the value of fracture skin factor, s_f, with the use of Table 10–1. This can be done by entering Table 10–1 in the abscissa with $k_{fD}w_{fD} = 2\pi = 6.29$ and reading in the ordinate

$$s_f + In\left(\frac{x_f}{r_w}\right) = 0.86$$

From Eq. 10–6a:

$$s_f = 0.86 - In\left(\frac{x_f}{r_w}\right) = 0.86 - In\left(\frac{150.3}{.4271}\right) = 0.86 - 5.86 = -5.02$$

This skin indicates that the fracture is large enough to provide an improvement in well productivity, in spite of the fact that the dimensionless fracture conductivity has an intermediate value.

Semilog Analysis

Figure 10–11 is a semilog graph, which shows a straight line through the last points with a slope $m = 218$ mmpsia^2/cP/cycle. A conventional permeability is calculated from Eq. 5–40:

$$k = \frac{57.920 \times 10^6 \times 12.225 \times 14.65 \times 710}{218 \times 10^6 \times 24 \times 520} = 2.71 \text{ mD}$$

which compares with 2.32 mD calculated by type curve matching. A conventional skin is calculated from Eq. 5–41:

$$s_f = 1.151\left[\frac{-87.5 \times 10^6}{218 \times 10^6} - \log\frac{2.71}{0.1004 \times 0.01925 \times 0.0002155 \times 0.4271^2} + 3.23\right]$$

$$= 1.151[-0.4014 - 7.55 + 3.23] = -5.43$$

which compares with $s_f = -5.13$ calculated by type curve matching.

If pressures for early times are not available, the remaining data could match any log-log curves presented in Figure 10–11. Consequently fracture geometry parameters could not be estimated, and the only possible way to evaluate the data quantitatively would be to use the conventional semilog plot.

Further, it is worthwhile to examine that only a few points should form the semilog straight line based on the match of Figure 10–10. The semilog plot of Figure 10–11 shows that a couple of points are in the straight line. This includes all points to the right of the arrow in Figure 10–11. The reason for the difference lies in the fact that the analytical solutions approach the semilog straight line asymptotically. Consequently, it may be possible in practice to stretch the rule with regard to beginning of the straight line based on type curves and still obtain some acceptable results. This must be done with caution and, if possible, using the two types of cross plots.

10.4 Pressure Transient Analysis in MHF Gas Wells

Figure 10–12 shows the log-log curve of Cinco-Ley and Samaniego[7] for a finite conductivity vertical fracture. An interesting finding was for a well with a low or intermediate conductivity fracture ($k_{fD}w_{fD} < 300$), the slope at early times did not exhibit the typical one-half slope straight line.

The following key assumptions were used in developing this curve:

1. An isotropic, homogeneous, horizontal infinite reservoir of constant thickness h, permeability k, and porosity ϕ which are independent of pressure.
2. Viscosity and compressibility (slightly compressible fluids only) are constant.
3. Production comes from a vertically fractured well intersected by a fully penetrating finite conductivity fracture of width w, half-length L_f, and permeability k_f.

The matching procedure is carried out as indicated previously. Quantitative reservoir evaluation is carried out with the use of Eqs. 10–4 through 10–7 as follows.

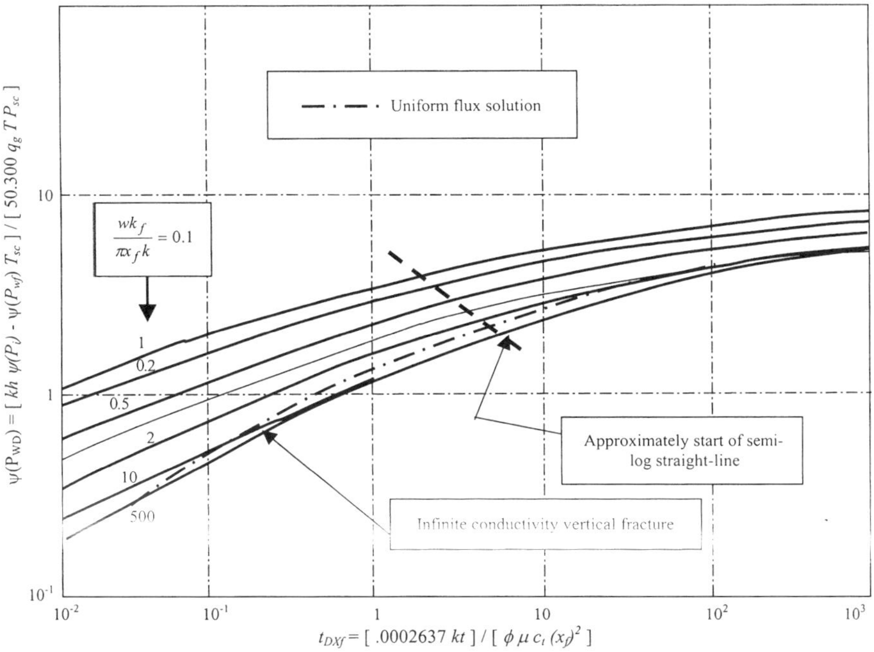

Figure 10–12. $\psi(P_{WD})$ versus t_D for a finite conductivity vertical fracture (after Cinco-Ley and Samaniego).[7]

Dimensionless time can be defined as a function of the half fracture length, x_f or L, as follows:

$$t_{Dxf} = \frac{0.0002637kt}{\phi\mu_i c_i x_f^2} = \frac{0.0002637kt}{\phi\mu_i c_i L^2} \tag{10–4}$$

Dimensionless pressure $\psi(p_{WD})$ is defined as function of real pressures by the relationship

$$\psi(p_{WD}) = \frac{kh\lfloor\psi(p_i) - \psi(p_{wf})\rfloor T_{sc}}{50{,}300\, q_g T P_{sc}} \tag{10–7}$$

where $\Delta\psi$ = incremental pressure, mmpsia^2/cP, equal to $\psi(p_i) - \psi(p_{wf})$ in a drawdown test and $\psi(p_{ws}) - \psi(p_{wf})$ in a buildup test; $\psi(p_i)$ = initial pressure, mmpsia^2/cP; $\psi(p_{wf})$ = flowing pressure, mmpsia^2/cP; $\psi(p_{ws})$ = shut-in pressure, mmpsia^2/cP; q_{sc} = flow rate, mmscfd; T = formation temperature, °R; T_{SC} = base temperature, °R; P_{SC} = base pressure, psi.

A new parameter called dimensionless fractured conductivity ($C_{fD}\eta_{fD}$) is introduced as

$$C_{fD}\eta_{fD} = \frac{k_f w}{\pi x_f k} \tag{10–7a}$$

where C_{fD} = dimensionless fracture storage capacity and η_{fD} = dimensionless fracture hydraulic diffusivity. Note that this parameter does not depend on porosity and total compressibility of the formation and fracture. The product of the following two dimensionless variables can also express the dimensionless fracture flow conductivity:

$$k_{fD} = \frac{k_f}{k} \tag{10–7b}$$

and

$$w_{fD} = \frac{w}{x_f} \tag{10–7c}$$

where k_{fD} = dimensionless fracture permeability and w_{fD} = dimensionless fracture width.

A decrease in the dimensionless fracture conductivity may be caused by a decrease in fracture permeability, an increase in fracture length, or both. Conversely, small values of dimensionless fracture conductivity may be caused by either low fracture permeability or large fracture length.

Combining Eqs. 10–7a and 10–7b leads to the relationship

$$k_{fd}w_{fD} = C_{fD}\eta_{fD} = \frac{k_f w}{k x_f} \tag{10–7d}$$

This equation is useful in the quantitative analysis of pressure behavior in wells with a finite conductivity vertical fracture. Table 10–1 shows a general correlation of the fracture skin factor (or pseudoskin factor) from which it is possible to evaluate s_f as a function of $w_{fD}k_{fD}$ and x_f/r_w. This is evaluated for finding whether the fracture is really contributing to well improvement. The application of Cinco-Ley and Samaniego's type curve for constant well rate case to analyze massive hydraulic fractured gas well is illustrated in Example 10–3.

Example 10–3[13] *Analyzing Drawdown Test Using Cinco-Ley and Samaniego Type Curves Long Enough to Reach and Pass the Start of the Semilog Straight Line*

The reservoir and drawdown data are given in Table 10–5. Table 10–6 shows the drawdown text data for conventional semilog analysis.

Table 10–5
Pressure Drawdown Data for a Gas Well Crossed by a Constant-Rate Finite Conductivity Vertical Fracture (T = 710°R; h = 22 ft; r_w = 0.42 ft; c_t = 0.00028 psi^{-1}; q_{sc} = 10.235 mmscfd; ϕ = 0.108 fraction; μ = 0.01725 cP)

Time t (hr)	$\psi(p_i)-\psi(p_{wf})$	Time t (hr)	$\psi(p_i)-\psi(p_{wf})$
1	96.12	34	298.12
2	115.18	30	322.19
3	143.32	40	348.43
4	159.45	50	371.54
5	172.05	60	393.57
6	185.42	70	411.42
7	197.88	80	426.12
8	207.92	90	439.16
9	216.46	100	454.19
10	221.39	120	474.26
12	238.24	150	499.29
14	251.28	200	537.35
16	262.39	250	569.47
20	282.42	300	586.45

Table 10–6
Drawdown Test Data for Conventional Semilog Analysis

Time t (hr)	$\psi(p_{wf})$ (mmpsia^2/cP)	Time t (hr)	$\psi(p_{wf})$ (mmpsia^2/cP)
1	800.00	40	742.11
2	794.24	50	734.18
3	789.05	60	732.76
4	784.75	70	730.55
5	781.67	80	725.13
6	782.72	90	722.23
7	778.00	100	720.14
8	775.22	120	712.25
9	774.19	150	705.00
10	770.20	200	690.17
15	765.15	250	684.12
20	755.25	300	677.12
30	750.35	350	673.14

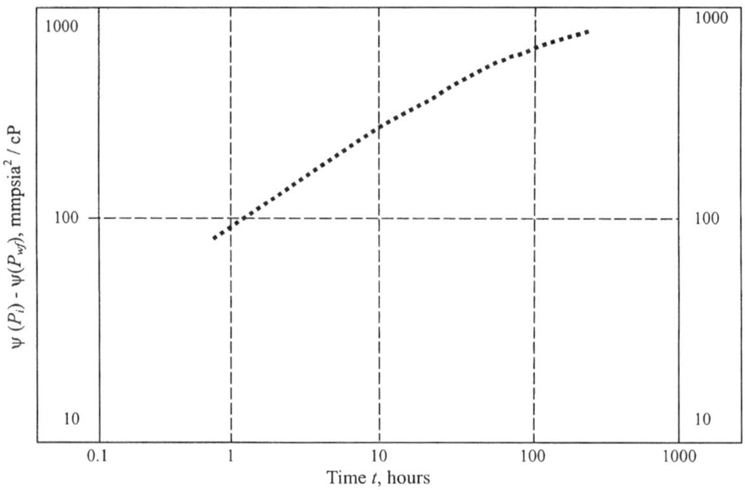

Figure 10–13. Data curve—Example 10–3.

Estimate the following:

1. Formation permeability k
2. Fracture half-length x_f
3. Fracture conductivity $k_f w$
4. Skin in the fracture s_f
5. Compare the result with that obtained using the semilog analysis technique

Solution A log-log cross plot of $\psi(p_i) - \psi(p_{wf})$ versus time was prepared on a sheet of tracing paper (Figure 10–13). Notice that early times the slope is smaller than 0.5. The data curve was placed over the type curve for a constant-rate finite conductivity vertical fracture (Figure 10–12) and was displaced until a match was obtained (Figure 10–14). This match was obtained for $F_{CD} = k_f w / k x_f = k_{fD} w_{fD} = 2\pi$ and indicates that the test was run long enough to reach the semilog straight line.

Match points from Figure 10–14: $t = 100$ hr, $\Delta\psi = \psi(p_i) - \psi(p_{wf}) = 100$ mmpsia²/cP, $t_D = 5.75$, and $\psi(p_{WD}) = 1.30$.

Calculate formation permeability k from Eq. 10–7:

$$k = \frac{50{,}300\, q_{sc} T P_{sc}}{h \psi(p_i) - \psi(p_{wf}) T_{sc}} [\psi(p_{WD})]_M$$

$$= \frac{50{,}300 \times 10.235 \times 10^3 \times 710 \times 14.65 \times 1.3}{22 \times 100 \times 10^6 \times 520} = 6.09 \text{ mD}$$

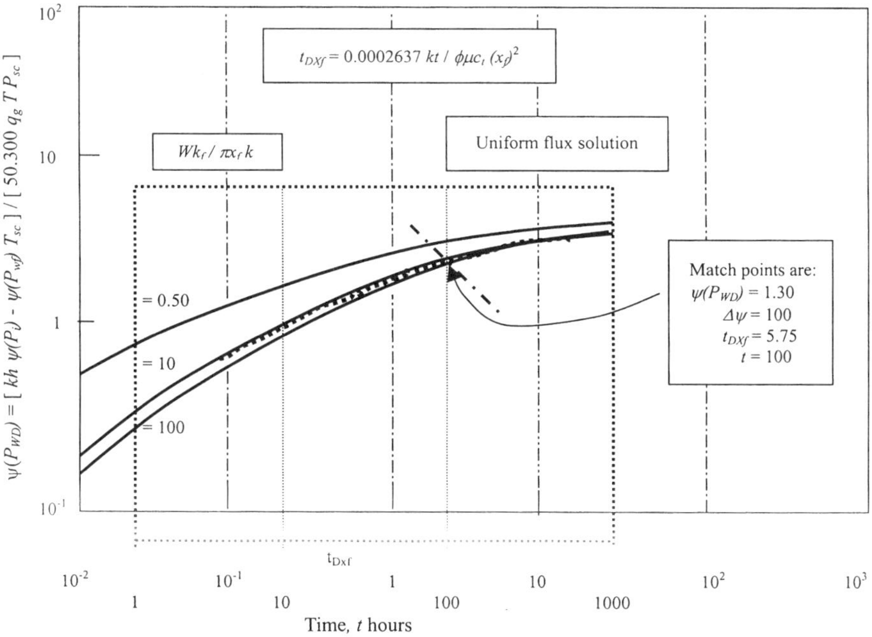

Figure 10–14. Type curve match—Example 10–3.

Calculate the half fracture length from Eq. 10–6:

$$x_f = \sqrt{\frac{0.0002637k}{\phi\mu_i c_i}\left[\frac{t}{t_{Dxf}}\right]_M}$$

$$= \sqrt{\frac{0.0002637(6.09)(100)}{(0.108)(0.01725)(0.00028)(5.75)}} = 231.4 \text{ ft}$$

Calculate fracture conductivity $k_f w$ from Eq. 10–5c:

$$k_f w = 2\pi(x_f)(k) = 2(22/7)(6.09)(231.4) = 8.858 \times 10^3 \text{ mD-ft}$$

Calculate the fracture skin factor with the use of Table 10–1. This can be done by entering Table 10–1 in the abscissa with $k_{fD}w_{fD} = 2\pi = 6.283$ and reading in the ordinate

$$\lfloor s_f + In(x_f/r_w)\rfloor_{Table 10-1} = 0.88$$

Then from Eq. 10–6a:

$$s_f = 0.88 - In\left(\frac{231.4}{.42}\right) = -5.43$$

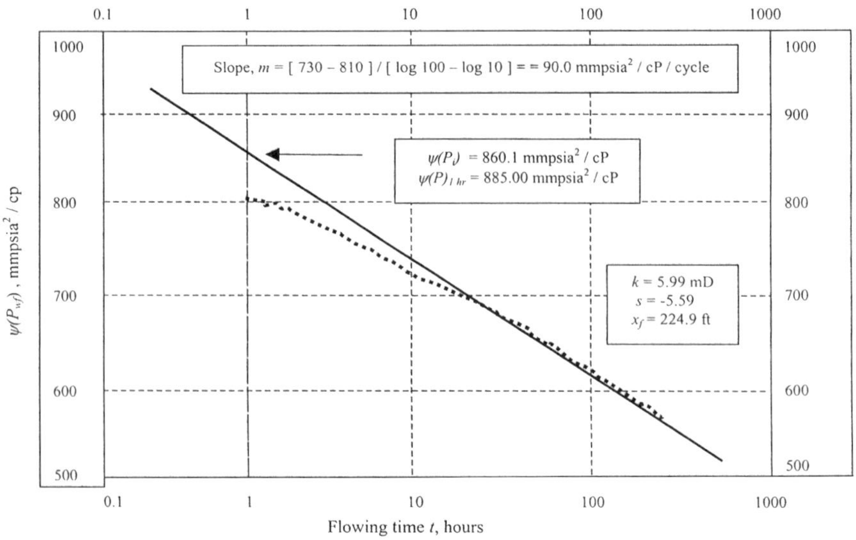

Figure 10–15. Conventional semilog drawdown test analysis for fractured gas well.

This skin indicates that the fracture is large enough to provide an improvement in well productivity, in spite of the fact that the dimensionless fracture conductivity has an intermediate value. Figure 10–15 shows a semilog plot of $\psi(p_{wf})$ versus time. From this figure the following information is obtained:

$$\text{Slope } m = \frac{730 - 810}{\log 100 - \log 10} = -90.0 \text{ mmpsia}^2/\text{cP/cycle}$$

$$p_{1hr} = 885.00 \text{ mmpsia}^2/\text{cP}$$

Conventional formation permeability calculations using Eq. 5–40:

$$k = \frac{57.921 \times 10^{-6} \times 10.235 \times 710 \times 14.65}{90.0 \times 10^6 \times 22 \times 520} = 5.99 \text{ mD}$$

A conventional skin calculation using Eq. 5–41:

$$s = 1.151\left[\frac{885.0 - 860.1}{-90.0} - \log\left(\frac{5.99}{0.108 \times 0.01725 \times 0.00028 \times 0.42^2}\right) + 3.23\right]$$

$$= 1.151[-0.277 - 7.81 + 3.23] = -5.59$$

which compares with $s_f = -5.43$ calculated by type curve matching.

Calculate fracture length x_f from the following equation:

$$r'_w = r_w e^{-(S)} = 0.42 \times e^{-(-5.59)} = 112.45 \text{ ft}$$

Hence,

$$x_f = 2 \times r'_w = 2(112.54) = 224.90 \text{ ft}$$

which compares with $x_f = 231.4$ ft calculated by type curve matching.

If pressures for early times are not available, the remaining data could match any of the curves presented in Figure 10–12. Consequently, fracture geometry parameters could not be estimated, and the only possibility to evaluate the data quantitatively would be to use the conventional semilog plot.

Final Remarks

The evaluations and prediction of performance of low-permeability gas wells stimulated by massive hydraulic fracturing warrant the following statements:

1. Pressure transient methods (both type-curve and square root graph), based on the concept of infinite flow-capacity fracture, are not adequate for evaluating MHF gas wells with finite flow capacity fracture.
2. Low-permeability MHF gas wells generally produce at a constant bottom-hole pressure rather than at a constant rate; the constant well pressure type curves appear more appropriate for analyzing the performance data. However, where production rates are reasonably constant or vary smoothly with bottom-hole pressure, constant-rate type curves should be used.
3. Finite flow-capacity type curves presented here for both the constant-pressure and constant-rate cases are intended for use with drawdown data. However, they may be used to analyze pressure buildup data, if producing time before shut-in is sufficiently long ($t_P + \Delta t \approx t_P$) so that the buildup data are least affected.
4. In low-permeability MHF wells, testing times may not be long enough because of practical limitations to permit semilog pressure analysis for determining formation.
5. The data of Cinco-Ley and Samaniego need to be extended to earlier times for MHF applications. Square root graphs with log-log type curves should be helpful.
6. The use of a mathematical model, such as the MHF simulator, with type curve analysis can provide a good prediction of well performance.

10.5 Fracture Characteristics Estimation Using Pressure Transient Testing

This section presents theoretical and practical aspects of methods used to determine formation permeability, fracture length, and fracture conductivity in low-permeability, hydraulically fractured gas reservoirs.

Horner Analysis

Plot buildup test data on a conventional Horner graph; determine the slope m, and thus estimate formation permeability k from

$$k = \frac{57.92 \times 10^6 q_g T P_{SC}}{m h T_{SC}}, \text{mD} \tag{10–8}$$

and fracture skin factor s_f from

$$s_f = 1.151\left[\frac{\psi(p)_{1hr} - \psi(p_{wf})}{m} - \log\left(\frac{k}{\phi\mu_g c_t r_w^2}\right) + 3.23\right] \tag{10–9}$$

Calculate fracture half-length from

$$x_f = 2r_w e^{-S_f} \tag{10–10}$$

Linear Flow Analysis: High Conductivity Fractures

When linear flow into a fracture dominates (at earliest times), a plot of $\psi(\Delta p)$ versus a square-root-of-time function will result in a straight line with slope m_{lf}, related to fracture half-length and formation permeability:

$$x_f = \frac{4.064 q_g \beta_g}{m_{lf} h} \cdot \left(\frac{\mu_g}{k\phi c_t}\right)^{0.5} \tag{10–11}$$

Calculate fracture skin factor s_f:

$$s_f = In\left(\frac{2r_w}{x_f}\right) \tag{10–12}$$

The assumptions on which Eq. 10–11 is based limit its applicability in many cases. These limiting assumptions include the following:

1. Formation permeability k must be available, if we wish to estimate x_f.
2. High fracture conductivity (but not infinite) and F_{CD} is greater than 100.
3. Earliest-time data is dominated by linear flow and no wellbore storage distortion.

Type Curve Analysis

Several type curves[3,4] have potential application to analysis of transient tests in low-permeability fracture gas reservoirs. Particularly important are the curves of Cinco-Ley and Samaniego[6,7], and Agarwal *et al.*[1] for finite conductivity fracture. These type curves have been discussed in this chapter.

Bilinear Flow Analysis: Low Conductivity Fractures

In the case of the bilinear flow regime, a Cartesian graph of $\psi(\Delta p)$ versus $t^{1/4}$ would yield a straight line. From the slope, the fracture permeability width, $k_f w$, may then be calculated using

$$k_f w = \left(\frac{44.1 q_g \beta_g \mu_g}{m_{bl} h}\right)^2 \cdot \left(\frac{1}{k\phi c_t}\right)^{0.5} \tag{10–13}$$

where q_g is in ft^3/bbl and β_g is in bbl/ft^3. Note that the reservoir permeability k is calculated from a semilogarithmic graph. For low conductivity fractures where only bilinear flow is evident, an additional step is necessary. Figure 10–3 is a graph between F_{CD} and $s_f + In(x_f/r_w)$. F_{CD} is defined by $F_{CD} = k_f w/kx_f$. Since k is known from conventional analysis technique and $k_f w$ from Eq. 10–5, assume an x_f and calculate F_{CD}. From Figure 10–3, obtain $s_f + In(x_f/r_w)$ and since s_f is known from buildup tests using Eq. 10–12 then solve for x_f and compare with the assumed value. Continue this trial-and-error process until the assumed and calculated values agree. Use and applications of these equations are illustrated in the following example.

Example 10–4[13] *Analyzing Buildup Data for Hydraulically Fractured Gas Well Using Log-Log, Horner, and Specialized Plots*

A post treatment test was conducted in a hydraulically fractured gas well and data are plotted in Figure 10–17. Gas well and reservoir data are as follows: $r_w = 0.39$ ft; $\phi = 0.12$ fraction; $h = 45$ ft; $T = 220°$F; $c_f = 0.000325$ psi^{-1}; $P_{SC} = 14.7$ psia; $T_{SC} = 520°$R; $q_{sc} = 75.0$ mscfd; $\gamma_g = 0.72$; and $\mu_g = 0.01652$ cP.

Solution Figure 10–16 is a log-log diagnostic plot and shows that nearly all of the data lie on a line of slope ½. The pressure derivative in this case is parallel to pressure change and offset by a factor of 2. This corresponds to the response labeled linear flow in Figure 10–17. Since there are no data showing a flat pressure derivative trend, an accurate value of permeability k can be determined from these data. Figure 10–17 is a Horner plot which shows that a value of permeability k equal to 0.35 mD and is determined from last

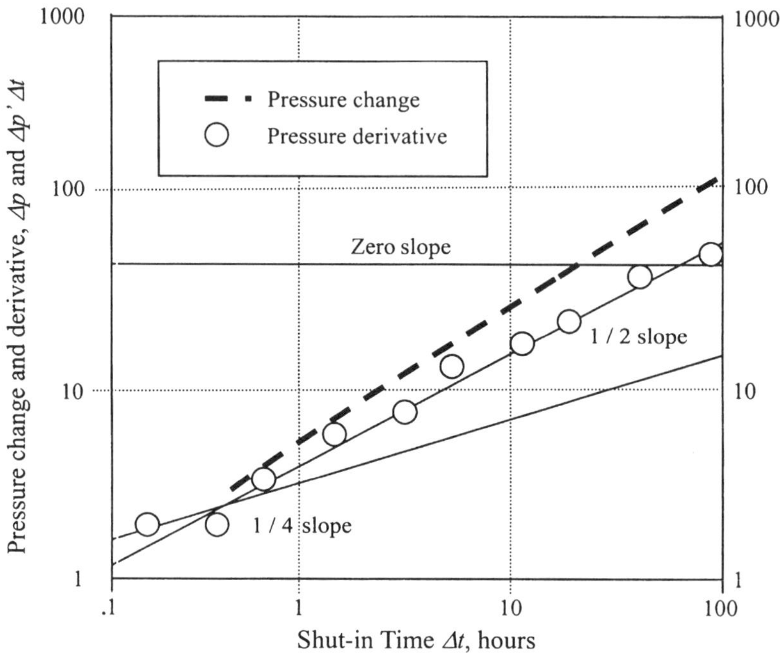

Figure 10–16. Log-log diagnostic plot—Hydraulically fractured gas well.

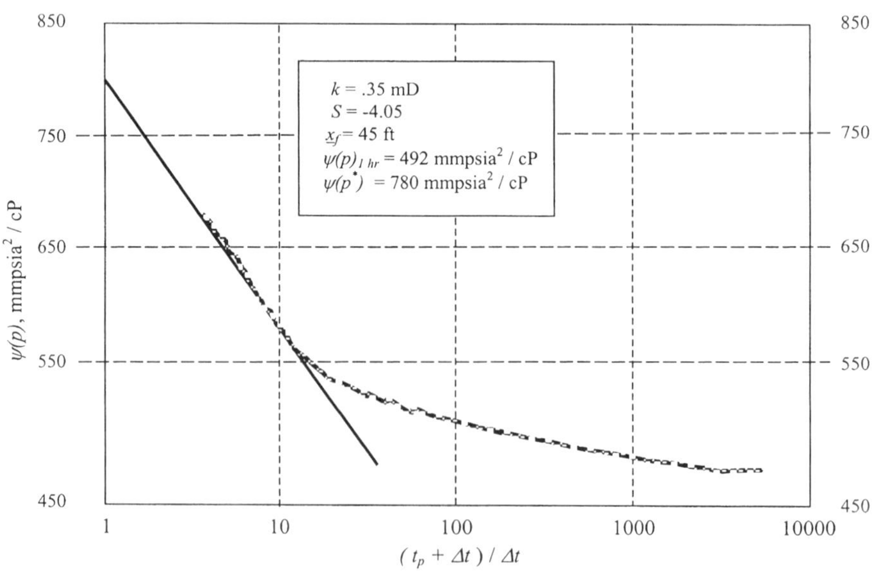

Figure 10–17. Horner plot—Hydraulically fractured gas well.

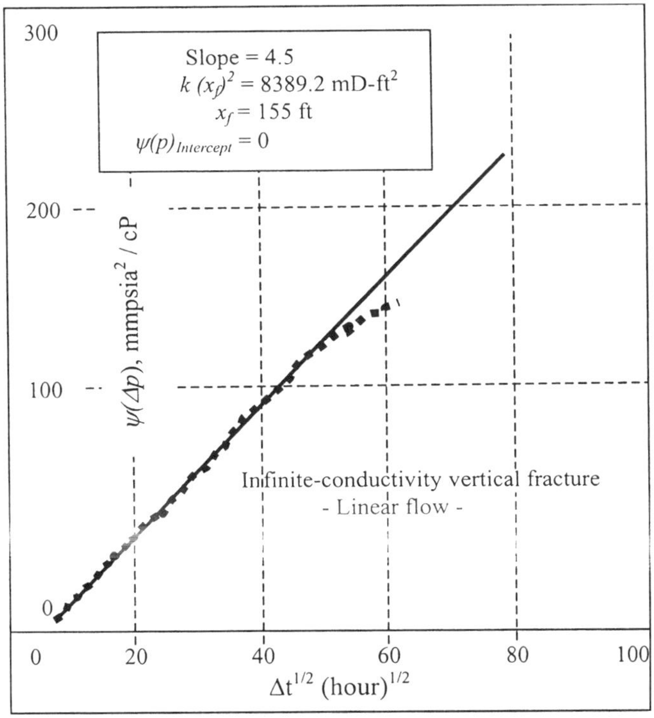

Figure 10–18. $\psi(\Delta p)$ versus $\Delta t^{1/2}$, $(\text{hr})^{1/2}$.

data points. Figure 10–18 is a specialized plot, $\lfloor \psi(\Delta p)$ versus $\Delta t^{1/2} \rfloor$; all but a few late-time data points fit on the same line, which has a slope of $m_{lf} = 4.5$. Using Eq. 10–11, the fracture half-length is computed as follows:

$$kx_f^2 = \left(\frac{\frac{4.064 \times 75.000 \times 10^3 \times 0.0525}{5.615}}{45 \times 4.50} \right) \cdot \frac{0.01652}{0.12 \times 0.00325} \text{ mD-ft}^2$$
$$= 14.073^2 \times 42.36 = 8389.2 \text{ mD-ft}^2$$

Using the maximum permeability k value estimated from the Horner plot,

$$x_f = \sqrt{\frac{8389.4}{0.35}} = 155 \text{ ft}$$

Using the maximum permeability value estimated from the Horner plot, $x_f = 45$ ft. Since the permeability could be less, this value is a minimum. The value for Δp_{int} determined from Figure 10–18 is approximately zero. Referring to Figure 10–18, this indicates that the fracture conductivity is effectively infinite (linear flow).

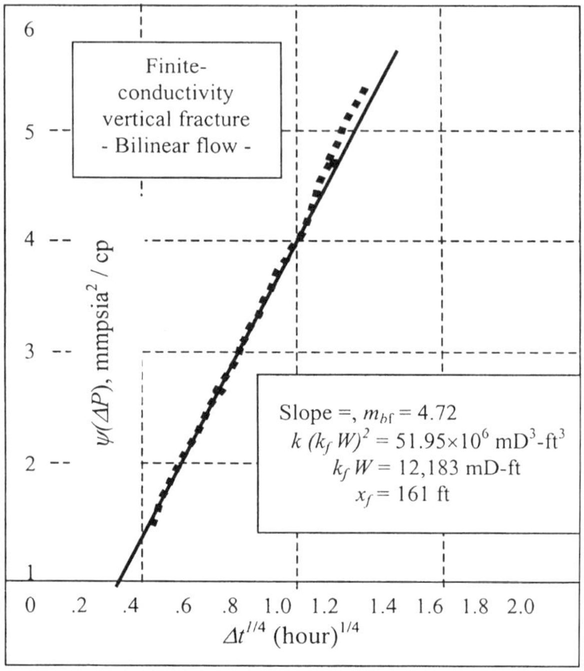

Figure 10–19. $\psi(\Delta p)$ versus $\Delta t^{1/4}$, $(\text{hr})^{1/4}$.

Figure 10–19 is $\Delta\psi(p)$ versus $\Delta t^{1/4}$—a specialized plot for the first few data points. From Eq. 10–13:

$$k(k_f w)^2 = \left(\frac{\frac{44.1 \times 57.0 \times 1000 \times 0.0525 \times 0.01652}{5.615}}{4.72 \times 45}\right)^4 \cdot \frac{1}{0.12 \times 0.01652 \times 0.000325}$$

$$= (2.405)^4 \times \frac{1}{0.12 \times 0.01652 \times 0.000325}$$

$$= 51.95 \times 10^6 \text{ mD}^3\text{-ft}^2$$

Therefore, $(k_f w)^2 = \frac{51.95 \times 10^6}{0.35} = 148.43 \times 10^6$ mD^3-ft^2 and $k_f w = 12{,}183$ mD-ft, which serves as a minimum.

To calculate fracture skin factor, s_f, follow these steps:

1. Using Eq. 10–5, find

$$F_{CD} = \frac{k_f w}{k x_f} = \frac{12{,}183}{0.35 \times 155} = 224.75$$

2. From Figure 10–3, find $s_f + In(x_f/r_w) = 0.76$.
3. Find fracture skin factor $s_f = 0.76 - In(\frac{155}{0.39}) = -5.33$.
4. Calculate fracture half-length $x_f = 2r_w e^{-S_f} = 2 \times 0.39e^{-(-5.33)} =$ 161 ft.

For fractures that have effectively infinite conductivity, a conventional technique for determining fracture half-length is from the skin value determined from Horner analysis. From Eq. 10–10:

$$x_f = 2r_w e^{-s} = 2 \times 0.39e^{-(-4.05)} = 45 \text{ ft}$$

Discussion

In this case, although the pressure buildup data were acquired for more than 450 hr, the pseudoradial regime has still not developed in late time. Massive hydraulic fractures are designed to reach up to 2000-ft fracture half-length. Such fractures should produce a transient response dominated by bilinear and linear flow for several months. Thus Eq. 10–10, $x_f = 2r_w e^{-s}$, can rarely be used in practice, because this equation applies only for infinite conductivity fractures, and the fracture half-length can be underestimated from finite-conductivity fractures using this approach. This example shows, however, that if the permeability k has been determined, perhaps from a pretreatment transient test, then fracture length and fracture conductivity can be computed from the bilinear and linear flow regimes.

10.6 Pretreatment Testing of Hydraulic Fractured Candidate

The well testing techniques described in previous chapters can be applied to gas wells that are candidates for hydraulic fracturing. However, low reservoir permeability prolongs wellbore storage, which delays the appearance of the recognizable and interpretable patterns from which reservoir permeability and other reservoir parameters may be determined. The time for the end of wellbore storage (in oilfield units) is

$$t(\text{hr}) = 3387(60 + 3.5s)\frac{\mu_g C}{kh} \tag{10–14}$$

Using typical values for a gas well, $\mu_g = 0.0245$ cP, but with $C = 10^{-1}$ bbl/psi, $h = 45$ ft, $k = 0.1$ mD, and $s = 0$ (best case), the corresponding time is, in hours,

$$t(\text{hr}) = 3387[60 + 3.5(0)]\frac{0.0245 \times 0.1}{0.1 \times 45} = 111$$

10.7 Pressure Transient Responses under Constant Rate

Low, Medium, and High Conductivity Hydraulic Fractured Gas Wells

Economides[8,9] gives comprehensive solutions for the pressure transient response under constant rate for a low ($F_{CD} = 1$), medium ($F_{CD} = 10$), and very high conductivity fracture ($F_{CD} = 100$), respectively. These figures, presented by Economides,[8] are given in terms of the dimensionless pressure p_D and the fracture dimensionless time t_{Dxf}, divided by the fracture dimensionless wellbore storage coefficient, C_{Df}, for a range of values of the C_{Df}. For the case of $F_{CD} = 1$ after a 45° straight line characteristic of wellbore storage, and after a transition period, there is a long log-log straight line with slope equal to $\frac{1}{4}$. Following an additional transition period, the well enters pseudoradial, infinite-acting reservoir flow. For the much larger conductivity fracture, a long linear flow is evident (with a log-log slope equal to $\frac{1}{2}$). Again, an early time 45° straight line and late-time pseudoradial flow are also present. The definition of the dimensionless variables for gas in oilfield units is

$$\psi(p_D) = \frac{kh\lfloor\psi(p_i) - \psi(p_{wf})\rfloor}{1424\, q_{sc} T} \tag{10–15}$$

Equation 10–15 includes no influence of the non-Darcy coefficient.

Including the influence of the non-Darcy coefficient, Eq. 10–15 becomes

$$\psi(p_i) - \psi(p_{wf}) = \frac{1424T\,\psi(p_D)}{kh} q_{sc} + \frac{1424TD}{kh} q_{sc}^2 \tag{10–16}$$

Equation 10–16 can be used to develop the transient IPR relationship for a fractured well.

Time:

$$t_{Dxf} = \frac{0.000264\Delta t}{\phi \mu c_t x_x^2} \tag{10–17}$$

Storage coefficient:

$$C_{Df} = \frac{5.615 C_S}{2\pi \phi c_t h x_x^2} \tag{10–18}$$

where

$$C_S = V_{Wb} \times C_{Wb} \tag{8–7}$$

Dimensionless fracture conductivity:

$$F_{CD} = \frac{k_f w}{k x_f} \tag{10–19}$$

A plot of pressure versus the square root of time on Cartesian coordinates forms a straight line from the slope of which the fracture half-length can be estimated if the reservoir permeability is known. For finite-conductivity fractures ($F_{CD} < 10$), the solutions by Cinco-Ley and Samaniego[6] suggest that while such a fracture controls the well response, bilinear flow (linear flow from the reservoir into the fracture and linear flow along the fracture into the well) will be dominant. During this time, a log-log plot of pressure versus time will form a straight line with a slope equal to 0.25. The slope of a Cartesian plot of pressure versus the quarter root of time yields the $k_f w$ product. This is also indicated in this chapter.

Example 10–5 *Calculating the Beginning and End of Bilinear Flow, Linear, and Pseudoradial Flow for a Low Conductivity Fractured Gas Well*

Well/reservoir data are as follows: $k = 10$ mD; $h = 55$ ft; $p_i = 4700$ psig; $T = 220°$F; $\mu_g = 0.02535$ cP; $r_w = 0.328$ ft; $c_t = 10^{-5}$ psi^{-1}; $C_S = 10^{-3}$ bbl/psi; $\phi = 0.12$; $x_f = 1000$ ft; $k_f w = 10{,}000$ mD-ft.

Solution From Eq. 10–18, the dimensionless storage coefficient is

$$C_{Df} = \frac{5.615 \times 10^{-3}}{2(3.14) \times 0.12 \times 10^{-5} \times 55 \times 1000^2} = 1.355 \times 10^{-5}$$

The required time in the well performance can be obtained from rearrangement of Eq. 10–17:

$$t = \frac{(t_{Dxf}/C_{Df})\phi \mu_g c_t x_f^2 C_{Df}}{0.000264k}$$

Since from References 9 and 14 and are at $C_{Df} = 1.355 \times 10^{-5}$, the beginning of bilinear flow is marked and is equal to $t_{Dxf}/C_{Df} = 6.5$, then

$$t = \frac{6.5 \times 0.12 \times 0.02535 \times 10^{-5} \times 1000^2 \times 1.355 \times 10^{-5}}{0.000264 \times 10}$$
$$= 0.00102 \text{ hr}$$

Also marked is the end of bilinear flow, which is at $t_{Dxf}/C_{Df} = 10^4$, leading to a real time of over 2.0 hr. From the same, the beginning of the infinite-acting pseudoradial flow is at $t_{Dxf}/C_{Df} = 1.355 \times 10^5$ and from Eq. 10–17 the real time would be

$$t = \frac{1.355 \times 10^5 \times 0.12 \times 0.02535 \times 10^{-5} \times 1000^2 \times 1.355 \times 10^{-5}}{0.000265 \times 10}$$
$$= 21.2 \cong 24 \text{ hr}$$

Therefore, the product $k_f w$ could be estimated through a test within a reasonable time (e.g., a 24-hr buildup and if the reservoir permeability is known); the fracture half-length could not (no infinite-conductivity behavior is evident).

Example 10–6 *Developing a Transient IPR Relationship for a Fractured Gas Well without and with Non-Darcy Coefficient Contribution*

Well/reservoir data are as follows: $k = 0.25$ mD; $h = 65$ ft; $p_i = 4700$ psig; $T = 220°$F; $\mu_g = 0.0245$ cP; $r_w = 0.328$ ft; $c_t = 1.08 \times 10^{-4}$ psi^{-1}; well depth = 6500 ft; tubing i.d. = 2.44 inch; $p_{wf} = 1000$ psig; $p_{tb} = 780$ psig; $\phi = 0.15$; $x_f = 1250$ ft; and $k_f w = 3125$ mD-ft.

Solution From Eq. 10–17, calculate the dimensionless fracture conductivity:

$$F_{CD} = \frac{k_f w}{k x_f} = \frac{3.125}{0.25 \times 1250} = 10$$

Calculate the wellbore storage coefficient from Eq. 8–7:

$$C_S = V_{Wb} \times C_{wb} = \frac{3.14 \times 2.44^2 \times 6500 \times 1.1 \times 10^{-3}}{4 \times 144 \times 5.615}$$
$$= 4.1 \times 10^{-2} \text{ bbl/psi}$$

where $C_{Wb} = 2/(p_{wf} + p_{tf}) = 2/(1000 + 780) = 1.1 \times 10^{-3}$ psi^{-1}, and from Eq. 10–18;

$$C_{Df} = \frac{5.615 C_S}{2\pi \phi c_t h x_f^2} = \frac{5.615 \times 4.1 \times 10^{-2}}{2 \times 3.14 \times 0.15 \times 1.08 \times 10^{-4} \times 65 \times 1250^2}$$
$$= 2.2 \times 10^{-3}$$

Time $t = 10$ days and from Eq. 10–17:

$$t_{Dxf} = \frac{0.000264 \times 10}{0.15 \times 0.0245 \times 1.08 \times 10^{-4} \times 1250^2} = 0.1022 = 1.02 \times 10^{-1}$$

and therefore

$$t_{Dxf}/C_{Df} = \frac{1.02 \times 10^{-1}}{2.2 \times 10^{-3}} = 46.4$$

Since from References 9 and 14 and are at $t_{Dxf}/C_{Df} = 46.4$ and $C_{Df} = 2.2 \times 10^{-3}$, $P_D = 0.105$; then using Eq. 10–15:

$$\psi(p_D) = \frac{kh\lfloor\psi(p_i) - \psi(p_{wf})\rfloor}{1424\,qT}$$

Substitution of the known variables and rearranging yields

$$0.105 = \frac{0.25 \times 65[1250 \times 10^6 - \psi(p_{wf})]}{0.105 \times 1424 \times (200 + 460)}$$

Therefore

$$q = \frac{0.25 \times 65[1250 \times 10^6 - \psi(p_{wf})]}{0.105 \times 1424 \times 660}$$

$$= 1.6467 \times 10^{-4}[1250 \times 10^6 - \psi(p_{wf})] \qquad (10\text{–}20)$$

This will be the transient IPR relationship without non-Darcy contribution. Equation 10–16 with non-Darcy contribution would become

$$\psi(p_i) - \psi(p_{wf}) = \frac{1424Tp_D}{kh}q + \frac{1424D}{kh}q^2$$

Assuming $D = 4.25 \times 10^{-5}$ (mscfd)$^{-1}$, then substitution of all known variables yields

$$\psi(p_i) - \psi(p_{wf}) = \frac{1424 \times 660 \times 0.105}{0.25 \times 65}q + \frac{1424 \times 4.25 \times 10^{-5}}{0.25 \times 65}q^2$$

$$= 6.073q + 3.7243 \times 10^{-3}q^2 \qquad (10\text{–}21)$$

Hence in quadratic form the above equation becomes

$$q = \frac{-6.073 + \{(6.073)^2 + 4(3.7245 \times 10^{-3}[\psi(p_i) - \psi(p_{wf})]\}^{0.5}}{2 \times 3.7243 \times 10^{-3}}$$

This will be the transient IPR relationship with non-Darcy coefficient contribution for a fractured gas well at $t = 10$ days (see Table 10–7). The results are shown in Figures 10–20 and 10–21.

Table 10–7
IPR Responses without and with Non-Darcy Flow Coefficient

Flowing pressure (psia)	$\psi(p_{wf})$ (mmpsia^2/cP) (1)	$\psi(P_i)-\psi(P_{wf})$ (mmpsia^2/cP) (2)	No turbulence Eq. 10–20 Flow rate (mmscfd) (3)	With turbulence Eq. 10–21 Flow rate (mmscfd) (4)
4700	1250.00	0	0	0
4000	1050.45	199.55	32.86	32.20
3000	640.22	609.78	100.40	94.92
2000	295.30	954.70	157.20	144.40
1000	135.15	1114.85	182.90	166.60
14.650			200.50	189.20

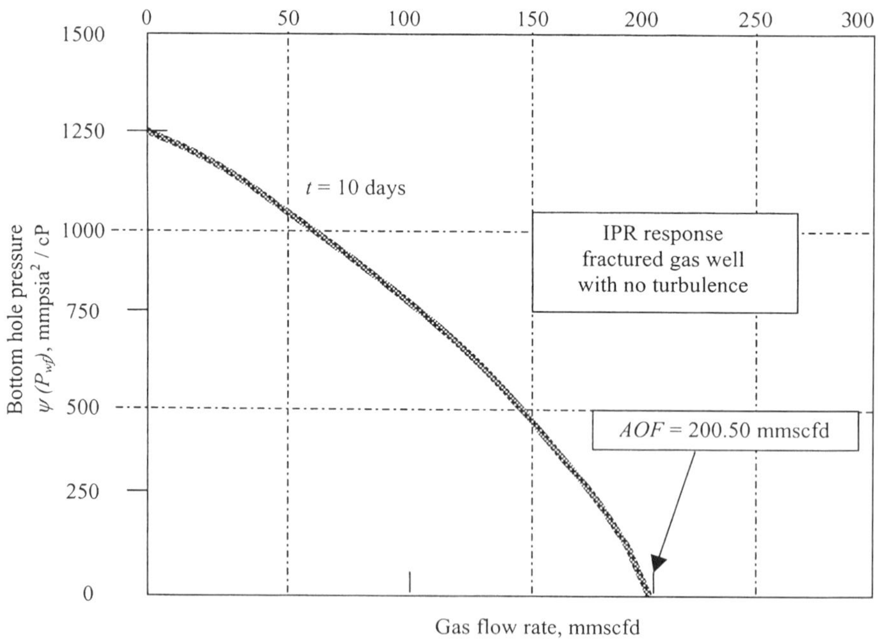

Figure 10–20. Inflow performance of fractured gas well with no turbulence.

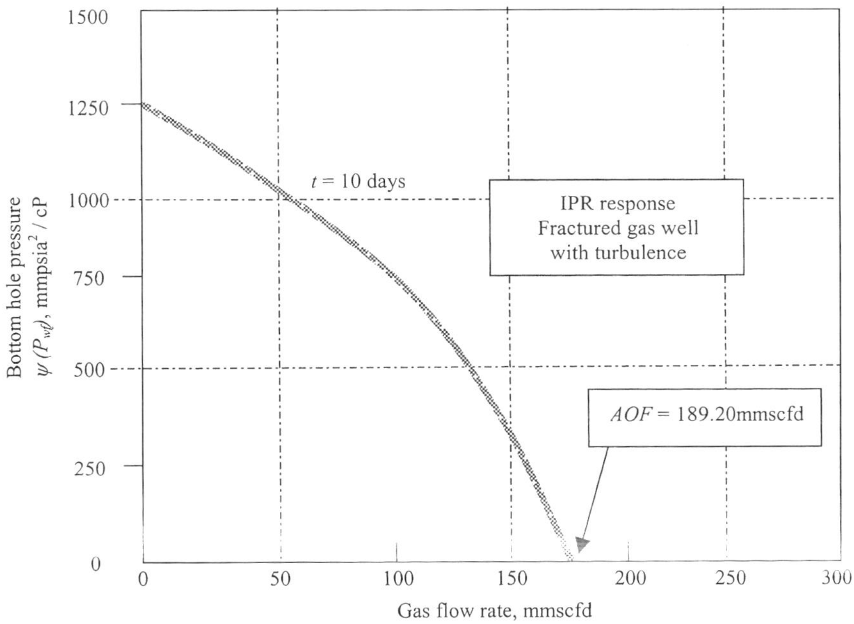

Figure 10–21. Inflow performance of fractured gas well with turbulence.

10.8 Summary

Based on the material presented in this chapter, the following remarks are pertinent:

- A new technique is presented to analyze data in the bilinear flow period. It is shown that, during this flow period, a graph of $\psi(p_{wf})$ versus $t^{1/4}$ yields a straight line when the slope is inversely proportional to $h_f(k_f b_f)^{1/2}$.
- New type curves are now available for pressure analysis of fractured gas wells, and the problem in the analysis is reduced considerably with the use of these type curves.
- Prefracture information about the reservoir is necessary to estimate fracture parameters.
- The type curve analysis method must be used simultaneously with the specific analysis methods $\psi(p_{wf})$ versus $t^{1/4}$, $\psi(p_{wf})$ versus $t^{1/2}$, and $\psi(p_{wf})$ versus $\log t$ to produce reliable results.

References and Additional Reading

1. Agarwal, R. G., Carter, R. D., and Pollock, C. B., "Evaluation and Prediction of Performance of Low-Permeability Gas Wells Stimulated by Massive Hydraulic Fracturing," *J. Petroleum Technol.* (March 1979) 362–372; *Trans. AIME* 67.
2. Cinco, H., and Samaniego, F. "Effect of Wellbore Storage and Damage on the Transient Pressure Behavior for a Well with a Finite-Conductivity Vertical Fracture," *Soc. Petroleum Eng. J.* (Aug. 1978) 253–264.
3. *Pressure Transient Testing Methods*, SPE Reprint Series No. 1, Society of Petroleum Engineers, Dallas, TX, 1980.
4. *Advances in Well Test Analysis*, SPE Monograph, Vol. 5, Society of Petroleum Engineers, Dallas, TX, 1977.
5. Prats, M., Hazebrock, P., and Sticker, W. R., "Effect of Vertical Fractures on Reservoir Behavior—Compressible Fluid Case," *Soc. Petroleum Eng. J.* (June 1962) 87–94; *Trans. AIME* 225, Vol. 8, Houston, TX, Oct. 1995.
6. Cinco-Ley, H., and Samaniego, F., "Transient Pressure Analysis for Fractured Wells," *J. Petroleum Technol.* (Sept. 1981) 1749–1766.
7. Cinco-Ley, H., and Samaniego, F., "Transient Pressure Analysis for Finite Conductivity Fracture Case versus Damage Fracture Case," SPE Paper 10179, 1981.
8. Economides, M. J., "Observations and Recommendations in the Evaluation of Tests of Hydraulically Fractured Wells," SPE Paper 16396, 1987.
9. Economides, M. J., and Nolte, K. G., *Reservoir Stimulation*, 2nd ed. Prentice-Hall, Englewood Cliffs, NJ, 1989.
10. Crawford, G. E., Hagedorn, A. R., and Pierce, A. E., "Analysis of Pressure Build-up Tests in a Naturally Fractured Reservoir," *J. Petroleum Technol.* (Nov. 1976) 1295–1300.
11. Kazemi, H., Seth, M. S., and Thomas, G. W., "The Interpretation of Interference Tests in Naturally Fractured Reservoirs with Uniform Fracture Distribution," *Soc. Petroleum Eng. J.* (Dec. 1969) 463–472.
12. Matthews, C. S., and Russel, D. G., *Pressure Build-up and Flow Tests in Wells*, SPE of AIME Monograph, Vol. 1, 25, Henry Doherty Series, 1967.
13. Amanat, U. C., *Pressure Transient Test Analysis User's Handbook.* Advanced TWPSOM Petroleum Systems Inc., Vol. 8, Houston, TX, Oct. 1995.
14. Michael, J. E., Daniel, H. A., and Christine, E. E., "Petroleum Production Systems," Prentice-Hall Inc; N.J. (1994). pp. 498–499.

Chapter 11

Fractured Gas Well Behavior Analysis Using Bilinear Flow Theory

11.1 Introduction

This chapter discusses the quantitative use and applications of type curves in gas well test analysis. The most generally useful type curves have been selected and are included herein. Fundamentals of type-curve use are presented and will allow the reader to understand and to apply newer type curves as they appear in the literature.

11.2 Special Type Curves for Pressure Analysis of Fractured Gas Wells

Transient behavior of a gas well with a finite conductivity vertical fracture has been simulated by Cinco and Samaniego.[9] Usually it is assumed that fractures have an infinite conductivity.

Finite-conductivity vertical fracture in an infinite slab is shown in Figure 11–1. Pressure data for each flow period should be analyzed using specific interpretation methods[8,9] such as

$$\Delta\psi \text{ versus } (t)^{1/4} \quad \text{for bilinear flow}$$

$$\Delta\psi \text{ versus } (t)^{1/2} \quad \text{for linear flow}$$

and

$$\Delta\psi \text{ versus } \log t \quad \text{for pseudoradial flow}$$

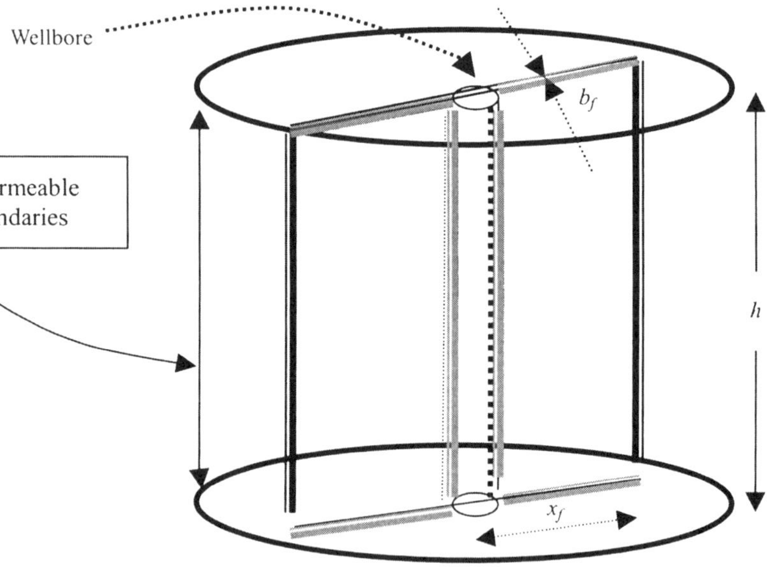

Figure 11–1. Finite conductivity vertical fracture in an infinite slab reservoir (after Cinco and Samaniego).[9]

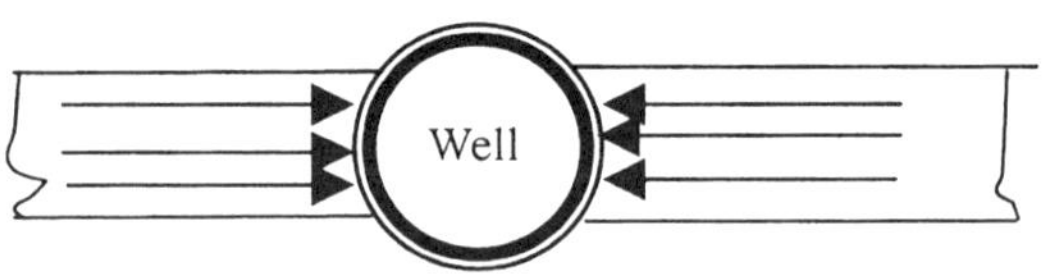

Figure 11–2. Fracture linear flow.[9]

11.3 Flow Regime Identification

There are four important flow regimes, which are discussed in the following sections.

Fracture Linear Flow

During this flow period, most of the fluid entering the wellbore comes from the expansion of the system within the fracture and the flow is essentially linear, as shown in Figure 11–2. Pressure response at the wellbore is given by

$$p_{WD} = \frac{3.546}{(k_f b_f)_D}\left[\frac{k_f \phi c_t}{k\phi_f c_{ft}} t_{Dxf}\right]^{0.5} \tag{11–1}$$

Hence

$$\Delta\psi = \psi(p_i) - \psi(p_{wf}) = \frac{0.275 q_g T}{b_f h}\left(\frac{t}{k_f \phi_f \mu_g c_t}\right)^{0.5} \tag{11–2}$$

Equation 11–2 indicates that a log-log graph of pressure difference against time yields a straight line whose slope is equal to one-half. A graph of pseudo-pressure versus the square root of time also gives a straight line whose slope depends on the fracture characteristics excluding the fracture half-length, x_f.

The fracture linear flow ends when

$$t_{Dxf} = \frac{0.01 (k_f b_f)_D^2}{\left(\frac{k_f \phi c_t}{k \phi c_t}\right)^2} \tag{11–3}$$

This flow period occurs at a time too early to be of practical use.

Bilinear Flow

This new type of flow behavior is called bilinear flow because two linear flows occur simultaneously. One flow is a linear within the fracture and the other is in the formation, as shown in Figure 11–3. The dimensionless wellbore pressure for the bilinear flow period is given by

$$p_{WD} = \frac{2.45}{[(k_f b_f)_D]^{0.5}}(t_{Dxf})^{1/4} \tag{11–4}$$

Equation 11–4 shows that a graph of p_{WD} versus $(t_{Dxf})^{1/4}$ produces a straight line whose slope is $2.45/[(k_f b_f)_D]^{0.5}$, intercepting the origin. Figure 11–6 presents that type of graph for different values of $(k_f b_f)_D$. Bilinear flow can be identified from a log-log plot of $\Delta\psi$ versus Δt, from which the pressure behavior for bilinear flow will exhibit a straight line whose slope is equal to

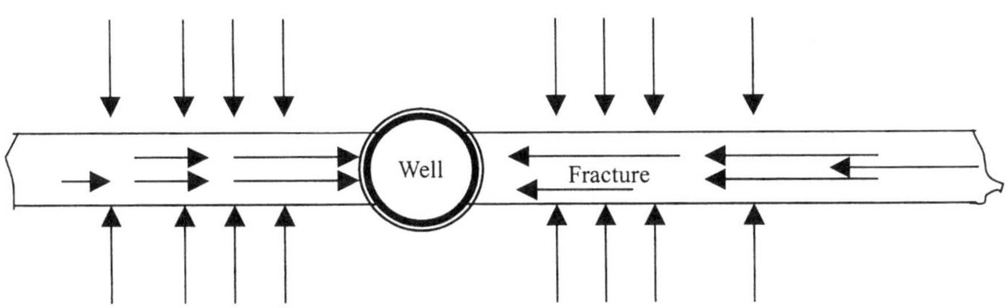

Figure 11–3. Bilinear flow.[9]

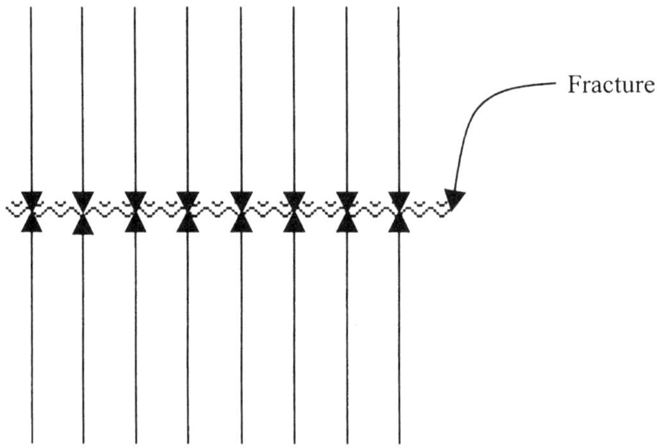

Figure 11–4. Formation linear flow.[9]

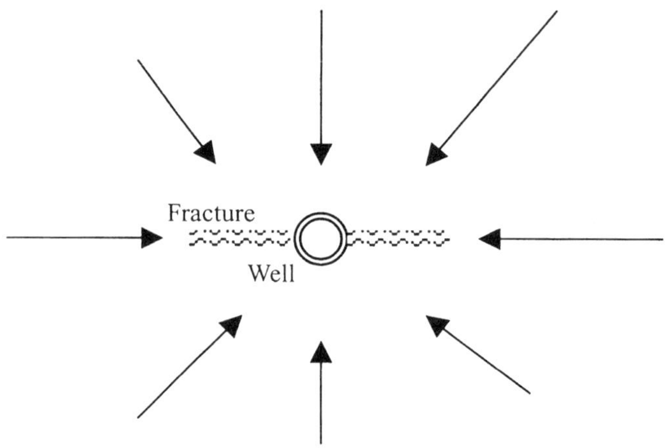

Figure 11–5. Pseudoradial flow.[9]

one-fourth to the linear flow period in which the slope is one-half. For buildup analysis of bilinear flow period, the pressure drop for gas may be expressed as

$$\psi(\Delta p) = \frac{444.75 q_g T}{h(k_f b_f)^{0.5}(\phi \mu_g c_t k)^{0.25}}(t)^{0.25} \tag{11–5}$$

Equation 11–5 indicates that a graph of $\psi(\Delta p)$ versus $t^{1/4}$ produces a straight line passing through the origin, whose slope m_{bf}, for gas, is given by

$$m_{bf} = \frac{444.75 q_g T}{h(k_f b_f)^{0.5}(\phi \mu_g c_t k)^{0.25}} \tag{11–6}$$

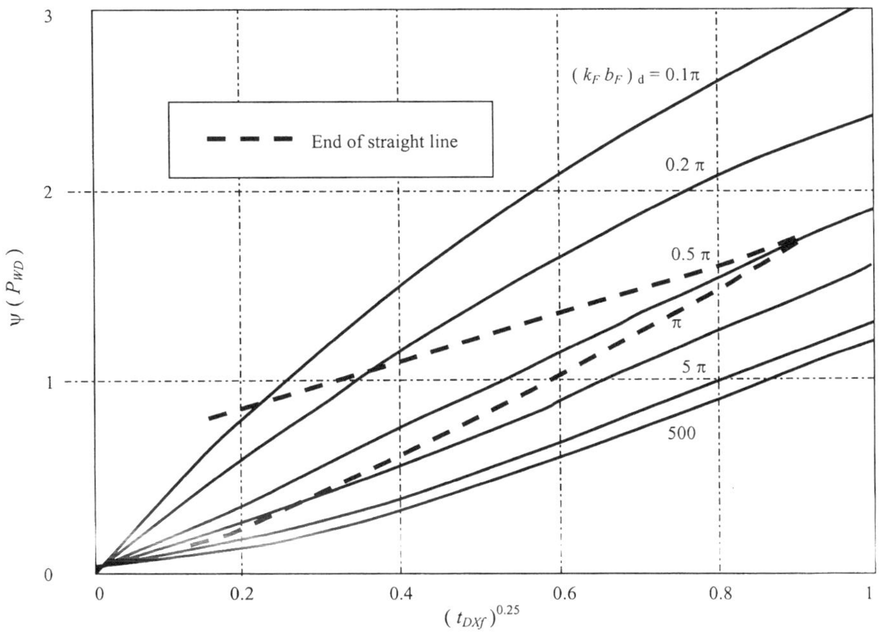

Figure 11–6. P_{WD} versus $[t_{DXF}]^{0.25}$ for a well with a finite-conductivity vertical fracture (after Cinco and Samaniego).[9]

Hence the product $h(k_f b_f)^{0.5}$ can be estimated by using the following equation:

$$h(k_f b_f)^{0.5} = \frac{444.75 q_g T}{m_{bf}(\phi \mu_g c_t k)^{0.25}} \tag{11–7}$$

Figure 11–7 shows a graph for analysis of pressure data of bilinear flow, while Figure 11–8 is a log-log graph of pressure data for bilinear flow. Figure 11–8 can be used as a diagnostic tool. The dimensionless time at the end of bilinear flow period is given by the following equations:

For $(k_f b_f)_D \leq 1.6$:

$$t_{Debf} = \left[\frac{4.55}{(k_f b_f)^{0.5}} - 2.5\right]^{-4} \tag{11–8}$$

For $(k_f b_f)_D \geq 3$:

$$t_{Debf} \approx \frac{0.10}{(k_f b_f)_D^2} \tag{11–9}$$

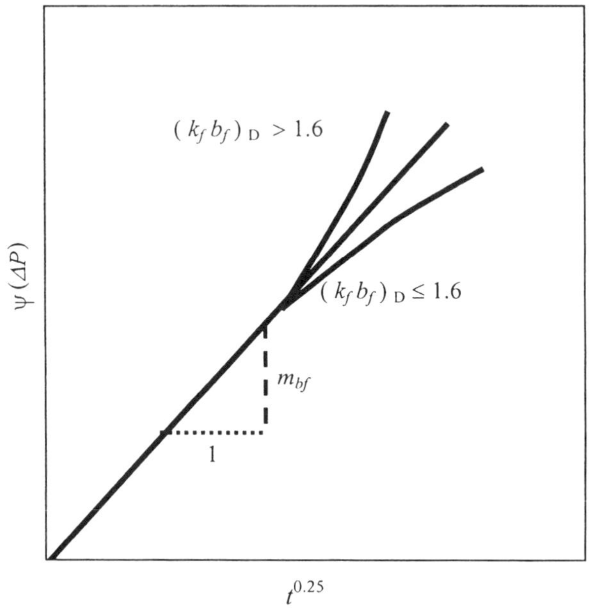

Figure 11–7. Graph for analysis of pressure data of bilinear flow (after Cinco and Samaniego).[9]

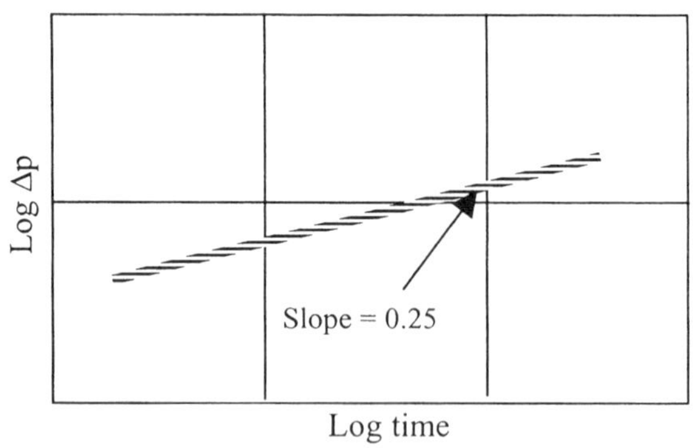

Figure 11–8. Log-log graph of pressure data for bilinear flow analysis (after Cinco and Samaniego).[9]

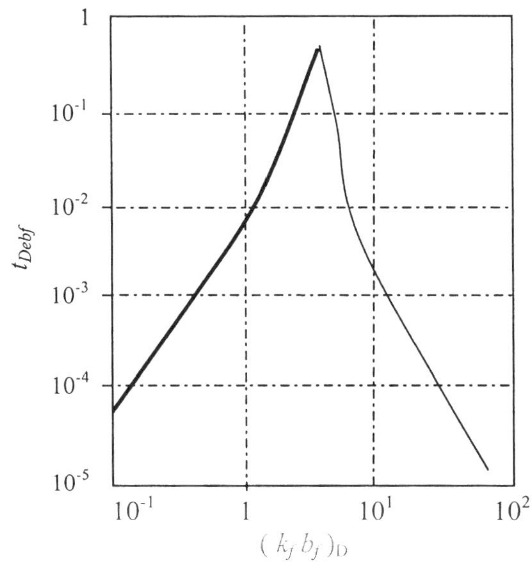

Figure 11–9. Dimensionless time for the end of the bilinear flow period versus dimensionless fracture conductivity.[9]

For $1.6 \leq (k_f b_f)_D \leq 3$:

$$t_{Debf} \approx 0.0205[(k_f b_f)_D - 1.5]^{-1.53} \tag{11–10}$$

Figure 11–9 shows a graphical representation of these equations. From Eqs. 11–4 and 11–8, if $(k_f b_f)_D \geq 3$, the dimensionless pressure drop at the end of the bilinear flow period is given by

$$(p_{WD})_{ebf} = \frac{1.38}{(k_f b_f)_D} \tag{11–11}$$

Hence the dimensionless fracture conductivity can be estimated using

$$(k_f b_f)_D \approx \frac{1.38}{(p_{Wd})_{ebf}} \tag{11–12}$$

$(p_{Wd})_{ebf}$ can be calculated using

$$(p_{WD})_{ebf} \approx \frac{kh\psi(\Delta p)}{1424\, q_g T} \tag{11–13}$$

where $\psi(\Delta p)$ is obtained from the bilinear flow graph. From Equation 11–5, a graph of $\log \psi(\Delta p)$ versus $\log t$ (see Figure 11–8) yields a quarter-slope straight line that can be used as a diagnostic tool for bilinear flow detection.

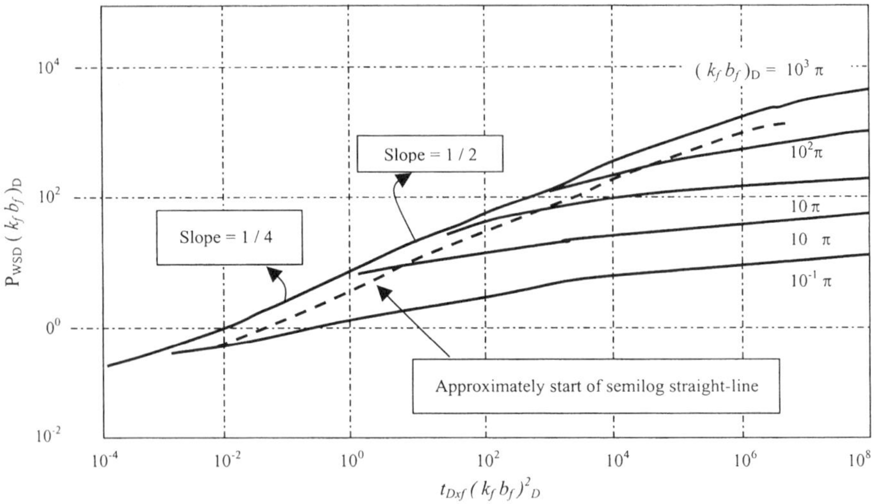

Figure 11–10. Type curve for vertically fractured gas wells (after Cinco and Samaniego).[9]

Formation Linear Flow

Figure 11–10 shows a graph of log $\lfloor p_{WD}(k_f b_f)_D \rfloor$ versus log $\lfloor t_{Dxf}(k_f b_f)^2_D \rfloor$. For all values of $(k_f b_f)_D$ the behavior of both bilinear flow (quarter-slope) and the formation linear flow (half-slope) is given by a single curve. Note that there is a transition period between bilinear and linear flows. Bilinear flow ends when fracture tip effects are felt at the wellbore. The beginning of the formation linear flow occurs at $(k_f b_f)^2_D \approx 10^2$, that is,

$$t_{Dblf} = \frac{100}{(k_f b_f)^2_D} \tag{11–14}$$

The end of this flow period is given by

$$t_{Delf} \approx 0.016$$

Hence, the fracture conductivity may be estimated as follows:

$$(k_f b_f)_D \approx \frac{10}{(t_{Dblf})^{0.5}} \tag{11–15}$$

or

$$(k_f b_f)_D \approx 1.25 \times 10^{-2} \left(\frac{t_{elf}}{t_{blf}}\right)^{0.5} \tag{11–16}$$

These equations apply when $(k_f b_f)_D \geq 100$.

Pseudoradial Flow

The dashed line in Figure 11–10 shows the approximate start of the pseudoradial flow period (semilog straight line). See Figure 11–5 on pseudoradial flow.[9]

11.4 Transient Pressure Behavior Analysis

This section describes type curve matching procedures.

Type Curve Matching Procedures

Figure 11–10 can be used as a type curve to analyze pressure data for a fractured well. Pressure data on a graph of log $\psi(\Delta p)$ versus log t are matched on a type curve to determine

$\lfloor \psi(\Delta p) \rfloor_M$, $\lfloor p_{WD}(k_f b_f)_D \rfloor_M$

$(t)_M$, $\lfloor t_{Dxf} \cdot (k_f b_f)_D^2 \rfloor_M$

$\lfloor t_{blf} \rfloor_M$, and $[t_{bssl}]_M$

From this information, we can estimate the following:

Dimensionless fracture conductivity:

$$\lfloor (k_f b_f)_D \rfloor_M$$

Formation permeability:

$$k_g = \frac{1424\, q_g T}{h[\psi(\Delta p)]_M} \cdot \frac{\lfloor p_{WD} \cdot (k_f b_f)_D \rfloor_M}{[(k_f b_f)_D]_M} \tag{11–17}$$

Fracture half length:

$$x_f = \frac{0.000264 k(t)_M}{\phi \mu g c_t} \cdot \frac{\lfloor (k_{fbf})_D^2 \rfloor_M}{[(k_{fbf})_D]_M} \tag{11–18}$$

Fracture conductivity:

$$k_f b_f = k x_f \lfloor (k_f b_f)_D \rfloor_M \tag{11–19}$$

End of bilinear flow:

$$\lfloor t_{lbf} \rfloor_M$$

Beginning of formation linear flow:

$$\lfloor t_{blf} \rfloor_M$$

Beginning of pseudoradial flow:

$$[t_{bsst}]_M$$

11.5 Specific Interpretation Methods

Reliable results can be obtained for using the specific analysis graphs. Now we will discuss various cases where all the pressure data fall on a very small portion of the type curve and a complete set of information may not be obtained.

Case 1: Bilinear Flow Type of Analysis

When a log-log graph of pressure data indicates that all of the test data are dominated by bilinear flow (quarter-slope), the minimum value of fracture half-length, x_f, can be estimated at the end of bilinear flow, i.e., for $(k_f b)_D \geq 3$, using the following equation:

$$x_f \geq \left(\frac{0.0002637 (k_f b_f)^2 t_{ebf}}{\phi \mu_g c_t k} \right)^{0.25} \tag{11–20}$$

By definition, the dimensionless fracture conductivity is

$$(k_f b_f)_D = \frac{k_f b_f}{k x_f} \tag{11–21}$$

where k_f b_f is calculated using Eq. 11–7 and slope m_{bf} can be found from the bilinear flow graph, which is a rectangular graph of real pseudopressure difference versus the quarter root of time. This graph will form a straight line passing through the origin. Deviations occur after some time depending on the fracture conductivity. The slope of this graph, m_{bf}, is used for the calculation of the fracture permeability–fracture width product $(k_f b_f)$.

The dimensionless fracture conductivity is correlated to the dimensionless effective wellbore radius, r'_w / r_f as shown in Table 11–1. Then the skin can be calculated from the following relationship:

$$s = In\left(\frac{r_w}{r'_w} \right) \tag{11–22}$$

Generally wellbore storage affects a test at early time. Thus it is expected to have pressure data distorted by this effect, causing deviation from the

Table 11–1
The Values of Effective Wellbore Radius as Function of Dimensionless Fracture Conductivity for a Vertical Fractured Well[9]

Dimensionless fracture conductivity $(k_f b_f)_D$	$\frac{r'_w}{x_f}$	Dimensionless fracture conductivity $(k_f b_f)_D$	$\frac{r'_w}{x_f}$
0.1	0.026	5.0	0.380
0.2	0.050	6.0	0.400
0.3	0.071	7.0	0.410
0.4	0.092	8.0	0.420
0.5	0.115	9.0	0.430
0.6	0.140	10.0	0.440
0.7	0.150	20.0	0.450
0.8	0.165	30.0	0.455
0.9	0.175	40.0	0.460
1.0	0.190	50.0	0.465
2.0	0.290	100.0	0.480
3.0	0.340	200.0	0.490
4.0	0.360	300.0	0.500

one-fourth-slope characteristic of this flow period. It is important to note that pressure behavior in Figure 11–11 for both wellbore storage dominated and bilinear flow portions is given by a single curve that completely eliminates the uniqueness-matching problem. Figure 11–11 is a new type curve and is used when pressure data exhibit one-fourth slope on a log-log graph. The end of wellbore storage effects occurs when $F_2(t_{Dxf}) = 2 \times 10^2$, yielding

$$t_{eWS} = 65,415.24\left[\frac{C^4}{(k_f b_f)^2 h^4 \phi c_t k}\right]^{1/3} \tag{11–23}$$

If Figure 11–11 is used as a type curve, the following information may be obtained:

$\lfloor F_1(p_{WD})\rfloor_M$, $\lfloor F_2(t_{Dxf})\rfloor_M$,
$(\Delta p)_M$, and $(t)_M$

Hence we can estimate the following:

Wellbore storage constant:

$$C = \frac{2.359 q_g T(t)_M}{[\psi(\Delta p)]_M} \cdot \frac{[F_1(p_{WD})]_M}{[F_2(t_{Dxf})]_M} \tag{11–24}$$

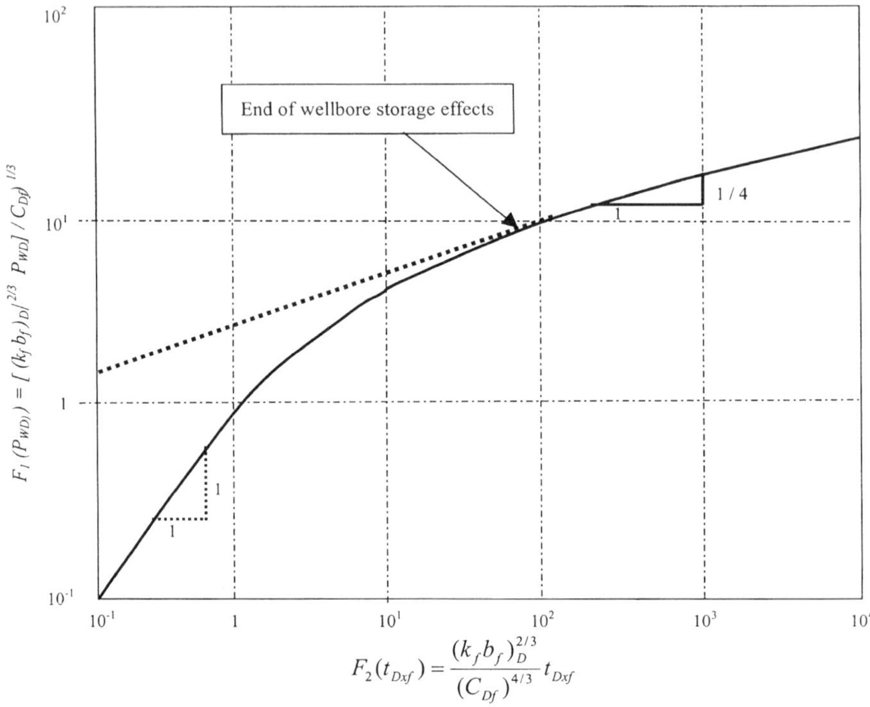

Figure 11–11. Type curve for wellbore storage under bilinear flow conditions (after Cinco and Samaniego).[9]

Fracture conductivity:

$$k_f b_f = \frac{0.4}{h^2}\sqrt{\frac{C}{\phi c_t k}\left\{\frac{1424 q_g T[F1(p_{WD})_M]}{[\psi(\Delta p)]_M}\right\}^3} \tag{11–25}$$

Case 2: Transition Period Analysis between Bilinear and Formation Flow

Cinco and Samaniego[9] have presented a new set of curves given in Figure 11–10. Figure 11–10 shows a graph of log $[p_{WD}(k_f b_f)_D]$ versus $\log\lfloor t_{Dxf}(k_f b_f)_D^2\rfloor$. From the type curve match of pressure data for this case in Figure 11–10, we obtain

$$[p_{WD}(k_f b_f)_D]_M, \lfloor t_{Dxf}(k_f b_f)_D^2 \rfloor_M,$$

$$(\Delta t)_M, \psi(\Delta p)_M, \quad \text{and} \quad \lfloor (k_f b_f)_D \rfloor_M$$

Hence

$$\left(\frac{k_f b_f}{x_f}\right) = \frac{50,300 q_g T P_{sc}}{h T_{sc}} \cdot \frac{\lfloor p_{WD}(k_f b_f)_D \rfloor_M}{[\psi(\Delta p)]_M} \tag{11–26}$$

Fracture half-length and fracture conductivity:

$$x_f = \left(\frac{k_f b_f}{x_f}\right)\left[\frac{0.0002637}{\phi \mu_g c_t k} \cdot \frac{(t)_M}{\left[t_{Dxf}(k_f b_f)_D^2\right]_M}\right]^{0.5} \tag{11–27}$$

and

$$k_f b_f = (x_f)\left[\frac{k_f b_f}{x_f}\right] \tag{11–28}$$

Since the formation permeability generally is known from prefracture tests, the dimensionless fracture conductivity can be estimated by using results from Eq. 11–26 or 11–27. If all pressure data fall on the transition period of the curve, type curve matching is the only analysis method available (see Figure 11–10).

Case 3: Pressure Data Exhibiting a Half-Slope Line on a Log-Log Graph

See Figure 11–12. There is no unique match with Figure 11–10; however, the linear flow analysis presented by Clark[2] can be applied to obtain fracture half-length if formation permeability is known, In addition, a minimum value for the dimensionless fracture conductivity, $(k_f b_f)_D$, can be estimated using Eq. 11–28. If the wellbore storage effects are present at early times in a test for this case, the analysis can be made using the type curve presented by Ramey and Gringarten:[1]

$$(k_f b_f)_D = 1.25 \times 10^{-2}\left(\frac{t_{elf}}{t_{blf}}\right)^{0.5} \tag{11–29}$$

Case 4: Pressure Data Partially Falling in the Pseudoradial Flow Period

Figure 11–13 is a graph of p_{Wd} versus $t_{Drw'}$, the dimensionless time defined by using $r_{w'}$ instead of x_f. This curve provides an excellent tool for type curve analysis of pressure data partially falling in the pseudoradial flow period because the remaining data must follow one of the curves for different fracture conductivity. Table 11–1 must be used as an auxiliary curve to determine

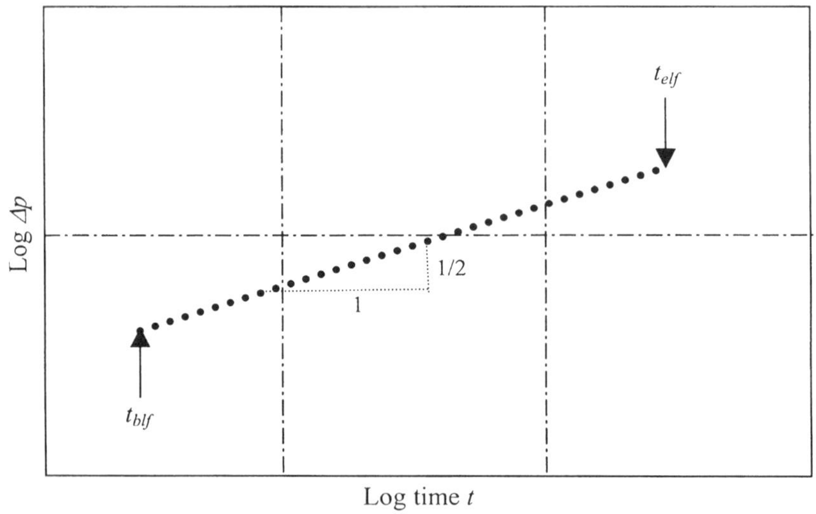

Figure 11–12. Pressure data for a half-slope straight line in a log-log graph (after Cinco and Samaniego).[9]

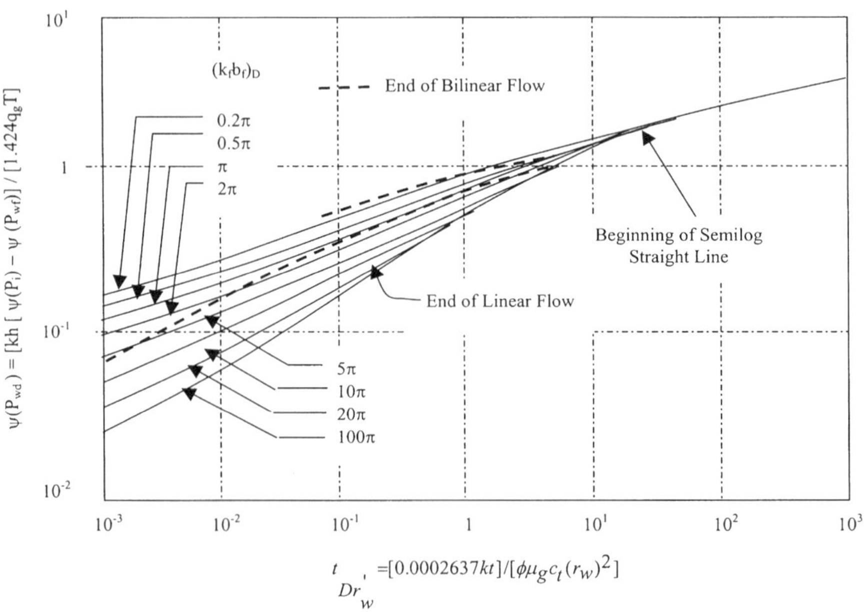

Figure 11–13. Type curve for a finite-conductivity vertical fracture (after Cinco and Samaniego).[9]

$(k_f b_f)_D$ when using Figure 11–13. The type curve of Figure 11–13 involves the following steps:

1. Plot a log-log graph of the pressure data to show that neither a one-fourth slope nor a half-slope is exhibited by the data.
2. Apply Figure 11–13 to match pressure data.
3. Estimate reservoir permeability from the pressure match point:

$$k = \frac{50{,}300 q_g T P_{sc}}{h T_{sc}} \cdot \frac{(p_{WD})_M}{(\Psi(\Delta p))_M} \tag{11–30}$$

4. Using information from the time match, estimate effective wellbore radius,

$$r'_w = \left[\frac{0.0002637k}{\phi \mu_g c_t} \cdot \frac{(\Delta t)_M}{(t_{Drw'})_M}\right]^{0.5} \tag{11–31}$$

5. By using $\lfloor (k_f b_f)_D \rfloor_M$ from Table 11–1, obtain $(r_{w'}/x_f)_{Table 11-1}$; hence

$$x_f = \frac{r'_w}{\left[\frac{r'_w}{r_f}\right]_{Table 11-1}} \tag{11–32}$$

6. Estimate the skin factor as follows:

$$s = In\left(\frac{r_w}{r'_w}\right) \tag{11–33}$$

7. Calculate fracture conductivity as follows:

$$k_f b_f = (k_f b_f)_D k x_f \tag{11–34}$$

8. The pressure data falling in the pseudoradial flow period also must be analyzed using the semilog method to estimate k, r'_w, and s.

The following three field examples illustrate the application of several of the methods and theory previously discussed.

Example 11–1[12] *Pressure Data Analysis for Bilinear Flow Period*

A buildup test was run after fracturing this gas well. The reservoir and pressure data are given in Table 11–2.

Table 11–2
Pressure Buildup Data (flowing wellbore pressure p_{wf} = 2550 psia; production rate q_q = 2.175 mmscfd; production time t_P = 1500 hr; h = 70 ft; k = 0.0075 mD; T = 710°R; c_t = 0.000145 psi^{-1}; ϕ = 0.11 fraction; μ_g = 0.02141 cP; r_w = 0.42 ft)

Δt (hr)	$\Delta t^{1/4}$ (hr$^{1/4}$)	$\Delta\psi$ (mmpsia^2/cP)
0.0016	0.20	18.80
0.0260	0.40	37.61
0.04	0.45	42.32
0.06	0.50	47.15
0.09	0.55	51.75
0.18	0.65	61.02
0.32	0.75	70.50
0.41	0.80	75.21
0.66	0.90	84.67
1.00	1.00	94.05
1.46	1.10	103.49
2.44	1.25	117.50
2.86	1.30	122.27
3.84	1.40	131.65
5.06	1.50	141.40
7.41	1.65	155.12
8.35	1.70	159.83
11.71	1.85	173.90
16.00	2.00	188.00
21.37	2.15	202.10
23.43	2.20	206.81
25.63	2.25	211.53
39.06	2.50	235.00
57.19	2.75	258.56
81.00	3.00	282.12

Determine the following:

1. Type of flow
2. Fracture conductivity $k_f b_f$
3. Fracture half-length x_f
4. Fracture skin factor s_f

Solution A log-log graph of pressure data (Figure 11–14) indicates that the buildup test was completely dominated by bilinear flow (quarter slope), corresponding in this example to case 1 in the type curve analysis section.

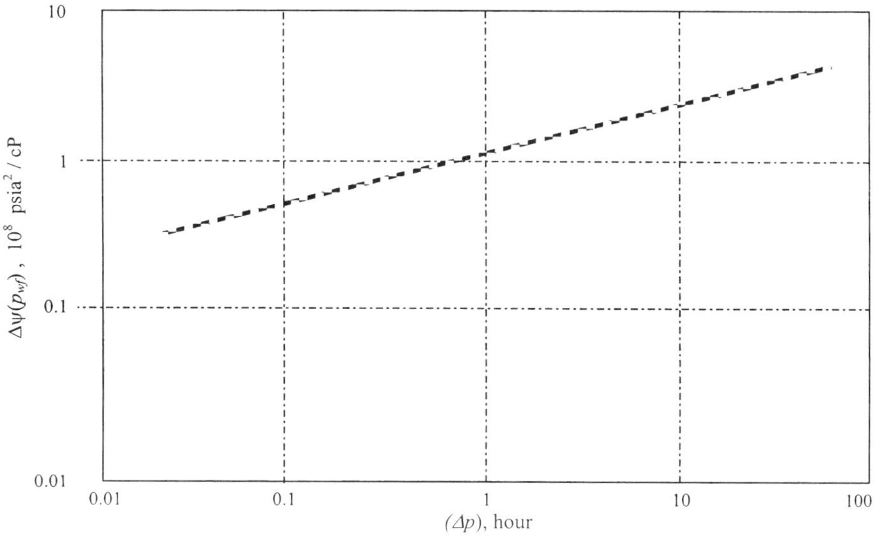

Figure 11–14. Log-log graph of pressure data for Example 11–1.

The bilinear flow graph (Figure 11–15) yields a straight line whose slope, m_{bf}, can be used to calculate $(k_f b_f)$ from Eq. 11–7:

$$(k_f b_f)^{0.5} = \frac{444.75 q_g T}{m_{bf}(\phi \mu_g c_t k)^{0.5}}$$

$$= \frac{444.75 \times 2.175 \times 10^3 \times 710}{94.0 \times 10^6 (0.11 \times 0.02141 \times 0.000145 \times 0.0075)^{0.5}}$$

$$= 379.96 \text{ mD-ft}$$

If we assume that the last data point corresponding to the end of the bilinear flow period and $(k_f b_f)_D \geq 3$, then from Eq. 11–20:

$$x_f \geq \left[\frac{0.0002637 (k_f b_f)^2 t_{ebf}}{\phi \mu_g c_t k}\right]^{0.25}$$

$$\geq \left[\frac{0.0002637(379.96)^2(81)}{0.11(0.02141)(0.000145)(0.0075)}\right]^{0.25} = 768.21 \text{ ft}$$

and from Eq. 11–21:

$$(k_f b_f)_D = \left(\frac{k_f b_f}{k x_f}\right) \leq \frac{379.96}{0.0075(768.21)} \leq 65.35$$

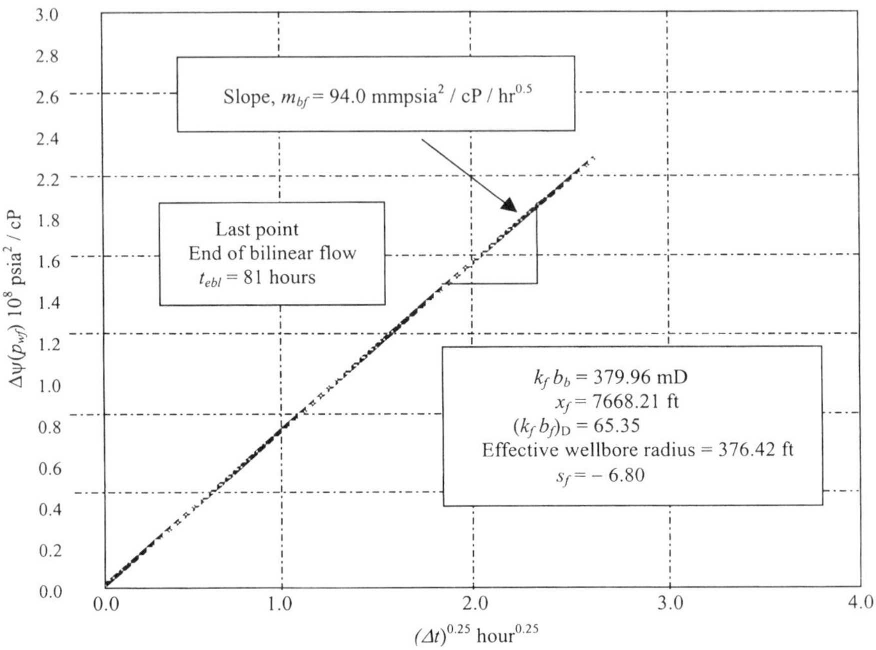

Figure 11–15. Bilinear flow graph for Example 11–1.

From Table 11–1, $r'_w/x_f \leq 0.49$. Hence $r'_w \leq (768.21 \times 0.49) = 376.42$, and from Eq. 11–22, calculate fracture skin factor s_f:

$$s_f = In\left(\frac{r_w}{r'_w}\right) = In\left(\frac{0.42}{376.42}\right) = -6.80$$

Example 11–2[12] *Pressure Data Analysis for Transition Period between Bilinear and Linear Flows*

A buildup test was run after fracturing this gas well. Information about the test and reservoir/well data is presented in Table 11–3.

Solution Figure 11–16 shows a log-log graph of pressure buildup data matching the type curve given in Figure 11–10. Notice that wellbore storage effects influence the first data points and the rest of the data fall in both the bilinear and the transition flow periods. The matching results also are presented in Figure 11–16. A minimum value for $(k_f b_f)_D$ can be estimated from the position of the last data point with respect to the type curves. For this case:

$$(k_f b_f)_D \min \approx 14.77$$

Table 11–3
Pressure and Reservoir Data (T = 250°R; c_t = 0.000135 psi^{-1}; p_{wf} = 1560 psia; k = 0.028 mD [prefracturing test]; h = 55 ft; R_w = 0.42 ft; ϕ = 0.1008 fraction; μ_g = 0.02152 cP; producing time t_P = 1750 hr; producing rate = 7.25 mmscfd; P_{sc} = 14.65; T_{sc} = 520°R)

Δt (hr)	$\Delta t^{1/4}$ (hr$^{1/4}$)	$\Delta\psi = \psi(p_{ws}) - \psi(p_{wf})$ (mmpsia^2/cP)
0.0001	0.10	1.01
0.00051	0.15	10.01
0.0016	0.20	25.21
0.0081	0.30	35.06
0.026	0.40	55.08
0.063	0.50	65.11
0.130	0.60	93.12
0.240	0.70	103.11
0.410	0.80	124.01
0.656	0.90	140.13
1.000	1.00	155.00
2.073	1.20	187.02
3.842	1.40	240.00
6.554	1.60	273.12
10.498	1.80	309.16
16.000	2.00	345.10
25.629	2.25	410.00
39.062	2.50	488.02
57.191	2.75	555.07
81.000	3.00	650.05
111.507	3.25	740.07
150.063	3.50	830.00

indicating that the dimensionless fracture conductivity is

$$(k_f b_f)_D \geq 5\pi$$

The end of wellbore storage occurs at approximately 0.35 hr and the end of bilinear flow is at 2.0 hr. We also see that the linear flow period was not reached in this test. Since the test was not long enough to match a curve for a specific value of $(k_f b_f)_D$, this example corresponds to case 2 in the type curve analysis section.

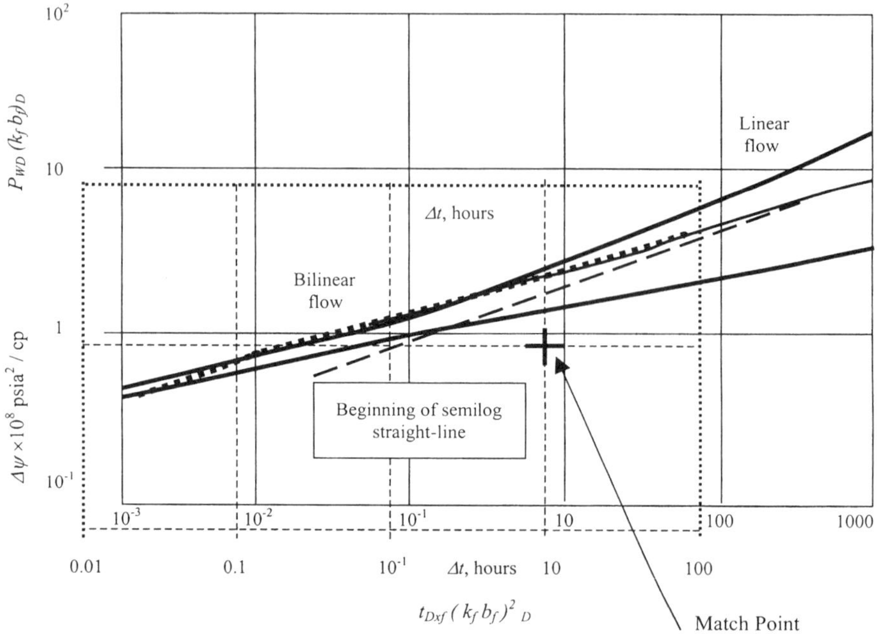

Figure 11–16. Type curve matching for Example 11–2.

Match points from Figure 11–16 are

$$\Delta\psi(p_{ws})_M = 100 \text{ mmpsia}^2/\text{cP} \quad \text{and} \quad \lfloor\psi(p_{WD})(k_f b_f)_D\rfloor_M = 0.79$$
$$(\Delta t)_M = 10 \text{ hr} \quad \text{and} \quad \lfloor t_{dxf}(k_f b_f)^2\rfloor_M = 0.82$$

Using the pressure data match points and from Eq. 11–26,

$$\left(\frac{k_f b_f}{x_f}\right) = \frac{50{,}300 q_g T P_{sc}}{h T_{sc}} \cdot \frac{[\Psi(p_{WD})(k_f b_f)_D]_M}{\Delta\psi(p_{ws})_M}$$

$$p_{WD}(k_f b_f)_D^2 = 0.79$$

$$\Delta\psi = 1.0 \times 10^8$$

$$t_{Dxf}(k_f b_f)_D^2 = 0.82$$

$$\Delta t = 10$$

$$= \frac{50{,}300 \times 7.25 \times 10^3 \times 710 \times 14.65}{55 \times 520} \cdot \frac{0.79}{100 \times 10^6}$$
$$= 1.048 \text{ mD-ft/ft}$$

From the time data match information and Eq. 11–27, we can calculate

$$x_f = \left(\frac{k_f b_f}{x_f}\right)\left[\frac{0.0002637}{\phi\mu_g c_t k}\cdot\frac{(\Delta t)_M}{[t_{dxf}(k_f b_f)^2]_M}\right]^{0.5}$$

$$= 1.2998\left[\frac{0.0002637}{0.1008\times 0.02152\times 0.000135\times 0.028}\cdot\frac{10}{0.82}\right]^{0.5}$$

$$= 814.0 \text{ ft}$$

Now application of Eq. 11–28 yields

$$k_f b_f = x_f\left[\frac{k_f b_f}{x_f}\right] = 814.0\,[1.048] = 895.40 \text{ mD-ft}$$

Match points are:

Estimate from Equation 11–1:

$$(k_f b_f)_D = \frac{k_f b_f}{x_f}\cdot\frac{1}{k} = \frac{1.048}{0.028} = 37.43$$

From Table 11–1:

$$\frac{r'_w}{x_f} = 0.49;\ \text{therefore } r'_w = 814.0\times 0.49 = 399.86 \text{ ft}$$

Calculate fracture skin factor s_f from Eq. 11–22:

$$s_f = In\left(\frac{r_w}{r'_w}\right) = In\left(\frac{0.42}{399.86}\right) = -6.84$$

Bilinear Flow Analysis

Figure 11–17 shows the bilinear flow graph $\lfloor\Delta\psi(p_{ws})$ versus $\Delta t^{1/4}\rfloor$ for this example. Based on the information provided by Figure 11–17, the correct straight line is drawn. The slope of this straight line is 155.0 mmpsia^2/cP/hr$^{1/4}$, and at end of bilinear flow, $\Delta\psi(p_{ws})_{ebl} \approx 200.0$ mmpsia^2/cP-hr$^{1/4}$. Notice that the pressure curve after the end of the bilinear flow period is concave upward, indicating that $(kb_f)_D \geq 1.6$.

From Equation 11–7:

$$(k_f b_f)^{1/2} = \frac{444.75 q_g T}{m_{bf}(\phi\mu_g c_t k)^{0.25}}$$

$$= \frac{444.75\times 7.25\times 10^3\times 710}{155.0\times 10^6(0.1008\times 0.02152\times 0.000135\times 0.028)^{0.25}}$$

$$= 28.21 \text{ mD-ft}$$

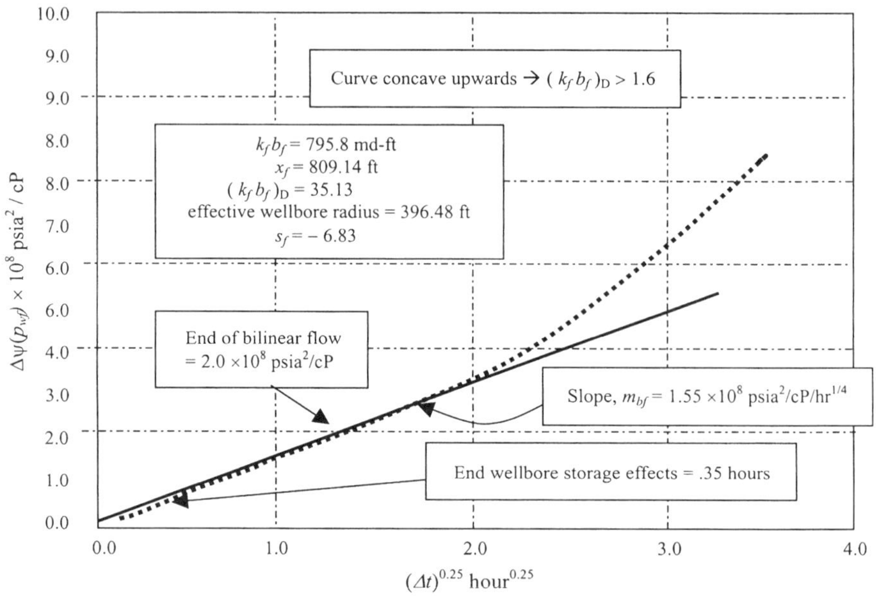

Figure 11–17. Bilinear flow graph for Example 11–2.

Therefore $k_f b_f = (28.21)^2 = 795.80$ mD-ft. From Eqs. 11–12 and 11–13,

$$(k_f b_f)_D = \frac{1.38(1424\, q_g T)}{kh\Delta\psi(p_{ws})_{ebl}}$$

$$= \frac{1.38(1424 \times 7.25 \times 10^3 \times 710)}{0.028 \times 55(187.0 \times 10^6)} = 35.126$$

Hence

$$x_f = \frac{kb_f}{k(k_f b_f)_D} = \frac{795.80}{0.028(35.126)} = 809.14 \text{ ft}$$

and

$$\frac{k_f b_f}{x_f} = 795.80/809.14 = 0.972$$

From Table 11–1, we can get $r'_w/x_f = 0.49$; therefore,

$$r'_w = 809.14 \times 0.49 = 396.48 \text{ ft}$$

Table 11–4
Summary of Analysis Results: Buildup Test for Example 11–2

Analysis results	Semilog solution	Type curve solution
$k_f b_f$, mD-ft	795.80	895.4
$(k_f b_f / x_f)$, mD-ft/ft	0.972	1.048
$(k_f b_f)_D$	35.126	37.43
r'_w, ft	396.48	399.86
x_f, ft	809.14	814.0
S_f	−6.83	−6.84

Calculate the fracture skin factor s_f using Eq. 11–22:

$$s_f = In\left(\frac{r_w}{r'_w}\right) = In\left(\frac{0.42}{396.48}\right) = -6.83$$

The results provided by both the type-curve analysis and semilog analysis methods are in good agreement as shown in Table 11–4.

Example 11–3[12] *Pressure Data Analysis for Pseudoradial Flow*
A buildup test was run on this fractured gas well after a flowing time of 1450 hr. Reservoir and test data are given in Table 11–5.

Identify the type of flow period and determine the following using type curve matching and semilog analysis techniques:

1. Reservoir permeability
2. Fracture wellbore radius
3. Fracture half-length
4. Fracture skin factor
5. Fracture conductivity

Solution Figure 11–18 shows a log-log graph of the pressure data; from this graph we can see that neither a one-fourth slope nor a half slope is exhibited by the data. Figure 11–19 shows that the pressure data match the curve for $(k_f b_f)_D = 2\pi$ given in Figure 11–10 and the last 14 points fall on the semilog straight line.

Match points from Figure 11–19:

$$[\psi(\Delta p)]_M = 100 \text{ mmpsia}^2/\text{cP} \quad \text{and} \quad (\Delta t)_M = 10 \text{ hr}$$

$$[\psi(p_{WD})]_M = 0.45 \quad \text{and} \quad \lfloor t_{Dr'_w} \rfloor_M = 1.95$$

Table 11–5
Pressure and Reservoir Data (q_g = 6.50 mmscfd; h = 52 ft; c_t = 0.000175 psi^{-1}; r_w = 0.39 ft; T = 690°R; ϕ = 0.11 (fraction); μ_g = 0.02035 cP; P_{sc} = 14.65 psia; T_{sc} = 520°R)

Δt (hr)	$\Delta\psi = \psi(p_{ws}) - \psi(p_{wf})$ (mmpsia^2/cP)	Δt (hr)	$\Delta\psi = \psi(p_{ws}) - \psi(p_{wf})$ (mmpsia^2/cP)
0.50	105.10	40.00	443.11
0.60	106.31	50.00	465.00
0.70	108.05	60.00	490.03
0.90	120.00	70.00	510.17
1.00	127.07	80.00	530.22
1.50	135.02	90.00	545.17
2.00	150.11	100.00	560.00
3.50	180.13	120.00	580.17
4.00	187.17	130.00	592.11
5.00	205.01	140.00	600.00
6.00	215.05	150.00	610.18
7.00	235.00	160.00	615.21
8.00	245.02	170.00	625.12
9.00	255.04	180.00	632.11
10.00	264.09	190.00	637.04
20.00	355.13	200.00	645.02
30.00	400.05		

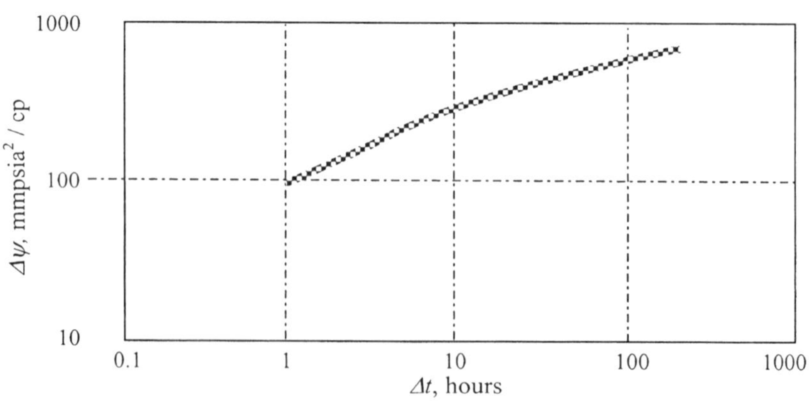

Figure 11–18. Log-log graph for Example 11–3.

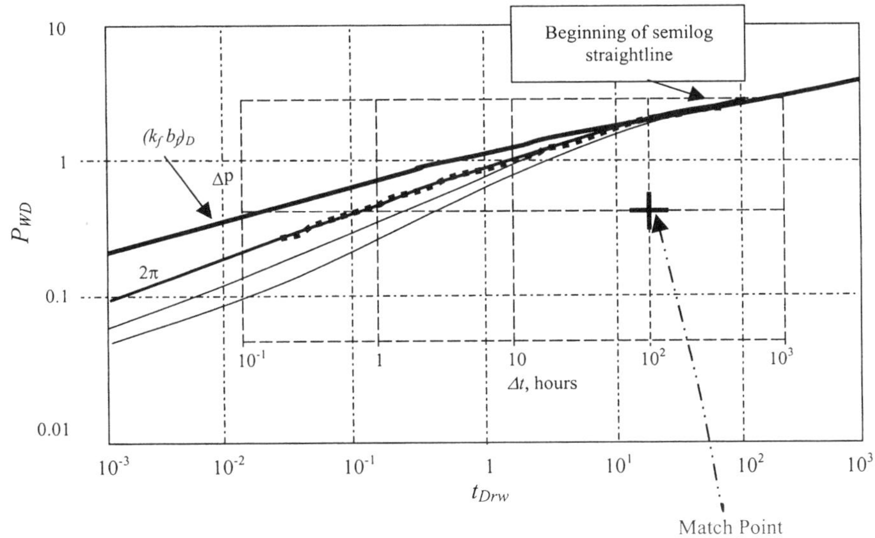

Figure 11–19. Type curve matching for Example 11–3.

From the pressure match, estimate reservoir permeability using Eq. 11–30:

$$k = \frac{50{,}300 q_g T P_{sc}}{h T_{sc}} \cdot \frac{[\psi(p_{WD})]_M}{[\psi(\Delta p)]_M}$$

$$= \frac{50{,}300 \times 6.25 \times 10^3 \times 690 \times 14.65}{52 \times 520} \cdot \frac{0.45}{100 \times 10^6} = 0.53 \text{ mD}$$

Using the information from time match in Eq. 11–31:

$$r'_w = \left[\frac{0.0002637k}{\phi \mu_g c_t} \cdot \frac{(\Delta t)_M}{(t_{Dr'_w})_M}\right]^{0.5}$$

$$= \left[\frac{0.0002637 \times 0.53}{0.11 \times 0.02035 \times 0.000175} \cdot \frac{10}{1.95}\right]^{0.5} = 42.77 \text{ ft}$$

From Table 11–1, $r'_w/x_f = 0.49$; hence, $x_f = 42.77/0.49 = 87.29$ ft.
The fracture skin factor is estimated by using Eq. 11–33:

$$s_f = In\left(\frac{r_w}{r'_w}\right) = In\left(\frac{0.39}{42.77}\right) = -4.70$$

From Eq. 11–34, the fracture conductivity is

$$k_f b_f = (k_f b_f)_D \cdot k x_f$$

$$= 2\pi(0.53)(87.29) = 1290.80 \text{ mD-ft}$$

Semilog Analysis

Figure 11–20 is a semilog graph for this example. The correct semilog straight line has a slope $m = 286.0$ mmpsia^2/cP/cycle and $\psi(\Delta p)_{1hr} = -65.0$ mmpsia^2/cP. The formation permeability can be calculated using Eq. 5–40:

$$k = \frac{57.920 \times 10^6 q_g T P_{sc}}{m h T_{sc}}$$

$$= \frac{57.920 \times 10^6 \times 6.25 \times 690 \times 14.65}{286.0 \times 10^6 \times 52 \times 520} = 0.47 \text{ mD}$$

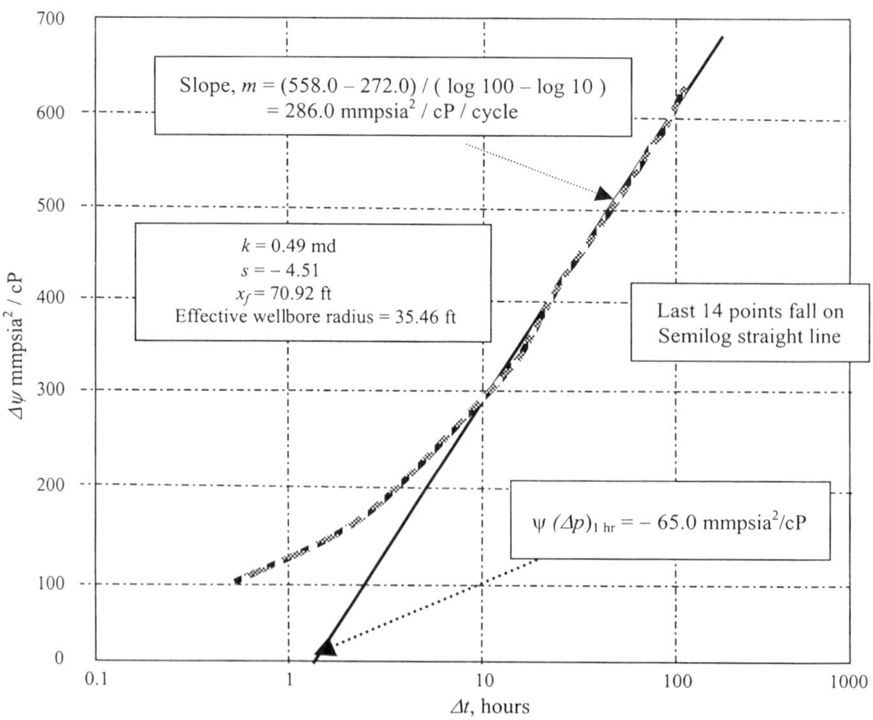

Figure 11–20. Semilog graph for Example 11–3.

Match Points are

$$\psi(\Delta p) = 100$$

$$\psi(p_{WD}) = 0.45$$

$$\Delta t = 10$$

$$t_{Dr'_w} = 1.95$$

$$s_f = 1.151\left[\frac{\psi(\Delta p)_{1hr}}{m} - \log\frac{k}{\phi\mu_g c_t r_w^2} + 3.23\right]$$

$$= 1.151\left[\frac{-65.0 \times 10^6}{286.0 \times 10^6} - \log\frac{0.47}{0.11 \times 0.02035 \times 0.000175 \times 0.39^2} + 3.23\right]$$

$$= 1.151[-0.227 - 6.92 + 3.23] = -4.51$$

Find the effective wellbore radius from rearranging Eq. 11–22:

$$r'_w = r_w e^{-s_f} = 0.39e^{-(-4.51)} = 35.46 \text{ ft}$$

Finally, the fracture half-length is calculated by using

$$In(x_f) = In(2r_w) - s_f = In(2 \times 0.39) - (-4.51) = 4.26$$

Hence, $x_f = e^{In(x_f)} = e^{4.26} = 70.92$ ft.

A summary of analysis results is given in Table 11–6. The results provided by both the type curve analysis and semilog analysis methods are reasonable.

Table 11–6
Summary of Analysis Results for Example 11–3

Analysis results	Type curve matching solution	Semilog solution
Permeability, mD	0.53	0.47
Fracture skin factor s_f	–4.70	–4.51
Effective wellbore radius r'_w, ft	42.77	35.46
Fracture half-length x_f, ft	87.29	70.92
Fracture conductivity mD-ft	46.26	33.33

From these examples it is demonstrated that type-curve matching analysis, when applied properly, provides an excellent diagnostic tool and a technique to estimate both reservoir and fracture parameters.

11.6 Summary

Based on the material presented in this chapter, the following remarks are pertinent:

- A new technique is presented to analyze data in the bilinear flow period. It is shown that, during this flow period, a graph of $\psi(p_{wf})$ versus $t^{1/4}$ yields a straight line when slope is inversely proportional to $h_f(k_f b_f)^{1/2}$.
- New type curves are now available for pressure analysis of fractured gas wells, and the problems in the analysis are reduced considerably with the use of these type curves.
- Prefracture information about the reservoir is necessary to estimate fracture parameters.
- The type curve analysis method must be used simultaneously with the specific analysis methods $\psi(p_{wf})$ versus $t^{1/4}$, $\psi(p_{wf})$ versus $t^{1/2}$, and $\Psi(p_{wf})$ versus log t to produce reliable results.

References and Additional Reading

1. Ramey, H. J., Jr., and Gringarten, A. C., "Effect of High-Volume Vertical Fractures in Geothermal Steam Well Behavior," Proc. Second United Nations Symposium on the Use and Development of Geothermal Energy, San Francisco, May 20–29, 1975.
2. Clark, K. K., "Transient Pressure Testing of Fractured Water Injection Wells," *J. Petroleum Technol.* (June 1968) 639–643; *Trans. AIME* 243.
3. Raghavan, R., Cady, G. V., and Ramey, H. J., Jr., "Well Test Analysis for Vertically Fractured Wells," *J. Petroleum Technol.* (1972) 24, 1014–1020.
4. *Pressure Transient Testing Methods*, SPE Reprint Series No. 1, Society of Petroleum Engineers, Dallas, TX, 1980.
5. Gringarten, A. C., Ramey, H. J., Jr., and Raghavan, R., "Applied Pressure Analysis for Fractured Wells," *J. Petroleum Technol.* (July 1975) 887–892; *Trans. AIME* 259.
6. Wattenbarger, R. A., and Ramey, H. J., Jr., "Well Test Interpretations of Vertically Fractured Gas Wells," *J. Petroleum Technol.* (May 1969) 625–633; *Trans. AIME* 246.
7. Raghavan, R., Cady, G. V., and Ramey, H. J., Jr., "Well Test Analysis for Vertically Fractured Wells," *J. Petroleum Technol.* (Aug. 1972) 1014–1020; *Trans. AIME* 253.

8. Scott, J., "The Effect of Vertical Fractures in Transient Pressure Behavior of Wells," *J. Petroleum Technol.* (Dec. 1963) 1365–1369; *Trans. AIME* 228.
9. Cinco, H., and Samaniego, F., "Effect of Wellbore Storage and Damage on the Transient Pressure Behavior for a Well with a Finite-Conductivity Vertical Fracture," *Soc. Petroleum Eng. J.* (Aug. 1978) 53–264.
10. Gringarten, A. C., Ramey, H. J., Jr., and Raghavan, R., "Applied Pressure Analysis for Fractured Wells," *J. Petroleum Technol.* (1975) 17, 887–892.
11. Earlougher, R. C., Jr., and Kerch, K. M., "Analysis of Short-Time Transient Test Data by Type-Curve Matching," *J. Petroleum Technol.* (1974) 26, 793–800.
12. Amanat U. C., *Pressure Transient Test Analysis User's Handbook*, Vol. 8, Advanced TWPSOM Systems Inc., Houston, TX, Oct. 1995.

Chapter 12

Practical Application of Interference and Pulse Tests

12.1 Introduction

This chapter discusses two well-testing techniques not yet discussed in the text: interference tests and pulse tests. This chapter also discusses various cross plotting techniques, the appearance of common flow regimes, log-log diagnostic, Horner, and specialized plots, and their field applications. Details and supporting materials are also presented in this chapter for the benefit of those who would like to learn more.

12.2 Interference Test Analysis Techniques

Interference tests are used to determine whether two or three wells are in pressure communication (i.e., in the same reservoir) and, when communication exists, to provide estimates of permeability k and porosity/compressibility product ϕc_t in the vicinity of the tested well. Convenient analysis techniques for interference tests are the use of type curves. Figure 12–1 shows these type curves, presented by Earlougher,[4] which are plot of the logarithm of $p_D(r_D, t_D)$ versus the logarithm of t_D/r_D^2. Using the following equations generate these type curves:

$$p_D(t_D, r_D) = -\frac{1}{2}E_i\left(\frac{-0.25\,r_D^2}{t_D}\right) \tag{12–1}$$

or

$$p_D(t_D, r_D) = \frac{1}{2}\left[In\left(t_D/r_D^2\right) + 0.80907\right] \tag{12–2}$$

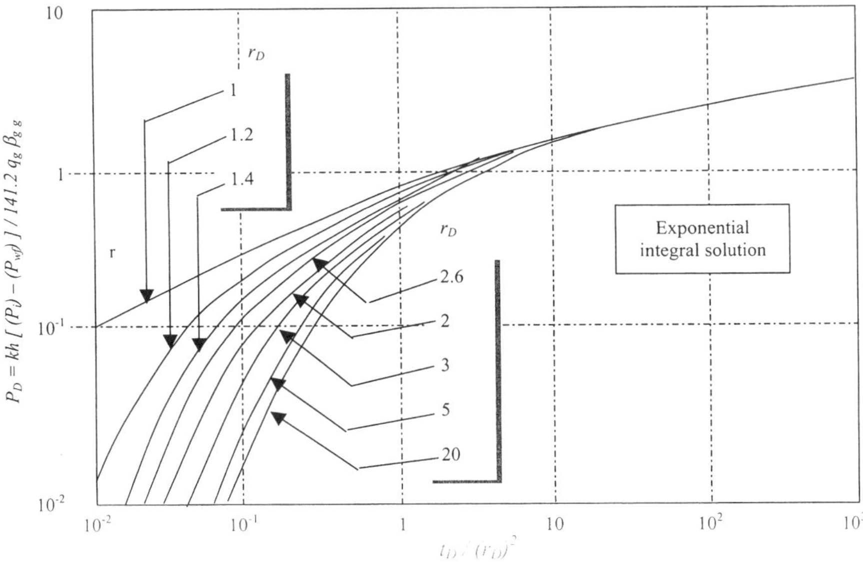

Figure 12–1. Type curves for interpretation of interference tests (after Mueller and Witherspoon *SPE*, *JPT*, April 1965).[1]

where

$$p_D = \text{dimensionless pressure} = \frac{kh\lfloor (p_i) - (p_{wf})\rfloor}{141.2 q_g \beta_{gi} \mu_{gi}} \tag{12–2a}$$

$$\beta_{gi} = 0.00504 \frac{z_i (T + 460)}{p_i} \times 10^6 \text{ (Rbbl/mmscfd)}$$

$$t_D = \text{dimensionless time} = \frac{0.0002637kt}{\phi \mu_i c_{ti} r_w^2} \tag{12–2b}$$

r_D = dimensionless distance between active and observation well, r/r_w

Evaluation of Eq. 12–1 gives the type curves shown in Figure 12–1. Figure 12–1 shows p_D as a function of t_d and r_D, the dimensionless radius distance from the well, for the infinite-acting system. When $r_D \geq 20$ and $t_D/r_D^2 \geq 0.5$, or when $t_D/r_D^2 \geq 25$, $r_d = 20$ and the "exponential–integral solution" lines on Figure 12–1 apply. Equation 12–2 may be used when

$$t_D/r_D^2 > 100 \tag{12–3}$$

but the difference between Eq. 12–1 and 12–2 is only about 2% when $t_D/r_D^2 > 5$. The exponential–integral solution is also called the line-source or the Theis solution to the flow equation. Figure 12–1 is useful for analyzing interference effects.

Interference Test Analysis by Type Curve Matching

The type curve analysis method is simple, fast, and accurate when the exponential integral P_D applies, that is, when $r_D = r/r_w > 20$ and $t_D/r_D^2 > 0.5$. Type curve matching can be performed as follows:

1. Plot pressure drawdown in an observation well, $\Delta p = p_i - p_r$, versus elapsed time t on the same size log-log paper as the full-scale type curve version of Figure 12–1 using an undistorted curve (the reader can prepare such a curve easily).
2. Slide the plotted test data over the type curve until a match is found. (Horizontal and vertical sliding both are required.)
3. Record pressure and time match points, $(p_D)_{MP}(\Delta p)_{MP}$ and $[(t_D/r_D 2)_{MP}, t_{MP}]$.
4. Calculate permeability k in the test region from the pressure match point:

$$k = \frac{141.2 q_g \beta_{gi} \mu_{gi}}{h} \cdot \frac{(p_D)_{MP}}{(\Delta p)_{MP}} \tag{12–4}$$

5. Calculate ϕc_t from the time match point using the following equation:

$$\phi c_t = \left(\frac{0.000264k}{\mu r^2}\right)\left[\frac{t_{MP}}{\left(t_D/r_D^2\right)_{MP}}\right] \tag{12–5}$$

where r is the distance between the two wells. The following example illustrates interference-test analysis by type-curve matching.

Example 12–1 *Analyzing Interference Test Data*

An interference test was run in gas reservoir. The producer well, well 2, produced 6.45 mmscfd gas. Pressure responses in shut-in wells 1 and 3 are plotted in Figures 12–3 and 12–4 and the locations of the producer and observation wells are shown in Figure 12–2. Addition reservoir/well data are as follows: $T = 250^0$ R; $P_i = 3700$ psia; $z_i = 0.9491$; $\mu_i = 0.0235$ cP; $\phi =$ 0.1004 fraction; $h = 41$ ft; $r_w = 0.4271$ ft.

Solution We assume that a gas reservoir is infinite-acting; we use the E_i function type curves to estimate k and the product of ϕc_t. The data fit the E_i function type curve well. A pair of match points is ($\Delta t = 160$ hr, $t_D/r_D^2 = 1.0$) and ($\Delta p = 2$ psi, $p_D = 0.1$) (see Figure 12–5).

Calculate reservoir permeability k from Eq. 12–4:

$$k = \frac{141.2 q_g \beta_{gi} \mu_{gi}}{h} \cdot \left[\frac{p_D}{\Delta p}\right]_{MP} = \frac{141.2 \times 6.45 \times 917.91 \times 0.0235}{41} \cdot \frac{0.1}{2}$$

$$= 23.96 \text{ mD}$$

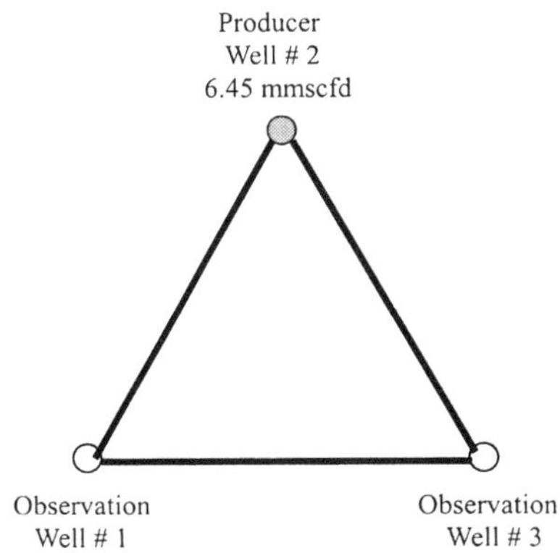

Figure 12–2. Location of producer and observation wells.

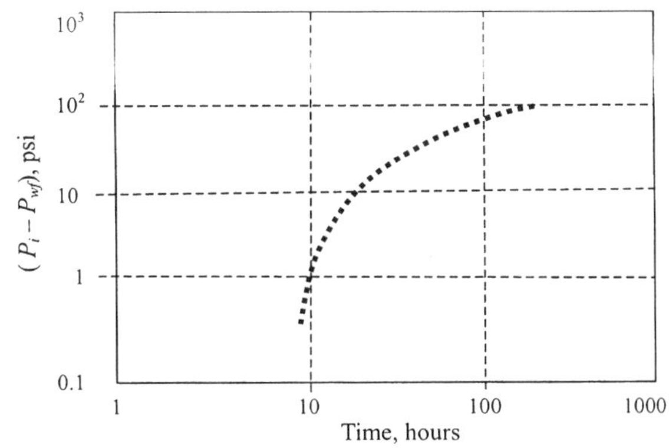

Figure 12–3. Interference data of well 3.

where

$$\beta_{gi} = .00504 \frac{0.9491 \times (460 + 250)}{3700} \times 10^6 = 917.91 \text{ rbbl/mmscfd}$$

Calculate product ϕc_t from Eq. 12–5:

$$\phi c_t = \left(\frac{0.000264k}{\mu_{gi} r^2}\right) \cdot \left[\frac{t}{t_D/r_D^2}\right]_{MP} = \frac{0.000264 \times 23.96}{0.0235 \times 1360^2} \cdot \frac{160}{1}$$

$$= 23.284 \times 10^{-6} \text{ psi}^{-1}$$

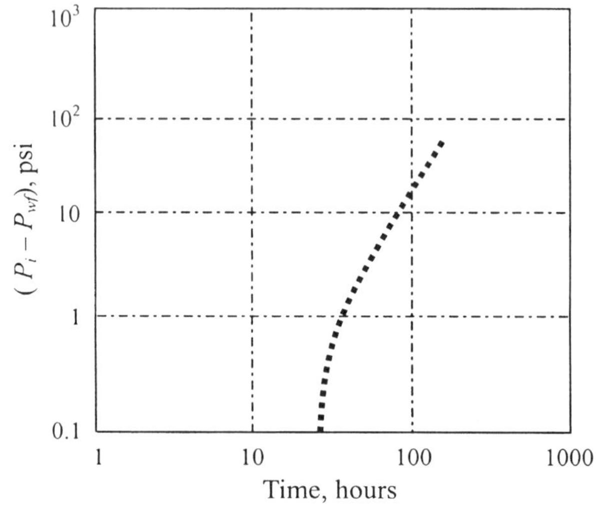

Figure 12–4. Interference data of well 1.

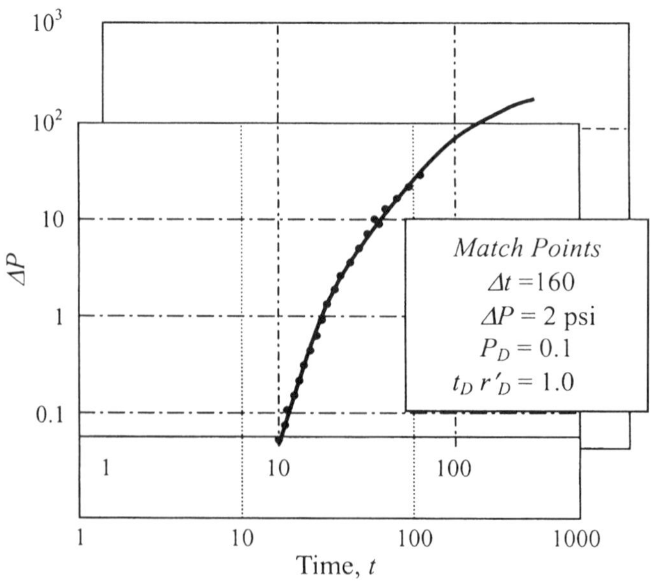

Figure 12–5. Match of interference data of well 3.

Therefore

$$c_t = \frac{\phi c_t}{\phi} = \frac{23.284 \times 10^{-6}}{0.1004} = 2.32 \times 10^{-4}\ \text{psi}^{-1}$$

12.3 Analysis of Pulse Test Pressure Response

Pulse tests have the same objective as conventional interference tests: to determine whether well pairs are in pressure communication and to determine reservoir permeability k and product of ϕc_t in the area of the tested wells. The tests are conducted by sending a coded signal or pulse sequence from an active well (producer or injector) to a shut-in observation well. The pulse sequence is created by producing from (or injecting into) the active well, then shutting it in, and repeating, that sequence in a regular pattern. An example is indicated in Figure 12–6. Highly sensitive pressure gauges usually are required to detect these small coded pulses, which may have magnitudes of less than 0.1 psi.

Analysis techniques for pulse tests usually are based on simulating the pressure response in an observation well with the familiar E_i function solution to the diffusivity equation, using superposition to model the rate changes in the pulsing sequence. From the simulations of pulse tests, Johnson *et al.*[3] have developed charts relating key characteristics of the tests to reservoir properties. Before we discuss these charts (Figures 12–9 through 12–16) and their application, it will be useful to introduce nomenclature used in pulse test analysis, using the system of Earlougher[4] and his schematic pulse-test rate and pressure-response history. Pulses can be analyzed for k and ϕc_t. It is good idea to analyze several pulses and compare the results.

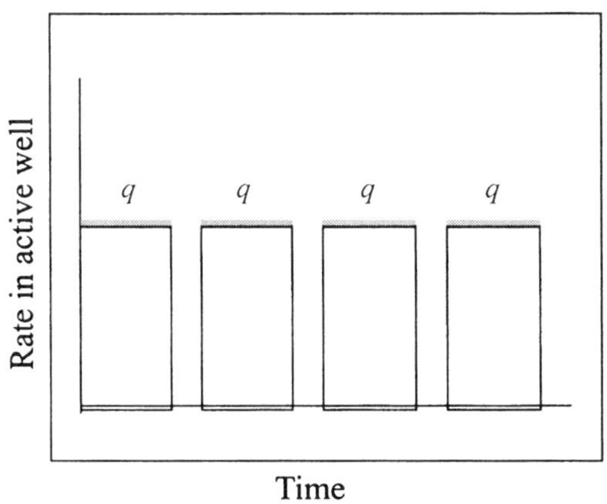

Figure 12–6. Typical rate schedules in pulse test.

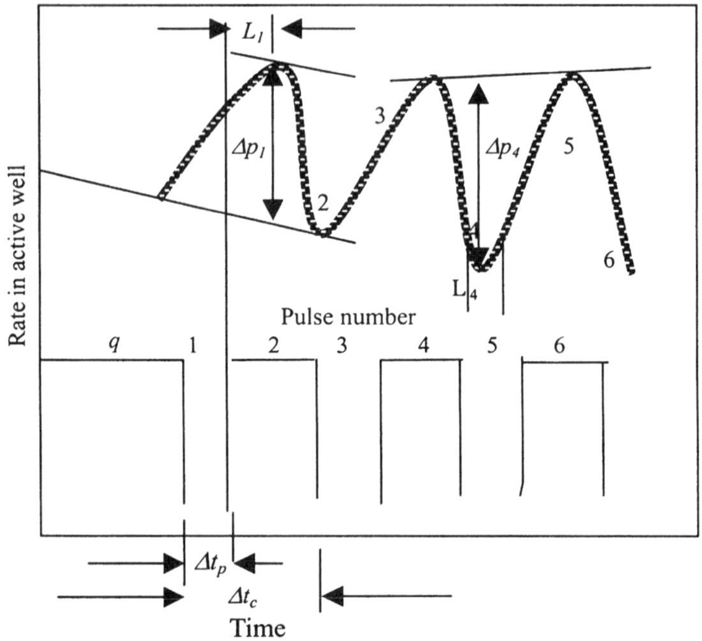

Figure 12–7. Pressure response in pulse test.

Characteristics of Pressure Response

For each pulse the pressure response (very small) at the observation well is recorded with a very sensitive pressure gauge. The pressure response in the pulse test is schematically illustrated in Figure 12–7. In pulse tests, pulse 1 and pulse 2 have characteristics that differ from those of all subsequent pulses. Following these pulses, all odd pulses have similar characteristics and all even pulses also have similar characteristics. Any one of the pulses can be analyzed for k and ϕc_t. It is good idea to analyze several pulses and compare the results.

Pulse Test Responses with Flow and Shut-in Time

Figure 12–8 shows pulse testing for a two-well system. The lower portion of the figure illustrates the pressure behavior at the observation well and correlates the pressure pulses with the rate pulses. The upper portion of the curve shows the constant production rate before the test and the rate pulses The flow time and shut-in time are equal as shown in Figure 12–8. Pulse testing can be done with unequal flow and shut-in times.

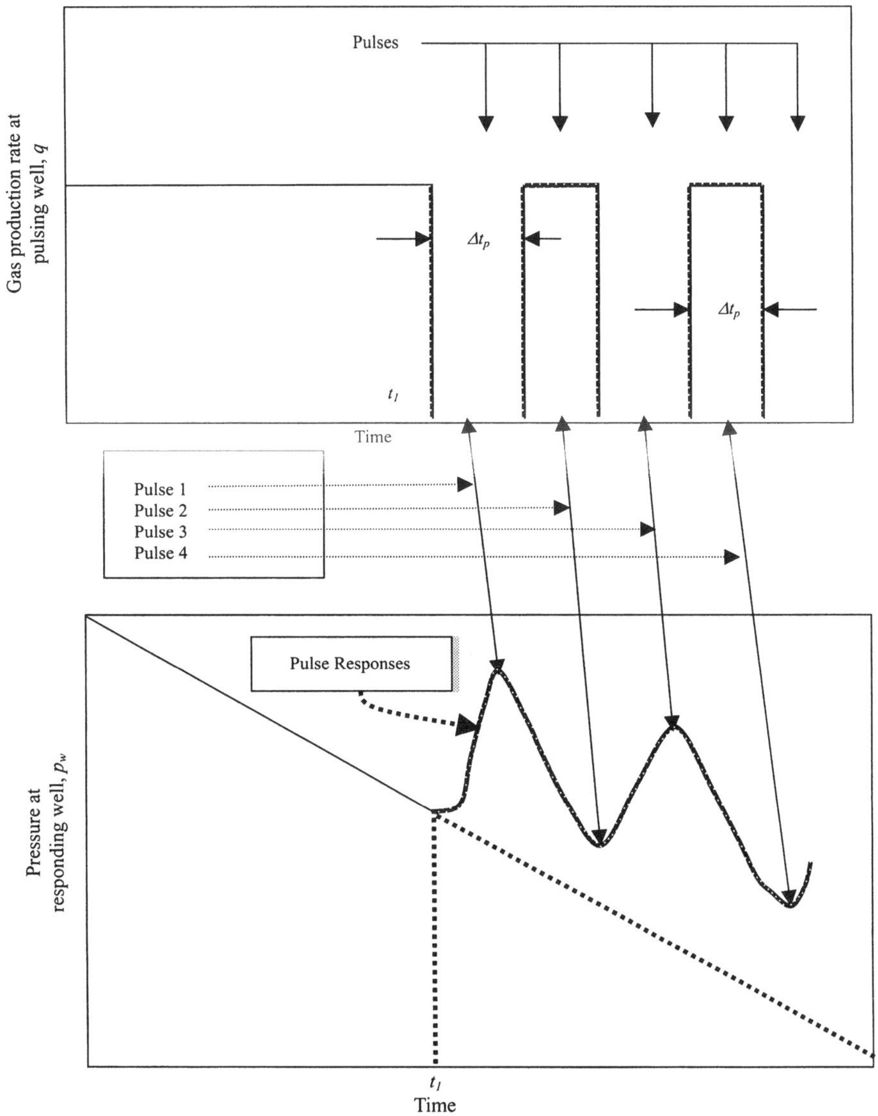

Figure 12–8. Rate history and pressure response for a pulse test (after Johnson *et al.*).[3]

Pulse Test Analysis Method

The following equations are used to calculate permeability and the porosity–compressibility product (ϕc_t):

$$k = \frac{141.2\, q_g \beta_{gi} \mu_{gi}}{h \Delta p} \cdot \left(\frac{\Delta p_D (t_L/\Delta t_c)^2}{(t_L/\Delta t_c)^2} \right) \tag{12–6}$$

$$\phi c_t = \frac{0.000264 k\, t_L}{\mu_g r^2 \left[(t_L)_D / r_D^2 \right]} \tag{12–7}$$

where Δp = amplitude of a pulse, Δt_c = total cycle length (including both shut-in and flow periods), t_L = time lag (time elapsed between the end of a pulse and the pressure peak caused by the pulse), Δp_D = dimensionless pressure response amplitude and is equal to

$$\Delta p_D = \frac{kh\Delta p}{141.2\, q_{sc} \mu_g B_{gi}},$$

where

$$B_{gi} = 0.00504 \frac{zT}{p} \times 10^6 \text{ rbbl/mmscfd}$$

$(t_L)_D$ = dimensionless time lag and is given by

$$(t_L)_D = \frac{0.0002637 k t_L}{\phi \mu_g c_t r_w^2}$$

$r_D = r/r_w$ = dimensionless distance between the tested wells (r_w is for the observation well). The values of the terms $\Delta p_D (t_L/\Delta t_c)^2$ and $\lfloor (t_L)_D / r_D^2 \rfloor$ are obtained from Figures 12–9 through 12–16. These figures use $t_L/\Delta t_c$ and $F' = \Delta t_P / \Delta t_c$, where Δt_P is the length of the pulse period. Example 12–2 illustrates how these figures are applied.

Pulse Test Design Procedure

A prior knowledge of the expected pressure response is important so that the range and sensitivity of the pressure gauge and the length of time needed for the test can be predetermined. To design a pulse test follow these steps:

1. The first step in designing a pulse test is to select the pulse ratio. If a specific pulse ratio is more convenient for gas field operations, this ratio should be used. Otherwise, a pulse ratio near 0.7 or 0.3 is recommended, depending on whether the odd pulses or the even pulses will be used to

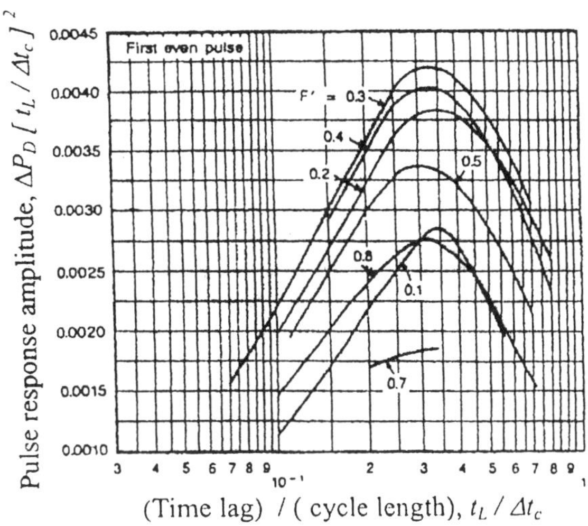

Figure 12–9. Time lag and response amplitude relationship for the first odd pulse.[2]

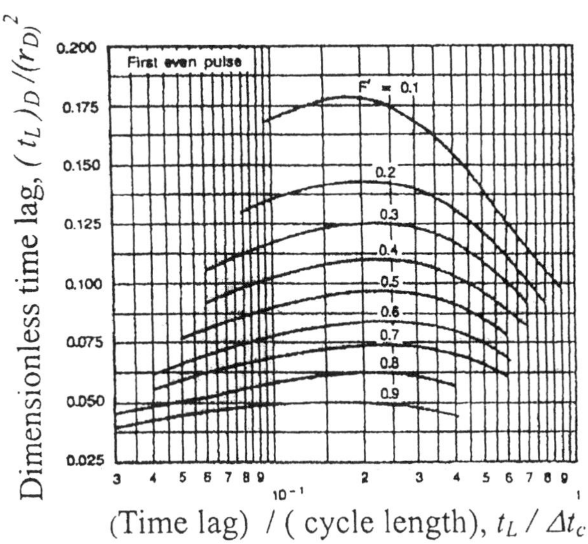

Figure 12–10. Time lag and cycle length relationship for the first odd pulse.[2]

analyze the results of the test. In no case should the ratio be below 0.2 or above 0.8.

2. Calculate the dimensionless time lag using the following equations:

$$(t_L)_D = 0.09 + 0.3R' \text{ (odd pulses)}$$

$$(t_L)_D = 0.09 \times 0.3(1 - R') \text{ (even pulses)}$$

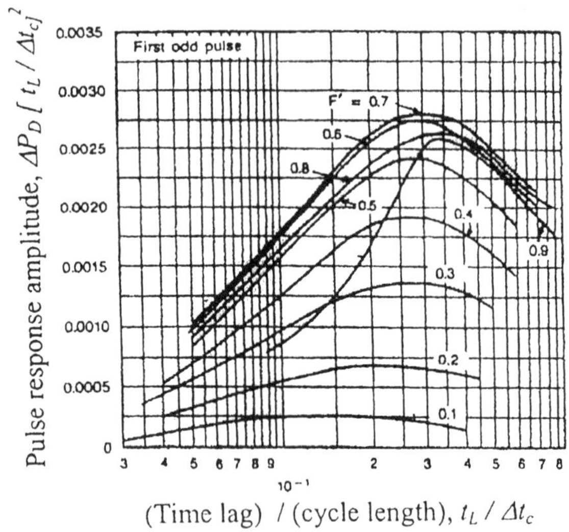

Figure 12–11. Time lag and response amplitude relationship for the first even pulse.[2]

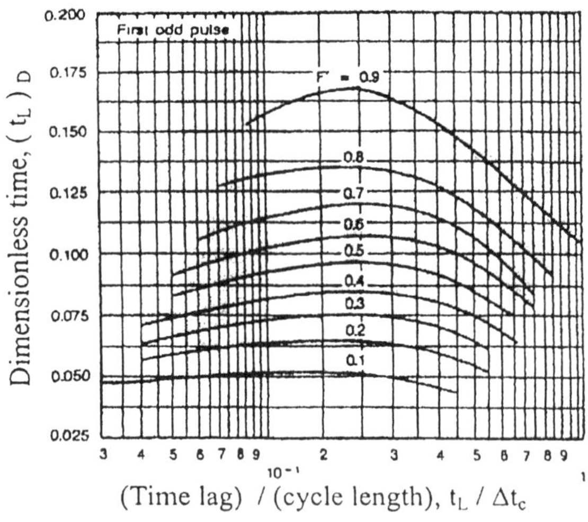

Figure 12–12. Time lag and cycle length for the first even pulse.[2]

3. Determine the dimensionless cycle period using the dimensionless time lag and the appropriate curve in Figures 12–10, 12–12, 12–14, or 12–16.
4. Determine the dimensionless response amplitude using the dimensionless time lag and the appropriate curve in Figures 12–9, 12–11, 12–13, or 12–15.

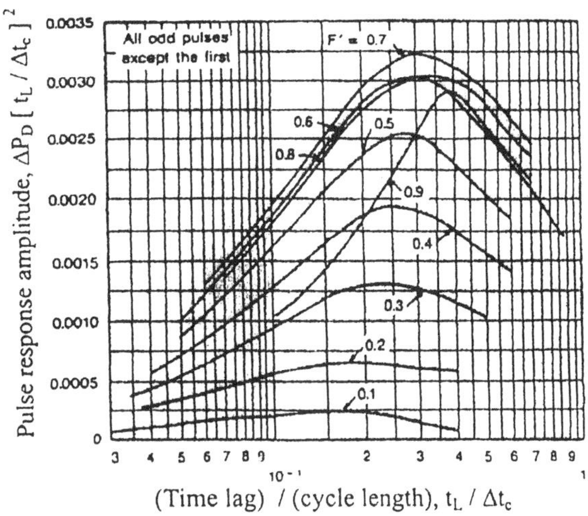

Figure 12–13. Time lag and response amplitude relationship for all odd pulses after the first.[2]

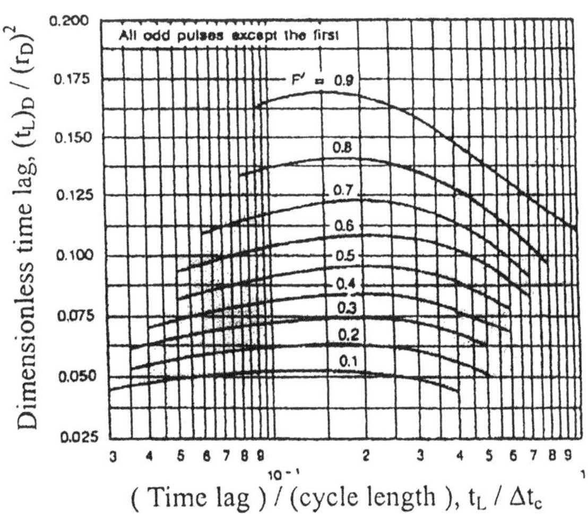

Figure 12–14. Time lag and cycle length relationship for all odd pulses after the first.[2]

5. Using approximate known values of the formation permeability, porosity, and thickness, the viscosity of the gas, and the total compressibility, together with the dimensionless cycle period, the dimensionless response amplitude, and Eqs. 12–8 and 12–9, calculate the cycle period and the response amplitude.

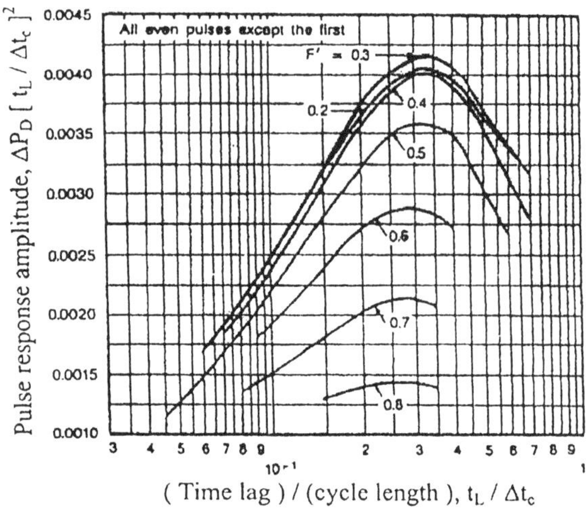

Figure 12–15. Time lag and response amplitude relationship for all even pulses after the first.[2]

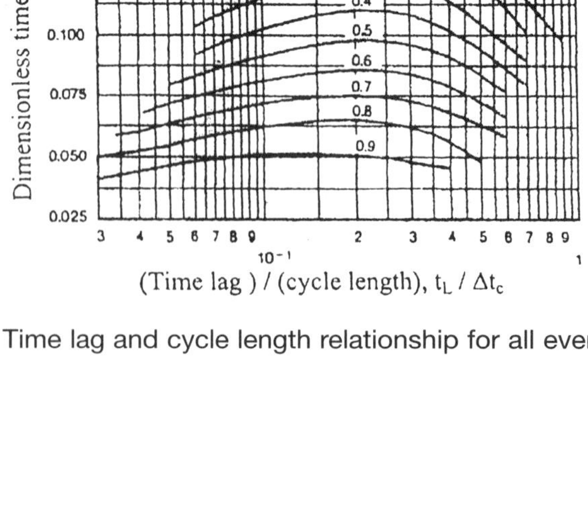

Figure 12–16. Time lag and cycle length relationship for all even pulses after the first.[2]

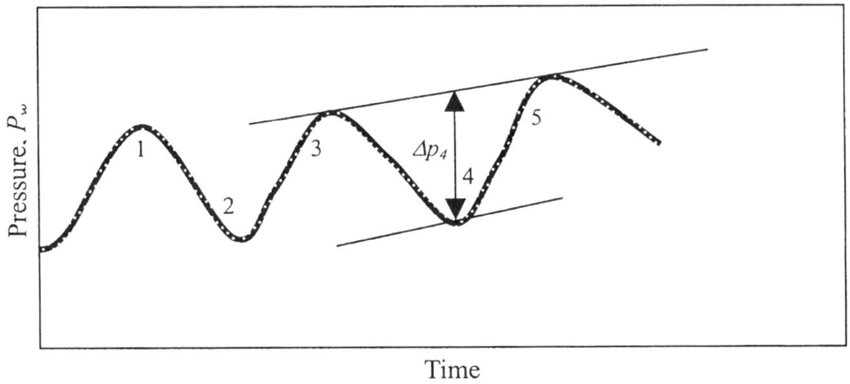

Figure 12–17. Schematic of pressure response in pulse test.

Dimensionless cycle period:

$$\Delta t_{cycD} = \frac{k\,\Delta t_{cyc}}{56{,}900\,\phi c_t \mu_g r_{bw}^2} \tag{12–8}$$

Dimensionless response amplitude:

$$\Delta P_D = \frac{kh\,\Delta P}{70.6\,\beta_g \mu_g q_{sc}} \tag{12–9}$$

6. Using the pulse ratio and the cycle period, calculate the pulsing period and the shut-in period.

The following example illustrates how to analyze a pulse test.

Example 12–2 *Analyzing Pulse Test Data*

A pulse test was run in a gas reservoir in which the distance between wells, r, was 660 ft. Formation gas viscosity, μ_g, was 0.0235 cP, formation thickness h 41 ft, and porosity ϕ 0.105. In the test following rate stabilization, the active well was shut in for 2 hr, and then produced for 2 hr, shut in for 2 hr, etc. Production rate q_g was 5.25 mmscfd, and formation volume factor β_g was 917.91 rbbl/mmscfd. The amplitude Δp of the fourth pulse (Figure 12–15) was 0.625 psi, and the time lag was 0.4 hr. From these data, estimate formation permeability k and ϕc_t.

Solution To analyze the fourth pulse, we use Figures 12–14 and 12–15. From these figures determine $\Delta p_D(t_L/\Delta t_c)^2$, and thus k. We note that

$$F' = \Delta t_P/\Delta t_c = 2/(2+2) = 0.5,\ t_L/\Delta t_c = 0.4/4 = 0.1$$

Then, from Figure 12–14:

$$\Delta p_D(t_L/\Delta t_c)^2 = 0.00221$$

and from Eq. 12–6:

$$\begin{aligned} k &= \frac{141.2\, q_g \beta_{gi} \mu_{gi}}{h \Delta p} \cdot \frac{\Delta p_D(t_L/\Delta t_c)^2}{(t_L/\Delta t_c)^2} \\ &= \frac{141.2 \times 5.25 \times 917.91 \times 0.0235}{41 \times 0.625} \cdot \frac{0.00221}{0.1^2} = 169.43\,\text{mD} \end{aligned}$$

From Figure 12–15: $(t_L)_D/r_D^2 = 0.091$. Thus using Eq. 12–7:

$$\begin{aligned} \phi c_t &= \frac{0.000264k}{\mu_g r^2} \cdot \frac{(t_L)}{\left[(t_L)_D/r_D^2\right]} \\ &= \frac{0.000264 \times 169.43}{0.0235 \times 660^2} \cdot \frac{0.4}{0.091} = 19.21 \times 10^{-6}\ \text{psi}^{-1} \end{aligned}$$

References and Additional Reading

1. Mueller, T. D., and Witherspoon, P. A., "Pressure Interference Effects within Reservoirs and Aquifers," *J. Petroleum Technol.* (April 1965) 471–474; *Trans. AIME* 234.
2. Kamal, M., and Brigham, W. E., "Pulse Testing Response for Unequal Pulse and Shut-In Periods," *Soc. Petroleum Eng. J.* (Oct. 1975) 399–410; *Trans. AIME* 259.
3. Johnson, C. R., Greenhorn, R. A., and Woods, E. G., "Pulse-Testing: A New Method for Describing Reservoir Flow Properties between Wells," *J. Petroleum Technol.* (Dec. 1966) 1599–1604; *Trans. AIME* 237.
4. Earlougher, R. C., Jr., *Advances in Well Test Analysis*. Society of Petroleum Engineers, Dallas, TX, 1977.
5. Wattenbarger, R. A., and Ramey, H. J., "Well Test Interpretation of Vertical Fractured Gas Wells," *J. Petroleum Technol.* (May 1969) 625–632.
6. Kamal, M. M., "Interference and Pulse Testing—A Review," *J. Petroleum Technol.* (Dec. 1983) 2257–2270.
7. Katsner, F. E., "Effects of Linear Boundaries on Pulse Testing." M.Sc. Thesis, Colorado School of Mines, 1970.
8. Matthews, C. S., and Russell, D. G., *Pressure Build-up and Flow Tests in Wells*. SPE of AIME Monograph, Vol. 1, 25, Henry Doherty Series, 1967.

Chapter 13

Well Testing Terminology in Multilayered Reservoir Systems

13.1 Introduction

This chapter discusses various types and testing of gas layered reservoir systems including multilayered responses in fractured reservoirs. It also describes crossflow identification and the nature and degree of communication between layers. Performance equations for cases of constant flowing pressure and constant producing rate are presented and discussed. The chapter also reviews "layer effect" on pressure and/or production behavior including economic aspects of interlayer crossflow.

13.2 Classification of Layered Reservoir Systems

Layered reservoirs can be classified into four groups.

Crossflow Reservoirs

Figure 13–1 shows a crossflow reservoir, which consists of four continuous layers that are communicating at the contact planes. These layers are not entirely separated by impervious layers, therefore interlayer crossflow could occur during the test. The crossflow would be directed from the layer of low permeability to the layer of higher permeability, as shown in the figure. If k_1 is greater than k_2 , then the pressure transients would travel faster in the layer of permeability k_1 than in the layer of permeability k_2. The duration of the crossflow periods depends on the storage of each layer. If the storage of the layer of permeability k_1 is negligible when compared to the storage of the layer of permeability k_2, then the crossflow will continue throughout the life of the

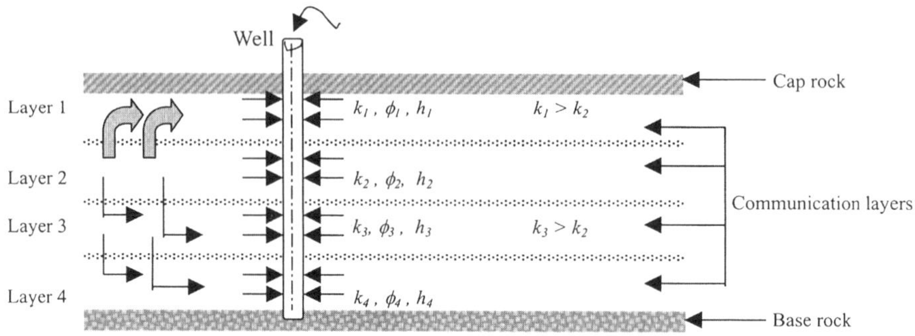

Figure 13–1. Four-layer crossflow reservoir.

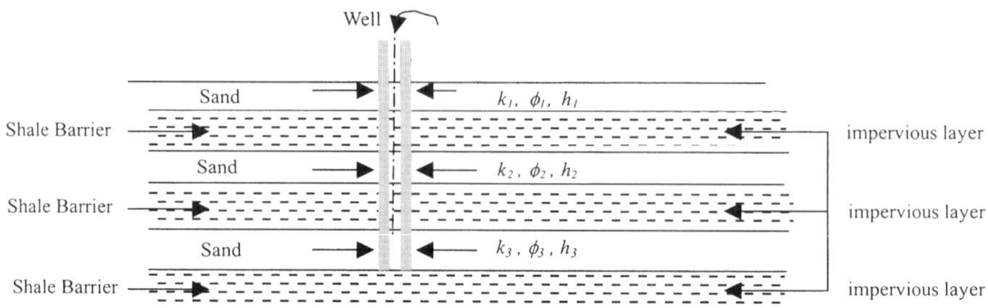

Figure 13–2. Three-layer without-crossflow reservoir.

well. If the opposite occurs, then the duration of the crossflow period will be short.

Without-Crossflow Reservoirs

This type of layered reservoir consists of two or more separate layers and is carried to the surface through a common wellbore. Each layer has different properties as shown in Figure 13–2. Production is commingled at the well, so layers communicate only through the wellbore.

Commingled Reservoirs

This type of reservoir is also known as a composite reservoir. Layers communicate only through the wellbore as shown in Figure 13–3. Investigators[1–3] have conducted studies on wells with commingled fluid production from two or more noncommunicating zones. In those cases, fluid is produced into the wellbore from two or more separate layers and is carried to the surface through a common wellbore.

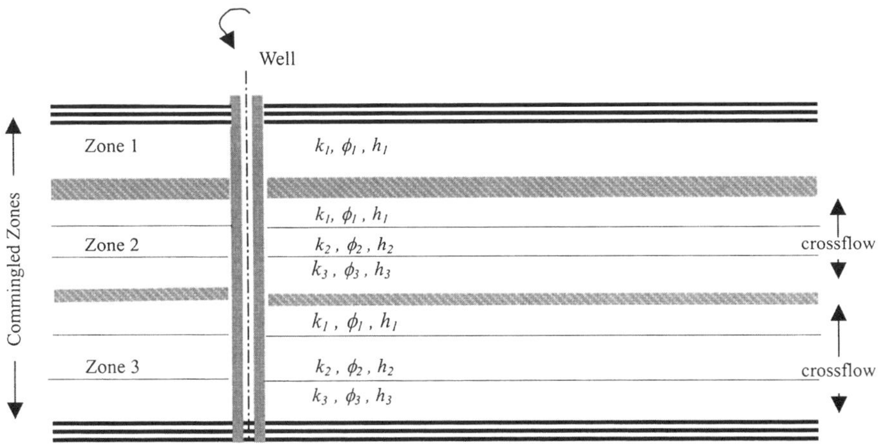

Figure 13–3. Reservoir consisting of commingled zones and crossflow layers.

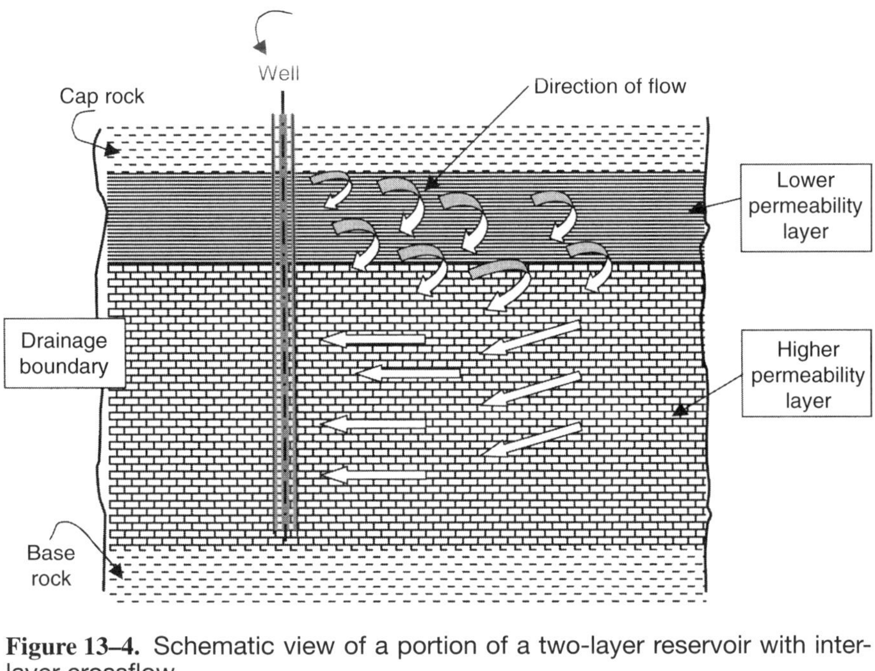

Figure 13–4. Schematic view of a portion of a two-layer reservoir with interlayer crossflow.

Interlayer Crossflow Reservoirs

Figure 13–4 shows that crossflows between the layers can occur; the pressure and production behavior of a gas well can be interpreted by the use of homogeneous reservoir theory. A gas well in a layered reservoir with crossflow behaves like a well in a homogeneous, single-layer reservoir that possesses the same dimensions and pore volume as the crossflow system and a permeability–thickness product (kh) equal to the total kh of the crossflow system. The occurrence of crossflow can be confirmed by the homogeneous-like appearance of the pressure and/or production behavior.

13.3 Pressure Analysis Methods in Layered Gas Reservoirs

Figure 13–5 shows pressure buildup for a layered gas reservoir with crossflow. Such a layered reservoir behaves like a homogeneous system. A semilog plot for any pressure transient test can be analyzed just as it can for homogeneous systems and should yield an estimate of $(kh)_t$. The following equations are applicable to analyze these tests:

$$(kh)_t = \sum_{j=1}^{n} (kh)_j \tag{13–1}$$

$$k_j = \frac{q_j}{q}\left[\frac{(kh_t)}{h_j}\right] \tag{13–2}$$

$$(\phi c_t h)_t = \sum_{j=1}^{n} [\phi c_t h]_j \tag{13–3}$$

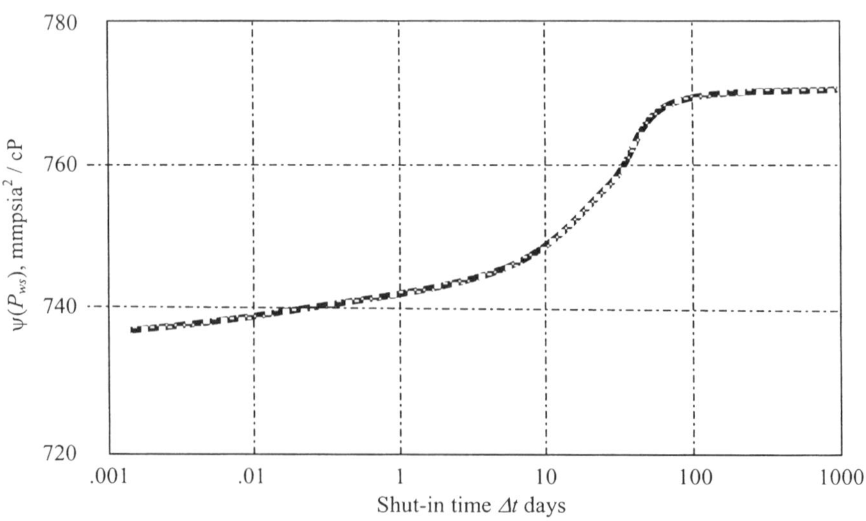

Figure 13–5. Pressure buildup for crossflow gas reservoir.

If $(kh)_t$ is known from a well test, individual layer permeabilities may be approximated from

$$k_j = \left(\frac{q_j}{q}[(kh_t)/h_j]\right), \quad j = 1, 2, \ldots, n \tag{13–4}$$

Pressure Buildup for Two-Layer Reservoir without Crossflow

Figure 13–6 shows a graph of dimensionless pressure P_D versus dimensionless time t_D for a two-layer reservoir with permeability ratios k_1/k_2 of 1, 2, 10, and 100. All four curves are for $r_e/r_w = 2000$. The dimensionless terms are

$$t_D = \frac{0.000264\,\bar{k}t}{\phi\mu_g c r_w^2} \tag{13–5}$$

$$t_{DA} = \frac{0.000264\,\bar{k}t}{\phi\mu_g cA} \tag{13–6}$$

$$P_D = \frac{\bar{k}h\,(P_i - P_{wf})}{141.2\,q\mu_g\beta_g} \tag{13–7}$$

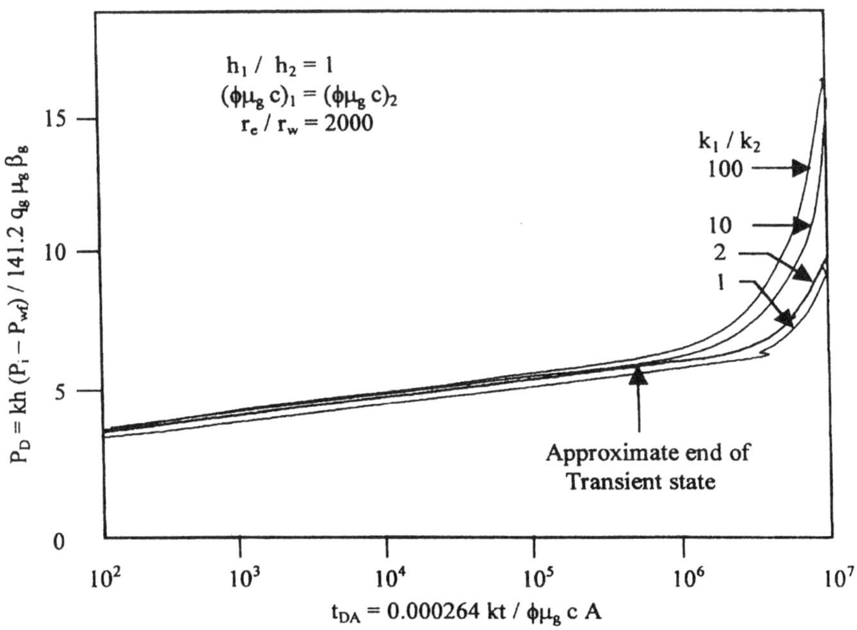

Figure 13–6. Muskat straight-line intercepts for two-layer reservoirs without crossflow (after Cobb, Ramey, and Miller).[4]

where $\beta_g = 0.00504(zTP_{sc})/(pT_{sc})$, bbl/scf, and

$$\bar{k} = \frac{k_1 h_1 + k_2 h_2}{h_{11} + h_2} \tag{13–8}$$

$$\bar{h} = h_1 + h_2 \tag{13–9}$$

$$\bar{\phi} = \frac{\phi_1 h_1 + \phi_2 h_2}{h_1 + h_2} \tag{13–10}$$

Figure 13–6 indicates that during the early transient period, the slope of the straight-line is 1.151 (2.303/2). The approximate semilogarithmic period ends at $t_D = 5 \times 10^5$ and behavior beyond the end of the semilogarithmic period is strongly influenced by permeability ratio.

Muskat Plot Characteristics

Figure 13–7 shows a Muskat plot for a well in the center of a closed two-layer reservoir with a permeability contrast of 2. Figure 13–8 illustrates Muskat straight-line intercepts for two-layer reservoirs as a function of producing time for selected permeability ratios. These plots can be used to estimate average kh and reservoir pore volume for a two-layer system produced to pseudo-steady-

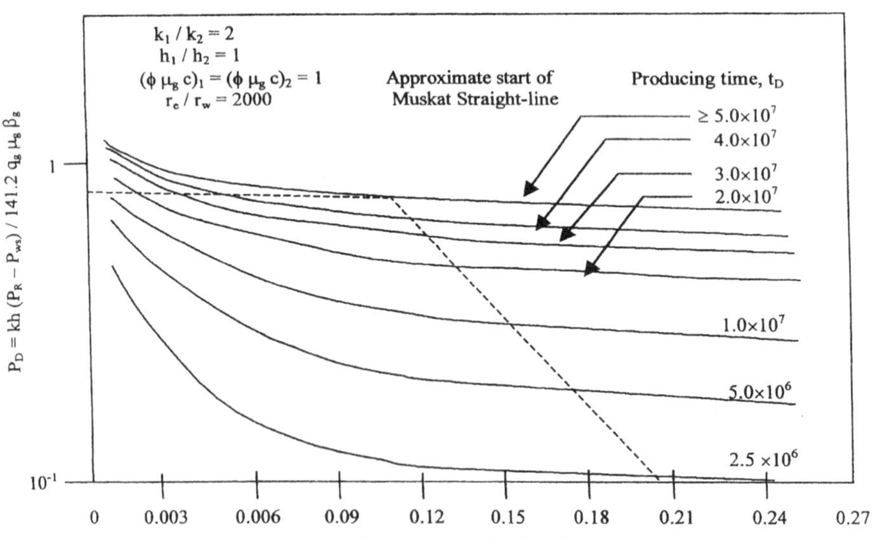

Figure 13–7. Muskat plot for-layer reservoir with a permeability contract of 2 (after Cobb, Ramey, and Miller).[4]

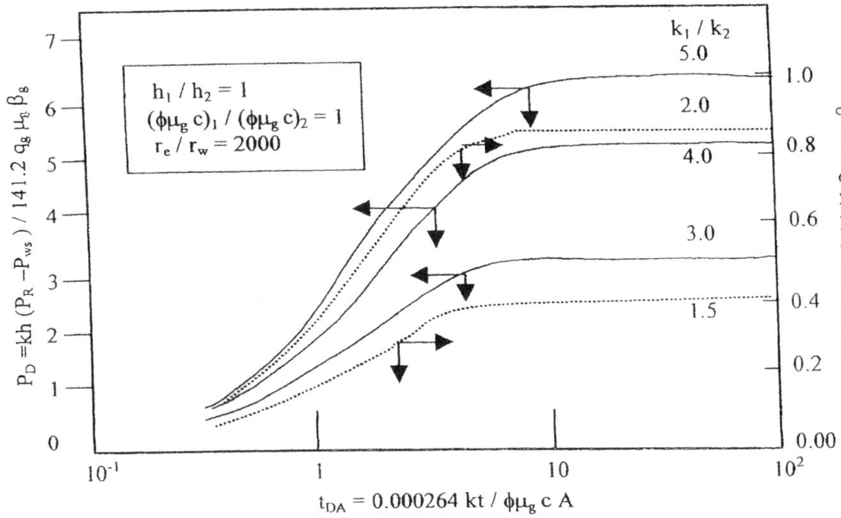

Figure 13–8. Muskat straight-line intercepts for two-layer reservoirs (after Cobb, Ramey, and Miller).[4]

state before shut-in by

$$\bar{k}\bar{h} = \frac{141.2\, q_g \mu_g \beta_g\, (0.87)}{(\bar{P}_R - P_{ws})_{\Delta t=0}} \tag{13–11}$$

The effective skin factor for a commingled reservoir system will be given by

$$s_t = \sum_{j=1}^{2} \frac{k_j h_j}{\bar{k}\bar{h}} \tag{13–12}$$

The reservoir volume may be obtained from

$$\bar{\phi} A c = 0.000264\, \frac{m\bar{k}}{\mu_g}\, (\text{slope, log-cycle/hr})^{-1} \tag{13–13}$$

where m is the slope of Muskat plot straight line.

MDH Method[2]

Figure 13–9 shows a MDH plot, which also provides a straight line for the early buildup, has an approximate slope for the early buildup, and has an approximate slope of 1.15 for a producing time of any length. Thus the method can be used to estimate average kh. This method can also be used to estimate static pressure, if theoretical buildup curves exist.

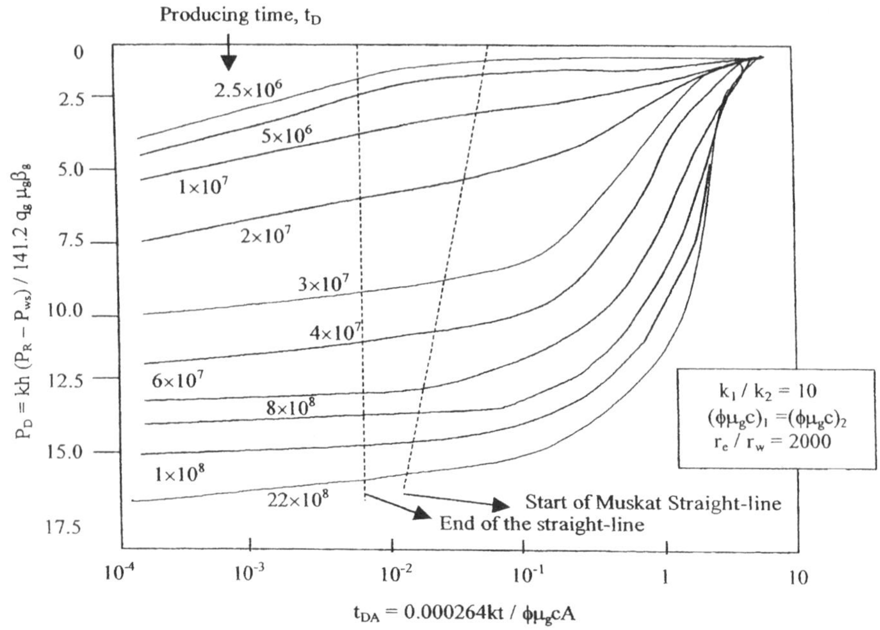

Figure 13–9. MDH buildup curves (after Raghavan *et al.*).[7]

13.4 Multilayered Responses in Fractured Gas Reservoirs

Reference 7 has introduced the concept of reservoir layer conductivity C_{RDj}, given by

$$C_{RDj} = \frac{k_j h_j}{\overline{k}\,\overline{h}} \sqrt{\overline{\eta}/\eta_j} \tag{13–14}$$

where η_j = diffusivity of layer j, $\overline{\eta} = \overline{k}/(\overline{\phi}\overline{c}_t \mu_g)$, $C_{RD} = \sum_{j=1}^{n} C_{RDj}$, and an equivalent fracture length and equivalent fracture conductivity are defined, respectively, by

$$\overline{x}_f = \sum_{j=1}^{n} C_{RDj} x_{fj} \tag{13–15}$$

and

$$\overline{k}_f \overline{w} = \frac{\left(\sum_{j=1}^{n} \sqrt{k_{fj} w_j h_j C_{RDj}}\right)^2}{h_t} \tag{13–16}$$

The dimensionless fracture conductivity is then defined by

$$C_{fD} = \frac{\overline{k}_f \overline{w}}{\overline{k}\overline{x}_f} \tag{13–17}$$

Camacho, Raghavan, and Reynolds[6] have studied the correlations of multiplayer responses with the single-layer solutions for a number of cases. They assume that the fractures are not in communication. If fractures are in communication, the values of C_{fD} are somewhat higher and the ration h_f/x_{fj} is an important factor in the performance of the fractured well. When layers are stimulated by fractures, then maximum productivity will be achieved if the fracture tip in each layer begins to affect the well response at approximately the same time. For a two-layer gas reservoir system, the criterion for maximum productivity is given by

$$\frac{\overline{C}_{fD2} h_2}{C_{fD1} h_1} = \frac{C_{RD2} x_{f2}}{C_{RD1} x_{f1}} \tag{13–18}$$

where

$$\overline{C}_f D_j = \frac{k_{fj} w_j}{\overline{k} x_{fj}}$$

The vertically dashed line in Figure 13–10 represents Eq. 13–18.

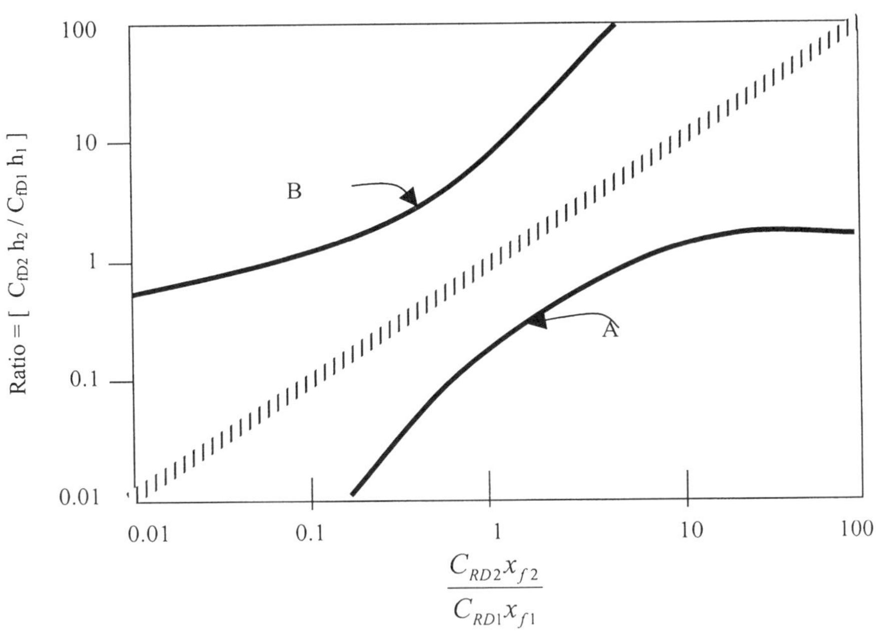

Figure 13–10. Criteria for maximum productivity (after Camacho, Raghavan, and Reynolds).[6]

These results are based on the assumption that boundary effects are negligible. The reservoir layer conductivity concept does not apply if boundary effects dominate the pressure response. During pseudo-steady-state flow the well response is given by a well in a circular reservoir.

13.5 Pressure–Production Performance Response Equations

During the period 1960 to 1962, seven papers[1–4] were published on the theoretical behavior of reservoir systems composed of intercommunicating layers. Russell and Prats[3] summarized the practical aspects of the finding of these papers in a later paper. Performance relationships are given here for two cases.

Constant Producing Rate

Pressure performance for transient flow period is given by

$$p_{wf}^2 = p_i^2 - \frac{57.920 \times 10^6 q_{sc} T \overline{\mu}_g \overline{z}\, p_{sc}}{(kh)_t T_{sc}} \left[\log \frac{(kh)_t t}{(\phi h)_t \mu_g c r_w^2} - 3.23 \right] \tag{13–19}$$

$$\psi(p_{wf}) = \psi(p_i) - \frac{57.920 \times 10^6 q_{sc} T p_{sc}}{(kh)_t T_{sc}} \left[\log \frac{(kh)_t t}{(\phi h)_t \mu_g c r_w^2} - 3.23 \right] \tag{13–20}$$

For larger times (semisteady state), the pressure behavior is described by

$$p_{wf}^2 = p_i^2 - \frac{141.2 q_{sc} \overline{\mu}_g \beta_{gi}}{(kh)_t} \left[\frac{0.000528 (kh)_t t}{(\phi h)_t \mu_g c r_e^2} + In(r_e/r_w) - 0.75 \right] \tag{13–21}$$

$$\psi(p_{wf}) = \psi(p_i) - \frac{141.2 q_{sc} T}{(kh)_t} \left[\frac{0.000528 (kh)_t t}{(\phi h)_t \mu_g c r_e^2} + In(r_e/r_w) - 0.75 \right] \tag{13–22}$$

where $(kh)_t = k_1 h_1 + k_2 h_2$; $(\phi h)_t = (\phi_1 h_1) + (\phi_2 h_2)$; $h_t = h_1 + h_2$; $t =$ time in hours; and $\beta_{gi} = 0.00504 zi T p_{sc}/p_i\ T_{sc}$ in rb/mmscfd.

The time at which the semisteady state starts is given by

$$t \cong 1136.4 \frac{(\phi h)_t \mu_g c r_e^2}{(kh)_t} \text{ hr} \tag{13–23}$$

For semisteady-state flow, the slope of the plot of flowing bottom-hole pressure versus time is given by

$$\text{Slope} = 0.07455 \frac{q_{sc} \beta_g}{(\phi h)_t c r_e^2} \text{ psi/hr} \tag{13–24}$$

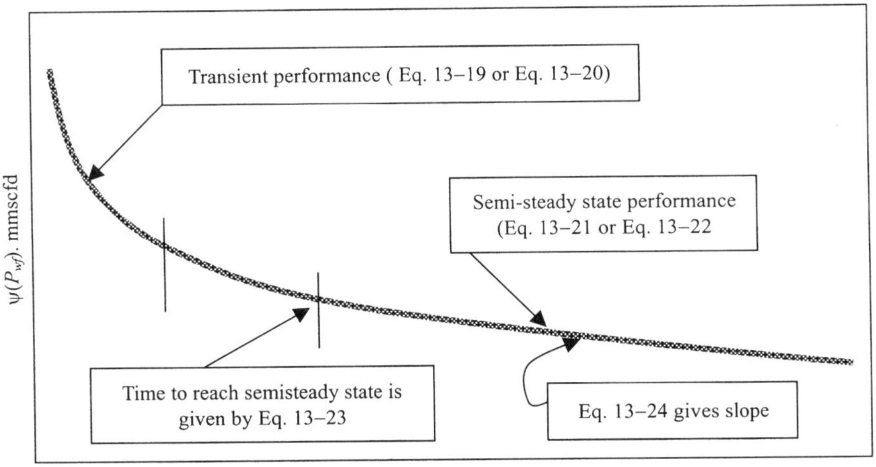

Figure 13–11. Idealized constant-rate pressure performance in two-layer reservoir with crossflow (after Russell and Prats).[3]

Figure 13-11 shows an idealized constant-rate flowing bottom-hole pressure performance curve, and reservoirs of the type should possess the properties shown on this plot.

Constant Producing Pressure

For the case of constant bottom-hole producing pressure, the following formula for producing rate was developed. Reference 5 has also provided the method to calculate cumulative production from multilayered reservoirs:

$$(kh)_t = k_1h_1 + k_2h_2, \quad \text{and} \quad (\phi h)_t = \phi_1h_1 + \phi_2h_2$$

$$q_g = \frac{(kh)_t\,(P_i - P_{wf})}{141.2\,\beta_g\mu_g\left(In\frac{r_e}{r_w} - 0.75\right)}e^{-AB} \tag{13–25}$$

where

$$AB = \left[\frac{0.0127\,(kh)_t}{(\phi h)_t c\,\mu_g\,r_e^2\left(In\frac{r_e}{r_w} - 0.75\right)}\right]$$

13.6 Flow Identification and Performance Analysis

Figure 13–12 presents the methods that can be used to identify the degree of communication between layers and type of crossflow.

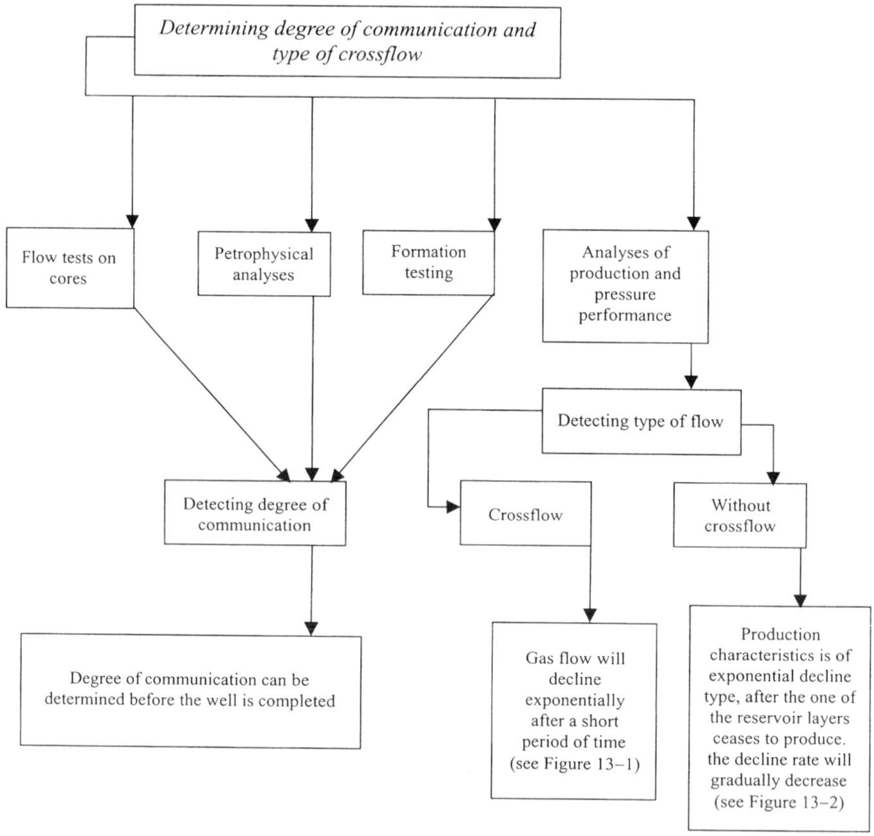

Figure 13–12. Systemic diagrams to determine degree of communication and type of crossflow.

13.7 Pressure Buildup Behavior in Layered Reservoir Systems

Figure 13–13 shows pressure buildup behavior in layered reservoir systems in a single-well, multiple-layer reservoir. Reference 3 has stated that, after the initial semilog straight line, the buildup curve flattens, then steepens and finally flattens toward the average reservoir pressure as indicated in the figure. This is not always correct. The C-D portion in Figure 13–13 can be insignificant for some systems. That is particularly true for large contrasts in porosity or thickness, for more than two layers, or for nonsymmetrical systems. Classifications and pressure response characteristics including detailed analysis of multiple-layered reservoir systems are described in the previous sections.

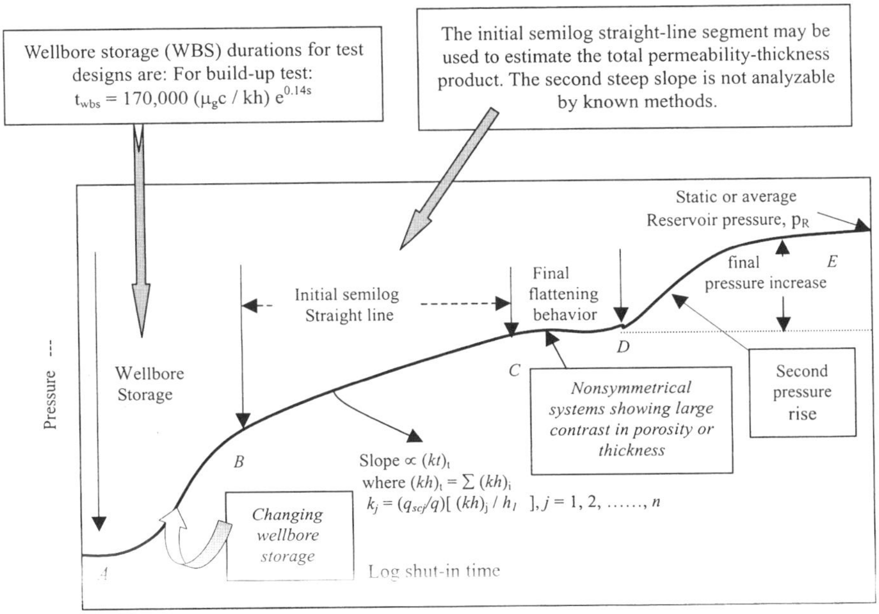

Figure 13–13. Schematic pressure buildup curve for a layered reservoir system.[5]

Buildup Behavior Curve in Two-Layer Gas Reservoir

The pressure builds up first in the more permeable layer, giving a straight-line section as shown in Figure 13–14. Then the less permeable layer, which is at a higher average pressure, begins to feed fluid into the more permeable layer. This causes the rise above the straight line. Finally, equalization will occur, and the curve will flatten as indicated by the dotted line. Figure 13–14 shows pressure buildup behavior in two-layer gas reservoir including the effects of wellbore damage and storage. Total flow capacity can be obtained from the shape of a field data plot by using the following equations:

Pressure squared case (Eq. 6–5):

$$(kh)_t = \frac{57.920 \times 10^6 q_{sc} T p_{sc} \overline{\mu}_g \overline{z}}{m\, T_{sc}}$$

Pseudopressure case (Eq. 6–10):

$$(kh)_t = \frac{57.920 \times 10^6 q_{sc} T p_{sc}}{m T_{sc}}$$

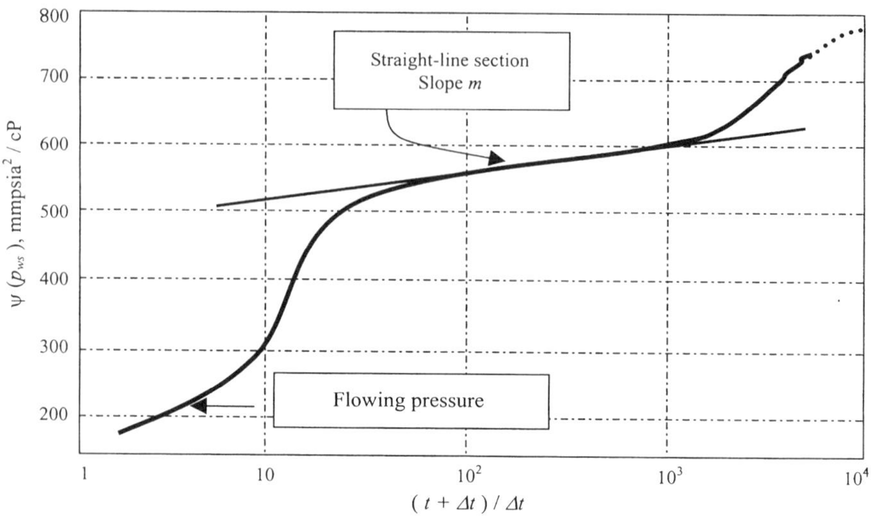

Figure 13–14. Pressure buildup behavior in a two-layer gas reservoir.

13.8 Determining Reservoir Characteristics in Commingled Systems

References 2 and 4 have provided techniques to determine average reservoir pressure in commingled systems. It requires some knowledge of the layer properties and correlations for specific systems. The following types of tests may be used to estimate individual zone properties for a two-layer reservoir with communication only at the wellbore:

- Single well test
- Pulse tests
- Flow meter surveys

Apparent kh/μ_g is always equal to or greater than the actual total $(kh/\mu_g)_t$ for the reservoir. Apparent $\phi c_t h$ is always equal to or less than the total $(\phi c_t h)_t$ for the reservoir. Deviation of apparent values from actual total values depends on the pulse duration.

13.9 Factors Affecting Performance

The following factors can affect the performance of multilayered reservoir systems:

- *Relative permeability.* If both layers have the same relative permeability characteristics, average gas saturation will be higher in the tighter layer than in the more permeable layer, because of the average pressure is always higher in the less permeable layer.

- *Pore size.* If pore size in the tight layer is smaller than that in the more permeable layer, then it will tend to reduce crossflow. This effect can be estimated from capillary pressure curves.
- *Reservoir geometry.* The geometrical nature and extent of interlayer communication have some effect on observed field performance.
- *Permeability anisotropy.* In most petroleum reservoirs, vertical permeability is significantly less than horizontal permeability.
- *Reservoir n-layer system.* Analysis of performance can be handled to acceptable accuracy merely by the previously presented formulas $\sum_j^n k_j h_j$ and $\sum_j^n \phi_j h_j$ for $(kh)_t$ and $(\phi h)_t$ respectively.

13.10 Economic Aspects of Interlayer Crossflow

The absence or presence of crossflow between interlayers can control the economic success of a gas production venture. Some of the advantages of interlayer crossflow are as follows:

- Shorter operating life
- Higher ultimate gas production
- Reduced perforating and completion costs
- Generally less engineering time required for interpretation of routine tests

Note: A without-crossflow reservoir can be converted into a crossflow reservoir by fracturing. Thus a vertical fracture can help to establish vertically adjacent gas-production strata, which were not in communication prior to the fracture job except at the wellbore.

References and Additional Reading

1. Lefkovits, H. C., Hazebrock, P., Allen, E., and Matthews, C. S., "A Study of the Behavior of Bounded Reservoirs Composed of Stratified Layers," *Soc. Petroleum Eng. J.* (March 1961) 43–58.
2. Cobb, W., "A Study of Transient Flow in Stratified Reservoirs with Commingled Fluid Production," Ph.D. dissertation, Stanford University, Stanford, CA, 1970.
3. Russell, D. G., and Prats, M., "The Practical Aspects of Inter Layer Cross Flow," *J. Petroleum Technol.* (June 1962) 589–594.
4. Cobb, W. M., Ramey, H. J., Jr., and Miller, F. G., "Well Test Analysis for Wells Producing Commingled Zones," *J. Petroleum Techol.* (Jan. 1972) 27–37; *Trans. AIME* 253.
5. Earlougher, R. C., Jr., Kersch, K. M., and Kunzman, W. J., "Some Characteristics of Pressure Buildup Behavior in Bounded Multiple Layer Reservoirs

without Crossflow," *J. Petroleum Technol.* (Oct. 1974) 1178–1186; *Trans. AIME* 257.

6. Camacho, V., Raghavan, R., and Reynolds, A. C., "Response of Wells Producing Layered Reservoirs, Unequal Fracture Length," *SPE Formation Evaluation* (Feb. 1987) 9–28.
7. Raghavan, R., Topaloglu, H. N., Cobb, W. M., and Ramey, H. J., Jr., "Well Test A Analysis for Wells Producing from Two Commingled Zones of Unequal Thickness," *J. Petroleum Technol.* (Sept. 1974) 1035–1043; *Trans. AIME* 257.

Chapter 14

Pressure Behavior Analysis in Heterogeneous Reservoir Systems

14.1 Introduction

This chapter discusses the effects of some common reservoir heterogeneities on pressure transient behavior. It is difficult to delineate specific heterogeneities from well tests. The difficulty occurs because many different conditions can cause the same or similar well test response. If we have an idea of the type of heterogeneity, it may be possible to determine some of the properties involved by pressure transient testing. Some knowledge of geological, seismic, fluid flow, and performance data is necessary before hypotheses are formed about the type and location of the heterogeneities. It may be possible to design a specific transient test to investigate the possibility of a particular type of heterogeneity. This chapter discusses and classifies reservoir heterogeneities, permeability, and anisotropy and describes how these heterogeneities affect transient testing. Several types of heterogeneities can cause similar transient test pressure response, but results should be supported by other data. The next section will illustrate a variety of situations such as faults, lateral changes in the hydraulic diffusivity such as occur at fluid contacts, and man-made heterogeneities.

14.2 Causes of Heterogeneities

Heterogeneities may have the following causes:

- Postdepositional changes in reservoir lithology
- Folding and faulting
- Changes in fluid type or properties
- Variations in rock and fluid properties from one location to another

- Physical barriers, gas–water contacts, thickness changes, lithology changes
- Different properties in each layer, etc.
- Man-made heterogeneities, including changes near the wellbore from hydraulic fracturing, acidzing, or gas injection

14.3 Pressure-Dependent Properties

It is well known from laboratory studies as well as from observed pressure behavior in some wells that both porosity and permeability decrease as reservoir pressure declines. For reservoir rocks, which are "normally" compacted, these effects are usually less than for those which have unusually high pore pressure, i.e., geopressured reservoirs. Carbonate rocks are more heterogeneous. Sandstone rocks are less complex than carbonate rocks. However, a quantitative evaluation of the porosity resulting from the interaction of the various factors is possible only by laboratory measurements. Sandstones and other classic rocks tend to be more elastic in their behavior than carbonate rocks. Limestones often are somewhat plastic in their behavior.

In general, it is expected to observe a decline in calculated permeability from successive transient pressure tests run throughout the life of a well in depleted reservoirs. Declines of 10% or so may be observed, but because of variations of other kinds such as two-phase flow effects, quantitative evaluation becomes difficult. Therefore laboratory-determined curves of porosity and permeability versus pressure should be used to predict pressure behavior.

References 1–5 concluded that neither permeability nor skin factor should be estimated from drawdown or buildup tests using techniques like those given in Chapters 5 and 6 in formation with pressure-dependent permeability. Figure 14–1 illustrates their findings.

14.4 Pressure Responses Near Flow Barriers

Linear sealing faults and barriers have been an interesting topic in the transient-testing literature.[6–8] Horner[7] considers pressure buildup and Russell[11] discusses two-rate flow testing in that system. Regardless of test type, the linear flow barrier affects the test in about the same way. To obtain the effect of the linear fault, the following interpretation formulas, which are needed in this particular instance, are given. A computed example of a buildup test in a well located 239.3 ft from a fault is shown in Figure 14–6. The data assumed for this example are given on the figure. It can be seen that the buildup test plot possesses two distinct straight-line slopes. As in the case with a pressure buildup, the second or "late-time" portion of the buildup test curve has a slope which is exactly double that of the "early-time" portion of the curve. The pressure

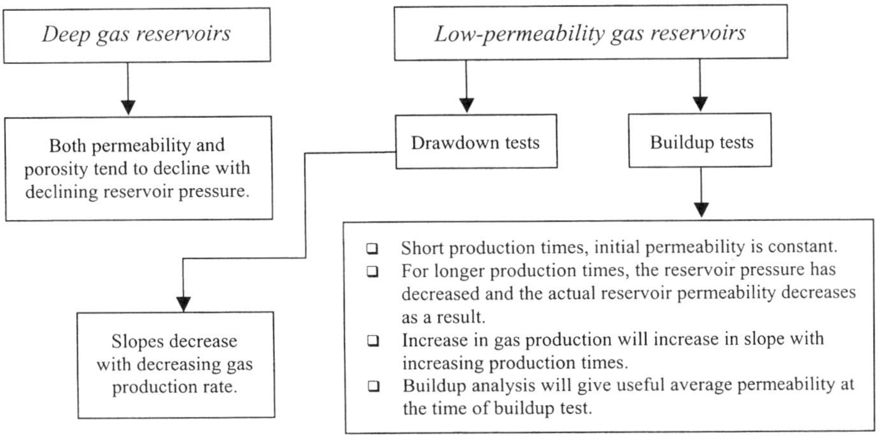

Figure 14–1. Effect of pressure-dependent permeability on drawdown and buildup tests.

response at a well near a sealing fault can be directly obtained from

$$[\psi(p_{WD}(t_D))] = -0.5\lfloor E_i(-1/4t_D) + E_i\left(-r_{dD}^2\right) + s\rfloor \qquad (14\text{–}1)$$

Here $r_{dD} = 2L/r_w$, where L is the distance to this fault. If times are small enough, then the second term in Eq. 14–1 can be assumed to be negligible compared with the first, and the line source solution (Eq. 14–2) can be used to analyze responses in the conventional way:

$$[\psi(p_D(r_D, t_D))] = 0.5E_i\left(-r_D^2/4t_D\right) \qquad (14\text{–}2)$$

If the logarithmic approximation to the exponential integral is used, then $\psi[p_D(r_D, t_D)]$ is given by

$$[\psi(p_D(r_D, t_D))] = 0.5\lfloor In\left(4t_D/r_D^2\right) - 0.5772\rfloor \qquad (14\text{–}3)$$

If flow times are long enough such that both exponential integrals can be approximated by the logarithmic approximation, then we have

$$[\psi(p_{WD})[t_D]] = In(4t_D/e^{0.5772}) - In(r_{dD}) + s \qquad (14\text{–}4)$$

This equation suggests that one should get a second straight line with a slope twice that of the first. In practice, the doubling of the slope on semilogarithmic coordinates is normally taken to be indicative of a sealing fault. If a fault exists, then the first straight line should exist for a time period given by

$$6 < t_D \leq 0.08r_{dD}^2 \qquad (14\text{–}5)$$

The second straight line should begin at $3r_{dD}^2$. In the time range $0.08r_{dD}^2 < t_D < 3r_{dD}^2$ it can be used to analyze pressure measurements or predict pressure responses. The distance to the fault can be obtained if we equate the semilog approximation of the line source solution Eq. 14–3 to the right-hand side of Eq. 14–4. If we denote this time by $0.8r_{dD}^2$, then $e^{0.5772}r_{dD}^2/4$; then the distance to the fault is given by

$$L = d = \sqrt{e^{-0.5772}(0.0002637kt_x/\phi c_t \mu_g)} \tag{14–6}$$

where t_x is the intersection time in hours. This procedure assumes that both straight lines are evident.

Estimating Techniques for Distance to the Discontinuity

The effect of a sealing fault or barrier in an infinite-acting reservoir is to cause the buildup plot to start off as a straight line with the proper slope, gradually bend over, and eventually become another straight line with twice the slope of the first. The first straight line gives the proper value of kh. The second straight line gives the proper extrapolation to $\Psi(p_i)$. The distance between the well and the fault may be obtained by using the expression given by Davis and Hawkins[9] for drawdown tests and seems to apply reasonably well to buildup tests. The approximation takes the final form

$$L = \sqrt{\left(\frac{0.000148k\Delta t_x}{\phi \mu_g c}\right)} \tag{14–7}$$

where Δt_x = value at the intersection of the two lines.

The distance to a barrier can also be calculated by using Eq. 14–8, developed by Van Poollen and coworkers:[10]

$$L = \sqrt{\left(\frac{0.000933kt_p}{\phi \mu_g c\left(\frac{t_p+\Delta t_x}{\Delta t_x}\right)}\right)} \tag{14–8}$$

where $(t_p + \Delta t_x)/\Delta t_x$ is the value at the point of deviation from the first straight line. The following equation is applicable to both buildup and drawdown tests and is known as Gray's equation:[8]

$$\Delta p = \left(\frac{-70.6q_{sc}\mu_g \beta_g}{kh}\right)\left[-E_i\left(-\frac{\phi \mu_g c L^2}{0.000264k\Delta t}\right)\right] \tag{14–9}$$

where $\beta_g = p_{sc}zT_R/T_{sc}p_R \times 10^6$ and q_{sc} is gas rate in mmscfd. This equation is most accurate if Δt is large. This is a trial-and-error procedure by assuming various values of L until the right-hand side of Eq. 14–9 is equal to the left-hand side. Gray also suggested that distance to the nearest boundary can be estimated approximately by

$$L \cong 0.01217\sqrt{k\Delta t_x/\phi\mu_g c} \tag{14–10}$$

where Δt_x is the time at which the buildup curve becomes nonlinear.

From pressure buildup testing, the intersection point of the two straight lines is related to the dimensionless pressure at the intersection line by

$$\{p_D[t_D/(2L/r_w)^2]\} = 0.5\, In\left(\frac{t_p + \Delta t}{\Delta t}\right)_x \tag{14–11}$$

Calculate p_D from Eq. 14–11. Then from Table 14–1, with the value of p_D, determine $t_D/(2L/r_w)^2$. Finally, use the following equation to estimate the distance to the fault:

$$L = \sqrt{\frac{0.0002637kt_p}{4\phi\mu_g c_t[t_D/(2L/r_w)^2]}} \tag{14–12}$$

Relationships between $p_D(t_D, r_D)$ and t_D/r_D^2 are given in Table 14–1. The detailed derivations of Eqs. 14–11 and 14–12 are given in Ref. 12.

Table 14–1
Dimensionless Pressure at Various Values of Dimensionless Time[13]

Dimensionless pressure p_D	Dimensionless time $t_D/(2L/r_W)^2$	Dimensionless pressure p_D	Dimensionless time $t_D(2L/r_W)^2$	Dimensionless pressure p_D	Dimensionless time $t_D/(2L/r_W)^2$
0.01	0.00	1.6	9.0	4.0	1500
0.02	0.00	1.7	12.0	4.1	1750
0.03	0.00	1.8	17.0	4.2	2000
0.04	0.15	1.9	20.0	4.3	2500
0.05	0.16	2.0	25.0	4.4	3000
0.06	0.18	2.1	27.5	4.5	3500
0.07	0.19	2.2	30.0	4.6	4200
0.08	0.20	2.3	45.0	4.7	5000
0.09	0.22	2.4	60.0	4.8	7000
0.10	0.24	2.5	70.0	4.9	9000
0.20	0.38	2.6	80.0	5.0	1.0×10^4
0.30	0.52	2.7	90.0	5.5	3.0×10^4
0.40	0.70	2.8	110.0	6.0	7.0×10^4
0.50	0.94	2.9	140.0	6.5	1.75×10^5
0.60	1.20	3.0	170.0	7.0	5.0×10^5
0.70	1.65	3.1	220.0	7.5	2.0×10^6
0.80	2.00	3.2	260.0	8.0	5.0×10^6
0.90	2.50	3.3	300.0	8.5	1.5×10^7
1.0	3.00	3.4	400.0	9.0	3.0×10^7
1.1	4.00	3.5	500.0	9.5	1.5×10^8
1.2	4.80	3.6	600.0	10.0	2.0×10^8
1.3	6.00	3.7	700.0	—	—
1.4	8.00	3.8	900.0	—	—
1.5	8.50	3.9	1200.0	—	—

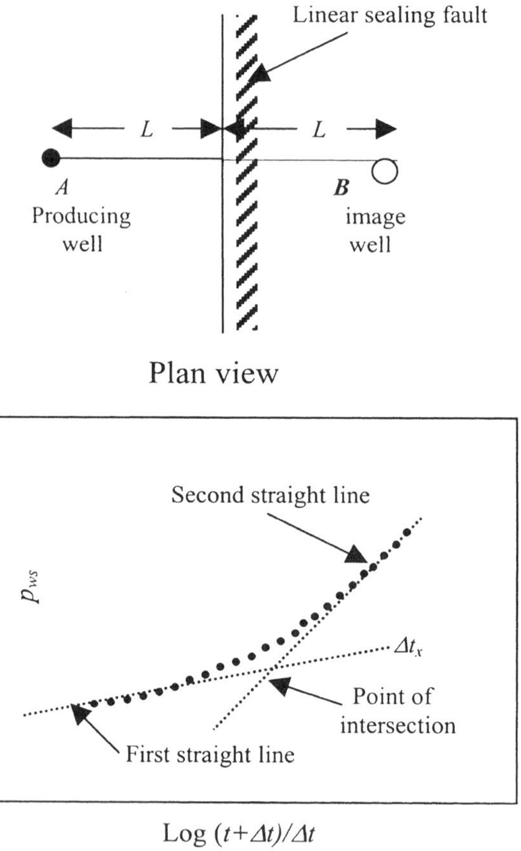

Figure 14–2. Fault near single boundary.

Figures 14–2 and 14–3 show various situations of linear discontinuities for single and multiple boundary cases. Figure 14–4 shows various methods to estimate the distance to a linear discontinuity and their limitations.

Example 14–1 *Estimating Distance to a No-Flow Boundary*

A pressure buildup test was run in a newly drilled gas well. Geologists suspect a fault. Data from the test are given in Table 14–2. Well and reservoir data include the following: $\phi = 0.088$ (fraction); $\mu_g = 0.01633$ cP; $h = 59$ ft; $r_w = 0.39$ ft; $c_t = 0.000255$ psi^{-1}; $q_{sc} = 5.20$ mmscfd; $T = 710°$R; pseudo-producing time $t_p = 819$ hr.

Calculate the distance to the linear fault using various methods.

Table 14–2
Analysis of Data from Well near Boundary

Δt (hr)	$\frac{t_p+\Delta t}{\Delta t}$	$\psi(p)$ (mmpsia^2/cP)	Δt (hr)	$\frac{t_p+\Delta t}{\Delta t}$	$\psi(p)$ (mmpsia^2/cP)
2.5	328.6	110.2	43.0	20.1	127.2
4.0	205.8	111.1	91.0	10.0	138.4
8.0	103.4	113.0	102.0	9.1	139.4
9.0	92.0	114.2	117.0	8.0	142.3
10.0	83.0	115.5	137.0	7.0	145.0
12.0	69.3	116.0	164.0	6.0	150.1
14.0	59.5	117.7	205.0	5.0	153.4
16.0	52.2	118.4	213.0	4.0	158.2
21.0	40.0	120.0	410.0	3.0	164.2
28.0	30.0	123.3			

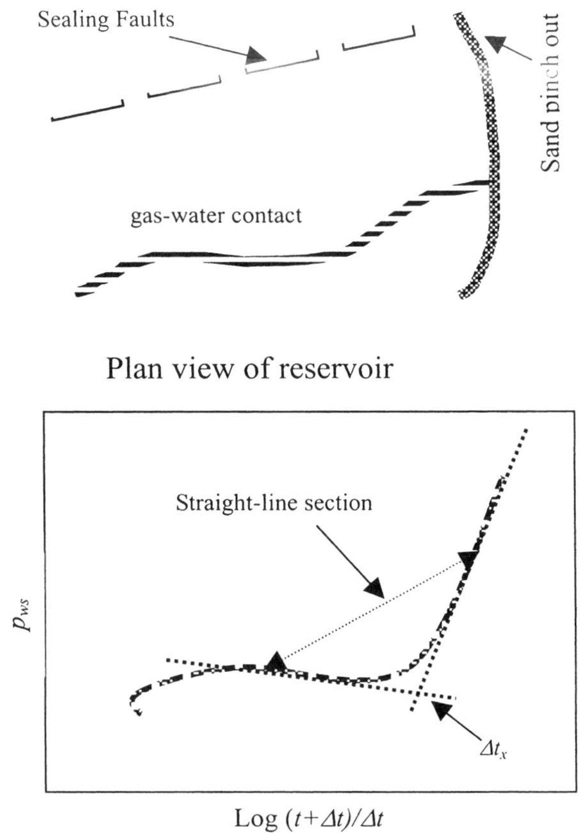

Figure 14–3. Fault near multiple boundaries.

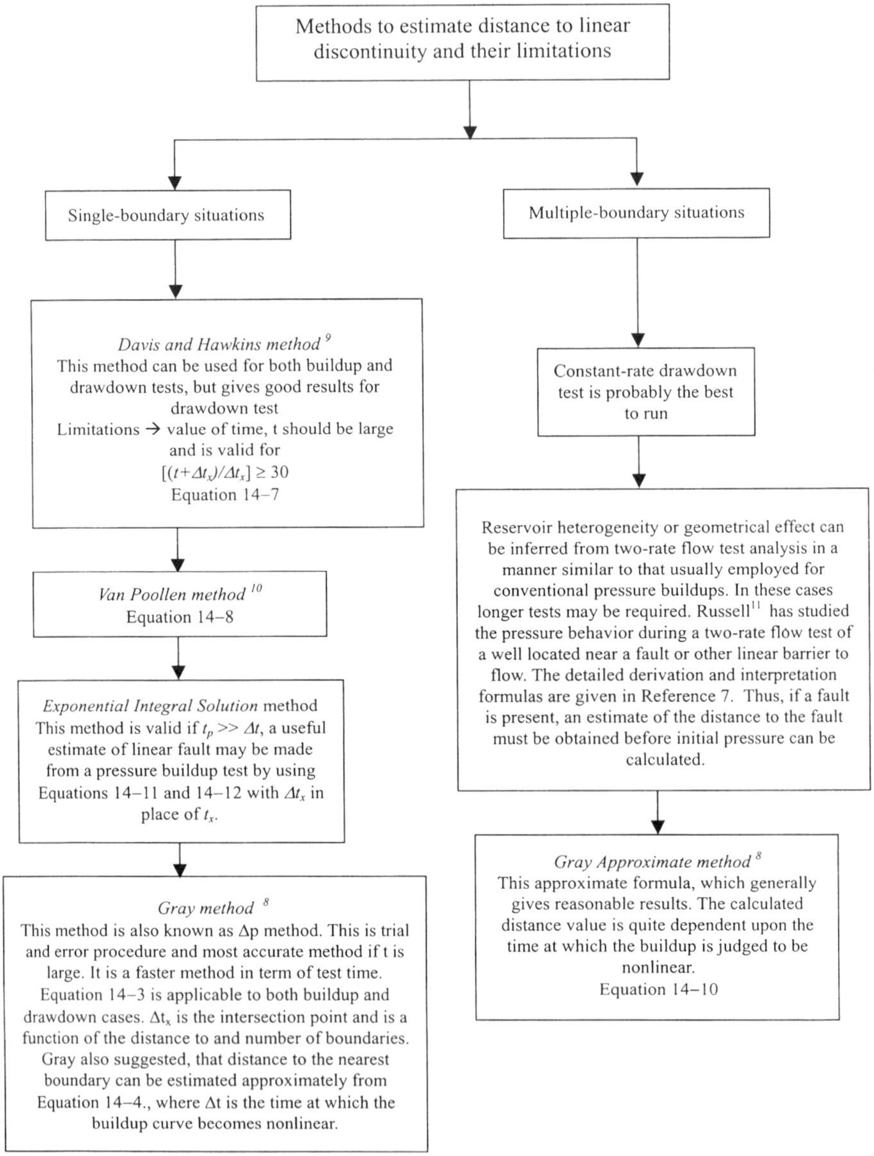

Figure 14–4. Various methods to determine the distance to a linear discontinuity.

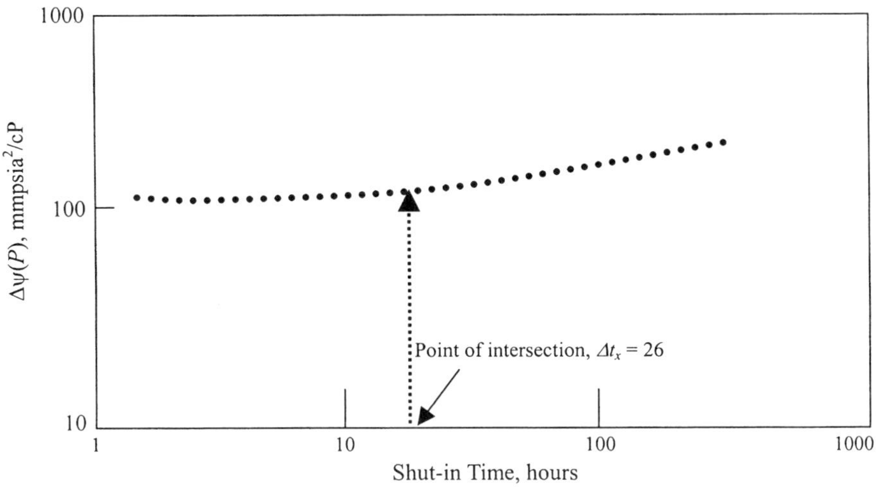

Figure 14–5. $\Delta\psi$(P) versus Δt for buildup test (log-log plot).

Solution Pressure buildup data are shown in Figures 14–5 and 14–6. The log-log plot of Figure 14–5 indicates that wellbore storage effects are not important, so the increase in slope in Figure 14–6 is probably caused by reservoir heterogeneity. The ratio of the two slopes is 2.20. Since the absolute value of the slopes is increasing with shut-in time, and since the slope ratio is about 2, a linear fault is suspected. Formation permeability k is estimated from the first straight line using Eq. 6–10. Recall that for a Horner plot the slope is $-m_1$, so $m_1 = 25.0$ mmpsia²/cP/cycle.

$$kh = \frac{57.92 \times 10^6 \times 5.20 \times 710 \times 14.65}{25 \times 520} = 240.98 \text{ mD-ft}$$

$$\therefore k = \frac{240.98}{59.0} = 4.08 \text{ mD}$$

If we wish, we may use Eq. 6–11 and data from the first straight line to estimate skin factor, s:

$$s = 1.51\left[\frac{\psi(p_{1hr}) - \psi(p_{wfo})}{m} - \log\frac{k}{\phi\mu_g c r_w^2} + 3.23\right]$$

$$= 1.51\left[\frac{112.4 - 101.5}{25} - \log\frac{4.08}{0.16 \times 0.0215 \times 0.000255 \times 0.42^2}\right]$$

$$= -3.75$$

To estimate the distance to the fault, we determine $(t_p + \Delta t_x)/\Delta t_x = 28.5$ and $\Delta t_x = 25.8$ (see Figure 14–6).

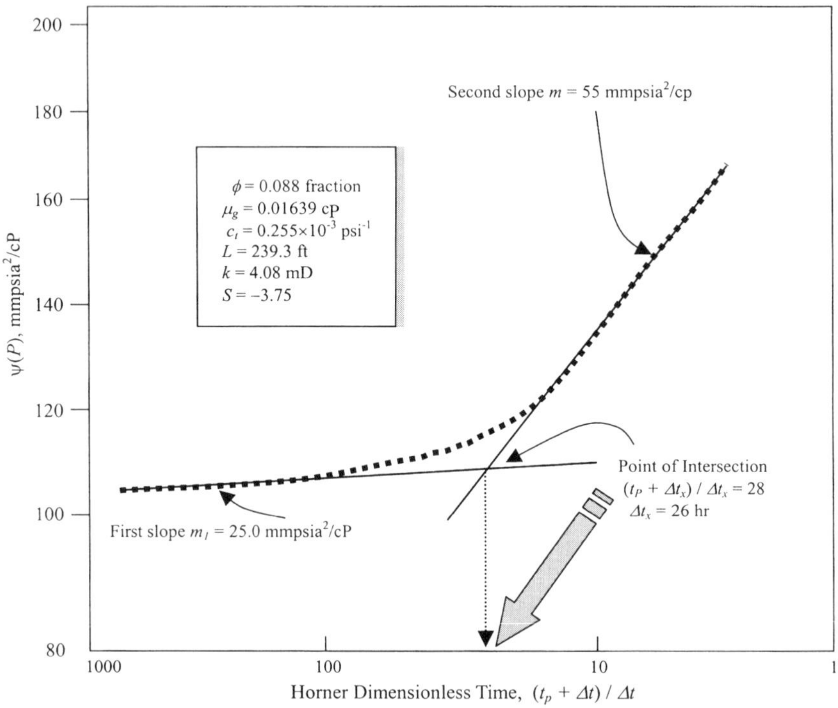

Figure 14–6. Estimating distance to a no-flow boundary.

1. Line source solution method (Eq. 14–6):

$$L = \sqrt{e^{-0.5772}\left(\frac{0.0002637kt_x}{\phi\mu_g c_t}\right)}$$

$$= \sqrt{0.5615\left(\frac{0.0002637 \times 4.08 \times 25.5}{0.088 \times 0.01633 \times 0.255 \times 10^{-3}}\right)} = 205.5 \text{ ft}$$

2. Davis and Hawkins method (Eq. 14–7):

$$L = \sqrt{\frac{0.000148k\Delta t_x}{\phi\mu_g c_t}} = \sqrt{\frac{0.000148 \times 4.08 \times 25.5}{0.088 \times 0.01633 \times 0.255 \times 10^{-3}}}$$

$$= 205.0 \text{ ft}$$

3. Van Poollen method (Eq. 14–8):

$$L = \sqrt{\frac{0.000933kt_p}{\phi\mu_g c\left(\frac{t_p+\Delta t_x}{\Delta t_x}\right)}} = \sqrt{\frac{0.00093(4.08)(819)}{0.088(0.01633)(0.000255)(28.5)}}$$

$$= 270.5 \text{ ft}$$

4. Gray Approximate method (Eq. 14–10):

$$L = 0.01217\sqrt{\frac{k\Delta t_x}{\phi\mu_g c_t}} = 0.01217\sqrt{\frac{4.08 \times 25.5}{0.088 \times 0.01633 \times 0.255 \times 10^{-3}}}$$

$$= 205.0 \text{ ft}$$

5. Gray method (Eq. 14–9):

$$\Delta p = \left(\frac{-70.6 q_{sc}\mu_g\beta_g}{kh}\right)\left[-E_i\left(-\frac{\phi\mu_g c L^2}{0.000264 k\Delta t}\right)\right]$$

6. Exponential integral solution method (Eqs. 14–11 and 14–12):

$$\{p_D[t_D/(2L/r_w)^2]\} = 0.5\, In\left(\frac{t_p + \Delta t}{\Delta t}\right)_x = 0.5\, In(28.5) = 1.675$$

When $p_D > 10$, the values of $[t_D/(2L/r_w)^2]$ can be calculated from

$$\left(\frac{t_D}{(2L/r_w)^2}\right) = e^{2(p_D - 0.4045)}$$

$$L = \sqrt{\frac{0.0002637 k t_p}{4\phi\mu_g c_t[t_D/(2L/r_w)^2]}}$$

$$= \sqrt{\frac{0.0002637 \times 4.08 \times 819}{4 \times 0.088 \times 0.01633 \times 0.255 \times 10^{-3} \times 10.5}}$$

$$= 239.3 \text{ ft}$$

Methods 1, 2, and 4 give reasonably close linear fault values (Table 14–3).

Table 14–3
Comparison of Linear Discontinuities by Six Methods

Methods	Equations used	Distance to fault L (ft)	Remarks
1. Line source solution	14–6	205.1	Low value
2. Davis and Hawkins	14–7	205.4	Low value
3. Van Poollen	14–8	270.5	Fairly good
4. Gray approximate equation	14–10	205.0	Low value
5. Gray Δp	14–9	296.0	Good
6. Exponential integral solution	14–11, 14–12	239.3	Fairly good

14.5 Effect of Lateral Changes on Pressure Behavior

Figure 14–7 shows idealized reservoir situation studies.[1,3,4,5] Changes in the hydraulic diffusivity occur at the boundary between differing geological depositional units due to changes in porosity and permeability. Reference 10 has investigated the effect of linear discontinuities in hydraulic diffusivity on pressure drawdown and buildup behavior. A brief summary follows in terms of these parameters: zone 1 hydraulic diffusivity, $\eta_1 = k_1/\phi_1\mu_1c_1$; zone 2 hydraulic diffusivity, $\eta_2 = k_2/\phi_2\mu_2c_2$; ratio $= \eta_2/\eta_1 = [(k_2/\phi_2\mu_2c_2)]/[(k_1/\phi_1\mu_1c_1)] = [k_2\mu_1/k_1\mu_2]/[\phi_2c_2/\phi_1c_1] = M_k/R_{pc} =$ permeability contrast/porosity contrast.

The ratio of hydraulic diffusivities η_2/η_1 is equal to M_k/R_{pc}. The greater the reduction in hydraulic diffusivity from the zone containing the well to the zone beyond the discontinuity, the move closely the slope change will approach a factor of 2, as with faults, gas–water contacts may not be distinguishable from a fault in practical cases. Large increases in diffusivity across the discontinuity will cause the pressure drop to arrest and become essentially constant.

For diffusivity contrast ratio $M_k/R_{pc} = 1$, homogeneous reservoir behavior results. If the diffusivity contrast ratio $M_k/R_{pc} > 1$, the buildup curve slope will flatten. If the diffusivity contrast ratio $M_k/R_{pc} = 1$, the buildup curve slope increases after the effect of discontinuities.

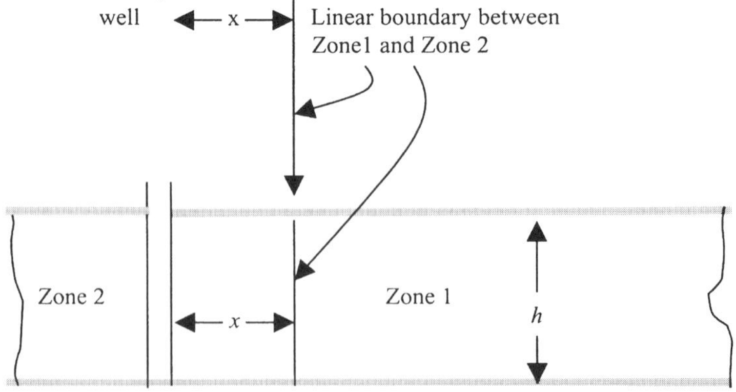

Figure 14–7. Schematic cross-section of some practical reservoir situations.

14.6 Evaluation of Heterogeneity of Reservoir Rock Porosity Systems

Evaluation of heterogeneity of the reservoir rock's porosity systems can be made using a wire-line formation tester. This a sample chamber of up to several gallons capacity combined with pressure gauges. The test chambers are forced against the borehole wall in a sealing pad, and firing a shaped charge perforates the formation. The signal to fire the charge is transmitted on logging cable. Fluid is collected during sampling, and pressure is recorded. Following sample collection, shut-in pressures are recorded as the buildup with time.

Reservoir Rock Porosity Distribution System Analysis

The pressure versus time records from the formation tester permit the evaluation of the heterogeneity of the reservoir rock's porosity system, whether it may be considered as a uniform and homogeneous porosity development or as a multiple porosity system made up of matrix porosity and of course porosity (vugs, fracture, fissures, joints, etc.). By a Pollard type plot[19] of $\log(p_s - p)$ versus time, it is possible to identify the type of prevailing porosity and the respective fraction of each; these data are important in the interpretation of fractured rocks. It has been shown by Pirson and Pirson[12] that the respective volumes of the coarse and fine pore systems may be evaluated by plotting the successive pressure differences versus time on semilog paper. Figure 14–8 is a representation of porosity partition in heterogeneous porous rock and

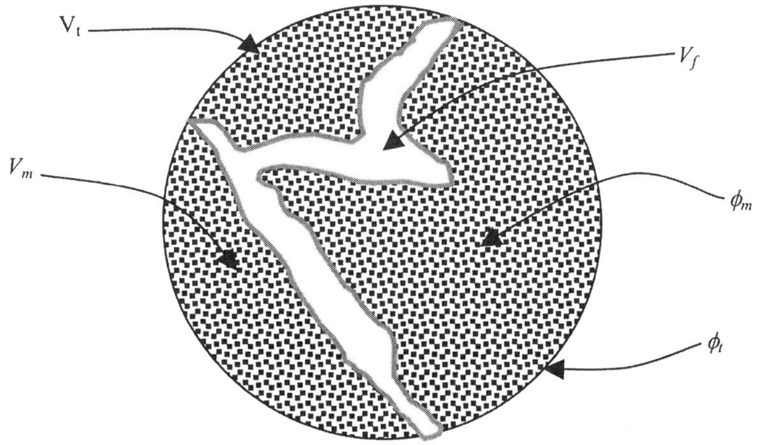

Figure 14–8. Porosity partition in heterogeneous porous rock.

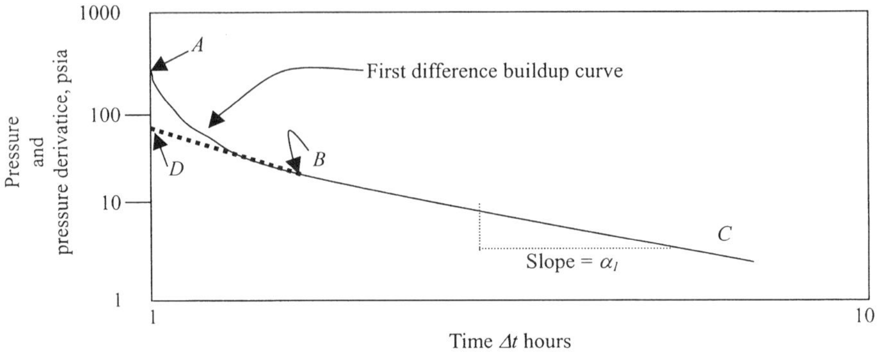

Figure 14–9. Log $(P_S - P_w)$ versus time—First difference curve.

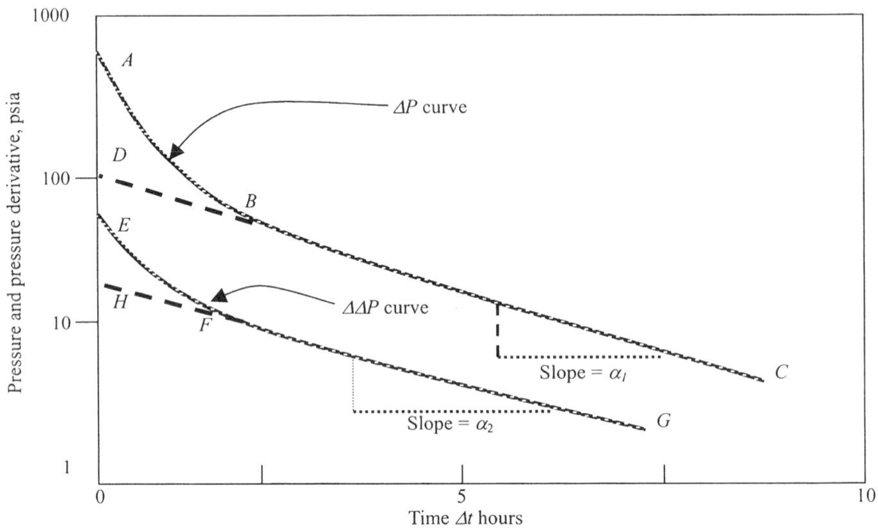

Figure 14–10. Log $(p_S - p_W)$ versus time—First and second difference curves.

shows how to develop the equation of partitioning coefficient. Notice that this approach interrelates buildup analysis with log interpretation. Figure 14–9 is a plot of $\log(p_s - p_w)$ versus time. The straight-line portion (BC) indicates the matrix porosity repressuring the fracture porosity, when Δp within the fractures and Δp between the coarse fissures and the wellbore has become negligible. Figure 14–10 is a log of pressure differential (average fracture pressure minus well pressure) across the "skin" near the wall of the well. This is represented by the difference plot (AB − DB) versus time. When the pressure drop due to skin becomes negligible, a straight line (FG) results. Pollard concluded that such plots of the log of pressure differential associated with any of the regions against time would result in a straight line from which it would be possible to determine properties such as volume of the fracture pore space system and well skin effect.

Matrix Pore Volume Calculation

With plots of Figures 14–9 and 14–10, it is possible to calculate the matrix pore volume, V_m from the following relationship as proposed by Pirson and Pirson:[12]

$$V_m = \frac{q}{\alpha_1 \phi_m (D + H) c_f} \tag{14–13}$$

where

q = flow rate at moment of shut-in, cm^3/sec
α_1 = slope of straight line (BC) of first differences, sec/cycle
ϕ_m = matrix porosity, fraction
D = intercept of first difference at time zero, psi
H = intercept of second difference at time zero, psi
C_f = compressibility of fluid in the fracture, psi

Fracture Pore Volume Calculation

From Figures 14–9 and 14–10, we can also evaluate the pore volume of the fracture V_f from the relationship

$$V_f = \frac{q}{\alpha_2 H c_f} \tag{14–14}$$

where α_2 is the slope of the straight line (FG) of second difference, sec/cycle.

Partitioning Coefficient Estimation

The partitioning coefficient concept, introduced by Pirson, has proven an important tool for the evaluation of fracture media. It is porosity breakdown between coarse (fracture) and fine (fracture) pore space. The partitioning coefficient υ can be estimated from the following relationship. Figure 14–9 shows how to develop the equation of the partitioning coefficient.

$$V_o = V_f + V_m \tag{14–15}$$

$$V_t = V_f + V_m\phi_m = \text{Total volume of a heterogeneous porous rock} \tag{14–16}$$

$$\upsilon = \frac{V_f}{V_t} = \frac{V_f}{V_f + V_M\phi_m} = \frac{\phi_t - \phi_m}{(1-\phi_m)\phi_t} \tag{14–17}$$

where

$$\phi_t = \frac{V_f + V_m \times \phi_m}{V_f + V_M} = \frac{V_t}{V_o} \tag{14–18}$$

Also

$$\upsilon = \frac{1}{1 + \frac{\alpha_2 H}{\alpha_1(D+H)}} \tag{14–19}$$

Well Skin Effects

Figure 14–10 indicates that the extrapolated E value is approximately the difference between the pressure in the fractures close to the wellbore and the average coarse fissure flowing pressure at shut-in. The pressure differential due to skin, Δp_s, is

$$\Delta p_s = (p_s - D - E), \text{ psi} \tag{14–20}$$

An example problem will illustrate the technique.

Example 14–2 *Analyzing Reservoir Rock's Porosity Distribution System*
A buildup test data for a well #1, Loving County, Texas believed to be fractured vertically. From these data, presented below and in Table 14–4 calculate the pressure differential. Table 14–5 shows the data for a formation tester.

Well depth = 15,545 ft; $T = 245°$F; Sp. gravity = 0.595; $T_c = 357°$R; $P_c = 672$ psia; $r_w = 0.3542$ ft; $re = 2980$ ft; $h = 16$ ft; $S_{wi} = 26.70\%$; $S_{gi} = 73.3\%$; $c_t = 64.75 \times 10^{-6}$ psi^{-1}; $z = 1.4924$; $\beta_{gi} = 0.001934$ ft^3/scf; $\mu_g = 0.03793$ cP; $\phi = 0.108$ fraction; and $\phi_{HC} = 0.0792$ fraction.

Solution Gas properties and necessary well/reservoir data were calculated from available literature and are listed in Table 14–2. Plot of real gas pseudopressure versus pressure is shown in Figure 14–11.

Table 14–4
Pressure Buildup Data (t_p = 42 hr; q_{sc} = 312.0 mscfd/d; p_{wfo} = 1306.2 psig ↔ $\psi(p_{wfo})$ = 132.20 mmpsia^2/cP; well depth = 15,545 ft; $\overline{p}_R$ = 12,360.0 psig ↔ $\psi(\overline{p}_R)$ = 4667.36 mmpsia^2/cP)

Δt (hr)	$\frac{t_p+\Delta t}{\Delta t}$	p_{wf} (psig)	$\psi(p_{wf})$ (mmpsia^2/cP)	$\Delta\psi$ (mmpsia^2/cP)
0.0000	∞	1306.2	132.200	0.000
0.0166	2532.12	1361.1	143.008	10.808
0.0333	1262.26	1352.9	141.369	9.169
0.0666	631.63	1358.2	141.428	10.227
0.1000	421.00	1382.7	147.366	15.166
0.1333	316.08	1443.7	161.048	28.848
0.1666	253.10	1482.0	168.159	35.958
0.2500	169.00	1602.3	194.980	62.780
0.3333	127.01	1712.7	221.122	88.922
0.5000	85.00	1930.0	276.664	144.464
0.6666	64.01	2109.3	326.363	194.162
0.8333	51.40	2299.1	382.532	250.332
1.00	43.00	2476.6	438.154	305.954
1.25	34.60	2724.9	520.572	388.371
1.50	29.00	2980.1	610.341	478.141
1.75	25.00	3206.7	693.871	561.671
2.00	22.00	3428.6	778.747	646.547
2.50	17.80	3878.9	958.886	826.685
3.00	15.00	4291.9	1131.523	999.327
3.50	13.00	4719.4	1315.728	1183.528
4.00	11.50	5103.1	1484.465	1352.265
5.00	9.40	5829.1	1808.993	1676.793
6.00	8.00	6499.5	2111.161	1978.961
7.00	7.00	7065.5	2365.996	2233.796
8.00	6.25	7512.5	2568.506	2436.306
10.00	5.20	8255.2	2896.115	2763.915
12.00	4.50	8786.6	3129.466	2997.265
14.00	4.00	9082.2	3258.259	3126.058
16.00	3.63	9312.3	3358.015	3225.815
18.00	3.33	9482.2	3431.397	3299.197
20.00	3.10	9662.7	3509.108	3376.908
24.00	2.75	9881.7	3603.060	3470.860
28.00	2.50	10049.9	3674.979	3542.778
29.67	2.42	10099.6	3696.191	3563.991
32.20	2.31	10089.6	3691.924	3559.724
36.00	2.17	10246.2	3758.662	3626.462
40.00	2.05	10366.9	3809.990	3677.789
44.00	1.95	10475.4	3856.050	3723.850

Table 14–4 (continued)

Δt (hr)	$\frac{t_p+\Delta t}{\Delta t}$	p_{wf} (psig)	$\psi(p_{wf})$ (mmpsia^2/cP)	$\Delta\psi$ (mmpsia^2/cP)
48.00	1.88	10582.3	3901.361	3769.161
52.00	1.81	10670.4	3938.653	3806.453
54.40	1.77	10720.9	3960.010	3827.810
95.40	1.44	11357.1	4227.999	4095.799
103.90	1.41	11385.0	4239.713	4107.513
108.00	1.39	11438.0	4261.959	4129.758
112.00	1.38	11474.0	4277.064	4144.863
116.00	1.36	11506.0	4290.487	4158.287
120.00	1.35	11539.0	4304.327	4172.126

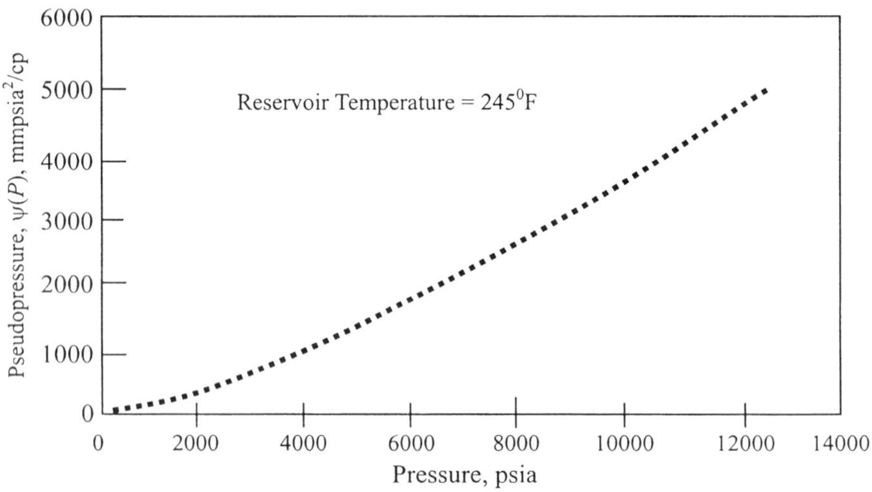

Figure 14–11. Pseudopressure $\psi(P)$ versus pressure ($\psi-P$ curve).

Pressure Buildup Analysis

The well was shut in for 120 hr during which the pressure buildup was monitored continuously. The pressure just prior to shut-in was 1321 psia. The pressure time data tabulated and the Horner plot (Figure 14–12), when extrapolated to $\log(\frac{t+\Delta t}{\Delta t}) = 1$, yielded a value of $\psi(\overset{*}{p}) = 4680$ mmpsia^2/cP. The average reservoir pressure $\psi(\overline{p})$ for a bounded reservoir could be calculated using the values of slope and $\psi(\overset{*}{p})$, obtained from the Horner plot and MBH curves.

Table 14–5 Pressure Record of Formation Tester

Shut-in time (min)	Pressure (psig)	Δp (psi)	$\Delta p^2 \times 10^6$	$\Delta\Delta p^2 \times 10^6$
15	1,602.3	9937.0	130.58	58.58
30	1,930.0	9609.0	129.42	57.42
60	2,476.6	9062.4	127.01	55.51
90	2,980.1	8558.9	124.27	53.27
120	3,428.6	8110.4	121.39	52.89
180	4,291.9	7247.1	114.73	47.73
240	5,103.1	6435.9	107.11	42.11
360	6,499.5	5039.5	90.91	28.91
420	7,065.5	4473.5	83.23	23.23
480	7,517.5	4021.5	76.64	17.92
600	8,255.2	3283.8	65.00	10.75
720	8,786.6	2752.4	55.94	5.14
840	9,082.2	2456.8	50.66	3.61
960	9,312.2	2226.8	46.43	—
1080	9,482.2	2056.8	43.24	—
1200	9,662.7	1876.3	39.78	—
1440	9,881.7	1657.3	35.50	—
1680	10,049.9	1489.1	32.15	—
1780	10,099.6	1439.4	31.15	—
1932	10,089.6	1449.4	31.35	—
2160	10,246.2	1292.8	28.16	—
2400	10,366.9	1172.1	25.68	—
2640	10,475.4	1063.6	23.41	—
2880	10,582.3	956.7	21.16	—
3120	10,670.4	869.0	19.29	—
3264	10,720.9	819.0	18.21	—
5724	11,357.1	181.9	4.16	—
6234	11,385.0	154.0	3.53	—
6480	11,438.0	101.0	2.32	—
6720	11,474.0	65.0	1.50	—
6960	11,506.0	33.0	0.76	—
7200	11,539.0	0.0	0.00	—

Examination of the log-log plot of the buildup data (Figure 14–12) shows a large increase on the predicted buildup over the interval of $\frac{1}{2}$ hr to 24 hr. A match of the above desuperposed buildup data plot with the type curve indicates the time of start of the middle-time (or Horner semilog straight-line) data is approximately 24 hr. The pressure behavior of the Warren and Root model is similar to one seen in the buildup test (Figure 14–14). The buildup, however, has only one distinct straight line. This situation arises when the ratio of matrix permeability thickness to fracture permeability thickness is small.

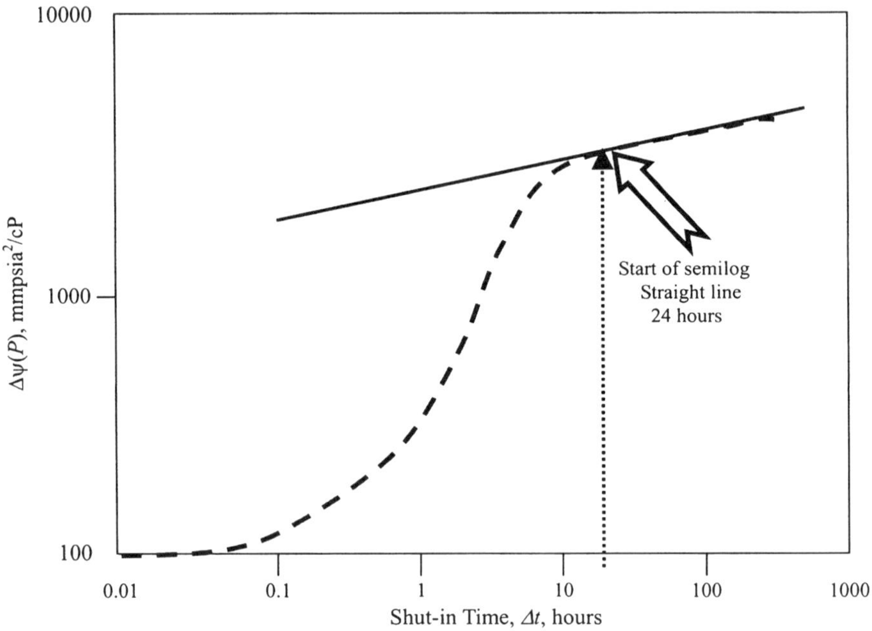

Figure 14–12. Log-log plot of fractured gas well.

Table 14–4 shows data for the buildup test. Figures 14–13 and 14–14 show that the pressure levels are consistently lower as flowing time previous to shut-in is increased. Further, it can be noticed that pressure levels tend to return to the initial pressure.

Using Eq. 6–10, calculate reservoir permeability k:

$$k = \frac{57.920 \times 10^6 q_{sc} T P_{SC}}{m T_{SC}} = \frac{57.920 \times 10^6 \times .312 \times 705}{2310 \times 10^6 \times 520} = 0.0097$$

$$\cong 0.01 \text{ mD}$$

Using Eq. 6–11, estimate apparent skin factor s':

$$s' = 1.151 \left[\frac{\psi(p)_{1hr} - \psi(p_{wfo})}{m} - \log\left(\frac{k}{\phi_{HC} \mu_g c_t r_w^2} \right) + 3.23 \right]$$

$$= 1.151 \left[\frac{(800 - 132.2)10^6}{2310 \times 10^6} \right.$$

$$\left. - \log\left(\frac{0.01}{0.0792 \times 0.037929 \times 64.75 \times 10^{-6} \times 0.3542^2} \right) + 3.23 \right]$$

$$= -2.41$$

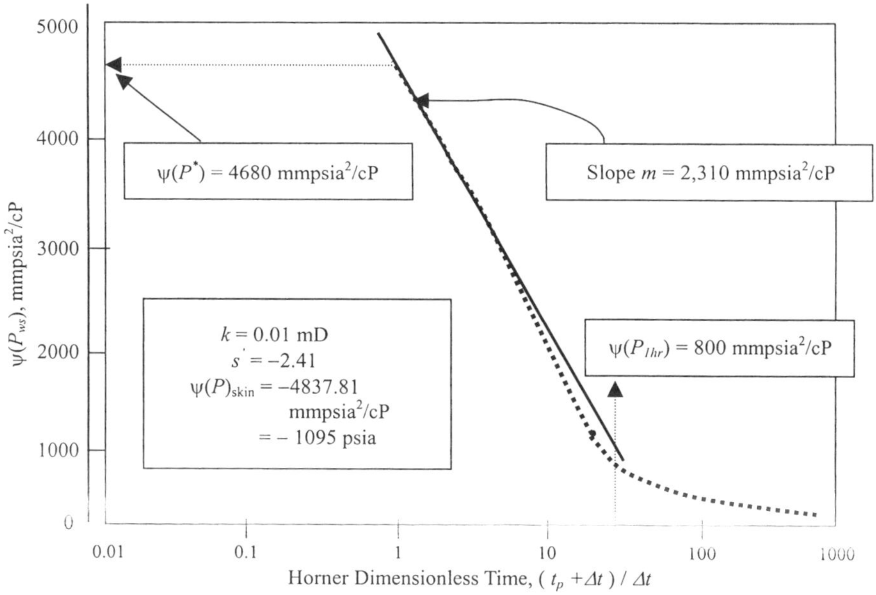

Figure 14–13. Semilog plot of buildup test of fractured gas well.

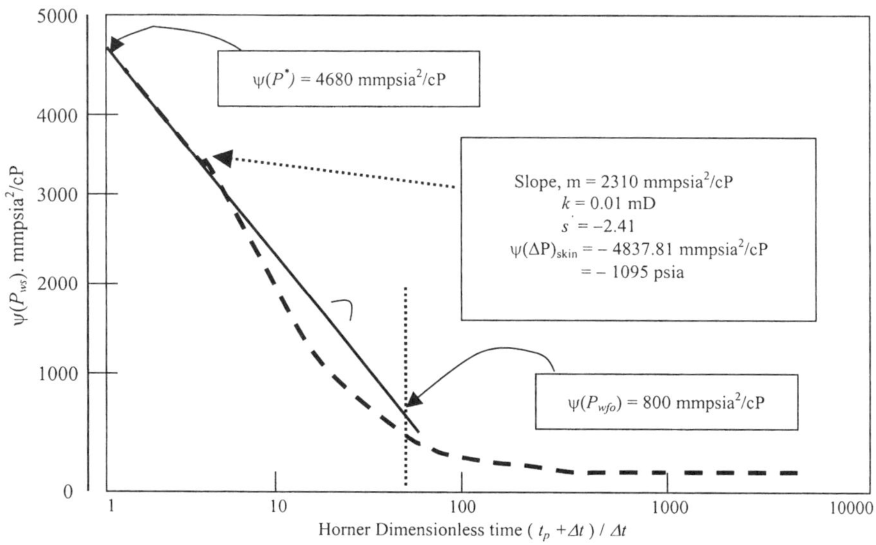

Figure 14–14. Semilog plots—Buildup test of fractured gas well.

Using Eq. 6–12, calculate the pressure drop due to the apparent skin factor, $\psi(p)_{skin}$:

$$\psi(p)_{skin} = 0.869ms' = 0.869 \times 2310 \times 10^6(-2.41)$$
$$= 4837.81 \text{ mmpsia}^2/\text{cP} \leftrightarrow -1095 \text{ psia}$$

A calculation after acidizing gives a negative value of additional pressure drop, which indicates that less pressure drop is required. Flow efficiency greater than 1, a negative skin effect, and a negative damage factor will all indicate successful stimulation. Percentage increase in flow efficiency can be determined by comparing the tests before acidizing.

Porosity Distribution

Since most fluid recovered was gas, the pressures were squared as shown in Table 14–5. Values of the first difference (Δp^2) and the second difference ($\Delta\Delta p^2$) were then calculated. These data were plotted in Figure 14–15 and resulted in a first straight line with a slope $\alpha_1 = 1/258{,}000$ cycle/sec. The straight line was extended to a time zero and yielded a D curve equal to 74×10^6 (Pirson). The second difference ($\Delta\Delta p^2$) resulted in a straight line

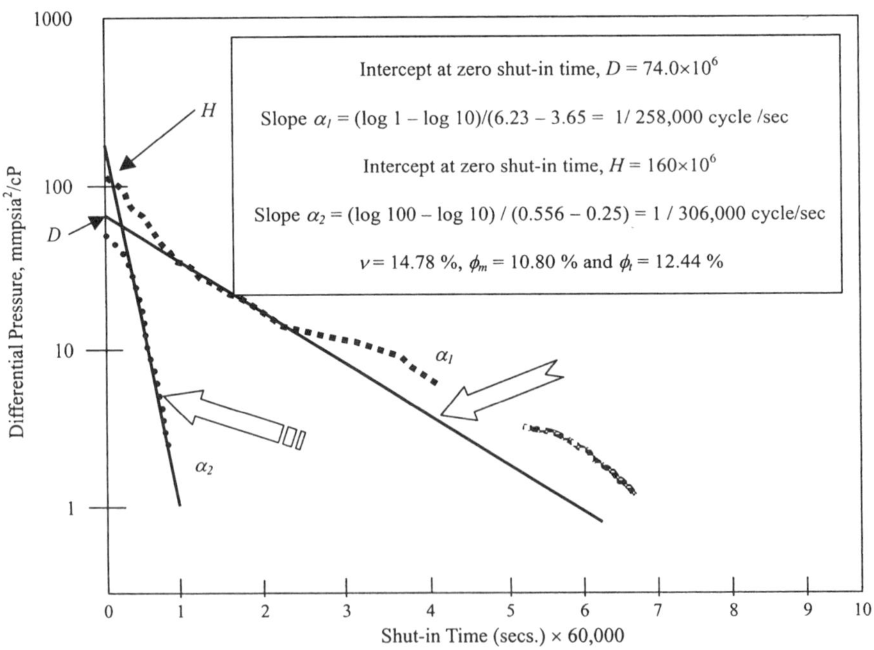

Figure 14–15. Interpretation of formation tester pressure buildup by the Pollard–Pirson method.

with a slope $\alpha_2 = 1/30{,}600$ cycle/sec. The intercept of this line at zero time yields $H = 160 \times 10^6$. Using Eq. 14–19 as follows makes use of the previous data to calculate the partitioning coefficient:

$$\upsilon = \frac{1}{1 + \frac{\alpha_2 H}{\alpha_1 (D+H)}}$$

$$\upsilon = \frac{1}{1 + \frac{(1/30{,}600)\times(160\times10^6)}{(1/258{,}000\times(74\times10^6+160\times10^6)}}$$

$$= \frac{1}{1 + \frac{5228.76}{906.98}} = \frac{1}{1 + 5.65} = 0.1478$$

This approach interrelates buildup analysis with log interpretation. With these plots, it is possible to determine porosity distribution from the matrix into the fracture system by using Eq. 14–17,

$$\upsilon = \frac{\phi_t - \phi_m}{(1 - \phi_m)\phi_t}$$

Knowing matrix porosity $\phi_m = 0.1080$, we can find total porosity as follows:

$$0.1478 = \frac{\phi_t - 0.1080}{(1 - 0.1080)\phi_t}$$

Therefore, $\phi_t = 0.1244$. The partitioning coefficient in formation producing gases was determined and can provide an insight into the percentage of total porosity made out of fracture. In this case, the fracture porosity comprises 12.44%.

14.7 Use of Pressure Transient Tests to Describe Reservoir Heterogeneity

Figure 14–16 presents the key steps involved in describing reservoir heterogeneities using single- and multiple-well tests as well as different models. Analysis techniques are described and guidelines are presented, including their advantages and limitations. The use of single- and multiple-well tests to determine their properties in different systems is discussed in the next section. Figures 14–16 (and 14–16a) also shows the different steps that should be used to analyze the results of multiple well tests to obtain a reservoir description. This figure also shows the criteria to be used when determining the validity of each method.

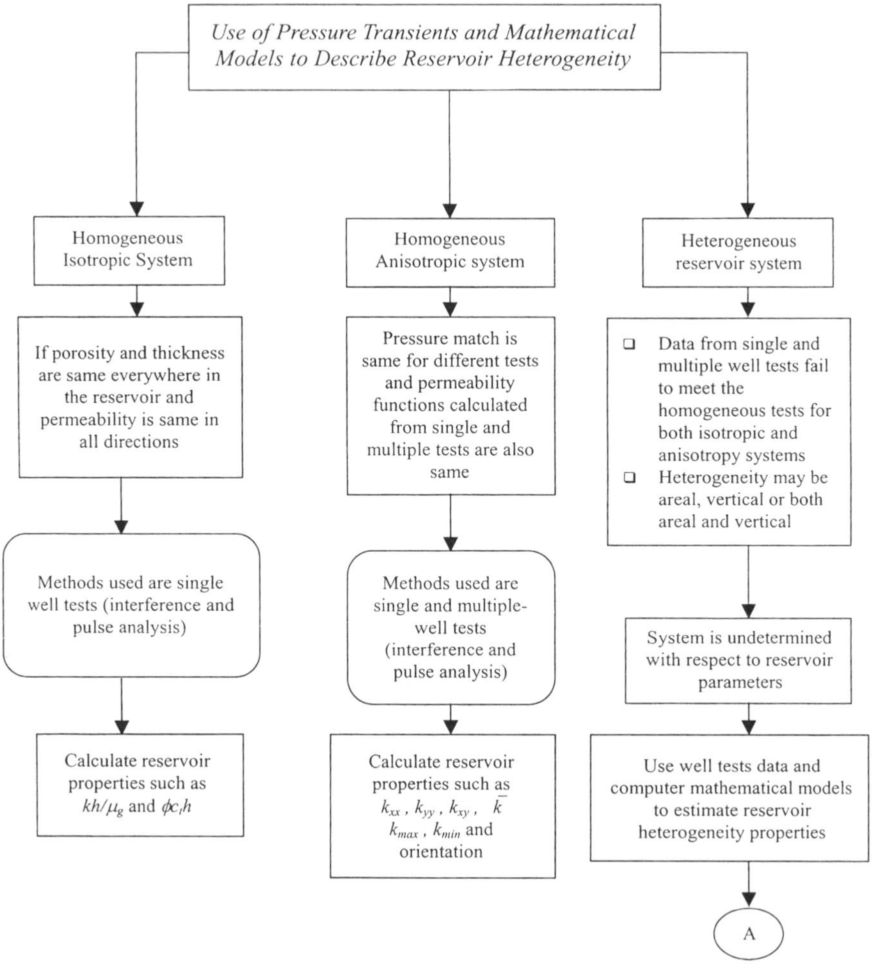

Figure 14–16. Systematic procedures and guidelines to describe reservoir heterogeneity.

14.8 Detecting Fracture Trends and Reservoir Heterogeneities

Kamal and Brigham[13] have proposed the following equation to investigate the presence of an isotropic reservoir without fracture or discontinuities:

$$[\Delta p]_{Calculate} = \frac{70.6 \times 10^6 q_{sc}\mu_g\beta_g}{kh} E_i\left(\frac{-56900\phi\mu_g cr^2}{kt}\right) \quad (14\text{–}21)$$

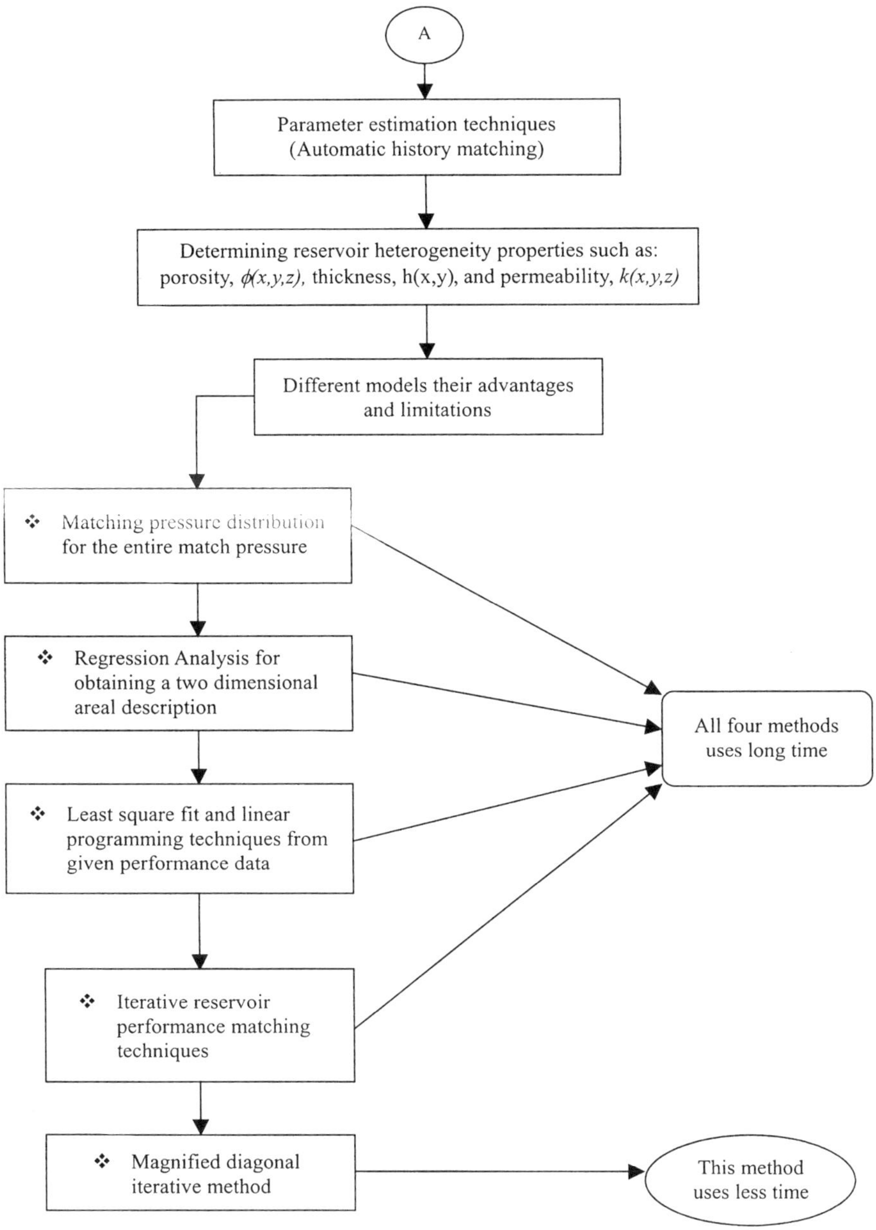

Figure 14–16a. Systematic procedures and guidelines to describe reservoir heterogeneity.

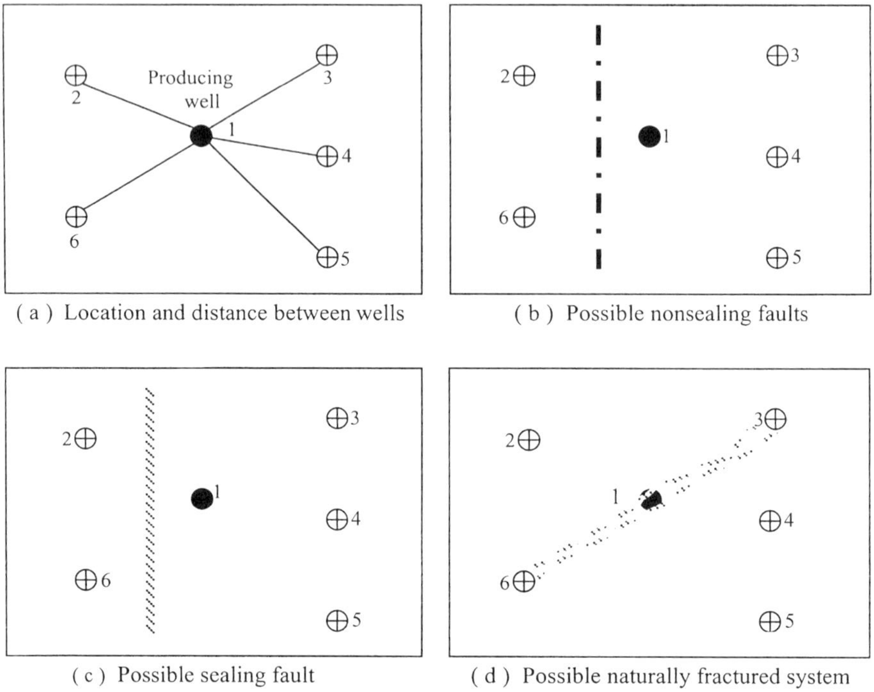

Figure 14–17. (a) Location and distance between wells. (b) Possible nonsealing faults. (c) Possible sealing fault. (d) Possible naturally fractured system.

where r is the distance between producing and observation well, ft, and t is the flowing time in minutes. Figures 14–17a through 14–17d show general trends of various reservoir heterogeneutities. Figure 14–18 can be used to confirm communication through the reservoir between producer and observation well or possible to determine general trends or possibilities.

Example 14–3 *Detecting Reservoir Heterogeneity or Fracture Trends*

Table 14–6 shows the location of six wells. All six wells were shut in until a stabilized pressure buildup was obtained. The middle well (1) was then placed on production for 110 days. Well/reservoir data are as follows: $P_i = 4450$ psi; $\mu_g = 0.025$ cP; $r_w = 0.550$ ft; $\beta_g = 6.06$ bbl/scf; $h = 45$ ft; gas injection rate $q_g = 5.0$ mmscfd; $T = 164°$F; $\phi = 8.5\%$; $c_t = 4.45 \times 10^{-3}$ psi^{-1}; well depth = 6340 ft.

Substituting the above values in Eq. 14–21 results in the values given in Table 14–7.

Table 14–7 shows as examples the match for well pairs A2–A1, A4–A1, A5–A1, and A6–A1. This will indicate an isotropic reservoir without fractures.

Table 14–6 Well Locations and Properties Related to Intereference Wells

Well number	Distance from well 1	k from buildup tests (mD)	kh product (mD-ft)
2	660	3.24	146
3	750	3.00	135
4	620	2.67	120
5	890	3.36	151
6	660	3.16	142

Table 14–7 Calculated Pressure Drop Values from Eq. 14–21

Well locations	r(ft)	$70.6q_g\mu_g\beta_g/kh$	$E_i(-x)$	Calculated Δp values (psi)
A2–A1	660	36.63	0.0669	2.45
A3–A1	750	39.61	3.0100	119.22
A4–A1	620	44.57	0.0660	2.94
A5–A1	890	35.42	1.4800	52.42
A6–A1	660	37.66	0.0728	2.74

Table 14–8 Comparison of Observed Well Pressures with Those Calculated Using Eq. 14–21

Well #	Calculated pressure (psi)	Observed pressure (psi)	Percent difference
2	4452	4450	0.05
3	4570	4685	2.45
4	4453	4456	0.06
5	4503	4493	0.22
6	4453	4450	0.07

The well pair A3–A1 shows possible nonsealing faults. Figure 14–18 can also help to interpret these results.

14.9 Determination of Reservoir Parameters and Fracture Orientations

For more accurate determination of reservoir anisotropic parameters and fracture orientations the methods proposed by Elkins and Skov[14] and by Ramey[15] are recommended. The following formula will permit estimation

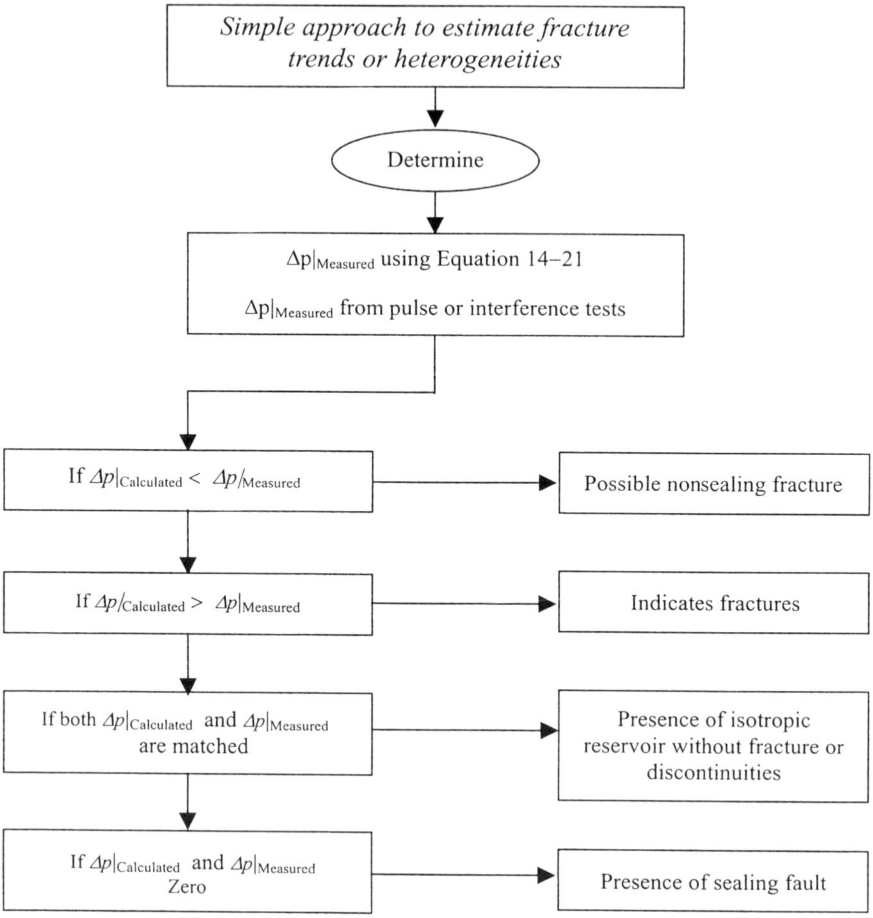

Figure 14–18. Systematic approaches to detect fracture trends and reservoir heterogeneity.

of the reservoir parameters in various directions based upon pressure drops measured in observation wells for conditions of single-phase flow:

$$\Delta p = (p_i - p) = \frac{q_g \mu_g \beta_g}{4\pi \sqrt{k_x k_y} h(1.127)} E_i \left[\frac{\frac{(x-x_0)^2}{k_x} + \frac{(y-y_0)^2}{k_y}}{\frac{4t}{\phi c \mu_g}(6.32)} \right] \tag{14–22}$$

where p_i = initial pressure, psi; p = pressure (psi) at x, y at time t in days; c = effective compressibility of gas, water and rock, psi^{-1}; k_x = effective permeability in x-direction, darcies; k_y = effective permeability in y-direction, darcies; $x - x_0$ = distance from producing to observation well in x-direction, ft;

$y - y_0$ = distance from producing to observation well in y-direction, ft; q_g = gas rate, mmscfd; β_g = gas formation volume factor, rbbl/mmscfd = $0.00504 z T_R \times 10^6/p_i$; and E_i is exponential integral, $-E_i(-x)$.

The pressure reductions at a point due to production of different wells are additive. For uniform permeability, Eq. 14–22 reduces to the simpler, well-known form involving r^2 and k. Reservoir parameters including effective compressibility and uniform or anisotropic permeability can be determined only by trial solutions until the set of values is found which gives the best match between calculated pressures and measured pressures. Fracture orientation, diffusivity parallel to the main fractures, and diffusivity perpendicular to the main fractures are related; $\sqrt{k_x k_y}$ and p_i are explicit. A sequence to determine the best set of these factors is given in Figure 14–19 and requires a computer (see also Figure 14–20).

14.10 Investigating Reservoir Heterogeneity by Multiple-Well Tests

Pressure transient tests can be used to investigate and obtain adequate reservoir descriptions for homogeneous (both isotropic and anisotropic) and heterogeneous systems. Type curves have proven very useful for evaluating pressure buildup, interference, and pulse tests in gas reservoirs influenced by reservoir boundaries. Multiple-well tests (interference and pulse tests) are used to establish communication between wells and to determine the interwell properties.[15]

The basic equations describing the pressure responses as well as pressure drop at some distance from a producing well are presented along with field examples in the next section to determine properties such as permeability $k(x, y, z)$, porosity $\phi(x, y, z)$, and thickness $h(x, y)$ in different systems.

Analysis of Homogeneous Isotropic Reservoir Systems

In these types of systems, the permeability is the same everywhere and in all directions. Porosity and thickness are also the same everywhere in the reservoir. The following analysis techniques (interference and pulse tests) can be used to determine reservoir properties in homogeneous isotropic formations.

Interference Tests

Interference tests can be analyzed by the type curve matching method, because it is simple, fast, and accurate when the exponential integral p_D applies: that is, when $r_D = r/r_w > 20$ and $t_D/r_D^2 > 0.5$. The reservoir properties such

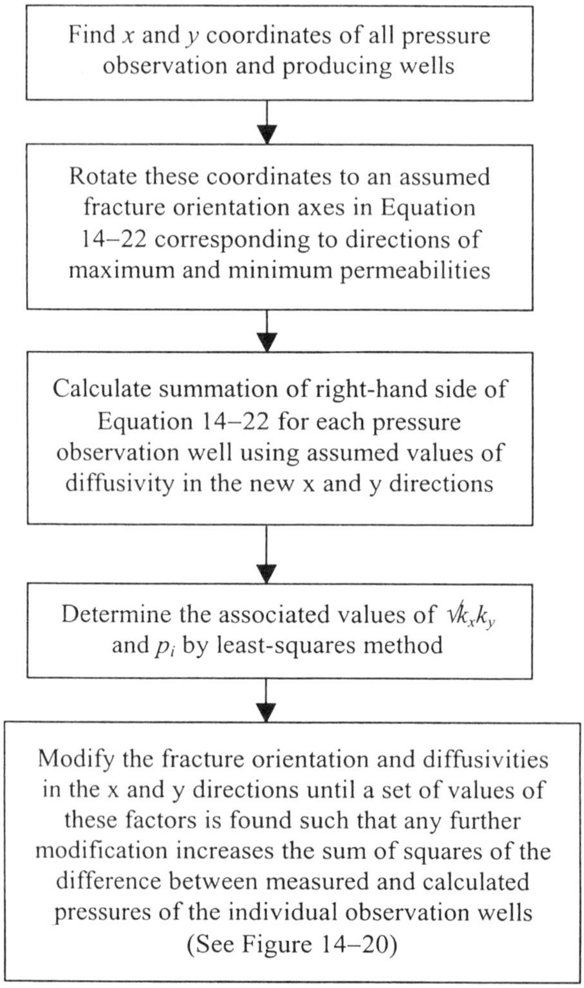

Figure 14–19. Systemic determination sequences of fracture orientations.

as the mobility–thickness product kh/μ_g and the porosity–compressibility–thickness product $\phi c_t h$ can be calculated from the following relationships:

$$\frac{kh}{\mu_g} = 141.2 q_g \beta_g \frac{|p_D|_M}{|\Delta p|_M} \tag{14–23}$$

$$\phi c_t h = \frac{0.0002637}{r^2} \cdot \frac{kh}{\mu_g} \cdot \frac{|t|_M}{|t_D/r_D^2|_M} \tag{14–24}$$

where q_g is gas flow rate in scfd and β_g is equal to 0.00504zT/p bbl/scf.

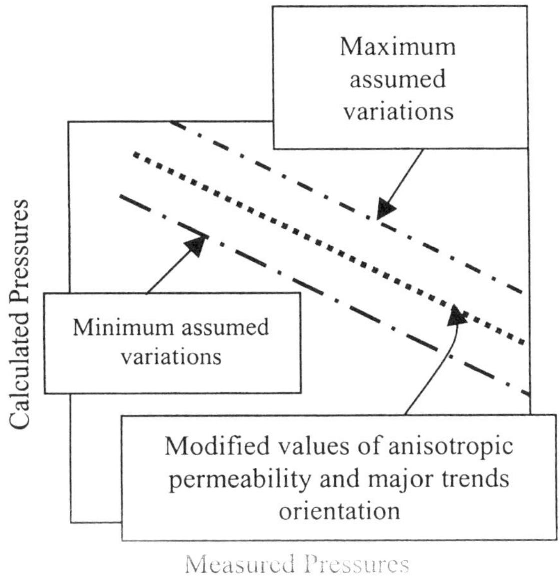

Figure 14–20. Calculated versus measured pressures.

Pulse Tests

Pulse tests can be used to determine the same information as interference tests. Pulse tests are not affected by unknown linear trends in reservoir pressures.[17] Therefore, conducting pulse tests rather than interference tests is preferable. Reference 19 has provided the relationships among dimensionless time lag, cycle period, and response amplitude in both graphical and analytical forms. The detailed discussion along with field examples can be found in Chapter 12. The following relationships can be applied to calculate the reservoir properties:

$$t_{tD} = \frac{t_l}{\Delta t_{cyc}} \tag{14–25}$$

$$\frac{kh}{\mu_g} = 70.6 q_g \beta_g \frac{\Delta p_D}{\Delta p} \tag{14–26}$$

and

$$\phi c_t h = \frac{kh}{\mu_g} \cdot \frac{1}{56{,}900\, r^2} \cdot \frac{\Delta t_{cyc}}{\Delta t_{cycD}} \tag{14–27}$$

General Remarks

If the reservoir is acting as a homogeneous isotropic system, reasonable identical values of kh/μ_g and $\phi c_t h$ can be calculated from several tests in the same areas. If there are big differences among the calculated values of kh/μ_g and those of $\phi c_t h$, then a homogeneous anisotropic system should be used.

Analysis Methods for Anisotropic Reservoir Systems

Porosity and thickness are uniform throughout the reservoir. Permeability is the same everywhere, but varies with direction. Figure 14–21 shows the major and minor axes of the permeability and axes of the well pattern. Many formations, such as channel sands, appear to exhibit simple ky–kx anisotropy. Directional permeability has an important effect on planning gas recovery by cycling. Ramey[5] presents a method for estimating anisotropic reservoir properties from interference data. At least three observation wells are required for analysis. Figure 14–21 defines the necessary nomenclature. The active well is located at the origin of the coordinate system and the observation wells are each located at coordinates indicated as (x, y). The anisotropic analysis requires pressure data from at least three observation wells. Assuming that the active well/observation well system is infinite acting and homogeneous

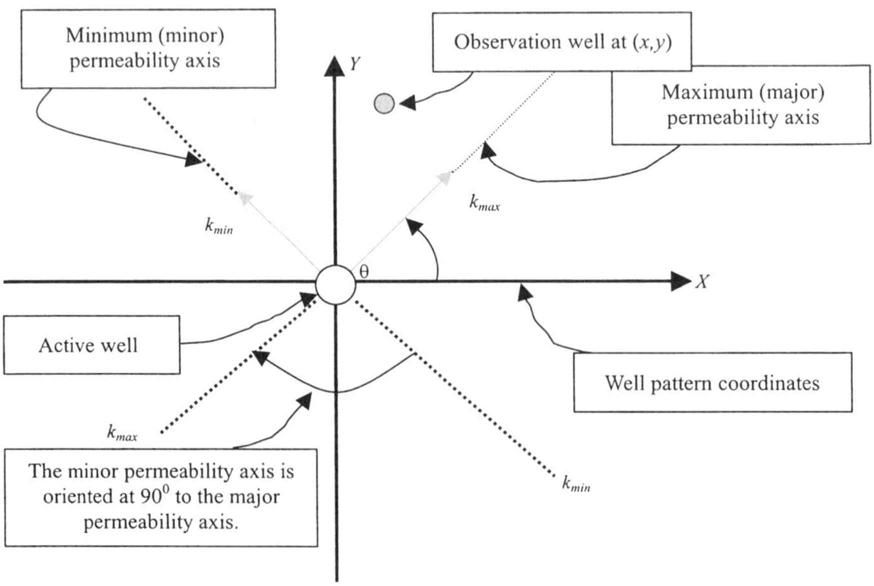

Figure 14–21. Nomenclatures for anisotropic permeability system.

with the exception of having anisotropic permeability, Ramey[5] shows that the pressure at an observation well is

$$p(t, x, y) = p_i - \frac{141.2 q_i \mu_w \beta_w}{\sqrt{k_{min} k_{max}} h} \cdot p_D \left(\left[\frac{t_D}{r_D^2} \right]_{direction} \right) \tag{14–28}$$

where

$$\left(t_D / r_D^2\right)_{direction} = \frac{0.0002637 t}{\varphi \mu_o c_t} \left[\frac{k_{min} k_{max}}{k_x y^2 + k_y x^2 - 2 k_{xy} x y} \right] \tag{14–29}$$

In Eq. 14–29:

k_x = principal permeability in x-direction, mD
k_y = principal permeability in y-direction, mD
k_{xy} = principal permeability in xy-direction, mD
k_{min} = minimum permeability in x-direction, mD
k_{max} = maximum permeability in x-direction, mD
θ = angle of orientation, degree

The following steps are used to analyze the interference test:

1. Observed pressure data from at least three wells are plotted and matched to the type curve of Figure 8–1. Each of the three data sets is matched so the pressure match point $[\Delta p_{MP}, (P_D)_{MP}]$ is the same for all three observation well responses. The time match point $\lfloor t_{MP}, (t_D/r_D^2)_{MP} \rfloor$ will be different for each set of observation data.
2. Rearrange Eq. 14–28 in the form

$$\sqrt{k_{min} k_{max}} = \frac{141.2 q_g \mu_w \beta_w}{h} \cdot \frac{(p_D)_{MP}}{(\Delta p)_{MP}} \tag{14–30}$$

$$k_{min} k_{max} = \left[\frac{141.2 q_g \mu_w \beta_w}{h} \frac{(p_D)_{MP}}{(\Delta p)_{MP}} \right]^2 \tag{14–31}$$

$k_{min} k_{max}$ = Average system permeability, mD

$$k_{min} k_{max} = \bar{k}^2 = k_x k_y - k_{xy}^2 \tag{14–32}$$

$$\bar{k} = \sqrt{k_{min} k_{max}} \text{ (average system permeability, mD)} \tag{14–33}$$

Rearrange Eq. 14–29 in the form

$$t_D / r_D^2 = \left(\frac{0.0002637 t}{\varphi \mu_g c_t} \right) \cdot \left(\frac{k_{xx} k_{yy} - k_{xy^2}}{k_{xx} y^2 + k_{yy} x^2 - 2 k_{xy} x y} \right) \tag{14–34}$$

or

$$t_D/r_D^2 = \left(\frac{0.0002637t}{\varphi\mu_g c_t}\right)\cdot\left(\frac{k_{max}k_{min}}{k_{xx}y^2 + k_{yy}x^2 - 2k_{xy}xy}\right) \tag{14–35}$$

Write the following equations for each observation well match:

$$\left[\frac{(t_D/r_D^2)_{MP}}{t_{MP}}\right]_{Well1} = \frac{0.0002637t}{\phi\mu_g c_t}\left(\frac{k_{max}k_{min}}{k_{xx}y^2 + k_{yy}x^2 - 2k_{xy}xy}\right) \tag{14–36}$$

$$\left[\frac{(t_D/r_D^2)_{MP}}{t_{MP}}\right]_{Well2} = \frac{0.0002637}{\phi\mu_g c_t}\left(\frac{k_{max}k_{min}}{k_{xx}y^2 + k_{yy}x^2 - 2k_{xy}xy}\right) \tag{14–37}$$

$$\left[\frac{(t_D/r_D^2)_{MP}}{t_{MP}}\right]_{Well3} = \frac{0.0002637}{\phi\mu_g c_t}\left(\frac{k_{max}k_{min}}{k_{xx}y^2 + k_{yy}x^2 - 2k_{xy}xy}\right) \tag{14–38}$$

where t_{MP} is the same for each well, and $(t_D/r_D^2)_{MP}$ is different for each well. Estimate average system permeability from Eq. 14–33. There are three equations (14–36, 14–37, 14–38) in four unknowns (k_{xx}, k_{yy}, k_{xy}, and $\phi\mu_o c_t$). They may be solved simultaneously to obtain such as k_{xx}, k_{yy}, and k_{xy}, each in terms of the unknown $\phi\mu_g c_t$. Then k_x, k_y, and k_{xy} (in terms of $\phi\mu_g c_t$) are substituted into

$$k_x k_y - k_{xy}^2 = k_{min}k_{max} = \bar{k}^2 \tag{14–32}$$

Since the right side of Eq. 14–32 is known from Eq. 14–30, it can be solved to estimate $\phi\mu_o c_t$. Then we estimate k_x, k_y, and k_{xy} from their relationship to $\phi\mu_o c_t$. Determine the minimum and maximum directional permeability and the angle of orientation by using the following equations:

$$k_{min} = k_{XX} = 0.5\left\lfloor (k_x + k_y) - \sqrt{(k_x - k_y)^2 + 4k_{xy}^2}\right\rfloor \tag{14–39}$$

$$k_{max} = k_{YY} = 0.5\left\lfloor (k_x + k_y) - \sqrt{(k_x - k_y)^2 + 4k_{xy}^2}\right\rfloor \tag{14–40}$$

$$\theta_{max} = \arctan\left(\frac{k_{max} - k_{XX}}{k_{xy}}\right) \tag{14–41}$$

$$\theta_{min} = \arctan\left(\frac{K_{min} - k_{YY}}{k_{xy}}\right) \tag{14–42}$$

where

β_g = gas formation volume factor, bbl/scf
c_t = total system effective compressibility, psi^{-1}

c_g = gas compressibility, psi^{-1}
c_w = water compressibility, psi^{-1}
c_f = pore space compressibility, psi^{-1}
h = net formation thickness, ft
$\bar{k}$ = average system permeability, mD
k_{XX} = maximum (major) principal permeability, mD
k_{YY} = minimum (minor) principal permeability, mD
k_{xx}, k_{yy}, k_{xy} = components of the permeability tensor, mD
θ_{max} = direction of maximum permeability, K_{max}
θ_{min} = direction of minimum permeability, K_{min}

Total system compressibility can be related to the pore space saturation of the two phases:

$$c_t = s_g c_g + s_w c_w + c_f \tag{14–43}$$

$$s_g = \frac{c_t - c_w - c_f}{(c_g - c_w)} \tag{14–44}$$

Equation 14–44 can be used to estimate in-place gas saturation using a transient test.

Important note: Analysis of more than one interference test in the same area should, therefore, provide information on the feasibility of using homogeneous anisotropic techniques. If the match of pressure is the same in different tests, the technique is applicable. If not, heterogeneous system analysis should be considered.

The following example will clarify the use of these equations to determine directional homogeneous anisotropic reservoir properties.

Example 14–4 *Analyzing Interference Tests Homogeneous Anisotropic Reservoirs*

An interference test was run in a 5-spot pattern. At the end of a gas cycling, before testing, all wells were shut in. Gas injecting at 7.25 mmscfd/d ran the test and the fluid levels were observed in five of the shut-in production wells, both during injection and during the subsequent falloff period. The test information and reservoir properties are as follows: $P_i = 265$ psi; $\mu_w = 1$ cP; $r_w = 0.550$ ft; $\beta_w = 1$ rbl/stb; $h = 30$ ft; injection rate $q_i = 120$ stb/d; $T = 72°$F; $\phi = 19\%$; $c_o = 7.5 \times 10^{-6}\ \text{psi}^{-1}$; $c_w = 3.3 \times 10^{-6}\ \text{psi}^{-1}$; $c_f = 3.7 \times 10^{-6}\ \text{psi}^{-1}$; API = 37°; $s_o = 0.25$; $s_w = 0.30$; well depth = 1200 ft: Figure 14–22 shows the well locations.

Tables 14–9, 14–10, and 14–10a give observation pressure data for wells 1, 2, and 3 during the gas injection period. Figure 14–22 shows the well locations.

Table 14–9 Observation Pressure Data for Well 1

t (hr)	$p_{x,y,t}$ (psi)	$\Delta p = p_i - p_{x,y,t}$ (psi)
34	272	−7
39	273	−8
50	279	−14
78	286	−21
98	291	−26
120	289	−24
188	280	−15

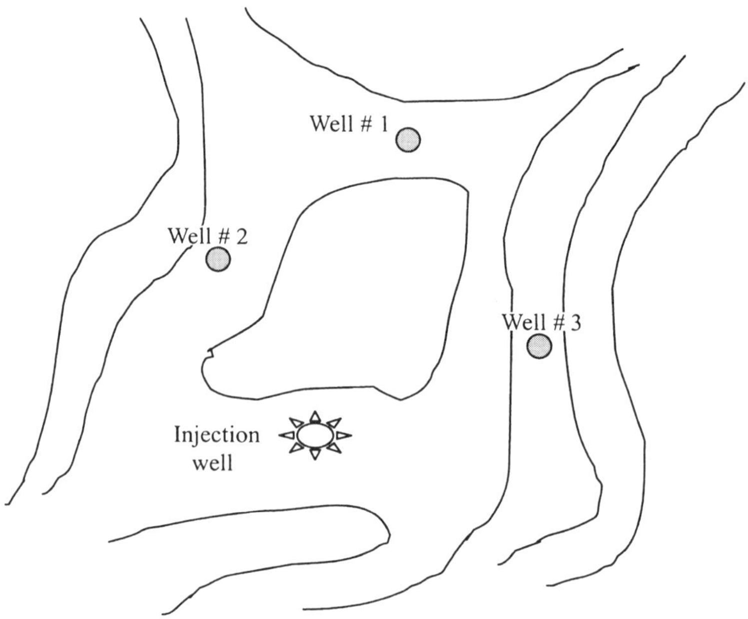

Figure 14–22. Well locations for Example 14–2.

Estimate the following homogeneous anisotropic reservoir parameters:

- Average system permeability $\overline{k}$
- Product of $\phi\mu c_t$
- Maximum directional permeability k_{max}
- Minimum directional permeability k_{min}
- Directions of k_{max} and k_{min}
- In-place gas saturation

Table 14–10 Observation Pressure Data for Well 2

t (hr)	$p_{x,y,t}$ (psi)	$\Delta p = p_i - p_{x,y,t}$ (psi)
22	270	−5
49	277	−12
71	281	−16
93	286	−21
116	288	−23
124	291	−26
210	284	−19
289	281	−16

Table 14–10a Observation Pressure Data for Well 3

t (hr)	$p_{x,y,t}$ (psi)	$\Delta p = p_i - p_{x,y,t}$ (psi)
28	269	−4
48	271	−6
70	275	−10
94	277	−12
117	282	−17
124	283	−18
190	276	−11
238	272	−7
297	271	−6

Table 14–10b

Well #	Match points $(\Delta t)_{MP}$	t_D/r_D^2	t_{MP} (Figure 14–24)	r(ft)
1	10.0	25	30	480
2	10.0	35	38	480
3	10.0	45	70	702

Solution Figure 14–22 shows well pattern, distances and coordinates in ft and Figure 14–23 is a net sand isopatch map. Figure 14–24 shows the match of the data in Tables 14–9 through 14–10b to the type curve of Figure 8–1. The match was made so the pressure match point $[(\Delta p)_{MP}, (P_D)_{MP}]$ is the same for all three responses, while the time match points vary.

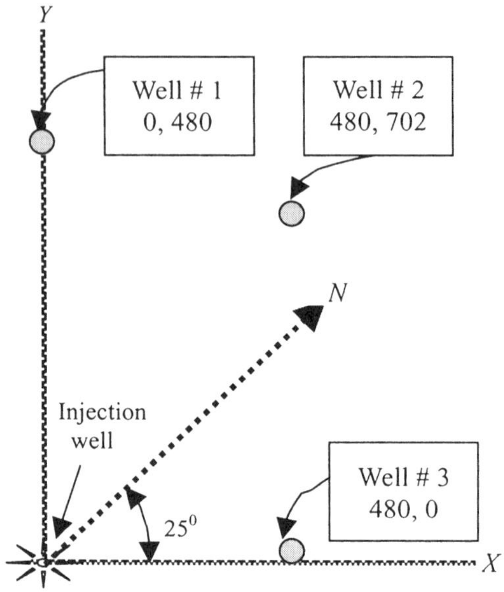

Figure 14–23. Net sand isopatch map—Example 14–2.

Well # location from injection well	Coordinates (x, y) (ft)	Dimentionless parameter t_D/r_D^2
1, injection well	480,0	25
2, injection well	0,480	35
3, injection well	480,702	45

Calculate System Permeability $\bar{k}$

From the pressure match point for all wells, $(p_D)_{MP} = 0.29$ and $(\Delta p)_{MP} = 10$. Rearranging Eqs. 14–31 and 14–3, we have

$$k_{min}k_{max} = \left[\frac{141.2q_i\mu_w\beta_w}{h} \cdot \frac{(p_D)_{MP}}{(\Delta p)_{MP}}\right]^2$$

$$= \left[\frac{141.2 \times 120 \times 1.0 \times 1.0}{30} \cdot \frac{0.29}{10}\right] = 268.3 \text{ mD}^2$$

$$\bar{k} = \sqrt{k_{min}k_{max}} = \sqrt{268.3} = 16.38 \text{ mD}$$

Also, $\bar{k} = \sqrt{(k_{xx}k_{yy} - k_{xy}^2)}$.

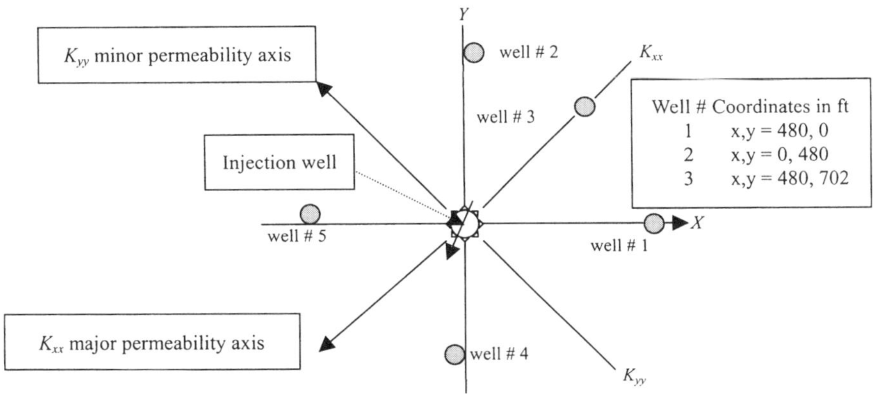

Figure 14–24. Coordinates for anisotropic permeability solution.

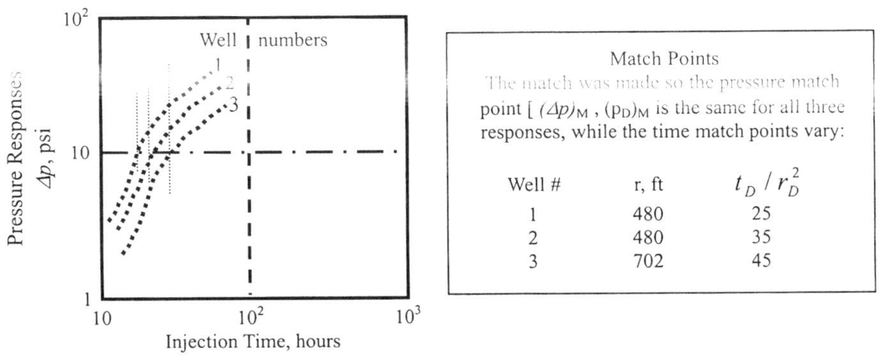

Figure 14–25. Interference data matched to Figure 8–1. Pressure match is the same for all curves.

Estimate Product $\phi\mu c_t$

Equations 14–36 through 14–38 now may be used with the time-match data to write three more equations. Match time was 10 hr and using the coordinate for each well from Figure 14–25 we have the following:

For well 1:

$$\left[\frac{(t_D/r_D^2)_{MP}}{t_{MP}}\right]_{Well1} \frac{0.0002637t}{\phi\mu_g c_t}\left(\frac{k_{max}k_{min}}{k_{xx}y^2 + k_{yy}x^2 - k_{xy}xy}\right)$$

Substituting the values, we get

$$\left[\frac{30}{25}\right]_{Well1} = \frac{0.0002637}{\phi\mu c_t}\left[\frac{268.3}{480^2k_x + (0)^2k_y - 2(0)(480)k_{xy}}\right]$$

Simplified and normalized, this equation becomes

$$k_x = \frac{3.685 \times 10^{-7}}{\phi \mu c_t} \tag{14–45}$$

For well 2:

$$\left[\frac{30}{25}\right]_{Well2} = \frac{0.0002637}{\phi \mu c_t}\left[\frac{268.3}{(0)^2 k_x + (480)^2 k_y - 2(0)(480)k_{xy}}\right]$$

Simplified and normalized this equation becomes

$$k_y = \frac{3.334 \times 10^{-7}}{\phi \mu c_t} \tag{14–46}$$

For well 3:

$$\left[\frac{70}{45}\right]_{Well2} = \frac{0.0002637}{\phi \mu c_t}\left[\frac{268.3}{530^2 k_x + (480)^2 k_y - 2(480)(530)k_{xy}}\right]$$

Simplified and normalized this equation becomes

$$0.5521 k_x + 0.4528 k_y - k_{xy} = \frac{1.101 \times 10^{-6}}{\phi \mu c_t}$$

Combining Eqs. 14–45 through 14–46 gives

$$k_{xy} = 0.5521 \frac{3.685 \times 10^{-7}}{\phi \mu c_t} + 0.4528 \frac{3.334 \times 10^{-7}}{\phi \mu c_t} - \frac{1.101 \times 10^{-6}}{\phi \mu c_t}$$

$$= \frac{3.434 \times 10^{-8}}{\phi \mu c_t} \tag{14–47}$$

Using Eqs. 14–45, 14–46, and 14–47 in Eq. 14–32 results in

$$k_x k_y - k_{xy}^2 = k_{min} k_{max} = \overline{k}^2$$

or

$$\frac{(3.685 \times 10^{-7})(3.334 \times 10^{-7})}{(\phi \mu c_t)(\phi \mu c_t)} - \left(\frac{3.434 \times 10^{-7}}{(\phi \mu c_t)^2}\right)^2 = 268.3$$

$$\frac{12.2858 \times 10^{-7}}{(\phi \mu c_t)^2} - \frac{22.7924 \times 10^{-7}}{(\phi \mu c_t)^2} = 268.3$$

Therefore

$$\phi \mu c_t = \sqrt{\frac{0.4934}{268.3}} = 3.24 \times 10^{-6} \text{cP/psi}^{-1}$$

Now Eqs. 14–45, 14–46, and 14–47 are solved using the computed $\phi\mu c_t$:

$$k_x = \frac{3.685 \times 10^{-7}}{3.24 \times 10^{-6}} = 11.373\,\text{mD}$$

$$k_y = \frac{3.334 \times 10^{-7}}{3.24 \times 10^{-6}} = 10.290\,\text{mD}$$

$$k_{xy} = \frac{3.434 \times 10^{-8}}{3.24 \times 10^{-6}} = 1.060\,\text{mD}$$

Now we can estimate maximum permeability k_{max} using Eq. 14–40:

$$k_{max} = k_{XX} = 0.5\left\lfloor k_x + k_y + \sqrt{(k_{XX} - k_{YY})^2 + 4k_{xy}^2}\right\rfloor$$

$$= 0.5\left\lfloor 11.373 + 10.290 + \sqrt{(11.373 - 10.290)^2 + 4(1.06)^2}\right\rfloor$$

$$0.5[21.663 + 2.3806] = 12.02\,\text{mD}$$

Estimate minimum permeability k_{min} using Eq. 14–39:

$$k_{min} = k_{YY} = 0.5[21.663 - 2.3806] = 9.64\,\text{mD}$$

We know $\sqrt{k_{max}k_{min}} = 16.38$ from Eq. 14–32, so we can check the computations:

$$\sqrt{k_{max}k_{min}} = \sqrt{(12.02)(9.64)} = 10.77(\text{ close enough})$$

Determine the direction of maximum permeability k_{max} from Eq. 14–41:

$$\theta_{max} = \arctan\left(\frac{k_{max} - k_{XX}}{k_{xy}}\right) = \arctan\left(\frac{12.02 - 11.373}{1.06}\right)$$

$$= \arctan\,(0.6104)$$

$$= \frac{0.5480(180)}{22/7} = 31.39^\circ \text{ from the } x\text{-axis}$$

Correcting for the orientation of the axes, the maximum permeability direction is

$$31.39^\circ - 25^\circ = \text{N } 6.38 \text{ W}.$$

Determine direction of minimum permeability k_{min} from Eq. 14–42:

$$\theta_{min} = \arctan\left(\frac{k_{max} - k_{YY}}{k_{xy}}\right) = \arctan\left(\frac{12.02 - 10.29}{1.06}\right)$$

$$= \arctan(0.6104) = \frac{1.021(180)}{22/7} = 58.48^\circ \text{ from the } x\text{-axis}$$

Correcting for the orientation of the axes, the minimum permeability direction is

$$58.48^\circ - 25^\circ = \text{N } 33.48 \text{ W}$$

As shown in Figure 14–23, the x-axis was chosen as a line through wells 1, 2, and 3. True north lies along the line through wells 2 and 3.

Estimate gas saturation using Eq. 14–44:

$$s_o = \frac{c_t - c_w - c_f}{c_o - c_w} = \frac{7.85 \times 10^{-6} - 3.30 \times 10^{-6} - 3.70 \times 10^{-6}}{7.40 \times 10^{-6} - 3.30 \times 10^{-6}}$$

$$= \frac{0.85 \times 10^{-6}}{4.1 \times 10^{-6}} = 0.21$$

Hence water saturation is $s_w = 1 - s_o = 1 - 0.21 = 0.79$.

If we check these saturation values with electric log and core data, they are in good agreement, hence it is possible to make a rough estimate of in-place saturation using a transient test.

Summary

Equations 14–28 through 14–47, coupled with the log-log type curve procedure, form a powerful tool for detecting reservoir anisotropy. The injection interference test described can be applied widely to aid in planning fluid injection programs.

Analysis of Heterogeneous Systems

If the data from multiwell tests fail to meet the homogenity requirement for both isotropic and anisotropic cases, numerical solutions must be used to analyze pressure transient data from heterogeneous systems. References 17 and 18 have suggested numerical solutions for performing the analysis by parameter estimation techniques to describe reservoir heterogeneities using pressure transient data. They consider the case of heterogeneous isotropic systems using the following diffusion equation:

$$\nabla \cdot \left(\frac{kh}{\mu} \nabla p \right) = \phi c_t \frac{\partial p}{\partial t} + Q \tag{14–48}$$

where Q is the diffusion equation source term and p is pressure, psi. In order to estimate the values of $kh(x, y, z)$ and $\phi c_t h(x, y, z)$ that minimize E:

Chavent *et al.*:[17]

$$E = \sum_{s=1}^{S} \int_0^t \left(p_S^{obs} - p_S^{calc} \right)^2 dT \tag{14–49}$$

Chen *et al.*:[18]

$$E = \sum_{n=1}^{N_s} \sum_{s=1}^{S} \left(p_{n,s}^{obs} - p_{n,s}^{calc}\right)^2 \quad (14\text{–}50)$$

where

E = sum of the squares of the difference between observed and calculated pressure, psi^2

S = number of observation wells

N = number of observations at well S

p_s^{obs} = observed pressure at well S, psi

p_s^{calc} = calculated pressure at well S, psi

$p_{n,s}^{obs}$ = observed pressure at well S and data point n

$p_{n,s}^{calc}$ = calculated pressure at well S and data point n

References and Additional Reading

1. Vairogs, J., Hearn, C. L., Dareing, D. W., and Rhoades, V. W., "Effect of Rock Stress on Gas Production from Low-Permeability Reservoirs," *J. Petroleum Technol.* (Sept. 1971) 1161–1167; *Trans. AIME* 251.
2. Thomas, R. D., and Ward, D. C., "Effect of Overburden Pressure and Water Saturation on the Gas Permeability of Tight Sandstone Cores," *J. Petroleum Technol.* (Feb. 1972) 120–124.
3. Vairogs, J., and Rhoades, V. W., " Pressure Transient Tests in Formations Having Stress-Sensitive Permeability," *J. Petroleum Technol.* (Aug. 1973) 965–970; *Trans. AIME* 255.
4. Raghavan, R., Scorer, J. D. T., and Miller. F. G., "An Investigation by Numerical Methods of the Effect of Pressure Dependent Rock and Fluid Properries on Well Flow Tests," *Soc. Petroleum Eng. J.* (June 1972) 267–275; *Trans. AIME* 253.
5. Ramey, H. J., Jr., "Non-Darcy Flow and Wellbore Storage Effects in Pressure Buildup and Drawdown of Gas Wells," *J. Petroleum Technol.* (Feb. 1965) 223–233; *Trans. AIME* 234. Also Reprint Series, No. 9—*Pressure Analysis Methods*, Society of Petrolem Engineers of AIME, Dallas, TX, 1967, pp. 233–243.
6. Matthews, C. S., and Russell, D. G., *Pressure Buildup and Flow Tests in Wells.* Monograph Series, Society of Petroleum Engineers of AIME, Dallas, TX, 1967.
7. Horner, D. R., "Pressure Build-up in Wells," Proc. Third World Pet. Cong., The Hague (1951), Sec. II, 503–523. Also Reprint Series, No. 9—*Pressure*

Analysis Methods, Society of Petroleum Engineers of AIME, Dallas, TX, 1967, pp. 25–43.

8. Gray, K. E., "Approximating Well-to-Fault Distance from Pressure Build-up Tests," *J. Petroleum Technol.* (July 1965) 761–767.
9. Davis, E. G., Jr., and Hawkins, M. F., Jr., "Linear Fluid-Barrier Detection by Well Pressure Measurements," *J. Petroleum Technol.* (Oct. 1963) 1077–1079.
10. Bixel, H. C., Larkin, B. K., and Van Poollen, H. K., "Effect of Linear Discontinuities on Pressure Build-Up and Drawdown Behavior," *J. Petroleum Technol.* (Aug. 1965) 885–895; *Trans. AIME* 228.
11. Russell, D. G., "Determination of Formation Chracteristics from Two Rate Flow Tests," *J. Petroleum Technol.* (Dec. 1963) 1347–1355; *Trans. AIME* 228. Also Reprint Series, No. 9—*Pressure Analysis Methods*. Society of Petroleum Engineers of AIME, Dallas, TX, 1967, pp. 136–144.
12. Pirson, R. S., and Pirson, S. J., " An Extension of the Pollard Analysis Method of Well Pressure Build-up and Drawdown Tests," Paper SPE 101 presented at the SPE-AIME 36th Annual Fall Meeting, Dallas, TX, Oct. 8–11, 1961.
13. Kamal, M., and Brigham, W. E., "The Effect of Linear Pressure Trends on Interference Tests," *J. Petroleum Technol.* (Nov. 1975) 1383–1384.
14. Elkins, L. F., and Skov, A. M., "Determination of Fracture Orientation from Pressure Intereference," *Trans. AIME* (1960) 219, 301–304. Also Reprint Series, No. 9—*Pressure Analysis Methods*. Society of Petroleum Engineers of AIME, Dallas, TX, 1967, pp. 97–100.
15. Ramey, H. J., Jr., "Intereference Analysis for Anisotropic Formations—A Case History," *J. Petroleum Technol.* (Oct. 1975) 1290–1298; *Trans. AIME* 259.
16. Brigham, W. E., "Planning and Analysis of Pulse Tests," *J. Petroleum Technol.* (May 1970) 618–624; *Trans. AIME* 249.
17. Chavent, C., Dupuy, M., and Lemonier, P., "History Matching by Use of Optimal Control Theory," *Soc. Petroleum Eng. J.* (Feb. 1975) 74–86; *Trans. AIME* 259.
18. Chen, W. H., Gavalas, G. R., and Seinfelds, J. H., "A New Alogrithm for Automatic History Matching," *Soc. Petroleum Eng. J.* (Dec. 1974) 593–608; *Trans. AIME* 257.
19. Pollard, P., "Evaluation of Acid Treatments from Pressure Build-up Analysis," *Trans. AIME* (1959) 216, 38–43.

Chapter 15

Gas Well Testing Field Case Studies

15.1 Introduction

This chapter presents various field case studies in low, high permeability, and fractured carbonate gas wells including summary, conclusions, and recommendations. It also includes a gas well test evaluation sheet, state report forms, and various cross plotting techniques before and after workovers.

15.2 Gas Well Test Evaluation Sheet

Well Data and Basic Parameters

Filed name		aaaa-
Well name		bbbb
Zone number		cccc-
Interval		feet
Reservoir datum		feet ss
Estimated reservoir pressure		psia
Reservoir temperature		^{o}R
Net hydrocarbon thickness		feet
Gas saturation		fraction
Porosity		fraction
Fluid viscosity		cP
Compressibility		psi^{-1}
Hydrocarbon porosity		fraction
Fluid gradient		psi/ft
Z-factor		—
Well radius		feet
Drainage radius		feet
Cumulative production prior to test		mmcf

Gas Composition

Gas composition	H_2S	CO_2	N_2	C_1	C_2	C_3	iC_4	nC_4	iC_5	nC_5	C_6	C_{7+}
Mol %												

Well Test Data

Choke size (- / 64")	Rate (mmcfd)	Duration (min)	Cumulative time (hr)	Final BHP (psia)	Final THP (psia)

Amerada Data

Amerada no.	Serial no.	Last calibrated data	Depth of Amerada

Interpreted data:

MBH correction: t_p (hr)

t_{pDA}(dimensionless time)

Buildup slope m(mmpsia^2/cP)

$\psi(P_{wfo})$(mmpsia^2/cP)

$\psi(p_{wf})_{\Delta t=1}$(mmpsia^2/cP)

Calculated data:

$$kh = \frac{1.632 \times 10^6 q_{sc} T}{m}, \; md - ft$$

$$k = kh/h$$

$$s' = 1.151 \left[\frac{\psi(p)_{1\,hr} - \psi(p_{wfo})}{m} - \log \frac{k}{\phi_h \mu\, c_t r_w^2} + 3.23 \right]$$

$$m(\Delta p)\, s' = 0.867\, m\, s'$$

$$\Delta p =$$

Pressure data:

$$\psi(p_i) =$$
$$F = 4\pi t_{DA}$$
$$\left[\frac{t + \Delta t}{\Delta t}\psi(p_{wf}) = \psi(\overline{p}_R) = e^F\right]$$

15.3 Shallow Low-Pressure and Highly Productive Gas Reservoirs

The following example illustrates how to determine the stabilized deliverability curve and *AOF*.

Example 15–1 *Determining Stabilized Deliverability Curve and AOF from the Test Data*

A gas well produces from a shallow low-pressure, highly productive reservoir. The well has been tested by a multirate test and the results are plotted in Figure 15–1. One and one-half durations of each flow period was enough to reach stabilization of flowing wellbore pressure. In fact, it was observed that pressures stabilized almost instantaneously after each rate change.

Solution The log-log backpressure plot gives a straight line which defines a backpressure exponent $n = 1/\text{slope} = 0.56$. The backpressure coefficient is

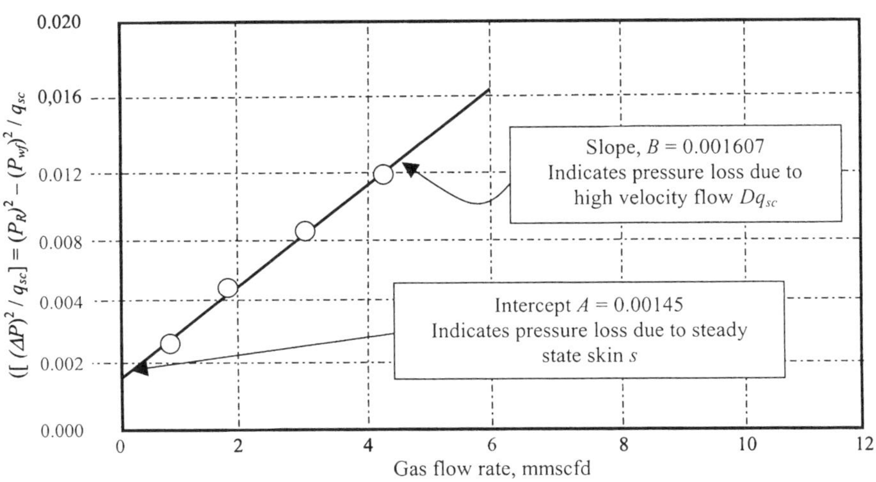

Figure 15–1. Linear plot for determining high-velocity effect on gas well performance.

calculated from the curve as

$$C = 8.0 \times 10^6 (11{,}200)^{0.56} = 43{,}204 \text{ scf/day/psia}^2$$

The backpressure equation then is $q_{sc} = 43{,}204(p_R^2 - p_{wf}^2)^{0.56}$ and the absolute open flow is 45.538 mmscfd. A Cartesian plot of $\Delta p^2/q_{sc}$ versus q_{sc} (Figure 15–1) gives a straight line (except for a small deviation and the low rate point). The intercept of the line is $A = 0.00145$ psia^2/scfd/D and the slope is

$$B = 1.607 \times 10^{-3} \frac{\text{pdia}^2/\text{scfd}}{\text{mmscfd}}$$

or, when expressed in scf/d,

$$B = 1.607 \times 10^{-9} \frac{\text{psia}^2/\text{scfd/d}}{\text{scfd/d}}$$

The low n value and the high B value indicate large rate-dependent skin. The slope B in Figure 15–1 indicates the significance of the high-velocity effect on the productivity of the well. A large slope implies large rate-dependent skin. The intercept A is related to steady-state skin factor. If the rate needs to be written in terms of flowing pressure, the quadratic equation can be solved as follows:

$$q_{sc} = \frac{\sqrt{A^2 + 4B\left(p_R^2 - p_{wf}^2\right)} - A}{2B}$$

$$= \frac{\sqrt{(0.00145)^2 + 4(1.607 \times 10^{-9})\left(p_R^2 - p_{wf}^2\right)} - 0.00145}{2(1.607 \times 10^{-9})}$$

This equation can be used to calculate the *AOF* for this example.

15.4 Recommended Form of Rules of Procedure for Backpressure Tests Required by State Regulatory Bodies

All backpressure tests required by a state regulatory body shall be conducted in according with the procedures set out by the state regulatory body except for those wells in pools where special testing procedures are applicable.[1–3] The calculations shall be made in the manner prescribed in the appropriate test examples. The observed data and calculations shall be reported on the prescribed forms. Gas produced from wells connected to a gas transportation facility should not be vented to the atmosphere during testing. When an accurate test can be obtained only under conditions requiring venting, the volume vented shall be the minimum required to obtain an accurate test. All surface pressure readings shall be taken with a dead weight gauge. Under special conditions where the use of a dead weight gauge is not practical, a properly calibrated spring gauge may be used when authorized by the state regulatory body. Subsurface pressures determined by the use of a properly calibrated pressure bomb are acceptable. The temperature of the gas column must be accurately known to obtain correct test results; therefore a thermometer well should be installed in the wellhead. Under shut-in or low-flow-rate conditions, the external temperature may distort the observed wellhead temperatures. Whenever this situation exists the mean annual temperature should be used.

15.5 Appropriate State Report Forms

The appropriate state report forms are as follows.

Texas Gas Well[1,2]

- Uses tubing pressures
- Square root chart entries for gas measurement
- GE system dialogue
- Answers transferred to G-1 form

New Mexico Gas Well[1,2]

- Uses tubing pressures
- Deviated well
- UCS system dialogue
- Answers transferred to preprinted state form C-122

Oklahoma Gas Well[3]

- Uses casing pressures
- Single-point test
- Case No. 1 assigned to input data
- GE system dialogue
- Answers presented in report form

Offshore Gas Well Using IOCC Procedure[3]

- Uses bottom-hole pressures
- UCS system dialogue
- Answers presented in report form for natural gas Oklahoma

15.6 Stimulation Efforts Evaluation, Summary, and Recommendations

This section presents theoretical and practical aspects of methods used to determine absolute open flow potential (*AOF*), formation permeability, overall skin factors, average reservoir pressure, and gas in place in low- and high-permeability gas reservoirs. Test analysis methods examined include deliverability, Horner, type curves, and reservoir limit test analysis. It also includes a brief summary, conclusions, and recommendations of two field case studies. One case is for a low-permeability gas reservoir; the other is for a high-permeability gas reservoir. These two cases demonstrate well test analysis applications in low- as well as high-permeability gas reservoirs.

Low-Permeability Gas Well, Nilam Gas Field, Indonesia

Case Studies: Nilam Gas Field, Well # N-38/gas, Zone G-50A

Nilam gas field is in Kalimantan, Indonesia, and is "offshore." The reservoir is 12,950 ft deep and consists of layers of clay and sandstone. The overall thickness is about 52 ft with average porosity of about 14 to 20%.

The empirical deliverability equations are

$$q_{sc} = 1.3152 \times 10^{-6} \left(\overline{p}_R^2 - p_{wh}^2\right) \text{ (wellhead conditions)}$$

$$q_{sc} = 0.5997 \times 10^{-6} \left(\overline{p}_R^2 - p_{wf}^2\right) \text{ (bottom-hole conditions)}$$

Stabilized flow equations are also developed using the LIT(ψ) approach to estimate deliverability potential of this gas well against any sandface pressure. The values of exponent $n = 1$ and formation permeability $= 8.274$ mD indicate, that it is a low-permeability gas reservoir (see Table 15–1 for a summary of results).

The laminar–inertial–turbulent (LIT) flow equations are

$$\psi(\overline{p}_R) - \psi(p_{wh}) = 45.5574\,_{sc}q + 2.1429\,q_{sc}^2 \text{ (wellhead conditions)}$$

$$\psi(\overline{p}_R) - \psi(p_{wf}) = 91.8273\,q_{sc} + 0.1785\,q_{sc}^2 \text{ (bottom-hole conditions)}$$

Returning again to the Forscheimer equation, $\overline{p}_R^2 - p_{wf}^2 = Aq_{sc} + Bq_{sc}^2$, kh is small (339.23 mD), $A\,q_{sc}$ becomes large, and the $B\,q_{sc}^2$ term can become negligible (not necessarily zero) when compared to the laminar pressure drop term. We could then write $q_{sc} \cong \frac{1}{A}(\overline{p}_R^2 - p_{wf}^2)^{1.0}$.

Calculate the following quantities

$$\sum_i^n s' = 76.145$$

$$\sum_i^n q = 27.087$$

Table 15–1
Summary of Results

	Wellhead pressure (psia)	Bottom-hole pressure (psia)	Flow rate (mmscf/d)	Choke size (inch)
Shut-in	2388	3700	—	
Rate 1	2015	3144	2.397	16
Rate 2	1640	2566	5.214	24
Rate 3	1365	2158	6.144	32
Rate 4	1015	1836	7.186	48
Extended rate	1015	1721	6.148	32
Final shut-in	2388	3700	—	
n	1.0	1.0		
C	1.3152×10^{-6}	0.5997×10^{-6}	mmscfd/psia	
AOF	7.50	8.21	mmscfd	

Table 15–2
Specific Results of Pressure Buildup Analysis Using Four Rate Tests

Parameters	Estimated values		Remarks
q_{sc}	6.148 mmscfd		
$\psi(P_{wf1})$	690×10^6 psia^2/cP		
$\psi(P_{wfo})$	669×10^6 psia^2/cP		
m	21.0×10^6 psia^2/cP		
kh	339.23 mD-ft		
k	8.274 mD		
s'	+16.869	Apparent skin	
s	+3.649	True skin	See Table 15–3
D	2.137511	Turbulent factor	See Table 15–3
$\psi(\Delta P)_{skin}$	64.44 mmpsia^2/cP	995 psia	True skin
$\psi(P_i)$	861.12 mmpsia^2/cP	3955 psia	From Horner plot
$\psi(P_R)$	772.0 mmpsia^2/ cP	3702 psia	
Static gradient	0.110 psi/ft		

$$\sum q^2 = 160.088$$

$$\sum s' \times q = 441.037$$

Solving by the least square method, we get

$$s = +3.649 \text{ (true skin)}$$
$$D = 2.137511 \text{ (turbulent factor)}$$

Figure 15–2 shows a graphical method to estimate true skin factor.

Table 15–3
Evaluation of True Skin and Turbulent Factor

Gas flow rate #	1	2	3	4	5
Gas flow rate q, mmscfd	2.397	5.214	6.144	7.186	6.148
$\psi(P_{wf1})$, mmpsia^2/cP	770	765	752	745	650
$\psi(P_{wfo})$, mmpsia^2/cP	592.45	418.12	306.21	227.88	201.25
m, mmpsia^2/cP/cycle	15	20	21	22	21
kh, mD-ft	185.163	302.078	339.00	378.480	339.229
K, mD	4.516	7.366	8.268	9.231	8.279
S', apparent skin	8.748	14.842	16.728	18.958	16.869

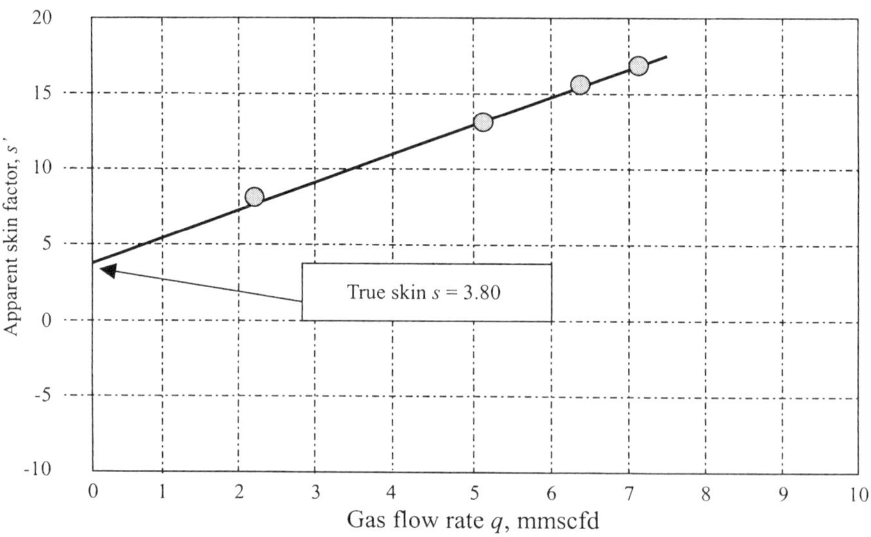

Figure 15–2. Apparent skin factor s' versus gas flow rate.

Radius of Investigation

At the beginning of the middle transient regime (MTR), $\Delta t = 1$ hr, 102 ft. At the end of the middle transient regime (MTR), $\Delta t = 10$ hr, 321 ft. Thus a significant fraction of the well's drainage area has been sampled and its permeability is 6.282 mD.

At $\Delta t = 147.12$ hr, 1233 ft.
At $\Delta t = 468.78$ hr, 2200 ft which is equal to the assumed r_e.

Reservoir limit:

$$m^* = 0.29412 \times 10^6 \text{ psia}^2\text{/cP/hr}$$

$$V_P = 2.59288 \times 10^{11} \text{ scf (gas filled pore volume of the reservoir)}$$

Basis on $p_i = 3965$ psia

Conclusions and Recommendations

A Horner plot using pseudopressure was used to obtain reservoir parameters. This completion has fair permeability to gas and positive true skin factor, indicating an undamaged well.

Overall, the results of analysis are reasonable and can be accepted as reliable. Based on this analysis, it can be concluded that:

- The completion would probably benefit from stimulation.
- Test procedures were suitable for this well.
- Production could continue from this reservoir at this gas well using a 32/64 inch choke size.

High-Permeability Gas Well: Batak Gas Field, Indonesia

Case Studies: Batak Gas Field, Well # B-9L/gas, Zone F-1

Batak gas field is in Kalimantan, Indonesia, and is "offshore." The reservoir is 5500 ft deep and consists of layers of sand and limestone. The overall thickness is about 68 ft with average porosity of about 15 to 22%.

The empirical deliverability equations are

$$q_{sc} = 0.00095\left(\overline{p}_R^2 - p_{wh}^2\right)^{0.69} \text{ (wellhead conditions)}$$

$$q_{sc} = 0.00473\left(\overline{p}_R^2 - p_{wf}^2\right)^{0.61} \text{ (bottom-hole conditions)}$$

Stabilized flow equations are also developed using the LIT(ψ) approach to estimate deliverability potential of this gas well against any sandface pressure. The values of exponent $n = 0.69$ (wellhead conditions), $n = 0.61$ (bottom-hole conditions), respectively, and permeability = 920 mD indicate that it is a high-permeability gas well. See Table 15–4 for a summary of results.

The laminar–inertial–turbulent (LIT) flow equations are

$$\psi(\overline{p}_R) - \psi(p_{wh}) = 1.1818q_{sc} + 0.2677q_{sc}^2 \text{ (wellhead conditions)}$$

$$\psi(\overline{p}_R) - \psi(p_{wf}) = 0.6430q_{sc} + 0.0597q_{sc}^2 \text{ (bottom-hole conditions)}$$

Table 15–4
Summary of Results

	Wellhead pressure (psia)	Bottom-hole pressure (psia)	Flow rate (mmscf/d)	Choke size (inch)
Shut-in	2602	3795	—	
Rate 1	2567	3774	1.446	16
Rate 2	2532	3757	6.681	24
Rate 3	2317	3723	15.790	32
Rate 4	2281	3717	16.566	48
Extended rate	2463	3754	10.822	32
Final shut-in	2602	3795	—	
n	0.69	0.61		
C	0.00095	0.00473	mmscfd/psia^2	
AOF	49.07	110.03	mmscfd	

Table 15–5
Specific Results of Pressure Buildup Analysis: Two-Rate Test

	Units	Buildup # 1	Buildup # 2
q_{sc}	mmscfd	16.566	10.822
$\psi(P_{wf1})$	mmpsia^2/cP	834.85	838.55
$\psi(P_{wfo})$	mmpsia^2/cP	816.51	827.90
m	mmpsia^2/cP	0.40	0.35
kh	mD-ft	45,284.82	33,809.17
k	mD	1053.14	786.26
s'	Apparent skin	+39.25	+27.57
s	True skin	+5.57	
D	Turbulent factor	2.03308	
$\psi(\Delta P)_{skin}$	mmpsia^2/cP	1.6975 psia (true skin)	
$\psi(P_i)$	mmpsia^2/cP	839.3	
		3786 psia from Horner Plot	
$\psi(P_R)$	mmpsia^2/cP		
Static gradient	psi/ft	0.116	

Returning again to the Forscheimer equation $\overline{p}_R^2 - p_{wf}^2 = Aq_{sc} + Bq_{sc}^2$, kh is large (920.0 mD), $A\,q_{sc}$ becomes small, and we would have

$$q_{sc} \cong \frac{1}{\sqrt{B}}\left(\overline{p}_R^2 - p_{wf}^2\right)^{0.61}$$

It is clear then that it is not necessary for flow to be completely turbulent throughout the reservoir for the slope (n) to be equal to 0.5.

Conclusions and Recommendations

A Horner plot using pseudopressure was used to obtain reservoir parameters. This completion has high permeability to gas and positive true skin factor, indicating an undamaged well.

Overall, the results of analysis are reasonable and can be accepted as reliable. Based on this analysis, it can be concluded that:

- The completion would benefit very little from stimulation.
- Test procedures were suitable for this gas well.
- Restrictions caused by turbulent effects are occurring in this well.
- Larger tubing size is recommended.
- Production could continue from this reservoir at this well using a 32/64 inch choke size.

15.7 Formation Characteristics from Fractured Carbonate Gas Reservoirs

Field Case Study for Analyzing Buildup Tests Having Two Slopes

Special pressure responses from well tests must be analyzed in light of all available information. Adams et al.[4] have presented a complete evaluation of a fractured carbonate gas reservoir. In a conventional buildup plot, two slopes were observed, with the first one having a higher value than the second one (Figure 15–2). After a detailed analysis, they concluded that the matrix permeability could be evaluated from the first slope, and the mean permeability of the matrix-fracture system could be evaluated from the second slope. Their results using this criterion were reasonable when compared with known geologic and core data. The use of special pressure responses for buildup tests in a fractured carbonate gas reservoir will be illustrated with an example using the method proposed by Adams *et al.*[4]

Example 15–2 *Analyzing Buildup Test with Two Slopes in a Fractured Carbonate (CO_3) Reservoir*

Figure 15–4 shows a buildup for well A, which is located in a fractured carbonate reservoir. The figure shows three straight lines. Figure 15–3 shows permeability variation of core data for this well. Gas properties are as follows: $T = 90°F$; $p_{SC} = 15.025$ psia; $T_{SC} = (60 + 460) = 520°R$; $q_g = 0.548$ mmscfd; $h = 160$ ft; $r_w = 0.3$ ft; $\mu_g = 0.0131$ cP; $\phi = 0.03$ fraction; $c_t = 0.0009\ psi^{-1}$.

Solution Figure 15–4 shows a buildup plot of $\log(\frac{t_p+\Delta t}{\Delta t})$ versus $\psi(p)$. This figure shows three straight lines, with their corresponding slopes, the

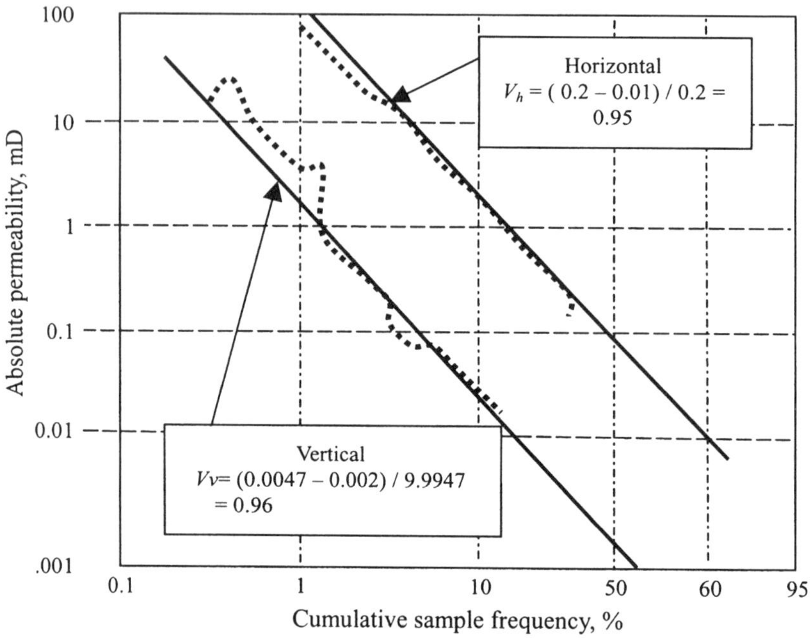

Figure 15–3. Core permeability variation for gas well (after Adams *et al.*).[4]

extrapolated value of $\psi(\dot{p})$ at infinite shut-in, and $\psi(p_{ws})_{1hr}$ at 1 hr. Adams *et al.* carried out the interpretation as follows (see Table 15–6).

1. The matrix permeability k_m was calculated from

$$\begin{aligned} k_m &= \frac{57.92 \times 10^6 q_{sc} P_{PS} T}{m_1 h T_{SC}} \\ &= \frac{57.92 \times .548 \times 15.025 \times (90 + 460)}{26.0 \times 10^6 \times 160} \times 520 \\ &= 0.121 \text{ mD} \end{aligned} \tag{15–1}$$

2. The skin factor s was calculated from

$$s = 1.151 \left[\frac{\psi(p)_{1hr} - \psi(p_{wf})}{m_1} - \log \frac{k_m}{\phi \mu c_t r_w^2} + 3.23 \right]$$

$$\begin{aligned} s = 1.151 &\left[\frac{(98.7 - 37.4) \times 10^6}{26.0 \times 10^6} \right. \\ &\left. - \log \frac{0.12}{0.03 \times 0.0131 \times 0.0009 \times 0.3^2} + 3.23 \right] \\ = 1.151\,&[2.358 - 7.019 - 3.23] = -1.131 \end{aligned}$$

Table 15–6
Calculated Gas Properties

Pressure (psia)	Z-factor	Gas viscosity μ (cP)	$\psi(p)$ (mmpsia^2/cP)
1800	0.792	0.0149	296.0
1600	0.805	0.0142	237.5
1400	0.821	0.0135	184.2
1200	0.841	0.0129	136.8
1000	0.861	0.0122	95.6
800	0.886	0.0188	61.3
600	0.912	0.0144	34.5
400	0.949	0.0111	15.4
200	0.969	0.0107	3.9
0	1.000	0.0105	0

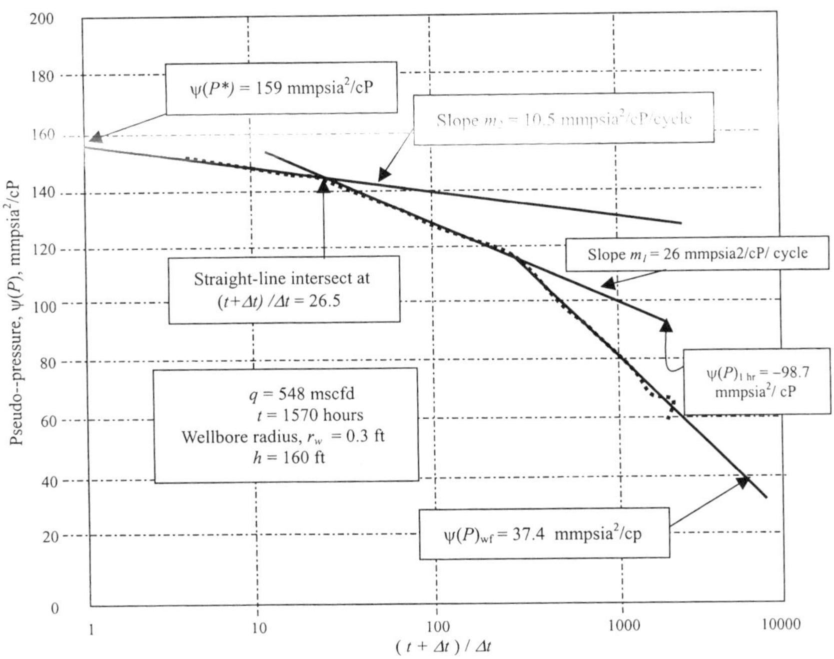

Figure 15–4. Pressure buildup curve for gas well.

3. Well A is located in the center of one-half of a 2:1 rectangle. The permeability of outer region k_2 was calculated from

$$k_2 = \frac{k_1 m_1}{m_2} \tag{15–2}$$

$$k_2 = \frac{0.12 \times 26.0 \times 10^6}{10.5 \times 10^6} = 0.3 \text{ mD}$$

The ratio k_2/k_1 equals 2.48. Since the drainage area is known to be 780,000 sq ft, the dimensionless producing time can be calculated from

$$t_{DA} = \frac{0.000264 k_2 t}{\phi \mu c_t A} = \frac{0.000264 \times 0.3 \times 1570}{0.03 \times 0.0131 \times 0.00063 \times 780{,}000} = 0.645 \tag{15–3}$$

The dimensionless correction to $\psi(\overset{*}{p})$ is 1.2 from curve III of Figure B–3 for the case of a 2:1 rectangle.

$$\psi(\overset{*}{p}) - \psi(\overline{p}) = \frac{m_2 \times \text{correction}}{2.303}$$

$$= \frac{10.5 \times 10^6 \times 1.2}{2.303} = 5.47 \times 10^6 \tag{15–4}$$

$$\psi(\overline{p}) = 159.0 \times 10^6 - 5.5 \times 10^6 = 153.5 \text{ mmpsia}^2/\text{cP}$$

4. The reservoir flow efficiency can be calculated from

$$FE = \frac{\psi(\overline{p}) - \psi(p_{wf}) - 0.869 m_1 s}{\psi(\overline{p}) - \psi(p_{wf})}$$

$$= \frac{153.5 - 37.5 - 0.869 \times 26.0(-1.131)}{153.5 - 37.4} = 1.25 \tag{15–5}$$

5. The distance x to the change in permeability can be calculated from

$$s = 1.151 \frac{k_1}{k_2} \left[\frac{\psi(p)_{1hr} - \psi(p_{wf})}{m_2} - \log \frac{k_2}{\phi \mu c_t r_w^2} \right.$$
$$\left. -2\left(\frac{k_2}{k_1} - 1\right) \log \frac{x}{r_w} \right] \tag{15–6}$$

$$-1.131 = 1.151 \frac{0.12}{0.3} \left[\frac{126.2 - 37.4}{10.5} \right.$$
$$- \log \frac{0.3}{0.03 \times 0.0131 \times 0.0009 \times 0.3^2} + 3.23$$
$$\left. -2\left(\frac{0.3}{0.12} - 1\right) \log \frac{x}{r_w} \right]$$

Consequently $x/r_w = 292$ and $x = 87.6$ ft. This distance to change in permeability can also be compared from[4]

$$\left(\frac{x}{r_w}\right)^2 = \frac{4t_D}{\gamma\left(\frac{t_P+\Delta t}{\Delta t}\right)} \times \left(\frac{k_2}{k_1}\right)^{\frac{1}{1-(k_2-k_1)}} \tag{15–7}$$

$$\left(\frac{x}{r_w}\right)^2 = \frac{4 \times 0.645 \times 0.78 \times 10^6}{1.78 \times 26.5 \times 0.3^2} \times (2.5)^{\frac{1}{1-(0.3-0.12)}} = 1.035 \times 10^5$$

$x/r_w = 322$ and $x = 96.6$ ft. The values of x calculated by two methods are of the same order of magnitude and consequently it can be concluded that the change of k is about 90 ft from the wellbore.

15.8 Buildup Interpretations Before and After Workovers

In practice, it is desirable to get as much information as possible from a pressure buildup test. Trying as many crossplotting techniques as possible can do this. The buildup data before and after workover were obtained in various zones, block III, Benuang Gas Field, and South Sumatra, Indonesia. Field examples are reproduced here because of their practical implications.

Buildup Data Completely Controlled by Afterflow

Figure 15–5 shows a conventional plot of shut-in pressure versus log of shut-in time. The pressure buildup data are completely controlled by afterflow.

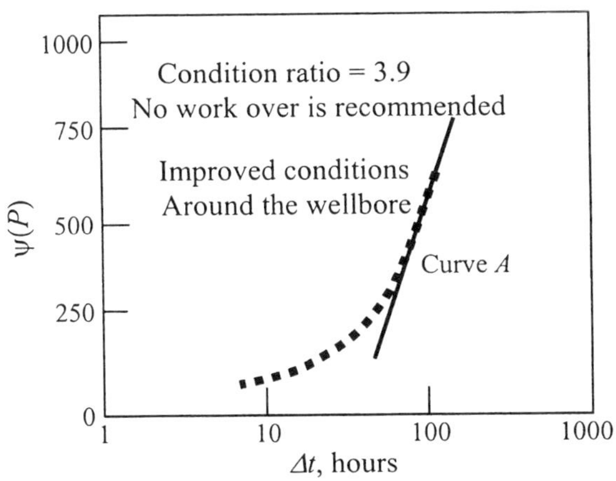

Figure 15–5. $\psi(P)$ versus Δt—Semilog plot.

Interpretation would be based on conventional radial flow equations, using what appears to be the straight-line portion of the curve A. The condition ratio using as a basis this straight line is 3.9 and indicates improved conditions around the wellbore. Under these conditions, no workover would be attempted in this well. Figure 15–6 shows a plot of shut-in pressure versus shut-in time in Cartesian coordinates. A straight line is obtained, which indicates that the buildup data are entirely dominated by wellbore storage (afterflow). Figure 15–7 shows a log-log plot of pressure differential versus time. A straight

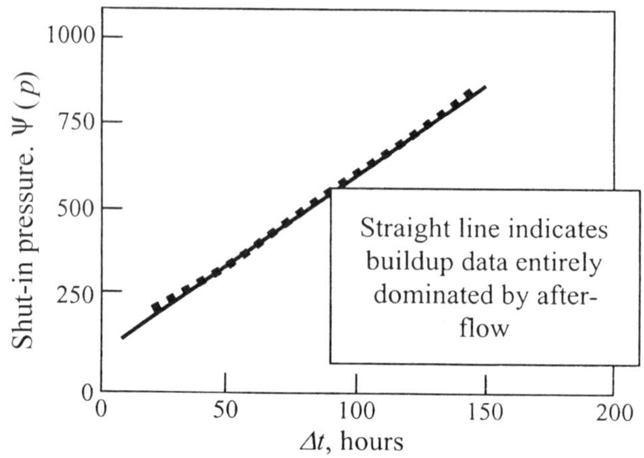

Figure 15–6. $\psi(P)$ versus Δt—Cartesian coordinate crossplot (fractured gas well).

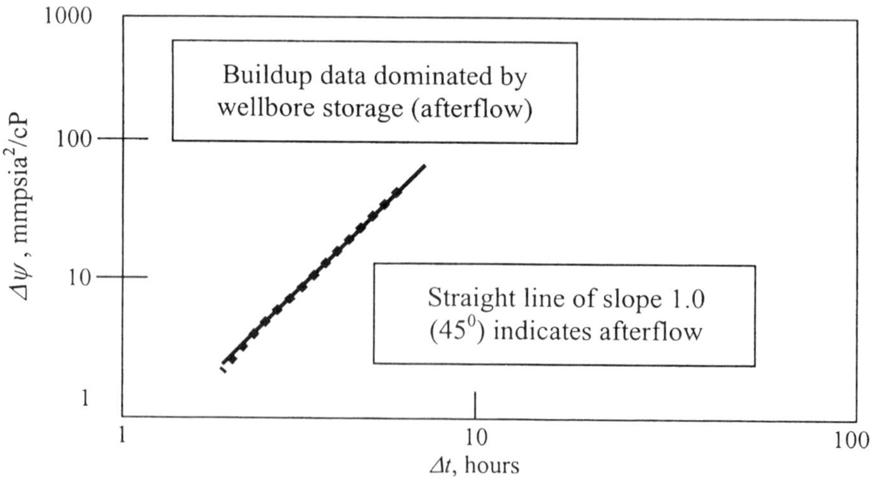

Figure 15–7. Log $\Delta\psi$ versus log Δt—Log-log plot (fractured gas well).

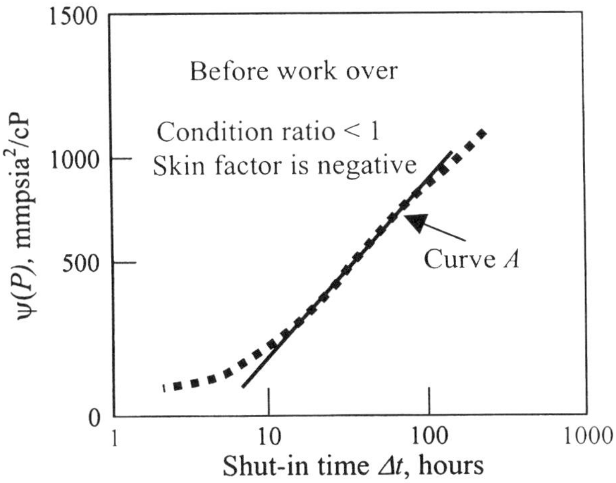

Figure 15–8. $\psi(P)$ versus log Δt—Semilog plot.

line with a unit slope is obtained, which also indicates that the buildup data are dominated by afterflow.

Pressure Buildup Data with Long Afterflow and Beginning of Linear Flow

Figure 15–8 shows a conventional semilog plot of shut-in pressure versus time. The apparent straight line A allows the calculation of a condition ratio greater than 1 and a negative skin. Under these conditions no workover should be attempted. The analysis, however, was followed by cross plotting the shut-in pressure versus time in Cartesian coordinates. The plot revealed a straight line for the first 25 hr (Figure 15–9), which was indicative of afterflow.

Figure 15–10 is a log-log cross plot of incremental pressure Δp (shut-in pressure − flowing pressure) versus time. A straight line of unit slope was not apparent during the first 40 hr of shut-in. Absence of such a straight line points to the possible presence of skin on the face of the fracture. Starting at 40 hr, there is a straight line of half-unit slope, which is indicative of linear flow. The interruption of the straight line is attributed to change of pumps and or skin damage on the surface of the fracture around the wellbore. Figure 15–11 is a plot of pressure differential versus square root of shut-in time on Cartesian coordinates. This type of cross plot results in a straight line, in which linear flow dominates. The intercept of the straight line at zero shut-in time equals the pressure drop due to skin. The slope of the straight line can calculate the length of the fracture or the formation permeability depending on which parameter can be reasonably assumed. In this case the combination of all plots

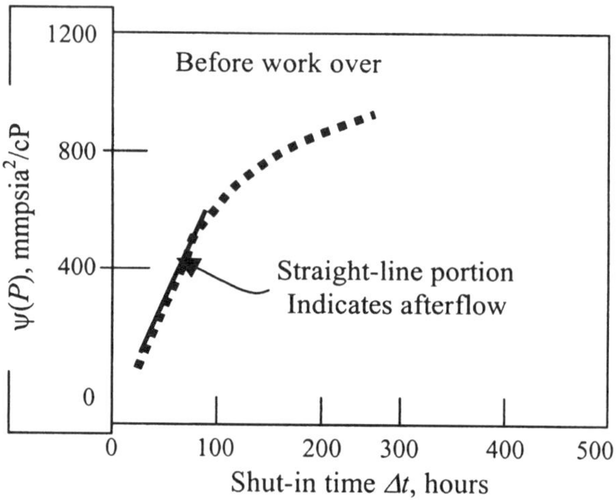

Figure 15–9. $\zeta(P)$ versus shut-in time—Cartesian coordinate plot (fractured gas well).

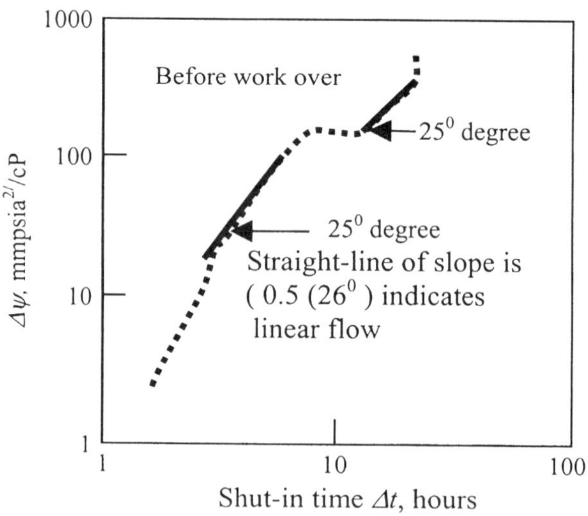

Figure 15–10. Log ΔP versus log Δt—log-log plot (fractured gas well).

provided valuable information and led to the recommendation of a stimulation job. Figure 15–12 shows the results of a pressure survey after the workover in semilog coordinates.

Figure 15–13 shows a straight-line portion on Cartesian coordinates, which is indicative of afterflow. This period ended after 16 hr shut-in. Figure 15–14 shows a log-log cross plot of pressure differential versus time. A straight line

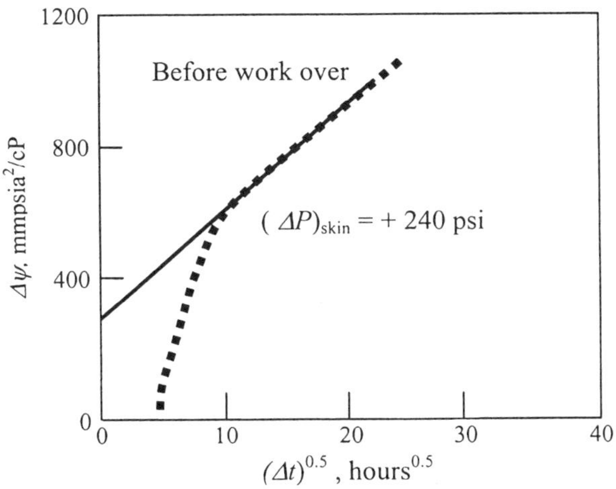

Figure 15–11. $\Delta\psi$ versus $\sqrt{\Delta t}$—Specialized plot (fractured gas well).

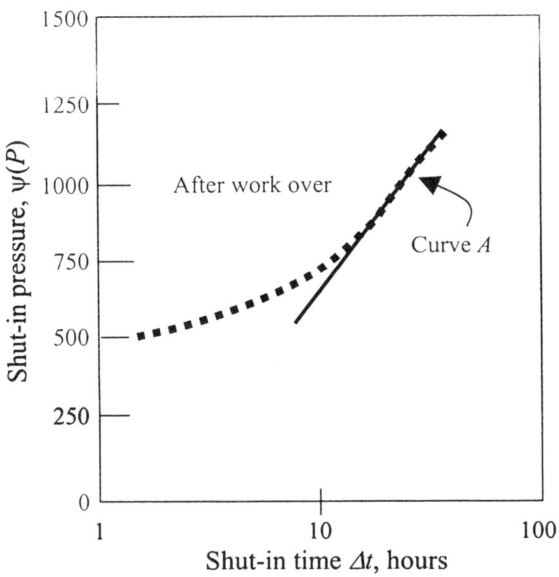

Figure 15–12. $\psi(P)$ versus log Δt—Semilog plot (gas well).

with unit slope is obtained which ends at 20 hr. This unit slope is indicative of afterflow. Another straight line is obtained after 20 hr. The slope of this line is 0.5 and indicates that flow becomes predominantly linear.

Figure 15–15 shows a Cartesian cross plot of pressure differential versus square root of shut-in time. The resulting straight line indicates the presence

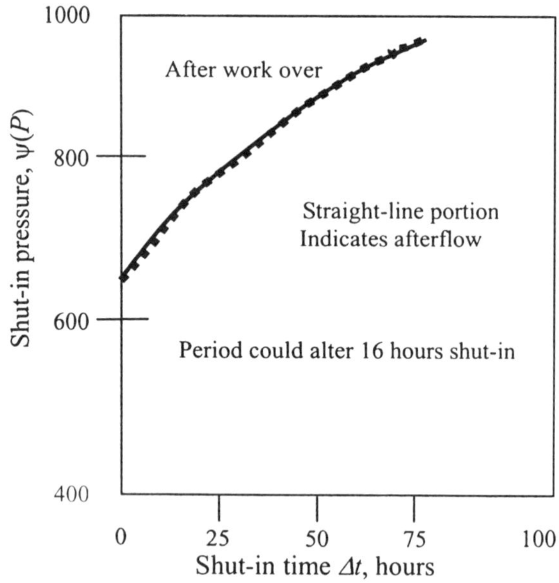

Figure 15–13. $\Delta\psi$ versus Δt—Cartesian coordinate plot (fractured gas well).

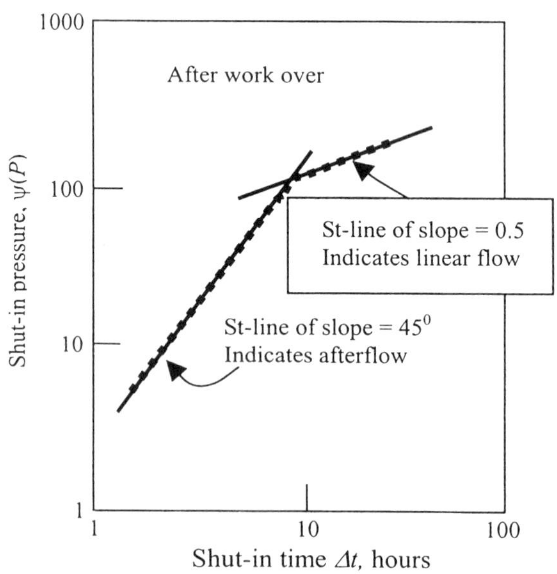

Figure 15–14. $\psi(P)$ versus log Δt—Semilog plot (gas well).

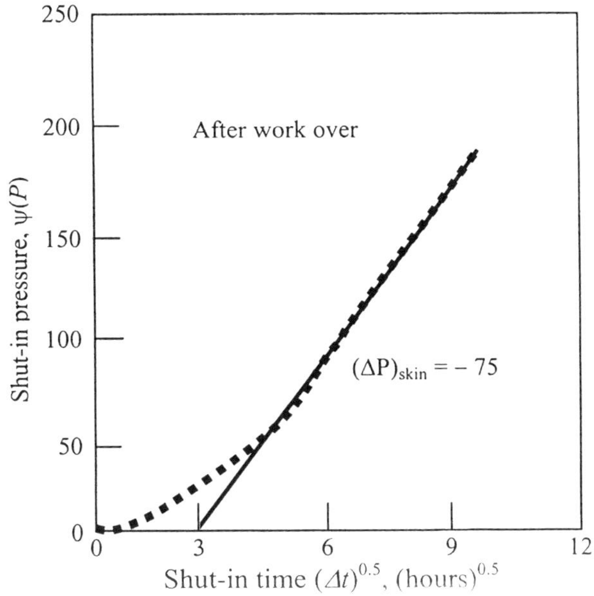

Figure 15–15. $\psi(P)$ versus Δt—Cartesian coordinate plot (fractured gas well).

of linear flow. The intercept at zero shut-in time is equivalent to the pressure change due to skin, and in this case is negative indicating that the stimulation job was successful. Note: The variety of cross plots presented in this case led to recommend stimulation of a well with damage and resulted in damage removal and improved conditions around the wellbore.

Pressure Buildup Data Controlled for a Short Period

Figure 15–16 shows the conventional semilog plot and the "first glance" straight line. Figure 15–17 shows a cross plot of shut-in pressure versus time in Cartesian coordinates. The lack of a straight line at early times indicates that the afterflow period dies very quickly. Figure 15–18 shows a log-log plot of pressure differential (shut-in pressure – flowing pressure) versus time. A straight line of half-unit slope is obtained which indicates linear flow. Also notice that the unit slope straight line, indicative of wellbore storage, is missing.

Figure 15–19 shows a Cartesian cross plot of pressure differential (shut-in pressure – flowing pressure) versus square root of shut-in time. A continuous straight line is obtained which indicates the predominance of linear flow. The intercept of this line at zero time is negative, indicating improved conditions around the wellbore.

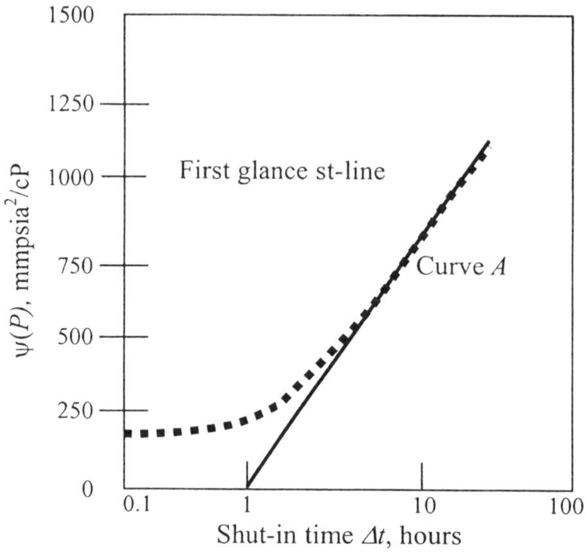

Figure 15–16. $\psi(P)$ versus Δt—Semilog plot (fractured gas well).

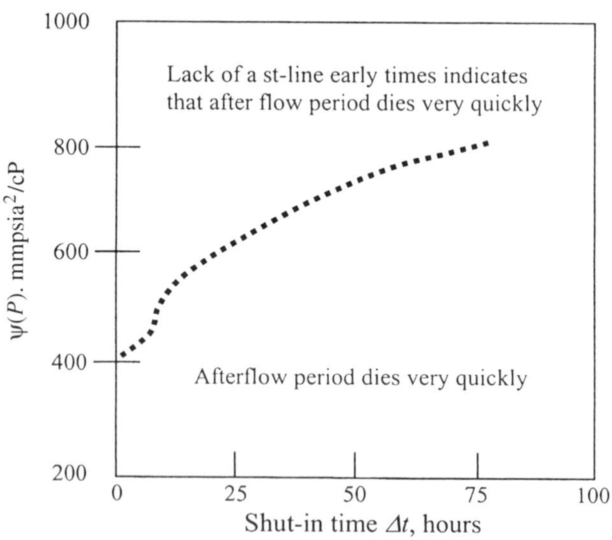

Figure 15–17. $\psi(P)$ versus Δt—Cartesian coordinate plot (fractured gas well).

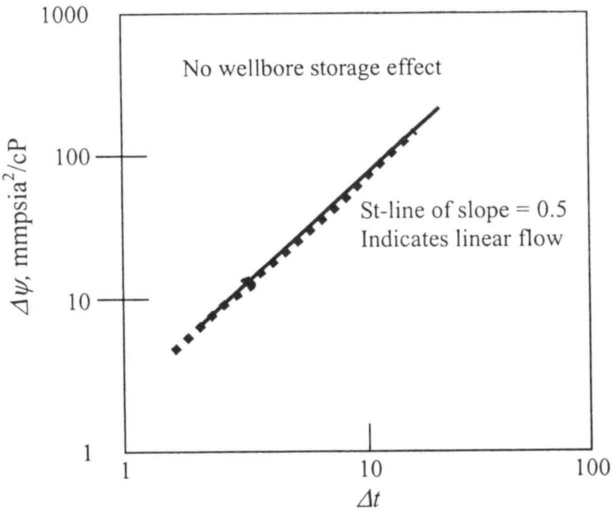

Figure 15–18. Log $\Delta\psi$ versus log Δt—Log-log plot (fractured gas well).

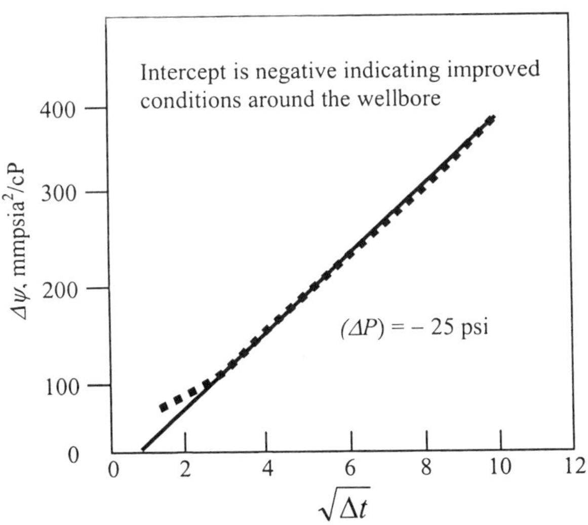

Figure 15–19. $\Delta\psi$ versus $\sqrt{\Delta t}$—Specialized plot (fractured gas well).

Pressure Buildup Data Showing a Small Afterflow

Figure 15–20 shows a conventional semilog cross plot which results in two straight lines named A and B. Figure 15–21 shows a Cartesian cross plot of shut-in pressure versus time which indicates no early straight line and,

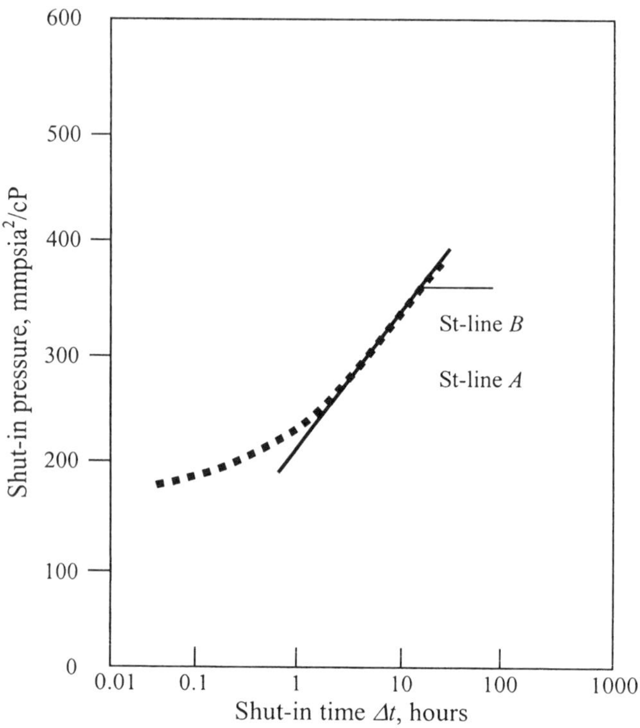

Figure 15–20. $\psi(P)$ versus log Δt—Semilog plot (fractured gas well).

consequently, that the afterflow period dies rapidly. Figure 15–22 shows a log-log cross plot of pressure differential (shut-in pressure − flowing pressure) versus shut-in time. There are a few early-scattered points and then a straight line in the half-unit slope, which indicates linear flow. In the late portion there is a slightly curved line, which indicates the presence of radial flow. Finally, a plot of pressure differential (Δp) versus square root of shut-in time is shown in Figure 15–23. There is a very clear straight line, which suggests linear flow, followed by a curved portion when radial flow is attained. The straight-line portion A of Figure 15–20 can calculate formation permeability using radial flow theory. With the slope of Figure 15–23, its intercept at zero shut-in, and the formation permeability determined from Figure 15–20, we can calculate the fracture length and pressure drop with a good degree of accuracy.

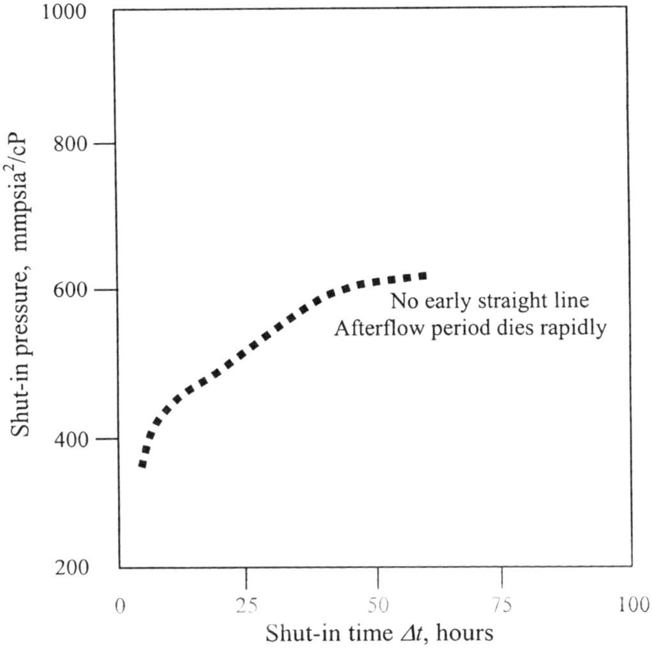

Figure 15–21. $\psi(P)$ versus t—Cartesian coordinate plot (fractured gas well).

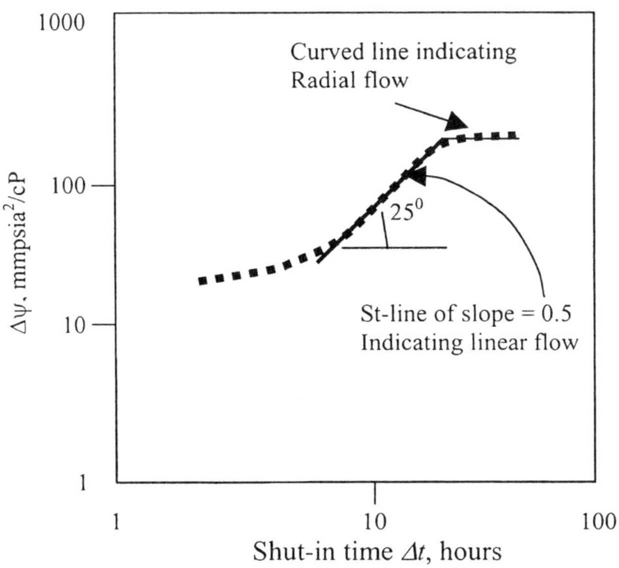

Figure 15–22. Log $\Delta\psi$ versus log Δt—Log-log plot (fractured gas well).

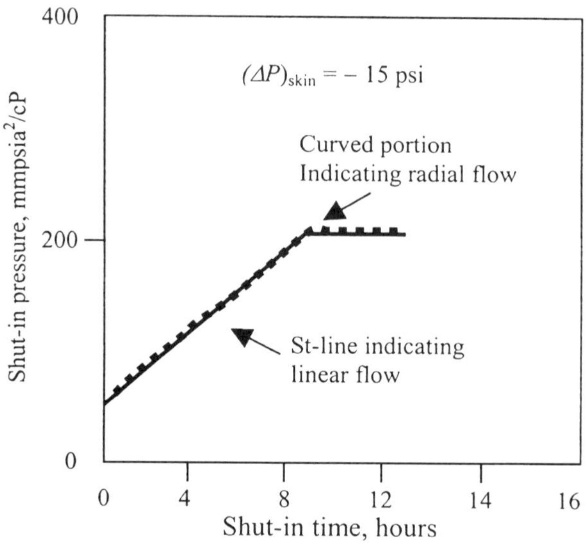

Figure 15–23. $\Delta\psi$ versus Δt—Cartesian coordinate plot (Fractured gas well).

References and Additional Reading

1. *Back Pressure Test for Natural Gas Wells*, revised edition. Railroad Commission of Texas, 1951.
2. Interstate Oil Compact Commission (1962). *Manual of Back Pressure Testing of Gas Wells.*
3. Kansas State Corporation Commission (1959). *Manual of Back Pressure Testing of Gas Wells.*
4. Adams, A. R., Ramey, H. J., and Burgass, R. J., "Gas Well Testing in a Fracture Carbonate Reservoir," *J. Petroleum Technol.* (Oct. 1988), 1187–1194.
5. Earlougher, R. C., and Ramey, H. J., Jr., Miller, F. G., and Mueller, T. D., "Pressure Distribution in Rectangular Reservoirs," *J. Petroleum Technol.* (1960) 20, 199–208.

Chapter 16

Application of Decline Curve Analysis Methods

16.1 Introduction

This chapter deals with analysis of decline curves during the transient and pseudo-steady-state flow periods. Classifications of production decline curves and their practical uses are discussed with field examples including methods to forecast performance of horizontal and vertical fracture gas reservoirs.

16.2 Transient Decline Behavior Analysis

The solution in Figure 16–1 presents the behavior of a well during the transient period, when it behaves as an infinite reservoir. Accordingly, this decline solution is referred to as "infinite acting." The terms "infinite-acting decline" and "transient decline" are used interchangeably. Note: The rapid rate decline is normal transient behavior. Figure 16–1 is useful for test analysis by type curve matching.

Transient Drainage Radius during Infinite-Acting Period

It is related to the dimensionless rate by

$$r_e(t) = r_{wa}\exp(-q_D) \qquad (16\text{–}1)$$

where

$$r_{wa} = r_w e^{-s} \qquad (16\text{–}2)$$

and solving for the skin factor;

$$s = -In\left(\frac{r_{wa}}{r_w}\right) \qquad (16\text{–}3)$$

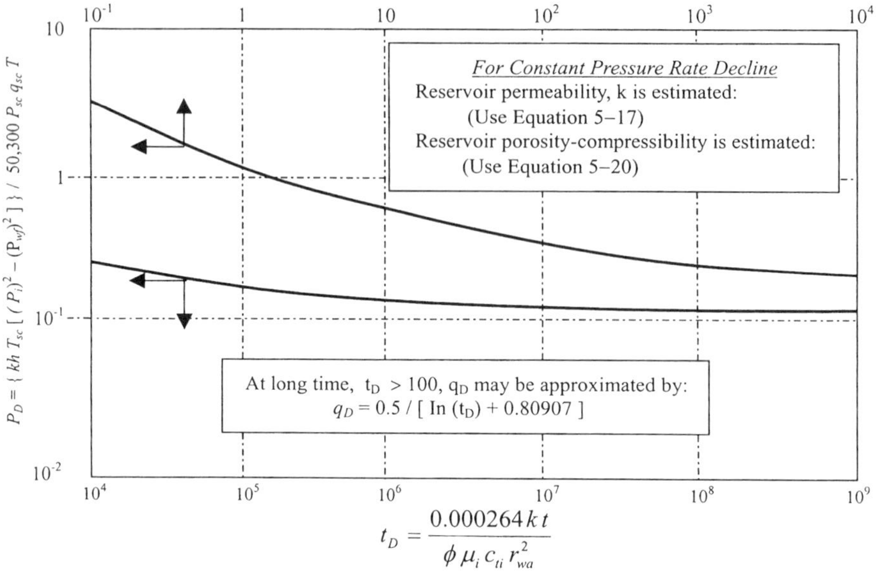

Figure 16–1. Infinite-acting dimensionless rate-solution (after Jacob and Lohman).[9]

Characteristics of Exponent *b* during Transient Flow

If rules during transient flow are used to compute the value of exponent b, then such measurements may suggest that the value of b is greater than unity and is given by

$$b = 2s + In\left(\frac{4t_D}{e^{\gamma}}\right)$$

In most cases the exponent b would be greater than unity, if transient responses were used to predict performance. Behavior in rate will follow the $b = 0$ curve only for the case $\overline{\lambda}_t/\overline{c}_t$ = constant as long as $\overline{c}_t / \overline{\lambda}_t$ is a linear function of time.

Production Characteristics during Transient Period

This section illustrates the use of the general infinite-acting solution and reservoir properties, to calculate wellbore conditions during the transient production period of a gas well.

Constant-Pressure Rate Decline

The following equations can be used to analyze a gas well producing from a low-permeability gas reservoir. Use the match points to calculate wellbore conditions, permeability, and skin factor for a gas well:

$$\left(\frac{q_D}{q_g}\right)_{match} = \frac{\mu_{gi}\, z_i T}{0.703\, kh\left(p_i^2 - p_{wf}^2\right)} \tag{16–4}$$

or

$$k = \frac{\mu_{gi}\, z_i T}{0.703\, kh\left(p_i^2 - p_{wf}^2\right)} \cdot \left(\frac{q_g}{q_D}\right)_{match} \tag{16–5}$$

$$\left(\frac{t_D}{t}\right)_{match} = \frac{0.000264\, k}{\phi\, \mu_{gi}\, c_i\, r_{wa}^2} \tag{16–6}$$

or

$$r_{wa}^2 = \frac{0.000264\, k}{\phi \mu_i\, c_i} \cdot \left(\frac{t}{t_D}\right)_{match} \tag{16–7}$$

where

$$r_{wa} = r_w e^{-s} \tag{16–8}$$

and, solving for the skin factor;

$$s = -In\left(\frac{r_{wa}}{r_w}\right) \tag{16–9}$$

To predict rate-time behavior, follow these steps:

1. Calculate dimensionless time for start of pseudo-steady-state flow by

$$t_{DApss} = 0.1 \tag{16–10}$$

or

$$t_{Dpss} = 0.1\,\pi \left(\frac{r_e}{r_{wa}}\right)^2 \tag{16–11}$$

In real time, this condition corresponds to

$$t_{pss} = \frac{\phi \mu_i\, c_{ti}\, r_w^2}{0.000264\, k} \cdot t_{Dpss} \tag{16–11a}$$

In terms of real time, the condition in Eqs. 16–11 and 16–11a becomes

$$t_{pss} = 379\frac{\phi\,\mu_i c_{ti} A}{k} \tag{16–12}$$

where

t_{pss} = start of pseudo-steady-state time, hrs
μ_I = gas viscosity, cP
c_{ti} = system compressibility, psi^{-1}
k = gas permeability, mD
A = gas field spacing, ft^2

2. List production time (in days) in 10-day increments.
3. Calculate the corresponding dimensionless time t_D.
4. Read q_D from Figure 16–1 at corresponding value of t_D.
5. Calculate the corresponding rates $q_g(t)$ from q_D using Eq. 16–13:

$$q_g(t) = \frac{0.703\,kh\left(p_i^2 - p_{wf}^2\right)}{\mu_i\, z_i\, T} \cdot q_D \tag{16–13}$$

Constant-Rate Production, Pressure Declining

To predict pressure–time behavior, follow these steps:

1. List production time in days, in day increments.
2. Calculate the corresponding dimensionless time t_D using

$$t_D = \frac{0.000634k}{\phi \mu_i c_i\, r_{wa}^2} t \tag{16–14}$$

3. Read P_D values from the type curve in Figure 16–1, corresponding to t_D values.
4. Calculate $\lfloor p_i^2 - p_{wf}^2(t) \rfloor$ by substituting P_D in Eq. 16–15:

$$\Delta p^2 = \left[p_i^2 - p_{wf}^2(t)\right] = \frac{\mu_i\, z_i\, T q_g}{0.703\,kh} P_D \tag{16–15}$$

5. Calculate $p_{wf}(t)$ by subtracting Δp^2 from p_i^2, which gives

$$p_{wf} = p_i^2 - \Delta p^2 \tag{16–16}$$

16.3 Pseudo-Steady-State Decline

A general expression for pseudo-steady-state decline for constant pressure production, according to the analytical solution, is

$$q_D = E_1\, e^{-E_2 t_D} \tag{16–17}$$

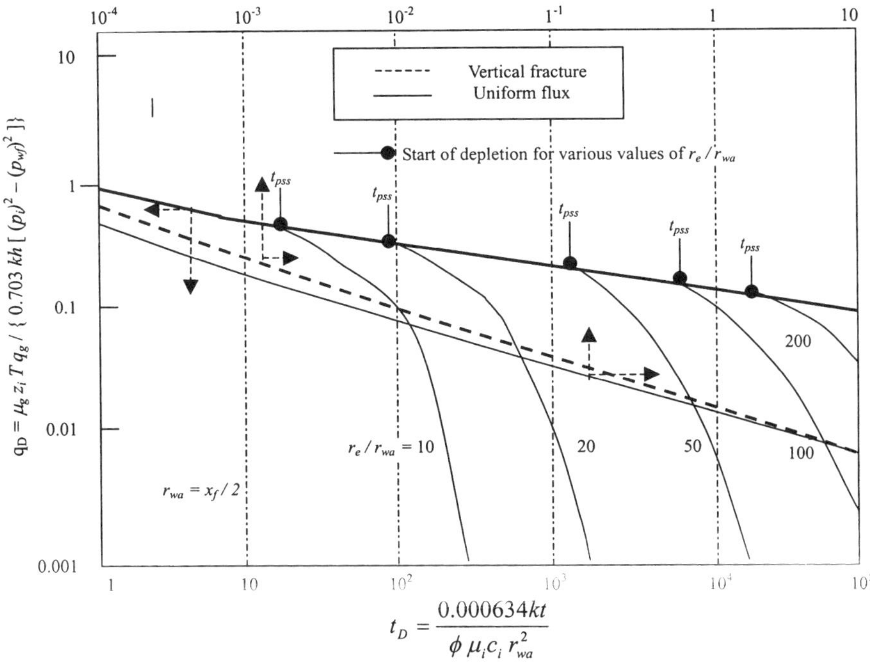

Figure 16–2. Full analytical constant pressure, dimensionless rate solution showing pseudo-steady-state depletion stems (after Fetkovich).[6]

where E_1 and E_2 are constants defined by the ratio r_e/r_{wa}. Fetkovich[6] developed expressions for E_1 and E_2 in Eq. 16–17 and stated that

$$E_1 = \frac{1}{\ln(r_e/r_{wa}) - 0.5} \tag{16–18}$$

$$E_2 = \frac{2E_1}{(r_e/r_{wa})^2 - 1} \tag{16–19}$$

The expressions for E_1 and E_2 reflect the observation that different ratios of r_e/r_{wa} give different stems as shown in Figure 16–2.

Forecasting Rate Decline

Figure 16–3 shows depletion decline characteristics after the start of the pseudo-steady-state as well as the transient or infinite-acting period prior to depletion. The instantaneous change occurs at t_{pss}, which can be estimated from Eq. 16–21. The dimensionless time for start of pseudo-steady-state-flow

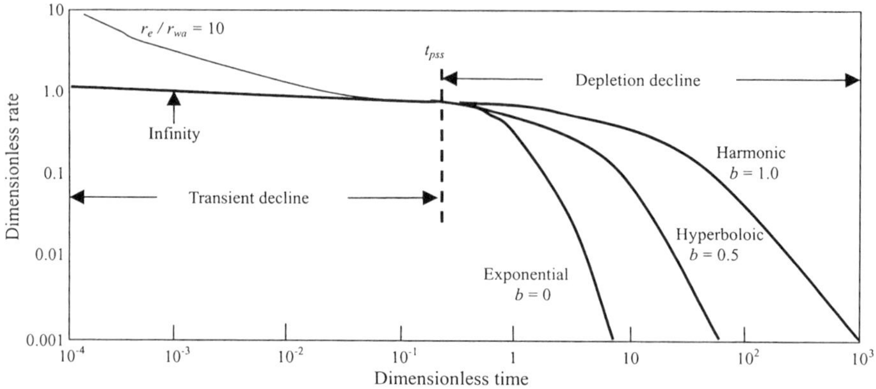

Figure 16–3. Complete rate solution plotted in terms of unit variables (after Fetkovich).[6]

is

$$t_{Dpss} = 0.1\,\pi\left(\frac{r_e}{r_{wa}}\right)^2 \tag{16–20}$$

Real time is

$$t_{pss} = \frac{\phi\,\mu_{gc}\,c_{ti}\left(\pi r_e^2\right)}{0.000264\,k}\cdot\left(t_{Dpss}\right)\text{ hr} \tag{16–21}$$

16.4 Characteristics and Classifications of Production Decline Curves

Development and the pertinent relationship for the three types of production decline curves on coordinate, semilog and, log-log graph paper can be found.[1,2] Decline curve analysis is a useful tool for reserves estimation and production forecasts. Decline curves also serve as diagnostic tools and may indicate the need for stimulation or remedial work. Production decline curves can be classified as follows:

- *Exponential decline.* Decline is constant ($b = 1$).
- *Hyperbolic decline.* Decline is proportional to a fractional power b of the production rate ($0 < b < 1$).
- *Harmonic decline.* Decline is proportional to production rate ($b = 1$). The use of type curves for decline curve analysis is demonstrated with examples.

Example 16–1 *Estimating Future Production Rates Using Decline Curve Method*

Using the production data from a gas field, estimate:

1. Future production down to a rate of 50 mmscfd
2. Decline rate
3. Effective monthly and annual decline rates
4. Remaining life

Production Data

Rate q_g (mmscfd)	Cumulative production G_p (mmscf)	Rate q_g (mmscfd)	Cumulative production G_p (mmscf)
200.00	10,000	130.00	190,000
210.00	20,000	133.00	220,000
190.00	30,000	115.00	230,000
193.00	60,000	110.00	240,000
170.00	100,000	115.00	250,000
155.00	150,000		

Solution A graph of q_g versus G_p is shown in Figure 16–4 on Cartesian coordinates. A straight line is obtained indicating constant percentage decline.

From graph $G_p = 398{,}000$ mmscf at $q_g = 50$ mmscfd. Future production = 398,000 − 250,000 = 144,000 mmscf.

The slope of the straight line gives the nominal decline rate. Picking two points on the straight line:

q_g(mmscfd)	G_p(mmscf)
250.00	0
100.00	276,000

The nominal daily decline rate is

$$D_d = \frac{215 - 100}{276{,}000} = 0.000417 \text{ day}^{-1}$$

The nominal monthly decline rate is

$$D_m = 30.4 \quad D_d = 0.0127 \text{ month}^{-1}$$

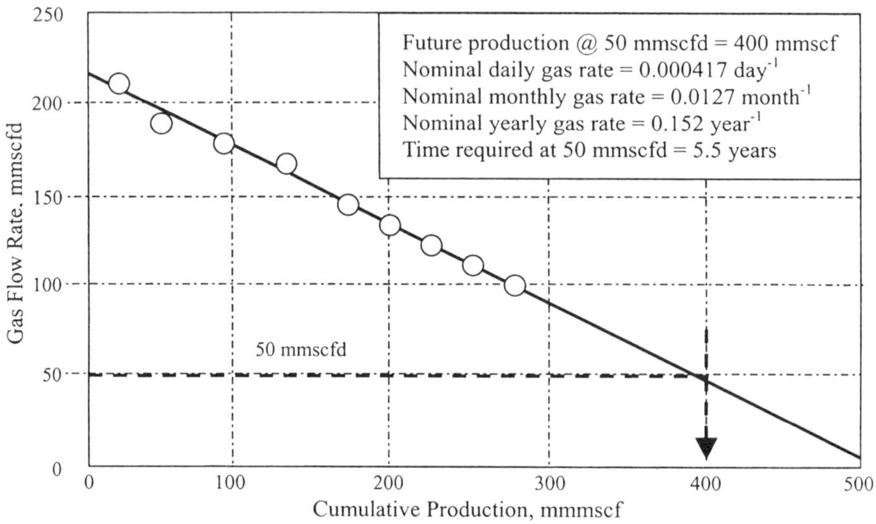

Figure 16–4. Gas flow rate versus cumulative production.

The nominal yearly decline rate is

$$D_a = 12 \quad D_m = 365 \quad D_d = 0.152 \text{ year}^{-1}$$

The effective monthly decline rate is

$$D'_m = 1 - e^{-D_m} = 1 - e^{-.0127} = 1.26\% \text{ per month}$$

The effective annual decline rate is

$$D'_a = 1 - e^{-12D_m} = 1 - e^{-12(.0127)} = 1 - e^{-0.1524} = 1 - 0.8586 = 14.14\%$$

The time to reach a production rate of 50.00 mmscfd or remaining life is obtained as follows:

At $t = 0$, $q_g = 115.00$ mmscfd:

$$t_a = \frac{In(q_1/q_n)}{-D_a} = \frac{In(50/115)}{-0.152} = 5.5 \text{ years}$$

Using Figure 16–4 requires that we calculate

$$\frac{G_g}{\overline{q}_o} = \frac{144{,}000}{(115)(365)} = 3.43$$

$$D'_a = 14.4\%$$

Table 16–1
Estimating Future Production History Using Decline Curve Method

Date	Production rate (mmscf/month)
1-1-95	1000
2-1-95	962
3-1-95	926
4-1-95	890
5-1-95	860
6-1-95	825
7-1-95	795
8-1-95	765
9-1-95	735
10-1-95	710
11-1-95	680
12-1-95	656
1-1-96	631

The remaining life from Figure 16–4 is 5.3 years, which is slightly less than the value calculated using decline curve equations.

Example 16–2 *Estimating Future Production History Using the Decline Curve Method*

Using the production history for the year 1995 given in Table 16–1, estimate the following:

1. Investigate the type of decline.
2. Calculate reserves at the end of 1985 to the economic limit of 25 mmscf/month.
3. When will the economic limit be reached?
4. Predict future rate and cumulative production until the economic limit is reached.

Solution The plot of q versus t on semilog graph paper (Figure 16–5) indicates a straight-line trend; therefore, constant percentage decline is assumed. The reserves at economic limit production rate can be calculated from

$$G_{pDa} = \frac{q_i - q_g}{D}$$

1. The nominal decline rate D can be determined from the rate–time equation or from the slope of the rate–time plot on semilog graph paper.

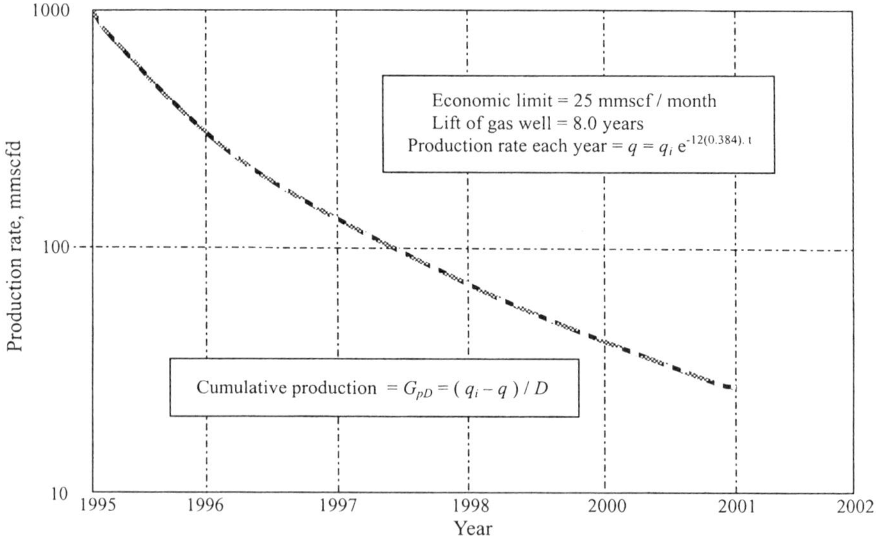

Figure 16–5. Plot of q versus time t on semilog paper.

Using two points on the straight line ($t = 0, q_i = 1000; t = 12, q_g = 631.0$ mmscfd/month):

$$-Dt = In\left(\frac{q}{q_i}\right) \quad \text{or} \quad D = \left(\frac{In(q/q_i)}{-t}\right)$$

$$= \left(\frac{In(631/1000)}{-12}\right) = 0.0384 \text{ per month}$$

2. Thus, reserves to be produced from 1995 to the economic limit of 25 mmscfd/month are

$$G_{pDa} = \frac{q_i - (E.L)}{D} = \frac{1000 - 25}{0.0384} = 25{,}391 \text{ mmscf}$$

3. The life of the gas well is given by

$$t = \frac{In[(E.L)/q_i]}{-D} = \frac{In(25/1000)}{-.0384} = 96 \text{ months or } 8.00 \text{ years}$$

4. The production rate each year is given by

$$q = q_i e^{-12Dt} = q_i e^{-12\,(.0384)t}$$

Table 16–2
Future Production Rate and Production

Year	q_i (mmscf/month)	q (mmscf/month)	Annual production G_{pD} (mmscf)	Cumulative production G_{pD} (mmscf)
1995	631	398	6068	6,068
1996	398	251	3828	9,896
1997	251	158	2422	12,318
1998	158	100	1510	13,828
1999	100	63	964	14,792
2000	63	40	599	15,391
2001	40	25	391	15,782

5. The cumulative production each year is given by

$$G_{pD} = \frac{q_i - q}{D} = \frac{q_i - q}{0.0384}$$

Note: q_i = rate at start of year and q = rate at end of year.

Table 16–2 shows the future production rate.

Characteristics and Decline Exponent b *Estimation Techniques*

Reference 5 has provided a graphical method to determine the value of b quickly at any time on decline, if the value of q_i/q is less than 100. These figures can also be used for extrapolating decline curves to some future point. The most important equations, which are applicable during depletion stage, are as follows. For hyperbolic decline ($b > 0$):

The rate–time relationship is

$$q = \frac{q_i}{(1 + bD_i t)^{1/b}} \tag{16–22}$$

where

q_i = gas rate at the beginning of depletion
D_i = initial decline rate, day^{-1}
t = time, days
b = decline exponent, dimensionless

The rate–cumulative relationship is

$$G_{pD} = \frac{q_i^b}{(1 - b)D_i}\left[q_i^{(1-b)} - q^{(1-b)}\right] \tag{16–23}$$

Cumulative production down to the economic limit is

$$G_{pDa} = \frac{q_i}{(1-b)D_i}\left[1 - (q_a/q_i)^{(b-1)}\right] \tag{16–24}$$

The remaining time on decline is

$$t_a = \frac{1-b}{b} \cdot \frac{G_{pDa}}{q_i}\left(\frac{q_i}{q_a}\right)^{1-b} \frac{(q_i/q_a)^b - 1}{(q_i/q_a)^{1-b} - 1} \tag{16–25}$$

It has been noted that for a bottom water-driven reservoir, $b = 0.5$.[1,6] The rate–time relationship becomes

$$q = \frac{q_i}{[1 + (D_i/2)t]^2} \tag{16–26}$$

and the rate–cumulative relationship,

$$G_{pD} = \frac{2\sqrt{q_i}}{D_i}\left(\sqrt{q_i} - \sqrt{q}\right) \tag{16–27}$$

The remaining life to abandonment for $b = 0.5$ is

$$t_a = \frac{2\left(\sqrt{q_i/q_a} - 1\right)}{D_i} \tag{16–28}$$

or

$$t_a = \frac{G_{pDa}}{q_i}\sqrt{\frac{q_i}{q_a}} \tag{16–29}$$

Gas wells usually produce at constant rates as prescribed by gas contracts. During this period the well pressure declines until it reaches a minimum level dictated by the line or compressor intake pressure. Therefore, the well will produce at a declining rate; the decline will be approximately hyperbolic with b equal to 0.5. The effective decline rate and nominal decline rate are related as follows:

$$D_i' = 1 - (1 + bD_i)^{-(1/b)} \tag{16–30}$$

$$D_i = \frac{1}{b}\left[(1 - D_i')^{-b} - 1\right] \tag{16–31}$$

Fetkovich[6] presents a refined decline curve analysis, which provides a tool with more diagnostic power. The use of type curves for decline curve analysis is demonstrated with examples. Some of the type curves are shown in Figures 16–6 through 16–8.

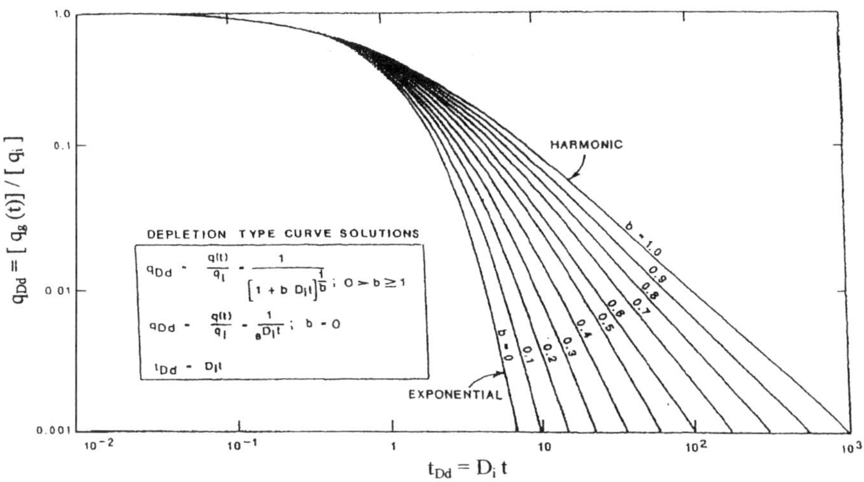

Figure 16–6. Type curves for Arps empirical rate time decline equations, unit solution ($D_i = 1$) (after Fetkovich).[6]

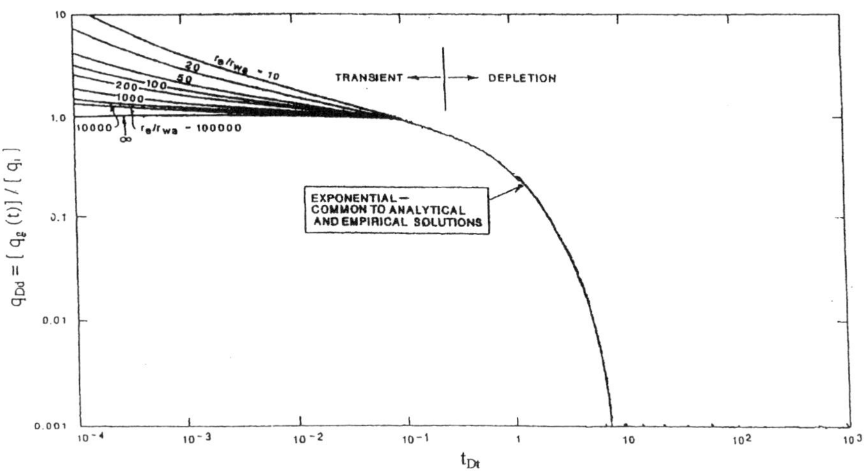

Figure 16–7. Dimensionless flow rate functions for plane radial system, infinite-acting and finite outer boundary, constant pressure at inner boundary (after Fetkovich).[6]

Table 16–3
Gas Well Production Data

Date	Time (year)	Daily production rate (mmscfd)	Cumulative production (mmmscf)
Jan. 1, 1987	0	10.0	0
July 1, 1987	0.5	8.45	1.70
Jan. 1, 1988	1.0	7.20	3.10
July 1, 1988	1.5	6.15	4.32
Jan. 1, 1989	2.0	5.40	5.38
July 1, 1989	2.5	4.75	6.30
Jan. 1, 1990	3.0	4.20	7.10
July 1, 1990	3.5	3.75	7.80
Jan. 1, 1991	4.0	3.40	8.50

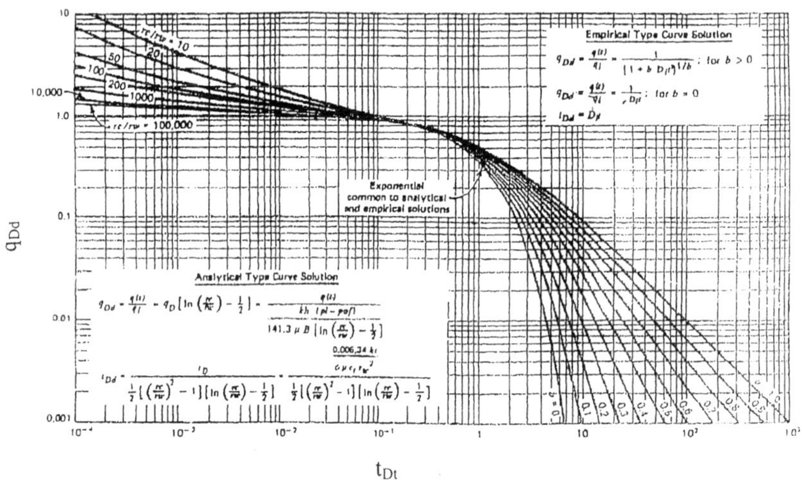

Figure 16–8. Composite of analytical and empirical type curves of Figures 16–6 and 16–7 (after Fetkovich).[6]

Example 16–3 *Estimating Future Production Down to an Economic Limit of 0.5 mmscfd*

The following production data are available for a gas well (see Table 16–3).

Solution First determine the values of b and D_i by using Table 16–4.

Figure 16–9 is a plot of $1/D$ versus t_{av}, which yields a straight line. From this plot we can find

$$\text{Slope } b = (5.0 - 2.60)/(4.0 - 0) = 0.60$$

Table 16–4

Time (years)	q (mmscfd)	$-\Delta q$ (mmscfd)	q_{av} (mmscfd)	$\frac{1}{D} = \frac{-q_{av}}{\Delta q/\Delta t}$	t_{av} (years)
0	10				
0.5	8.45	1.55	9.23	2.98	0.25
1.0	7.20	1.25	7.83	3.13	0.75
1.5	6.15	1.05	6.68	3.18	1.25
2.0	5.40	0.75	5.78	3.85	1.75
2.5	4.25	0.65	5.08	3.91	2.25
3.0	4.20	0.55	4.48	4.07	2.75
3.5	3.75	0.45	3.98	4.42	3.25
4.0	3.40	0.35	3.58	5.11	3.75

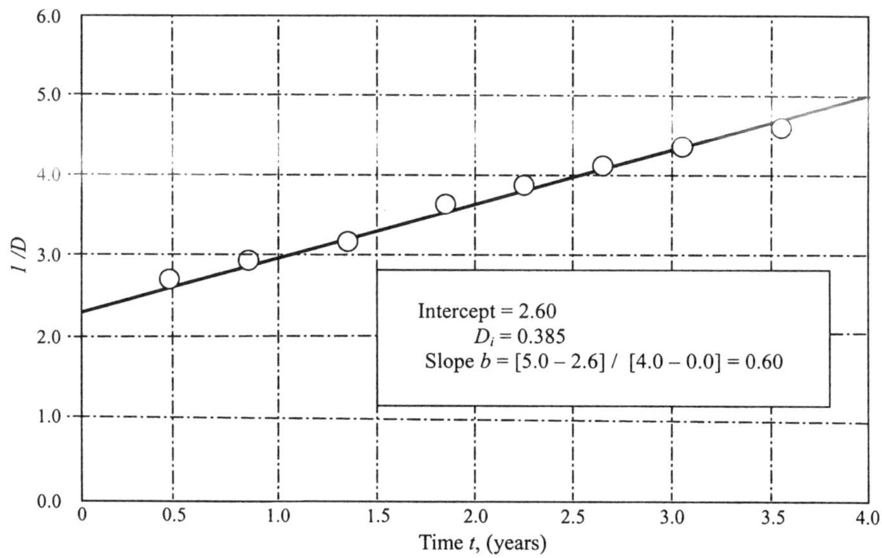

Figure 16–9. $1/D$ versus time.

Intercept = 2.60 and D_i = 0.385 per year.

$$G_{pD}/tq_i = \frac{8.500}{4(365)(10)} = 0.582$$

$$q_i/q = \frac{10}{3.40} = 2.94$$

To check the value of linearity, plot q^{-b} versus t (see Figure 16–10). The correct value of b will yield the best straight line. See Table 16–5.

Table 16–5
Graphical Determination of Value of b

t_{av}	q_{av}	$b = 0.40$ $q^{-0.40}$	$b = 0.60$ $q^{0.60}$ Correct	$b = 0.80$ $q^{-0.80}$
0.25	9.25	0.4107	0.2632	0.1687
0.75	7.83	0.4390	0.2909	0.1927
1.25	6.68	0.4678	0.3200	0.2189
1.75	5.78	0.5140	0.3490	0.2457
2.25	5.08	0.5220	0.3771	0.2725
2.75	4.48	0.5489	0.4067	0.3013
3.25	3.98	0.5755	0.4366	0.3312
3.75	3.58	0.6004	0.4652	0.3605

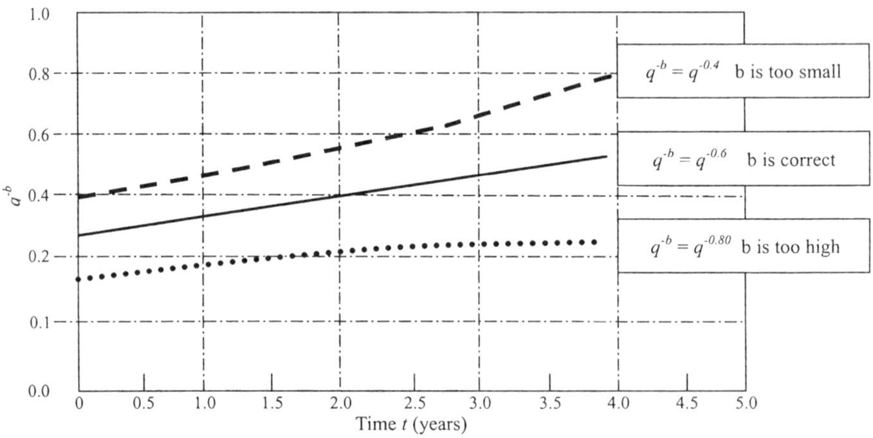

Figure 16–10. Graph of q^{-b} versus time.

Calculate the gas rate from Equation 16–22:

$$q = \frac{q_i}{(1 + bD_i t)^{1/b}} = \frac{10.00}{(1 + 0.6 \times 0.385 \times 5)^{1/0.6}} = 2.78 \text{ mmmscfd}$$

Calculate gas cumulative production from Eq. 16–23:

$$G_{pD} = \frac{q_i^b}{(1 - b)D_i}\left[q_i^{(1-b)} - q^{(1-b)}\right]$$

$$= \frac{3.36^{.6}}{(1 - 0.6) \times 0.385}\left[3.36^{1-.6} - 2.78^{1-.6}\right] = 10.019 \text{ mmmscfd}$$

Estimate cumulative production down to the economic limit from Eq. 16–24:

$$
\begin{aligned}
G_{pDa} &= \frac{q_i}{(1-b)D_i}\left[1-(q_a/q_i)^{(b-1)}\right] \\
&= \frac{3.36}{(1-0.6)\times 0.001}[1-(0.5/3.36)^{1-.6}] = 4480 \text{ mmscf}
\end{aligned}
$$

Estimate remaining time on decline from Eq. 16–25:

$$
\begin{aligned}
t_a &= \frac{1-b}{b}\cdot\frac{G_{pDa}}{q_i}\left(\frac{q_i}{q_a}\right)^{1-b}\frac{(q_i/q_a)^b-1}{(q_i/q_a)^{1-b}-1} \\
&= \frac{1-0.6}{0.6}\left(\frac{4480}{3.36}\right)\left(\frac{3.36}{0.5}\right)^{1-.6}\frac{(3.36/0.5)^{.6}-1}{(3.36/0.5)^{1-.6}-1} \\
&= 9.76 \text{ years}
\end{aligned}
$$

The calculated values are shown in Table 16–6. Results are plotted in Figures 16–11 and 16–12.

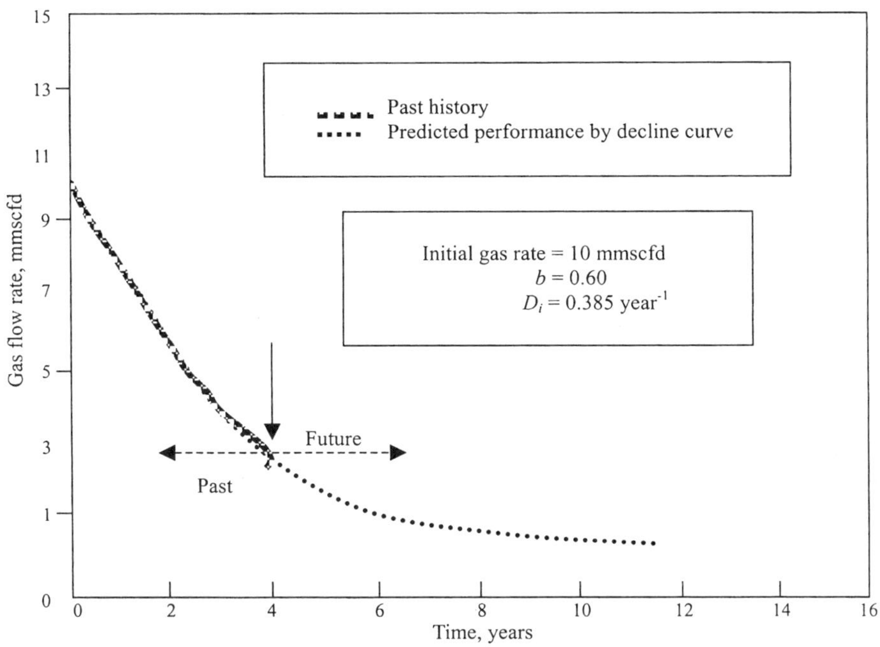

Figure 16–11. Final history match and prediction.

Table 16–6
Field History Match and Prediction of Future Gas Rate and Production

Time (years)	Actual gas rate (mmscfd)	Predicted gas rate (mmscfd)	Actual cumulative production (mmmscf)	Predicted cumulative production (mmmscfd)
0	10.00	10.00	0	0
0.5	8.45	8.33	1.70	1.762
1.0	7.20	7.07	3.10	3.238
1.5	6.15	6.09	4.38	4.498
2.0	5.40	5.31	5.40	5.592
2.5	4.75	4.68	6.35	6.548
3.0	4.20	4.16	7.19	7.398
3.5	3.75	3.72	7.80	8.167
4.0	3.40	3.36	8.57	8.839
		Future performance		**Future performance**
4.5		3.05		9.453
5.0		2.78		10.019
6.0		2.35		10.993
7.0		2.01		11.841
7.0		1.75		12.551
8.0		1.53		13.202
9.0		1.36		13.745
10.0		1.22		14.125
11.0		1.09		15.572
12.0		.99		15.984
14.0		.50		18.495

16.5 Horizontal Gas Reservoir Performance Using Production Type Curves

The following example illustrates the use of production type curves in horizontal gas wells.[7] These type curves represent the constant-pressure solution, i.e., the bottom-hole pressure remains constant during the well's life.

Example 16–4 *Forecasting Cumulative Production from the Production Type Curve for a Horizontal Gas Well*

Given the following data, calculate cumulative gas production of a horizontal gas reservoir: $T = 645°R$; $p = 3400$ psia; $p_{wf} = 2500$ psia; $T_c = 354.5°R$;

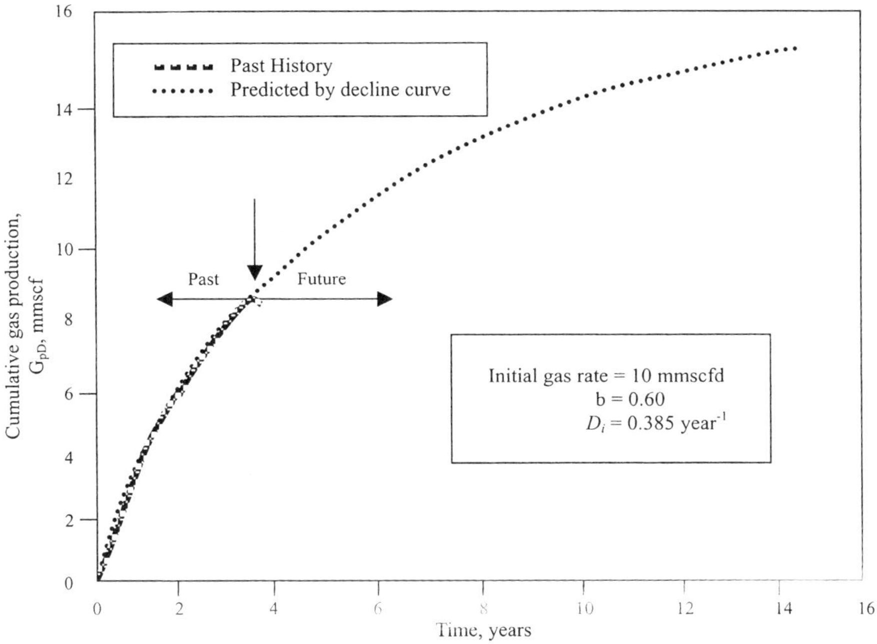

Figure 16–12. Final history match and prediction.

Table 16–7
Gas PVT Properties

Pressure (psia)	Z-factor	Gas viscosity (cP)	Real gas pseudopressure (mmpsia^2/cP)
3400	0.9250	0.01959	765.79
3000	0.9111	0.01847	619.26
2500	0.9011	0.01711	448.47
2000	0.9109	0.01585	296.92
1500	0.9115	0.01486	171.12
1000	0.9327	0.01401	76.99

$P_c = 664.5$ psia; $G = 0.80$; $\overline{\mu} = 0.01835$ cP; $r_w = 0.2385$ ft; $L = 2000$ ft; $h = 60$ ft; $k_h = 6$ mD; $k_v = 0.7$ mD; $\phi = 0.14$ fraction; $c_t = 0.0002941$ psi^{-1}. (Assume well spacing is 640 acres.)

Table 16–7 shows gas PVT properties.

Solution These three steps are illustrated below for time $t = 1$ day

1. Calculate t_D for different time steps using Eq. 3–118.
2. Determine G_{pD} from type curves.
3. Calculate q for different time steps from G_{pD} using Eq. 3–119.

These three steps are illustrated below for time $t = 1$ day.

1. Calculate t_D corresponding to $t = 1$ day from Eq. 3–118:

$$t_D = \frac{0.001055kt}{\phi\mu c_t L^2}$$

$$= \frac{0.001055 \times 6 \times 1}{0.14 \times 0.0835 \times 0.002941 \times 2000^2} = 0.0021$$

2. Determine Q_{pD} from type curves. To choose the correct type curve for calculating Q_{pD}, it is necessary to calculate L_D and r_{wD} from Eqs. 3–117 and 3–116, respectively.

$$\text{Dimensionless length } L_D = [L/(2h)]/(k_v/k_h)^{0.5} \quad \text{(Eq. 3–117)}$$

$$= 2000/(2 \times 60)/(0.7/6)^{0.5} = 48.8 \approx 50$$

For a square drainage area,

$$r_{wD} = \sqrt{4r_w^2/L^2} = \sqrt{\frac{4 \times 0.2385^2}{2000^2}} = 0.000239 \approx 2.5 \times 10^{-4}$$

(Eq. 3–116)

3. From Figure 3–26, the dimensionless production G_{pD} on curve $L_D = 50$ corresponding to $t_D = 0.0021$ and $r_{wD} = 2.5 \times 10^{-4}$ is $G_{pD} = 0.0178$. Then find G_P from Eq. 3–119:

$$G_{pD} = \frac{9.009 G_P T}{h\phi\mu c_t L^2[\psi(p_i) - \psi(p_{wf})]}$$

$$0.018 = \frac{9.009\, G_P \times 645}{60 \times 0.14 \times 0.01835 \times 0.0002941 \times 2000^2[(765.79 - 448.470) \times 10^6]}$$

Therefore, $G_P = 178.0$ mmscf. This procedure is repeated to calculate the cumulative gas production over time as shown in Table 16–8. Results are plotted in Figures 16–13 and 16–14.

Table 16–8
Summary of Results, Production Forecasting Horizontal Gas Well

Time (days)	Time (years)	Dimensionless time t_D	Dimensionless production function G_{pD}	Cumulative gas production G_p (scf×10^9)
0	0	—	—	0
1	0.0027	0.0021	0.018	0.178
30	0.0822	0.0630	0.148	1.465
60	0.1644	0.1260	0.195	1.931
90	0.2466	0.1890	0.220	2.178
120	0.3288	0.2520	0.270	2.673
180	0.4932	0.3780	0.350	3.465
250	0.6849	0.5250	0.480	4.753
300	0.8219	0.6300	0.560	5.545
365	1.0000	0.7665	0.800	7.921

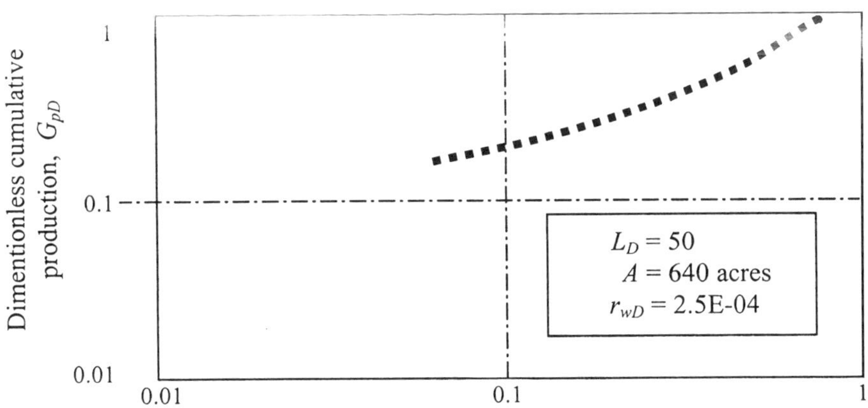

Figure 16–13. G_{pD} versus t_D.

16.6 Horizontal and Fractured Vertical Gas Reservoir Production Forecasting

Arps[1] and Fetkovich[6] have provided decline analysis equations. Equations 16–22 through 16–29 are applicable to predict gas rate and production with modification of Eq. 16–31 as

$$D_i = \frac{0.00634k/\left(\phi\mu_g c_t r_w^2\right)}{0.5\left[\left(\frac{r_e}{r_w}\right)^2 - 1\right]\left[In(r_e/r_w') - 0.5\right]}$$

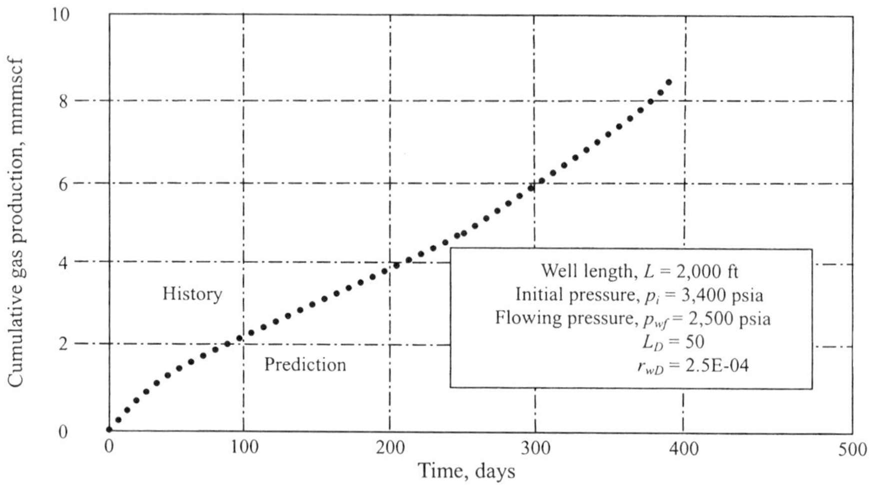

Figure 16–14. Predicted productions—Horizontal gas well.

For a horizontal gas well (Eq. 3–7):

$$r'_w = \frac{0.5 r_{eh} L/a}{\left[1+\sqrt{1-(0.5L/a)^2}\right][0.5\beta h/r_w]^{\beta h/L}}$$

where variables a and β are defined by Eqs. 3–8 and 3–8a, respectively.

For a fractured vertical gas well:

$$r'_w = x_f/2 \text{ for infinite-conductivity fracture}$$
$$= x_f/e \text{ for uniform flux fracture} (e = 2.718)$$

For hyperbolic decline ($b > 0$), the rate–time relationship is

$$q = \frac{q_i}{(1+bD_i t)^{1/b}} \tag{16–32}$$

where

q_i = gas rate at the beginning of depletion
D_i = initial decline rate, day^{-1}
t = time, days
b = decline exponent, dimensionless

The rate–cumulative relationship is

$$G_{pD} = \frac{q_i^b}{(1-b)D_i}\left[q_i^{(1-b)} - q^{(1-b)}\right] \tag{16–33}$$

Example 16–5 *Predicting Performance of Horizontal and Fractured Vertical Gas Well Using Decline Analysis Equations*

Given the following data, calculate performance of a horizontal gas well; $A = 100$ acres; $T = 110°F$; $p_i = 2000$ psia; $\mu = 0.0250$ cP; $z = 0.95$; $k_h = 0.1$; $k_v = 0.1$; $q_i = 25.60$ mmscfd; $r_w = 0.396$ ft; $x_f = 250$ ft; $h = 45$ ft; $c_t = 0.000375$ psi^{-1}; $\phi = 0.14$; well length $L = 2000$ ft; $b = 0.5$.

Solution First calculate the following parameters:

$$r_{ev} = \left(\frac{A \times 43{,}560}{\pi}\right)^{0.5} = \left(\frac{100 \times 43{,}560}{22/7}\right) = 1177.29 \text{ ft}$$

From Eq. 3–6,

$$r_{eh} = [(L/2 + r_{ev})\,(r_{ev})]^{0.5} = [(2000/2 + 1177.29)\,(1177.29)]^{0.5}$$
$$= 1601 \text{ ft}$$

From Equation 3–8,

$$a = 0.5L\left[0.5 + \sqrt{0.25 + (2r_{eh}/L)^4}\right]^{0.5}$$
$$= 0.5 \times 2000\left[0.5 + \sqrt{2(1601/2000)^4}\right]^{0.5} = 1763.95$$

For a horizontal well, r'_w from Eq. 3–7 is

$$r'_w = \frac{0.5r_{eh}L/a}{\left[1 + \sqrt{1 - (0.5L/a)^2}\right][0.5\beta h/r_w]^{\beta h/L}}$$
$$= \frac{0.5 \times 1601 \times 2000/1763.95}{\left[1 + \sqrt{1 - (0.5 \times 2000/1763.95)^2}\right]\left[\frac{0.5 \times 1 \times 45}{0.396}\right]^{(1 \times 45/2000)}}$$
$$= 327.22 \text{ ft}$$

For a fractured vertical well,

$$r'_w = x_f/2 = 250/2 = 125 \text{ ft}$$

Calculate D_i for a horizontal well from Eq. 16–29:

$$D_i = \frac{0.00634 \times 0.1/\,(0.14 \times 0.025 \times 0.000375 \times 0.396^2)}{0.5[\,(1177.29/0.396)^2 - 1]\,[\,In\,(1177.29/327.22) - 0.5]}$$

$$= 0.0008933 \text{ day}^{-1}$$
$$= 0.0268 \text{ month}^{-1}$$
$$= 0.3261 \text{ year}^{-1}$$

Calculate D_i for a fractured vertical well as:

$$D_i = \frac{0.00634 \times 0.1/(0.14 \times 0.025 \times 0.000375 \times 0.396^2)}{0.5[(1177.29/.396)^2 - 1]\,[In(1177.29/125) - 0.5]}$$

$$= 0.0004 \text{ day}^{-1}$$

$$= 0.0126 \text{ month}^{-1}$$

$$= 0.1460 \text{ year}^{-1}$$

Calculate gas production rate, $q(t)$, corresponding to $t = 0.25$ year from Eq. 16–22:

$$q(t) = \frac{q_i}{[1 + bD_i t]^{1/b}}$$

$$= \frac{25.60}{[1 + 0.5 \times 0.3261 \times t]^2}$$

$$= \frac{25.60}{(1 + 0.1632 \times 0.25)^2}$$

$$= 23.63 \text{ mmscfd}$$

Calculate cumulative gas production, G_P, corresponding to $q(t) = 23.63$ from Eq. 16–23:

$$G_P = \frac{q_i}{(1-b)D_i}\left[q_i^{1-b} - q^{1-b}\right] = \frac{25.60}{(1-0.5) \times 0.3262}[25.60^{1-0.5} - q^{.5}]$$

$$= 156.959\,[5.0596 - q^{0.5}] = 156.959\,[5.0596 - 4.8611]$$

$$= 31.161 \text{ mmscf}$$

The foregoing procedure is repeated to calculate the well rates and cumulative production over time as shown in Table 16–9. A field history match is shown using Arps[1] and Fetkovich[6] decline curves (see Figure 16–15). The example provides a 10-year forecast obtained from the type curve for $b = 0.5$. A horizontal well's gas rate and cumulative production versus time forecast are plotted in Figures 16–16 and 16–17.

16.7 Estimating In-Place Gas Reserves

Knowing the drainage radius r_e, in-place gas reserves drained by the well are calculated by a simple volumetric equation:

$$N(\text{well}) = \frac{43{,}560(\pi r_e^2)h\phi\,(1 - S_{wi} - S_{gr})}{\beta_{gi}} \text{ scf} \tag{16–34}$$

Table 16–9
Summary of Results, Forecasting Horizontal Gas Well Performance

Time (days)	Time (years)	Gas rate $q(t)$ (mmscfd)	Dimensionless time $t_{Dd} = D_i t$	Dimensionless rate $q_{Dd} = q(t)/q_i$	Cumulative gas production G_P (mmscf)
0	0	25.60	—	1.0000	0
36.50	0.10	24.78	0.0326	0.9680	12.816
91.25	0.25	23.63	0.0815	0.9230	31.161*
182.50	0.50	21.89	0.1630	0.8551	59.790
273.75	0.75	20.32	0.2445	0.7938	86.614
365	1.00	18.92	0.3261	0.7391	111.423
547.5	1.50	16.53	0.4891	0.6457	156.000
730	2.00	14.56	0.6521	0.5688	195.232
1095	3.00	11.54	0.9782	0.4508	260.951
1460	4.00	9.38	1.3042	0.3664	313.435
1825	5.00	7.77	1.6303	0.3035	356.631
2190	6.00	6.54	1.9563	0.2555	392.752
2555	7.00	5.58	2.2824	0.2180	423.381
2920	8.00	4.82	2.6084	0.1883	449.554
3285	9.00	4.20	2.9345	0.1641	472.479
3650	10.00	3.70	3.2605	0.1445	492.233

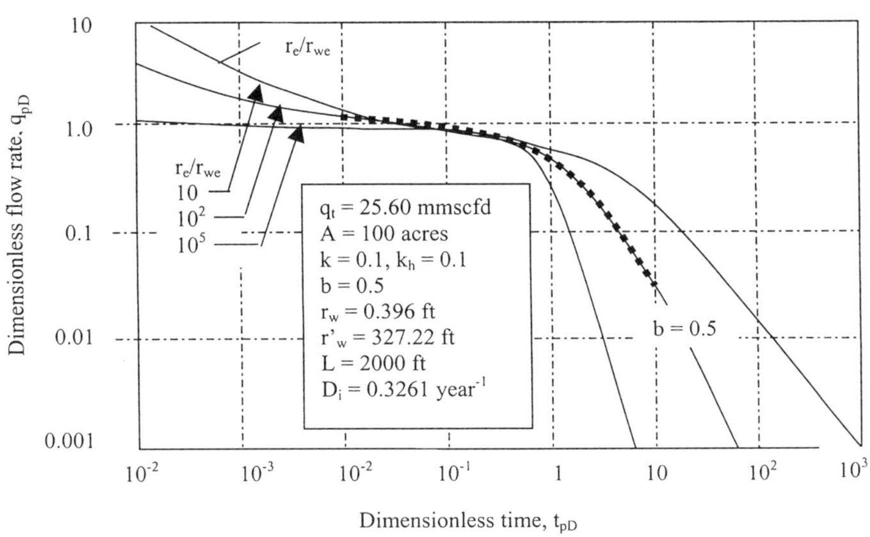

Figure 16–15. History match and prediction of horizontal gas reservoir (Arps and Fetkovich decline curve).

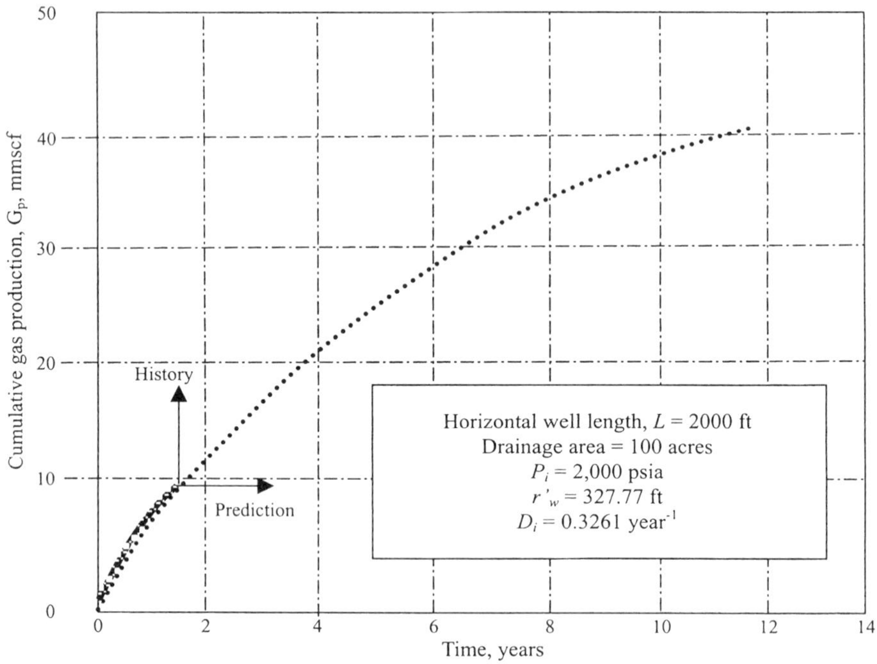

Figure 16–16. History match and prediction, G_p versus time.

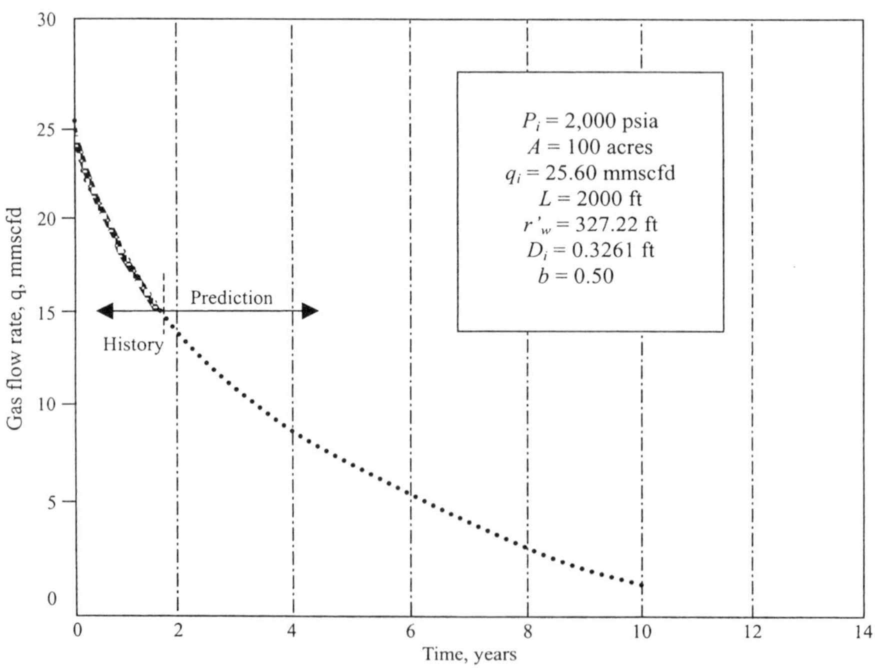

Figure 16–17. History match and gas flow rate versus time—Example 16–5.

where

$$\beta_{gi} = \frac{z\, P_{scT}}{P\, T_{sc}\, z_{sc}} = 0.0283 \frac{zT}{P_i} \text{ bbl/scf} \tag{16–35}$$

16.8 Determination of Economic Limit

The following example illustrates the method of estimating the economic limit.

Example 16–6 *Estimating Economic Limit*
Given: Natural gas price, mscf = \$3.0;
severance tax = 5%; additional (valoren) tax = 3%; royalty = 12.5%;
direct operating cost = \$2800/month. Estimate the economic limit.

Solution Net income/mscf = 7/8(1 − 0.05)(1 − 0.3)(\$3.0) = \$2.42/mscf
Economic limit = (\$2800)/(\$2.42) (30.4 days/month) = 38 gross mscfd
= 38 × 30.4 ≅ 1160 gross mscfd/month
= 38 × 30.4 × 12 ≅ 13,870 gross mscfd/year

References and Additional Reading

1. Arps, J. J., "Analysis of Decline Curve." *Trans. AIME* (1945) 160, 228.
2. Arps, J. J., "Estimation of Primary Oil and Gas Reserves," Chapter 37, *Petroleum Production Handbook* (T. C. Frick, ed.). McGraw-Hill, New York, 1962.
3. Seba, R. D., "Estimation of Economically Recoverable Oil from Decline Curve Analysis." Lecture Notes, Stanford University, 1976.
4. Shoemaker, R. P., "Graphical Method for Solving Decline Problems." *World Oil* (Oct. 1967) 123.
5. Gentry, R. W., "Decline Curve Analysis." *J. Petroleum Technol.* (Jan. 1972) 38.
6. Fetkovich, M. J., "Decline Curve Analysis Using Type Curves," *J. Petroleum Technol.* (June 1980) 1065–1077.
7. Aminian, K., and Ameri, S., "Predicting Horizontal Well Production Performance Using Type Curves," paper SPE 19342 presented at the SPE Eastern Regional Meeting, Morgantown, WV, Oct. 24–27, 1989, a: Oct 8–11, 1989.
8. Duda, J. R., "Type Curves for Predicting Production Performance from Horizontal Wells in Low Permeability Gas Reservoirs," paper SPE 18993, Richardson, TX.
9. Jacob, C. E., and Lohman, S. W., "No Steady Flow to a Well of Constant Drawdown in an Extensive Aquifer," *Trans. AGU* (Aug. 1952) 559–569.

Chapter 17

Overall Skin Effects and Impact on Gas Well Performance

17.1 Introduction

Various types of skin factor evaluation techniques and impact on gas well performance with examples are described in this chapter; more discussion including equations is given, and the accuracy, data requirements, and results of these techniques are presented in detail.

17.2 Rate-Dependent Skin Factor

In many gas wells, the observed flow rate is different from that calculated theoretically. As noted in Chapter 1, the concept of skin factor was developed to account for the deviation from the theoretical rate. For example a gas well located centrally in the drainage plane, during pseudo-steady-state flow, the gas flow rate can be calculated as follows.

In term of pressure-squared treatment:

$$q_g = \frac{0.0007027kh\left(\overline{p}_R^2 - p_{wf}^2\right)}{T\overline{z}\,\overline{\mu}\,[In(r_e/r_w) - 0.75 + s_t]} \tag{17–1}$$

In term of pseudopressure treatment:

$$q_g = \frac{0.0007027kh\lfloor\psi(\overline{p}_R) - \psi(p_{wf})\rfloor}{T[In(r_e/r_w) - 0.75 + s_t]} \tag{17–2}$$

All parameters in Eqs. 17–1 and 17–2 are in U.S. gas field units and s_t is total skin factor, which includes the effects of partial penetration, perforation density, mechanical skin damage due to drilling and completion, etc. Positive skin effects can be created by "mechanical" causes such as partial completion (i.e., a perforated height that is less than the reservoir height) and inadequate number of perforations (again, causing a distortion of flow lines), by phase changes (relative permeability reduction to the main fluid), by turbulence, and, of course, by damage to the natural reservoir permeability.

Negative skin effects would result in a flow enhancement. The pressure drop in the near-wellbore zone is less than it would have been from the normal, undisturbed, reservoir flow mechanisms. Such negative skin effects, or a negative contribution to the total skin effect, may be the result of matrix stimulation (near-wellbore permeability exceeds the natural value), hydraulic fracturing, or a highly inclined wellbore. Damaged wells have positive skin factors and stimulated wells have negative skin factors. The mechanical skin factor s_m represents the damage caused by drilling and completion fluids. Most of the drilled wells when completed show mechanical skin damage, and hence, the well is normally acidized before it is put on production. The mechanical damage, denoted as a positive skin factor, would cause a loss in well productivity if it is not removed. In addition to mechanical skin damage, many other parameters cause either loss or gain in well productivity. These parameters include:

1. Wells completed in part of the pay zone, i.e., partially penetrated wells
2. Near-wellbore turbulence
3. Perforated density
4. Slant wells

The damage in well productivity due to these parameters is described by assigning an equivalent skin factor called pseudoskin factors. The skin factor of a gas well is estimated by either drawdown or buildup tests. The skin factor calculated from well test analysis is usually a linear combination of a mechanical skin factor and various pseudoskin factors; for example, for a partially penetrating well,

$$s_t = s + Dq_{sc} \tag{17–3}$$

where

s_t = total skin factor
s = true skin factor
D = rate-dependent skin factor, mmscfd^{-1}
q_{sc} = gas flow rate, mmscfd

Dq_{sc} is an important term in Eq. 17–3. This is called turbulence skin, or rate-dependent skin factor.[2,3] The term accounts for additional pressure drop in the wellbore region due to high gas velocity. The total skin factor s_t or s can be obtained from well test analysis test performed at several different rates. They can be used to isolate the skin effect, s. A plot such as the one shown in Figure 17–1 of apparent skin factor s' versus flow rate q_{sc} suggests that s is the intercept and D is the slope. This is the proper manner for the field determination of D and the forecast of the impact of the rate-dependent skin on future well production.

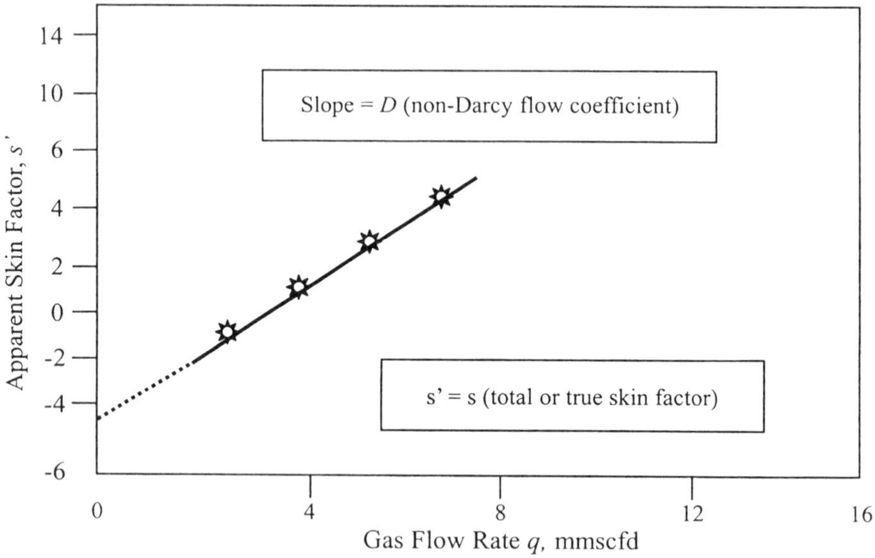

Figure 17–1. Field determination of skin factor and non-Darcy coefficient from multiple gas well tests.

D is also called turbulence coefficient (rate-dependent skin factor) and is defined as

$$s_{rds} = Dq_{sc} \tag{17–4}$$

where q_{sc} is gas flow rate in mscfd and *D* is turbulence coefficient (1/mscfd), given by[1–3]

$$D = \frac{2.226 \times 10^{-15} K_a \beta' \gamma_g h}{\mu_{pwf} r_w h_P^2} \tag{17–5}$$

where

$$\beta' = \frac{2.73 \times 10^{10}}{K_a^{1.1045}} \tag{17–6}$$

$$\beta' = \frac{2.33 \times 10^{10}}{K_a^{1.201}} \tag{17–7}$$

β' = high velocity flow coefficient, 1/ft; k_a = effective gas permeability near wellbore, mD; μ_{pwf} = gas viscosity at wellbore conditions, cP; γ_g = specific gas gravity (air = 1.000); r_w = wellbore radius, ft; and h_P = thickness of perforated interval, ft.

Note that in Eq. 17–5 μ_{pwf} is a function of pressure, thus the turbulence coefficient D is pressure dependent. Equations 17–6 and 17–7 for turbulent coefficient β' are from Refs. 14 and 1, respectively. References 2 and 3 include detailed discussions on the effect of turbulent flow in porous media. Chapter 3, 5, and 6 include methods to estimate pseudoskin factors due to horizontal and fractured vertical wells.

17.3 Skin Factor Due to Partial Penetration

Many gas wells are completed in a part of the pay zone. This is normally referred to a partially penetrated well. The following correlations[4–7] are available to calculate skin factors due to a partial penetration.

Papatzacos's correlation[4] showed that for a single-layer, infinite reservoir the skin factor, s_P, for partial penetration can be determined as follows.

Well producing from the top or bottom of the formation (see Figure 17–2a):

$$s_P = \left(\frac{1}{b'} - 1\right) In\left(\frac{\pi \frac{h}{r_w}\sqrt{\frac{k_h}{k_v}}}{2}\right) + \frac{1}{b'} In\left[\frac{b'}{z + b'}\left(\frac{A-1}{B-1}\right)^{0.5}\right] \quad (17\text{–}8)$$

Well producing only from the central section (see Figure 17–2b):

$$s_P = \left(\frac{1}{b'} - 1\right) In\left(\frac{\pi \frac{h}{2r_w}\sqrt{\frac{k_h}{k_v}}}{2}\right) + \frac{1}{b'} In\left[\frac{b'}{z + b'}\left(\frac{A-1}{B-1}\right)^{0.5}\right] \quad (17\text{–}9)$$

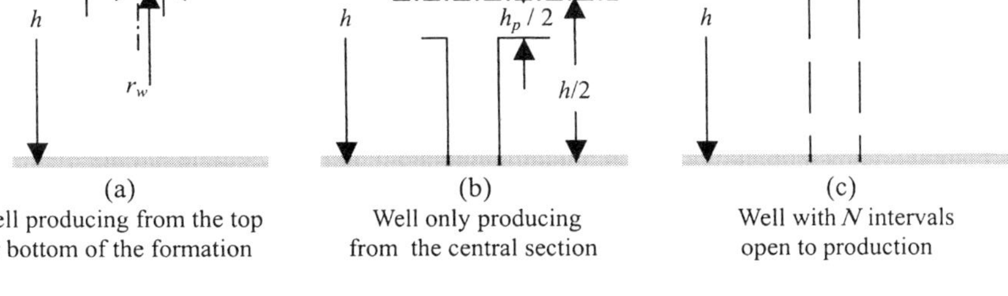

Figure 17–2. Effect of restricted fluid entry on well productivity.

Well with N intervals open to production (see Figure 17–2c):

$$s_P = \left(\frac{1}{b'} - 1\right) In\left(\frac{\pi \frac{h}{2Nr_w}\sqrt{\frac{k_h}{k_v}}}{2}\right) + \frac{1}{b'} In\left[\frac{b'}{z + b'}\left(\frac{A-1}{B-1}\right)^{0.5}\right] \quad (17\text{–}10)$$

where

$$A = \frac{h}{h_1 + .25h_P} \quad (17\text{–}11)$$

and

$$B = \frac{h}{h_1 + .75h_P} \quad (17\text{–}12)$$

where

h_1 = distance from top of the reservoir to the top of the open interval, ft (see Figure 17–3)
h_2 = distance from bottom of the reservoir to the bottom of open interval, ft (see Figure 17–3)
h_P = perforated interval, ft
$b' = \frac{h_P}{h}$, penetration ratio

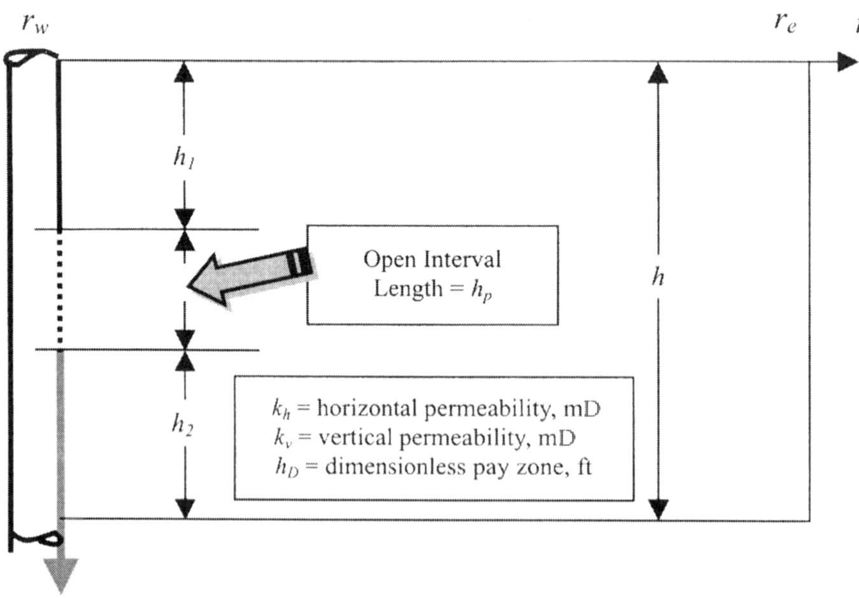

Figure 17–3. A schematic view of a restricted-entry well.

h_D = dimensionless pay zone thickness, defined in Figures 17–2a through 17–2c
h = total pay zone thickness, ft
k_H = horizontal permeability, mD
k_V = vertical permeability, mD

Odeh's correlation[5] presents an equation for calculating skin factor for an arbitrary location of perforated interval, h_P:

$$s_P = 1.35\left(\frac{1}{b'} - 1\right)^{0.825} [In(r_w h_D + 7) - 1.95$$
$$-In(r_{we})\{0.49 + 0.1\, In(r_w h_D\}] \tag{17–13}$$

where

$$r_{wc} = r_w \exp\left[0.2126\left(2.753 + \frac{z_m}{h}\right)\right] \tag{17–14}$$

for $0 < z_m/h < 0.5$, and

$$z_m = h_1 + (h_P/2) \tag{17–15}$$

If $h_1 = 0$, i.e., if the well perforates at the top of the formation, then use

$$r_{wc} = r_w$$

The Brons and Marting method[6] has suggested the following correlation to calculate an additional pressure drop due to restricted fluid entry: If $z_m/h >$ 0.5, replace z_m/h in Eq. 17–14 by $[1 - (z'_m/h)]$:

$$s_P = \left(\frac{1}{b'} - 1\right)[In(h_D) - F(b')] \tag{17–16}$$

where

$h_D = \dfrac{h}{r_w\sqrt{\frac{k_H}{k_v}}}$, dimensionless pay zone thickness
h_P = perforated interval, ft
h = total pay zone thickness, ft
k_H = horizontal permeability, mD
k_V = vertical permeability, mD
$b' = h_P/h$, penetration ratio

$$F(b') = 2.948 - 7.363b' + 11.45(b')^2 - 4.67(b')^3 \tag{17–16a}$$

Reference 6 has also presented a plot of s_P versus penetration ratio b' for several h_D values (see Figure 17–4). For each well configuration, the

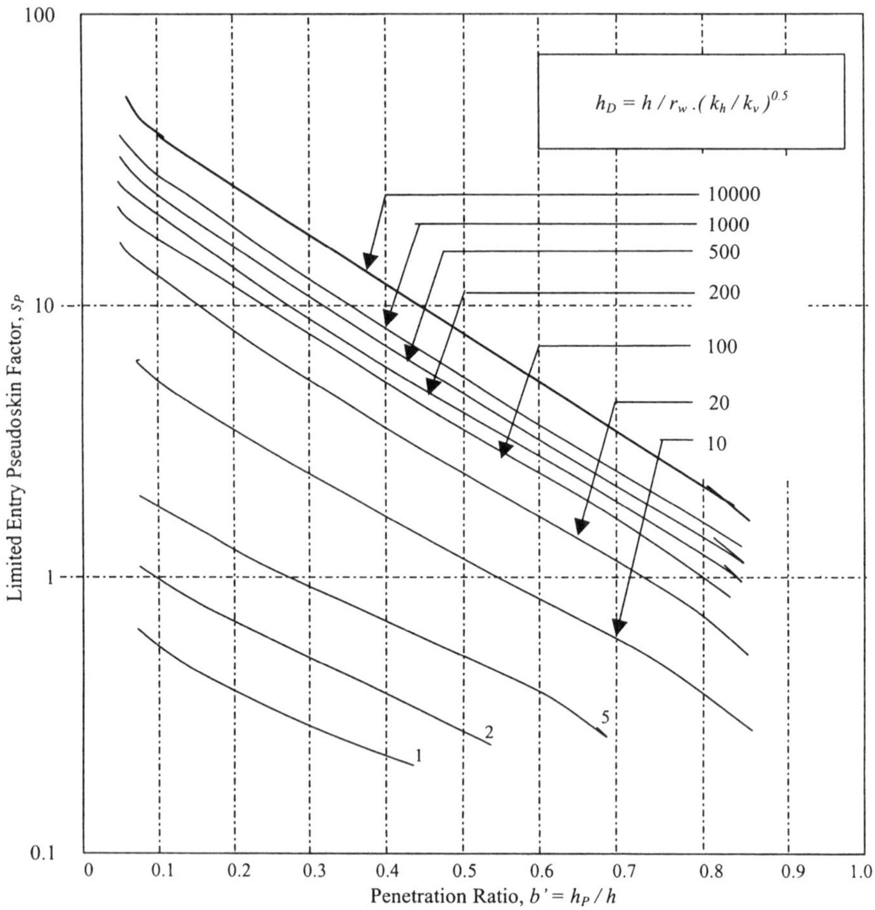

Figure 17–4. A correlation of pseudoskin factor due to partial penetration.[6]

dimensionless pay zone thickness h_D is defined differently (see Figures 17–2a through 17–2c).

The Yeh and Reynolds method[7] has presented an equation to calculate skin factor s_P caused by partial well completion:

$$s_P = \left(\frac{1 - b'}{b'}\right) In(h_{WD}) \tag{17–17}$$

where

$$h_{WD} = \frac{c'b'(1 - b')\, h_D}{\exp(c_1)} \tag{17–18}$$

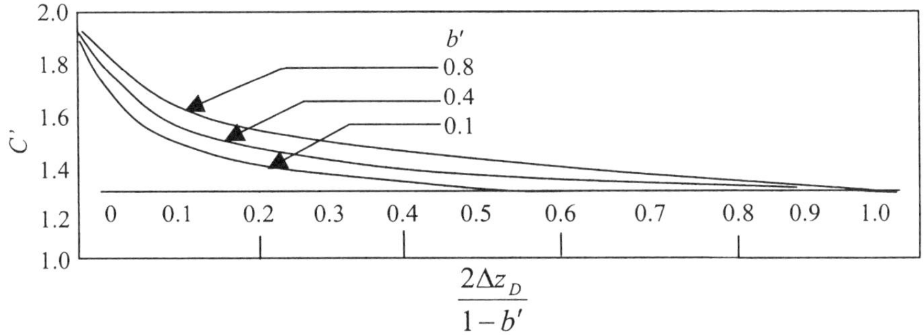

Figure 17–5. A graphical correlation of parameter C' to be used in Eq. 17–18 (after Yeh and Reynolds).[7]

The parameter c' accounts for the location of the open interval and is obtained from Figure 17–5, where

$$c_1 = 0.481 + 1.01\,(b') - 0.838\,(b')^2$$

$$\Delta z_D = \min[h_1/h,\ h_2/h_1] \tag{17–18a}$$

The definitions of h_1, h_2, and h are shown in Figure 17–3. If the open interval is at the top or bottom of the reservoir, then $c' = 2$. This correlation can also be used for a multilayer reservoir.

17.4 Skin Factor Due to Perforation

Karakas and Tariq[8] have presented a semianalytical solution for the estimation of the perforation skin effect. The majority of the wells are cemented and perforated. Perforations, depending upon their short density, offer flow restrictions to the wellbore, resulting in a reduced production rate. Loss of productivity due to perforations can also be expressed as a skin factor s_P and depends upon perforation geometry and perforation quality. The total perforation skin effect is then

$$s_{P_f} = s_H + s_V + s_{wb} \tag{17–19}$$

where

s_H = plane flow effect
s_V = vertical skin effect
s_{wb} = wellbore effects

Calculation of s_H:

$$s_H = In\left(\frac{r_w}{r'_w(\theta)}\right) \tag{17–20}$$

where $r'_w(\theta)$ is the effective wellbore radius and is a function of the phasing angle θ:

$$r'_w(\theta) = \{a_\theta(r_w + L_{Perf})\} \quad \text{for} \quad \theta \neq 0 \tag{17–20a}$$

$$r'_w(\theta) = \left\{a_\theta\left(r_w + \frac{L_{Perf}}{4}\right)\right\} \quad \text{for} \quad \theta = 0 \tag{17–20b}$$

Calculation of s_V:

$$s_V = 10^a h_D^{b-1} r_D^b \tag{17–21}$$

where h_D and r_D are defined by

$$h_D = \frac{h_P}{L_P}\sqrt{\frac{k_H}{k_V}} \tag{17–22}$$

$$r_D = \frac{r_P}{2h_P}\left(1 + \sqrt{\frac{k_V}{k_H}}\right) \tag{17–23}$$

$$a = a_1 \log r_D + a_2 \tag{17–24}$$

$$b = b_1 r_D + b_2 \tag{17–25}$$

The constants a_1, a_2, b_1, and b_2 are also functions of the perforation phasing and can be obtained from Table 17–1.

Table 17–1
Constants for Perforation Skin Effect Calculation[8]

Perforation phasing	a	a_1	a_2	b_1	b_2	c_1	c_2
0° (360°)	0.250	−2.091	0.0453	5.1313	1.8672	1.6E-01	2.675
180°	0.500	−2.025	0.0943	3.0373	1.8115	2.6E-02	4.532
120°	0.648	−2.018	0.0634	1.6136	1.7770	6.6E-03	5.320
90°	0.726	−1.905	0.1038	1.5674	1.6935	1.9E-03	6.155
60°	0.813	−1.898	0.1023	1.3654	1.6490	3.0E-04	7.509
45°	0.860	−1.788	0.2398	1.1915	1.6392	4.6E-05	8.791

Calculation of s_{Wb}:

$$s_{Wb} = c_1 \, e^{c_2 r_{WD}} \tag{17–26}$$

where

$$r_{WD} = \frac{r_w}{L_P + r_w} \tag{17–27}$$

The constants c_1 and c_2 also can be obtained from Table 17–1.

Example 17–1 *Calculating Total Perforation Skin Factor*

Given: $r_w = 0.39$ ft; SPF $= 4$; $r_P = 0.25$ inches (0.0208 ft); $L_P =$ 8 inches (0.667 ft); and $\theta = 168$ and 0. Calculate the perforation skin effect if $k_H/k_V = 10$ and 1.

Solution From Eq. 17–20a and Table 17–1 ($\theta = 120°$),

$$r'_w(\theta) = (0.648)(0.39 + 0.667) = 0.685$$

Then, from Eq. 17–20,

$$s_H = In\left(\frac{0.39}{0.685}\right) = -0.563$$

From Eq. 17–22 and remembering that

$$h_P = 1/SPF = 1/4 = 0.25$$

$$h_D = \frac{0.25}{0.667}\sqrt{10} = 1.185$$

and from Eq. 17–23:

$$r_D = \frac{0.0208}{2 \times .25}(1 + \sqrt{0.1}) = 0.0548$$

Using Eqs. 17–24 and 17–25 and the constants in Table 17–1,

$$a = -2.018 \log(0.0548) + 0.0634 = 2.4817 \quad \text{and}$$
$$b = 1.6136(0.0548) + 1.777 = 1.8654$$

From Eq. 17–21:

$$\begin{aligned} s_V &= 10^{2.4817} 1.185^{(1.8654-1)} 0.0548^{1.8654} \\ &= (303.18)(1.1582)(0.0044) = 1.5589 \end{aligned}$$

Finally, from Eq. 17–27:

$$r_{WD} = \frac{0.39}{0.667 + 0.39} = 0.369$$

and with the constants in Table 17–1 and Eq. 17–26:

$$s_{Wb} = 6.6 \times 10^{-3} e^{(5.320)(0.369)} = 0.047$$

Then total perforation skin effect is

$$s_{Pf} = -0.563 + 1.5589 + 0.047 = 1.043$$

For $\theta = 120^\circ$ and $k_H/k_V = 1$, s_H and s_{Wb} do not change; s_V, though, is only 0.3443, leading to

$$s_{Pf} = -0.563 + 0.3443 + 0.047 = -0.1717$$

reflecting the beneficial effects of good vertical permeability even with relatively unfavorable perforation density (SPF = 4).

17.5 Skin Factor from Partial Completion and Slant

Cinco-Ley *et al.*[10] presented tables of these skin effects for various combinations of partial completion, completion elevation, and well deviation. Figure 17–6 shows the relevant variables.

h_w = perforated height, ft
z_w = elevation of the perforation midpoint from the base of the reservoir
h = reservoir height, ft
θ = angle of well deviation
r_w = well radius, ft
$h_D = h/r_w$ = dimensionless thickness
z_w/h = elevation ratio
h_w/h = completion ratio
$s_{c+\theta}$ = composite skin effect

Tables 17–2 through 17–5 give the results for reservoir dimensionless thickness $h_D (= h/r_w)$ equal to 100 and 1000, respectively. Relevant ratios are z_w/h (elevation ratio) and h_w/h (completion ratio). The composite skin effect, $s_{c+\theta}$, and the individual parts, s_c and s_θ, are listed.

Example 17–2 *Calculating Composite Skin Effect for a Slanting Well*

Given: Well radius $r_w = 0.39$ ft; reservoir height = 40 ft; perforated completed height $h_w = 10$ ft; and midpoint elevation $z_m = 35$ ft. Calculate the skin effect due to partial completion for a vertical gas well (if $\theta = 0^\circ$ and $\theta = 60^\circ$ slant).

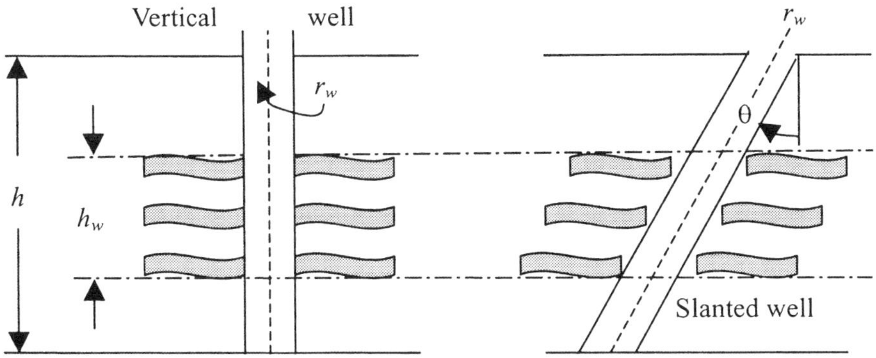

Figure 17–6. Inclined partially completed and skewed well configuration (after Cinco-Ley *et al.*).[10]

Solution Calculate the following parameters:

Dimensionless reservoir thickness $= h/r_w = 40/.39 = 102.56 \cong 100$

Elevation ratio $= z_m/h = 35/40 = 0.875$ and

Completion ratio $= h_w/h = 10/40 = 0.25$

From Table 17–3 for $\theta = 0$ slant, $s_{C+\theta} = s_C + s_\theta = 8.641 + 0 = +8.64$

From Table 17–3 for $\theta = 60°$ slant, $s_{C+\theta} = s_C + s_\theta = 8.641 - 4.924 = 3.717$

17.6 Skin Factor Due to Reduced Crushed-Zone Permeability

The effect of a crushed zone can be expressed as a mechanical skin factor[1,9] and is given by the following equations known as "McLeod's equations" for steady-state skin factor due to reduced crushed-zone permeability:

If perforated overbalanced,

$$s_c = \left(10k - \frac{k}{k_d}\right)\frac{12h_P}{NL_P} In\left(\frac{r_P + 0.5}{r_P}\right) \tag{17–28}$$

If perforated underbalanced,

$$s_c = \left(2.5k - \frac{k}{k_d}\right)\frac{12h_P}{NL_P} In\left(\frac{r_P + 0.5}{r_P}\right) \tag{17–29}$$

where k_{dp} = crushed-zone permeability, mD; k = formation permeability, mD; N = total number of perforations, k_d = damaged-zone permeability near

(text continued on page 680)

Table 17–2
Skins from Partially Penetrating Slanted Wells (h_D = 100)[a]

θ	h_D	z_w/h	h_w/h	$s_{\theta+r}$	s_r	s_θ
0	100	0.95	0.1	20.810	20.810	0
15				20.385	20.810	−0.425
30				18.948	20.810	−1.861
45				16.510	20.810	−4.299
60				12.662	20.810	−8.147
75				6.735	20.810	−14.074
0	100	0.80	0.1	15.809	15.809	0
15				15.449	15.809	−0.360
30				14.185	15.809	−1.623
45				12.127	15.809	−3.682
60				8.944	15.809	−6.864
75				4.214	15.809	−11.594
0	100	0.60	0.1	15.257	15.257	0
15				14.898	15.257	−0.359
30				13.636	15.257	−1.621
45				11.583	15.257	−3.674
60				8.415	15.257	−6.842
75				3.739	15.257	−11.517
0	100	0.50	0.1	15.213	15.213	0
15				14.854	15.213	−0.359
30				13.592	15.213	−1.620
45				11.540	15.213	−3.673
60				8.372	15.213	−6.841
75				3.699	15.213	−11.514
0	100	0.875	0.1	8.641	8.641	0
15				8.359	8.641	−0.282
30				7.487	8.641	−1.154
45				5.968	8.641	−2.673
60				3.717	8.641	−4.024
75				0.464	8.641	−8.177
0	100	0.75	0.1	7.002	7.002	0
15				6.750	7.002	−0.251
30				5.969	7.002	−1.032
45				4.613	7.002	−2.388
60				2.629	7.002	−4.372
75				−0.203	7.002	−7.206
0	100	0.60	0.1	6.658	6.658	0
15				6.403	6.658	−0.249
30				5.633	6.658	−1.024
45				4.290	6.658	−2.447
60				2.337	6.658	−4.320
75				0.418	6.658	−7.076

[a] From Cinco-Ley *et al.*[10]

Table 17–3
Skins from Partially Penetrating Slanted Wells (h_D = 100)[a]

θ	h_D	z_w/h	h_w/h	$s_{\theta+r}$	s_r	s_θ
0	100	0.75	0.50	3.067	3.067	0
15				2.878	3.067	−0.189
30				2.308	3.067	−0.759
45				1.338	3.067	−1.729
60				−0.082	3.067	−3.150
75				−2.119	3.067	−5.187
0	100	0.60	0.50	2.430	2.430	0
15				2.254	2.430	−0.176
30				1.730	2.430	−0.700
45				0.838	2.430	−1.592
60				−0.466	2.430	−2.807
75				−2.341	2.430	−4.772
0	100	0.50	0.50	2.369	2.369	0
15				2.149	2.369	−0.175
30				1.672	2.369	−0.697
45				0.785	2.369	−1.584
60				−0.509	2.369	−2.879
75				−2.368	2.369	−4.738
0	100	0.625	0.75	0.934	0.924	0
15				0.778	0.924	−0.145
30				0.337	0.924	−0.587
45				−0.411	0.924	−1.336
60				−1.507	0.924	−2.432
75				−3.099	0.924	−4.024
0	100	0.50	0.75	0.694	0.694	0
15				0.554	0.694	−0.139
30				0.134	0.694	−0.560
45				−0.581	0.694	−1.275
60				−1.632	0.694	−2.336
75				−3.170	0.694	−3.864
0	100	0.50	1.0	0	0	0
15				−0.128	0	−0.128
30				−0.517	0	−0.517
45				−1.178	0	−1.178
60				−2.149	0	−2.149
75				−3.577	0	−3.577
0	100	0.50	0.25	6.611	6.611	0
15				6.361	6.611	−0.249
30				5.587	6.611	−1.023
45				4.245	6.611	−2.365
60				2.295	6.611	−4.315
75				−0.451	6.611	−7.062

[a] From Cinco-Ley *et al.*[10]

Table 17–4
Skins from Partially Penetrating Slanted Wells ($h_D = 1000$)[a]

θ	h_D	z_w/h	h_w/h	$s_{\theta+r}$	s_r	s_θ
0	1000	0.95	0.1	41.521	41.521	0
15				40.343	41.521	−1.178
30				36.798	41.521	−4.722
45				30.844	41.521	−10.677
60				22.334	41.521	−19.187
75				10.755	41.521	−30.766
0	1000	0.80	0.1	35.840	35.840	0
15				34.744	35.840	−1.095
30				31.457	35.840	−4.382
45				25.973	35.840	−9.867
60				18.261	35.840	−17.599
75				8.003	35.840	−27.837
0	1000	0.60	0.1	35.290	35.290	0
15				34.195	35.290	−1.095
30				30.910	35.290	−4.380
45				25.430	35.290	−9.860
60				17.710	35.290	−17.580
75				7.522	35.290	−27.768
0	1000	0.50	0.1	35.246	35.246	0
15				34.151	35.246	−1.095
30				30.866	35.246	−4.380
45				25.386	35.246	−9.860
60				17.667	35.246	−17.579
75				7.481	35.246	−27.765
0	1000	0.875	0.25	15.733	15.733	0
15				15.136	15.733	−0.597
30				13.344	15.733	−2.389
45				10.366	15.733	−5.367
60				6.183	15.733	−9.550
75				0.632	15.733	−15.101
0	1000	0.75	0.25	14.040	14.040	0
15				13.471	14.040	−0.569
30				11.770	14.040	−2.270
45				8.959	14.040	−5.081
60				5.047	14.040	−8.993
75				−0.069	14.040	−14.109
0	1000	0.60	0.25	13.701	13.701	0
15				13.133	13.701	−0.568
30				11.437	13.701	−2.264
45				8.638	13.701	−5.063
60				4.753	13.701	−8.948
75				−0.288	13.701	−13.989

[a] From Cinco-Ley *et al.*[10]

Table 17–5
Skins from Partially Penetrating Slanted Wells (h_D = 1000)[a]

θ	h_D	z_w/h	h_w/h	$s_{\theta+r}$	s_r	s_θ
0	1000	0.50	0.25	13.655	13.655	0.000
15				13.087	13.655	−0.568
30				11.391	13.655	−2.264
45				8.593	13.655	−5.063
60				4.711	13.655	−8.944
75				−0.321	13.655	−13.976
0	1000	0.75	0.50	5.467	5.467	0
15				5.119	5.467	−0.348
30				4.080	5.467	−1.387
45				2.363	5.467	−3.104
60				−0.031	5.467	−5.498
75				−3.203	5.467	−8.670
0	1000	0.60	0.5	4.837	4.837	0
15				4.502	4.837	−0.335
30				3.503	4.837	−1.334
45				1.858	4.837	−2.979
60				−0.424	4.837	−5.261
75				−0.431	4.837	−8.268
0	1000	0.50	0.5	4.777	4.777	0
15				4.443	4.777	−0.334
30				3.446	4.777	−1.331
45				1.806	4.777	−2.971
60				−0.467	4.777	−5.244
75				−3.458	4.777	−8.235
0	1000	0.625	0.75	1.735	1.735	0
15				1.483	1.735	−0.252
30				0.731	1.735	−1.004
45				−0.512	1.735	−2.247
60				−2.253	1.735	−3.988
75				−4.595	1.735	−6.330
0	1000	0.50	0.75	1.508	1.508	0
15				1.262	1.508	−0.246
30				0.528	1.508	−0.980
45				−0.683	1.508	−2.191
60				−2.380	1.508	−3.888
75				−4.665	1.508	−6.173
0	1000	0.50	1.00	0.000	0	0
15				−0.206	0	−0.206
30				−0.824	0	−0.824
45				−1.850	0	−1.850
60				−3.298	0	−3.298
75				−5.282	0	−5.282

[a]From Cinco-Ley *et al.*[10]

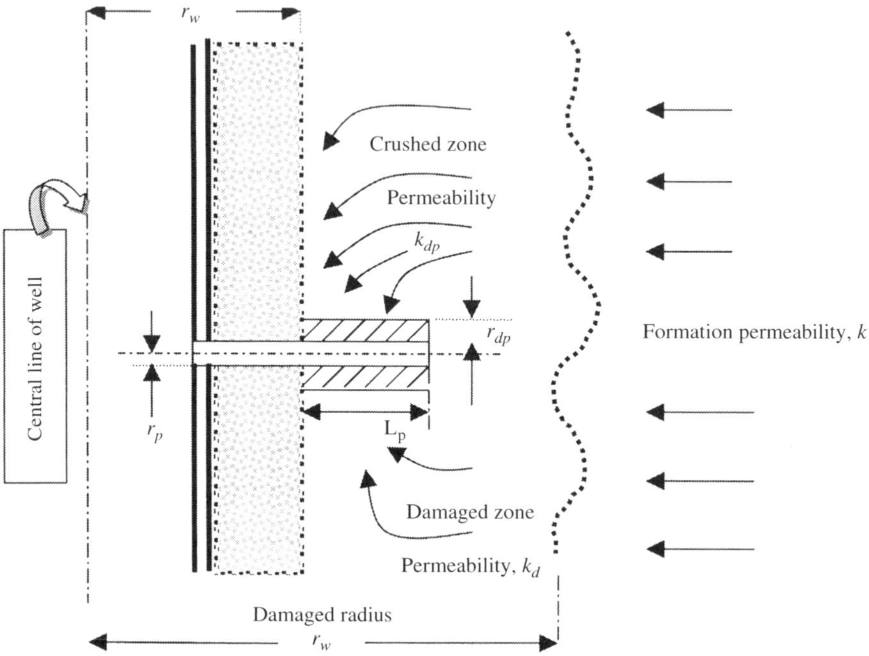

Figure 17–7. Geometry of perforation with crushed zone (after McLeod).[9]

(text continued from page 675)

the wellbore, mD; r_P = perforation radius, ft; $r_{dp} = r_P + 0.5$ = crushed-zone radius, ft; L_P = depth of perforation, ft and h_P = perforated interval, ft.

It is believed that these equations are difficult to use with certainty and thus are primarily useful for sensitivity analysis. Figure 17–7 shows the geometry of perforation with a crushed zone.

17.7 Slant Well Damage Skin Effect on Well Productivty

Figure 17–8 shows a schematic diagram of a slant well and Figure 17–9 presents pseudoskin factors of slant wells as a function of h_D for different slant angles θ. Reference 10 presented an equation to calculate the skin factor due to a slant well, which depends on the well geometry:

$$s_s = -(\theta'/41)^{2.06} - (\theta'/56)^{1.865} \log(h_D/100) \qquad (17\text{–}30)$$

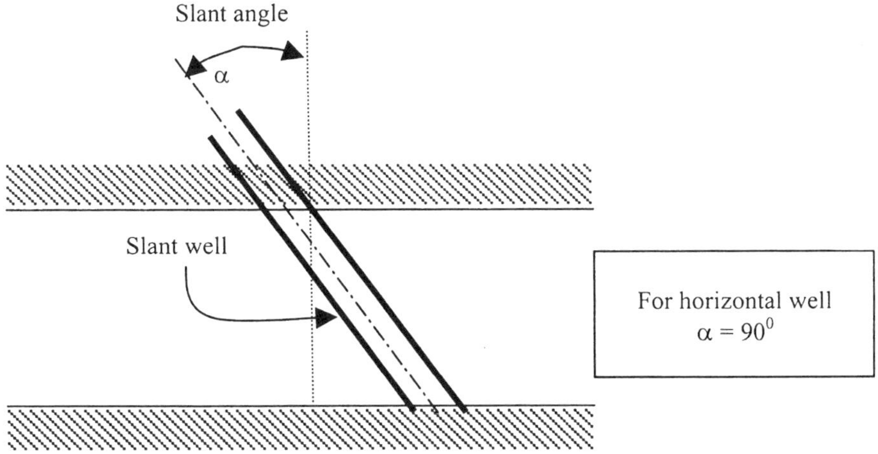

Figure 17–8. A schematic of a slant well (after Cinco *et al.*).[10]

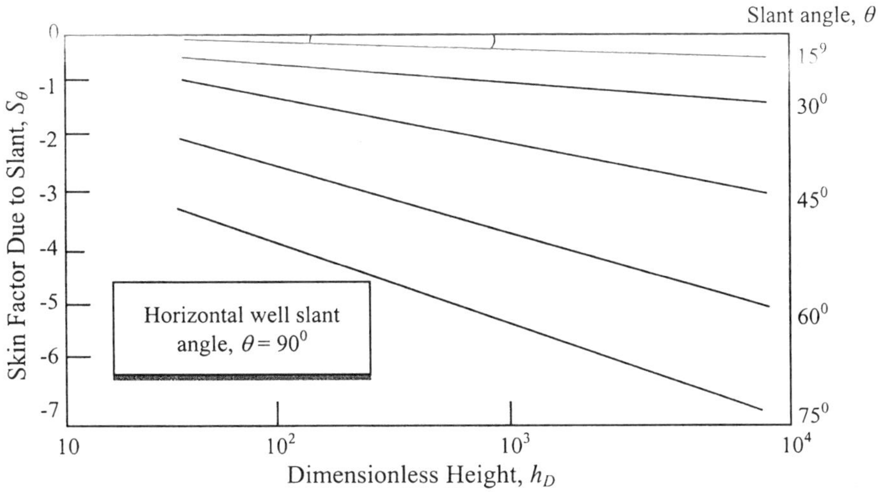

Figure 17–9. Slant well skin factor (after Schechter).[11]

for $t_D \geq t_{D1}$ and $\theta' \leq 75^\circ$, where

$$\theta' = \tan^{-1}\left[\sqrt{\frac{k_V}{k_H}}\tan\theta\right] \tag{17–31}$$

$$h_D = h/r_w\left(\frac{k_H}{k_V}\right)^{0.5} \tag{17–32}$$

$$t_D = 0.000264 k_H t/\left(\phi\mu c_t r_w^2\right) \tag{17–33}$$

and

$$t_{D1} = \max \begin{vmatrix} 70r_D^2 \\ (25/3)\ [r_D \cos\theta + (h_D/2)\tan\theta')]^2 \\ (25/3)\ [r_D \cos\theta - (h_D/2)\tan\theta')]^2 \end{vmatrix} \tag{17–34}$$

θ is slant angle. Note that in Eq. 17–33, t is in hours. The effective wellbore radius is given by

$$r'_w = r_w \exp^{(-S_s)} \tag{17–35}$$

and the productivity index of a slant well can be compared to an unstimulated vertical well by using the following relationship:

$$J_S/J_V = \frac{In(r_e/r_w)}{In(r_e/r'_w)} \tag{17–36}$$

$$r'_w = (L/4)[0.454\, Sin\,(360^\circ \times r_w/h)]^{h/L}$$

where

$$L = h/\cos\theta$$

It has been reported that the results of effective wellbore radius and skin factors calculated from the Cinco-Ley *et al.* equation, Eq. 17–30, and from the equation of Van Der Vlis *et al.*, Eq. 17–36, are in fairly good agreement with each other, and therefore either one of them could be used for calculation purposes.

Example 17–3 *Calculating Productivity Improvement of a Slant Well over a Vertical Well*

Given: Well spacing = 80 acres; $k_V/k_H = 1.0$; $r_w = 0.39$ ft; $h = 550$ ft; slant angles = 30°, 45°, 60°, and 70°. Calculate ratio J_S/J_V using the Cinco-Ley *et al.* and Van Der Vlis *et al.* correlations.

Solution The calculations are illustrated for a slant angle of 30°.

1. Cinco *et al.* method (Eqs. 17–30, 17–32, 17–35, and 17–36):

$$\theta = 30^\circ$$

$$h_D = \frac{h}{r_w}\sqrt{\frac{k_H}{k_V}} = \frac{550}{.39}\sqrt{1.0} = 1410.3$$

$$s_{s_s} = -(30/41)^{2.06} - (30/56)^{1.865}\log(1410.3/100) = -0.8843$$

$$r'_w = 0.39\exp^{-(-0.8843)} = 0.944$$

$$r_e = \sqrt{80 \times 43{,}560/\pi} = 1053\ \text{ft}$$

Table 17–6
Comparison of Cinco *et al.* and Van Der Vlis *et al.* Methods

Slant angle θ	Cinco *et al.* method			Van Der Vlis *et al.* method		
	L(ft)	S_S	J_S/J_V	r'_w	S_S	J_S/J_V
30	635	−0.884	1.125	0.737	−0.636	1.088
45	778	−1.795	1.332	2.504	−1.860	1.291
60	1100	−1.174	1.816	13.268	−3.527	1.780
70	1608	−4.315	2.606	53.262	−4.917	2.631

$$\frac{J_S}{J_V} = In(1053/0.39) \Big/ In(1053/0.944) = 1.125$$

2. Van Der Vlis *et al.* method (Eq. 17–36):

$$L = h/\cos\theta = 550/0.866 = 635 \text{ ft}$$

$$r'_w = (635/4)[0.454 \sin(360^\circ \times .39/550)]^{550/635}$$
$$= 158.75[0.454 \sin(0.2553)]^{.8661} = 0.737$$

$$s_s = -In\left(\frac{r'_w}{r_w}\right) = -In\left(\frac{0.737}{0.39}\right) = -0.636$$

$$\frac{J_S}{J_V} = In(1053/0.39) \Big/ In(1053/0.737) = 1.088$$

Calculated results using these two methods are shown in Table 17–6, which indicates that the calculated results using these two methods are in good agreement with each other. Therefore, either equation could be used for predicting productivity improvements.

A more detailed discussion on estimation of perforation depth, L_P, is provided in Refs. 1 and 14. Table 17–7 presents the penetration depths for different perforation diameters.[1]

Example 17–4 *Calculating Flowing Pressure Drops Due to Skin Effect*
Flow Path (Reservoir, Laminar Skin, and Turbulent Skin) for a Perforated Well

Given: $re = 1177$ ft; $k = 49.71$ mD; $r_w = 0.39$ ft; $h = 50$ ft; $ct = 0.0003329$ psi^{-1}; $z = 0.895$; $\mu_g = 0.0275$ cP; $\phi = 0.15$; $q_g = 40$ mmscfd; and from two-rate build-up test, $s = 6.13$, $D = 0.09850$ mmscfd^{-1}. Determine pressure drops due to reservoir, laminar, and skin factors.

Table 17–7
Perforating Gun Data[1]

Gun size (inch)	Tubing/casing (inch)	Perforation diameter, averge (in)	Perforation* average (inch)	Perforation* longest (inch)
Retrievable through tubing				
$1\frac{3}{8}$	$4\frac{1}{2}$	0.21	3.03	3.30
$1\frac{9}{16}$	$5\frac{1}{2}$	0.24	4.70	5.48
$1\frac{11}{16}$	$4\frac{1}{2}$ to $5\frac{1}{2}$ csg	0.24	4.80	5.50
2	$4\frac{1}{2}$ to $5\frac{1}{2}$ csg	0.32	6.50	8.15
$2\frac{1}{8}$	$2\frac{7}{8}$ tbg to $4\frac{1}{2}$ csg	0.33	7.20	8.15
$2\frac{5}{8}$	$4\frac{1}{2}$ csg	0.36	10.36	10.36
Expendable through tubing				
$1\frac{1}{8}$	$4\frac{1}{2}$ csg	0.19	3.15	3.15
$1\frac{1}{4}$	$2\frac{3}{8}$ tbg	0.30	3.91	3.91
$1\frac{3}{8}$		0.30	5.10	5.35
$1\frac{11}{16}$	$2\frac{7}{8}$ tbg $5\frac{1}{2}$ csg	0.34	6.00	8.19
$2\frac{1}{16}$	$5\frac{1}{2}$ to 7 csg	0.42	8.20	8.60
$2\frac{1}{8}$	$2\frac{7}{8}$ tbg to $5\frac{1}{2}$ csg	0.39	7.70	8.60
Retrievable casing guns				
$2\frac{1}{4}$	$4\frac{1}{2}$ csg	0.38	10.55	10.5
$2\frac{7}{8}$	$4\frac{1}{2}$ csg	0.37	10.63	10.6
$3\frac{1}{8}$	$4\frac{1}{2}$ csg	0.42	8.60	11.1
$3\frac{3}{8}$	$4\frac{1}{2}$ csg	0.36	9.10	10.8
$3\frac{5}{16}$	$4\frac{1}{2}$ & $5\frac{1}{2}$ csg	0.39	8.90	12.8
4	$5\frac{1}{2}$ to $9\frac{5}{8}$ csg	0.51	10.60	13.5
5	$6\frac{3}{4}$ to $9\frac{3}{8}$ csg	0.73	12.33	13.6

*Penetration length measured from casing ID.

Solution After rearranging Eqs. 17–1 and 17–3, the pressure drop is

$$\Delta p^2 = \overline{p}_R^2 - p_{wf}^2 = \frac{T\overline{z}\,\overline{\mu}q_g\lfloor In(r_e/r_w) - 0.75 + s + Dq_g\rfloor}{0.0007027kh}$$

$$\Delta p^2 = \frac{669 \times 0.895 \times 0.0275 \times 40.0\,[In(1177/0.39) - 0.75 + 6.13 + 0.0985 \times 40.0]}{0.0007027 \times 49.71 \times 50}$$

$$= 377.102\,[0.7262 + 6.13 + 3.94]$$

Therefore,

Pressure drop due to reservoir = $[377.102(0.7262)]^{1/2} = 17$ psi
Pressure drop due to laminar skin = $[377.102(6.13)]^{1/2} = 48$ psi
Pressure drop due to turbulent skin = $[377.102(3.94)]^{1/2} = 39$ psi
Total pressure drop due (reservoir + skin + turbulent skin) = 104 psi

17.8 Horizontal Well Damage Skin Effects

Frick and Economides[20] developed equations for the skin effect that reflect the damage around a horizontal well. Figure 17–10 describes the shape of damage along and normal to a horizontal well. The shape of damage depends on the permeability anisotropy index I_{ani}, is a measurement of vertical to horizontal permeability anisotropy, and is given by Eq. 17–37. Figure 17–11 shows simulated responses for three different values of I_{ani}.

$$I_{ani} = \sqrt{\frac{k_H}{k_V}} \tag{17–37}$$

The geometry of the shape of damage resulted in a skin effect analogous to Hawkins's formula for a vertical well:

$$s'_{eq} = \left(\frac{k}{k_s} - 1\right) In\left[\frac{1}{(I_{ani}+1)}\sqrt{4/3}\left(\frac{\alpha^2_{H,\max}}{r^2_w} + \frac{\alpha_{H,\max}}{r_w} + 1\right)\right] \tag{17–38}$$

Equation 17–38 assumes no damage at the end of the well. This skin effect can be added to the denominator of the horizontal gas well production rate Eqs. 17–39 and 17–40, but it must be multiplied by $I_{ani}h/L$, called the anisotropic scaled aspect ratio, as shown in Eqs. 17–39 and 17–40.

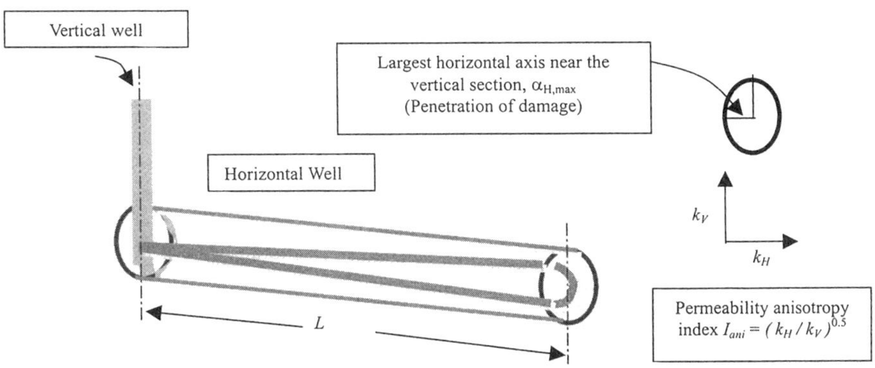

Figure 17–10. Distribution of damage along and normal to horizontal well.[20]

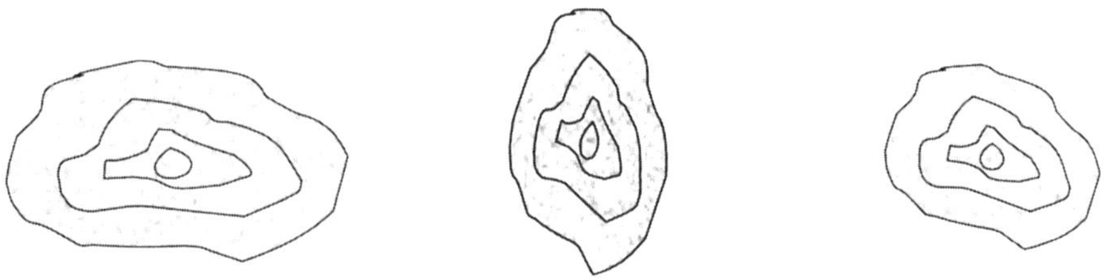

Figure 17–11. Effect of vertical to horizontal permeability anisotropy.[20]

Impact of Skin Effect on Horizontal Gas Well Performance

The impact of this skin effect on the gas production rate can be very large. Joshi[21] presented a horizontal well deliverability relationship that was augmented by Frick and Economides.[20] The relationships for gas flow in pressure squared and pseudopressure (mixed steady state in the horizontal plane and pseudo-steady state in the vertical plane) are

$$q_g = \frac{0.0007027\, k_H h \left(\overline{p}_R^2 - p_{wf}^2\right)}{\mu_g z T \left(In\left\{ \frac{[a+\sqrt{a^2-(L/2)^2}]}{L/2} \right\} + \left(\frac{I_{ani} h}{L}\right) \left\{ In \frac{I_{ani}}{[r_w(I_{ani}+1)]} + s'_{eq} \right\} \right)} \qquad (17\text{–}39)$$

$$q_g = \frac{0.0007027 k_H h (\psi(\overline{p}_R) - \psi(p_{wf}))}{T \left(In\left\{ \frac{[a+\sqrt{a^2-(L/2)^2}]}{L/2} \right\} + \left(\frac{I_{ani} h}{L}\right) \left\{ In \frac{I_{ani}}{[r_w(I_{ani}+1)]} + s'_{eq} \right\} \right)} \qquad (17\text{–}40)$$

where q_g is gas flow rate in Mscfd at 14.65 psia and 60°F and a is the large half-axis of the drainage ellipsoid formed by a horizontal well of length L. The equation for this ellipsoid is

$$a = \frac{L}{2}\left\{0.5 + \left[0.25 + \left(\frac{r_{eH}}{L/2}\right)^4\right]^{0.5}\right\}^{0.5} \quad \text{for} \quad \frac{L}{2} < 0.9 r_{eH} \qquad (17\text{–}41)$$

The productivity index ratio J_H/J_V in a specific reservoir may be large, assuming that an appropriate candidate is selected, the well is drilled in the optimum direction, and it is stimulated effectively. This productivity index ratio can be manifested by an increase in the gas production rate, a decrease in the pressure drawdown, or both. Therefore, horizontal wells can be excellent means of reservoir management where problems of water or gas coning or sand production are present.

Sandstone Reservoirs

Stimulation fluids will penetrate the pore space, eliminating foreign damage, and the posttreatment skin effect equation is

$$s'_{eq} = \frac{1}{2}\left(\frac{k}{k_s} - 1\right) In\left[\frac{\alpha_{sH,\max}}{r_w^2} + \frac{\alpha_{s,H,\max}}{r_w} + 1\right] + \frac{1}{2}\left(\frac{k}{k_i} - \frac{k}{k_s}\right) In\left[\frac{\alpha_{iH,\max}^2}{r_w^2} + \frac{\alpha_{iH,\max}}{r_w} + 1\right] - \left(\frac{k}{k_i} - 1\right) In[0.866\,(I_{ani} + 1)] \tag{17–42}$$

Limestone Reservoirs

The shape of the stimulated zone, affected by reaction kinetics and not by flow in the porous medium, is cylindrical. For the skin effect with elliptical damage but cylindrical stimulated zone, the equation is

$$s'_{eq} = \frac{1}{2}\left(\frac{k}{k_s} - 1\right) In\left[\frac{\alpha_{sH,\max}}{r_w^2} + \frac{\alpha_{s,H,\max}}{r_w} + 1\right] + \frac{1}{2}\left(\frac{k}{k_i} - \frac{k}{k_s}\right) In\left[\frac{r_{i,\max}^2}{r_w^2} + \frac{r_{i,\max}}{r_w} + 1\right] - \left(\frac{k}{k_i} - 1\right) In[0.866\,(I_{ani} + 1)] \tag{17–43}$$

where $r_{i,max}$ is the largest radius. The productivity index of a horizontal well can be compared to an unstimulated vertical well by using Eq. 17–39 or 17–40 and dividing by Eq. 3–2 or 3–4. The productivity index ratio equation is

$$\frac{J_H}{J_V} = \frac{\left[In\left(\frac{r_e}{r_w}\right) + s\right]}{In - \left[\frac{\left[a + \sqrt{a^2 - (L/2)^2}\right]}{L/2} + \left(\frac{I_{ani}h}{L}\right)\left\{In\frac{I_{ani}}{r_w(I_{ani}+1)} + s'_{eq}\right\}\right]} \tag{17–44}$$

Example 17–5 *Analyzing Horizontal Well Skin Effect and Impact on Gas Well Performance*

Use the data for gas well in Chapter 3, Examples 3–1 and 3–2. Calculate the following:

- Productivity index ratio J_H/J_V for various values of $\alpha_{i\,H,max}$
- Skin effect versus penetration of damage ($\alpha_{H,max}$) for different permeability impairment ratios such as $k/k_s = 20$, 10, 5, and 1

Also plot the equivalent skin factor versus k/k_s ratio for different values of $\alpha_{i\,H,max}$.

Solution Productivity index ratios can be calculated using Eqs. 17–39 and 3–2. The results are reported in Table 17–8. Calculate horizontal well equivalent skin effect from Eq. 17–42 for various values of permeability impairments (k/k_s). The calculated results are reported in Table 17–9.

Table 17–8
Productivity Index Ratio versus s'_{eq}

$\alpha_{s\,H\max}(ft)$	s'_{eq}	J_H/J_V
1.0	0.15	4.10
1.5	8.04	3.80
2.0	12.40	3.50
2.5	16.03	2.78
3.0	18.51	2.25
3.5	22.21	2.00
4.0	24.50	1.95
4.5	26.45	1.75
5.0	28.02	1.32

Table 17–9
Horizontal Gas Well Skin Effects on a Range of Penetration Damage and Permeability Impairments

Impairement ratio $\alpha_{s,H,max}, ft$	$k/k_s = 20$ s'_{eq}	$k/k_s = 10$ s'_{eq}	$k/k_s = 5$ s'_{eq}
1.0	0.15	0.1	0.05
1.5	8.38	4.13	2.18
2.0	15.25	8.15	3.85
2.5	19.21	10.33	4.48
3.0	23.15	12.50	5.10
3.5	24.20	13.90	5.38
4.0	25.26	15.30	5.65
4.5	26.64	16.78	5.84
5.0	28.02	18.25	6.0

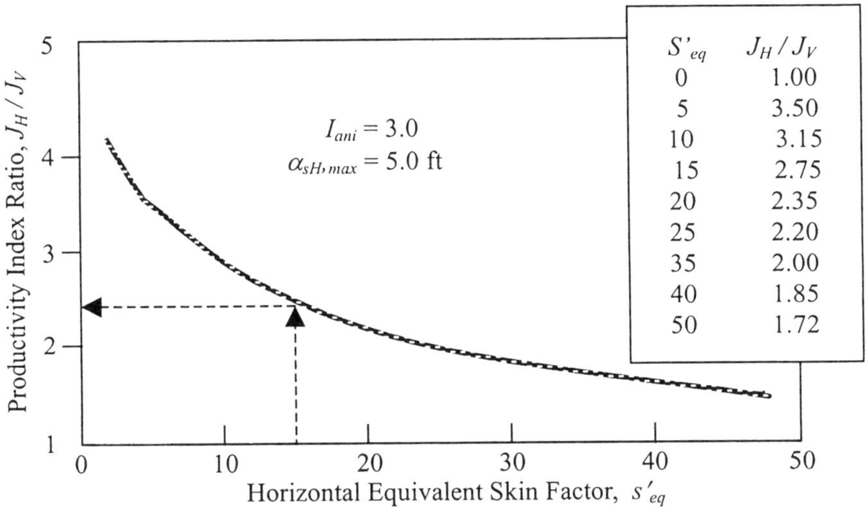

S'_{eq}	J_H / J_V
0	1.00
5	3.50
10	3.15
15	2.75
20	2.35
25	2.20
35	2.00
40	1.85
50	1.72

Figure 17–12. J_H / J_V versus s'_{eq}.

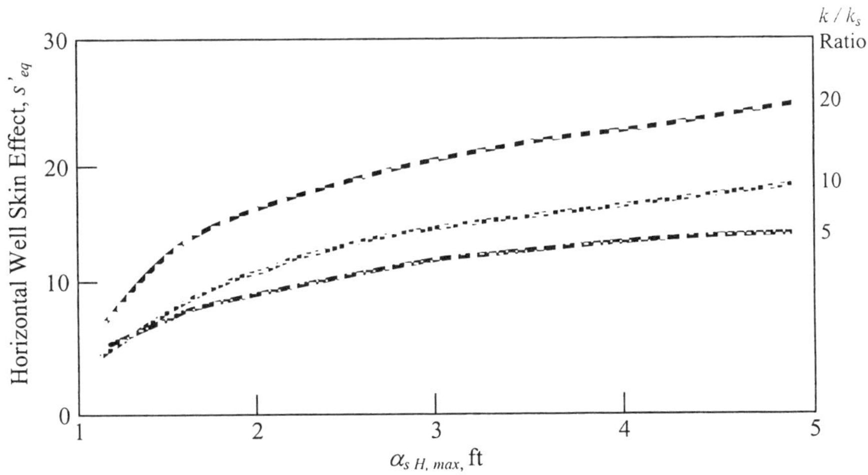

Figure 17–13. Horizontal well equivalent skin effect for range of penetration of damage and k/k_s as parameters.

Figure 17–12 shows a plot of productivity index ratio J_H / J_V versus s_{eq}. Horizontal gas well equivalent skin effects are significant, because they denote that the productivity index benefits (either increased production rate or decreased drawdown) may be affected by a horizontal well that is not removed. Figure 17–13 presents a plot of horizontal well equivalent skin effect versus

Table 17–10
Horizontal Well Equivalent Effects for Maximum Penetration Damage with k/k_s as Parameters

Permeability impairment ratio k/k_s	$\alpha_{sH,\,max}$, (ft) = 1	$\alpha_{sH,\,max}$, (ft) = 2	$\alpha_{sH,\,max}$ (ft) = 3	$\alpha_{sH,\,max}$ (ft) = 4	$\alpha_{sH,\,max}$ (ft) = 5
1	0.025	0.045	0.15	0.210	0.23
5	0.105	3.2	4.3	5.75	6.42
10	0.210	6.7	9.7	11.26	13.50
20	0.254	13.5	19.5	24.5	25.15

Table 17–11
Stimulated Horizontal Well and Associated Productivity Index Ratio for Various Values of Maximum Penetration of Damage

Maximum penetration of damaged $\alpha_{iH,max}$,(ft)	Horizontal well equivalent skin effect s'_{eq}	Productivity index ratio J_H/J_V
1	29.50	1.15
2	31.32	1.95
3	12.85	2.98
4	7.75	3.92
5	3.65	4.14

penetration damage $\alpha_{sH,max}$ with permeability impairment ratio k/k_s as a parameter. From this figure the equivalent skin effect can be calculated.

Table 17–10 is prepared by using Eq. 17–42 for various values of permeability impairment ratio of 5, 10, and 20 with maximum penetration damage $\alpha_{sH,max}$ as a parameter. Figure 17–14 is prepared using these data.

Figure 17–15 is a summary of the results of this example. The posttreatment skin effect is obtained with Eq. 17–42, assuming that the shape of stimulation imitates the shape of damage. This shape would require the appropriate distribution of the stimulation fluids. When $\alpha_{iH,max} = \alpha_{sH,max}$ then $s_{eq} = 0$. The corresponding productivity index ratio J_H/J_V can be used in an economic evaluation of the benefits versus costs of the treatment. Table 17–11 shows various values of maximum penetration of stimulation $\alpha_{iH,max}$; the original maximum penetration of damage $\alpha_{sH,max}$ is 5.

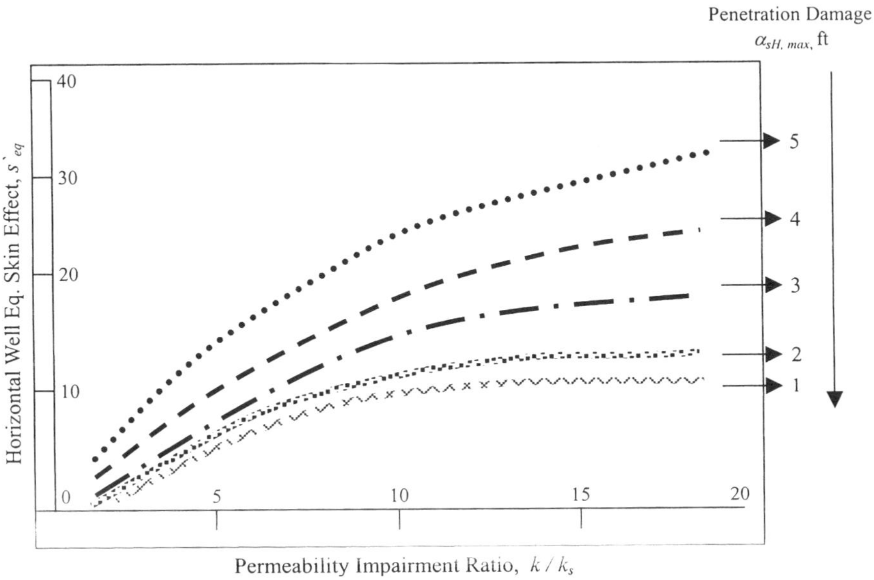

Figure 17–14. Horizontal well equivalent skin effect versus permeability impairment ratio for various values of penetration of damage—Sandstone reservoir.

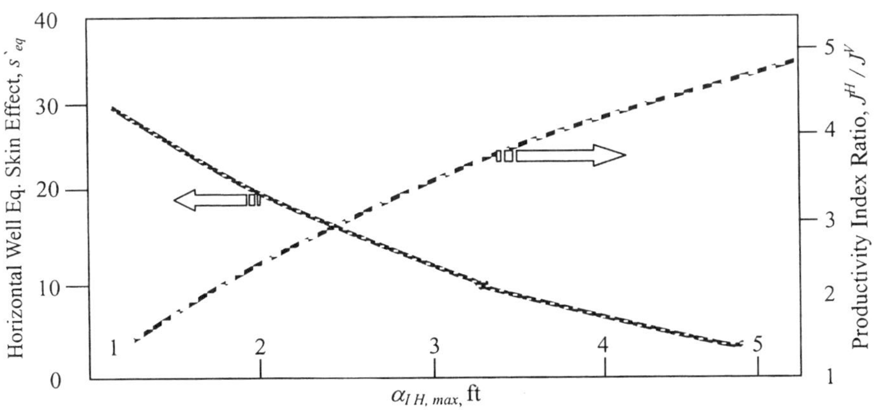

Figure 17–15. Stimulated horizontal well and associated productivity index ratio increase (over that of a vertical well).

References and Additional Reading

1. Brown, K. E., *The Technology of Artificial Lift Methods*, Vol. 4. Penn Well Publishing Co., Tulsa, OK, 1984.
2. Wittenberg, R. A., and Ramey, H. J., Jr., "Gas Well Testing with Turbulence," *J. Petroleum Technol.* (Jan. 1983) 31–39.
3. Ding, W., "Gas Well Test Analysis," M.S. Thesis, University of Tulsa, Tulsa, OK, 1986.
4. Papatzacos, P., "Approximate Partial-Penetration Pseudo-Skin for Infinite-Conductivity Wells," *SPE Reservoir Eng.* (May 1988) 227–234; *Trans. AIME* 283.
5. Odeh, A.S., "An Equation for Calculating Skin Factor Due to Restricted-Entry," *J. Petroleum Technol.* (June 1980) 964–965.
6. Brons, F., and Marting, V. E., "The Effect of Restricted Fluid Entry on Well Productivity," *J. Petroleum Technol.* (Feb. 1961) 172–174.
7. Yeh, N. S., and Reynolds, A. C., "Computation of the Pseudo-Skin Caused by a Restricted-Entry Well Completed in a Multilayer Reservoir," *SPE Formation Eval.* (June 1989) 253–263.
8. Karakas, M., and Tariq, S., "Semi-Analytical Production Model for Perforated Completion," SPE Paper 18247, 1988.
9. McLeod, H. O., Jr., "The Effect of Perforating Conditions on Well Performance," *J. Petroleum Technol.* (Jan. 1983) 31–39.
10. Cinco-Ley, H., Ramey, H. J., Jr., and Miller, F. G., "Pseudo-Skin Factors for Partially Penetrating Directionally Drilled Wells," SPE paper 5589, 1975.
11. Schechter, R. S., *Oil Well Stimulation.* Prentice-Hall, Englewood Cliffs, NJ, 1982.
12. Schechter, R. S., and Gidley, J. L., "The Change in Pore Size Distribution from Surface Reactions in Porous Media," *AICHEJ* (May 1969) 339–350.
13. Tyler, T. N., Metzger, R. R., and Twyford, L. R., "Analysis and Treatment of Formation Damage at Prudhoe Bay, AK," SPE Paper 12471, 1984.
14. Golan, M., and Whitson, C. H., *Well Performance.* International Human Resources Corporation, Boston, 1986.
15. Cinco, H., Miller, F. G., and Ramey, H. J., Jr., "Unsteady-State Pressure Distribution Created by a Directionally Drilled Well," *J. Petroleum Technol.* (Nov. 1975) 1392–1402.
16. Jones, L. G., Blount, E. A., and Glaze, O. H., "Use of Short-Term Multiple-Rate Flow Tests to Predict Performance of Wells Having Turbulence," Paper SPE 6144 presented at the SPE Annual Technical Conference and Exhibition, New Orleans, Oct. 3–6, 1976.
17. Mach, J., Proano, E., and Brown, K. F., "Application of Production Systems Analysis to Determine Completion Sensitivity on Gas Well Completion," Paper 8113 presented at the ASME Energy Sources Technical Conference, Houston, Jan. 18–22, 1981.

18. Klotz, J. A., Krueger, R. F., and Pyle, H., "Effect of Perforation Damage on Well Productivity," *J. Petroleum Technol.* (Nov. 1974) 1303–1314; *Trans. AIME* 257.
19. Firoozabadii, A., and Katz, D. L., "An Analysis of High-Velocity Gas Flow through Porous Media." *J. Petroleum Technol.* (Feb. 1979) 211–216.
20. Frick, T. P., and Economides, M. J., "Horizontal Well Damage Characterization and Removal," SPE Paper 21795, 1991.
21. Joshi, S. D., "A Review of Horizontal Well and Drainhole Technology," Paper SPE 16868, presented at the 1987 Annual Technical Conference, Dallas, TX. A revised version was presented at the SPE Rocky Mountain Regional Meeting, Casper, WY, May 1988.

Chapter 18

Selection of Gas Wells for Production Stimulation

18.1 Introduction

Before a well is selected for stimulation treatment, it must be determined that the reservoir contains sufficient gas-in-place and has adequate potential gradients or formation pressure available to produce gas at higher rates following the creation of a high-permeability fracture. The cause for low productivity must also be determined so that the right type of treatment job can be applied.

18.2 Major Causes of Low-Productivity Gas Wells

Figure 18–1 illustrates causes of low productivity of a well.

18.3 Formation Condition Evaluation Techniques

Formation damage may be indicated by production tests, pressure buildup and drawdown tests, comparisons with offset wells, careful analyses of production history, including prior completion, and workover. Figure 18–2 shows the difference in pressure drawdown in a normal well compared with a well with serious "skin damage." In a relatively undamaged well with low reservoir permeability, days or weeks may be required for reservoir pressure to stabilize. In a relatively high-permeability well with severe skin damage, reservoir pressure measured in the well may stabilize within a few hours. "Skin" damage calculations using productivity tests and buildup and drawdown analyses are carried out many areas prior to planning well stimulation. Equations 18–1 and 18–2 can be applied to calculate average formation permeability k_{BU} and interwell permeability plus wellbore effects k_{PI}. Pressure drop in the skin or damaged zone near the well by the amount of the skin effect is shown in Figure 18–2.

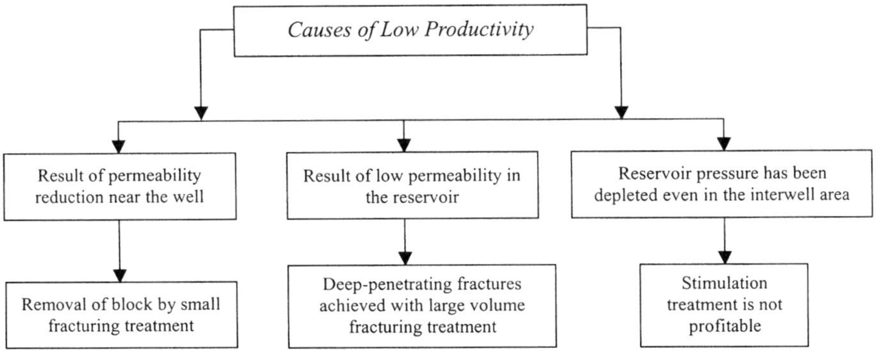

Figure 18–1. Causes of low permeability.

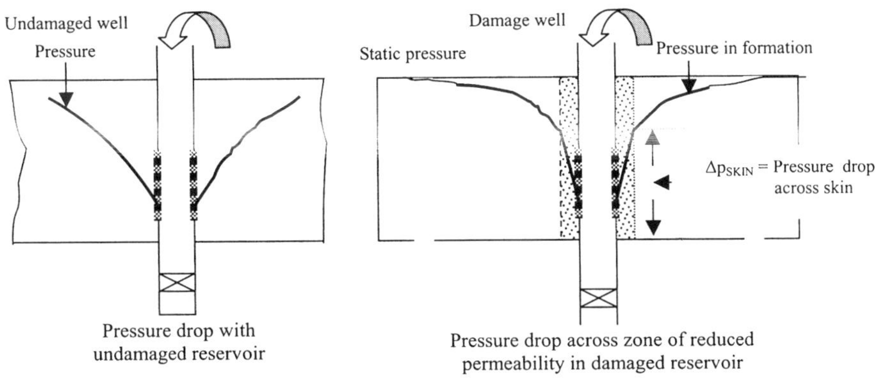

Figure 18–2. Pressure drop difference between pressure drawdown in damaged and undamaged reservoirs.

Figure 18–3 can be used to find causes of low productivity.

$$k_{BU} = \frac{162.6 q_g \mu_g \beta_g}{mh} \tag{18–1}$$

$$k_{PI} = \frac{141.2 q_g \mu_g \beta_g In(r_e/r_w)}{h(p_{ws} - p_{wf})} \tag{18–2}$$

where k_{BU} is the average formation permeability and can be calculated from buildup test data measured during the first 5 hr after the well is shut in, and k_{PI} may be determined from the productivity index test by producing a well at a nearly constant rate as possible for as long as possible, or until the producing bottom-hole pressure has ceased to change significantly. Interwell permeability k plus wellbore effects can be approximated using Eq. 18–2.

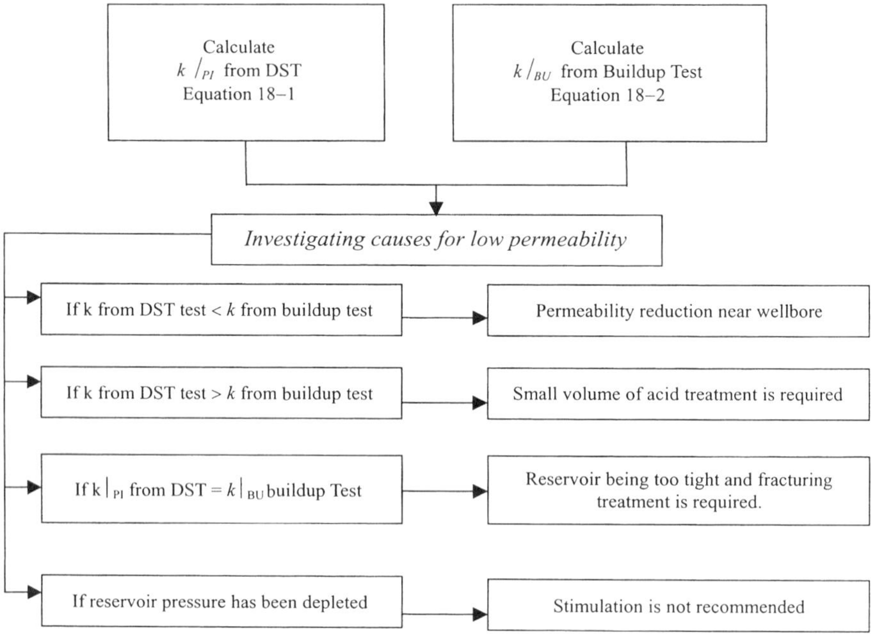

Figure 18–3. Causes of low permeability and stimulation treatment.

18.4 Relative Indicators of Wellbore Conditions

The efficiency of the completion can be determined by comparing the actual productivity index, J, and the ideal (no skin). The ratio of these two quantities is

$$\text{Flow efficiency} = [J_{actual}/J_{ideal}] = (P_i - P_{wf} - \Delta p_{skin})/(P_i - P_{wf})$$

This ratio is quite similar to the condition ratio of Gladfelter *et al.*[4] Find flow efficiency or condition ratio from Eqs. 18–3 and 18–4. Well problems can be interpreted using Figure 18–3. Condition ratio (CR) $= k|_{PI}/k|_{BU}$. Figure 18–4 shows various values of condition ratio and their effect on fracturing treatment jobs.

18.5 Skin Factor Concepts, Relationships, and Equations

The skin concept was originally introduced by Van Everdingen[2] and Hurst *et al.*[3] to describe the behavior of damaged wells[2] and is illustrated in Figure 18–5. Equation 18–6 shows that the pressure should rise by an amount Δp_{skin} immediately after shut-in to calculate the skin factor; it is necessary to measure the well pressure both before and after closing in. The skin factor s

$C.R < 1$ →	Permeability reduction near wellbore and required small acidizing job.
$C.R > 1$ →	Formation permeability is low and fracturing treatment job is required.
$C.R = 1$ →	Reservoir is too tight and fracturing treatment is required.
$C.R = 2$ →	Moderately high permeability formation and stimulation treatment is required.
$C.R = 5$ →	Low permeability formation and fracture treatment is required.

Figure 18–4. Various values of CR and treatment jobs.

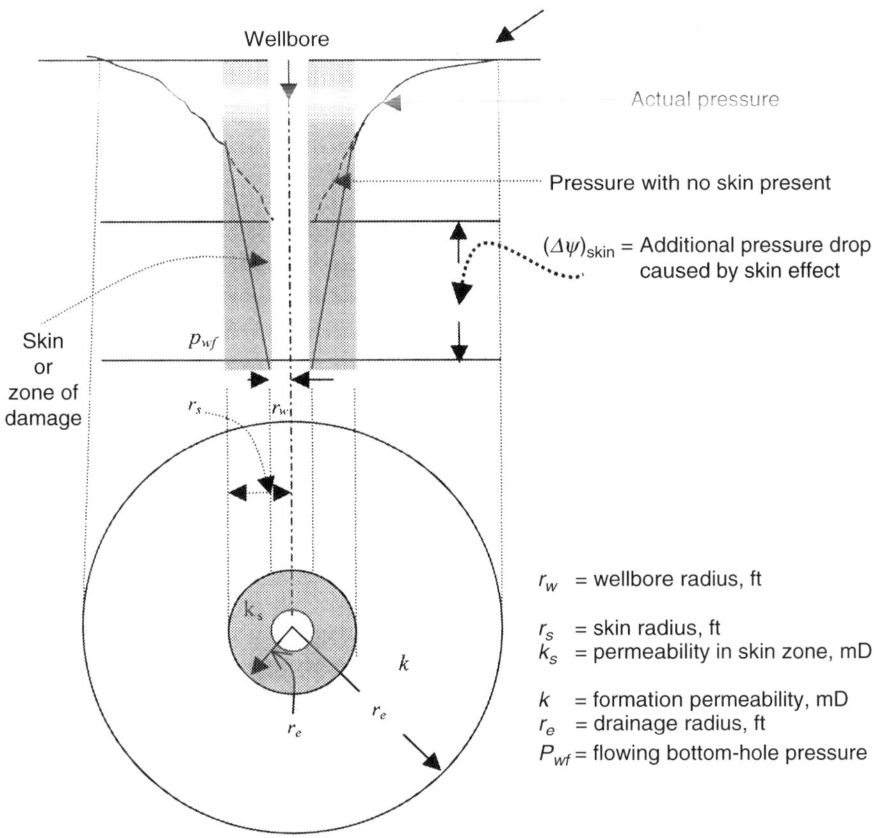

Figure 18–5. Pressure responses in a gas reservoir with a skin.

can be expressed as a function of k_s and r_s by the relation

$$s = (k/k_s - 1)In(r_s/r_w) \tag{18–3}$$

Equation 18–3 indicates that if $k_s < k$, then s is positive; if $k_s = k$, then s is zero; and finally, if the permeability k_s is greater than that in the formation (k), such as from fracturing or acidizing, s will be negative. Hydraulically fractured wells often show values of s ranging from −3 to −5.

$$s' = s + Dq_{sc} = 1.151\left[\frac{\psi(p_i) - \psi(p_{1hr})}{m'} - \log\left(\frac{k}{\phi\mu_{gi}c_{ti}r_w^2}\right) + 3.23\right] \tag{18–4}$$

$$s = 1.151\left[\frac{\psi(p_{1hr}) - \psi(p_{wf})}{m} - \log\left(\frac{k}{\phi\mu_g c_t r_w^2}\right) + 3.23\right] \tag{18–5}$$

$$(\Delta p)_{skin} = 0.869ms \quad \text{where } s = \text{skin} \tag{18–6}$$

Important Skin Factor Relationships and Equations

This section summaries analysis equation for single-well tests and their relationships with skin factor. These equations listed below are also presented elsewhere in the text.

$r'_w = r_w e^{-S}$, $x_f = 2r'_w = 2r_w e^{-S}$; and fracture penetration $= x_f/r_e$

Pressure Derivative Type Curves with Double Porosity Behavior

$$\omega = \frac{C_D e^{2S}}{(C_D e^{2S})_f} \tag{18–7}$$

$$k_f h = \frac{50{,}300\, q_g T_R p_{sc}}{T_{sc}} \tag{18–8}$$

$$C = \frac{0.000295\, k_f h}{\mu_g} \cdot \left[\frac{(\Delta t)_{MP}}{\left(\frac{t_D}{C_D}\right)_{MP}}\right] \tag{18–9}$$

$$C_{D_{f+m}} = \frac{0.8936C}{\phi c_t h r_w^2} \tag{18–10}$$

$$s = 0.5\, In\left[(C_D e^{2S})_{fm}/C_{D_{f+m}}\right] \tag{18–11}$$

$$\lambda = \frac{(1-\omega)^2}{C_{D_{f+m}}} \cdot \left[\lambda C_{D_{f+m}}/(1-\omega)^2\right]_{MP} \tag{18–12}$$

Linear Flow Analysis

$$x_f = \frac{4.014 q_g \beta_g}{m_{lf} h} \cdot \left(\frac{\mu_g}{k\phi c_t}\right)^{0.5} \tag{18–13}$$

$$s_f = In\left(\frac{2r_w}{x_f}\right) \tag{18–14}$$

Bilinear Flow Analysis

$$k_f w = \left(\frac{44.1 q_g \beta_g \mu_g}{m_{bl}}\right)^2 \cdot \left(\frac{1}{k\phi c_t}\right)^{0.5} \tag{18–15}$$

$$s_f = (s_f + In(x_f/r_w))_{Table\ 7\text{–}13} - In(x_f/r_w) \tag{18–16}$$

$$x_f = 3r_w e^{-S_f} \tag{18–17}$$

18.6 Completion Types and Related Skin Factors

Table 18–1 shows various types of stimulation treatments to improve skin factor.

Table 18–1
For Wellbore Radius r_w = 0.29 ft and 0.39 ft

Type of stimulation treatment	Skin factor s	Apparent wellbore radius (ft)	Fracture radius x_f (ft)	Apparent wellbore radius (ft)	Fracture radius x_f (ft)
Natural completion	0	0.2900	0.5800	0.3900	0.7800
Light acid	−0.5	0.4781	0.9562	0.6430	1.2860
Medium acid or light fracture	−1.0	0.7883	1.5766	1.0601	2.1220
Heavy acid or medium fracture	−2.0	2.1428	4.2856	2.8817	5.7634
Heavy fracture	−3.0	5.8248	11.6496	7.8334	15.6668
Heavy fracture in low permeability	−4.0	15.8335	31.6670	21.2933	42.5866
Very large fracture in low permeability	−5.0	43.0398	86.0796	57.8811	115.7622

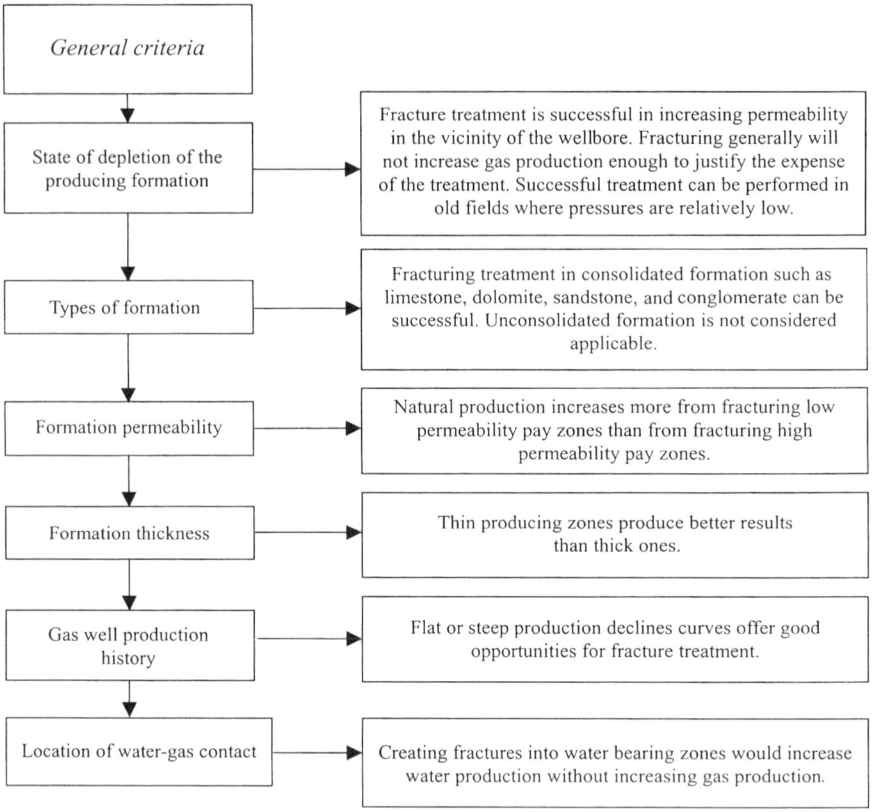

Figure 18–6. General criteria for selecting gas well for fracturing treatment.

18.7 Selecting Gas Wells for Fracturing Treatment

Kaufman[7] has published general criteria for selecting wells for fracturing treatment. The criteria in Figure 18–6 are believed to be applicable in most cases.

18.8 Productivity Improvement and Treatment Variables

The optimum fracturing treatment requires that the productivity increase resulting from various fracture radii be determined. The ratio of productivity after fracturing to that before fracturing is a function of fracture radius, fracture capacity, and formation characteristics. These variables are related to well production by the discontinuous permeability formula for steady-state flow.[4]

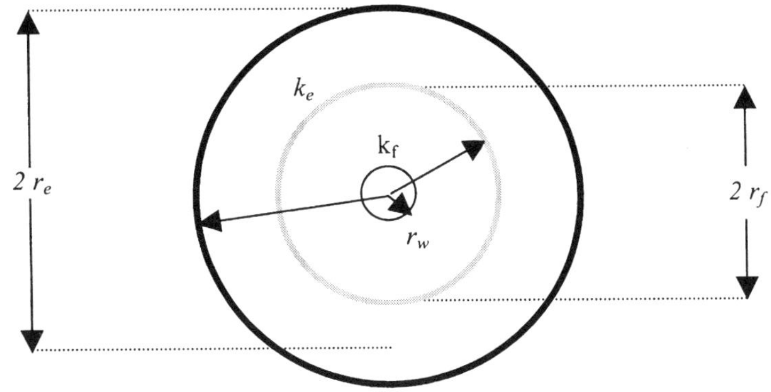

Figure 18–7. Showing various variables.

With this method it is assumed that, because of the creation of a fracture, the permeability in a zone around the wellbore differs from that at a distance. It may be shown that

$$k_{avg} = \frac{\log(r_e/r_w)}{\frac{1}{k_f}\log[(r_f/r_w) + 1/k_e] \cdot \log(r_e/r_f)} \tag{18–18}$$

where k_e is the original permeability of the formation before treatment and k_f is the permeability of the formation from the wellbore to the fracture radius. Figure 18–7 illustrates this system. In applying Eq. 18–17, k_f is equal to the effective horizontal permeability of the formation lying within the radius of fracture. The value that should be assigned to this effective horizontal permeability is somewhat indefinite because the height of formation, vertical permeability, thickness of fracture, etc., all influence it. For the purpose of these calculations, however, it is believed that a sufficiently accurate estimate of its value may be determined from

$$k_f = \frac{k_e h + k_f w}{h} \tag{18–19}$$

When all factors in Eq. 18–18 have been estimated as explained earlier, the average permeability of the whole producing zone (k_{avg}) is calculated. After this average permeability is obtained, the stabilized production gas rate following hydraulic fracturing may be estimated as follows.

Pressure-squared case:

$$q_{sc} = \frac{0.000305 k_{avg} h \left(\overline{p}_R^2 - p_w^2\right)}{\overline{\mu}_g \overline{z} T \log(r_e/r_w)} \tag{18–20}$$

Pseudopressure case:

$$q_{sc} = \frac{0.000305 k_{avg} h \lfloor \psi(\overline{p}_R) - \psi(p_w) \rfloor}{T \log(r_e/r_w)} \tag{18–21}$$

18.9 IPR Modification to Different Hydraulic Fracture Designs

Hydraulic fracture design depends on a large number of variables. The cost of the treatment and the associated costs of testing are only one component of the net present value (NPV) design approach. In the parametric studies shown here, four of the most important variables affecting fracture design are as follows.

- Permeability
- Fracture height
- Optimum fracture half-length
- Net present value (NPV)

Figure 18–8 shows the effect of the reservoir permeability on NPV. In this study the optimum fracture half-length is 350 ft, corresponding to NPV values of $\$1.86 \times 10^6$, $\$1.28 \times 10^6$, and $\$0.62 \times 10^6$ for permeabilities of 0.15, 0.75, and 1.5 mD, respectively. Figure 18–9 shows a parametric study with IPR modification (e.g., corresponding to different hydraulic fracture design).

Figure 18–9 indicates the benefit of well stimulation. Once again, it should be remembered that stimulation costs money. Therefore, the gain in production

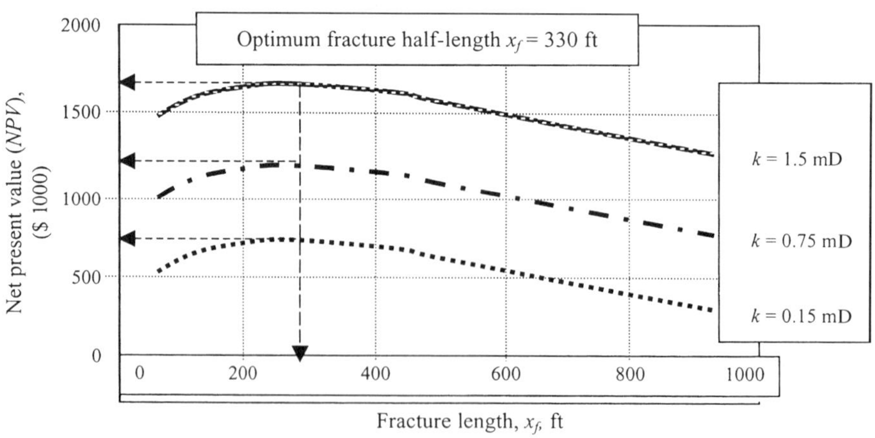

Figure 18–8. Effect of permeability on net present value (NPV).

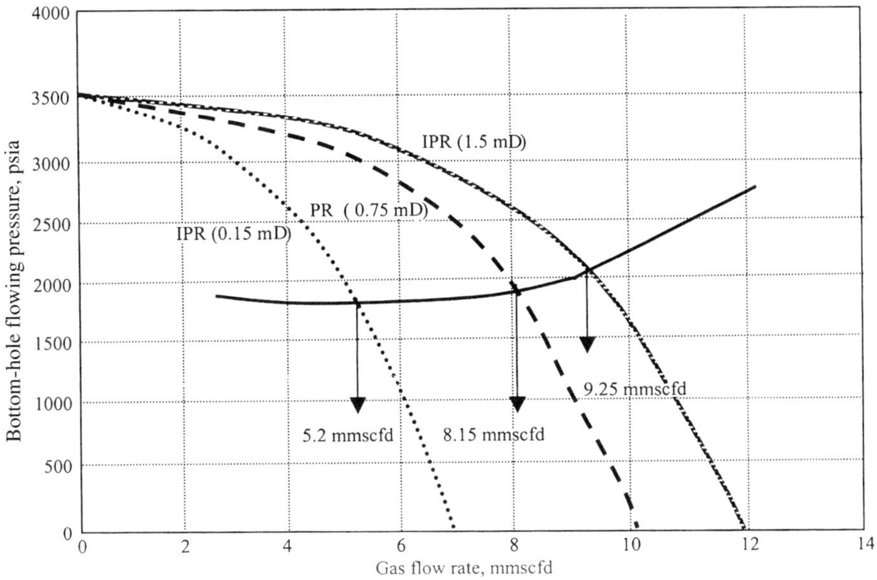

Figure 18–9. Parametric study with IPR modifications (corresponding to different hydraulic fracture design).

from the stimulation should be sufficient to pay for the stimulation in a reasonable time to make the stimulation job economically justified.

References and Additional Reading

1. Hawkins, M. F., Jr., "A Note on the Skin Effect," *Trans. AIME* (1956) 207, 356–357.
2. Van Everdingen, A. F., "The Skin Effect and Its Influence on the Productive Capacity of a Well," *Trans. AIME* (1953) 198, 171–176.
3. Hurst, W., Clark, D., and Brauer, E. B., "The Skin Effect in Producing Wells," *J. Petroleum Technol.* (1969) 21, 1483–1489.
4. Gladfelter, R. E., Tracy, G. W., and Wiley, L. E.,"Selecting Wells Which Will Respond to Production Stimulation Treatment," *Drill Prod. Practice API* (1955) 117–129.
5. Howard, G. C.,"Driving Maximum Profit from Hydraulic Fracturing," *Drill Prod. Practice API* (1958) 91.
6. Muskat, M., *Physical Principles of Oil Production.* McGraw-Hill, New York, 1949, p. 242.
7. Kaufman, M. J.,"Well Stimulation by Fracturing," *Petroleum Eng.* (Sept. 1956) B-53.

8. Clark, J. B., Fast, C. R., and Howard, G. C., "A Multiple-Fracturing Process for Increasing Productivity of Wells," *Drill Prod. Practice API* (1952) 104.
9. Maly, J. W., and Morton, T. E., "Selection and Evaluation of Wells for Hydrafrac Treatment," *Oil Gas J.* (May 1951) 52, 126.
10. Clark, R. F., Freedman, H. G., Bolstead, J. H., and Coffer, H. F., "Application of Hydraulic Fracturing to the Stimulation of Oil and Gas Production," *Drill Prod. Practice API* (1953) 113.

Chapter 19

Design Criteria of Flow and Pressure Transient Tests

19.1 Introduction

Specific test data are required for each transient test analysis technique; adequate data are essential for satisfactory transient test results. Thus, an important part of preparation for a transient well test is deciding which data are needed and how they will be obtained. This chapter discusses the design of deliverability and transient tests and data required and describes characteristics of suitable equipment.[1]

19.2 Deliverability Tests

The accuracy of the results calculated depends on the accuracy of data used. Obtaining accurate field data can be accomplished only if the field personnel follow established procedures for data collection. Detailed procedures can be found in Refs. 1–3. Brief procedures are summarized below.

Equipment and Procedure for Field Tests

The wells were shut in preparatory to running the initial drawdown test. In Appendix D, Figure D–1 shows the equipment used for these tests. A bottom-hole pressure bomb recorded bottom-hole pressures. A dead-weight tester periodically measured surface pressures. For all tests, the bomb was lowered on a wire line to the same pressure datum.

The initial flow rate was selected with the aid of well capacity information made available from the *AOF* data, considering that two or three added flow rates were to be used within the span of the well's indicated capacity. Once the rate was selected, the pressure controller was set to provide a constant pressure downstream of the choke, thus maintaining a constant rate throughout the test. The specific rate was obtained by using the appropriate orifice plate and critical flow proper pressure.

The bomb was pulled after a period of 24 to 72 hr, depending on the clock used. Pressures and flow rates were determined immediately. Plots were prepared. By flowing pressure performance it could be determined whether the test should be continued or terminated. Rerunning the pressure bomb did not affect flow rate or pressure data.

Guidelines for Designing Deliverability Tests

In designing a deliverability test, collect and utilize all information, which may include logs, drill-stem tests, previous deliverability tests conducted on that well, production history, fluid composition and temperature, cores, and geological studies. Knowledge of the time required for stabilization is a very important factor in deciding the type of test to be used for determining the deliverability of a well. This may be known directly from previous tests, such as drill-stem or deliverability tests, conducted on the well or from the production characteristics of the well. If such information is not available, it may be assumed that the well will behave in a manner similar to neighboring wells in the same pool, for which the data are available. When the approximate time to stabilization is not known, it may be estimated from Eq. 19–3. If the time to stabilization is of the order of a few hours, a conventional test may be conducted. Otherwise one of the isochronal tests is preferable. The isochronal test is more accurate than the modified isochronal test and should be used if the greater accuracy is required. A single-point test is appropriate when the deliverability relationship of the well is known from previous tests, and only updating of this relationship is desired.

Some of the factors affecting the choice of equipment are the expected flow rates and pressures and the composition of gas and liquid. The possibility and location of hydrate formation must be investigated. Production of liquid, water or condensate, causes fluctuations in the rate and pressure measurements. Long flow times of at least 6 to 8 hr are needed before the liquid-to-gas ratio stabilizes.

Some idea of the flow rates at which a well is capable of flowing may be obtained from the drill-stem test or from the preliminary well cleanup flows. In the absence of any data whatsoever, the *AOF* may be estimated from the equation given below by assuming stabilized, purely laminar flow in the reservoir.

$$AOF \cong \frac{kh(\overline{\Psi}_R - \Psi_{wf} = 0)}{3.274 \times 10^6 T \left[\log\left(0.472\frac{r_e}{r_w}\right) + \frac{s}{2.303}\right]} \qquad (19\text{–}1)$$

or

$$AOF \cong \frac{kh(\overline{\Psi}_R - \Psi_{wf} = 0)}{1.422 \times 10^6 T\left[\text{In}\left(0.472\frac{r_e}{r_w}\right) + s\right]} \quad (19\text{–}2)$$

s may be estimated from similar stimulation treatment performed on approximately similar wells in the formation. The duration of the isochronal periods is determined by two considerations, namely, (a) wellbore storage time and (b) the radius of investigation.

The wellbore storage time t_{ws} is the approximate time required for the wellbore storage effects to become negligible. This can be calculated by

$$t_{ws} = \frac{36{,}177\ \overline{\mu}_g\ V_{ws} C_{ws}}{kh} \quad (19\text{–}3)$$

where

V_{ws} = volume of the wellbore tubing (and annulus, if there is no packer)
C_{ws} = compressibility of the wellbore fluid evaluated at the wellbore at the mean wellbore pressure and temperature

For wells with no damage or improvement an approximate time to investigate 100 ft is obtained from Eq. 19–4:

$$t_{100} \cong 10^7 \frac{\phi\, \overline{\mu}_g}{k\overline{P}_R} \quad (19\text{–}4)$$

Equations 19–3 and 19–4 give the minimum duration of flow that will yield data representative of the bulk formation rather than the wellbore area. Duration equal to about four times this value is recommended for the isochronal periods. In conducting a multipoint test, the minimum flow rate used should be at least equal to that required lifting the liquids, if any, from the well. It should also be sufficient to maintain a wellhead temperature above the hydrate point. When these considerations do not apply, the minimum and maximum flow rates are chosen, whenever practical such that the pressure drops cause at the well are approximately 10%, respectively of the shut-in pressure. Alternatively, they may be taken to be about 10%, respectively, of the *AOF*.

Designing Suitable Deliverability Tests

The following example illustrates how to design a suitable deliverability test.

Example 19–1 *Designing a Suitable Deliverability Test*

A gas well was completed in a new pool and no deliverability tests have, so far, been performed on it. It has been cored, logged and drill-stem tested, acidized, and cleaned. Design a suitable deliverability test. Given: P_R = 2200 psig; $r_e = 2640$ ft; $\phi = 0.20$ (from logs); $s_g = 0.733$ (from logs); k = 125 mD (from drill-stem tests); $h = 10$ ft (from logs); $\mu = 0.0159$ cP; length tubing = 5000 ft; $T = 580°$R; diameter of tubing = 0.50 ft; $r_w = 0.29$ ft; and $C_{ws} = 0.0006$ psi^{-1}.

Solution No data are available. Assume $s = 0.0$ and $\phi_{HC} = 0.20 \times 0.733 = 0.15$. Calculate time of stabilization using the following equation:

$$t_S = 1000\frac{\phi_{HC}\overline{\mu}r_e^2}{k\overline{p}_R} = \frac{1000 \times 0.15 \times 0.0159 \times 2640^2}{125 \times 2200} = 60.5 \text{ hr}$$

The time of stabilization is considered to be too long to conduct the four rates of a conventional test. The isochronal procedures will be considered instead. The permeability and the buildup characteristics experienced during drill-stem testing suggest that a modified isochronal test will be chosen to determine the deliverability relationship. Determine the time necessary to investigate 100 ft into the reservoir by using the following equation:

$$t_{100} \cong 10^7\frac{\phi_{HC}\overline{\mu}}{k\overline{p}_R} = \frac{10^7 \times 0.15 \times 0.0159}{125 \times 2200} = 0.09 = 0.10 \text{ hr}$$

Flow Periods and Rates

$$V_{ws} = \pi \times 0.25^2 \times 5000 = 982.14 \text{ cu ft}$$

Calculate time required for wellbore storage effects using Eq. 19–3:

$$t_{WS} = \frac{36{,}177\overline{\mu}V_{WS}C_{WS}}{kh} = \frac{36{,}177 \times 0.0159 \times 982.14 \times 0.0006}{125 \times 10}$$
$$= 0.27 \text{ hr}$$

since $t_{WS} > t_{100}$.

The duration of the isochronal period is

$$= 4\ t_{WS} = 0.27 \times 4 = 1.08 \text{ hr} = 1.5 \text{ hr (say)}$$

The duration of the extended flow period is

$$t_S = 60 \text{ hr} = 72 \text{ hr (say)}$$

Flow rates during well cleanup are not available, therefore approximate, estimate of the *AOF* will be made from Eq. 19–2. Find $\Psi(\overline{p}_R) = 335 \times 10^6$ psia^2/cP from the $\Psi - p$ curve.

$$AOF \cong \frac{kh\Psi(\overline{p}_R)}{1.422 \times 10^6 T\left[In\frac{0.472r_e}{r_w} + s\right]}$$

$$= \frac{125 \times 10 \times 335 \times 10^6}{1.422 \times 10^6 \times 580\left[In\left(\frac{0.472\times 2640}{0.29}\right) + 0\right]} = 60.69 \text{ mmscfd}$$

10% of $AOF = 6$ mmscfd

75% of $AOF = 45$ mmscfd

A suitable range of approximate flow rates would be

First rate = 6 mmscfd, for 1.5 hr

Second rate = 12 mmscfd, for 1.5 hr

Third rate = 24 mmscfd, 1.5 hr

Fourth rate = 48 mmscfd, for 1.5 hr

An extended flow rate of about 25 mmscfd for 72 hr is recommended.

19.3 Procedures for Conducting Deliverability Tests

Next section will illustrate to conduct deliverability tests such as conventional backpressure, isochronal, and modified isochronal tests.

Conventional Backpressure Tests

Gas well deliverability tests have conventionally been called backpressure tests because flowing against particular pipeline backpressure greater than atmospheric pressure tests wells. The conventional backpressure test is also referred to as a flow-after-flow test, or a multipoint test. A test is referred to as a multipoint test if the rates of flow are imposed in succession without allowing a shut-in period in between the flow rates. If the well is shut in between the flow rates, the test is isochronal. A multipoint test is also referred to as a three-, four-, or five-point test, depending on the number of flow-rate changes during the test. The flow-rate and pressure histories for such a test are depicted in Figure 19–1.

The following steps are used for successfully conducting and analyzing a backpressure test:

1. Produce a well for sufficient length of time at a flow rate large enough to clear the wellbore of accumulated liquids prior to the shut-in period.

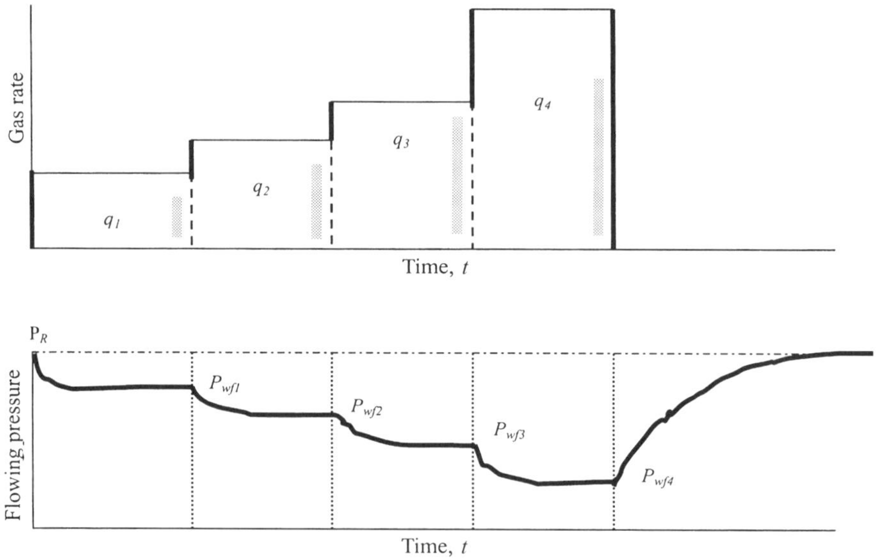

Figure 19–1. Conventional test—Flow rate and pressure diagrams.

2. Shut in the well until the rate of pressure *stabilizes*, i.e., pseudo-steady state is reached. The Railroad Commission of Texas defines "stabilized" as when two consecutive pressure readings over a period of 15 min agree within 0.1 psi. The IOCC defines "stabilized" as when two consecutive pressure readings over a period of 30 min agree within 1% of the previous recorded pressure.
3. A series of at least four stabilized flow rates and the pressure corresponding to each flow rate are recorded. Any shut-in time between flow rates shall be held to a minimum. The flow rates are normally in increasing sequence; a decreasing sequence may be used in case of high-liquid-ratio wells or unusual temperature conditions. A decreasing sequence will result in higher wellbore temperature, thus avoiding hydrate formation. A decreasing sequence may be required if liquid holdup in the wellbore will be a problem. A good spread of flow rates is recommended.
4. If the well produced hydrocarbon liquids, using the specific gravity of separator gas, calculate the specific gravity of the flowing fluid.
5. Calculate the shut-in pressure, $\overline{p}_R$, and the flowing bottom-hole pressure p_{wf} at each rate of flow.
6. Calculate the difference between the shut-in pressure and the flowing pressure for each rate of flow, $(\overline{p}_R^2 - p_{wf}^2)$.
7. Plot the $\overline{p}_R^2 - p_{wf}^2 = \Delta p^2$ versus the corresponding rate of flow on 3×3 log-log graph paper.

8. Draw a best-fit line through at least three of the total points. The line is referred to as the stabilized deliverability curve. Calculate the exponent, n using the straight line and the following equation. The $(\overline{p}_R^2 - p_{wf}^2)_{q_{sc1}}$ and $(\overline{p}_R^2 - p_{wf}^2)_{q_{sc2}}$ should be read on the straight line corresponding to q_{sc1} and q_{sc2}, respectively, exactly one log cycle apart. The value of n may also be obtained from the angle the straight line makes with the vertical, in which case $1/n = \tan\theta$.

$$n = \log\left[\frac{(\overline{p}_R^2 - p_{wf}^2)_{q_{sc1}}}{(\overline{p}_R^2 - p_{wf}^2)_{q_{sc2}}}\right]$$

9. Determine the value of performance coefficient C by extrapolating the straight line until the value of $(\overline{p}_R^2 - p_{wf}^2)$ is equal to 1.0.
10. Determine the *AOF* from the straight line (or its extrapolation) at $\overline{p}_R^2$, if $p_{wf}^2 = 0$ psi, or at $(\overline{p}_R^2 - p_{wf}^2)$ when p_{wf} is the atmospheric pressure.
11. The following equation represents the straight-line deliverability curve:

$$q_{sc} = C\left(\overline{p}_R^2 - p_{wf}^2\right)^n \tag{19–5}$$

Generally, the value of n ranges from 0.5 to 1.0. Exponents $n < 0.5$ may be caused by liquid accumulation in the wellbore. Exponents apparently greater than 1.0 may be caused by fluid removal during testing. When a test is conducted using decreasing rate sequence in slow stabilizing reservoirs, an exponent greater than 1.0 may be experienced. If n is outside the range of 0.5 to 1.0, the test data may be in error because of insufficient cleanup or liquid loading in the gas well.

Note that the bottom-hole static and flowing pressures are determined by Amerada-type downhole pressure gauges or by converting the stabilized static and flowing tubing pressures (determined at the surface) to bottom-hole conditions using the Cullender and Smith method (see Appendix E).

If the value of n is known or if it can be assumed, only a one-point test will provide the stabilized deliverability curve. The Oklahoma Corporation Commission allows the operator to use one-point test and assume the value of $n = 0.85$. The one-point test is conducted by shutting in the well until a stabilized static reservoir pressure is obtained. The well is then flowed at a constant rate for 1 to 3 days, and the stabilized bottom-hole flowing pressure is recorded.

Isochronal Tests

When dealing with low-permeability reservoirs or when flaring has to be minimized, the time required to obtain stabilized flow conditions in conventional backpressure tests may be very long. In an isochronal test, a series of one-point tests is performed, each starting with a well shut in and the shut-in pressure constant or nearly constant with time. The flow rates are for a predetermined and fixed period of time. The basic principle is to establish a deliverability curve for a smaller portion of the drainage area. The procedure normally followed in this test is as follows:

1. Shut in well until stabilized static reservoir pressure $\overline{p}$ is obtained.
2. Open the well at the first rate q_{sc}; flow for 6 hr.
3. Shut in well again until the same static pressure as in step 1 is obtained.
4. Repeat steps 2 and 3 two or three additional times at different flow rates.
5. After the last flow period, one flow test is conducted for an extended time period to attain stabilized flow conditions.

The stabilized flow data that are obtained above are analyzed as follows:

1. Plot the three or four isochronal points on log-log paper.
2. Draw a best-fit line through the points.
3. Obtain the value of the exponent n from the slope of this line, n = 1/slope.
4. Plot the point of extended flow rate and the corresponding $(\overline{p}_R^2 - p_{wf}^2)$ at the stabilized pressure P_{wf} at this rate. Draw a line through this point parallel to the best straight line plotted in step 2. This line will represent the stabilized deliverability curve. Once the stabilized deliverability curve is determined. *AOF* is established in the usual way as previously discussed. The behavior of the flow rate and pressure with time is illustrated in Figure 19–2.

Modified Isochronal Tests

In extremely low-permeability gas formations, an isochronal test may not always be practical since it is very difficult to attain a completely stabilized static reservoir pressure before the first flow period and during each subsequent shut-in period. Modified isochronal tests are used widely because they conserve time and money. However, the method is an approximation of the regular isochronal test. The only difference is in actually conducting the test and not in the analysis of the test data. In modified isochronal test, the flowing and shut-in periods are of equal duration, and the final shut-in BHP (p_{ws}) before the beginning of a new flow period is used as an approximation to $\overline{p}_R$ in the test analysis procedure. Note that the initial static reservoir pressure is used to calculate $(\overline{p}_R^2 - p_{wf}^2)$ for the flowing pressure obtained during the extended flow period. The flow-rate pressure behavior with time is shown in Figure 19–3.

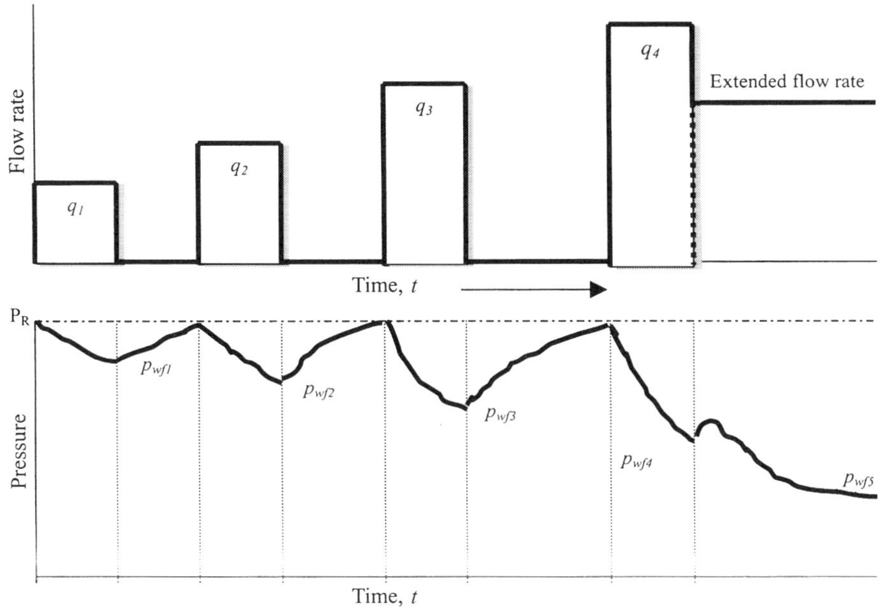

Figure 19–2. Isochronal test—Flow and pressure diagrams.

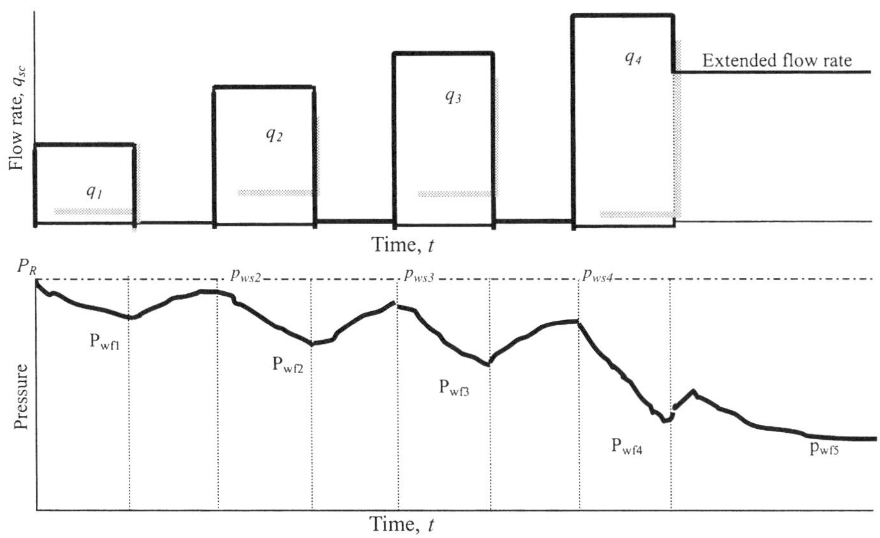

Figure 19–3. Modified isochronal test—Flow rate and pressure diagrams.

19.4 General Concepts for Designing Transient Pressure Tests

Figure 19–4 describes the steps in designing a transient test.

Choice of Test Type

Production Well Transient Tests

Choose between buildup, drawdown, and multiple rate testing

Design Calculations

1. Estimate the complete expected pressure response using assumed formation properties.
2. Determine key factors in test response, such as the end of wellbore storage effects, the end of the semilog straight line, the semilog straight line slope, and the general magnitude of the pressure response.
3. Run the test without design calculations.

Interference Test Design

It is best to estimate the pressure response at the observation well as a function of time.

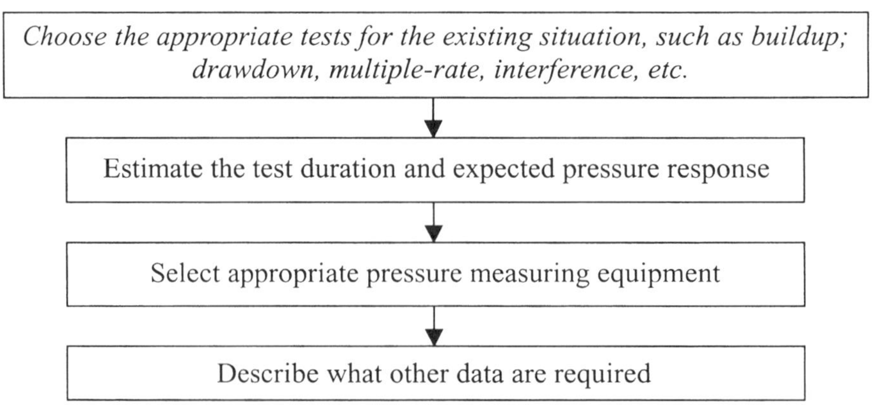

Figure 19–4. Important steps in designing a pressure transient test.

Pulse Test Design

Reference 2 has developed a method of interference determination by pulse testing. In this method a production well near the observation well is alternately produced and then closed in to give series of pressure pulses. These pressure pulses are detected at the observation well by a very accurate (0.001 psi) pressure gauge. Use of this gauge allows the interference pressure pulses to be detected much more rapidly.

Design of Flow and Buildup Tests

Test design should include the following:

1. A set of calculations are performed similar to those made during analysis of results using likely assumed values, so that the flow rates and time are chosen properly.
2. Be certain that test results can be adequately analyzed before spending the time and money, including risk in the field.

Guidelines for Designing Drawdown Tests

1. Wells that have not been produced or wells that have been shut in long enough to permit pressure stabilization are good candidate for drawdown testing and analysis.
2. In the isochronal deliverability tests discussed in the previous section, a series of shut-in periods is required to attain pressure stabilization; in the conventional deliverability tests, each flow period must be continued to pressure stabilization. In either case, the time required for stabilization may be very large and will limit the application of such tests.
3. A drawdown test may prove to be particularly attractive, since such a test minimizes the loss of production associated with a shut-in.
4. Drawdown tests may also be conducted to supplement information obtained from other tests such as buildup or deliverability tests.
5. When it is difficult to achieve constant flow rates because of slugging of the well, a drawdown test is not recommended.
6. Tests utilizing early-time data and their interpretations are only recommended, if, after testing, it is found that they are the only data amenable to analysis.
7. Tests utilizing transient flow data are recommended when an accurate knowledge of reservoir characteristics and skin effects is desired.
8. A single-rate test is acceptable when IT flow effects are negligible; otherwise two single-rate tests should be conducted to evaluate the skin and IT flow components of the apparent skin factor.

9. In some situations, a long shut-in period is necessary between two single-rate tests. When this is not practical, two-rate tests may be more appropriate. A two-rate test, with a declining rate, is particularly suitable when wellbore storage effects are to be minimized or phase redistribution in the wellbore during shut-in is to be eliminated.
10. Variable-rate transient flow tests are rarely designed. Single-rate tests in which the flow rate cannot be maintained at a constant value lend themselves to a variable-rate analysis.
11. To obtain information on reservoir limits, a conventional deliverability test with a continuous recording of the flowing well pressure is recommended. Economic limits tests should be used whenever possible to minimize flaring and wastage of gas.
12. In a fractured well during stimulation, early-time data might provide a good match on the type curve and where transient flow tests economically prohibitive, tests may be designed with a view to utilizing early-time data. However, it is advisable to confirm the results, whenever possible, by an appropriate transient flow analysis.
13. For tests of long duration, longer than 1 week, a surface-recording bottom-hole pressure bomb is recommended. When conducting a two-rate test in which the first flow rate is the production rate itself and only the second rate is being analyzed, the pressure bomb should be lowered into the well, preferably without stopping the first flow rate.
14. Duration of flow rates must last for at least 5, but preferably 10, times this wellbore storage time, t_{ws}:

$$t_{ws} = \frac{36,177\,\overline{\mu}\,v_{ws}\,C_{ws}}{kh} \tag{19–6}$$

Duration of flow applies to the single-rate test, the two single-rate tests, and the first rate of all multirate tests. In the case of reservoir limits tests, the time required for a limit is given by

$$t_s \cong 1000\frac{\phi\,\overline{\mu}_g\,r_e^2}{k\,\overline{p}_R} \tag{19–7}$$

If the reservoir is noncircular, then a reservoir limits test should be run for a time equal to at least three times that given by the value of t_{DA} for the appropriate shape.
15. For a fractured well, the time of departure from the straight line of slope on half has been given by Wattenbarger[13] as

$$t \cong \frac{\phi\mu_i c_i x_f^2}{25\,k} \tag{19–8}$$

Approximately 10 to 20 times this value of t should be sufficient to match the data plot and type curves.

Procedures for Estimating Reservoir Drainage Volume and System Shape Factor

Drawdown tests run specifically to determine the reservoir volume communicating with the well are called *reservoir limit tests*. Such tests, introduced by Jones,[2,3] use the pseudo-steady-state part of the drawdown data. The following test procedures should be used to determine reservoir drainage volume and system shape factor.

1. Prepare a plot of $\psi(P_{wf})$ versus time t on Cartesian coordinate graph paper as shown in Figure 19–5. From this plot find slope m^* and intercept $\psi(P_{int})$ (intercept of straight line). The equation of the straight line is

$$\psi(P_{wf}) = m^* t + \psi(P_{int}) \tag{19–9}$$

where

$$m^* = -\frac{0.23395 q_{sc} \beta_g}{\phi c_t h A} \tag{19–10}$$

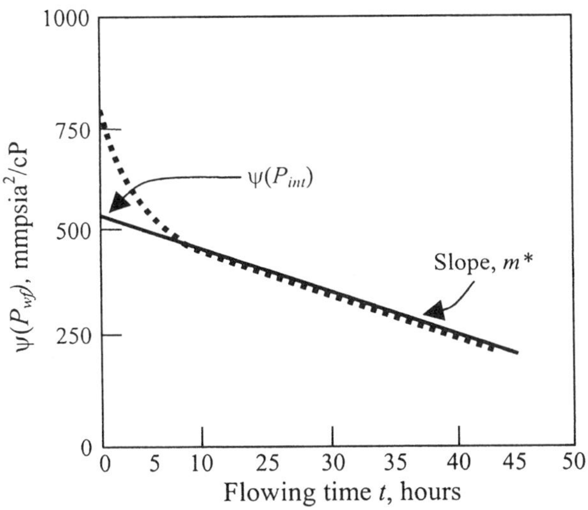

Figure 19–5. Cartesian plot of the drawdown test.

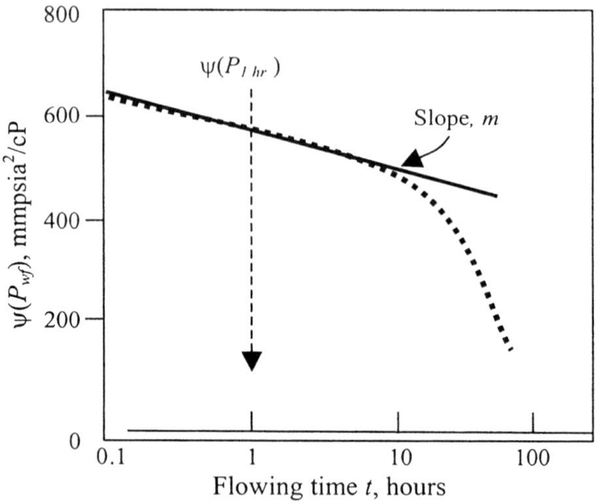

Figure 19–6. Semilog curve.

and

$$\psi(P_{int}) = \psi(P_i) - \frac{70.60 q_{sc} \beta_g \mu_g}{kh}\left[In\left(\frac{A}{r_w^2}\right) + In\left(\frac{2.2458}{C_A} + 2s\right)\right] \quad (19\text{–}11)$$

Equation 19–9 indicates that a Cartesian plot (Figure 19–5) of bottom-hole flowing pressure versus flowing time should be a straight line during pseudo-steady-state flow, with slope m^* given by Eq. 19–10 and intercept by Eq. 19–11.

2. The system shape factor is estimated from

$$C_A = 5.456 \frac{m}{m^*} Exp[2.303(\psi(P_{1hr}) - \psi(P_{int}))/m] \quad (19\text{–}12)$$

where m and $\psi(P_{1hr})$ can be determined from a semilog plot (see Figure 19–6).

Important Factors Influencing Pressure Buildup and Drawdown Tests

Buildup Tests

1. Pressure buildup tests are difficult to conduct and many factors can influence the shape of a pressure buildup curve.

2. In addition to wellbore storage effects, hydraulic fractures, particularly in low-permeability formations, can have a major effect on buildup curve shape and analysis. Chapter 6 gives a more detailed discussion of both these factors.

Drawdown Tests

1. The test may be hard to control since it is a flowing gas well.
2. The early part of the drawdown data is influenced by wellbore storage. Sometimes it is possible to draw a straight line through the semilog plot of data taken during this time. The slope of that line gives incorrect values of permeability and skin. A log-log data plot of the drawdown data must be made to select the correct semilog straight line.

19.5 Test Planning and Data Acquisition

The important parts of test operation and planning include the following:

1. Good and complete rate stabilization
2. Placement of the pressure instrument before the test begins
3. Careful documentation of what happens during the test, both at the test well and at nearby operating wells

The general data checklist in Figure 19–7 is an aid to complete data acquisition.

19.6 Guidelines for Gas Well Testing

Accurate field data can be obtained if the field personnel follow established procedures for data collection.

Choice of Testing Equipment

It depends upon the nature of the produced fluids and the type of test being conducted. Various wellhead-testing facilities are necessitated by the presence of condensate, water, or acid gases in the natural gas being produced.

Sweet Dry Gas

Figure 19–8 may be used. As shown in this figure, if valve A is closed gradually while valve B is being opened, maintaining a constant pressure in the flow string, the flow rate being measured by the flow prover will be the same as the production rate. The deliverability of constant flow rates during tests is very important. Figure 19–9 illustrates the wellhead rigging that may provide constant flow rates. Flow downstream from the flow prover is usually

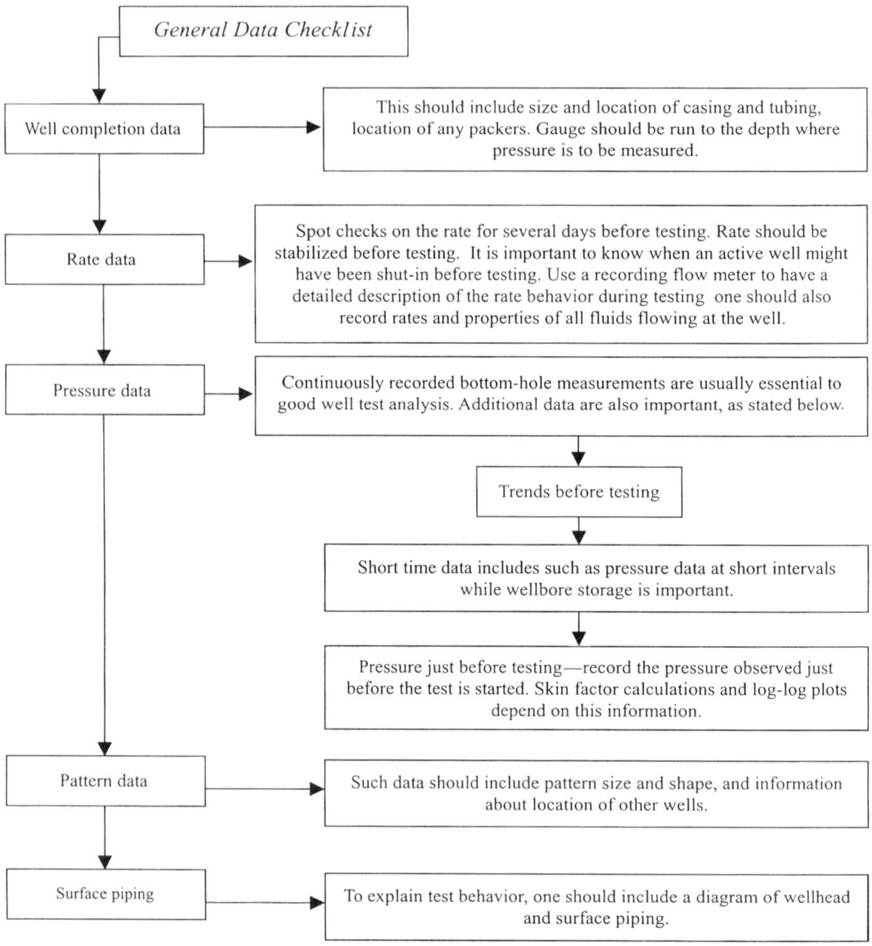

Figure 19–7. General data check list.

vented in the atmosphere. If, however, the produced gas must be flared, care must be taken to ensure that critical flow conditions are maintained in the flow prover. The theory and application of the metering devices are discussed in Appendix E.

Sweet Wet Gas

A natural gas containing heavier hydrocarbons appears as a condensate in the produced gas. In some instances, water may also be produced but it is not included in the definition of a wet gas. The presence of condensate in produced

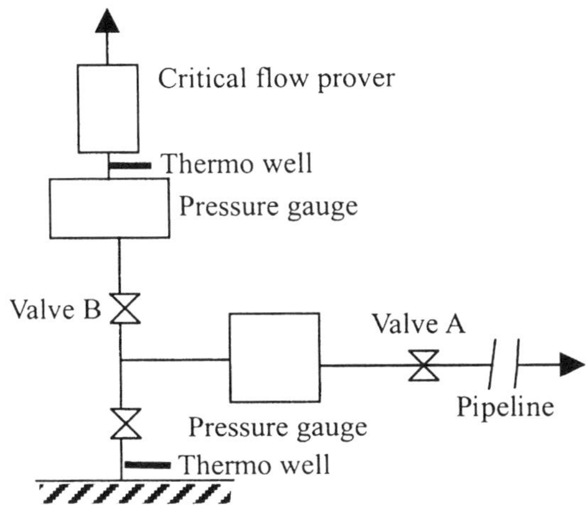

Figure 19–8. Flow diagram for measuring flow rates not equipped with flow-rate measuring equipment.

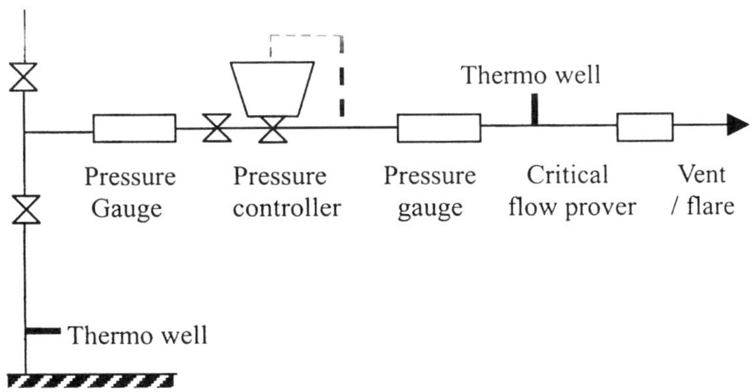

Figure 19–9. Flow diagram of wellhead rigging for constant-rate tests.

gas creates requirements for more complex testing facilities than those required for sweet, dry gas wells. A typical facility includes the following:

- Flow rate measurement devices
- Pressure measurement devices
- Thermometers
- Gas and condensate sampling equipment
- Line heaters
- Separation facilities

Several stages and a combination of measurements may be required for highly productive wells, but the most commonly used configurations involve

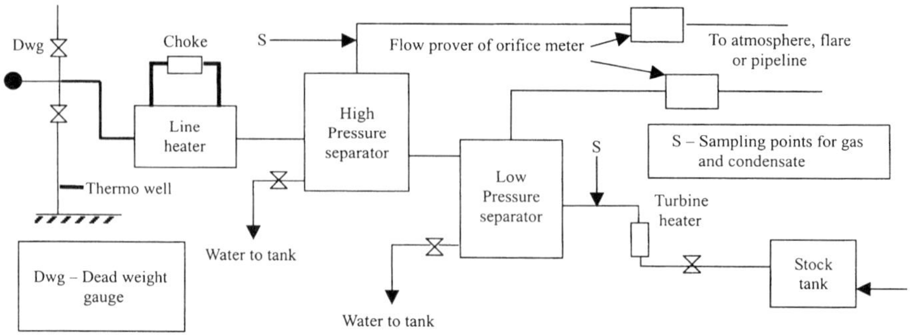

Figure 19–10. Flow diagram of surface well testing facilities for wet gas—Two-stage separation.

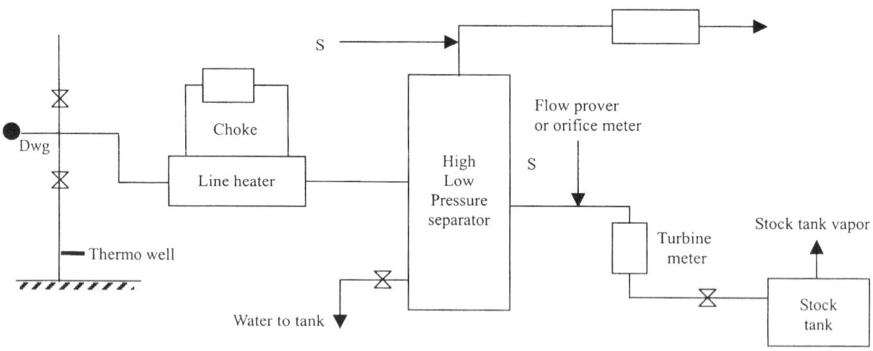

Figure 19–11. Flow diagram of surface well testing facilities for wet gas—Single-stage separation.

either a single separator or two separators in series. Figures 19–10 and 19–11 are a guide in the selection of test equipment. The requirement for line heaters is necessitated by the possibility of hydrate formation within the flow line and testing equipment.

Sour Gas

For testing sour gas wells, more elaborate facilities are required. In addition to the standard equipment, depending on whether the gas is dry or wet a gas meter and a flow line to an appropriate flow stack are required. In addition, liquid seals may also be necessary to protect the gas meter and pressure-measuring device from H_2S gas.

Flow Measurement

The accurate measurement of gas and liquid production rates is essential to the proper conduct and analysis of gas well tests. Correct sampling procedures are also necessary in order to obtain representative samples of the produced fluids and an accurate estimate of the constituents of the serving gas.

Pressure Measurement

The accurate measurement of static pressures and the pressures corresponding to flow rates measured during the flow periods of various tests is of great importance in gas well testing, since interpretation of deliverability drawdown and buildup test results must be based on the theory of flow in the reservoir sandface pressure (in the wellbore). Ideally this pressure is measured directly through use of an accurate, carefully calibrated bottom-hole pressure gauge.

19.7 Problems in Gas Well Testing

The following problems can result when testing gas wells. These problems are discussed from the viewpoint of the types of errors that can result in the test data obtained.

Hydrate Formation

This kind of problem normally occurs in high-pressure gas wells. Maintaining the well stream temperature above the hydrate formation temperature can eliminate this problem.

Liquid Loading

This problem occasionally occurs when testing low-productivity gas wells with high liquid gas ratios. Wide variations of surface pressures may indicate liquid loading.

Sour (H_2S) Gas

This hazardous substance is highly toxic and at certain concentrations can cause illness and death. Special precautions should be taken when testing wells where hydrogen sulfide is present to ensure that exposure will not exceed the safe maximum allowable concentration for the work period required. In testing sour gas wells, care must be taken to not allow any unflared gas to escape to the atmosphere, especially in populated areas. Extensive testing should be delayed until the gas can be sent to a pipeline or gas plant. All gas produced during preliminary tests should be flared. Testing equipment such as separators, pressure gauges, and meters should be thoroughly tested before beginning a sour gas well test, especially if the equipment has previously been used in sour gas service.

Wet Gas Streams

Such streams will often deposit liquid in the flow line downstream of the point where the orifice was installed. Therefore it is necessary to meter a gas stream at the wellhead. If the gas gravity is measured on the gas sample obtained at such a downstream point; its value will not represent the gravity of the gas that flowed through the orifice. In this instance, the measured gas gravity must be adjusted to give the gravity of the full stream this can be accomplished by using

$$\gamma_{gas} = \frac{\gamma_g + 4584\,\gamma_o/R}{1 + V_o/R} \qquad (19\text{–}13)$$

where

γ_{gas} = specific gravity of the mixture (air = 1.00)
γ_g = specific gravity of the separator gas
γ_o = specific gravity of the condensate

$$\gamma_o = \frac{141.5}{131.5 +^0 API}$$

R = producing gas-condensate ratio
V_o = condensate vaporizing volume, ft^3/bbl

19.8 Reporting Gas Well Test Data

Rather than attempt to list data for a particular purpose, a complete list of well test data is given in Figure 19–12.

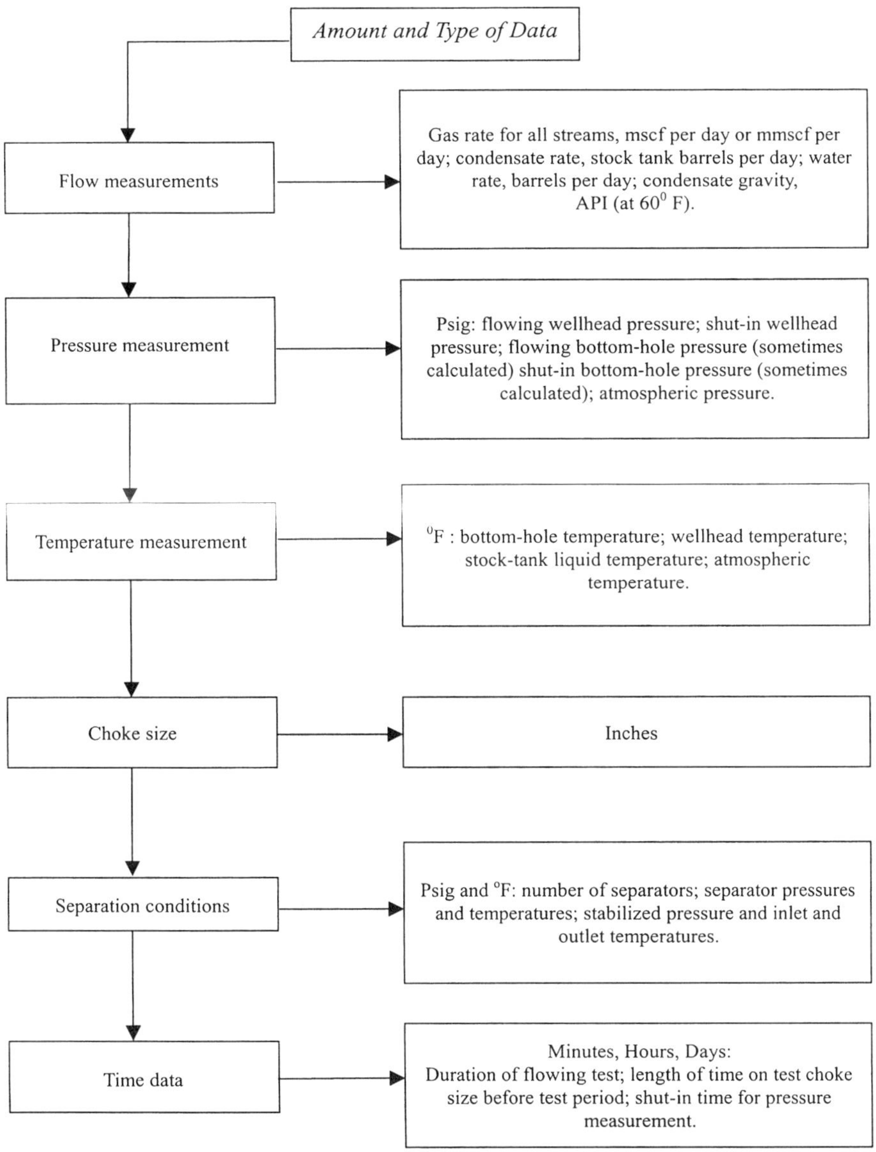

Figure 19–12. List of necessary well test data.

References and Additional Reading

1. *The Theory and Practice of Testing of Gas Wells*, 3rd ed. Energy Resources Conservation Board, Calgary, Alta. 1975.
2. Jones, P., "Reservoir Limit Test," *Oil Gas J.* (June 1956) 184–196.
3. Jones, P., "Drawdown Exploration Reservoir Limit, Well and Formation Evaluation," Paper 824-G presented at the *SPE, AIME*, Midland, TX, April 18–19, 1957.
4. Dale Beggs H., *Gas Production Operations*. Oil and Gas Consultants International Inc., Tulsa, OK, 1975.
5. Shiu, K. C., and Beggs, H. D., "Predicting Temperature in Flowing Wells," *J. Energy Res. Technol.* (March 1980), *Trans. AIME.*
6. *Field Handling of Natural Gas*, Petroleum Extension Service, University of Texas at Austin, TX, 1972.
7. *Engineering Data Book*, 9th ed. Natural Gas Processors Associations, Tulsa, OK, 1972.
8. Campbell, L. M., *Gas Conditioning and Processing*. Campbell Petroleum Series, Norman, OK, 1976.
9. *Manual of Back Pressure Testing of Gas Wells*. Interstate Oil Compact Commission, 1962.
10. *Manual of Back Pressure Testing of Gas Wells*. Kansas State Corporation Commission, 1959.
11. *Back Pressure Test for Natural Gas Wells*. Railroad Commission of Texas, 1950. Revised edition, 1951.
12. Rawlins, E. L., and Schellhardt, M. A., *Backpressure Data on Natural Gas Wells and Their Application to Production Practices*. U.S. Bureau of Mines, Monograph 7, 1936.
13. Wattenbarger, R. A., and Ramey, H. J., Jr., "Gas Well Testing with Turbulence, Damage and Wellbore Storage," *J. Petroleum Technol.* (Aug. 1968) 877–877., *Trans., AIME*, 243.

Appendix A

Use of SI Units in Gas Well Testing Equations

Example A–1 *Converting Metric to English Gas Field Units*

Equation A–1 describes the gradual decline in flowing bottom-hole pressure (atmospheres) in a gas or water well as related to time in seconds, when the rate of production is q (reservoir cc/sec). This liquid flow equation is useful generally even when gas is present. The following conversion factors and arithmetic will convert Eq. A–1 to A–2 from metric to English gas field units (however, leaving the t_D parameter in Darcy units).

$$
\begin{aligned}
1\ \text{foot} &= 30.48\ \text{cm} \\
1\ \text{day} &= 86{,}400\ \text{sec} \\
1\ \text{B/D} &= 1.84\ \text{cm}^3/\text{sec} \\
1\ \text{mD} &= 0.001\ \text{D} \\
1\ \text{psi} &= 1/14.65\ \text{atm} \\
\text{in}\, x &= 2.303 \log x
\end{aligned}
$$

Change q to $(q\beta_o)$, so that q = tank barrels

	Darcy units	English units
h	cm	ft
r_w	cm	ft
t	sec	day
q	cm^3/sec	bbl/day
μ	cP	cP
k	D	mD
p	atm	psi

$$\Delta p = p_i - p_{wf} = \frac{q\mu}{4\pi kh}\left[In\left(\frac{kt}{\phi c_t \mu r_w^2}\right) + 0.809 + 2s\right] \tag{A–1}$$

$$\frac{p_i - p_{wf}}{14.65} = \frac{(1.84)q\mu\beta_o}{4 \times 22/7k(0.001)h(30.48)}$$

$$\times\left[2.303\log\left(\frac{kt}{\phi c_t \mu r_w^2}\right) + 0.809 + 2s\right]$$

$$p_i - p_{wf} = \frac{162.6q\mu\beta_o}{kh}\left[\log\left(\frac{kt}{\phi c_t \mu r_w^2}\right) + 0.35 + 0.869s\right] \tag{A–2}$$

Table A–1
Conversion of Common Field Units to Metric (SI) Units (Base conditions: Field 60°F, 14.65 psia; Metric (SI) 15°C, 101.325 kPa)

Field unit	Multiplication factor	Metric (SI) unit	Symbol
Acre	4.046 856 E+03	Square meter	m^2
Acre	4.046 856 E-01	Hectare	ha
Acre-foot	1.233 482 E+03	Cubic meter	m^3
Atmosphere	1.013 25 E+02	Kilopascal	kPa
Barrel (35 imp. gal.)	1.589 873 E-01	Cubic meter	m^3
Btu per standard cubic foot (60°F, 14.65 psia)	8.799 136 E-01	Kilojoule per mole	kJ/mol
Centipoise	1.0 E+00	Millipascal	mPa*s
Cubic foot	2.831 685 E-01	Cubic meter	m^3
Cubic foot gas per gallon (60°F, 14.65 psia)	7.494 773 E+00	Mole per cubic meter	mol/m^3
Darcy	9.869 233 E-01	Square micrometer	μm^2
Degree Fahrenheit	(°F-32)5/9 E+00	Degree Celsius	°C
Degree Rankine	5/9 E+00	Kelvin	K
Gallon (Cdn.)	4.546 09 E-03	Cubic meter	m^3
Gallon (U.S.)	3.785 412 E-03	Cubic meter	m^3
Gas constant	8.314 32 E+00	Joule per mole kelvin	J/(mol*K)
Mcf (thousand cubic foot)	1.191 574 E+00	Kilomole	kmol
(60°F, 14.65 psia)	2.826 231 E+01	cubic meter (API)	m^3 API
Millidarcy	9.869 233 E-04	Square micrometer	μm^2
MMcf (million cubic foot)	1.191 574 E+00	Megamole	mmol
(60°F, 14.65 psia)	2.826 231 E+01	cubic meter (API)	m^3 API
Pound-force per square inch (psi)	6.894 757 E+00	Kilopascal	kPa
Pound-mass	4.535 924 E-01	Kilogram	kg
Psi per foot	2.262 059 E+01	Kilopascal per meter	kPa/m
Section (540 acres)	2.589 988 E+06	Square meter	m^2
Section (640 acres)	2.589 988 E+02	Hectare	ha
Standard cubic foot	1.191 574 E+00	Mole	mol
(60°F, 14.65 psia – ideal gas)	2.826 231 E-02	cubic meter (API)	m^3 API
Tcf (trillion cubic foot)	1.191 574 E+00	Teramole	Tmol
(60°F, 14.65 psia)	2.826 231 E-02	cubic meter (API)	m^3 API
Ton (U.S. short—2000 lb)	9.071 847 E-01	Tonne	t
Ton (U.K. long—2240 lb)	1.016 047 E+00	Tonne	t

Appendix B

Correlation Tables and Charts for Dimensionless Functions

This appendix presents correlations tables and charts for dimensionless functions for single-well systems producing at constant rate. Some data from the literature have been modified to be consistent with the nomenclature used in this text.

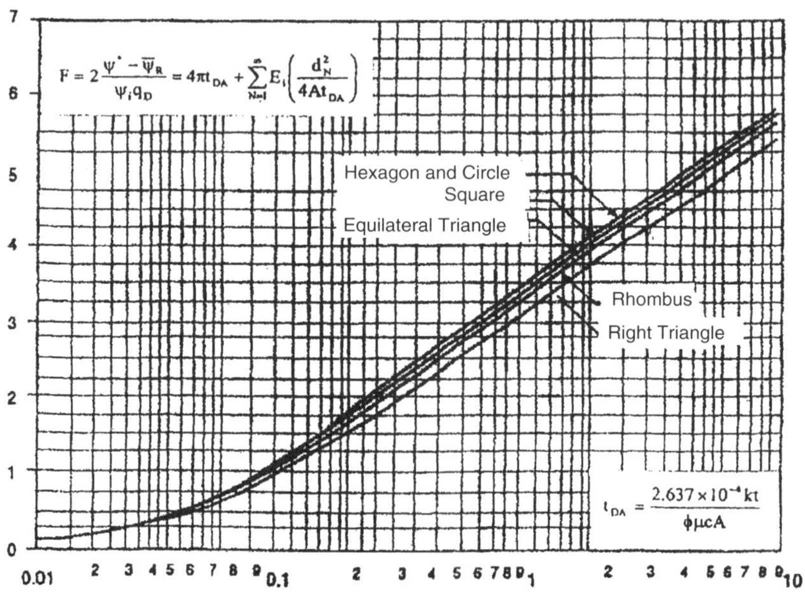

Figure B–1. MBH curves for a well at the center of a regular shaped drainage area (after Matthews, Brons, and Hazebrock).[3]

Table B–1
MBH Dimensionless Pressure Functions for Various Closed Shaped Reservoirs (after Earlougher, Ramey, Miller, and Mueller)[1]

F = MBH DIMENSIONLESS PRESSURE FUNCTION

DIMENSIONLESS TIME t_{DA}	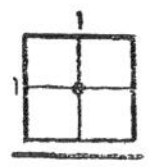	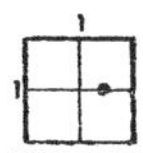	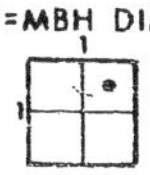	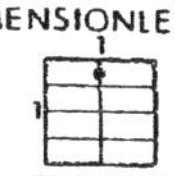	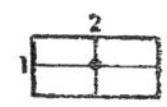	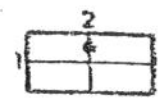	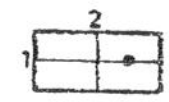	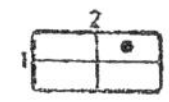
0.0010	0.0126	0.0126	0.0126	0.0126	0.0126	0.0126	0.0126	0.0126
0.0015	0.0188	0.0188	0.0188	0.0188	0.0188	0.0188	0.0188	0.0188
0.0020	0.0251	0.0251	0.0251	0.0251	0.0251	0.0251	0.0251	0.0251
0.0025	0.0314	0.0314	0.0314	0.0311	0.0314	0.0314	0.0314	0.0314
0.0030	0.0377	0.0377	0.0377	0.0368	0.0377	0.0377	0.0377	0.0377
0.0040	0.0503	0.0503	0.0503	0.0460	0.0503	0.0502	0.0503	0.0502
0.0050	0.0628	0.0628	0.0628	0.0517	0.0628	0.0626	0.0628	0.0626
0.0060	0.0754	0.0754	0.0754	0.0537	0.0754	0.0745	0.0754	0.0745
0.0070	0.0880	0.0880	0.0879	0.0524	0.0880	0.0858	0.0880	0.0858
0.0080	0.1005	0.1005	0.1004	0.0483	0.1005	0.0963	0.1005	0.0963
0.0090	0.1131	0.1130	0.1128	0.0422	0.1131	0.1059	0.1131	0.1059
0.0100	0.1257	0.1254	0.1251	0.0345	0.1257	0.1145	0.1257	0.1145
0.0150	0.1885	0.1854	0.1823	−0.0162	0.1884	0.1449	0.1884	0.1449
0.0200	0.2513	0.2402	0.2287	−0.0700	0.2508	0.1601	0.2505	0.1598
0.0250	0.3141	0.2892	0.2630	−0.1181	0.3119	0.1678	0.3107	0.1663
0.0300	0.3769	0.3333	0.2864	−0.1584	0.3708	0.1723	0.3676	0.1683
0.0400	0.5016	0.4108	0.3087	−0.2166	0.4804	0.1812	0.4686	0.1657
0.0500	0.6237	0.4791	0.3099	−0.2512	0.5785	0.1955	0.5511	0.1593
0.0600	0.7415	0.5413	0.3002	−0.2694	0.6667	0.2169	0.6167	0.1513
0.0700	0.8537	0.5991	0.2856	−0.2766	0.7471	0.2450	0.6683	0.1428
0.0800	0.9597	0.6531	0.2700	−0.2765	0.8217	0.2787	0.7085	0.1346
0.0900	1.0592	0.7038	0.2553	−0.2716	0.8917	0.3167	0.7398	0.1269
0.1000	1.1524	0.7516	0.2427	−0.2633	0.9581	0.3581	0.7642	0.1198
0.1500	1.5364	0.9583	0.2226	−0.1951	1.2524	0.5864	0.8261	0.0954
0.2000	1.8212	1.1314	0.2637	−0.1027	1.4987	0.8106	0.8476	0.0873
0.2500	2.0439	1.2854	0.3412	0.0025	1.7064	1.0104	0.8629	0.0923
0.3000	2.2262	1.4257	0.4365	0.1129	1.8830	1.1841	0.8829	0.1087
0.4000	2.5139	1.6720	0.6400	0.3300	2.1678	1.4674	0.9447	0.1687
0.5000	2.7370	1.8797	0.8321	0.5268	2.3905	1.6899	1.0299	0.2537
0.6000	2.9193	2.0563	1.0028	0.6994	2.5728	1.8722	1.1280	0.3518
0.7000	3.0735	2.2083	1.1527	0.8499	2.7269	2.0263	1.2307	0.4545
0.8000	3.2070	2.3411	1.2847	0.9821	2.8605	2.1599	1.3329	0.5567
0.9000	3.3249	2.4586	1.4019	1.0994	2.9782	2.2776	1.4315	0.6552
1.0000	3.4302	2.5638	1.5070	1.2045	3.0836	2.3830	1.5252	0.7489
2.0000	4.1234	3.2569	2.2000	1.8976	3.7768	3.0762	2.2002	1.4239
4.0000	4.8166	3.9501	2.8933	2.5908	4.4701	3.7695	2.8932	2.1170
8.0000	5.5099	4.6435	3.5867	3.2842	5.1633	4.4627	3.5866	2.8103
10.0000	5.7331	4.8667	3.8098	3.5073	5.3865	4.6859	3.8098	3.0335

F = MBH DIMENSIONLESS PRESSURE FUNCTION

DIMENSIONLESS TIME t_{DA}	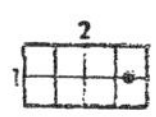	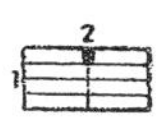	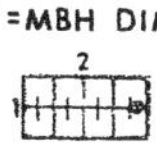					
0.0010	0.0125	0.0126	0.0125	0.0126	0.0126	0.0126	0.0126	0.0126
0.0015	0.0179	0.0188	0.0179	0.0188	0.0188	0.0188	0.0188	0.0188
0.0020	0.0209	0.0251	0.0208	0.0251	0.0251	0.0251	0.0251	0.0251
0.0025	0.0203	0.0314	0.0200	0.0314	0.0311	0.0314	0.0311	0.0314
0.0030	0.0160	0.0377	0.0155	0.0377	0.0368	0.0377	0.0368	0.0377
0.0040	−0.0019	0.0502	−0.0027	0.0503	0.0460	0.0503	0.0460	0.0503
0.0050	−0.0284	0.0626	−0.0295	0.0628	0.0517	0.0628	0.0517	0.0628
0.0060	−0.0596	0.0745	−0.0612	0.0754	0.0537	0.0754	0.0537	0.0753
0.0070	−0.0932	0.0858	−0.0951	0.0879	0.0524	0.0879	0.0524	0.0878
0.0080	−0.1277	0.0962	−0.1298	0.1004	0.0483	0.1004	0.0483	0.1000
0.0090	−0.1620	0.1058	−0.1644	0.1129	0.0422	0.1129	0.0422	0.1119
0.0100	−0.1957	0.1144	−0.1983	0.1251	0.0345	0.1251	0.0345	0.1234
0.0150	−0.3468	0.1445	−0.3502	0.1823	−0.0162	0.1823	−0.0162	0.1713
0.0200	−0.4670	0.1589	−0.4718	0.2291	−0.0701	0.2291	−0.0701	0.2015
0.0250	−0.5615	0.1641	−0.5695	0.2643	−0.1186	0.2643	−0.1187	0.2163
0.0300	−0.6357	0.1633	−0.6507	0.2897	−0.1600	0.2897	−0.1600	0.2200
0.0400	−0.7395	0.1492	−0.7839	0.3197	−0.2231	0.3194	−0.2235	0.2075
0.0500	−0.8012	0.1224	−0.8965	0.3332	−0.2662	0.3315	−0.2682	0.1820
0.0600	−0.8339	0.0862	−0.9989	0.3385	−0.2957	0.3335	−0.3013	0.1516
0.0700	−0.8457	0.0437	−1.0949	0.3399	−0.3158	0.3290	−0.3278	0.1203
0.0800	−0.8422	−0.0028	−1.1859	0.3401	−0.3291	0.3199	−0.3510	0.0899
0.0900	−0.8272	−0.0512	−1.2723	0.3403	−0.3375	0.3072	−0.3727	0.0613
0.1000	−0.8038	−0.1004	−1.3542	0.3412	−0.3421	0.2915	−0.3942	0.0351
0.1500	−0.6223	−0.3322	−1.7021	0.3663	−0.3257	0.1826	−0.5128	−0.0580
0.2000	−0.4138	−0.5189	−1.9613	0.4269	−0.2661	0.0468	−0.6499	−0.0935
0.2500	−0.2196	−0.6580	−2.1508	0.5120	−0.1811	−0.0959	−0.7928	−0.0855
0.3000	−0.0479	−0.7555	−2.2854	0.6102	−0.0829	−0.2344	−0.9312	−0.0477
0.4000	0.2343	−0.8547	−2.4344	0.8152	0.1220	−0.4789	−1.1758	0.0773
0.5000	0.4567	−0.8671	−2.4768	1.0075	0.3143	−0.6712	−1.3681	0.2266
0.6000	0.6389	−0.8284	−2.4564	1.1783	0.4852	−0.8134	−1.5103	0.3753
0.7000	0.7931	−0.7620	−2.4011	1.3282	0.6351	−0.9129	−1.6098	0.5143
0.8000	0.9267	−0.6820	−2.3278	1.4602	0.7670	−0.9775	−1.6744	0.6409
0.9000	1.0444	−0.5969	−2.2469	1.5774	0.8843	−1.0145	−1.7114	0.7555
1.0000	1.1497	−0.5115	−2.1640	1.6825	0.9894	−1.0301	−1.7270	0.8595
2.0000	1.8430	0.1507	−1.5058	2.3755	1.6824	−0.7325	−1.4294	1.5516
4.0000	2.5363	0.8436	−0.8129	3.0688	2.3757	−0.0756	−0.7725	2.2448
8.0000	3.2295	1.5370	−0.1195	3.7623	3.0691	0.6173	−0.0796	2.9381
10.0000	3.4527	1.7601	0.1036	3.9854	3.2922	0.8406	0.1438	3.1615

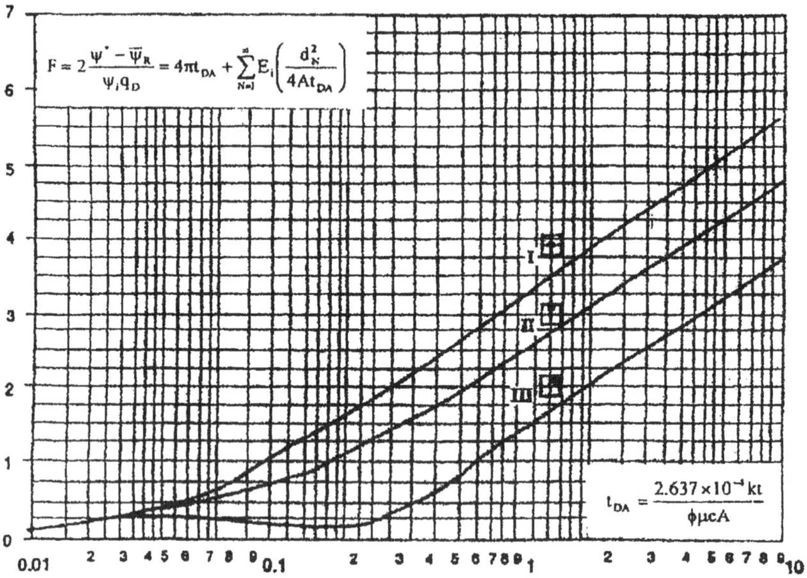

Figure B–2. MBH curves for a well situated within a square (after Matthews, Brons, and Hazebrock).[3]

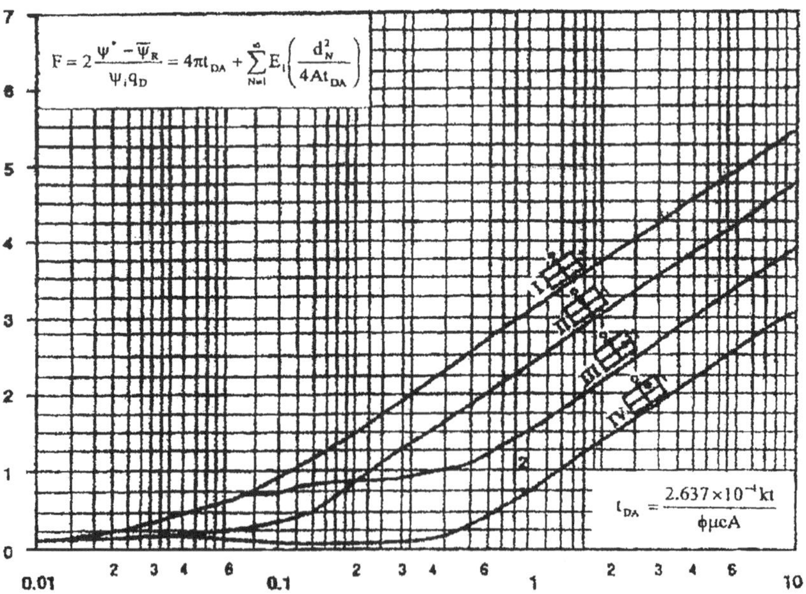

Figure B–3. MBH curves for a well situated within a 2:1 rectangular (after Matthews, Brons, and Hazebrock).[3]

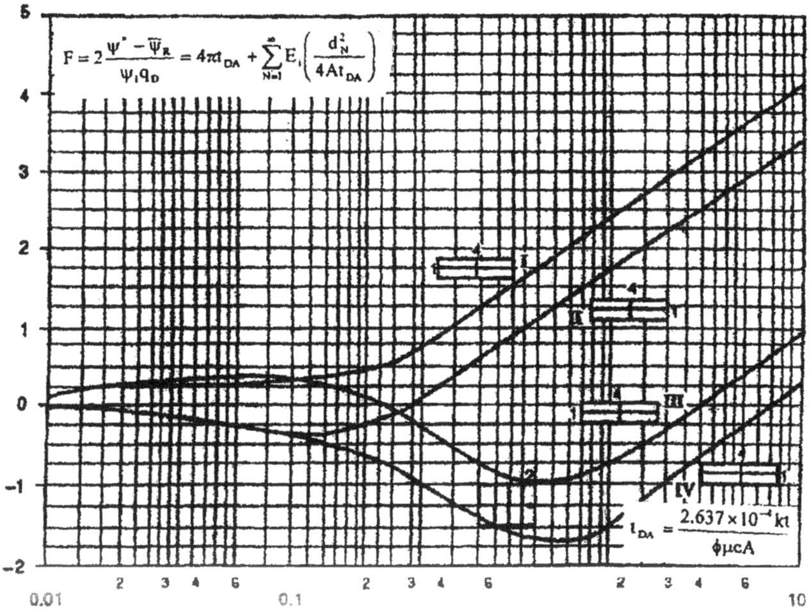

Figure B–4. MBH curves for a well situated within a 4:1 rectangle (after Matthews, Brons, and Hazebrock).[3]

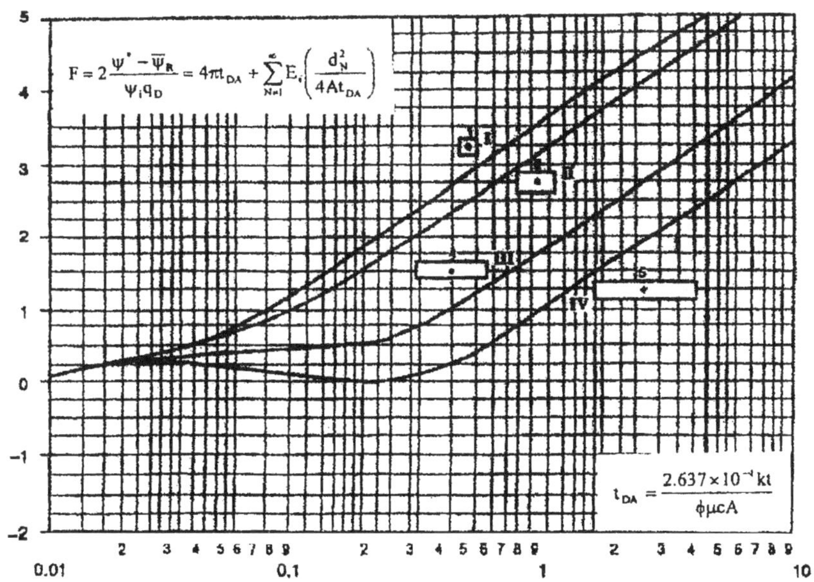

Figure B–5. MBH curves for a well situated in various rectangular geometries (after Matthews, Brons, and Hazebrock).[3]

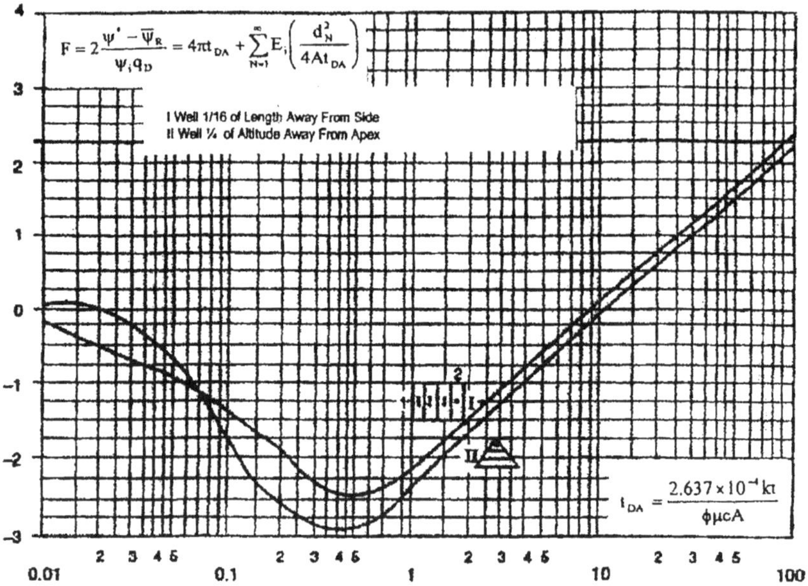

Figure B–6. MBH curves for a well situated within a square and in 2:1 rectangle (after Matthews, Brons, and Hazebrock).[3]

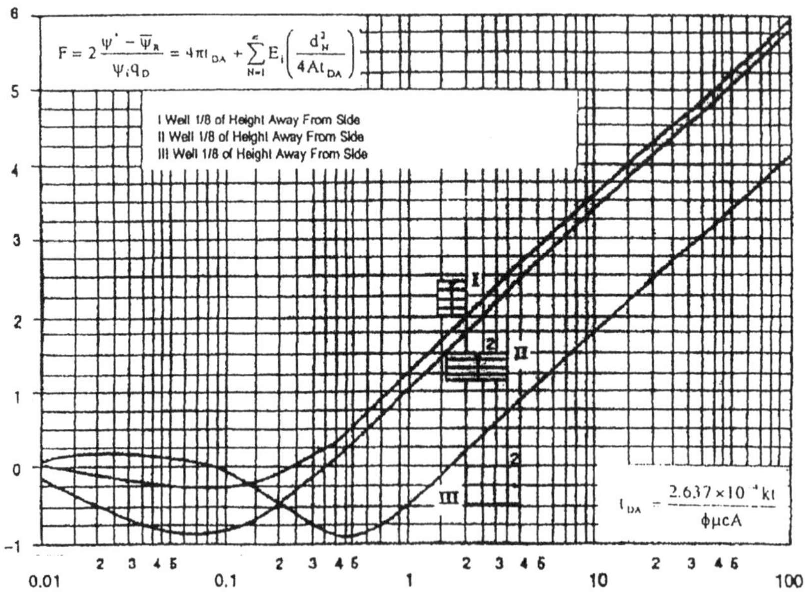

Figure B–7. MBH curves for a well situated in 2:1 rectangle and in an equilateral triangle (after Matthews, Brons, and Hazebrock).[3]

Table B–2
Pseudo-Steady-State Shape Factors for Various Reservoirs (from Dietz)[2]

In Bounded Reservoirs	ln C_A	C_A	Stabilized Conditions for t_{DA} >		ln C_A	C_A	Stabilized Conditions for t_{DA} >
	3.45	31.6	0.1		2.38	10.8	0.3
	3.43	30.9	0.1		1.58	4.86	1.0
	3.45	31.6	0.1		0.73	2.07	0.8
	3.32	27.6	0.2		1.00	2.72	0.8
	3.30	2.71	0.2		−1.46	0.232	2.5
	3.09	21.9	0.4		−2.16	0.115	3.0
	3.12	22.6	0.2		1.22	3.39	0.6
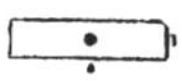	1.68	5.38	0.7		1.4	3.13	0.3
	0.86	2.36	0.7	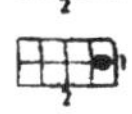	−0.50	0.607	1.0
	2.56	12.9	0.6	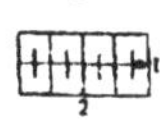	−2.20	0.111	1.2
	1.52	4.57	0.5		−2.32	0.098	0.9
				In water drive reservoirs			
					2.05	19.1	0.1
				In reservoirs of unknown production character			
				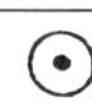	2.22	2.5	0.1

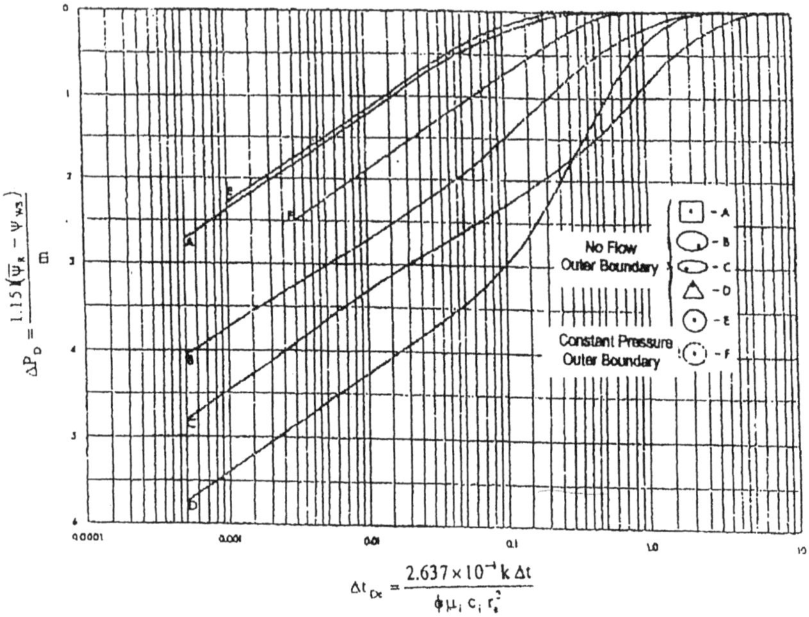

Figure B–8. MBH dimensionless pressures ABCDEF (after Pitzer).[4]

References and Additional Reading

1. Earlougher, R. C., Jr., Ramey, H. J., Jr., Miller, F. G., and Mueller, T. D., "Pressure Distributions in Rectangular Reservoirs," *J. Petroleum Technol.* (1968) 20, 199–208.
2. Dietz, D. N., "Determination of Average Reservoir Pressure from Build-up Surveys," *Trans. AIME* (1965) 234, 935–959.
3. Matthews, C, S., Brons, F., and Hazebrock, P., "A Method for Determination of Average Pressure in a Bounded Reservoir," *Trans. AIME* (1954) 209, 182–189.
4. Pitzer, S. C., "Evaluation of Acid Treatments from Pressure Buildup-up Analysis," *Trans. AIME* (1964) 216, 38–43.

Appendix C

Estimation of Formation Characteristics from Drill-Stem Test

The normal Horner and/or Miller, Dyes, and Hutchinson methods are applicable to drill-stem tests. At times, shortcut methods as used by service companies in field analysis are reliable. The drill-stem test often uses two bombs, and one or more flow and shut-in sequences are recorded, as illustrated in Figures C–1 and C–2. To illustrate how a typical DST is performed, we will examine a schematic chart (Figure C–1) of pressure versus time from a test with two flow periods and two shut-in periods.

Point A: Tool is lowered into the hole.
Point B: Tool is on bottom.
Point C: Packers are set, the mud column is compressed, and a still higher pressure is recorded.
Point D: Tool is opened for an initial flow period and the pressure drop as shown.
Point E to point F: Fluid accumulates in the drill stem above the pressure gauge, the well is shut in, and pressure rises to point F.
Point G to point H: After a suitable shut-in period, the well is reopened for a second final flow period from point G to point H.
Point H to point I: Final shut-in period.
Point J to point K: Packers are then released; the testing device is then removed from the hole.

C.1 Normal Routine Drill-Stem Test

The first flow is very short and is designed to remove any excess pressure, which may have resulted from setting the packers. The first buildup is rather long since a reliable value for the initial reservoir pressure is desired.

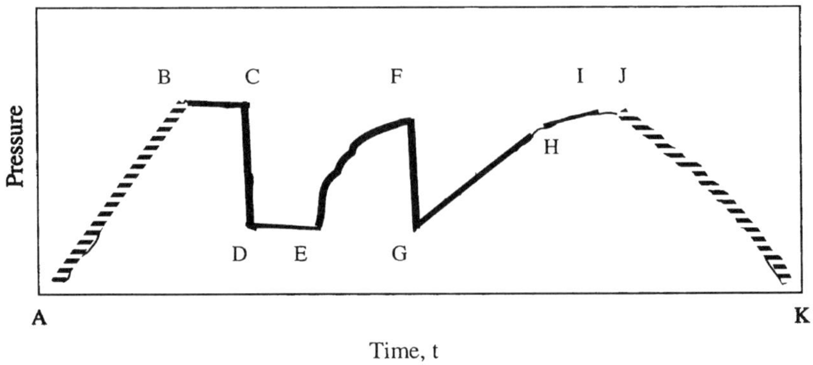

Figure C–1. Schematic of drill-stem test pressure chart.

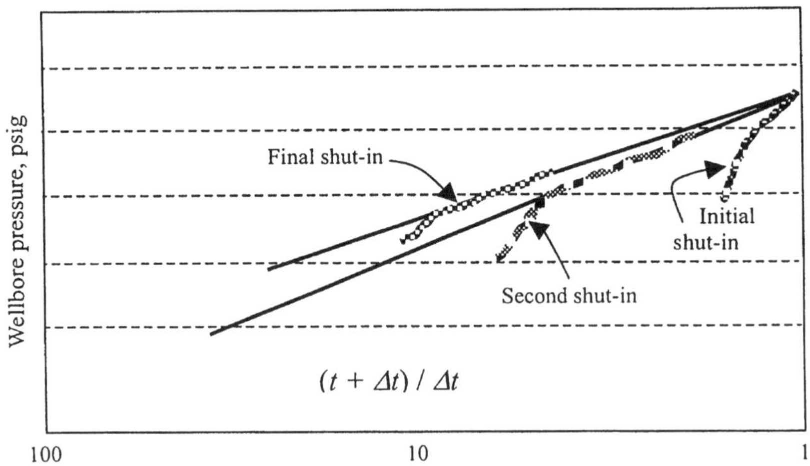

Figure C–2. Interpretation method for pressure buildup: Horner plot with minimum after flow obtainable with packer.

The second flow is somewhat longer and is designed to evaluate the formation for some distance from the well. The second shut-in is used to calculate transmissibility and other characteristics of the reservoir. If the second extrapolated pressure is less than the pressure of the first shut-in, depletion of the small reservoir should be suspected.

C.2 Determination of Effective Permeability, Skin Factor, and Damage Ratio

A drill-stem test (DST) is a short-duration test. Drill-stem test pressure buildup data are analyzed much like any other pressure buildup data; the techniques of Chapter 6 apply. In a DST, the flow period is about the same duration as the shut-in period, so pressure buildup data must be analyzed with the Horner plot,

$$\Psi(p_{ws}) \text{ versus } \log\left[\frac{t_p + \Delta t}{\Delta t}\right].$$

The value used for t_p is usually the length of the preceding flow period. If the shut-in period is long enough and if wellbore storage is not dominant, a Horner plot of the buildup data should have a straight-line section with slope m, as indicated in Chapter 6. The value of m may be used to estimate reservoir permeability k from Eq. 6–5:

$$k = \frac{57.920 \times 10^6 q_{sc} T P_{sc}}{m h T_{sc}} \tag{C–1}$$

If μ_g and h are not known, kh/μ_g may be estimated by rearranging Eq. 8–4. The flow rate normally used is the average over t_p. The skin factor is estimated from

$$s' = 1.151\left[\frac{\psi(p_{1hr}) - \psi(p_{wfo})}{m} + \log\left(\frac{t_p + 1}{t_p}\right) - \log\left(\frac{k}{\phi \mu_{gi} c_i r_w^2}\right) + 3.23\right] \tag{C–2}$$

DST analyses commonly report damage ratio:

$$DR = \frac{\psi(\overline{p}_R) - \psi(p_{wf})}{\psi(\overline{p}_R) - \psi(p_{wf}) - \psi(\Delta p)_{skin}} \tag{C–3}$$

where pressure drop across the skin is computed from

$$\psi(\Delta p)_{skin} = 0.869 m s' \tag{C–4}$$

C.3 Initial Reservoir Pressure Estimation Technique

Initial or average reservoir pressure is estimated by extrapolating the Horner straight line to infinite shut-in time, $(\frac{t_p+\Delta t}{\Delta t}) = 1$. If the rate varies during the flow period, then the multiple analysis technique is used. Odeh and Selig[2] proposed a simplified analysis technique that is useful for large rate variation when t_p is less than shut-in time. The rate and t_p are modified by

$$q^* = \frac{1}{t_p^*} \sum_{j=1}^{N} q_j (t_j - t_{j-1}) \tag{C–5}$$

and

$$t_p^* = 2\left[t_p - \frac{\sum q_j(t_j^2 - t_{j-1}^2)}{2\sum_{j=1}^{N} q_j(t_j - t_{j-1})}\right] \qquad \text{(C–6)}$$

C.4 Radius of Investigation

The modified values t_p^* and q^* are used in the Horner plot. For practical purpose, the radius of investigation during DST is equivalent to the radius of drainage given by

$$r_i = \sqrt{\frac{k\Delta t_{max}}{948\phi\mu_g c_t}} \qquad \text{(C–7)}$$

References and Additional Reading

1. Van Poollen, H. K., "Status of Drill-Stem Testing Techniques and Analysis." *J. Petroleum Technol.* (April 1961) 333–339. Also Reprint Series, No. 9—*Pressure Analysis Methods*, Society of Petroleum Engineers of AIME, Dallas, TX, 1967, pp. 104–110.
2. Odeh, A. S. and Selig, F., "Pressure Buildup Analysis Variable-Rate Case," *J. Petroleum Tech.* (July 1963) 790–794. *Trans. AIME*, 228. Also Reprint Series, No. 9—*Pressure Analysis Methods*, Society of Petroleum Engineers of AIME, Dallas, TX, 1967, pp. 131–135.

Appendix D

Gas Flow Rate Measurement Techniques

D.1 Gas Flow Rate Calculations

The natural gas is measured by volume in standard cubic feet at an operating pressure and temperature and is corrected to some reference or base pressure and temperature, generally 60° F and atmospheric pressure. The two most commonly used gas measurement devices are the orifice meters and critical flow provers.

D.2 Determining Orifice Meter Constants and Factors

Orifice Meters

An orifice meter is utilized for gas measurement if environmental constraints prohibit gas venting to the atmosphere. A typical closed-orifice metering system consists of orifice plates, meter tubes, flange taps, and pipe taps. The gas flow rate through a closed-orifice metering system is determined using

$$q_{sc} = C'\sqrt{h_w p_f} \tag{D–1}$$

where

$$C' = F_b F_{Pb} F_{tb} F_g F_{tf} F_r F_{Pv} F_m Y \tag{D–2}$$

The term C' is known as the orifice constant, the value of which depends primarily on the basic orifice factor, F_b. Values for most of these constants are tabulated for various orifice sizes and flowing conditions in Tables D–1 and D–2. The $F_b F_{Pb} F_{tb} F_g F_{tf} F_r F_{Pv} F_m Y$ factors are determined empirically and are periodically updated by the AGA.

Table D–1
Basic Critical Flow Prover Factors (P_{sc} = 14.65 psia; T_{sc} = 520°R; T_b = 520°R; γ_b = 1.000)

2-inch prover orifice diameter (inch)	2-inch prover factor F_p (mscfd)	4-inch prover orifice diameter (inch)	4-inch prover factor F_p (mscfd)
1/16	0.06560	1/4	1.74
3/32	0.1446	3/8	2.414
1/8	0.2716	1/2	4.319
3/16	0.6237	5/8	6.729
7/32	0.8608	3/4	9.643
1/4	1.115	7/8	13.11
5/16	1.714	1	17.08
3/8	2.439	1-1/8	21.57
7/16	3.495	1-1/4	26.57
1/2	0.06560	1/4	1.74

Orifice Constants

The values of the constants in Eq. D–2 depend on the points between which the differential pressure h_w is measured. Two standards are provided in gas measurement flange taps and pipe taps. With the former, the flange or orifice holder is tapped so that the center of the upstream and downstream taps is 1 inch from the respective orifice-plate surface. For standard pipe taps the upstream tap is located $2\frac{1}{2}$ inch pipe diameters upstream and 8-inch pipe diameter downstream. The location of the taps makes an obvious difference in the values obtained. Table D–1 is provided for both configurations. The relative locations of the taps are shown in Figure D–1.

Basic Orifice Factor F_b

The charts show values of F_b for both flange and pipe taps.

Pressure Base Factor F_{pb}

The F_{pb} factor corrects the value of F_b for cases where the pressure base used is not 14.73 psia. It may be determined by the equation $F_{pb} = 14.73/P_b$.

Temperature-Base Factor F_{tb}

The F_{tb} factor corrects for any contract wherein the base temperature is not 520°R (60°F). This factor may be computed by the formula $F_{tb} = T_b/520$.

Table D–2
Orifice Coefficient for Critical Flow Provers[a]

Size of orifice (inches)	2-inch prover	4-inch prover
1/16	1.524	
3/32	3.355	
1/8	6.301	
5/16	14.47	
7/32	19.97	
1/4	25.86	24.92
5/16	39.77	
3/8	56.68	56.01
7/16	81.09	
1/2	101.8	100.2
5/8	154.0	156.1
3/4	224.9	223.7
7/8	309.3	304.2
1	406.7	396.3
1-1/8	520.8	499.2
1-1/4	657.5	616.4
1-3/8	807.8	742.1
1-1/2	1002.0	884.3
1-3/4		1208
2		1596
2-1/2		2566
3		3904

[a] Adopted from Bureau of Mines Monograph 7 by the Interstate Oil Compact Commission.

Specific-Gravity Factor F_g

The F_g factor is to correct the basic orifice equation for those cases where the specific gravity of the gas is other than 1.000. The equation is $F_g = \sqrt{1/\gamma_g}$.

Flowing-Temperature Factor F_{tf}

The F_{tf} factor corrects for those cases where the flowing temperature of the gas is other than 60°F. The equation is $F_{tb} = \sqrt{(520/T_f)}$.

Reynolds-Number Factor F_r

The F_r factor takes into account the variation of the discharge coefficient with Reynolds number. In gas measurement the variation is slight and is often

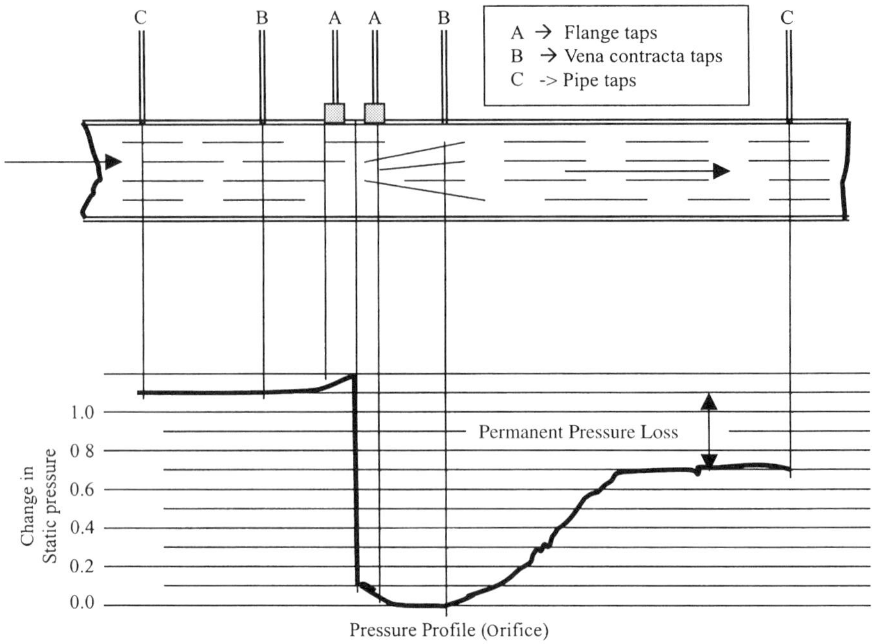

Figure D–1. Relative locations of taps. Courtesy John M. Cambell.

ignored in production operations. Values are shown in the charts. It has been assumed in these charts that gas viscosity is substantially constant. The constant b shown in the charts is then primarily a function of pipe diameter, orifice diameter, and the location of the differential-pressure taps.

Expansion Factor Y

The Y factor accounts for the change in gas density as the pressure changes across the orifice. This correction is small and often ignored; the value used depends on which of the differential-pressure taps is used to measure static pressure and the location of the tap. The additional primary variables involved are (1) β, (2) ratio of differential pressure to absolute pressure, and (3) the specific-heat ratio C_p/C_v. In the standard chart the last variable is taken as constant and equal to 1.3. Tables of this factor are shown in the charts.

Supercompressibility Factor F_{pv}

The variation from the ideal gas laws of an actual gas is corrected by the F_{pv} factor. It may be estimated from the equation $F_{pv} = \sqrt{1/z}$ where z is equal to the compressibility factor obtained from standard correlations.

Manometer Factor F_m

The F_m factor is used only with mercury-type meters, to correct for the slight error in measurement caused by having different heads of gas above the two legs of the manometer. For all practical purposes it is insignificant. A thorough discussion of all types of fluid meters may be found in Ref. 2. A limited selection of tables for determining the constants for use in orifice meters is included in this appendix. A more complete set of tables and charts may be found in Ref. 1, 2, and 3.

Example D–1 *Calculating Gas Flow Rate*

A meter run that is equipped with flange taps and a 2.5-inch orifice has an inside diameter of 7.625 inches. The static pressure, obtained from the downstream tap, reads 795 psia and the average differential pressure is 22 inches of water. If the pressure and temperature bases are 15.05 psia and 60°F, respectively, calculate the daily flow rate of gas through an orifice meter. The gas specific gravity is 0.734 and the flowing temperature is 88°F.

Solution

1. Calculate $C' = F_b F_{Pb} F_{tb} F_g F_{tf} F_r F_{Pv} F_m Y$ by using Eq. D–2. From the tables, for $d = 2.50$ and $D = 7.625$,

$$F_b = 1272.3, \quad F_{pb} = \frac{14.73}{p_b} = \frac{14.73}{15.05} = 0.9787, \quad F_{tb} = 1.000,$$

$$F_g = (\gamma_g)^{-0.5} = (0.734)^{-0.5} = 1.1672$$

From Table D–1, $b = 0.0239$; therefore

$$F_r = 1 + \frac{b}{\sqrt{h_w p_f}} = 1 + \frac{0.0239}{\sqrt{(33)(795)}} = 1.00015$$

For $\gamma_g = 0.734$, $T = 88°F$, $P_f = 795$ psia, the value calculated for z is 0.8448.

$$F_{PV} = (z)^{-0.5} = (0.8448)^{-0.5} = 1.0880; \quad F_m = 1.000$$

To determine Y,

$$B = d/D = 2.50/7.625 = 0.3279$$

$$h_w/p_f = 33/795 = 0.04151$$

From the tables, $Y = 0.9988$ (requires interpolation).

$$C' = 1272.3(0.9787)(1.000)(1.1672)(1.00015)(1.088)(0.9988)$$
$$= 1579.64$$

2. Calculate q_{sc}:

$$q_{sc} = C'\sqrt{h_w p_f} = 1579.64\sqrt{(33)(795)} = 255{,}857.80 \text{ scf/hr}$$
$$= 24 \times 255{,}857.80 = 6.141 \text{ mmscfd}$$

Example D–2 *Determining Orifice Plate Size*
A metering system is required to measure approximately 2.316 mmscf/d of 0.732 gravity gas at a line pressure of 815 psia. The meter run is to be made of 8-inch pipe (7.625 inch i.d.). Determine the size of orifice plate to give a differential of about 77 inches. Flowing temperature averages about 88°F.

Solution For $h_w = 77$ and using Eq. C–1,

$$C' = \frac{2.316 \times 10^6}{24\sqrt{77(815)}} = 359.3$$

For an approximation all of the terms in Eq. C–2 except F_g and F_{tf} can be ignored in this case. Therefore,

$$F_b = \frac{C'}{F_g F_{tf}} = \frac{359.3}{(1.1688)(0.9741)} = 315.58$$

From the F_b tables for large taps, for $D = 7.625$ one obtains:

d	F_b
1.25	313.19
1.376	380.25

Therefore, a 1.25-inch orifice plate would be selected to obtain an h_w reading of approximately 77 inches at the design flow rate.

D.3 Critical-Flow Prover

The critical-flow prover is a device that also exhausts the gas to the atmosphere. It is also a special pipe nipple with a flange for holding special plates to the end. The equation for calculating the rate of flow through a critical prover is

$$q_{sc} = \frac{Cp}{(\gamma_g T)^{0.5}} \tag{D–3}$$

where

q_{sc} = rate of flow, mcf/d, measured at 14.4 psia and 60°F
γ_g = specific gravity of gas (air = 1.000)
T = absolute temperature, °R,
p = pressure on prover, psia
C = orifice coefficient for prover

The critical-flow prover is one of the basic devices used for determining the gas flow rate in the open-flow testing of gas wells. In the open orifice metering system, gas is measured using the critical-flow prover in which gas flows directly to the atmosphere. This type of gas metering is quick and easy to install for well testing; however, when the gas is vented, a large pressure drop across the orifice may cause hydrates or ice to form. The equation for calculating the rate of flow through a critical-flow prover is

$$q_{sc} = F_p \times p_m \times \sqrt{\frac{520}{T + 460}} \cdot \sqrt{\frac{1}{\gamma_g}} \cdot \sqrt{\frac{1}{z}} \qquad \text{(D–4)}$$

where F_p is the basic orifice prover factor and p_m is the upstream pressure of the orifice plate in psi.

References and Additional Reading

1. American Gas Association; "Orifice Metering of Natural Gas," Gas Measurement Committee Report No. 3, 1969, revision.
2. *Fluid Meters—Their Theory and Application: Report of ASME Research Committee on Fluid Meters*, 6th ed. The American Society of Mechanical Engineers, New York, 1971.
3. *Orifice Meter Constants: Handbook E-2*. Singer American Meter Division, 1973.

Appendix E

Computing Flowing Bottom-Hole Pressure from Wellhead Pressure

Ideally, pressure should be recorded continuously during a transient test. Best results are obtained when the bottom-hole pressure is measured, although surface pressures often can be converted to bottom-hole values if adequate information is available about the wellbore system. It is usually beneficial to record bottom-hole, tubing-head, and casing-head pressures during a well test. That combination of data can provide information about wellbore effects, such as storage, and leaking packers or tubing. Such surface pressures data may be valuable in verifying correct operation of the down-hole pressure gauge. This appendix will describe a method to determine bottom-hole pressure from wellhead pressure.

E.1 Cullender and Smith Method

The procedure for this method is as follows:

1. Calculate the Reynolds number:

$$N_{\text{Re}} = \frac{20{,}011 \gamma_g q_{sc}}{\overline{\mu}\, d} \tag{E–1}$$

2. Estimate friction factor f:

$$\frac{1}{\sqrt{f}} 1.14 - 2 \log \left(\frac{\epsilon}{d} + \frac{21.25}{N_{\text{Re}}^{0.9}} \right) \tag{E–2}$$

3. Calculate F^2:

$$F = \frac{0.667 f\, q_{sc}^2}{d^5} \tag{E–3}$$

4. Find the ratio

$$\frac{TVD}{MD} = \frac{\text{True vertical depth}}{\text{Measured depth}} = \cos\theta \tag{E–4}$$

5. Estimate I_{tf}:

$$I_{tf} = \frac{\frac{P}{Tz}}{0.001\left(\frac{P}{Tz}\right)^2 \frac{TVD}{MD} + F^2} \tag{E–5}$$

6. Calculate p^*_{mf}:

$$p^*_{mf} = p_{tf}(1 + 2.5 \times 10^{-5} \times \frac{MD}{2} \tag{E–6}$$

7. Estimate I_{mf}:

$$I = \frac{P/TZ}{0.001\left(\frac{P}{TZ}\right)^2 \frac{TVD}{MD} + F^2} \tag{E–7}$$

8. Calculate P_{mf}:

$$P_{mf} = P_{tf} + \frac{18.75\gamma_g H}{I_{mf} + T_{tf}} \tag{E–8}$$

9. Compare P_{mf} and P^*_{wf}. If not close enough, set $P^*_{wf} = P_{wf}$ and go to 4.
10. Repeat until abs $(P_{wf} - P^*_{wf}) < 0.001$ or any other tolerance preferred. When the tolerance is met, the pressure calculated in step 8 is correct.

Example E–1 *Calculating the Flowing Bottom-Hole Pressure Using Cullendar and Smith Method*

Given: Well depth = 10,000 ft; gas gravity = 0.75; formation temperature = 245°F; wellhead temperature = 110°F; roughness = 0.0006 inches; μ = 0.012 cP; d = 2.441 inches; P_{tf} = 2000 psia; q_{scx} = 4.915 mmscfd.

Solution From Eq. E–1, $N_{\text{Re}} = 2.518 \times 10^6$, and find from Eq. E–2 friction factor $f = 0.015$.

Using Eq. E–3:

$$F^2 = 0.00279$$

First Trial

From Eq. E–5:

$$I_{tf} = 181.60$$

Using Eq. E–6:

$$P_{wf}^{*} = 2250\,\text{psia}$$

From Eq. E–7:

$$I_{mf} = 197.81$$

Using Eq. E–8:

$$P_{mf} = 2371\,(\text{not close enough to } P_{wf}^{*})$$

Second Trial

From Eq. E–5:

$$I_{tf} = 189.88$$

Using Eq. E–8:

$$P_{wf} = 2379$$

Third Trial

From Eq. E–5:

$$I_{tf} = 189.41$$

Using Eq. E–8:

$$P_{wf} = 2379\,\text{psia}$$

Therefore the pressure at the midpoint of the gas well is 2379 psia. The value of P_{wf} is now calculated.

Estimate from Eq. E–6:

$$P_{wf}^{*} = 2676$$

Calculate from Eq. E–7:

$$I_{mf} = 199.39$$

First Trial

Calculate from Eq. E–8:

$$P_{wf} = 2741$$

Second Trial

Calculate from Eq. E–7:

$$I_{mf} = 196.00$$

Calculate from Eq. E–8:

$$P_{wf} = 2744 \text{ psia}$$

This is close enough to the previously calculated value of 2741 psia. Therefore, the flowing bottom-hole pressure is 2744 psia.

References and Additional Reading

1. Cullender, M. H., and Smith, R. V., "Practical Solution of Gas Flow Equation for Wells and Pipelines with Large Temperature Gradients," *Trans. AIME* (1956) 207.

Appendix F

Fluid and Rock Property Correlations

This appendix presents concepts and applications of fluid and rock properties usually required for solving reservoir engineering and transient well test analysis problems. The engineering equations and correlations presented in this appendix represent technical papers well known to the petroleum engineers. For most of these properties, laboratory analysis provides the most accurate answer; however, in many cases, laboratory results are not available, and the test analyst must use the following two approaches, which are adopted for computing or finding the various properties:

1. Equation approach and
2. Figure, chart, or table approach

When laboratory results are not available, the test analyst must use empirical correlations of experimental data. This appendix provides a summary of correlations that have proved useful for test analysis. The appendix is divided into the following sections:

- Gas properties and correlations
- Reservoir rock properties
- Reservoir PVT water properties

For the properties where the equations require simple mathematical manipulations, both the equations and the charts are presented. You may use either the equations or the charts and tables. Each property computation and its use are illustrated by a solved example.

F.1 Gas Properties and Correlations

Pseudocritical Properties

Each component of a gas mixture has its own critical temperature T_c and critical pressure P_c. When the individual critical property is multiplied by the mole fraction of the whole gas mixture, we get pseudocritical temperature T_{pc} and pseudocritical pressure P_{pc}. If gas composition is available, more accurate properties will be calculated using the composition. T_{pc} and P_{pc} will be estimated from empirical correlations developed by Brown *et al.*[1] using equations or charts.[1]

For Condensate Fluids

$$T_{pc} = 187 + 330\gamma_g - 71.5\gamma_g^2 \tag{F–1}$$

$$P_{pc} = 706 - 51.7\gamma_g - 11.1\gamma_g^2 \tag{F–2}$$

For Miscellaneous Gases

$$T_{pc} = 168 + 325\gamma_g - 12.5\gamma_g^2 \tag{F–3}$$

$$P_{pc} = 677 + 15.0\gamma_g - 37.5\gamma_g^2 \tag{F–4}$$

where γ_g is the specific gravity of whole gas mixture (air = 1.000). Equations F–1 and F–2 are used for calculations when gas is in equilibrium with crude oil or condensate in the reservoir. Equations F–3 and F–4 are used for dry surface gases. These equations were developed for gases with no contaminants, such as H_2S, CO_2, and N_2. If any or all of these contaminants are present, the values obtained above must be corrected using the Wichert–Aziz correction.[2]

In Eqs. F–1 through F–4, γ_g is replaced by γ_{hc} and is given by

$$\gamma_{hc} = \frac{\gamma_g - 0.967Y_{N_2} - 1.5195Y_{CO_2} - 1.1765Y_{H_2S}}{A} \tag{F–5}$$

and the pseudocritical temperature and pressure for the whole gas mixture are

$$T_{pc1} = AT_{pc} + 227.3Y_{N_2} + 547.6Y_{CO_2} + 672.4Y_{H_2S} \tag{F–6}$$

$$P_{pc1} = AP_{pc} + 493.0Y_{N_2} + 1071Y_{CO_2} + 1306Y_{H_2S} \tag{F–7}$$

where

Y_{CO_2} = mole fraction of carbon dioxide in gas mixture

Y_{N_2} = mole fraction of nitrogen in gas mixture

Y_{H_2S} = mole fraction of hydrogen sulfide in gas mixture

$A = \left(1 - Y_{N_2} - Y_{CO_2} - Y_{H_2S}\right)$

The pseudocritical properties T_{pc1} and P_{pc1} are then corrected for acid gases (CO_2 and H_2S) by means of the Wichert–Aziz correction factor. The Wichert–Aziz correction factor ε is given by

$$\varepsilon = 120\left(Y_{CO_2} + Y_{H_2S}\right)^{0.90} - \left(Y_{CO_2} + Y_{H_2S}\right)^{1.6} + 15\left(Y_{H_2S}\right)^{0.5} - \left(Y_{H_2S}\right)^{4} \quad \text{(F–8)}$$

The corrected pseudocritical temperature and pressure are

$$T^*_{pc} = T_{pcx1} - \varepsilon \quad \text{(F–9)}$$

$$P^*_{pc} = \frac{P_{pc1} \times T^*_{pc}}{T_{pc1} + Y_{H_2S}\,(1 - Y_{H_2S})\varepsilon} \quad \text{(F–10)}$$

The pseudocritical temperature and pressure may also be determined graphically using gas gravity and Figure F–1. Figure F–1 represents Eqs. F–1 through F–4. If there are contaminants present, the pseudocritical properties are corrected by determining the Wichert–Aziz temperature correction factor from Figure F–2 and using Eqs. F–9 and F–10.

Pseudoreduced Properties

These pseudoreduced properties are related to the pseudocritical properties by the following equations:

$$T_{pr} = \frac{T_R + 460}{T_{pc}} \quad \text{(F–11)}$$

$$P_{pr} = \frac{P_R}{P_{pc}} \quad \text{(F–12)}$$

where T_{pc} is the pseudoreduced temperature in °R and P_{pc} is the pseudoreduced pressure in psia; T_R is the reservoir temperature in °F; and P_R is the reservoir pressure in psia. If T_{pc} and P_{pr} are the same for two gases, their physical properties will be equivalent even if their absolute pressure and temperature are not the same. These properties are widely used in gas properties determination. If contaminants are present, then T^*_{pc} and P^*_{pc} are used in calculating pseudoreduced properties.

Example F–1 *Computing Pseudocritical Pressure and Temperature*

Given: Dry gas gravity = 0.732 (air = 1.000); reservoir temperature = 240°F; reservoir pressure = 3700 psia; and gas contains H_2S = 5%, N_2 = 2%, and CO_2 = 3%. Determine pseudocritical and pseudoreduced temperature using charts.

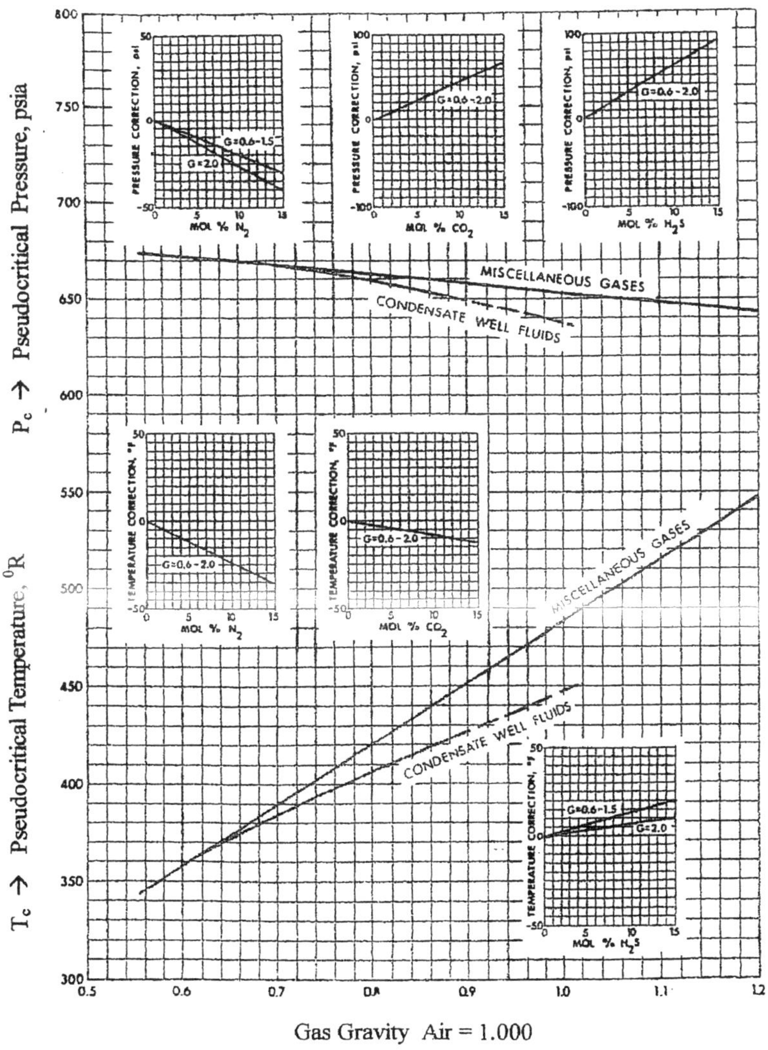

Figure F–1. Pseudocritical properties of natural gases (after Brown *et al.*;[1] inserts from Carr *et al.*).[5]

Solution From Figure F–1 using dry gas gravity = 0.732, determine T_{pc} and P_{pc}.

$$T_{pc} = 401^\circ\text{R} \quad \text{and} \quad P_{pc} = 671 \text{ psia}$$

Compute A.

$$\begin{aligned} A &= 1 - Y_{N_2} - Y_{CO_2} - Y_{H_2S} \\ &= 1 - 0.02 - 0.03 - 0.05 = 0.90 \end{aligned}$$

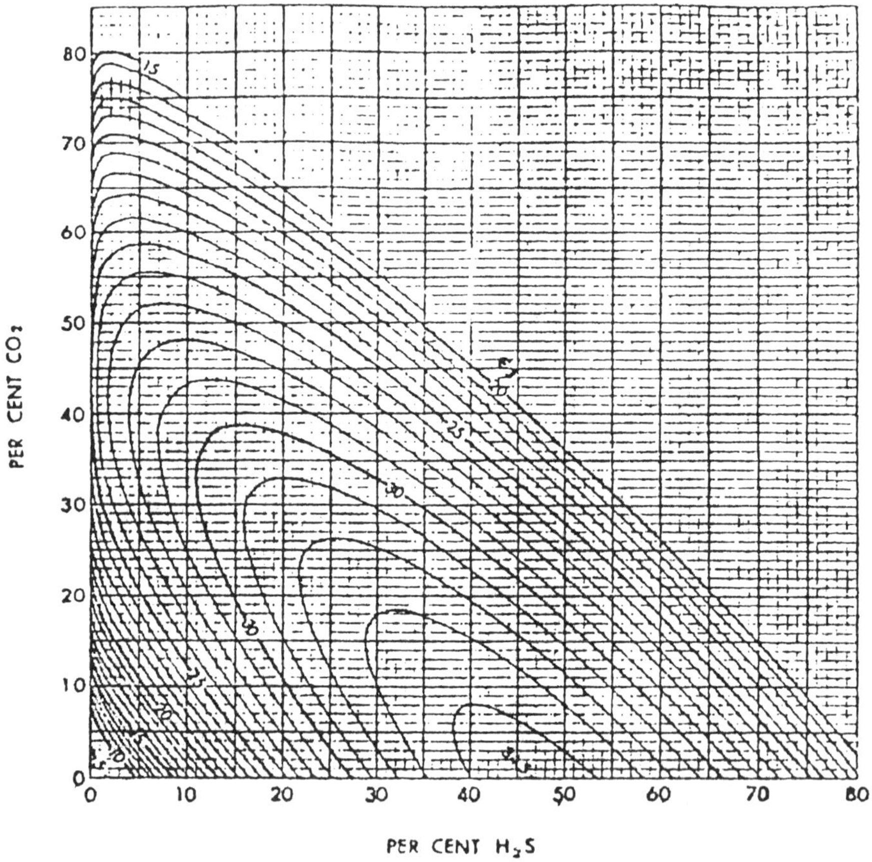

Figure F–2. Pseudocritical temperature adjustment factor *e* (from Wichert and Aziz).[2]

Using Eqs. F–6 and F–7:

$$T_{pc1} = 0.9\,(401) + 227.3(0.02) + 547.6(0.03) + 672.4(0.05) = 420.5^{\circ}\text{R}$$

$$P_{pc1} = 0.9(671) + 493.0(0.02) + 1071(0.03) + 1306(0.05) = 711.19 \text{ psia}$$

Using Figure F–2, compute the Wichert–Aziz correction factor ε:

$$\varepsilon = 13.2$$

Using Eqs. F–9 and F–10, compute corrected T^*_{pc} and P^*_{pc}:

$$T^*_{pc} = 420.5 - 13.2 = 407.3^{\circ}\text{R}$$

$$P^*_{pc} = \frac{711.19 \times 407.3}{420.5 + 0.05(1 - 0.05) \times 13.2} = 687.84 \text{ psia}$$

Using Eqs. F–11 and F–12, determine pseudoreduced properties as follows:

$$T_{pr} = \frac{240 + 460}{407.3} = 1.72$$

$$P_{pr} = \frac{3700}{687.84} = 5.38$$

Gas Deviation Factor

Gas compressibility factors account for the deviation of a real gas from the ideal gas behavior. In order to facilitate computations of PVT properties of natural gases using ideal gas laws, a gas deviation or compressibility factor, symbol z, is used. A natural gas with a gas compressibility factor of 1 will behave in the same way as an ideal gas would.

The deviation factor in most cases is determined from empirical correlation developed by Standing and Katz as a function of pseudoreduced pressure and temperature.[4] The Standing and Katz z-factor chart and the extended Standing and Katz z-factor chart developed by Yarborough are presented in Figures F–3 and F–4, respectively.[15] These charts will be utilized in finding the gas deviation factor. Equations have been developed to fit this correlation, but they are nonlinear, and therefore iterative solutions are required. Solving the equations by hand would be very time consuming.

Example F–2 *Determining Gas Deviation Factor (z-Factor)*
Given: Gas gravity = 0.732 (dry gas); reservoir temperature $T_R = 240°F$; reservoir pressure = 3700 psia.

Solution Using Figure F–1 or equations, calculate pseudocritical pressure and temperature:

$T_{pc} = 401°R$ (from Figure F–1) and $P_{pc} = 671$ psia (from Figure F–1)

Determine pseudoreduced pressure and temperature using Eqs. F–11 and F–12:

$$T_{pr} = \frac{240 + 460}{401} = 1.746$$

$$P_{pr} = \frac{3700}{671} = 5.514$$

Read the gas deviation factor z from Figure F–3 or Figure F–4 as $z = 0.915$.

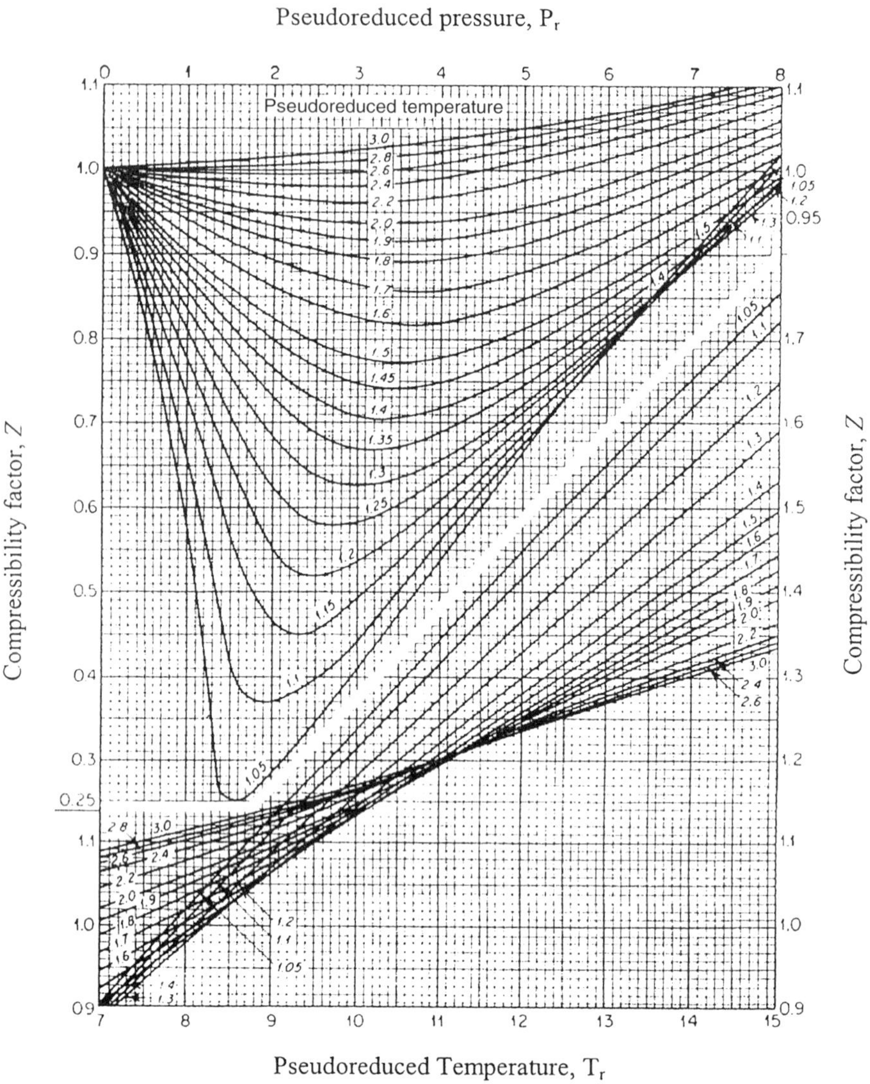

Figure F–3. Gas deviation factor for natural gases (from Standing and Katz).[4]

Gas Formation Volume Factor

The gas formation volume factor, symbol β_g, is used to convert the reservoir gas volume (at reservoir temperature and pressure) to gas volume at standard conditions, P_{sc} and T_{sc}. The factor is generally expressed in either cubic feet or

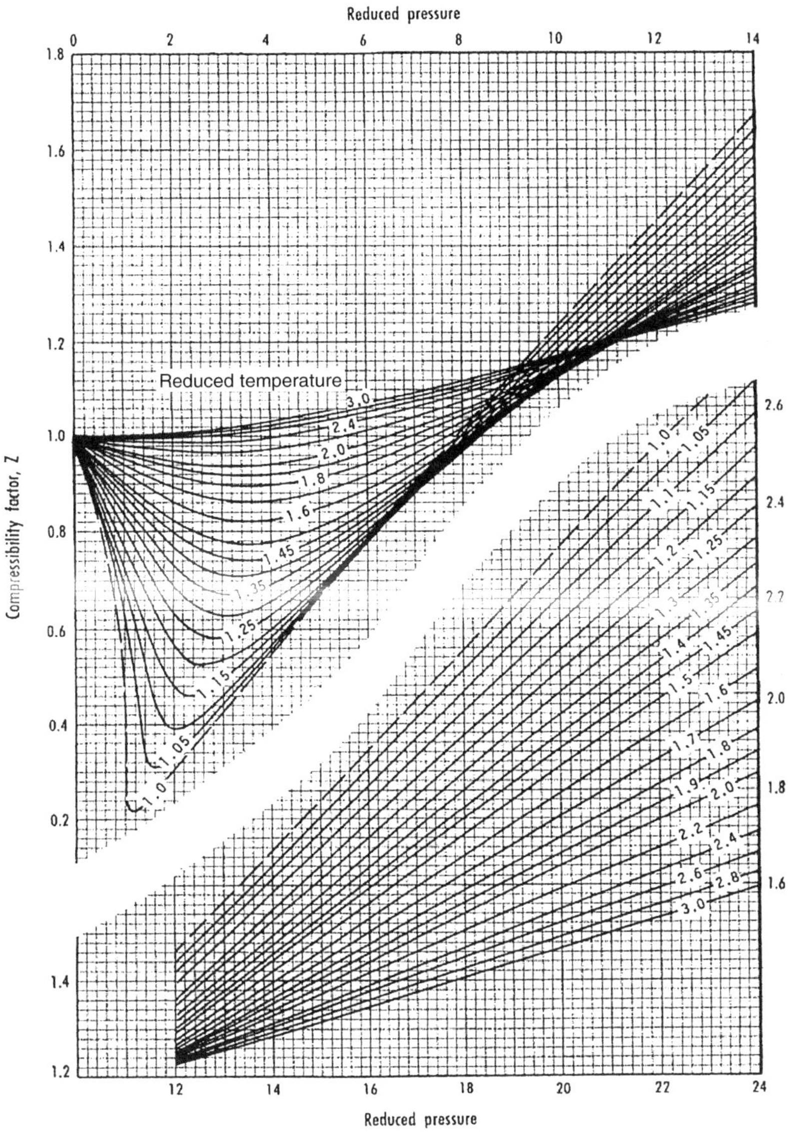

Figure F–4. Extended Standing and Katz Z-factor chart (after Yarborough).[15]

Table F–1
Standard Pressures by Location

Location	P_{sc} (psia)	Location	P_{sc} (psia)
California	14.73	Texas	14.65
Arkansas	14.65	Oklahoma	14.65
Colorado	15.025	Utah	15.025
Illinois	14.65	West Virginia	14.85
Kansas	14.65	Wyoming	15.025
Louisiana	15.025	New Mexico	15.025
Michigan	14.73	U.S. federal lands	14.73
Mississippi	15.025	Canada	14.696

barrels of reservoir volume per standard cubic foot of gas, or as the reciprocals of these in standard cubic feet per cubic foot or per barrel of reservoir volume. The following equations are used to compute β_g:

$$\beta_g = \frac{P_{sc} z T}{T_{sc} p} \tag{F–13}$$

Where $P_{sc} = 14.7$ psia and T_{sc} is 60°F,

$$\begin{aligned} \beta_g &= 0.02829 \frac{zT}{p} \text{ cu ft/scf} \\ &= 0.00504 \frac{zT}{p} \text{ bbl/scf} \\ &= 35.35 \frac{p}{zT} \text{ scf/cu ft} \\ &= 198.4 \frac{p}{zT} \text{ scf/bbl} \end{aligned} \tag{F–14}$$

The constants in Eqs. F–14 are only for 14.7 psia and 60°F, and different constants must be calculated for other standards. The standard pressures used in various locations in the United States are presented in Table F–1.

Example F–3 *Computing Gas Formation Volume Factor, β_g*

Given: z-Factor $= 0.915$; reservoir temperature $T_R = 240$°F; reservoir and pressure $= 3700$ psia; standard temperature and pressure are 60°F and 14.65 psia.

Solution Using Eq. F–13,

$$\beta_g = \frac{(60 + 460) \times 3700}{14.65 \times 0.915 \times (240 + 460)} = 205.045 \text{ scf/cu ft}$$

Gas Viscosity

Gas viscosity of natural gases, symbol μ_g, at reservoir temperature and pressure can be determined from correlations developed by Carr *et al.*[5] and presented in Figures F–5 and F–6. The correlations require gas gravity or molecular weight, pseudoreduced temperature and pressure, and reservoir temperature and pressure. If the gas contains any contaminant gases, the gas viscosity read from Figure F–5 must be corrected using the correction factors from the insets. The correction factors may be calculated using Eqs. F–16 through F–18. The gas viscosity is reported in cP. The following steps are required to obtain gas viscosity:

1. Find molecular weight from gas gravity (MW $= \gamma_g \times 28.968$). Using Figure F–5, determine gas viscosity μ_1 at atmospheric pressure and reservoir temperature.
2. Correct μ_1 for contaminant gases, if any, using insets in Figure F–5 or Eqs. F–16 through F–18.

$$\text{Corrected } \mu_1 = \mu_1(\text{uncorrected}) + N_2 \text{ correction} + H_2S \text{ correction} + CO_2 \text{ correction}$$

3. Read viscosity ratio (μ/μ_1) from Figure F–6 and convert gas gravity at atmospheric pressure to reservoir pressure using the equation

$$\mu_g = (\mu/\mu_1) \times \mu_1 \tag{F–15}$$

where μ_g is the gas viscosity at reservoir conditions in cP and μ_1 is the gas viscosity at one atmospheric in cP. The equations used to compute correction factors are

$$H_2S \text{ correction} = Y_{H_2S}\, 8.49 \times 10^{-3} \log \gamma_g + 3.73 \times 10^{-3} \tag{F–16}$$

$$CO_2 \text{ correction} = Y_{CO_2} 9.08 \times 10^{-3} \log \gamma_g + 6.24 \times 10^{-3} \tag{F–17}$$

$$N_2 \text{ correction} = Y_{N_2} 8.48 \times 10^{-3} \log \gamma_g + 9.59 \times 10^{-3} \tag{F–18}$$

Example F–4 *Estimating Gas Viscosity for the Gas in Example F–1*
Given: Reservoir temperature $T_R = 240°\text{F}$; reservoir pressure = 3700 psia; $T_{pr} = 1.72$; $P_{pr} = 5.38$; gas gravity = 0.732; H_2S = 5%; N_2 = 2%; and CO_2 = 3%.

Solution Find molecular weight (MW = $0.732 \times 28.869 = 21.13$). Read μ_1 from Figure F–5, using gas gravity $\gamma_g = 0.732$ and $T_R = 240°\text{F}$. $\mu_1 =$ 0.0126 cP. Using insets in Figure F–5,

$$\text{Corrected } \mu_1 = 0.0126 + 0.0002 + 0.00048 + 0.00015 = 0.01343 \text{ cP}$$

Read $\mu/\mu_1 = 1.70$ (from Figure F–6). $\mu_g = 1.70 \times 0.01343 = 0.02283$ cP.

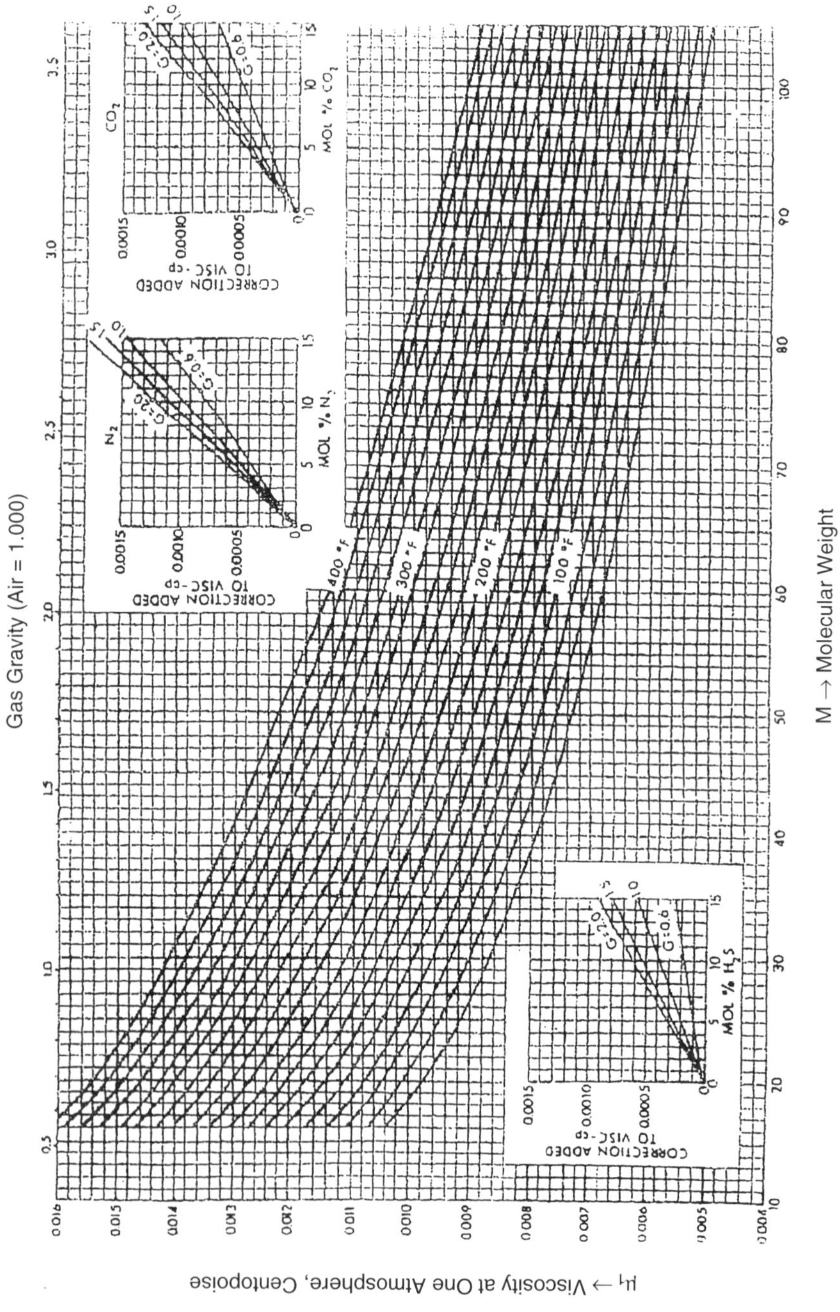

Figure F–5. Viscosity of natural gases at 1 atm (© SPE *Trans. AIME* 1954).[5]

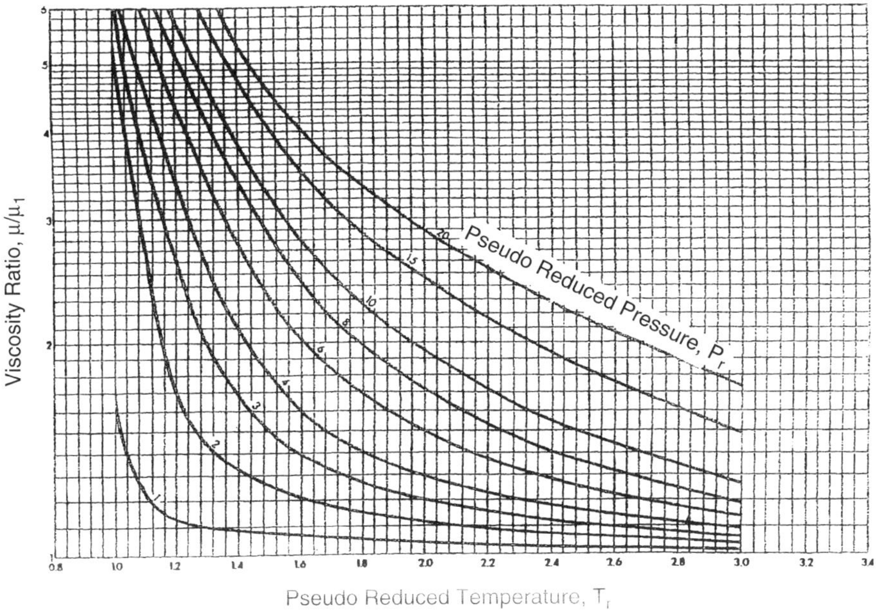

Figure F–6. Viscosity ratio at elevated pressures and temperatures (from Carr *et al.*).[5]

Gas Isothermal Compressibility

Gas isothermal compressibility c_g is extensively used in well test analysis and other reservoir engineering problems and is defined as the change in pressure.

$$c_g = -\frac{1}{V}\left(\frac{\partial V}{\partial p}\right)_T \qquad \text{(F–19)}$$

For an ideal gas,

$$c_g = 1/p \qquad \text{(F–20)}$$

For a real gas,

$$c_g = \frac{1}{p} - \frac{1}{z}\left(\frac{\partial z}{\partial p}\right)_T \qquad \text{(F–21)}$$

If z values are known as a function of pressure, Eq. F–21 can be used to estimate c_g by plotting z against p:

$$\text{Slope of plot} = \frac{\Delta z}{\Delta p} = \left(\frac{\partial z}{\partial p}\right)$$

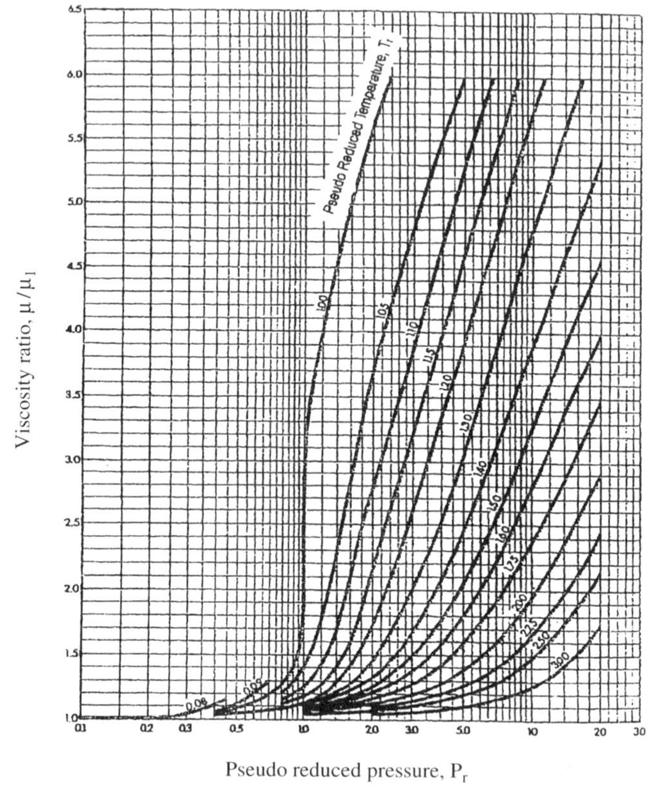

Figure F–7. Viscosity ratio versus pseudoreduced pressure (from Carr *et al.*).[5]

The slope of the plot can be substituted in Eq. F–21 to calculate gas compressibility.

Correlations

Correlations developed by Trube as presented in Figure F–8 are utilized.[9] The required parameters are T_R, p_R, T_{pc}, P_{pc}, T_{pr}, and P_{pr}. Figure F–8 is used to determine pseudoreduced compressibility c_{pr} as a function of T_{pr} and P_{pr}. c_g is then estimated using

$$c_g = \frac{c_{pr}}{P_{pc}} \qquad \text{(F–22)}$$

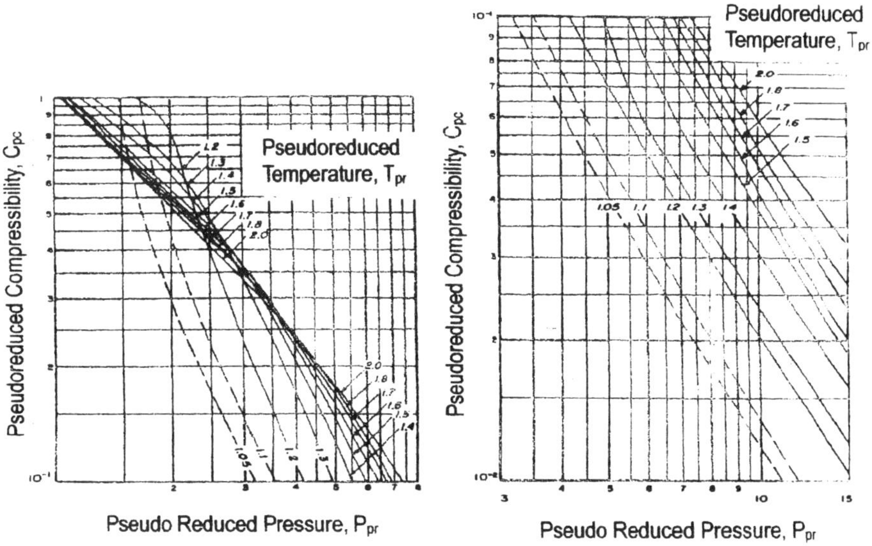

Figure F–8. Correlation of pseudoreduced compressibility for natural gases (© SPE, *Trans. AIME*).[9]

Example F–5 *Estimating Gas Isothermal Compressibility—Example F–1*
Given: Reservoir temperature $T_R = 240°F$; reservoir pressure $= 3700$ psia; $T_{pc} = 420.5°R$; $P_{pc} = 711.19$ psia; $T_{pr} = 1.72$; $P_{pr} = 5.38$; gas gravity $= 0.732$; $H_2S = 5\%$; $N_2 = 2\%$; $CO_2 = 3\%$.

Solution Determine pseudoreduced compressibility from Figure F–7, and using Equation F–22, compute c_g.

$c_{pr} = 0.148$ (from Figure F– 7)

F.2 Reservoir Rock Properties

Naturally occurring reservoir rocks contain fluid, water, gas, or a combination of these fluids. The petroleum engineer is obviously concerned with the properties of reservoir rocks. These properties constitute a set of fundamental parameters by which the rock can be quantitatively described.

Table F–2
Formation Compressibility as a Function of Formation Porosity

Porosity (%)	$C_f \times 10^{-6}$ (psi^{-1})	Porosity (%)	$C_f \times 10^{-6}$ (psi^{-1})
2	9.48	14	4.23
4	7.11	16	4.00
6	6.01	18	3.81
8	5.33	20	3.65
10	4.86	22	3.51
12	4.51	24	3.38

Formation Compressibility

Formation compressibility c_f in most cases is estimated from the correlation developed by Hall[10] presented in equation form as well as in Table F–2:

$$c_f = \frac{1.87}{10^6} \times \phi^{-0.415} \tag{F–23}$$

where ϕ is the formation porosity in fraction and c_f is the formation isothermal compressibility in psi^{-1}.

Total isothermal compressibility above dew-point pressure:

$$c_t = c_g S_g + c_w S_w + c_f \tag{F–24}$$

Total isothermal compressibility below dew-point pressure:

$$c_t = c_o S_o + c_g S_g + c_w S_w + c_f \tag{F–25}$$

Effective compressibility above dewpoint pressure:

$$c_e = c_g S_g + \frac{S_w}{1 - S_w}(c_w) + \frac{c_f}{1 - S_w} \tag{F–26}$$

F.3 Reservoir PVT Water Properties

This section presents Tables F–3 through F–10 for computing reservoir formation PVT properties such as gas solubility and then compressibility, formation volume factor, density, and density gradient for both gas-free and gas-saturated conditions. Viscosity is calculated which is gas-free. The author of this book using the data supplied by previous authors[12–14] developed

(text continued on page 783)

Table F–3

Reservoir Formation Water PVT Properties[11] (reservoir temperature T = 100°F; total dissolved solids TDS = 0.10; base pressure P_{sc} = 14.70 psia; base temperature, T_{sc} = 60°F; specific gravity, Sp. Gr. = 1.721; density of pure water at standard conditions = 62.42 lb/ft^3; density of brine water at standard conditions = 66.92 lb/ft^3)

Gas free conditions							
Pressure (psia)	Compressibility (psi^{-1})	FVF (bbl/bbl)	RSW (ft^3/bbl)	Density (lb/ft^3)	Sp. Gr. (air = 1.000)	Gradient (psi/ft)	Viscosity (cP)
8000	0.210535E-05	0.9814	20.11	68.1923	1.0924	0.4736	0.9218
7000	0.220900E-05	0.9843	17.96	67.9884	1.0892	0.4722	0.9231
6000	0.227536E-05	0.9876	15.94	67.7648	1.0856	0.4706	0.9235
5500	0.230103E-05	0.9892	14.97	67.6495	1.0837	0.4698	0.9234
5000	0.232466E-05	0.9909	14.03	67.5334	1.0819	0.4690	0.9233
4500	0.234816E-05	0.9926	13.10	67.4177	1.0800	0.4682	0.9231
4000	0.237318E-05	0.9943	12.15	67.3030	1.0782	0.4674	0.9230
3500	0.240115E-05	0.9960	11.18	67.1900	1.0764	0.4666	0.9228
3000	0.243322E-05	0.9977	10.14	67.0792	1.0746	0.4658	0.9226
2500	0.247034E-05	0.9993	9.02	66.9708	1.0729	0.4651	0.9225
2000	0.251315E-05	1.0009	7.77	66.8648	1.0712	0.4644	0.9225
1500	0.256211E-05	1.0024	6.35	66.7612	1.0695	0.4636	0.9224
1000	0.261738E-05	1.0039	4.74	66.6597	1.0679	0.4629	0.9224
750	0.264737E-05	1.0047	3.84	66.6096	1.0671	0.4626	0.9224
500	0.267889E-05	1.0054	2.87	66.5599	1.0663	0.4622	0.9223
250	0.271191E-05	1.0062	1.82	66.5104	1.0655	0.4619	0.9223
15	0.274424E-05	1.0069	0.76	66.4641	1.0648	0.4616	0.9223

Table F–3 (continued)

Gas saturated conditions						
Pressure (psia)	Compressibility (psi^{-1})	FVF (bbl/bbl)	RSW (ft^3/bbl)	Density (lb/ft^3)	Sp. Gr. (air = 1.000)	Gradient (psi/ft)
8000	0.253267E-05	0.9921	35.35	67.4558	1.0806	0.4685
7000	0.255487E-05	0.9941	31.56	67.3182	1.0784	0.4675
6000	0.257268E-05	0.9959	28.01	67.1968	1.0765	0.4667
5500	0.258336E-05	0.9968	26.32	67.1383	1.0756	0.4663
5000	0.259601E-05	0.9976	24.66	67.0800	1.0746	0.4658
4500	0.21080E-05	0.9985	23.02	67.0209	1.0737	0.4654
4000	0.262766E-05	0.9994	21.36	66.9606	1.0727	0.4650
3500	0.264828E-05	1.0003	19.64	66.8989	1.0717	0.4646
3000	0.266612E-05	1.0013	17.82	66.8358	1.0707	0.4642
2500	0.268637E-05	1.0023	15.84	66.7714	1.0680	0.4637
2000	0.270598E-05	1.0032	13.65	66.7064	1.0686	0.4633
1500	0.272354E-05	1.0042	11.17	66.6415	1.0676	0.4628
1000	0.273737E-05	1.0052	8.33	66.5777	1.0666	0.4624
750	0.274227E-05	1.0056	6.74	66.5467	1.0661	0.4621
500	0.274545E-05	1.0061	5.04	66.5163	1.0656	0.4619
250	0.274562E-05	1.0065	3.20	66.4870	1.0651	0.4617
15	0.274561E-05	1.0069	1.34	66.4605	1.0647	0.4615

Table F–4

Reservoir Formation Water PVT Properties[11] (reservoir temperature T = 100°F; total dissolved solids TDS = 0.20; base pressure P_{sc} = 14.70 psia; base temperature T_{sc} = 60°F; specific gravity, Sp. Gr. = 1.1311; density of pure water at standard conditions = 62.42 lb/ft^3; density of brine water at standard conditions = 70.61 lb/ft^3)

Gas free conditions							
Pressure (psia)	Compressibility (psi^{-1})	FVF (bbl/bbl)	RSW (ft^3/bbl)	Density (lb/ft^3)	Sp. Gr. (air = 1.000)	Gradient (psi/ft)	Viscosity (cP)
8000	0.194705E-05	0.9886	12.43	71.4250	1.1442	0.4960	1.0574
7000	0.201171E-05	0.9912	11.09	71.2309	1.1411	0.4947	1.0600
6000	0.204519E-05	0.9940	9.85	71.0330	1.1379	0.4933	1.0607
5500	0.205627E-05	0.9954	9.25	70.9333	1.1363	0.4926	1.0606
5000	0.206628E-05	0.9968	8.67	70.8336	1.1348	0.4919	1.0604
4500	0.207698E-05	0.9982	8.09	70.7342	1.1332	0.4912	1.0601
4000	0.208983E-05	0.9996	7.51	70.6352	1.1316	0.4905	1.0597
3500	0.210608E-05	1.0010	6.90	70.5369	1.1300	0.4909	1.0594
3000	0.212672E-05	1.0024	6.26	70.4397	1.1284	0.4892	1.0591
2500	0.215249E-05	1.0038	5.57	70.3436	1.1269	0.4885	1.0589
2000	0.218389E-05	1.0051	4.80	70.2489	1.1254	0.4879	1.0587
1500	0.222116E-05	1.0064	3.93	70.1558	1.1239	0.4872	1.0586
1000	0.226432E-05	1.0071	2.93	70.0646	1.1224	0.4866	1.0585
750	0.228803E-05	1.0084	2.37	70.0197	1.1217	0.4863	1.0585
500	0.231310E-05	1.0090	1.77	69.9753	1.1210	0.4860	1.0585
250	0.233946E-05	1.0097	1.12	69.9315	1.1203	0.4856	1.0584
15	0.236534E-05	1.0130	0.47	69.8907	1.1196	0.4854	1.0583

Table F–4 (continued)

Gas saturated conditions						
Pressure (psia)	Compressibility (psi^{-1})	FVF (bbl/bbl)	RSW (ft^3/bbl)	Density (lb/ft^3)	Sp. Gr. (air = 1.000)	Gradient (psi/ft)
8000	0.219652E-05	0.9952	35.35	70.9499	1.1366	0.4927
7000	0.219763E-05	0.9973	31.56	70.7986	1.1342	0.4917
6000	0.219689E-05	0.9992	28.01	70.6663	1.1321	0.4908
5500	0.219866E-05	1.0001	26.32	70.6033	1.1311	0.4903
5000	0.220264E-05	1.0009	24.66	70.5408	1.1301	0.4899
4500	0.220914E-05	1.0018	23.02	70.4778	1.1291	0.4894
4000	0.221830E-05	1.0027	21.36	70.4140	1.1280	0.4890
3500	0.223008E-05	1.0037	19.64	70.3488	1.1270	0.4885
3000	0.224427E-05	1.0046	17.82	70.2823	1.1259	0.4881
2500	0.226042E-05	1.0056	15.84	70.2147	1.1248	0.4876
2000	0.227790E-05	1.0066	13.65	70.1465	1.1237	0.4871
1500	0.229580E-05	1.0076	11.17	70.0784	1.1227	0.4867
1000	0.231300E-05	1.0085	8.33	70.0115	1.1216	0.4862
750	0.232091E-05	1.0090	6.74	69.9789	1.1211	0.4860
500	0.232810E-05	1.0094	5.04	69.9471	1.1206	0.4858
250	0.233436E-05	1.0099	3.20	69.9163	1.1201	0.4855
15	0.233920E-05	1.0103	1.34	69.8884	1.1196	0.4854

Table F–5

Reservoir Formation Water PVT Properties[11] (reservoir temperature T = 100°F; total dissolved solids TDS = 0.30; base pressure P_{sc} = 14.70 psia; base temperature T_{sc} = 60°F; specific gravity, Sp. Gr. = 1.1690; density of pure water at standard conditions = 62.42 lb/ft^3; density of brine water at standard conditions = 72.97 lb/ft^3)

	Gas free conditions						
Pressure (psia)	**Compressibility (psi^{-1})**	**FVF (bbl/bbl)**	**RSW (ft^3/bbl)**	**Density (lb/ft^3)**	**Sp. Gr. (air = 1.000)**	**Gradient (psi/ft)**	**Viscosity (cP)**
8000	0.161694	0.9902	7.06	73.6941	1.1806	0.5118	1.2172
7000	0.169788	0.9927	6.30	73.5047	1.1775	0.5105	1.2213
6000	0.174782	0.9952	5.59	73.3239	1.1746	0.5092	1.2224
5500	0.176674	0.9964	5.25	73.2351	1.1732	0.5086	1.2222
5000	0.178413	0.9976	4.92	73.1470	1.1718	0.5080	1.2219
4500	0.180151	0.9988	4.60	73.0592	1.1704	0.5074	1.2213
4000	0.182022	1.0000	4.26	72.9714	1.1690	0.5068	1.2208
3500	0.184130	1.0012	3.92	72.8838	1.1676	0.5062	1.2202
3000	0.186556	1.0024	3.56	72.7963	1.1662	0.5055	1.2198
2500	0.189357	1.0036	3.16	72.7092	1.1648	0.5049	1.2194
2000	0.192565	1.0048	2.72	72.6227	1.1634	0.5043	1.2192
1500	0.196187	1.0060	2.23	72.5372	1.1620	0.5037	1.2190
1000	0.200205	1.0071	1.66	72.4533	1.1607	0.5032	1.2189
750	0.202350	1.0077	1.35	72.4122	1.1600	0.5029	1.2189
500	0.204577	1.0083	1.01	72.3716	1.1594	0.5026	1.2188
250	0.206875	1.0088	0.64	72.3318	1.1588	0.5023	1.2187
15	0.209092	1.0093	0.27	72.2951	1.1582	0.5021	1.2186

Table F–5 (continued)

Gas saturated conditions						
Pressure (psia)	**Compressibility (psi^{-1})**	**FVF (bbl / bbl)**	**RSW (ft^3/bbl)**	**Density (lb/ft^3)**	**Sp. Gr. (air=1.000)**	**Gradient (psi/ft)**
8000	0.174606E-05	0.9939	35.35	73.4153	1.1761	0.5098
7000	0.178667E-05	0.9961	31.56	73.2510	1.1735	0.5087
6000	0.181628E-05	0.9981	28.01	73.1087	1.1712	0.5077
5500	0.183038E-05	0.9990	26.32	73.0414	1.1701	0.5072
5000	0.184514E-05	0.9999	24.66	72.9751	1.1691	0.5068
4500	0.186110E-05	1.0008	23.02	72.9096	1.1680	0.5063
4000	0.187865E-05	1.0017	21.36	72.8415	1.1669	0.5059
3500	0.189794E-05	1.0027	19.64	72.7732	1.1658	0.5054
3000	0.191899E-05	1.0036	17.82	72.7038	1.1647	0.5049
2500	0.194158E-05	1.0046	15.84	72.6334	1.1636	0.5044
2000	0.196533E-05	1.0056	13.65	72.5625	1.1624	0.5039
1500	0.198963E-05	1.0066	11.17	72.4917	1.1613	0.5034
1000	0.201368E-05	1.0076	8.33	72.4221	1.1602	0.5029
750	0.202532E-05	1.0080	6.74	72.3882	1.1597	0.5027
500	0.203651E-05	1.0086	5.04	72.3550	1.1591	0.5025
250	0.204712E-05	1.0089	3.20	72.3229	1.1586	0.5023
15	0.205642E-05	1.0093	1.34	72.2938	1.1581	0.5021

Table F–6

Reservoir Formation Water PVT Properties[11] (reservoir temperature T = 200°F; total dissolved solids TDS = 0.10; base pressure P_{sc} = 14.70 psia; base temperature T_{sc} = 60°F; specific gravity, Sp. Gr. = 1.0721; density of pure water at standard conditions = 62.42 lb/ft^3; density of brine water at standard conditions = 66.92 lb/ft^3)

				Gas free conditions			
Pressure (psia)	Compressibility (psi^{-1})	FVF (bbl/bbl)	RSW (ft^3/bbl)	Density (lb/ft^3)	Sp. Gr. (air = 1.000)	Gradient (psi/ft)	Viscosity (cP)
8000	9.230275E-05	1.0099	40.59	66.2674	1.0616	0.4602	0.3695
7000	0.240640E-05	1.0133	33.47	66.0404	1.0580	0.4586	0.3599
6000	0.247276E-05	1.0168	27.13	65.8152	1.0544	0.4571	0.3699
5500	0.249844E-05	1.0186	24.27	65.7054	1.0526	0.4563	0.3698
5000	0.252207E-05	1.0202	21.59	65.5980	1.0509	0.4556	0.3697
4500	0.254556E-05	1.0218	19.10	65.4932	1.0492	0.4548	0.3695
4000	0.257058E-05	1.0234	16.77	65.3908	1.0476	0.4541	0.3694
3500	0.259855E-05	1.0250	14.59	65.2906	1.0460	0.4534	0.3693
3000	0.263063E-05	1.0286	12.52	65.1920	1.0444	0.4527	0.3692
2500	0.266774E-05	1.0281	10.54	66.0941	1.0428	0.4521	0.3691
2000	0.271056E-05	1.0296	8.61	64.9952	1.0412	0.4514	0.3690
1500	0.275951E-05	1.0312	6.69	64.8958	1.0396	0.4507	0.3690
1000	0.281478E-05	1.0328	4.73	64.7945	1.0380	0.4500	0.3690
750	0.284477E-05	1.0337	3.73	64.7418	1.0372	0.4496	0.3689
500	0.287629E-05	1.0346	2.69	64.6877	1.0363	0.5592	0.3689
250	0.290931E-05	1.0354	1.61	64.6321	1.0354	0.4488	0.3689
15	0.294164E-05	1.0363	0.57	64.5781	1.0345	0.4485	03689

Table F–6 (continued)

Gas saturated conditions						
Pressure (psia)	Compressibility (psi^{-1})	FVF (bbl/bbl)	RSW (ft^3/bbl)	Density (lb/ft^3)	Sp. Gr. (air = 1.000)	Gradient (psi/ft)
8000	0.268492E-05	1.0237	60.99	65.3742	1.0473	0.4540
7000	0.273038E-05	1.0258	50.29	65.2393	1.0451	0.4531
6000	0.276266E-05	1.0274	40.77	65.1363	1.0435	0.4523
5500	0.277716E-05	1.0281	36.46	65.0916	1.0428	0.4520
5000	0.279154E-05	1.0288	32.45	65.0492	1.0421	0.4517
4500	0.280622E-05	1.0294	28.70	65.0079	1.0414	0.4515
4000	0.282139E-05	1.0301	25.20	64.9664	1.0408	0.4512
3500	0.283705E-05	1.0308	21.92	64.9239	1.0401	0.4509
3000	0.285298E-05	1.0315	18.82	64.8798	1.0394	0.4506
2500	0.286876E-05	1.0322	15.84	64.8337	1.0386	0.4502
2000	0.288371E-05	1.0330	12.94	64.7855	1.0379	0.4499
1500	0.289690E-05	1.0338	10.06	64.7353	1.0371	0.4496
1000	0.290715E-05	1.0346	7.11	64.6835	1.0362	0.4492
750	0.291074E-05	1.0350	5.60	64.6572	1.0358	0.4490
500	0.291305E-05	1.0355	4.04	64.6306	1.0354	0.4488
250	0.291386E-05	1.0359	2.43	64.6040	1.0350	0.4487
15	0.291307E-05	1.0363	0.85	64.5791	1.0346	0.4485

Table F–7

Reservoir Formation Water PVT Properties[11] (reservoir temperature T = 200°F; total dissolved solids TDS = 0.20; base pressure P_{sc} = 14.70 psia; base temperature T_{sc} = 60°F; specific gravity, Sp. Gr. = 1.1311; density of pure water at standard conditions = 62.42 lb/ft^3; density of brine water at standard conditions = 70.6 lb/ft^3)

Gas free conditions							
Pressure (psia)	Compressibility (psi^{-1})	FVF (bbl/bbl)	RSW (ft^3/bbl)	Density (lb/ft^3)	Sp. Gr. (air = 1.000)	Gradient (psi/ft)	Viscosity (cP)
8000	0.218482E-05	1.0165	27.38	69.4615	1.1128	0.4824	0.4275
7000	0.224948E-05	1.0196	22.57	69.2483	1.1094	0.4839	0.4281
6000	0.228296E-05	1.0226	18.30	69.-495	1.1062	0.4795	0.4281
5500	0.229403E-05	1.0240	16.37	68.9550	1.1047	0.4789	0.4279
5000	0.230405E-05	1.0253	14.56	68.8634	1.1032	0.4782	0.4277
4500	0.231474E-05	1.0267	12.88	68.7743	1.1018	0.4776	0.4275
4000	0.232760E-05	1.0280	11.31	68.6871	1.1004	0.4770	0.4272
3500	0.234385E-05	1.0292	9.84	68.6012	1.0990	0.4764	0.4270
3000	0.236449E-05	1.0305	8.45	68.5161	1.0976	0.4758	0.4268
2500	0.239026E-05	1.0318	7.11	68.4311	1.0963	0.4752	0.4267
2000	0.242166E-05	1.0331	5.81	68.3452	1.0949	0.4746	0.4265
1500	0.245893E-05	1.0344	4.51	68.2577	1.0935	0.4740	0.4264
1000	0.250208E-05	1.0358	3.19	68.1676	1.0920	0.4734	0.4263
750	0.252580E-05	1.0365	2.51	68.1212	1.0913	0.4731	0.4263
500	0.255087E-05	1.0372	1.81	68.0738	1.0905	0.4727	0.4263
250	0.257723E-05	1.0380	1.09	68.0252	1.0898	0.4724	0.4262
15	0.260310E-05	1.0387	0.38	67.9784	1.0890	0.4721	0.4261

Table F–7 (continued)

Gas saturated conditions

Pressure (psia)	Compressibility (psi^{-1})	FVF (bbl/bbl)	RSW (ft^3/bbl)	Density (lb/ft^3)	Sp. Gr. (air = 1.000)	Gradient (psi/ft)
8000	0.238565E-05	1.0258	60.99	68.8313	1.1027	0.4780
7000	0.240981E-05	1.0280	50.29	68.6830	1.1003	0.4770
6000	0.242345E-05	1.0297	40.77	68.5701	1.0985	0.4762
5500	0.242917E-05	1.0304	36.46	68.5216	1.0977	0.4750
5000	0.243518E-05	1.0311	32.45	68.4758	1.0970	0.4755
4500	0.244201E-05	1.0318	28.70	68.4314	1.0963	0.4752
4000	0.245005E-05	1.0325	26.20	68.3871	1.0956	0.4749
3500	0.245956E-05	1.0332	21.92	68.3420	1.0948	0.4746
3000	0.247061E-05	1.0339	18.82	68.2954	1.0941	0.4743
2500	0.248310E-05	1.0346	15.84	68.2468	1.0933	0.4740
2000	0.249673E-05	1.0354	12.94	68.1962	1.0925	0.4736
1500	0.251102E-05	1.0362	10.06	68.1434	1.0917	0.4732
1000	0.252528E-05	1.0370	7.11	68.0890	1.0908	0.4729
750	0.253214E-05	1.0374	5.60	68.0613	1.0903	0.4727
500	0.253822E-05	1.0378	4.04	68.0334	1.0899	0.4725
250	0.254470E-05	1.0363	2.43	68.0054	1.0894	0.4723
15	0.254984E-05	1.0387	0.85	67.9791	1.0890	0.4721

Table F–8

Reservoir Formation Water PVT Properties[11] (reservoir temperature T = 200°F; total dissolved solids TDS = 0.30; base pressure P_{sc} = 14.70 psia; base temperature T_{sc} = 60°F; specific gravity, Sp. Gr. = 1.721; density of pure water at standard conditions = 62.42 lb/ft^3; density of brine water at standard conditions = 72.97 lb/ft^3)

Gas free conditions							
Pressure (psia)	Compressibility (psi^{-1})	FVF (bbl/bbl)	RSW (ft^3/bbl)	Density (lb/ft^3)	Sp. Gr. (air = 1.000)	Gradient (psi/ft)	Viscosity (cP)
8000	0.186768E-05	1.0176	17.87	71.7131	1.1488	0.4980	0.5070
7000	0.194862E-05	1.0204	14.73	71.5082	1.1456	0.4966	0.5082
6000	0.199856E-05	1.0230	11.94	71.3282	1.1427	0.4953	0.5081
5500	0.201748E-05	1.0242	10.68	71.2449	1.1413	0.4949	0.5078
5000	0.203487E-05	1.0253	9.50	71.1650	1.1401	0.4942	0.5074
4500	0.205226E-05	1.0265	8.41	71.0876	1.1388	0.4937	0.5070
4000	0.207096E-05	1.0276	7.38	71.0117	1.1376	0.4932	0.5066
3500	0.209204E-05	1.0286	6.42	70.9367	1.1364	0.4926	0.5063
3000	0.211630E-05	1.0297	5.51	70.8619	1.1352	0.4921	0.5060
2500	0.214431E-05	1.0308	4.64	70.7864	1.1340	0.4916	0.5057
2000	0.217639E-05	1.0319	3.79	70.7098	1.1328	0.4911	0.5056
1500	0.221261E-05	1.0331	2.96	70.6313	1.1315	0.4905	0.5053
1000	0.225279E-05	1.0343	2.08	70.5503	1.1302	0.4899	0.5052
750	0.227424E-05	1.0349	1.64	70.5088	1.1295	0.4897	0.5051
500	0.229651E-05	1.0355	1.18	70.4665	1.1289	0.4894	0.5050
250	0.231949E-05	1.0361	0.71	70.4233	1.1282	0.4891	0.5049
15	0.234166E-05	1.0368	0.25	70.3819	1.1275	0.4888	0.5048

Table F–8 (continued)

Gas saturated conditions

Pressure (psia)	Compressibility (psi^{-1})	FVF (bbl/bbl)	RSW (ft^3/bbl)	Density (lb/ft^3)	Sp. Gr. (air = 1.000)	Gradient (psi/ft)
8000	0.194914E-05	1.0236	60.99	71.2876	1.1420	0.4951
7000	0.200925E-05	1.0259	50.29	71.1265	1.1394	0.4939
6000	0.205152E-05	1.0277	40.77	71.0044	1.1375	0.4931
5500	0.206923E-05	1.0284	36.46	70.9521	1.1366	0.4927
5000	0.208591E-05	1.0291	32.45	70.9031	1.1359	0.4924
4500	0.210231E-05	1.0298	28.70	70.8558	1.1351	0.4921
4000	0.211901E-05	1.0305	25.20	70.8089	1.1344	0.4917
3500	0.213643E-05	1.0312	21.92	70.7614	1.1336	0.4914
3000	0.215485E-05	1.0319	18.82	70.7126	1.1328	0.4911
2500	0.217437E-05	1.0326	15.84	70.6518	1.1320	0.4907
2000	0.219494E-05	1.0334	12.94	70.6089	1.1312	0.4904
1500	0.221635E-05	1.0342	10.06	70.5540	1.1303	0.4900
1000	0.223825E-05	1.0351	7.11	70.4972	1.1294	0.4896
750	0.224923E-05	1.0356	5.60	70.4683	1.1289	0.4894
500	0.226015E-05	1.0359	4.04	70.4391	1.1284	0.4892
250	0.227094E-05	1.0363	2.43	70.4099	1.1280	0.4890
15	0.228090E-05	1.0368	0.86	70.3842	1.1275	0.4888

Table F–9

Reservoir Formation Water PVT Properties[11] (reservoir temperature T = 300°F; total dissolved solids TDS = 0.10; base pressure P_{sc} = 14.70 psia; base temperature T_{sc} = 60°F; specific gravity, Sp. Gr. = 1.721; density of pure water at standard conditions = 62.42 lb/ft^3; density of brine water at standard conditions = 66.92 lb/ft^3)

Gas free conditions							
Pressure (psia)	Compressibility (psi^{-1})	FVF (bbl/bbl)	RSW (ft^3/bbl)	Density (lb/ft^3)	Sp. Gr. (air = 1.000)	Gradient (psi/ft)	Viscosity (cP)
8000	0.278618E-05	1.0417	18.57	64.2415	1.0291	0.4461	0.2497
7000	0.288982E-05	1.0471	17.70	63.9103	1.0238	0.4438	0.2498
6000	0.295619E-05	1.0520	16.57	63.6167	1.0191	0.4418	0.2497
5500	0.298286E-05	1.0542	15.90	63.4833	1.0170	0.4409	0.2496
5000	0.300549E-05	1.0562	15.17	63.3580	1.0150	0.4400	0.2494
4500	0.302899E-05	1.0582	14.36	63.2400	1.0131	0.4392	0.2493
4000	0.305401E-05	1.0601	13.47	63.1282	1.0113	0.4384	0.2491
3500	0.308198E-05	1.0619	12.46	63.0212	1.0096	0.4377	0.2490
3000	0.311405E-05	1.0636	11.33	62.9175	1.0079	0.4369	0.2489
2500	0.315116E-05	1.0654	10.05	62.8155	1.0063	0.4362	0.2488
2000	0.319398E-05	1.0671	8.58	62.7131	1.0047	0.4355	0.2488
1500	0.324294E-05	1.0689	6.91	62.6083	1.0030	0.4348	0.2487
1000	0.329821E-05	1.0708	4.98	62.4986	1.0012	0.4340	0.2487
750	0.332820E-05	1.0718	3.91	62.4413	1.0003	0.4336	0.2486
500	0.335972E-05	1.0728	2.76	62.3817	0.9994	0.4332	0.2486
250	0.339273E-05	1.0738	1.53	62.3196	0.9984	0.4328	0.2486
15	0.342507E-05	1.0749	0.28	62.2587	0.9974	0.4324	0.2486

Table F–9 (continued)

Gas saturated conditions						
Pressure (psia)	**Compressibility (psi^{-1})**	**FVF (bbl/bbl)**	**RSW (ft^3/bbl)**	**Density (lb/ft^3)**	**Sp. Gr. (air = 1.000)**	**Gradient (psi/ft)**
8000	0.340631	1.0619	99.02	63.0216	1.0096	0.4377
7000	0.349258	1.0641	87.37	62.8899	1.0075	0.4367
6000	0.354600	1.0657	68.19	62.7976	1.0060	0.4361
5500	0.356373	1.0663	59.54	62.7600	1.0054	0.4358
5000	0.357674	1.0669	51.52	62.7252	1.0049	0.4356
4500	0.358576	1.0675	44.12	62.6915	1.0043	0.4354
4000	0.359132	1.0681	37.31	62.6572	1.0038	0.4351
3500	0.359375	1.0687	31.07	62.6209	1.0032	0.4349
3000	0.359317	1.0694	25.38	62.5815	1.0026	0.4346
2500	0.358949	1.0701	20.18	62.5381	1.0019	0.4343
2000	0.358235	1.0709	15.43	62.4900	1.0011	0.4340
1500	0.357121	1.0718	11.08	62.4367	1.0002	0.4336
1000	0.355527	1.0728	7.06	62.3780	0.9993	0.4332
750	0.354521	1.0734	5.15	62.3467	0.9988	0.4330
500	0.353357	1.0739	3.30	62.3139	0.9983	0.4327
250	0.352022	1.0745	1.50	62.2799	0.9977	0.4326
15	0.350599	1.0751	0.16	62.2468	0.9972	0.4323

Table F–10

Reservoir Formation Water PVT Properties[11] (reservoir temperature T = 300°F; total dissolved solids TDS = 0.20; base pressure P_{sc} = 14.70 psia; base temperature T_{sc} = 60°F; specific gravity, Sp. Gr. = 1.721; density of pure water at standard conditions = 62.42 lb/ft^3; density of brine water at standard conditions = 70.61 lb/ft^3)

Gas free conditions							
Pressure (psia)	Compressibility (psi^{-1})	FVF (bbl/bbl)	RSW (ft^3/bbl)	Density (lb/ft^3)	Sp. Gr. (air = 1.000)	Gradient (psi/ft)	Viscosity (cP)
8000	0.270918E-05	1.0485	11.99	67.3418	1.0788	0.4677	0.2928
7000	0.277384E-05	1.0529	11.43	67.0620	1.0743	0.4657	0.2931
6000	0.280732E-05	1.0586	10.70	66.8244	1.0705	0.4641	0.2928
5500	0.281839E-05	1.0583	10.27	66.7183	1.0688	0.4633	0.2926
5000	0.282841E-05	1.0599	9.80	66.6192	1.0572	0.4626	0.2924
4500	0.283911E-05	1.0614	9.28	66.5257	1.0657	0.4620	0.2922
4000	0.285196E-05	1.0628	8.70	66.4362	1.0643	0.4614	0.2919
3500	0.286821E-05	1.0642	8.05	66.3496	1.0629	0.4608	0.2917
3000	0.288885E-05	1.0655	7.32	66.2641	1.0615	0.4602	0.2915
2500	0.291462E-05	1.0669	6.49	66.1783	1.0602	0.4596	0.2914
2000	0.294602E-05	1.0683	5.54	66.0906	1.0588	0.4590	0.2913
1500	0.298329E-05	1.0698	4.46	65.9993	1.0573	0.4583	0.2911
1000	0.302644E-05	1.0714	3.22	65.9026	1.0558	0.4577	0.2910
750	0.305016E-05	1.0722	2.52	65.6518	1.0549	0.4573	0.2910
500	0.307523E-05	1.0731	1.78	66.7989	1.0541	0.4570	0.2909
250	0.310159E-05	1.0740	0.99	65.7439	1.0532	0.4566	0.2909
15	0.312747E-05	1.0749	0.18	65.6899	1.0523	0.4562	0.2908

Table F–10 (continued)

Gas saturated conditions						
Pressure (psia)	Compressibility (psi^{-1})	FVF (bbl/bbl)	RSW (ft^3/bbl)	Density (lb/ft^3)	Sp. Gr. (air = 1.000)	Gradient (psi/ft)
8000	0.313009E-05	1.0615	99.95	66.5156	1.0656	0.4519
7000	0.319562E-05	1.0638	87.37	66.3703	1.0632	0.4509
6000	0.323-26E-05	1.0655	68.10	66.2685	1.0616	0.4602
5500	0.323923E-05	1.0661	59.54	66.2273	1.0610	0.4599
5000	0.324409E-05	1.0667	51.52	66.1894	1.0604	0.4597
4500	0.324575E-05	1.0673	44.12	66.1529	1.0598	0.4594
4000	0.324499E-05	1.0679	37.31	66.1161	1.0592	0.4592
3500	0.324244E-05	1.0686	31.07	66.0773	1.0586	0.4589
3000	0.323853E-05	1.0692	26.38	66.0355	1.0579	0.4586
2500	0.323354E-05	1.0700	20.18	65.9895	1.0572	0.4583
2000	0.322751E-05	1.0708	15.43	65.9387	1.0563	0.4579
1500	0.322034E-05	1.0717	11.08	65.8824	1.0554	0.4575
1000	0.321175E-05	1.0727	7.05	65.8205	1.0544	0.4571
750	0.320681E-05	1.0733	5.15	65.7873	1.0539	0.4569
500	0.320136E-05	1.0738	3.30	65.7528	1.0534	0.4566
250	0.319534E-05	1.0744	1.50	65.7168	1.0528	0.4564
15	0.318910E-05	1.0750	0.16	65.6818	1.0522	0.4561

(text continued from page 766)

the graphical and empirical correlations.[11] New tables showing water PVT properties as functions of pressure–temperature and total dissolved solids are presented which improve accuracy of prediction and save time.

Tables F–3 through F–10 show the following formation water PVT properties:

1. Formation water compressibility—gas free
2. Formation water compressibility—gas-saturated
3. Gas solubility in brine
4. Water formation volume factor—gas free
5. Water formation volume factor—gas saturated
6. Formation water density—gas free
7. Formation water density—gas saturated
8. Density gradient—gas free
9. Density gradient—gas saturated
10. Water viscosity—gas free

References and Additional Reading

1. Brown, G. G., Katz, D. L., Oberfell, C. G., and Alden, R. C., *Natural Gasoline and the Volatile Hydrocarbons*. NGAA, Tulsa, OK, 1948.
2. Wichert, E., and Aziz, K., "Calculation of Z's for Sour Gases," *Hydrocarbon Processing* (1972) 51(5).
3. *Engineering Data Book,* 9th ed. Gas Processors Suppliers Association, Tulsa, OK, 1977.
4. Standing, M. B., and Katz, D. L., "Density of Natural Gases," *Trans. AIME* (1942) 146, 144.
5. Carr, N. L., Kobayashi, R., and Burrows, D. B., "Viscosity of Hydrocarbon Gases under Pressure," *Trans. AIME* (1954) 201, 264–272.
6. Weber, J. H., "Predicting Properties of Gas Mixtures," *Chem. Eng.* (May 1980).
7. Lee, A., Gonzalez, M. H., and Eakin, B. E., "The Viscosity of Natural Gases," *J. Petroleum Technol.* (Aug. 1966) 18, 997–1000.
8. Katz, D. L., Cornell, D., Kobayashi, R., Poettmann, F. H., Vary, J. A., Elenbaas, J. R., and Weinary, C. F., *Handbook of Natural Gas Engineering*. McGraw-Hill, New York, 1959.
9. Trube, A. S., *et al.,* "Compressibility of Under-Saturated Hydrocarbon Reservoir Fluids," *Trans. AIME* (1957) 341.
10. Hall, H. N., "Compressibility of Reservoir Rocks," *Trans. AIME* (1953) 198, 309–311.

11. Amanat, U. C., "New Generalized Correlations for Predicting Reservoir Formation Water Properties as Functions of Pressure, Temperature and Total Dissolved Solids, Gives Best Accuracy, Computer Application, Chart or Table Use, Unpublished development (June 1998).
12. Ulbertson, O. L., and McKetta, J. J., "Solubility of Methene in Water at Pressures to 10,000 psia," *Trans. AIME* (1951) 223.
13. McCain, W. D., Jr., *The Properties of Petroleum Fluids.* Petroleum Publishing Co., Tulsa, OK, 1990.
14. Standing, M. B., and Dodson, C. R., "Pressure–Volume–Temperature and Solubility Relations for Natural Gas Water Mixtures," *Drill Prod. Proc. API* (1944) 1973.
15. Yarbrough, L., and Hall, K. R., "How to Solve Equation of State for z-Factors," *Oil & Gas Journal* (Feb. 18, 1974) 8688.

Appendix G

Substantial Set of Problems without Solutions

G–1 A gas reservoir was produced at a constant rate q_{sc} of 4.525 mmscfd for a time t of 48 hr. The sandface pressure p_{wf} at that time was 1750 psia. General data are as follows: $\overline{p} = 3100$ psia; $p_i = 3375$ psia; $z_I = 0.895$; $c_i = 0.000296$ psi^{-1}; $\overline{c} = 0.000323$ psi^{-1}; $k = 7.5$ mD; $T = 605°$R; $r_w = 0.42$ ft; $h = 35$ ft; $\overline{\mu}_i = 0.01425$; $\mu_i = 0.01402$; $\phi = 0.142$. Calculate dimensionless quantities t_D, P_D, and q_D using the p, p^2, and ψ treatments. (Use $\psi - P$ curve for Example 2–1.)

G–2 A gas well in an infinite-acting reservoir was produced at a constant rate q_{sc} of 6.85 mmscfd for a period of 45 hr. The flowing bottom-hole pressure p_{wf1} at that time was 1850 psia. The same well was produced at a constant rate q_{sc2} of 10.75 mmscfd for a time of 35 hr. The flowing bottom-hole pressure p_{wf2} at that time was 1560 psia. The stabilization shut-in pressure p_R prior to each of the two flowing periods was 2100 psia.

Other data pertinent to the test are as follows: $k = 21.5$ mD; $r_w = 0.33$ ft; $T = 604°$R; $\phi = 0.145$; $\mu_i = 0.01465$ cP; $c_i = 0.00054$ psi^{-1}; $\psi_i = 320.12$ mmpsia^2/cP; $t_1 = 45$ hr; $q_{sc1} = 6.85$ mmscfd; $p_{wf1} = 1850$ psia; $t_2 = 35$ hr; $q_{sc2} = 10.75$ mmscfd; $p_{wf2} = 1560$ psia.

Calculate the skin and IT flow effects, s and D, respectively. Also calculate, for the second flow rate,

(a) The pressure drop due to laminar flow effect
(b) The pressure drop due to skin effect
(c) The pressure drop due to IT flow effect
(d) Total pressure drop

(Use the same ψ–p curve given in Example 2–1.)

G–3 Estimate steady-state gas flow rate from a gas well with a 200-ft-long conductivity fracture. Given: $r_w = 0.4271$ ft; $r_e = 2106$ ft; $P_{sc} = 14.65$ psia; $T_{sc} = 520°$R; $T = 686°$R; $z = 0.9148$; $p_R = 1660$ psia; $\mu_g = 0.01639$ cP;

Table G–1
Calculated Gas PVT Properties

Pressure (psia)	Compressibility factor Z	Gas viscosity (cP)	Pseudopressure $\psi(p)$ (mmpsia^2/cP)
4000	0.9647	0.024580	872.92
3600	0.9445	0.023151	739.56
3200	0.9282	0.021721	610.28
2800	0.9169	0.020329	486.72
2400	0.9113	0.019008	371.18
2000	0.9120	0.017784	266.41
1600	0.9189	0.016681	175.33
1200	0.9317	0.015723	100.82
800	0.9503	0.014932	45.51
400	0.9733	0.014337	11.47
14.65	0.9995	0.013978	0.52

Table G–2
Gas Well Deliverability Stabilized Flow Data

Initial shut-in	Duration (hr)	Surface pressure (psia)	Bottom-hole pressure (psia)	Choke size (inches)	Flow rate (mmscfd)
Initial shut-in	147.2	2388	3700	—	—
Flow 1	6	2015	3144	16	2.397
Shut-in	6	2388	3700	—	—
Flow 2	6	1640	2566	24	5.214
Shut-in	6	2388	3700		—
Flow 3	6	1365	2158	32	6.144
Shut-in	6	2368	3698		—
Flow 4	6	1015	1836	48	7.186
Extended flow	24	1015	3690	32	6.148
Final shut-in	22.75	2388	1727		

$h = 69$ ft; reservoir permeability $k_h = 10.0$ mD; vertical permeability $k_v = 10.0$ mD; well flowing pressure 1250 psia.

G–4 Calculate inflow performance responses for vertical and horizontal gas wells. The reservoir and gas properties are as follows: Reservoir = sandstone; depth = 12,550 ft; $p_R = 3700$ psis; reservoir thickness $h = 54$ ft; average reservoir permeability $k = 6.282$ mD; vertical permeability $k_v = 6.282$ mD (assume); well spacing = 640 acres; average porosity $\phi = 0.179$; $T = 710°$R; $\gamma = 0.665$; $T_c = 380.16°$R; $P_c = 645.08$ psia; $T_{sc} = 520°$R; $P_{sc} = 14.7$ psia.

G–5 The data in Table G–2 were reported for a flow-after-flow test. At each rate, pseudosteady state was reached. Initial (i.e., before the test) shut-in wellhead and bottom-hole pressures were 2388 psia and 3200 psia, respectively.

Gas properties are given in Example 4–1. Estimate the following:

(a) Gas well deliverability at wellhead and bottom-hole conditions using simplified and LIT(ψ) analysis approaches.
(b) Inflow performance response using both approaches.

G–6 A flow-after-flow test was performed on a well located in a low pressure reservoir in which the permeability was high. Using the following test data, determine:

(a) The values of n and C for the deliverability equation
(b) The *AOF*
(c) The flow rate for $P_{wf} = 160$ psia

Test	P_{wf} **(psia)**	q_{sc} **(mmscfd)**
	201 P_R	0
1	196	2.730
2	195	3.970
3	193	4.440
4	190	5.550

G–7 An isochronal test was conducted on a well located in a reservoir that had an average pressure of 1952 psia. The well was flowed on four choke sizes and the flow rate and the flowing bottom-hole pressure were measured at 3 hr and 6 hr for each choke size. An extended test was conducted for a period of 72 hr at a rate of 6.0 mmscfd, at which time P_{wf} was measured at 1151 psia. Using the preceding data, find the following:

(a) Stabilized deliverability equation
(b) *AOF*
(c) Generate an inflow performance curve

q_{sc} **(mmscfd)**	$t = 3$ **hr** P_{wf} **(psia)**	$t = 6$ **hr** P_{wf} **(psia)**
2600	1793	1761
3300	1757	1657
5000	1623	1510
6300	1505	1320
6000	Extended flow, $t = 72$ hr	1151

G–8 A four-point test was conducted on a gas well that had a perforated zone of 20 ft. Static reservoir pressure is 5250 psia. Determine the following:

(a) Laminar and turbulent coefficients A and B
(b) Absolute open flow potential AOF
(c) New AOF if the perforated interval is increased to 30 ft

Test Data

Test no.	q_{sc} (mmscfd)	p_{wf} (psia)
1	9.300	5130
2	6.000	5190
3	5.200	5203
4	3.300	5225

G–9 A stabilized four-point drawdown test is run on a gas well when the average pressure in the drainage area is 2800 psia. The last flow rate of 4.5 mmscfd gave a stabilized flowing bottom-hole pressure of 2445 psia. The well drains 2250 acres. Assume that turbulence effects are negligible although this is not realistic. Other reservoir data are as follows: $k = 74$ mD; $\phi = 15\%$; $\mu_g = 0.021$ cP; $z_{avg} = 0.87$; $c = 3.57 \times 10^{-4}$ psi^{-1}; $h = 4.0$ ft.

Calculate the following:

(a) What must the bottom-hole pressure be in this well to produce at a rate of 5 mmscfd if P_e has declined to 2000 psia and the other reservoir parameters remain unchanged?
(b) How long must each rate be maintained to reach stabilized conditions during the flow test?
(c) During the flow test at 4.5 mmscfd where $P_i = 2800$ psia, how rapidly does the average reservoir pressure decline once the pseudo-steady state is reached ?
(d) A pressure drawdown test was performed on an exploratory gas well. The data obtained are shown in Table G–3. Other test data, reservoir, and fluid properties are as follows: Pressure prior to test = 6966 psia; constant gas-producing rate during test $q_g = 1.800$ mmscfd; $h = 15$ ft; $\phi = 0.19$ fraction; $s_w = 25\%$; $\mu = 0.0385$ cP; gas gravity = 0.920 (air = 1.00); $T = 120°$F; $T_{sc} = 520°$R; $P_{sc} = 14.70$ psia; $c_g = 0.000065$ psi^{-1}; $c_w = 0.000004$ psi^{-1}; $c_f = 0.0000039$ psi^{-1}; z at 6966 psia = 1.223; z at 6546 psia = 1.171; $p_{av} = 6756$ psia; and $\beta_g = 0.5174$ res bbl/mscf.

Find:

- Formation permeability, k_g
- Sketch of drainage area
- Estimate initial gas in-place

Table G–3
Pressure Drawdown Test Data

Time t (hr)	Time t (days)	Flowing well pressure P_{wf} (psia)
0.0331	0.00138	6928
0.0533	0.00222	6919
0.0660	0.00275	6915
0.0984	0.00410	6908
0.125	0.00520	6901
0.247	0.00665	6895
0.353	0.0103	6885
0.487	0.0147	6875
0.648	0.0203	6860
1.044	0.0270	6845
1.584	0.0435	6820
2.448	0.0660	6800
3.216	0.102	6776
4.896	0.134	6758
6.384	0.204	6730
10.44	0.266	6712
15.72	0.435	6670
20.22	0.655	6617
22.80	0.950	6546

G–10 Determine k, s, and C_s from the data below and in Table G–4, which were obtained in a pressure drawdown test on a gas well. Given: $P_I = 3000$; $s_{wi} = 0.22$; $V_W = 290$ cu ft; $h = 12$ ft; $T = 210°$F; $r_w = 0.39$ ft; $q_g = 1000$ mcf/d; $C_{ti} = 0.000245$ psi^{-1}; $\phi = 0.20$; $\mu_i = 0.01925$ cP; $\gamma_g = 0.655$; drainage area $= 640$ acres (square); well centered in drainage area.

G–11 A two-rate test was conducted on a well in gas reservoir, $p_i = 2550$ psia. The pressure time data for the first flow rate, $q_{sc1} = 17.016$ mmscfd, were not recorded. The flow rate was changed at $t_0 = 4$ hr at which time the flowing bottom-hole pressure, p_{wfo}, was 2250 psia. The second flow rate, $q_{sc2} = 11.463$ mmscfd, was continued for 8 hr, during which time the following bottom-hole pressures were recorded continuously. These pressure time data are given in Table G–5.

Gas properties and well/reservoir data were as follows: $p_c = 380.16$; $T_c = 645.08°$R; $G = 0.666$; $CO_2 = 7.84\%$; $N_2 = 0.11\%$; $H_2S = 0.00\%$; $\mu_I = 0.0186$ cP; $c_i = 0.000274$ psi^{-1}; $T = 632°$R; $h = 59$ ft; $r_w = 0.3542$; $\phi = 0.272$ fraction; $\phi_{HC} = 0.1801$; $s_g = 0.662$.

Table G–4
Single-Rate Drawdown Test Data for Ramey's Type Curve Analysis

Time t (hr)	Flowing pressure p_{wf} (psia)	$\psi(p_{wf})$ (mmpsia^2/cP)	$\Delta\psi = \psi(p_i) - \psi(p_{wf})$
0.02	1810.65	221.41	639.71
0.03	1807.45	220.68	640.44
0.07	1798.95	218.74	642.38
0.10	1786.35	215.87	645.24
0.17	1775.75	213.47	647.64
0.25	1768.05	211.74	649.38
0.33	1764.75	211.00	650.12
0.50	1757.45	209.36	651.76
0.75	1754.65	208.73	652.38
1.00	1755.45	208.91	652.20
1.50	1757.85	209.45	651.67
2.00	1754.65	208.73	652.38
2.50	1754.65	208.73	652.38
3.00	1751.35	208.00	653.12
3.50	1748.95	207.46	653.66
4.00	1747.35	207.10	654.01
5.00	1745.25	206.64	654.48
5.50	1742.05	205.92	655.19
6.00	1740.45	205.57	655.55

Calculate the following:

- The formation permeability k
- Apparent skin factors s'_1 and s'_2
- Inertial-turbulent flow factor D
- True skin factor s
- Pressure drop due to actual skin
- Rate-dependent pressure drop

The ψ–p curve, developed in Example 4.1, is applicable to this problem.

$$\psi(p_i) = 424.50 \text{ mmpsia}^2/\text{cP} \leftrightarrow 2550 \text{ psia and}$$
$$\psi(p_{wfo}) = 365.12 \text{ mmpsia}^2/\text{cP} \leftrightarrow 2250 \text{ psia}$$

$t_1 = 6$ hr, Two-rate drawdown test data are given in Table G–5.

G–12 An isochronal flow test is performed on a gas well at two different rates. Given the reservoir data and fluid properties below, determine the following using pressure-squared and real pseudopressure approaches:

Table G–5
Two-Rate Drawdown Test Data for Gas Well (q_{sc1} = 17.016 mmscfd; q_{sc2} = 11.463 mmscfd)

Time t (hr) (1)	Drawdown pressure P_{wf} (psig) (2)	Drawdown pressure P_{wf} (psia) (3)
0.02	2355.0	2357.0
0.03	2352.0	2355.5
0.07	2350.0	2355.3
0.10	2350.0	2355.0
0.13	2349.8	2354.4
0.17	2349.8	2353.8
0.25	2349.3	2353.0
0.33	2349.6	2352.0
0.50	2349.6	2351.9
0.75	2349.6	2351.9
1.00	2350.0	2351.9
1.50	2350.0	2351.9
2.00	2349.6	2351.9
3.00	2350.0	2351.9
4.00	2350.0	2351.9

Table G–6
PVT Gas Properties and Pseudopressure

P_{wf} (psia)	Z	μ (cP)	$z(p/\mu z)$	Mean $z(p/\mu z)$	Δp (psia)	$z(p/\mu z) \times \Delta p$	$\psi(P_{wf})$ (mmpsia^2/cP)
400	0.95	0.0117	71.975	35.988	499	14.4×10^6	14.4
800	0.90	0.0125	142.222	107.099	400	42.9×10^6	57.3
1200	0.86	0.0132	211.416	176.819	400	70.7×10^6	128.0
1600	0.81	0.0146	270.590	241.003	400	96.5×10^6	224.5
2000	0.80	0.0163	306.748	288.669	400	115.5×10^6	340.0
2400	0.81	0.0180	329.218	319.000	400	127.6×10^6	467.6

- Flow capacity kh
- The apparent skin factors s_1' and s_2'
- Non-Darcy flow coefficient D for the well
- True skin factors

Reservoir and well data are as follows $p_i = 2300$ psia; $r_w = 0.5$ ft; $r_e = 2980$ ft (640-acre spacing); $T = 590°$R; $z_i = 0.805$; $\mu_I = 0.0176$ cP; $\phi = 0.077$; and $c_i = 0.00041$ psi^{-1}.

Table G–7
Two-Rate Drawdown Test Data

Flowing time t (hr)	Flow no. 1 $q_{sc1} = 2.65$ mmscfd p_{wf} (psia)	Flow no. 2 $q_{sc2} = 4.23$ mmscfd p_{wf} (psia)
0.232	1855	1105
0.4	1836	1020
0.6	1814	954
0.8	1806	906
1.0	1797	860
2.0	1758	700
4.0	1723	539
6.0	1703	387

Table G–8
Two-Rate Drawdown Test Data for Gas Well (q_{sc1} = 17.016 mmscfd; q_{sc2} = 11.463 mmscfd)

Time t (hr) (1)	Drawdown pressure P_{wf} (psig) (2)	Drawdown pressure P_{wf} (psia) (3)
0.02	2355.0	2357.0
0.03	2352.0	2355.5
0.07	2350.0	2355.3
0.10	2350.0	2355.0
0.13	2349.8	2354.4
0.17	2349.8	2353.8
0.25	2349.3	2353.0
0.33	2349.6	2352.0
0.50	2349.6	2351.9
0.75	2349.6	2351.9
1.00	2350.0	2351.9
1.50	2350.0	2351.9
2.00	2349.6	2351.9
3.00	2350.0	2351.9
4.00	2350.0	2351.9

G–13 A two-rate drawdown test was conducted on a gas well. Given the following reservoir data and fluid properties: $T = 632°\text{R}$; $r_w = 0.3542$ ft; $r_e = 2200$ ft; $h = 59$ ft; $\phi = 0.1801$ fraction; $\mu_{gi} = 0.018017$ cP; $c_t = 0.00028$ psi^{-1}; initial pressure prior to test = 2550 psia ↔ 424.00 mmpsia^2/cP; first flow rate $q_{sc1} = 11.463$ mmscfd; time t_1 at which first flow rate changed = 4 hr;

flowing bottom-hole pressure at that time = 2355 psia ↔ 410.70 mmpsia^2/cP; second flow rate $q_{sc2} = 17.016$ mmscfd. The ψ–p curve in Example 4.1 is applicable to this problem.

Calculate the following:

- Formation permeability k
- Apparent skin factor s'_1 related to flow rate q_{sc1}
- Apparent skin factor s'_2 related to flow rate q_{sc2}
- Non-Darcy flow coefficient D
- True skin factor s
- Pressure drop across the skin related to flow rate q_{sc1}
- Pressure drop across the skin related to flow rate q_{sc2}
- Reservoir pressure p_R
- The values of deliverability constants A and B
- Absolute open flow potential AOF
- Inflow performance response

G–14 A modified isochronal test was made on a well believed to be producing from a 640-acre drainage area. The bottom-hole pressure–time data are given in Table G–9. During the first flow and buildup periods, the pressure buildup was monitored and also presented in Table G–9. Determine k and s; both transient and semisteady-state deliverability; and the deliverability at a wellhead pressure of 1200 psia, given the following reservoir, fluid, and tubing characteristics: initial pressure $P_i = 3200$ psia; $\psi(P_i) = 639.00$ mmpsia^2/cP; $\gamma_g = 0.876$; $T = 271$°F; $\mu_I = 0.02052$ cP; $N_2 = 0.0$; $H_2S = 0.0$; $CO_2 = 0.0$; $Z_i = 0.9192$; $C_t = 3.2 \times 10^{-4}$ psi^{-1}; $r_w = 0.25$ ft; $\phi = 0.17$; $h = 60$ ft; depth = 10,000 ft; and tubing i.d. = 2.441 inches.

G–15 The well described in Example 6–2 was flowed at a rate of 565.0 mscfd for a period of 120.5 hr and then shut in for a buildup test. The flowing pressure at shut-in was 3295 psia. Calculate k, s', and $\overline{p}_R$ if the well is producing from the center of a square drainage area containing 22×10^6 sq ft. The pressure versus time data are tabulated below ($t_P = 120.5$ hr).

Δt (hr)	P_{ws} (psi)
0	3295
0.53	3296
1.60	3385
2.67	3547
3.73	3573
4.80	3591
5.87	3605
6.93	3614
8.00	3623
9.87	3634

(continued)

Δt (hr)	P_{ws} (psi)
12.00	3644
14.67	3654
18.67	3664
24.53	3672
29.33	3676
35.73	3684
45.87	3688
49.87	3691

Table G–9
Modified Isochronal Deliverability Test

	Duration (hr)	Bottom-hole pressure (psia)	Flow rate q (mmscfd)
1	12	3041	4.00
Shut-in	12	3193	
2	12	2997	5.00
Shut-in	12	3188	
3	12	2954	6.00
Shut-in	12	3183	
4	12	2911	7.00
5	96	2878	7.00
Shut-in	60	3183	

Buildup data, first flow period

Δt (hr)	Pressure (psia)	ψ (P) (mmpsia^2/cP)
0	3041	586.349
0.5	3166	627.399
1	3173	629.699
1.5	3177	630.994
2	3179	631.829
3	3183	633.069
4	3185	633.849
5	3187	634.429
6	3188	634.869
7	3189	635.229
8	3190	635.519
9	3191	635.749
10	3192	635.959
11	3192	636.139
12	3193	636.299

G–16 A pressure-buildup test was performed on a well located in a gas field on 640-acre spacing. Data obtained were as follows:

Δt (hr)	$\frac{t_p+\Delta t}{\Delta t}$	P_{ws} (psig)
0	—	1727
1	2401	1850
3	801	1964
6	401	2008
10	241	2039
15	161	2064
22	110	2087
34	71.6	2113
45	54.3	2130
65	37.9	2155
126	20.0	2175

$t_P = 100$ days; $h = 54$ ft; $T = 605°$R; $z = 0.85$; $\phi = 18.0\%$; $r_w = 0.33$ ft; $P_R = 2320$ psia; $P_i = 2390$ psi; $\mu_g = 0.12$ cP.

(a) Compute the reservoir pressure in the drainage area of the well assuming finite boundary conditions, using the Horner and Matthews *et al.* methods.
(b) If the well is completed across the entire formation thickness, calculate the effective permeability.
(c) Calculate the value of the mechanical skin factor.
(d) What is the additional pressure drop in the wellbore due to the skin?
(e) If it is initially assumed that the well is draining from the center of a circle, is it valid to equate P_i to P^*?

G–17 A modified isochronal test was made on a well believed to be producing from a 640-acre draining area ($r_e = 2980$ ft). The bottom-hole pressure–time data are given in Table G–10. During the first flow and buildup periods, the pressure buildup was monitored and also presented in Table G–10. The objective is to determine k and s; both transient and semisteady-state deliverability; and the deliverability at a wellhead pressure of 1200 psia, given the following reservoir, fluid, and tubing characteristics: $h = 60$ ft; $C_t = 3.2 \times 10^{-4}$ psi^{-1}; $P_i = 3200$ psia; $\phi = 0.17$; $\gamma_g = 0.876$ (air = 1.00); depth = 10,000 ft; $N_2 = 0$; $CO_2 = 0$; $H_2S = 0$; $\psi(P_i) = 638.739 \times 10^6$ psia^2/cP; $\mu_I = 0.02052$ cP; tubing i.d. = 2.441 inches.

G–18 A modified isochronal test was made on a well believed to be producing from a 640-acre draining area. The bottom hole pressure–time data are

Table G–10
Modified Isochronal Deliverability Test

Duration (hr)	End pressure (psia)	Gas flow rate (mmscfd)
4	3041	12
Shut-in	3193	12
5	2997	12
Shut-in	3188	12
6	2954	12
Shut-in	3183	12
7	2911	12
7	2878	96
Shut-in	3183	60

Buildup data, first flow period

t (hr)	Δt (hr)	Pressure (psia)	Δ(P) (mmpsia^2/cP)
12		3041	586.349
12.5	0.5	3166	627.399
13	1.0	3173	629.699
13.5	1.5	3177	630.999
14	2.0	3179	631.879
15	3.0	3183	633.879
16	4.0	3185	633.849
17	5.0	3187	634.429
18	6.0	3188	634.869
19	7.0	3189	635.229
20	8.0	3190	635.519
21	9.0	3191	635.749
22	10.0	3191	635.959
23	11.0	3192	636.139
24	12.0	3193	636.299

given in Table G–11. During the first flow and buildup periods, the pressure buildup was monitored and also presented in Table G–11.

Determine k and s; both transient and semisteady-state deliverability; and the deliverability at a wellhead pressure of 1200 psia, given the following well/reservoir characteristics: $P_i = 3200$ psia; $\gamma_g = 0.876$; $\psi(P_i) = 638.739 \times 10^6$ psia^2/cP; $\mu_I = 0.02052$ cP; $z_i = 0.9192$; $\phi = 0.17$ fraction; $c_t = 3.2 \times 10^{-4}$ psi^{-1}; $h = 60$ ft; depth $= 10{,}000$ ft.

G–19 The well was shut-in at a constant rate of 25.794 mmscfd for 41.67 hr, during which time the pressure buildup was monitored continuously.

Table G–11
Modified Isochronal Deliverability Test

Q (mmscfd)	Duration (hr)	End pressure (psia)
4	12	3041
Shut-in	12	3193
5	12	2997
Shut-in	12	3188
6	12	2954
Shut-in	12	3183
7	12	2911
7	12	2878
Shut-in	12	3183

Buildup data, first flow period

t (hr)	Δt (hr)	Pressure (psia)	$\Delta(P)$ (mmpsia^2/cP)	$(t_P + \Delta t)/\Delta t$
12		3041	586.349	—
12.5	0.5	3166	627.399	25.0
13	1	3173	629.699	13.0
13.5	1.5	3177	630.999	9.0
14	2	3179	631.879	7.0
15	3	3183	633.069	5.0
16	4	3185	633.849	4.0
17	5	3187	634.429	3.4
18	6	3188	634.869	3.0
19	7	3189	635.229	2.71
20	8	3190	635.519	2.5
21	9	3191	635.749	2.33
22	10	3192	635.959	2.2
23	11	3192	636.139	2.09
24	12	3193	636.299	2.0

The pressure just prior to shut-in was 2362.65 psia. General data pertinent to the test are given below. The pressure–time data are also tabulated in Table G–12.

From a recombined gas analysis, $N_2 = 0.18\%$, $CO_2 = 2.83\%$, $H_2S = 0.0\%$, $C_1 = 88.51\%$, $C_2 = 4.02\%$, $C_3 = 2.45\%$, $iC_4 = 0.52\%$, $nC_4 = 0.55\%$, $iC_5 = 0.24\%$, $nC_5 = 0.14\%$, $C_6 = 0.56\%$, $C7^+ = 0.00$.

Well/reservoir data: well depth = 12,000 ft; $T = 172°F$; $h = 59$ ft; $\phi = 0.272$; $c_g = 0.00041$ psi^{-1}; $c_w = 3.10 \times 10^{-6}$ psi^{-1}; $c_O = 3.30 \times 10^{-6}$ psi^{-1}; $c_f = 4.0 \times 10^{-6}$ psi^{-1}; $\overline{z} = 0.8620$; $\overline{\mu}_g = 0.018017$ cP; $s_w = 0.338$;

Table G–12
PVT Gas Properties and Pseudopressure Calculations

Pressure (psia)	Z-Factor —	Gas viscosity (cP)	Pseudopressure (mmpsia^2/cP)
4000	0.9226	0.02433	956.79
3600	0.8988	0.02278	815.17
3200	0.8798	0.02122	676.27
2800	0.8671	0.01966	542.01
2400	0.8620	0.01817	415.02
2000	0.8657	0.01677	298.62
1600	0.8783	0.01554	196.63
1200	0.8993	0.01449	112.92
800	0.9273	0.01366	50.82
400	0.9608	0.01303	12.78
14.65	0.9970	0.01266	0.17

$s_g = 0.662$; $c_t = 0.00028\ \text{psi}^{-1}$; $p_R = 2374$ psia; production rate at shut-in time = 27.497 mmscfd; cumulative production prior to test = 36.5245 mmscf;

$$\begin{aligned}\beta_g &= 0.00646\ \text{ft}^3/\text{scf}\\ &= 0.001151\ \text{bbl/scf}\\ &= 154.7567\ \text{scf/ft}^3\\ &= 869\ \text{scf/bbl}\\ &= 0.8689\ \text{mscf/bbl}\\ &= 1.1508\ \text{bbl/mscf}\end{aligned}$$

From gas compositional analysis, the gas properties are as follows Mol. wt. = 19.29; $G = 0.666$; $P_c = 658.73$ psia; $T_c = 370.62°\text{R}$; H_2S = 0.00%; CO_2 = 2.83%; N_2 = 0.18%; $P_i = 2400$ psia; $\mu_I = 0.01817$ cP; $c_i = 0.0002195$.

Using the Horner method, determine the following: permeability k, skin factor s, pressure drop due to skin Δp_{skin}, flow efficiency using p^*, and effective wellbore radius

(a) Using the MDH method
(b) Using the Ramey and Cobb method
(c) Using the Dietz method

G–20 A gas well was flowed at a rate of 3.543 mmscfd. The stabilized sandface pressure at the end of the flow test was 2566 psia, and the current average reservoir pressure was estimated to be 1660 psia. For a gas gravity

Table G–13
Pressure Build-up Test Data

Time Δt (hr) (1)	$\frac{t_p+\Delta t}{\Delta t}$ (2)	P_{ws} (psia) (3)	P_{ws} $\psi(P_{ws})$ (mmpsia^2/cP) (4)	$\Delta(P_{wf})$ (mmpsia^2/cP) (5)
0.0	—	1735	204.35	0.00
0.02	2666.92	1738	204.96	0.00
0.03	1333.96	1747	207.12	0.059
0.07	667.48	1788	216.16	14.89
0.10	445.32	1818	223.10	21.83
0.13	334.24	1869	234.89	33.62
0.17	267.59	1925	248.28	47.01
0.25	178.73	2028	273.44	72.17
0.33	134.30	2135	300.34	99.07
0.50	89.86	2312	347.09	145.82
0.75	60.24	2615	432.19	230.92
1.00	45.43	2819	492.58	291.31
1.50	32.62	3146	593.35	392.07
2.00	23.22	3310	645.38	444.11
2.50	18.77	3350	658.17	456.89
3.00	15.81	3366	663.42	462.14
3.50	13.69	3382	668.65	467.37
4.00	12.11	3385	669.42	468.14
4.83	10.19	3391	671.52	470.24
5.00	9.89	3397	673.23	471.96
5.50	9.08	3403	675.21	473.93
6.00	8.41	3407	676.66	475.38
6.50	7.84	3411	677.96	476.68
7.00	7.35	3415	679.15	477.88
7.50	6.92	3418	680.32	479.04
8.00	6.55	3421	681.13	479.85
8.50	6.23	3425	682.30	481.02
9.00	5.94	3428	683.37	482.09
9.50	5.68	3432	684.66	483.39
10.00	5.44	3436	685.99	484.72
10.50	5.23	3440	687.32	486.05
11.00	5.04	3443	688.36	487.08
11.50	4.86	3447	689.43	488.16
12.00	4.70	3448	689.95	488.68
12.50	4.55	3451	690.86	489.59
13.00	4.42	3453	691.55	490.27
13.50	4.29	3456	692.46	491.18
14.00	4.17	3459	693.40	492.12

Table G–13 (continued)

Time Δt (hr) (1)	$\frac{t_p+\Delta t}{\Delta t}$ (2)	P_{ws} (psia) (3)	P_{ws} $\psi(P_{ws})$ (mmpsia^2/cP) (4)	$\Delta(P_{wf})$ (mmpsia^2/cP) (5)
14.50	4.06	3461	694.18	492.90
15.00	3.96	3464	694.96	493.68
15.50	3.87	3466	695.88	494.62
16.00	3.78	3468	696.53	495.27
16.50	3.69	3471	697.31	496.06
17.00	3.61	3473	697.86	496.61
17.50	3.54	3475	698.77	497.52
18.00	3.47	3477	699.29	498.04
18.50	3.40	3479	700.11	498.85
19.00	3.34	3481	700.76	499.51
19.50	3.28	3483	701.41	500.16
20.00	3.22	3486	702.22	500.97
20.50	3.17	3488	702.75	501.49
21.00	3.12	3489	703.14	501.88
21.50	3.07	3491	703.92	502.67
22.00	3.02	3493	704.47	503.22
22.50	2.97	3494	704.99	503.74
23.00	2.93	3496	705.52	504.26
23.50	2.89	3497	705.78	504.53
24.00	2.85	3499	706.59	505.34
24.50	2.81	3501	706.98	505.73
25.00	2.78	3502	707.38	506.12
26.00	2.71	3505	708.58	507.33
27.00	2.65	3509	709.76	508.51
28.00	2.59	3512	710.71	509.45
29.00	2.53	3515	711.62	510.37
30.00	2.48	3518	712.53	511.28
31.00	2.43	3521	713.61	512.36
32.00	2.39	3524	714.66	513.41
33.00	2.35	3526	715.21	513.96
34.00	2.31	3529	716.26	515.01
35.00	2.27	3532	717.21	515.96
36.00	2.23	3533	717.73	516.48
37.00	2.20	3536	718.65	517.40
38.00	2.17	3538	719.17	517.92
39.00	2.14	3541	720.12	518.87
40.00	2.22	3544	721.17	519.92
41.00	2.08	3546	721.72	520.47

Table G–13 (continued)

Time Δt (hr) (1)	$\frac{t_p+\Delta t}{\Delta t}$ (2)	P_{ws} (psia) (3)	P_{ws} $\psi(P_{ws})$ (mmpsia²/cP) (4)	$\Delta(P_{wf})$ (mmpsia²/cP) (5)
42.00	2.06	3548	722.64	521.39
43.00	2.03	3550	723.30	522.04
44.00	2.01	3553	724.25	522.99
45.00	1.99	3555	724.77	523.52
46.00	1.97	3557	725.56	524.31
47.00	1.95	3560	726.38	525.12

Table G–14 PVT Gas Property and Pseudopressure

P (psia)	Z	μ (cP)	$\Psi(P)$ (mmpsia²/cP)
4000	0.9598	0.023689	903.57
3750	0.9470	0.022859	816.26
3500	0.9354	0.022018	730.52
3250	0.9256	0.021176	646.66
3000	0.9177	0.020345	565.11
2750	0.9119	0.019533	486.41
2500	0.9085	0.018748	411.18
2250	0.9074	0.017997	340.12
2000	0.9089	0.017285	273.93
1750	0.9128	0.016618	213.36
1500	0.9192	0.016002	159.12
1250	0.9279	0.015441	111.91
1000	0.9389	0.014940	72.35
750	0.9518	0.014507	41.00
500	0.9665	0.014147	18.31
250	0.9825	0.013868	4.60
14.65	0.9985	0.013687	0.53

of 0.681 and a bottom-hole temperature of 686°R, the P–$\Psi(P)$ table was calculated and tabulated in Table G–14.

Deliverability coefficients are $a = 28.5136$ psia²/cP-mmscfd, and $b = 0.34783$ psia²/cP-mmscfd². The objective is to simplify the method suggested in this section by calculating the following parameters:

Table G–15 Constant-Rate Drawdown Test Data

Time t (hr)	Flowing pressure (psia)	Pseudopressure (mmpsia^2/cP)
0.02	3608.95	742.50
0.03	3544.35	721.30
0.07	3480.25	700.37
0.10	3440.05	687.30
0.13	3385.15	669.52
0.17	3346.65	657.11
0.25	3270.25	632.62
0.33	3224.35	618.00
0.50	3172.75	601.65
0.75	3141.55	591.82
1.00	3130.15	588.23
1.50	3145.15	592.95
2.00	3127.75	587.48
2.50	3129.75	588.10
3.00	3133.85	589.39
4.00	3136.65	590.27
5.00	3139.85	591.28
6.00	3143.55	592.45

(a) $(AOF)_{current}$ at current conditions ($\overline{p}_R = 1660$ psia)
(b) Deliverability at a flowing bottom-hole pressure $P_{wf} = 1000$ psia
(c) $(AOF)_{future}$ at a future average pressure $(P_R)_{future} = 1500$ psia
(d) Deliverability at a future bottom-hole pressure $P_{wf} = 1250$ psia

G–21 Determine wellbore storage coefficients C_S, C_{SD}, s, and formation permeability k from the following data and those in Table G–15, which were obtained in a pressure drawdown test on a gas well: $q_{sc} = 2.397$ mmscfd; $h = 41$ ft; $r_w = 0.4271$ ft; $\phi = 0.1004$; $\mu_i = 0.02441$ cP; $c_{ti} = 0.0002295$ psi^{-1}; $p_i = 3700$ psia $\leftrightarrow$ $\psi(p_i) = 861.12$ mmpsia^2/cP; $T_{SC} = 520°$R; $P_{SC} = 14.65$ psia; well depth = 12,550 ft; $C_{WS} = 0.000292$ psi^{-1} at $P_{WS} =$ 3420 psi.

Nomenclature

Symbol	Description	Field units	Metric (SI) units
A	Drainage area	ft^2	m
A	Gross cross-sectional area	—	—
a	Coefficient in the stabilized deliverability equation	$\frac{psia^2}{(cP)(mmscfd)}$	$\frac{kPa^2}{(\mu Pa.s)(kmol/d)}$
a_t	Coefficient in the transient form of deliverability equation	$\frac{psia^2}{(cP)(mmscfd)}$	$\frac{kPa^2}{(mPas)(kmol/d)}$
a^1	Coefficient in the stabilized deliverability equation	$\frac{psia^2}{mmscfd}$	$\frac{kPa^2}{kmol/d}$
a^{11}	Coefficient in the stabilized deliverability equation	$\frac{psia}{mmscfd}$	$\frac{kPa}{kmol/d}$
A	Gross cross-sectional area	—	—
A	Drainage area	ft^2	m^2
AOF	Absolute open flow potential of a well or the deliverability against a zero sandface pressure (Ch. 3)	mmscfd	kmol/d
b	Coefficient in the stabilized deliverability equation	$\frac{psia^2}{(cP)(mmscfd)}$	$\frac{kPa^2}{(\mu Pa.s)(kmol/d)}$

(continued)

Symbol	Description	Field units	Metric (SI) units
b_1	Coefficient in the stabilized deliverability equation	$\frac{\text{psia}}{\text{mmscfd}}$	$\frac{\text{kPa}}{\text{kmol/d}}$
b_{11}	Coefficient in the stabilized deliverability equation	$\frac{\text{psia}}{\text{mmscfd}}$	$\frac{\text{kPa}}{\text{kmol/d}}$
c	Compressibility of gas at average conditions	psia^{-1}	kPa^{-1}
c_f	Formation compressibility	psia^{-1}	kPa^{-1}
c_g	Gas compressibility	psia^{-1}	kPa^{-1}
c_0	Oil compressibility	psia^{-1}	kPa^{-1}
c_t	Effective total compressibility	psia^{-1}	kPa^{-1}
c_w	Water compressibility	psia^{-1}	kpa^{-1}
c_{wf}	Gas compressibility corresponding to p_{wf}	psia^{-1}	kPa^{-1}
C_{ws}	Compressibility of wellbore fluids evaluated at the mean wellbore temperature and pressure	psia^{-1}	kPa^{-1}
C	Coefficient in the deliverability equation	—	—
C_A	Shape factor	—	—
C_S	Wellbore storage constant	ft^3/psia	m^3/kPa
C_{SD}	Dimensionless wellbore storage constant	—	—
C_W	Coefficient in the wellhead deliverability equation	—	—
D	IT flow factor	mmscfd^{-1}	$(\text{kmol/d})^{-1}$
e	2.718, base of natural logarithms	—	—
E_i	Exponential integral	—	—
FE	Flow efficiency	—	—
F_r	A factor defined by Equation 6–29		
F	MBH dimensionless pressure function	—	—
G	Specific gravity of a gas	—	—
h	Net formation thickness	ft	m
h_i	Thickness of ith layer in a multilayer reservoir	ft	m
h_D	Dimensionless formation thickness	—	—

(continued)

Symbol	Description	Field units	Metric (SI) units
k	Permeability of a medium	—	—
k	Reservoir permeability	md	μm^2
kh	Permeability thickness of a formation	md-ft	$\mu m^2 \times m$
$\frac{kh}{\mu}$	Transmissibility of a formation	$\frac{md-ft}{cp}$	$\frac{\mu m^2 \times m}{\mu Pa \times m}$
k_i	Permeability of the ith layer of a multilayer reservoir	md	μm^2
k_g, k_o, k_w	The permeability of a medium to gas, oil and water, respectively, in multiphase systems	mD	μm^2
in	Logarithm to the base e	—	—
log	Logarithm to the base 10	—	—
L	Length of the flow string	ft	m
m	Slope of the semilog straight line	$\frac{psia^2/cP}{cycle}$	$\frac{kPa^2/\mu Pa's}{cycle}$
m'	Slope of the straight line plot Equation		
m''	Slope of the straight line plot Equation		
M	Molecular weight		
M	Match point	—	—
MBH	Abbreviated form of Matthews, Brons, and Hazebrock (1954)	—	—
MDH	Abbreviated form of Miller, Dyes, and Hutchison (1950)	—	—
M_i	Molecular weight of any pure component i		
n	Reciprocal slope of the deliverability line	—	—
N	Number of data points for least-square curve fit	—	—
p	Pressure	psi	kPa
p_a, p_b, p_c, p_d	Pressure drawdown calculated by different methods	—	—
p_c	Pseudocritical pressure	psia	kPa
p_{ci}	Critical pressure of any pure component i	psia	kPa
p_e	Pressure at the external boundary of the reservoir	psia	kPa

(continued)

Symbol	Description	Field units	Metric (SI) units
p_f	Flowing reservoir pressure at any position in the reservoir, except at the well and the external boundary	psia	kPa
P_i	Initial stabilized shut-in pressure in a new reservoir	psia	kPa
p_i	Stabilized shut-in pressure prior to a flow test	psia	kPa
p_{wf}	Flowing well midpoint pressure	psia	kPa
p_r	Pseudoreduced pressure	—	—
$\overline{p}_R$	Stabilized shut-in reservoir pressure	psia	kPa
p_{sc}	Standard pressure	psia	kPa
p_{wf}	Flowing bottom-hole pressure	psia	kPa
p_{ws}	Static bottom-hole pressure	psia	kPa
p_R^*	Arbitrary reservoir pressure	psia	kPa
p_t	Dimensionless pressure drop at the well excluding skin and inertial–turbulent flow effects	psia	kPa
Δp	Pressure difference	psi	kPa
Δp_D	Dimensionless pressure drop	—	—
$\Delta p_{D\ HOR}$	Horner dimemsionless pressure drop	—	—
$(\Delta p_D)_{IT}$	Dimensionless pressure drop due to IT flow	—	—
$\Delta p_{D\ MDH}$	MDH dimensionless pressure drop	—	—
$(\Delta p_D)_{skin}$	Dimensionless pressure drop due to skin	—	—
q	Flow rate	mmscfd	kmol/d
q_A	Gas flow rate at well A	mmscfd	kmol/d
q_B	Gas flow rate at well B		
q_D	Production rate of gas in a multiphase system	mmscfd	kmol/d
q_{sc}	Volumetric flow rate at standard conditions of temperature and pressure	mmscfd	kmol/d
q_{total}	Sum of the flow rates of two wells creating a no-flow boundary between them	mmscfd	kmol/d

(continued)

Symbol	Description	Field units	Metric (SI) units
q_w	Production rate of water in a multiphase system	Bbl/d	M^3/d
q^*	Modified flow rate	mmscfd	kmol/d
q	Production rate	mmscfd	kmol/d
Q_t	Dimensionless total production number	—	—
Q_T	Cumulative production	scf	mol
r	Radius	ft	mol
r_A	Distance of well A from a point P in the reservoir	ft	mol
r_{AD}	Dimensionless value of r_A, $/r_W$	—	—
r_B	Distance of well B from a point P in the reservoir	ft	m
r_{BD}	Dimensionless value of r_B, r_B/r_W	—	—
r_d	Effective drainage radius	ft	m
r_D	Dimensionless radius, r/r_w	—	—
r_e	Radius of external boundary	ft	m
r_{eD}	Dimensionless external (boundary) radius, r_e/r_W	—	—
r_f	Fracture radius	ft	m
r_{inv}	Radius of investigation	ft	m
r_{skin}	Radius of a hypothetical permeability k_{skin}	ft	m
r_w	Well radius	ft	m
$r_{weffective}$	Effective well radius	ft	m
R	Gas constant	$10.7\frac{ft^3.psia}{lbmol°R}$	$8.3\frac{m^3.kPa}{mol\ K}$
s	Skin factor	—	—
s_c	Condensate skin effect	—	—
s_K	Skin due to altered permeability	—	—
s'	Apparent skin factor	—	—
s'_n	Apparent skin factor associated with the flow rate q_n	—	—
S	Parameter	—	—
S_c	Hydrocarbon liquid saturation required to reach mobility (fraction of pore volume)	—	—
SF	Stabilized factor	—	—
t	Time	hr	hr
t_c	Corrected time of flow	hr	hr
t_D	Dimensionless time for various systems	—	—

(continued)

Symbol	Description	Field units	Metric (SI) units
t_{DA}	Dimensionless time based on drainage area	—	—
t_s	Time to stabilization	hr	hr
t_{ws}	Time for wellbore storage effects to become negligible	hr	hr
t_{WSD}	Dimensionless time for wellbore storage effects to become negligible	—	—
T	Reservoir temperature	°R	K
T_c	Pseudocritical temperature	°R	K
T_{ci}	Critical temperature of any pure component i	°R	K
T_{mf}	Flowing well midpoint temperature	°R	K
T_r	Pseudoreduced temperature	—	—
T_{SC}	Standard temperature	60°F	15°C
T_{tf}	Flowing wellhead (top-hole) temperature	°R	K
T_{wf}	Flowing bottom-hole temperature	°R	K
T_{ws}	Static bottom-hole temperature	°R	K
t_{100}	Time required to investigate 100 ft of reservoir	hr	hr
t^*	Modified time	hr	hr
Δt	Shut-in time	hr	hr
Δt_{DA}	Dimensionless time based on drainage area	—	—
Δt_{De}	Dimensionless shut-in time based on external radius	—	—
Δt_i	Time of intersection of semilog straight lines for a well near a fault	—	—
V_P	In-place gas volume of a reservoir	mmscfd	kmol
V_{pm}	Minimum in-place gas volume	mmscfd	kmol
V_{ws}	Volume of wellbore	ft^3	m^3
x	Distance	ft	m
x, y, z	Rectangular coordinate	—	—
x_D	Dimensionless distance defined for various systems	—	—
x_e	Distance from well to external boundary	ft	m
x_f	Fracture half-length	ft	m
x_i	Mole fraction of component i in a mixture	—	—

(continued)

Symbol	Description	Field units	Metric (SI) units
X	Parameter in the viscosity correlation		
X	Boltzmann transformation	—	—
Δx	Distance in the x-direction	ft	m
y_D	Dimensionless location of a well in a rectangular drainage area	—	—
Y	Parameter used in the viscosity correlation	—	—
z	Vertical downward direction	ft	m
z_i	Compressibility factor at initial conditions	—	—
α	$=\frac{kp}{\phi\mu}$, diffusivity constant (Ch. 2, Eq. 2–35)	—	—
α'	Slant angle	degree	—
α_{ij}	Mass fraction of jth component in ith phase	—	—
β_g	Gas formation volume factor	—	—
β_w	Water formation volume factor	—	—
β	Turbulent coefficient, ft^{-1} (Ch. 2, Eq. 2–14a)	—	—
β	Ratio diameter orifice to inside diameter	—	—
β	Ratio of horizontal and vertical permeability defined in Ch. 3 (Eq. 3–8a)	—	—
β'	High velocity flow coefficient	1/ft	1/m
γ_g	Gas gravity (air = 1.000)	—	—
γ_{hc}	Wichert–Aziz correction term	—	—
	2.637×10^{-4} constant term	—	—
λ_t	Total mobility	md/cp	$\frac{\mu m}{\mu pa^*s}$
ε	Correction factor (Appendix F, Eq. F–8)	—	—
η	=0.159 (constant term)	—	—
$\delta(x)$	Delta function (Ch 2, Eq. 2–129)	—	—
θ	Angle of well deviation	degree	—
∞	Infinity	—	—
μ_1	Gas viscosity at atmospheric pressure	cP	μPa*s
μ	Gas viscosity	cP	μPa*s
μ_g	Gas viscosity	cP	μPa*s
$\overline{\mu}$	Average gas viscosity	cP	μPa*s

(continued)

Symbol	Description	Field units	Metric (SI) units
μ_i	Gas viscosity at initial conditions	cP	μPa*s
μ_w	Water viscosity	cP	μPa*s
π	A constant, 3.1415	—	—
ρ	Fluid density	lb_m/ft^3	kg/m^3
ϕ	Porosity of the medium	—	—
ϕ_t	Total porosity	—	—
ψ	Pseudopressure as defined by Al-Hussainy	$psia^2/cP$	$kPa^2/\mu Pa^*s$
ψ_i	Pseudopressure corresponding to p_i	$psia^2/cP$	$kPa^2/\mu Pa^*s$
ψ_{wf}	Pseudopressure corresponding to p_{wf}	$psia^2/cP$	$kPa^2/\mu Pa^*s$
ψ_{wfo}	Pseudopressure corresponding tp p_{wfo}	$psia^2/cP$	$kPa^2/\mu Pa^*s$
ψ_{wf1}	Pseudopressure corresponding to p_{wf1}	$psia^2/cP$	$kPa^2/\mu Pa^*s$
ψ_{ws}	Pseudopressure corresponding to p_{ws}	$psia^2/cP$	$kPa^2/\mu Pa^*s$
ψ_{ws1}	Pseudopressure corresponding to p_{wf1}	$psia^2/cP$	$kPa^2/\mu Pa^*s$
$\overline{\psi}_R$	Pseudopressure corresponding to p_R	$psia^2/cP$	$kPa^2/\mu Pa^*s$
$\psi*$	Pseudopressure corresponding to p^*	$psia^2/cP$	$kPa^2/\mu Pa^*s$
$\Delta\psi_D$	Dimensionless pseudopressure	—	—
$(\Delta\psi)_{IT}$	Inertial turbulent pseudopressure drop	$psia^2/cP$	$kPa^2/\mu Pa^*s$
$(\Delta\psi)_{skin}$	Skin pseudopressure drop	$psia^2/cP$	$kPa^2/\mu Pa^*s$
$(\Delta\psi)'_s$	Apparent skin pseudopressure drop	$psia^2/cP$	$kPa^2/\mu Pa^*s$
∇, ∇^2, ∇'	Gradient operators (Ch. 2)	—	—

Bibliography

Abel, W., R. F. Jackson, and R. A. Wattenbarger. "Simulation of a Partial Pressure Maintenance Gas Cycling Project with a Compositional Model, Carson Creek Field, Alberta," *J. Petroleum Technol.* (Jan. 1970), 38–46.

Abramowitz, M., and I. A. Stegun (ed.). *Handbook of Mathematical Functions with Formulas, Graphs and Mathematical Tables*, National Bureau of Standards Applied Mathematics Series-55 (June 1964) 227–253.

Adams, A. R., H. J. Ramey, and R. J. Burgess, "Gas Well Testing in a Fracture Carbonate Reservoir," *J. of Petroleum Technol.*, October 1968, 1187–1194.

Agarwal, R. G., R. Al-Hussainy, and H. J. Ramey, Jr. "The Importance of Water Flux in Gas Reservoirs," *Trans*, AIME, v. 234, p. 1336, 1965.

Agarwal, R. G., R. Al-Hussainy, and H. J. Ramey, Jr. (1970). An Investigation of Wellbore Storage and Skin Effect in Unsteady Liquid Flow: I. Analytical Treatment, *Society of Petroleum Eng. J*; 10, 279–290.

Agarwall, R. G., R. D. Carter, and C. B. Pollock. "Evaluation and Prediction of Performance of Low-Permeability Gas Wells Stimulated by Massive Hydraulic Fracturing," *J. Petroleum Technol.* (March 1979) 362–372; *Trans; AIME*, **267**.

Aguilera, R. "Well Test Analysis of Naturally Fractured Reservoirs," APEEJ, Sept. 1987, pp. 239–252.

Aguilera, R. and M. C. Ng. "Transient Pressure Analysis of Horizontal Wells in Anisotropic Naturally Fractured Reservoirs," *paper SPE 19002*, presented at the SPE Joint Rocky Mountain Regional/Low Permeability Reservoirs Symposium and Exhibition, Denver, Colorado, March 6–8, 1989.

Al-Hussainy, R. (1967). Transient Flow of Ideal and Real Gases Through Porous Media, Ph. D. Thesis, Texas A. and M. University.

Al-Hussainy, R., and H. J. Ramey, Jr. (1966). "Application of Real Gas Flow Theory to Well Testing and Deliverability Forecasting," *J. of Petroleum Technol.* pp. 18, 637–642.

Al-Hussainy, R., H. J. Ramey, Jr., and P. B. Crawford. "The Flow of Real Gases Through Porous Media," *J. of Petroleum Technol.* (May 1966), pp. 624–636; Trans; AIME; 237.

Alagoa, A., D. Bourdet, and J. A. Ayoub, "How to Simplify the Analysis of Fractured Well Tests," *World Oil*, Oct. 1985.

Amanat U. C. "Pressure Transient Test Analysis User's Handbook," © Advanced TWPSOM Petroleum Systems Inc; Houston, TX, Vol. 8, Oct. 1992.

American Gas Association Natural Gas Dept; Gas Measurement Committee, Report No. 3 (1955). Orifice Metering of Natural Gas. Revised 1969, reprinted 1972.

Aminian, K., and S. Ameri. "Predicting Horizontal Well Production Performance Using Type Curves," paper SPE 19342 presented at the SPE Eastern Regional Meeting, Morgantown, WV, Oct. 24–27, 1989.

Amyx, J. W., D. M. Bass, Jr., and P. L. Whiting. *Petroleum Reservoir Engineering*, New York: McGraw-Hill, 1960.

Aronofsky, J. A., and R. Jenkins. "A Simplified Analysis of Unsteady Radial Gas Flow," *Trans; AIME* (1954) **201**, 149–154.

Arps, J. J. "Analysis of Decline Curve." *Trans. AIME* **160**. p. 228, 1945.

Arps, J. J. "Estimation of Primary Oil and Gas Reserves," Chapter 37 of *Petroleum Production Handbook*, edited by T. C. Frick. New York: McGraw-Hill, 1962.

Aziz, K., and D. L. Flock (1963). Unsteady State Gas Flow – Use of Draw Down Data in the Prediction of Gas Well Behavior, *J. Can. Pet. Tech*; 2(1), 9–15.

Back Pressure Test for Natural Gas Wells, Revised edition, Railroad Commission of Texas (1951).

Beggs, D. H.: "Gas Production Operations, *OGCI Publications*, Oil & Gas Consultants International Inc. Tulsa, OK, Dec. 1984.

Bill, J. P., and D. H. Beggs. *Two Phase Flow in Pipes, 2th Edition*, University of Tulsa, Tulsa, OK, Dec. 1988.

Bixel, H. C., B. K. Larkin, and H. K. van Poolan. "Effect of Linear Discontinuities on Pressure Build-up and Drawdown Behavior," *J. Petroleum Technol.* (Aug. 1963) 885–895; *Trans; AIME*, **228**.

Bourdet, D., A. Alagoa, J. A. Ayoub, and Y. M. Pirard, "New Type Curves Aid Analysis of Fissured Zone Well Tests," *World Oil*, April 1984.

Bourdet, D., J. A. Ayoub, T. M. Whittle, Y. M. Pirard, and Y. Kniazeff, "Interpreting Well Tests in Fractured Reservoirs," *World Oil*, Oct. 1983.

Bordet, D., T. M. Whittle, A. A. Douglas, and Y. M. Pirard, "A New Set of Type Curves Simplifies Well Test Analysis," *World Oil*, May 1983.

Bourdet, D, J. A. Ayoub, and Y. M. Pirard. "Use of Pressure Derivative in Well-Test Interpretation," *SPEFE*, 293–302, June 1989; *Trans. AIME* 293.

Brar, G. S., and K., Aziz. "The Analysis of Modified Isochronal Tests to Predict the Stabilized Deliverability of Gas Wells without Using Stabilized Flow Data," paper SPE 6134, presented at the SPE 51st Annual Meeting, New Orleans, Oct. 3–6, 1976.

Brauser, E. B. (1965). Simplification of the Superposition Principle for Pressure Analysis at Variable Rates, paper SPE 1184, 40th Fall Meeting of *AIME*, Denver, CO.

Brigham, W. E. "Estimating Reservoir Parameters From the Gas backpressure Equation," SPE Reservoir Engineers, May 1988, pp. 649–650.

Brons, F., and V. E. Marting. "The Effect of Restricted Fluid Entry on Well Productivity," *J. of Petroleum Technol.*, pp. 172–174, February 1961.

Brown, G. G., D. L. Katz, C. G. Oberfell, and R. C. Alden. "*Natural Gasoline and the Volatile Hydrocarbons*," NGAA, Tulsa, OK, 1948.

Brown, K. E. *The Technology of Artificial Methods*, Tulsa, OK, PennWell Publishing Co; 1984.

Bruce, G. H., D. W. Peaceman, A. A. Rachford, Jr., and J. D. Rice. "Calculations of Unsteady-State Gas Flow Through Porous Media:" *Trans. AIME*. Vol 198, 1953, pp 79–92.

Burns, William A., Jr.: "New Single Well Test for Determining Vertical Permeability," *J. Petroleum Technol.* (June 1969) 743–752; *Trans; AIME*, **246**.

Cannon, J. R., and A. H. Dogru. "Estimation of Permeability and Porosity from Well Test Data," *paper SPE* 5345 presented at the SPE. *AIME 45th Annual California Regional Meeting*, Ventura, April 2–4, 1975.

Carr, N. L., R. Kobayashi, and D. B. Burrows. "Viscosity of Hydrocarbon Gases Under Pressure," *Trans. AIME*. 201, 264–272, 1954.

Carroll, J. A., III, and R. N. Horne. "Multivariate Optimization of Production Systems," *J. Petroleum Technol.*, pp. 782–789, July 1992.

Carter, R. D. (1962). Solutions of Unsteady-State Radial Gas Flow, *J. Petroleum Technol.*, 14, 549–554.

Carter, R. D. (1966). Performance Predictions for Gas Reservoirs Considering Two-Dimensional Unsteady-State Flow, *Society of Petroleum Engineers Journal*, 6, 35–43.

Carter, R. D., S. C. Millers, Jr., and H. G. Riley. "Determination of Stabilized Gas Well Performance from Short Flow Tests," *J. Petroleum Technology* (June 1963) 651–653.

Celier, G. C. M. R., P. Jouault, O. A. M. C. de Montigny. "Zuidwal: A Gas Field Development with Horizontal Wells," paper SPE 19826, presented at *the SPE 64th Annual Technical Conference and Exhibition of the Society of Petroleum Engineers*, San Antonio, TX, Oct. 8–11, 1989.

Chatas, Angelos T. "A Practical Treatment of Non-Steady State Flow Problems in Reservoirs Systems,"

Cinco, H., F. G. Miller, and H. J. Ramey, Jr. "Unsteady-State Pressure Distribution Created by a Directionally Drilled Well," *J. Petroleum Technol.*, pp. 1392–1402, November 1975.

Cinco, H., and F. Samaniego. "Effect of Wellbore Storage and Damage on the Transient Pressure Behavior for a Well with a Finite-Conductivity Vertical Fracture," *Soc. Pet. Eng. J.* (Aug. 1978) 253–264.

Cinco-Ley, H., H. J. Ramey, Jr., and F. G., Miller. "Pseudo-skin Factors for Partially Penetrating Directionally Drilled Wells," SPE paper 5589, 1975.

Cinco-Ley, H., and F. Samaniego. "Transient Pressure Analysis for Finite Conductivity Fracture Case versus Damage Fracture Case," *SPE Paper 10179*, 1981b.

Cobb, W. M., H. J. Ramey, Jr., and F. G. Miller (1972). Well-Test Analysis for Wells Producing Commingled Zones, *J. Petroleum Technol.* 29, 28–37.

Collins, R. E. (1961). *Flow of Fluids Through Porous Materials*, Reinhold Publishing Corporation, New York.

Compressed Air and Gas Data, Ingersoll-Rand Co; Woodcliff Lake, NJ.

Conversion of Operational and Process Measurement Units to the Metric (SI) System, *Manual of Petroleum Measurement Standards*, Pub. API 2564, API (March 1974) Chap. 15, Sec. 2.

Cornell, D. "*How to Determine Gas Well Interference Graphically*" *World Oil*, Nov. 1952, p. 187–188.

Cornell, D., and D. L. Katz. "*Pressure Gradients in Natural Gas Reservoirs*" Trans. AIME, vol. 198, 1953, pp. 61–70.

Cornelson, D. W. "Analytical Prediction of Natural Gas Reservoir Recovery Factors." *J. Can. Petroleum Technology* (1974) 13 (4), 17024.

Cornett, J. E. (1961). How to Locate Reservoir Limits, *Pet. Eng. J*; 33, B19-B24.

Craft, B. C., and M. F. Hawkins. "*Applied Petroleum Reservoir Engineering*" Prentice-Hall. Inc; 1959, Chapter 6.

Crafton, J. W., and C. D. Harris. "Direct Finite Difference Simulation of a Gas Well with a Finite Capacity Vertical Fracture." Paper SPE 5736 presented at the 4th Symposium of Numerical Simulation of Reservoir Performance, SPE of AIME, Los Angeles (February 19–20, 1979).

Crawford, G. E., A. R. Hagedorn, and A. E. Pierce (1973). *Analysis of Pressure Buildup Tests in a Naturally Fractured Reservoir*, Paper SPE 4558, 48th Fall Meeting of AIME, Las Vegas, NV.

Cregg, M. W. (1968). The Flow of Real Gases in Porous Media, Paper SPE 2091, 43rd Fall Meeting of *AIME*, Houston, Texas.

Cullender, M. H. "The Isochronal Performance Method of Determining the Flow Characteristics of Gas Wells," *Trans.* AIME (1955) **204**, 137–142.

Cullender, M. H., and R. V. Smith (1956). Practical Solution of Gas-Flow Equations for Well and Pipelines with Large Temperature Gradients, *Trans. AIME*, 207, 281–287.

Culbertson, O. L., and J. J. Mcketta. "Solubility of Methane in Water at Pressures to 10,000 psia," *Trans. AIME* (1951), 223.

Dake, L. P. *Fundamentals of Reservoir Engineering,* Elsevier Scientific Pub. Co. (1978).

De Swaan, A. O. "Analytical Solutions for Determining Naturally Fractured Reservoirs Properties by Well Testing." *Society of Petroleum Engineers Journal*, June 1976, 117–122.

De Wiest, R. J. M. ed. (1969). *Flow Through Porous Media*, Academic Press, Inc., New York.

Deaton, W. M., and E. M. Frost. "Gas Hydrates," USBM Monograph (1946) 8.

Denson, A. H., J. T. Smith, and W. M. Cobb. "Determining Well Drainage Pore Volume and Porosity from Pressure Build-up Tests," *Soc. Pet. Eng. J.* (Aug. 1976) 209–216; *Trans. AIME*, **261**.

Derradii S. *Bessel Functions, Laplace Transforms and Their Application*, M.S. Report, University of Tulsa, Tulsa, OK, (1983).

Dietz, D. N. "Determination of Average Reservoir Pressure from Build-Up Surveys," (1965) *Trans. AIME.*

Ding, W. *Gas Well Test Analysis*, MS Thesis, University of Tulsa, Tulsa, OK, 1986.

Dodson, C. R., and M. B. Standing. "Pressure-Volume-Temperature and Solubility Relations for Natural-Gas-Water Mixtures," *Drill and Prod. Prac*; API (1944) 173–179.

Dranchuk, P. M., and J. G. Flores (1973). Non-Darcy Transient Radial Gas Flow Through Porous Media, Paper SPE 4595. 48th Fall Meeting of *AIME*, Las Vegas, NV.

Duda, J. R. "Type Curves for Predicting Production Performance from Horizontal Wells in Low Permeability Gas Reservoirs," *paper SPE 18993*, Richardson, TX.

Dykstra, H. (1961). Calculated Pressure Buildup for a Low-Permeability Gas-Condensate Well, *J. Petroleum Technol.*, 13, 1131–1134.

Earlougher, R. C., Jr. (1971). Estimating Drainage Shapes from Reservoir Limit Tests, *J. Petroleum Technol.*, 23, 1266–1268.

Earlougher, Robert C., Jr.: "Variable Flow Rate Reservoir Limit Testing," *J. Petroleum Technol.* (Dec. 1972) 1423–1429.

Earlougher, R. C., Jr., *Advances in Well Test Analysis*," Monograph Vol. 5 of the Henry L. Doherty Series in Society of Petroleum Engineers of AIME, 1977.

Earlougher, R. C., Jr., and K. M. Kerch (1974). Analysis of Short-Time Transient Test Data by Type-Curve Matching, *J. Petroleum Technol.*, 26, 793–800.

Earlougher, R. C., Jr., and H. J. Ramey, Jr. (1968). The Use of Interpolation to Obtain Shape Factors for Pressure Buildup Calculations, *J. Petroleum Technol.*, 20, 449–450.

Earlougher, R. C., Jr., and H. J. Ramey, Jr. (1973). Interference Analysis in Bounded Systems, *J. Can. Pet. Tech.*, 12(4), 33–45.

Earlougher, R. C., Jr., H. J. Ramey, Jr., F. G. Miller, and T. D. Mueller (1968). Pressure Distributions in Rectangular Reservoirs, *J. Petroleum Technol.*, 20, 199–208.

Economides, C. E. "Use of the Pressure Derivative for Diagnosing Pressure-Transient Behavior," *J. Petroleum Technol.*, Oct. 1988, pp. 1280–1282.

Economides, M. J. "Observations and Recommendations in the Evaluation of Tests of Hydraulically Fractured Wells." *SPE Paper 16396*, 1987.

Economodes, M. J. *et al. Petroleum Production Systems*, Prentice-Hall, Englewood Cliffs, NJ, 1994.

Economides, M. J., and K. G. Nolte. *Reservoir Stimulation*, 2nd ed; Prentice-Hall, Englewood Cliffs, NJ, 1989.

Edwards, A. G., and R. H. Winn. "A Summary of Modern Tools and Techniques Used in Drillstem Testing," Pub. T-4069, Halliburton Co; Duncan, OK (Sept. 1973).

Eilerts, C. K. "Methods for Estimating Deliverability After Massive Fracture Completions in Tight Formations," *paper SPE* 5112 presented at the *SPE-AIME Deep Drilling and Production Symposium*, Amarillo, TX; Sept. 8–10, 1974.

Energy Resources Conservation Board (1974). *Guide for the Planning, Conducting and Reporting of Subsurface Pressure Tests.*

Engineering Data Book, 9th ed; Gas Processors Suppliers Association; Tulsa, OK (1972) Sec. 1.

Fetkovich, M. J. (1973). Decline Curve Analysis Using Type Curves, Paper SPE 4629, 48th Fall Meeting of *AIME*, Las Vegas, NV.

Fetkovich, M. J., and M. E. Vienot. "Shape Factors, C_A, Expressed as a Skin, s_{CA}," *J. Petroleum Technol.*, pp. 321–322, February 1985.

Fetkovich, M. J. "Multi-point Testing of Gas Wells," Continuing Education Course, SPE Mid-Continent Section, March 17, 1975.

Firoozabadii. A., and D. L. Katz. "An Analysis of High-Velocity Gas Flow Through Porous Media. *J. Petroleum Technol.* (Feb. 1979) 211–16.

Flopetrol Johnston Schlumberger, *Course Manual*, 100 Macco Boulevard, Sugarland, TX 77478.

Fluid Meters—Their Theory and Application: Report of ASME Research Committee on Fluid Meters, 6th edition, The American Society of Mechanical Engineers, New York (1971).

Gas Processors Suppliers Association, *Engineering Data Book.* 9th edition 1972, Revised 1974.

Gas Technology, Vol 1, SPE Reprint Series No. 13, Society of Petroleum Engineers of AIME, Dallas, TX, 1977 Edition.

Gentry, R. W. "Decline Curve Analysis." *J. Petroleum Technol.*, p. 38, January 1972.

Gilbert. W. E. "Flowing and Gas-Lift Well Performance," *Drilling and Prod. Prac*; API (1954) 16–43.

Golan, M., and C. H. Whitson. *Well Performance* International Human Resources Corporation, Boston, MA, 1986.

Goode P. A., and R. K. M. Thambynayagam. "Pressure Drawdown and Buildup Analysis for Horizontal Wells in Anisotropic Media," *SPE Formation Evaluation*, pp. 683–697, December 1987.

Govier. G. W. (1961). Interpretation of the Results of Back Pressure Testing of Gas Wells, *Trans. AIME*, LXIV, 511–514.

Govier, G. W. *Theory and Practice of the Testing of Gas Wells*, Energy Resources Conservation Board, Calgary, Alberta, Canada, 1975.

Govier. G. W., and M. Forgarasi (1975). Pressure Drop in Wells Producing Gas and Condensate, Paper Presented at 26th Technical Meeting of Petroleum. Soc. of CIM, Banff, Alta.

Gray, H. E. "Vertical Flow Correlation in Gas Wells," User Manual for API 14B. Subsurface Controlled Safety Vales String Computer Program. App. B. API, Dallas, TX (June 1974).

Gray, K. E. (1965). "Approximating Well-to-Fault Distance from Pressure Build-Up Tests," *J. Petroleum Technol.*, 17, 761–767.

Gringarten, A. C. "Reservoir Limit Testing for Fractured Wells," paper SPE 7452, presented at the *SPE 53rd Annual Fall Technical Conference and Exhibition*, Houston, TX, Oct. 1–3, 1978.

Gringarten, A. C., H. J., Ramey, Jr., and R. Raghavan. "Unsteady-State Pressure Distribution Created by a Well with a Single Infinite-Conductivity Vertical Fracture," *Society of Petroleum Engineer's Journal*, pp. 347–360, August 1974.

Gringarten, A. C., H. J. Ramey, Jr., and R. Raghvan (1975). Applied Pressure Analysis for Fractured Wells, *J. Petroleum Technol.*, 17, 887–892.

Hadinoto, N., R. Raghavan, and G.W. Thomas. "Determination of Gas Well Deliverability of Vertically Fractured Wells." Paper SPE 6136 presented at the 51st Annual Fall Technical Conference and Exhibition of SPE of *AIME*, New Orleans (October 3–6, 1976).

Hall, H. N. "Compressibility of Reservoir Rocks," *Trans. AIME*, Vol. 198, 1953, pp. 309–311.

Hammerschmidt, E. G. "Formation of Gas Hydrates in Natural Gas Transmission Lines," *Ind. And Eng. Chem.* (1934) 26, 851.

Hankinson, R. W., L. K. Thomas, and K. A. Phillips. "Predict Natural Gas Properties," *Hydrocarbon Processing,* pp. 106–108, April 1969.

Havlena, D., and A. S. Odeh. "The Material Balance as an Equation of Straight Line," *J. Petroleum Technol.* (August 1963): 896–900.

Havlena, D., and A. S. Odeh. "The Material Balance as an Equation of Straight Line – Part II, Field Cases," *J. Petroleum Technol.* (July 1964) : 815–822.

Horner, D. R. (1951). Pressure Build-Up in Wells, *Proceedings*, Third World Pet. Congress—Sect. II, 503–521.

Houpeurt, A.: "Analog Study of Radial Circular Transient Flow of Gas in porous media," *Revue IFP* (1953) **8**. 129, 193, 248 (in French).

Hurst, W., J. D. Clark, and E. B. Brauer (1969). The Skin Effect in Producing Wells, *J. Petroleum Technol.*, 21, 1483–1489.

Hurst, W., W. C. Goodson, and R. E. Leeser (1963). Aspects of Gas Deliverability, *J. Petroleum Technol.*, 15, 568–676.

Interstate Oil Compact Commission (1962*). Manual of Back Pressure Testing of Gas Wells.*

Ishteiwy, A. A., and H. K. Van Poolen (1967). *Radius-of-Drainage Equation for Pressure Buil*dup, Paper Presented at Libyan Assoc. of Pet. Technologies' Meeting, Tripoli, Libya, Jan. 25–26.

Jacoby, R. H., R. C. Koeller, and V. J. Berry, Jr. "Effect of Composition and Temperature on Phase Behavior and Depletion Performance of Gas-Condensate Systems," Paper presented before SPE of AIME, Houston, TX, Oct. 5–8, 1958.

Jahnke, E., and F. Emde (1945). *Tables of Functions with Formulae and Curves*, 4th edition, Dover Publications, New York.

Janicek, J., and D. L. Katz (1955). *Applications of Unsteady-State Gas Flow Calculations*, Preprint, University. Of Michigan Publishing Services, Ann Arbor, MI.

Johnson, C. R., R. A. Greenhorn, and E. G. Woods. "Pulse-Testing: A New Method for Describing Reservoir Flow Properties Between Wells," *J. Petroleum Technol.* (Dec. 1966) 1599–1604; *Trans.* AIME, 237.

Jones, L. G. (1961). "An Approximate Method for Computing Non-steady-State Flow of Gases in Porous Media," *Society of Petroleum Engineers J.* 1, 264–276.

Jones, L. G., E. M. Blount, and O. H. Glaze, "Use of Short Term Multiple Rate Flow Tests to Predict Performance of Wells Having Turbulence," paper SPE 6133 presented at the SPE 51st Annual Meeting, New Orleans, Oct. 3–6, 1976.

Jones, P. (1963). Reservoir Limits Test on Gas Wells, *J. Petroleum Technol.*, 14, 613–619.

Joshi, S. D. "A Review of Horizontal Well and Drain Hole Technology," Paper No. SPE 16868, SPE Annual Technical Conference, Dallas, 1987, and revised version SPE Rocky Mountain Regional Meeting, Casper, WY, May 1988.

Joshi, S. D. "Horizontal Well Technology," *PennWell Books*, Tulsa, OK, 1991.

Kamal, M. M. "Interference and Pulse Testing—A Review, *J. Petroleum Technol.*, 2257–2270, Dec., 1983.

Kansas State Corporation Commission (1959). *Manual of Back Pressure Testing of Gas Wells.*

Karakas, M., and S. Tariq. "Semi-Analytical Production Model for Perforated Completion," *SPE Paper* 18247, 1988.

Katsner, F. E. "Effects of Linear Boundaries on Pulse Testing." M. Sc. Thesis, Colorado School of Mines, 1970.

Katz, D. L., D. Cornell, R. Kobayashi, F. H. Poettmann, J. A. Vary, J. R. Elenbaas, and C. F. Weinaug (1959). *Handbook of Natural* Gas *Engineering,* McGraw-Hill, New York.

Kazemi, H. (1969). "Pressure Transient Analysis of Naturally Fractured Reservoirs with Uniform Fracture Distribution," *Soc. Pet. Eng. J*; 69, 451–462.

Kazemi, H. (1974). "Determining Average Reservoir Pressure from Pressure Buildup Tests," *Soc. Pet. Eng. J*; 14, 55–62.

Kazemi. H., M. S. Seth, and G. W. Thomas. "The Interpretation of Interference Tests in Naturally Fractured Reservoirs with Uniform Fracture Distribution." *Society of Petroleum Engineers Journal*, Dec. 1969, 463–472.

Klotz. J. A., R. F. Krueger, and H. Pyle, "Effect of Perforation Damage on Well Productivity," *J. Petroleum Technol.* (Nov. 1974) 1303–14; Trans; AIME. 257.

Kulczycki, W. (1955). "New Method of Determination of the Output and the Absolute Open-Flow of Gas Wells," *Nafta*, 11(10), 233–237.

Larson, V. C. (1963). Understanding the Muskat Method of Analyzing Pressure Build-Up Curves, *J. Can. Pet. Tech.*, 2(3), 136–141.

Lee, A., M. H. Gonzales, and B. E. Eakin (August, 1966). "The Viscosity of Natural Gases," *J. Petroleum Technol.*, 18, 997–1000.

Lee, W. J. "Wellbore Storage: How It Effects Pressure Build-up and Pressure Drawdown Tests," paper Presented at the SPWLA 12th Annual Logging Symposium, Dallas, TX, May 2–5, 1971.

Lee. W. J. "Well Testing," vol. 1. SPE, Textbook Series, *Society of Petroleum Engineers of AIME*, Dallas, TX, 1982.

Lee, W. J., R. R. Harrell, and W. D. McCain, Jr. (1972). "Evaluation of a Gas Well Testing Method," paper *SPE* 3872, N. Plains Sect, Regional Meeting of *AIME*, Omaha, NE.

Lee, W. J., Jr. "Analysis of Hydraulically Fractured Wells with Pressure Build-up Tests," paper SPE 1820 Presented at the *SPE-AIME* 42nd Annual Fall Meeting, Houston, TX, Oct. 1–4, 1967.

Letkeman, J. P., and R. L. Ridings (1970). "A Numerical Coning Model," *Trans. AIME*, 249, 418–424.

Livak, B. I. Texaco E & P Technology Department: Personal Contact.

Mach. J., E. Proano, and K. F. Brown. "Application of Production Systems Analysis to Determine Completion Sensitivity on Gas Well Completion," paper 8113 presented at the *ASME Energy Sources Technical Conference*, Houston, TX, Jan. 18–22, 1981.

Maer, N. K., Jr. (1974). "Type Curves for Analysis of Afterflow-Dominated Gas Well Build-up Data," Paper SPE 5134, 49th Fall Meeting of *AIME*, Houston, TX.

Martin, J. C. (1959). Simplified Equations of Flow in Gas Drive Reservoirs and the Theoretical Foundation of Multiphase Pressure Buildup Analyses, *Trans. AIME*, 216, 309–311.

Mathews, C. S., F. Brons, and P. Hazebrock. "A Method for Determination of Average Pressure in a Bounded Reservoir," (1954) *Trans.* AIME.

Mathews, C. S., and D. G. Russell (1967). *Pressure Buildup and Flow Tests in Wells*, AIME, Monograph.

Mattews, L., G. S. F. Brons, and P. Hazebroek (1954). "A Method for Determination of Average Pressure in a Bounded Reservoir," *Trans. AIME*, 201, 182–191.

McCain W. D., Jr. *The Properties of Petroleum Fluids*, Petroleum Publishing Co. Tulsa, OK (1990).

McKinley, R. M. (1970). Wellbore Transmissibility from Afterflow-Dominated Pressure Buildup Data, Paper SPE 2416, 45th Fall Meeting of AIME, Houston, TX.

McKinley, R. M. (1974). Estimating Flow Efficiency from Afterflow-Distorted Pressure Buildup Data, *J. Petroleum Technol.*, 26(6), 696–697.

McLeod, H. O. "The Effect of Perforating Conditions on Well Performance," *J. Petroleum Technol.* (Jan. 1983).

McMahon, J. J. (1961). "Determination of Gas Well Stabilization Factors from Surface Flow Tests and Build-Up Tests," Paper SPE 114, 36th Fall Meeting of AIME, Dallas, TX.

Meng. H. Z. *et al.* "Production Systems Analysis of Vertically Fractured Wells," paper SPE/DOE 10842 presented at the SPE/DOE Unconventional Gas Recovery Symposium, Pittsburgh, PA, May 16–18, 1982.

Mishra, S. "Deliverability Testing of Gas Wells Using Dimensionless IPR Curves," M.S. Thesis, The University of Texas at Austin, May 1983.

Mueller, T. D., and Paul A. Witherspoon. "Pressure Interference Effects Within Reservoirs and Aquifers," *J. Petroleum Technol.* (April 1965) 471–474; *Trans*. AIME, 234.

Muskat, M. (1936). Use of Data on the Build-Up of Bottom-Hole Pressures, Paper presented Fort Worth Meeting, Fort Worth, TX, Oct. 1966.

Muskat, M. *The Flow of Homogeneous Fluids Through Porous Media*, McGraw-Hill, New York. (1973).

Mutalik, P. N., S. P. Godbole, and S. D. Joshi. "Effect of Drainage Area Shapes on Horizontal Well Productivity," paper SPE 18301, presented in *the SPE 63rd Annual Technical Conference*, Houston, TX, Oct. 2–5, 1988.

Najurieta, H. L. "A Theory for the Pressure Transient Analysis in Naturally Fractured Reservoirs," Paper SPE 6017 presented at the *SPE-AIME* 51st Annual Fall Technical Conference and Exhibition, New Orleans, Oct. 3–6, 1976.

Nemeth, L. K., and H. T. Kennedy, "A Correlation of Dew Point Pressure with Fluid Composition and Temperature," *J. Petroleum Technol.*, June 1967, p. 99.

NGPSA Data Book, 9th ed., Tulsa, OK (1972).

Nisle, R. G. (1956). "The Effect of a Short Term Shut-in on a Subsequent Pressure Build-up Test on an Oil Well," *Trans. AIME*, 207, 320–321.

Odeh, A. S. (1965). Unsteady-State Behavior of Naturally Fractured Reservoirs, *Soc. Pet. Eng*. J., 5, 60–66.

Odeh, A. S. (1969). "Flow Test Analysis for a well with Radial Discontinuity," *J. Petroleum Technol.* 21, 207–210.

Odeh, A. S. "An Equation for Calculating Skin Factor Due to Restricted-Entry," *J. Petroleum Technol.*, pp. 964–965, June 1980.

Odeh, A. S., and R. Al-Hussainy (1971). "A Method for Determining the Static Pressure of a Well from Build-up Data," *J. Petroleum Technol.*, 23, 621–624.

Odeh, A. S., and D. K. Babu. "Transient Flow Behavior of Horizontal Wells, Pressure Drawdown and Buildup Analysis," SPE Formation Evaluation, pp. 7–15, March 1990.

Odeh, A. S., and L. G. Jones (1965). "Pressure Drawdown Analysis Variable-Rate Case," *J. Petroleum Technol.*, 17, 960–964.

Odeh, A. S., and G. W. Nabor (1966). "The Effect of Production History on Determination of Formation Characteristics from Flow Tests," *J. Petroleum Technol.*, 18, 1343–1350.

Odeh, A. S., and F. Selig (1963). "Pressure Buildup Analysis, Variable-Rate Case," *J. Petroleum Technol.*, 15, 790–794.

Omana, R., C. Houssiere, K. E. Jt. Brown, J. O. Brill, and R. E. Thompson. "Multiphase Flow Through Chokes," *SPE* Paper **2682**, 1969.

Orifice Meter Constants: Handbook E-2, Singer American Meter Division (1973).

Overbey, W. K., Jr., A. B. Yost II, and D. A. Wilkins. "Inducing Multiple Hydraulic Fractures from a Horizontal Wellbore," paper SPE 18249, presented at the SPE 63rd Technical Conference and Exhibit Annual Meeting, Houston, TX, Oct. 2–5, 1988.

Ozkan, E. "Performance of Horizontal Wells," Ph.D. Dissertation, The University of Tulsa, Tulsa, OK, 1988.

Ozkan, E., R. Raghavan, and S. D. Joshi. "Horizontal Well Pressure Analysis," *SPE Formation Evaluation*, pp. 567–575, Dec. 1989.

Papatzacos, P. "Approximate Partial-Penetration Pseudo-Skin for Infinite-Conductivity Wells," *SPE Reservoir Engineering*, pp. 227–234, May 1988, *Trans. AIME*, vol. 283.

Perrine, R. L. (1956). "Analysis of Pressure-Buildup Curves," *API Drill. and Prod. Practice*, 482.

Pierce, H. R., and E. L. Rawlins (1929). *The Study of a Fundamental Basis for Controlling and Gauging Natural-Gas Wells*, U.S. Dept. of Commerce-Bureau of Mines, Series 2929.

Pirson, R. S., and S. J. Pirson (1961). "An Extension of the Pollard Analysis Method of Well Pressure Build-up and Drawdown Tests," paper SPE 101, 36th Fall Meeting of *AIME*, Dallas, TX.

Pitzer, S. C. (1964). "Uses of Transient Pressure Tests," *API Drill. and Prod. Practice*; 115–130.

Pollard, P. (1959). "Evaluation of Acid Treatments from Pressure Build-up Analysis," *Trans. AIME*, 216, 38–43.

Prasad, R. K. (1973). "Pressure Transient Analysis in the Presence of Two Intersecting Boundaries," paper SPE 4560, 48th Fall Meeting of *AIME*, Las Vegas, NV.

Quon, D., P. M. Dranchuk, S. R. Allada, and P. K. Leung (1966). "Application of the Alternating Directional Explicit Procedure to Two-Dimensional Natural Gas Reservoirs," *Soc. Pet. Eng. J*; 6, 137–142.

Raghavan, R., G. V. Cady, and H. J. Ramey, Jr. (1972). "Well Test Analysis for Vertically Fractured Wells," *J. Petroleum Technol.*, 24, 1014–1020.

Railroad Commission of Texas (1950). *Back Pressure Test for Natural Gas Wells*. Revised Edition, 1951.

Ramey, H. J., Jr. "Non-Darcy Flow and Wellbore Storage Effects in Pressure Build-up and Drawdown of Gas Wells." *J. Petroleum Technol.*, (1965) 223–233.

Ramey H. J., Jr. (1967). "Application of the Line Source Solution to Flow in Porous Media—A Review," *Producers Monthly*, 31 (5), 4–7 and 25–27.

Ramey, H. J., Jr. (1970). "Short-Time Well Test Data Interpretation in the Presence of Skin Effect and Wellbore Storage," *J. Petroleum Technol.*, 22, 97–104.

Ramey, H. J., Jr. A. Kumar, and M. S. Gulati (1973). *Gas Well Test Analysis Under Water-Drive Conditions*, American Gas Association; VA.

Ramey, H. J., Jr. A. Kumar, and M. Gulati. *Gas Well Test Analysis Under Water Drive Conditions*, Monograph, American Gas Association Project 61–51 (1975).

Rawlins, E. L., and M. A. Schellhardt. "Back-Pressure Data on Natural Gas Wells and their Application To Production Practices," *Monograph 7. USBM*, 1936.

Roebuck, I. F., G. E. Henderson, J. Douglas, Jr., and W. T. Ford, "The Compositional Reservoir Simulator: The Linear Model," *Trans. AIME* (1969), 246, 115.

Russell, D. G. (1963). "Determination of Formation Characteristics from Two-Rate Flow Tests," *J. Petroleum Technol.*, 15, 1317–1355.

Russell, D. G. (1966). "Extensions of Pressure Build-up Analysis Methods," *J. Petroleum Technol.*, 18, 1624–1636.

Russell, D. G., J. H. Goodrich, G. E. Perry, and J. F. Bruskotter. "Methods for Predicting Gas Well Performance," *J. Petroleum Technol.*, (Jan. 1966) 99–108.

Russell, D. G., and N. E. Truitt. "Transient Pressure Behavior in Vertically Fractured Reservoirs," Society of Petroleum Engineers, August 1964.

Saidikowski, R. M. "Numerical Simulation of the Combined Effects of Wellbore Damage and Partial Penetration," paper SPE 8204, Sept. 23–26, 1979.

Schechter, R. S. *Oil Well Stimulation*, Prentice-Hall, Englewood Cliffs, NJ, 1982.

Schechter, R. S., and J. L. Gidley. "The Change in Pore Size Distribution from Surface Reactions in Porous Media." *AICHEJ*; May 1969, 339–350.

Schlumberger Educational Services, repeat Formation Tester, SMP-9070, Houston, TX, 1986b.

Seba, R. D. "Estimation of Economically Recoverable Oil from Decline Curve Analysis," Lecture Notes, Stanford University, 1976.

Shoemaker, R. P. "Graphical Method for Solving Decline Problems." *World Oil*, p. 123, October 1967.

Slider, H. C. "Application of Pseudo-Steady-State Flow to Pressure Build-up Analysis," paper SPE 1403 Presented at the SPE-AIME Regional Symposium, Amarillo, TX, Oct. 27–28, 1966.

Slider, H. C. (1971). "A Simplified Method of Pressure Analysis for a Stabilized Well," *J. Petroleum Technol.*, 23, 1155–11160.

Smith, R. V. (1961). "Unsteady-State Gas Flow into Gas Wells," *J. Petroleum Technol.*, 13, 1151–1159.

Smith, R.V. *Practical Natural Gas Engineering*, Tulsa, OK, Penn-Well Publishing Co; 1983.

Smolen, J. J., and L. R. Litsey. "Formation Evaluation Using Wireline Formation Tester Pressure Data," *J. Petroleum Technol.*, (Jan. 1979) 25–32.

Standing, M. B. *Volumetric and Phase Behavior of Oil Field Hydrocarbon Systems*, Reinhold Publishing Corp; New York (1952).

Standing, M. B. "Concerning the Calculation of Inflow Performance of Wells Producing Solution Gas Drive Reservoirs," *J. Petroleum Technol.* (Sept. 1971) 1141–1142.

Standing, M. B., and C. R. Dodson. "Pressure-Volume-Temperature and Solubility Relations for Natural Gas Water Mixtures," *Drill. And Prod. Proc*; API (1944), 1973.

Standing, M. B., and D. L. Katz. "Density of Natural Gases," *Trans. AIME* (1942) **146**, 140–149.

Strobel, C. J., M. S. Gulati, H. J. Ramey, Jr. "Reservoir Limit Tests in a Naturally Fractured Reservoir—A Field Case Study Using Type Curves," *J. Petroleum Technol.*, Sept. 1976, pp. 1097–1106.

Swift, G. W., and O. G. Kiel (1962). "The Prediction of Gas Well Performance Including the Effect of Non-Darcy Flow," *J. Petroleum Technol.*, 14, 791–798.

Szilas, A. P. *Production and Transport of Oil and Gas*, Elevier, Amstrdam, 1975.

Theory and Practice of the Testing of Gas Wells, Energy Resources Conservation Board, Calgary, Alberta, Canada, 1975.

Tiab, D., and A. Kumar. "Application of p_D Function to Interference Analysis," paper SPE 6053 Presented at the SPE-AIME 51st Annual Fall Technical Conference and Exhibition, New Orleans, Oct. 3–6, 1976.

Timmerman, E. H., and H. K. van Poolen. "Practical Use of Drill-Stem Tests," *J. Cdn Pet. Tech.* (April-June 1972) 31–41.

Trube, A. S. "Compressibility of Natural Gases," *Trans. AIME* (1957) **210**, 355–357.

Trube, A. S. *et al.* "Compressibility of Under-saturated Hydrocarbon Reservoir Fluids," *Trans. AIME* (1957), 341.

Tyler, T. N., R. R. Metzger, and L. R. Twyford. "Analysis and Treatment of Formation Damage at Prudhoe Bay, AK," *SPE Paper* 12471, 1984.

U.S. Bureau of Mines, Monograph 7 (1936). *Backpressure Data on Natural Gas Wells and Their Application to Production Practices.*

Uldrich, D. O., and I. Ershaghi. "A Method for Estimating the Interporosity Flow Parameter in Naturally Fractured Reservoirs," *Society of Petroleum Eng, J*; Oct. 1979, pp. 324–332.

Van Everdingen, A. F. (1953). "The Skin Effect and Its Influence on the Productive Capacity of a Well," *Trans. AIME*, 198, 171–176.

Van Everdingen, A. F., and W. Hurst (1949). "The Application of the Laplace Transformation to Flow Problems in Reservoirs," *Trans. AIME*, 179, 305–324.

Van Everdingen, A. F., I. Meyer, and L. Joffre. "Analysis of Buildup Curves Obtained After Well Treatment," *J. Petroleum Technol.*, (April 1971) 513–524; *Trans. AIME*, **251**.

Van Poollen, H. K. (1964). "Radius-of-Drainage and Stabilization-Time Equations," *Oil and Gas J*; 62, 138–146.

Van Poollen, H. K., H. C. Bixel, and J. R. Jargon. "Reservoir Modeling—4: Explicit Finite-Difference Technique," *Oil and Gas J.* (Nov. 3, 1969) 81–87.

Van Poollen, H. K., H. C. Bixel, and J. R. Jargon. "Reservoir Modeling—5: Implicit Finite-Difference Technique," *Oil and Gas J.* (Jan. 3, 1970) 88–92.

Van Poollen, H. K., H. C. Bixel, and J. R. Jargon. "Reservoir Modeling—8: Single-Phase Gas Flow," *Oil and Gas J.* (March 30, 1970) 106–107.

Van Poollen, H. K., H. C. Bixel, and J. R. Jargon. "Reservoir Modeling—9: Here Are Fundamental Equations for Multiphase Fluid Flow," *Oil and Gas J.* (May 11, 1970) 72–78.

Vogel, J. L. "Inflow Performance Relationships for Solution-Gas Drive Wells," *J. Petroleum Technol.* (Jan. 1968) 83–92.

Wang, B., and T. S. Teasdale. "GASWAT-PC: A Microcomputer Program for Gas Material Balance with Water Influx." SPE Paper 16484, Petroleum Industry Applications of Microcomputer, Del Lago on Lake Conroe, TX, June 23–26, 1987.

Warren, J. E., and J. H. Hartsock. "Well Interference," *Trans. AIME* (1960) **218**, 89–91.

Warren, J. E., and P. J. Root (1963). "The Behavior of Naturally Fractured Reservoirs," *Soc. Pet. Eng. J*; 3, 245–255.

Warren, J. E., and P. J. Root. "Discussion of Unsteady-State Behavior of Naturally Fractured Reservoirs," *Soc. Pet. Eng. J.* (March 1965) 64–65; *Trans. AIME*, **234**.

Watson, E. J. *Laplace Transforms and Applications*, van Nostrand Reinhold Company, New York (1981) 89.

Watson, G. N. *Theory of Bessel Functions*, Cambridge University Press, London (1944) 44.

Wattenbarger, R. A. (1967). Effects of Turbulence, Wellbore Damage, Wellbore Storage and Vertical Fractures on Gas Well Testing, Ph.D. Thesis, Stanford University; Stanford, CA.

Wattenbarger, R. A., and H. J. Ramey, Jr. "Gas Well Testing with Turbulence Damage, and Wellbore Storage," *J. Petroleum Technol.* (Aug. 1968) 877–887; *Trans.* AIME, **243**.

Wattenbarger, R. A., and H. J. Ramey, Jr. "Well Test Interpretations of Vertically Fractured Gas Wells," *J. Petroleum Technol.* (May 1969) 625–632; *Trans. AIME*, 246.

Weber, J. H. "Predicting Properties of Gas Mixtures," *Chemical Eng.* (May 19, 1980).

Wichert, E., and K. Aziz. "Calculation of Z's for Sour Gases," *Hydrocarbon Processing*, 51 (5), 1972.

Willis, R. B.: "How to Simply Gas-Well Test Analysis," Petroleum Technology J. (1965) 37, 95–98.

Yarborough, L., and K. R. Hall. "How to Solve Equation of State for z-factors," *Oil and Gas J.* (Feb. 18 1974) 86–88.

Yeh, N. S., and A. C. Reynolds. "Computation of the Pseudo-Skin Caused by a Restricted-Entry Well Completed in a Multilayer Reservoir," *SPE Formation Evaluation*, pp. 253–263, June 1989.

Yost II, A. B., W. K. Overbey, Jr., D. A. Wilkens, and C. D. Locke. "Hydraulic Fracturing of a Horizontal Well in a Naturally Fractured Reservoir: Case Study for Multiple-Fracture Design," *paper SPE* 17759, presented at the SPE Gas Technology Symposium, Dallas, TX, June 13–15, 1988.

Zana, E. T., and G. W. Thomas (1970). "Some Effects of Contaminants on Real Gas Flow," *J. Petroleum Technol.*, 22(9), 1157–1168.

Index

E

J

K

L

M

N

V

W

Z